# Historical Geology

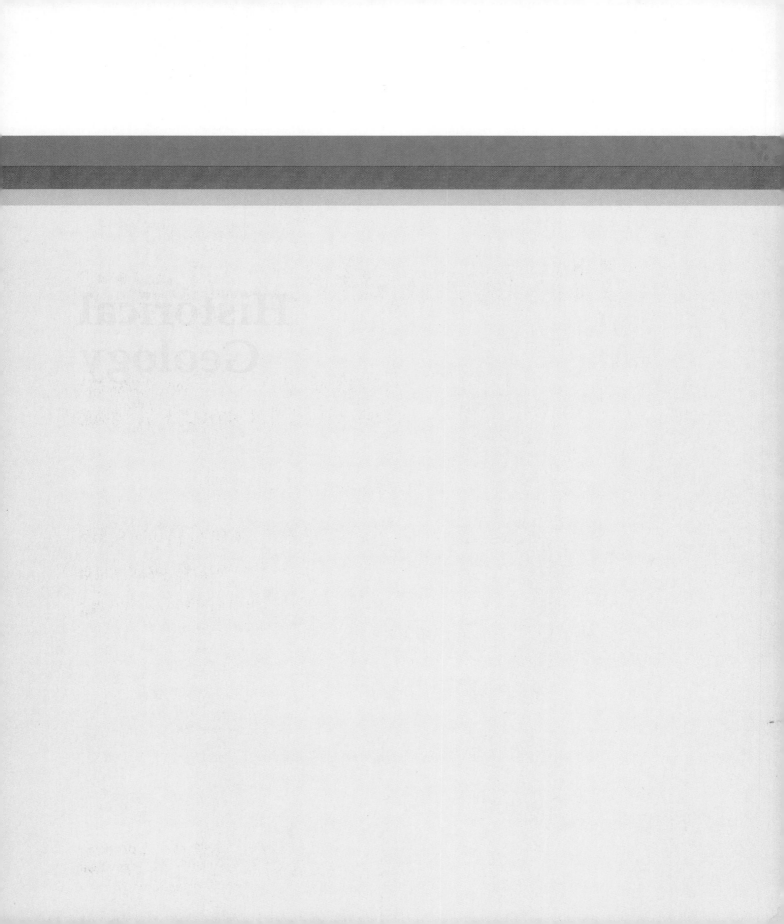

# Historical Geology

## Evolution of the Earth and Life through Time

**Reed Wicander**

**James S. Monroe**

Central Michigan University

**West Publishing Company**

St. Paul New York Los Angeles San Francisco

***Copyediting:*** Margaret Jarpey
***Text Design:*** John Edeen
***Text Illustrations:*** Rolin Graphics
Michael Thomas Associates (Jim Kiehne, Dan Thorson, Toshiharu Maeda, Michelle Recke)
Erin O'Brien
***Composition:*** Rolin Graphics
***Cover Design:*** Roslyn M. Stendahl
***Cover Image:*** "The Tower of Time" by John Gurche.
Copyright 1980. Smithsonian Institution.


50 West Kellogg Boulevard
P.O. Box 64526
St. Paul, MN 55164-1003



Printed in the United States of America

96 95 94 93 92 91 90 89 8 7 6 5 4 3 2 1

**Library of Congress Cataloging-in-Publication Data**

Wicander, Reed, 1946–
Historical geology : evolution of the earth and life through time /Reed Wicander, James S. Monroe.
p. cm.
Bibliography: p.
Includes index.
ISBN 0-314-46336-4
1. Historical geology I. Monroe, James S. (James Stewart), 1938- II. Title.
OE28.3.W53 1989
551.7–dc19

88-28127
CIP

## CREDITS

**CHAPTER 1** p. 3 Photo courtesy of NASA; p. 14 Modified from W. Hamilton, U.S. Geol. Survey.

**CHAPTER 2** p. 22 Modified from H.L. Levin, *The Earth through Time,* 3rd Ed., 1988, p. 111, fig. 4–20; p. 24 Photo courtesy of National Park Services, U.S. Department of Interior; p. 25 Biblioteca Reale, Torino, Italy; p. 26 Modified from R.H. Dott, Jr. and R.L. Batten, *Evolution of the Earth,* 4th Ed., p. 34, fig. 2–13; p. 27 Reprinted from *The Birth and Development of the Geological Sciences,* F. D. Adams, Dover Publications, NY, 1954; p. 28 From Craig, McIntyre, and Waterson, *James Hutton's Theory of the Earth: The Lost Drawing,* 1978, Scottish Academic Press Ltd. Reprinted with permission; p. 30 Courtesy of the Geological Society, London; p. 34 Don L. Eicher, *Geologic Time,* 2d ed., ©1976, p. 120 Reprinted by permission of Prentice Hall, Inc., Englewood Cliffs, New Jersey; p. 37 Reprinted by permission from *Nature* vol. 197, p. 949, copyright © 1963 Macmillan Magazines Ltd.; p. 39 Stokes and Smiley, *An Introduction to Tree-Ring Dating,* 1968, the University of Chicago Press. Reprinted with permission; p. 39 Modified from B.J. Skinner and S.C. Porter, 1987, *Physical Geology,* p. 193, fig. 8.11.

**CHAPTER 3** p. 42 Reprinted from *Fossils Magazine,* v. 1, Issue 1, May, 1976; p. 55 M.F. Miller and L.W. Knox, 1985, reprinted with permission of the Society of Economic Paleontologists and Mineralogists; p. 55 Neg. #125158, Courtesy Dept. of Library Services, American Museum of Natural History; p. 55 Reprinted with permission from Martin and Bennett, *Palaeogeography, Palaeoclimatogoy, Palaeoecology,* ©1977, Elsevier Science Publishers; p. 56 Neg #39442, Courtesy Department of Library Services, American Museum of Natural History; p. 56 Photograph from the Rancho La Brea Archives, George C. Page Museum, Los Angeles County Museum of Natural History; p. 57 Neg. # 35606, Courtesy Department of Library Services, American Museum of Natural History; p. 57 Neg. #35608, Courtesy Department of Library Services, American Museum of Natural History; p. 57 Photo courtesy of Soufoto, by V. Khristoforov. p. 59 From Morris S. Peterson and J. Keith Rigby, *Interpreting Earth History,* 2d ed. Copyright © 1978 Wm. C. Brown Publishers, Dubuque, IA. All Rights Reserved. Reprinted by permission; p. 60 Photo courtesy of Badlands National Park; p. 61 Photograph from the Rancho La Brea Archives, George C. Page Museum, Los Angeles County Museum of Natural History; p. 61 A.L. Karowe and T.H. Jefferson, "Burial of Trees by Eruptions of Mount St. Helens, Washington: Implications for the Interpretations of Fossil Forests," *Geological Magazine* 124, no. 3f (1987). Re-

# Contents

## 6 Plate Tectonics—A Unifying Theory



## 7 The Origin of the Universe, Solar System, and Planet Earth



## 8 Precambrian History—The Archean Eon



## 9 Precambrian History—The Proterozoic Eon



## 10 Geology of the Early Paleozoic Era

## 11 Geology of the Late Paleozoic Era



## 12 Life of the Paleozoic Era



## 13 Geology of the Mesozoic Era



## 14 Life of the Mesozoic Era



## 15 Cenozoic History—Tertiary Period

## 16 Life of the Tertiary Period



## 17 Geology of the Quaternary Period and the Evolution of Humans

# Preface

The Earth as a dynamic planet has changed continuously during its 4.6 billion years of existence. The size, shape, and geographic distribution of continents have changed through time, the atmosphere has evolved, and life forms existing today differ from those that lived during the past. Historical geologists are concerned with all aspects of the Earth and its life history. They seek to determine what events occurred during the past, place those events into an orderly chronologic sequence, and provide conceptual frameworks for explaining such events. All of this makes the study of the history of the Earth and its life such a fascinating endeavor.

Some broad similarities exist in all historical geology textbooks. Most begin with several chapters on concepts and principles, followed by a chronologic discussion of Earth history. In this respect we have not departed from convention. We have, however, attempted to place greater emphasis on basic concepts and principles, their historical development, and their importance in deciphering the history of the Earth. Stratigraphic terminology is used sparingly with the idea that understanding the causes and consequences of events is more important than knowing the names of numerous formations. Additionally, we have attempted to present a balanced coverage of both physical and biological events because we feel that such balance is lacking in many textbooks.

## THEMATIC APPROACH

Our primary goal in this textbook is to develop three major themes that are essential to the interpretation and appreciation of historical geology, introduce these themes early, and reinforce them throughout the book. These themes are *time*, the dimension that sets historical geology apart from the other sciences; *evolutionary theory*, the explanation for inferred relationships among living and fossil organisms; and *plate tectonics*, a unifying theory for interpreting much of the Earth's physical history, and to a large extent, its biological history. In addition, we have emphasized the intimate interrelationship that exists between physical and biological events.

The concept of geologic time and its historical development is presented in Chapter 2. The concept is more fully developed in the following chapter on Rocks, Fossils, and Time. In Chapter 5, we discuss the development of evolutionary theory in the context of the historical period during which it was formulated. Included in this discussion are sections in which the evidence for evolution and the importance of the fossil record in evolutionary studies are emphasized. Evolutionary theory serves as the unifying theme for subsequent discussions of life history, and is constantly reinforced throughout the book.

Particular emphasis is given to the evidence that substantiates plate tectonic theory, why this theory is one of the cornerstones of geology, and why plate tectonic theory serves as a unifying paradigm in explaining many apparently unrelated geologic phenomena. Several examples of how plate tectonics is responsible for seemingly disparate aspects of geology are given in this context, events are viewed as parts of an ever-changing continuum, rather than as discrete events separated from one another in time and space. While we emphasize the geologic evolution of North America, each chapter on geologic history contains an overview of global tectonic events so that the history of North America is discussed in a global context.

## TEXT ORGANIZATION

This book was written for a one semester course in historical geology, to serve both majors and nonmajors in geology and in the Earth sciences. It may also be used in a one quarter course by selecting the appropriate material the instructor wishes to cover or emphasize. Many students taking a historical geology course will have had an introductory physical geology course. In this case, Chapter 1 can be used simply as a review of the principles and concepts of physical geology. For those students taking historical geology with no prerequisite, the text is appropriate, but the instructor may have to spend more time explaining some of the concepts and terminology in Chapter 1. Time, organic evolution, and plate tectonics are established in Chapter 1 as the integrative themes for the book. In chapters 2, 3, 4, 5, and 6, the three themes are developed more fully.

Depositional environments are covered rather superficially (perhaps little more than a summary table) in most historical geology textbooks. Chapter 4, "The Origin and Interpretation of Sedimentary Rocks," is devoted to this topic, and contains sufficient detail to be meaningful but avoids an overly detailed review more appropriate for advanced courses.

Chapter 7, "The Origin of the Universe, Solar System, and Planet Earth," provides a more comprehensive treatment of this topic than appears in other historical geology textbooks. The chapter includes discussions of the origin of the universe from the time when physics as we know it began, the theories for the origin of the solar system, the significance of meteorites, and the competing theories for the formation and early history of the Earth.

Precambrian time, fully 87 percent of all geologic time, is typically considered in a single chapter. Chapters 8 and 9 are devoted to the geologic and biologic histories of the Archean and Proterozoic eons respectively.

Chapters 10 through 17 constitute our chronologic treatment of the Phanerozoic geologic and biologic history of the Earth. These chapters are arranged such that the geologic history of an era is followed by a chapter in which the biologic history of that era is considered. We believe this format allows easier integration of life history with geologic history.

## CHAPTER ORGANIZATION

All chapters have the same organizational format. Each begins with a chapter outline followed by a prologue and an introduction giving a brief overview of the chapter contents. Special interest topics are discussed at the beginning of each chapter (prologues) and within chapters (perspectives). Each is relevant to its chapter and expands further on a particular topic. Examples include "The Evolutionary Significance of Vestigial Structures," "The California Gold Rush," "Dinosaur National Monument," "The Search for the Elusive Eve—A Controversial Theory Concerning Human Origins," "The Lysenko Affair," and "Submarine Hydrothermal Vents and the Origin of Life."

End of chapter materials begin with a concise review of important concepts and conclusions in the *Chapter Summary*. A list of *Important Terms* also appears at the end of each chapter for easy review, and a full glossary of important terms appears at the end of the text. A set of *Review Questions* to serve as a study guide is also provided. The last section of each chapter has a list of *Additional Readings*, most of which are written at a level appropriate for beginning students interested in further investigation of particular topics.

## SPECIAL FEATURES

### Figures

Many of the figures are original artwork especially designed for this book. Our paleogeographic maps are designed to illustrate clearly and accurately the geography for the various geologic periods. Many of the illustrations depicting geologic processes or events are block diagrams rather than cross sections so that students can more easily visualize the salient features of these processes and events. Full color scenes showing associations of plants and animals are based on the most current interpretations.

### Summary Tables

A number of summary tables appear in the book and especially in the chapters on geologic and biologic history. These tables are designed to give an overall perspective of geologic and biologic events for a particular time interval, and by summarizing the characteristics and times of appearance of organisms in the fossil record. The emphasis in these tables is on the geologic evolution of North America. However, global tectonic events and sea-level changes are incorporated to provide a global perspective.

### Appendixes

With a few exceptions, units of length, mass, and volume are given in metric units in the text. Appendix A is an *English-Metric Conversion Chart*. The terminology of biological classification is complex, and while minimized in the text, some is necessary. A *Classification of Organisms* appears in Appendix B.

### Glossary

It is important for students to comprehend geologic terminology if they are to understand geologic principles and concepts. Accordingly, an alphabetical and comprehensive glossary of all important terms appears at the back of the textbook.

## ANCILLARY MATERIALS

A set of 100 of the important figures and tables have been reproduced as transparency masters. These transparency masters have been designed such that they may be effectively used on overhead projectors in large lecture halls.

An instructor's manual, written by the authors, is available containing chapter outlines, sets of test questions, lecture hints and suggestions, sample transparency masters, a complete list of transparency masters and their corresponding figures and tables, and an expanded list of references.

## ACKNOWLEDGMENTS

As the authors, we are of course responsible for the organization, style, and accuracy of the text, and any mistakes, omissions, or errors are our responsibility. The finished product is the culmination of a two-year project during which time we received numerous comments and advice from many geologists who reviewed parts of the text. We wish to express our sincere appreciation to the following reviewers whose contributions were invaluable.

Leonard Alberstadt
Vanderbilt University

Tony Arnold
Florida State University

William I. Ausich
Ohio State University

Roger Bain
University of Akron

Kennard B. Bork
Denison University

Scott Brande
University of Alabama–Birmingham

Thomas W. Broadhead
University of Tennessee–Knoxville

J. Allan Cain
University of Rhode Island

Allen Cichanski
Eastern Michigan University

Robert DeMar
University of Illinois at Chicago

George F. Engelmann
University of Nebraska at Omaha

David E. Fastovsky
University of Rhode Island

Richard H. Fluegeman
Ball State University

Andrew N. Genes
University of Massachusetts–Boston

Bryan Gregor
Wright State University

Edward A. Hay
De Anza College

David R. Hickey
Lansing Community College

John Howe
Bowling Green State University

Roger L. Kaesler
University of Kansas

Patricia Kelley
University of Mississippi

James Kirkland
University of Nebraska–Lincoln

R.L. Langenheim, Jr.
University of Illinois at Urbana–Champaign

Barbara J. Leitner
University of Montevallo

David Lumsden
Memphis State University

Hulon M. Madeley
North Harris County College

William J. Neal
Grand Valley State College

Anne Noland
University of Louisville

Anthony Randazzo
University of Florida

James A. Shea
University of Wisconsin–Parkside

Hubert C. Skinner
Tulane University

Paula J. Steinker
Bowling Green State University

Thomas Straw
Western Michigan University

Jim W. Teeter
University of Akron

Michael J. Tevesz
Cleveland State University

Jamie L. Webb
California State University at Dominguez Hills

We also wish to thank Professors Emeriti Richard V. Dietrich and Wayne E. Moore, Central Michigan University, with whom we had numerous discussions regarding various aspects of the text. We are grateful for the generosity of the various agencies and individuals from many countries who provided photographs. Special thanks must go to Mrs. Martha Brian whose word processing skills and general efficiency were indispensable during the preparation of the manuscript. Jerry Westby, acquisitions editor for West Publishing Company, made many suggestions and patiently guided us through the project. The developmental editor, Elizabeth Lee, assisted during the review process and was instrumental in coordinating our efforts on the ancillary materials. Cheryl Wilms, the production editor, was especially helpful in responding to our last-minute concerns as well as guiding the book through final production.

Our families were particularly patient and encouraging throughout the two years during which most of our spare time and energy was devoted to this book. We thank them for their support and understanding.

# 1

## CHAPTER OUTLINE

PROLOGUE
INTRODUCTION
HISTORICAL GEOLOGY AND THE FORMULATION OF THEORIES
EARTH MATERIALS
The Rock Cycle
THE EARTH'S SURFACE AND INTERIOR
PLATE TECTONICS
The Rock Cycle and Plate Tectonics
FOSSILS AND EVOLUTION
THE GEOLOGIC TIME SCALE
CHAPTER SUMMARY

## PROLOGUE

*If we could film the history of the Earth from its beginning 4.6 billion years ago to the present, and speed up that film so it could be shown as a feature movie, we would see a planet undergoing remarkable change. Its geography would change as continents moved about its surface, and as a result of these movements, ocean basins would open and close. Mountain ranges would form along continental margins or when continents collided with each other. Oceanic and atmospheric circulation patterns would shift in response to the movement of continents. Massive ice-sheets would form, grow, and then melt away. Our film would show other times when there were extensive swamps or vast interior deserts.*

*If we focused our imaginary camera closer to the Earth's surface, we would view a breathtaking panorama of different organisms. We would witness the first living cells evolving from a primordial organic soup sometime between 4.6 and 3.5 billion years ago. After about 2 billion years, cells that produce their own food would appear, followed by cells with a nucleus. Next, (about 700 million years ago) the first multicelled soft-bodied organisms would evolve in the oceans, then animals with hard parts, and then animals with backbones. As we scanned the various continents, we would see an essentially bare landscape until about 420 million years ago. At that point in time the landscape would seem to come to life as plants and animals moved out of the water and onto the land. From then on, the landscape would be dominated successively by amphibians, reptiles, and mammals. In our film, humans and their history would occupy only the last second or so of the movie.*

*Three interrelated themes run through the preceding discussion. The first is that the Earth is composed of a series of moving plates whose interactions have affected the physical and biological history of the Earth. The second is that the Earth's biota has evolved. The third is that the physical and biological changes that occurred took place over long periods of time.*

*One of these themes,* plate tectonics, *provides geologists with a unifying model that explains the Earth's internal workings and accounts for many seemingly unrelated geologic phenomena. The second theme, the theory of* organic evolution, *explains how life has changed through time, based on the idea that all living organisms are the evolutionary descendants of life-forms*

# The Dynamic Earth

*that existed in the past. Finally, the third theme, the concept of* geologic time, *allows geologists to show how small, almost imperceptible changes have over vast lengths of time resulted in significant changes. For example, the formation of the Grand Canyon is the result of erosion by the Colorado River over the past 2 to 3 million years.*

*These three interrelated themes, plate tectonics, organic evolution, and geologic time, are central to our understanding and appreciation of the history of our planet. Each will be discussed in depth in later chapters.*

The Apollo 17 view of the Earth. In a view extending from the Mediterranean Sea area to the Antarctic south polar ice cap, almost the entire coastline of Africa is clearly shown. The Arabian Peninsula can be seen at the northeastern edge of Africa, while the Asian mainland is on the horizon toward the northeast. The present location of continents and ocean basins is the result of plate movement. The interaction of plates through time has affected the physical and biological history of the Earth.

## INTRODUCTION

**Geology,** the study of the Earth, has traditionally been divided into two broad areas—physical geology and historical geology. Physical geology is the study of the origin, classification, and composition of Earth materials as well as the processes that operate beneath and upon the Earth's surface. Historical geology is concerned with the origin and evolution of the Earth, its continents, atmosphere, oceans, and life.

Historical geology is, however, more than just a recitation of past events. It is the study of a dynamic planet that has changed continuously over the past 4.6 billion years—although not always at the same rate. Geologists seek not only to place events in an orderly chronologic arrangement but, more importantly, to explain how and why past events happened. It is one thing to observe in the fossil record that dinosaurs went extinct but quite another to ask why they became extinct. Geologists have long noted evidence in the rock record of numerous mountain-building episodes. However, it has only been in the last few decades that a theory has emerged to provide geologists with a comprehensive model explaining how mountain building occurs and how individual episodes of mountain building may be related to each other.

In addition to aiding in the reconstruction of the Earth's history, the basic principles of historical geology have practical application. William Smith, an English surveyor and engineer, recognized that by studying the sequences of rocks and the fossils they contained, he could predict the kinds and thicknesses of rocks that would have to be excavated in the construction of canals. The same principles Smith used in the late eighteenth and early nineteenth century are still used today in mineral and oil exploration.

## HISTORICAL GEOLOGY AND THE FORMULATION OF THEORIES

In colloquial usage **theory** commonly means a conjecture or guess, hence the widespread belief that scientific theories are little more than wild guesses. In scientific usage, however, a theory is an explanation for some natural phenomenon, and it has a large body of supporting evidence. The law of universal gravitation, for example, is a theory describing the attraction between masses—an apple and the Earth in the popularized story of Newton and his discovery. Newton's theory is a predictive statement that can be tested so its validity can be assessed (Fig. 1–1).

This testability criterion is probably the one element of the so-called scientific method upon which all agree. Obviously, if an idea cannot be tested by any conceivable means it is not scientifically useful. And, in fact, it is this aspect of science that separates it from other forms of human inquiry. For example, religious concepts are held to be absolute truths; there is no conceivable observation or test that could verify or falsify such beliefs. Scientific theories, however, can be tested and, being testable, have the potential of being supported or proven wrong. Accordingly, science must proceed with no appeal to supernatural explanations, not because such explanations are necessarily untrue, but because there is no way to investigate them.

Since scientific theories must be testable, how is it possible to formulate theories to explain events that occurred before there were human observers to record such events? In order to deal with such difficulties, geologists must make two assumptions. One assumption is that the evidence preserved in rocks, the geologic record, is an accurate record of prehistoric events. That is, rocks contain evidence that can be used to determine the processes whereby they formed. The second assumption is that the physical, chemical, and biological processes operating today have operated the same way in the past, that is, the laws of nature have not changed.

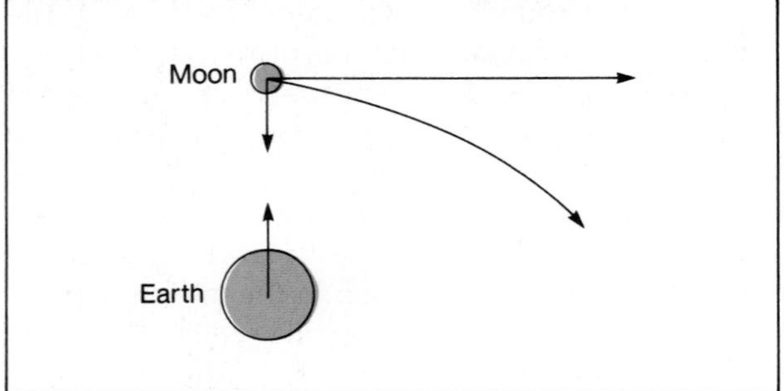


**Figure 1–1** Newton's law of universal gravitation. If there were no gravitational attraction the moon would travel in a straight path. The moon deviates from a straight path because of gravity and has a nearly circular orbit around the Earth.

The following example demonstrates that these two assumptions are justified. Figure 1–2 shows some common features observed in modern nearshore sediments and similar features in ancient sedimentary rocks. Having examined such modern features, geologists have determined that they form in sand subjected to water currents. In short, these current-formed ripple marks are the products of a physical process. The ancient features in Figure 1–2 were also produced by a physical process, and it seems reasonable to infer that that process was current action.

As a matter of fact, people commonly interpret events that they did not witness. Although they may not propose a formal theory or go through a testing process, most people would have little difficulty interpreting a shattered, charred tree in the forest as the product of a lightning strike. Geologists use similar reasoning in their investigations of Earth history. However, they do propose formal theories, some of which are discussed at greater length in other parts of this book. For example, the theory of evolution (Chapter 5) serves as the theoretical framework to explain changes in life-forms through time, and the differences and similarities among modern organisms. Plate tectonic theory is a well-supported explanation for a number of ancient and modern aspects of the Earth (Chapter 6).

## EARTH MATERIALS

The only evidence available for interpreting Earth history is preserved in the rock record. Consequently, the origin, distribution, and interrelationships among rocks are fundamental to an understanding of Earth history. We will review the three major rock groups with an emphasis on sedimentary rocks, since they comprise the majority (over 75 percent) of rocks exposed on the Earth's surface and tell us about physical conditions at the time of their deposition.

A **rock** is an aggregate of minerals. **Minerals** are naturally occurring, inorganic, crystalline solids that have definite physical and chemical properties. Minerals are composed of elements such as oxygen or aluminum,

a.

b.

**Figure 1–2** Sedimentary features of a nearshore environment and their counterparts in the rock record. a. Modern wave-formed ripples at Luddington, Michigan. b. Ripples preserved in Jurassic rocks at Gull Hill Park, South Dakota.

and elements are made up of atoms, the smallest particles of matter that still retain the characteristics of an element. More than 2,000 different minerals are known, but only about a dozen make up the bulk of the rocks in the Earth's crust (Table 1–1).

**Table 1–1** Common rock-forming minerals.

| **Relative Abundances of Minerals in the Earth's Crust** | |
|---|---|
| Plagioclase | 39% |
| Quartz | 12% |
| Orthoclase | 12% |
| Pyroxenes | 11% |
| Micas | 5% |
| Amphiboles | 5% |
| Clay Minerals | 5% |
| Olivine | 3% |
| Others (Mainly Non-Silicates) | 8% |

## The Rock Cycle

The **rock cycle** is one way of viewing the interaction between the Earth's internal and external processes (Fig. 1–3). It relates the three major rock types—igneous, sedimentary, and metamorphic—to each other; to surficial processes such as weathering, transportation, and deposition; and to internal processes such as magma generation and mountain building.

**Igneous rocks** result from the solidification of molten material called *magma*. As magma cools, minerals crystallize, and the resulting rock is characterized by interlocking mineral crystals. If magma solidifies slowly beneath the Earth's surface, the resulting rock is said to

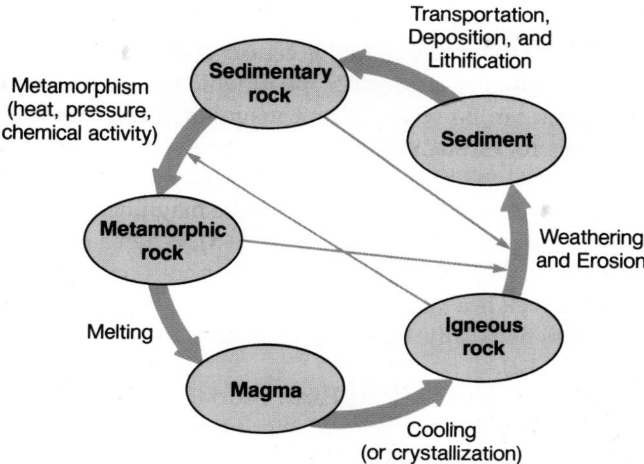


**Figure 1–3** The rock cycle.

a.

b.

**Figure 1–4** Coarse-grained texture characteristic of intrusive igneous rocks. a. Hand specimen of granite in which the individual crystals can be seen without magnification. b. The same rock seen through a polarizing microscope. Note the interlocking crystal grains of individual minerals.

a.

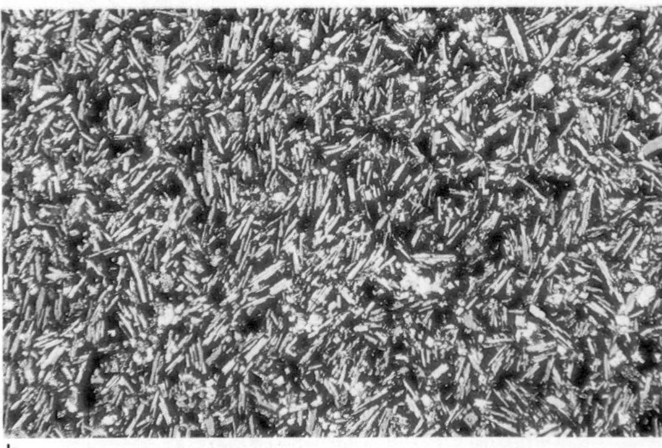
b.

**Figure 1–5** Fine-grained texture characteristic of extrusive igneous rocks. a. Hand specimen of a basalt from a lava flow in which the individual grains cannot be seen without magnification. b. The same rock seen through a polarizing microscope. The individual grains can be seen, but they are microscopic in size.

be intrusive, and its texture is coarse-grained (Fig. 1–4). A coarse-grained texture is one in which the individual minerals can be seen without magnification. Extrusive igneous rocks, such as lava, have a fine-grained texture that results from magma cooling rapidly at the Earth's surface. Fine-grained textures must be magnified for one to see the individual crystals (Fig. 1–5). When a magma cools so rapidly that crystallization cannot occur, the resulting rock is a volcanic glass (Fig. 1–6). Igneous rocks are classified on the basis of their texture and mineralogy (Table 1–2).

Rocks exposed at the Earth's surface are broken up and dissolved by various weathering processes. The resulting particles and dissolved material may be transported by wind, water, or ice and eventually deposited. The process of converting this material, called *sediment*, into a sedimentary rock is termed *lithification* (Fig. 1–7).

Two major categories of **sedimentary rocks** are recognized (Table 1–3). Detrital sedimentary rocks are composed of solid particles that were once part of other rocks and are classified according to particle size. For example, conglomerate is composed of rounded particles larger than 2 millimeters (mm), while shale is made up of clay-sized particles less than 1/256 mm in diameter (Fig. 1–8).

Chemical sedimentary rocks are derived from the dissolved material carried in solution (Fig. 1–9) and are classified on the basis of their chemical composition. Some chemical rocks are the result of precipitation of minerals by organisms or are accumulations of organic material such as shells and plants (Fig. 1–10).

Sedimentary rocks are very useful for interpreting Earth history. These rocks form at the Earth's surface;

a.

b.

**Figure 1–6** Fine-grained texture in which the magma cools so rapidly that crystallization cannot occur. Such rocks are referred to as volcanic glass. a. Obsidian. b. Pumice.

**Table 1–2** Classification of igneous rocks.

| | | Light | COLOR | | Dark | | |
|---|---|---|---|---|---|---|---|
| | Feldspar | Orthoclase Dominant | Plagioclase Dominant | | No Orthoclase Very Minor or No Plagioclase | MINERALOGY | |
| | Quartz | Quartz | Quartz Rare to Absent | | No Quartz | MINERALOGY | |
| | Accessory Minerals | Biotite Mainly | Hornblende Mainly | Augite Mainly | Augite/Hornblende Olivine | MINERALOGY | |
| TEXTURE | Fine-grained | Rhyolite | Andesite | Basalt | - - - - | Extrusive | ORIGIN |
| TEXTURE | Coarse-grained | Granite | Diorite | Gabbro | Peridotite | Intrusive | ORIGIN |

Obsidian: Glassy texture–dark colored glassy rock  Pumice: A very frothy variety of volcanic glass
Tuff: Volcanic rock formed from consolidation of volcanic ash (Pyroclastic)

**Table 1–3** Classification of sedimentary rocks.

**DETRITAL ROCKS**

| Sediment Name and Particle Size | Description | Rock Name |
|---|---|---|
| Gravel > 2 mm | Rounded rock fragments | Conglomerate |
| | Angular rock fragments | Breccia |
| Sand 1/16–2 mm | Quartz predominant | Quartz sandstone |
| | Quartz with > 25% feldspar | Arkose |
| | Dark color; quartz with considerable feldspar, clay, and rock fragments | Graywacke |
| Mud < 1/16 mm | Splits into thin layers | Shale |
| | Breaks into clumps or blocks | Mudstone |

**CHEMICAL ROCKS**

| Texture | Composition | Rock Name |
|---|---|---|
| Clastic or nonclastic | Calcite, $CaCO_3$ | Limestone |
| Clastic or nonclastic | Dolomite, $CaMg(CO_3)_2$ | Dolomite (dolostone) |
| Nonclastic | Halite, NaCl | Rock salt |
| Nonclastic | Gypsum, $CaSO_4 \cdot 2H_2O$ | Rock gypsum |
| Nonclastic | Microcrystalline quartz, $SiO_2$ | Chert |
| Nonclastic | Altered plant remains | Coal |

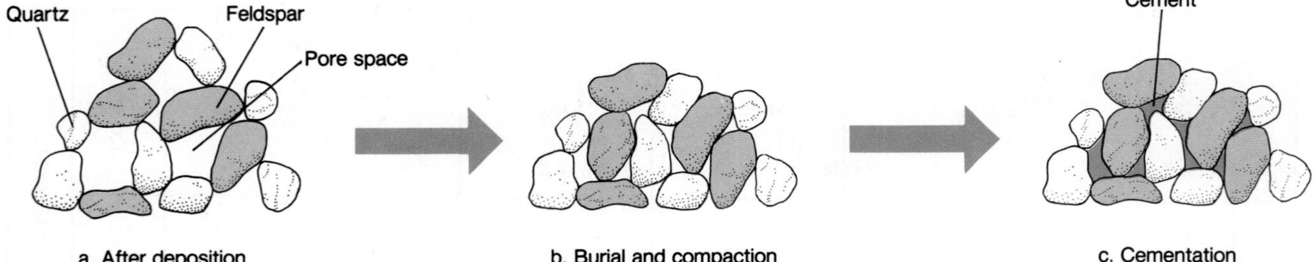


**Figure 1–7** Lithification of sand grains into a sandstone. a. Deposition of loose sand grains with large amounts of pore space between the grains. b. The weight of the overlying sediments (overburden) compacts the sand grains into a tighter arrangement and reduces pore space. c. Groundwater, moving through the pores between the sand grains precipitates cement, thus converting the sand grains into the rock sandstone.

**Figure 1–8** Common detrital sedimentary rocks. a. Shale. b. Sandstones. c. Conglomerate. d. Breccia.

**Figure 1–9** Two common chemical sedimentary rocks. a. Limestone. b. Chert.

and as layers of sediment accumulate, they record something about the environment of deposition, the type of transporting agent, and perhaps even something about the source from which the particles were derived (Fig. 1–11). If the sedimentary layers contain fossils, geologists can use these to reconstruct past environmental conditions, identify the different types of organisms that were living at that time, and also perhaps be able to date the age of the rocks and correlate them with rocks of the same age in different places (Fig. 1–12).

Some rock layers contain sedimentary structures such as cross-beds that allow geologists to reconstruct

a.

b.

**Figure 1–10** Two common chemical sedimentary rocks that are the result of organic processes. a. Fossiliferous limestone. b. Coal.

**Figure 1–11** Sedimentary rocks preserve evidence of the processes that were responsible for deposition of the sediment. This rock specimen was most likely deposited in a stream system. Evidence for this conclusion is its reddish color which is more typical of sediments deposited on land rather than in the sea, and the small ripple marks showing that the current flowed from left to right. Furthermore, it is associated with other rocks that contain desiccation cracks (mudcracks) and footprints of land animals.

**Figure 1–12** Although limestone may be deposited in lakes, most forms in warm, shallow, tropical seas. A marine origin for this limestone is indicated by the fossil crinoid (sea lily) stems, and its association with other rocks interpreted as marine.

the environment at the time the cross-beds formed (Fig. 1–13). Furthermore, some rocks by their very nature are good indicators of past environmental conditions. Tillites, for example, indicate glacial conditions (Fig. 1–14), while coals signify swampy environments (Fig. 1–10). In Chapters 3 and 4 we will discuss in more detail how sedimentary rocks and sedimentary structures can be used to interpret Earth history.

**Metamorphic rocks,** the third major rock group, result from the alteration of other rocks beneath the Earth's surface by heat, pressure, and the chemical activity of fluids (Table 1–4). Metamorphic rocks are either foliated or nonfoliated (Fig. 1–15). Foliation is the parallel alignment of minerals due to pressure. It gives the rock a layered or banded appearance. If the metamorphic rock is foliated, the type of foliation determines its name, while nonfoliated metamorphic rocks are named on the basis of their composition (Fig. 1–16).

The sequence of processes shown by the thick dark lines in the rock cycle (Fig. 1–3) illustrates how the three major rock groups are interrelated and typifies an idealized cycle. However, modifications of this idealized cycle are common. For example, an igneous rock may never be exposed at the surface. After forming, it may be subjected to heat and pressure and converted directly to metamorphic rock without ever going through the intermediate sedimentary phase. Sedimentary rocks may never be buried deeply enough to be converted to meta-

**Figure 1–13** Cross-bedding in this sandstone in Zion National Park, Utah, indicates it was once a sand dune.

**Figure 1–14** Glacial till is an unsorted mixture of many different sediment sizes deposited at the front of a melting glacier. The rock tillite (which is lithified till) provides evidence that glaciation once occurred at this location. (Photo courtesy of Wayne E. Moore, Central Michigan University)

**Table 1–4** Classification of metamorphic rocks.

**Nonfoliated Metamorphic Rocks**

| Rock Name | Parent Rock | Predominant Minerals | Characteristics |
|---|---|---|---|
| Marble<br>Dolomitic marble | Limestone<br>Dolomite | Calcite<br>Dolomite | Coarse interlocking grains of calcite or dolomite. |
| Quartzite | Quartzose sandstone | Quartz | Rock composed of interlocking small quartz grains. |

**Foliated Metamorphic Rocks**

| Type of Foliation | Rock Name | Characteristics |
|---|---|---|
| Slaty | Slate | Fine-grained rock. Splits into thin sheets. |
| Schistose | Schist | Composed of platy or elongated minerals showing parallel alignment. A wide variety of minerals occur in schists. |
| Gneissic | Gneiss | Separate layers of light and dark minerals. |

morphic rock or melted to form magma. Instead, the sedimentary rock may undergo several cycles of weathering, erosion, and deposition such that its mineral composition changes from the effects of repeated weathering, erosion, and deposition.

The concept of the rock cycle was first outlined in the late eighteenth century by James Hutton as part of his overall theory of the Earth (See Fig. 2–5 in Chapter 2). At that time, he believed that some type of internal heat source provided the mechanism for magma generation, uplift, and mountain building. Today, plate tectonic theory offers an explanation for how and where magma is generated, and how and why mountain building occurs. Following a brief discussion of the Earth's interior and plate tectonics, we will return to the interrelationship between the rock cycle and plate tectonics.

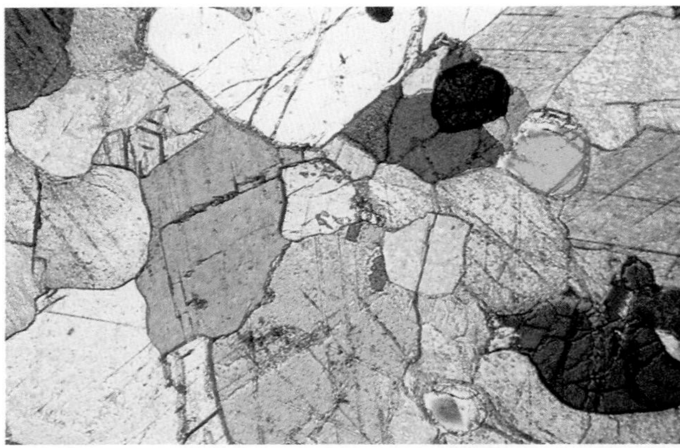
a.

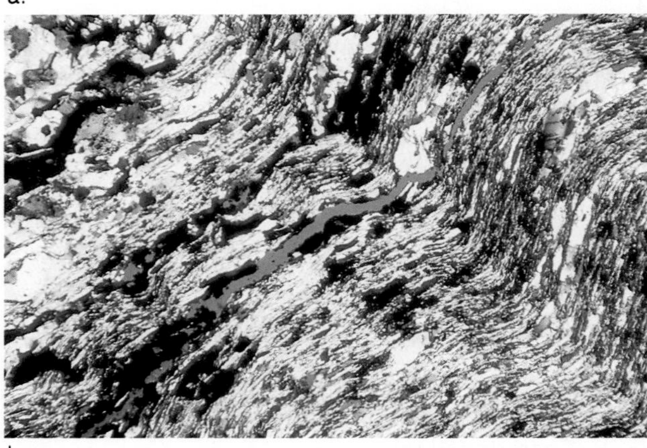
b.

**Figure 1–15** Photomicrographs of metamorphic textures. a. Nonfoliated texture of marble. b. Schist showing schistose foliation.

a.

b.

c.

**Figure 1–16** Common metamorphic rocks. a. View of schist parallel to foliation. b. View of schist perpendicular to foliation. The large, spherical minerals are garnets. c. Marble, a nonfoliated metamorphic rock, results from metamorphism of limestone, a sedimentary rock composed of the mineral calcite.

## THE EARTH'S SURFACE AND INTERIOR

Figure 1–17 is a cross section through the Earth. The **core** of the Earth is divided into two parts which are usually termed the outer core and the inner core. It is believed that the inner core is solid while the outer core is liquid. Both are apparently composed largely of iron and nickel. Above the core is the **mantle**. It comprises about 80 percent of the Earth's volume and is composed of a dense rock called *peridotite* that contains mainly the iron and magnesium silicate mineral olivine. The mantle can be divided into three distinct zones based on physical characteristics. The lower mantle is solid and forms most of the volume of the Earth's interior. Above the lower mantle, between approximately 100 to 250 kilometers (km) in depth, is the **asthenosphere**. It has the same composition as the lower mantle but behaves plastically. It is in this zone that partial melting takes place, generating magma that migrates to the Earth's surface. Above the asthenosphere is the solid upper mantle. The upper mantle and the overlying crust form the **lithosphere**. The lithosphere is broken into several individual pieces called *plates*. These plates are believed to move over the

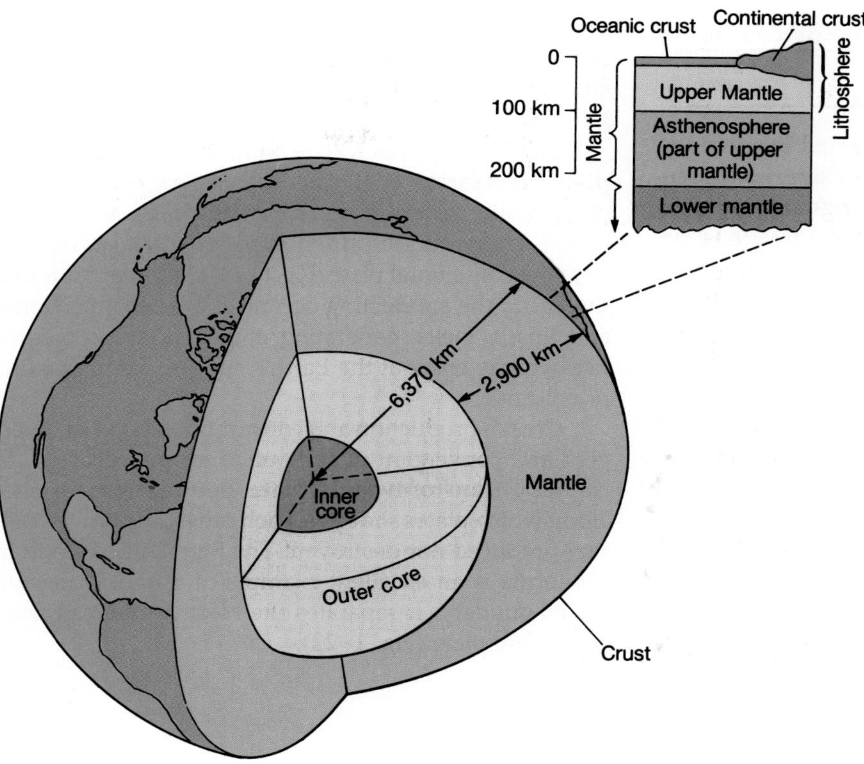


**Figure 1–17** Cross section of the Earth. The expanded section shows the relationship between the two types of crust, the lithosphere, the asthenosphere, and the mantle.

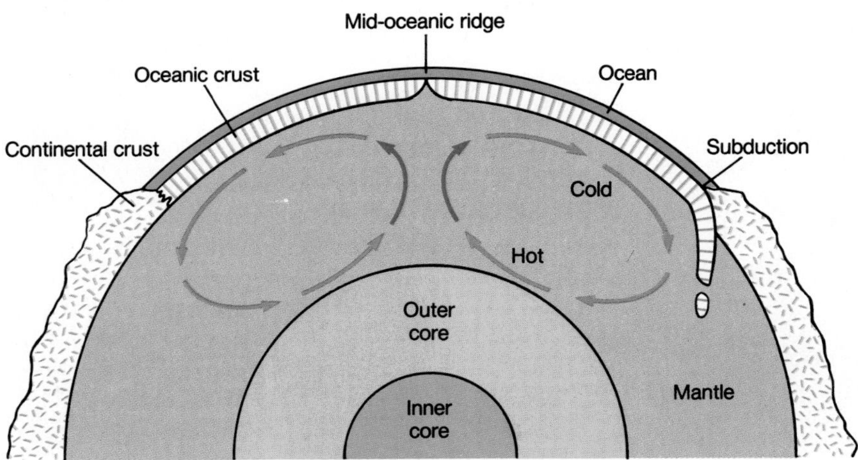


**Figure 1–18** The Earth's plates are believed to move as a result of underlying mantle convection cells.

asthenosphere as a result of underlying mantle convection cells (Fig. 1–18).

The **crust** is of two kinds. The continental crust is thick (20 to 70 km), has an average density of 2.7 grams per cubic centimeter ($g/cm^3$), and is commonly referred to as *sialic* in composition, a term that means it contains considerable silicon, oxygen, and aluminum. The oceanic crust is thin (5 to 10 km), denser than continental crust (3.0 $g/cm^3$), and is composed of the extrusive igneous rock *basalt*.

## PLATE TECTONICS

**Plate tectonic theory** is generally accepted by most geologists because it provides a comprehensive model that accounts for the Earth's internal workings and explains the relationships among many seemingly unrelated geological phenomena. Although the concept of continental movement goes back to the early 1900s, the theory of plate tectonics was first seriously proposed in the 1960s. We will present here a brief overview of plate tectonics and provide a more thorough treatment of it and its significance in reconstructing Earth history in Chapter 6.

According to plate tectonic theory, the lithosphere (crust and upper mantle) is divided into plates that move over the asthenosphere (Fig. 1–19). Zones of volcanic or earthquake activity mark most plate boundaries. Along these boundaries, plates diverge, converge, or slide past each other.

**Divergent plate boundaries** are those where magma rises to the surface from the asthenosphere as plates move apart (Fig. 1– 20). Divergent boundaries in oceanic crust are marked by mid-oceanic ridges. Beneath continental crust, divergent boundaries are recognized by rift valleys.

Plates move toward one another along **convergent plate boundaries**. When an oceanic plate meets a continental plate, for example, the oceanic plate sinks along what is known as a **subduction zone** because it is denser than the continental plate (Fig.1–21). As it descends into the Earth, the subducting oceanic plate becomes hotter and finally melts, generating magma. As this magma rises, it may erupt at the Earth's surface, forming a line of volcanoes.

Crustal production and consumption occur at divergent and convergent plate boundaries, respectively. In contrast, **transform-fault plate boundaries** are sites along which plates slide past each other, and crust is neither produced nor destroyed. The San Andreas fault in California is an excellent example of a transform-fault plate boundary. It separates the North American from the Pacific plate (Fig. 1–22).

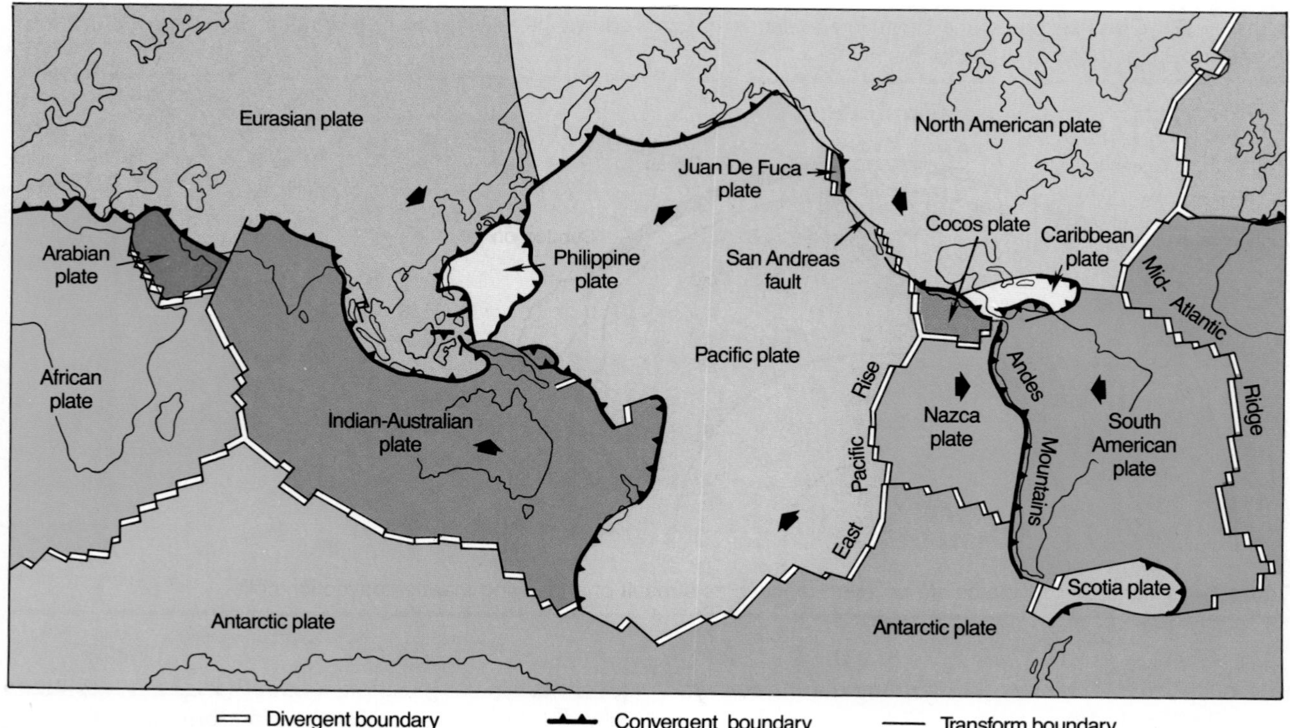


**Figure 1–19** Plates of the world. Double lines are divergent boundaries, lines with barbs are converging boundaries, and single lines are transform-fault boundaries.

While geologists can describe and explain what is happening at plate boundaries, there is still debate about a possible driving mechanism that causes plate movement. Most geologists, however, accept a mechanism involving some type of a convection system in which mantle rock is heated and flows like a very viscous liquid.

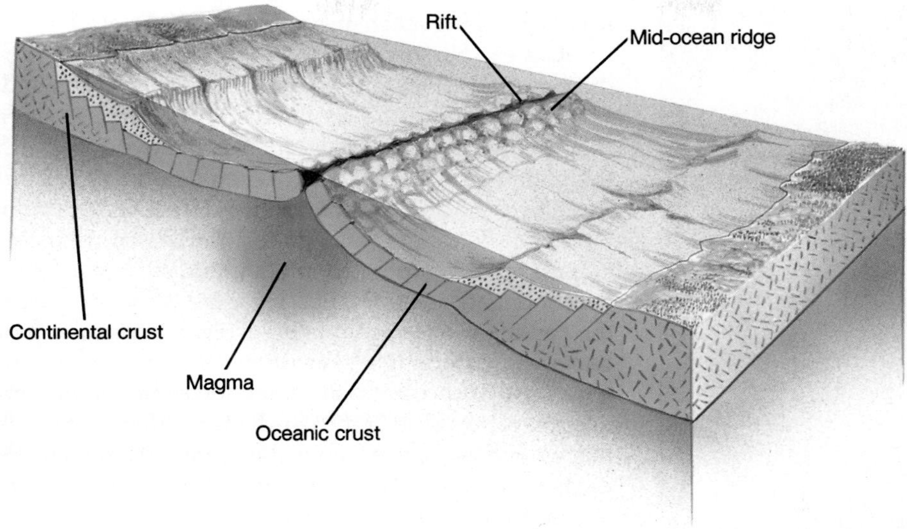


**Figure 1–20** A divergent plate boundary forms when magma wells up beneath oceanic crust and forms a mid-oceanic ridge.

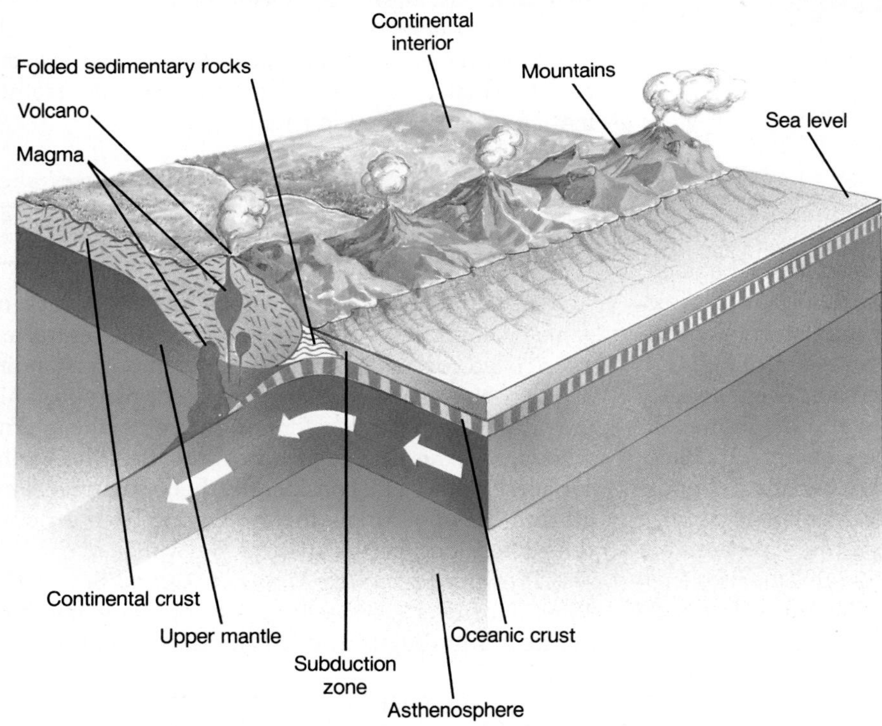


**Figure 1–21** Convergent plate boundary between oceanic crust and continental crust. Because the oceanic crust is denser, it is subducted under the continental crust.

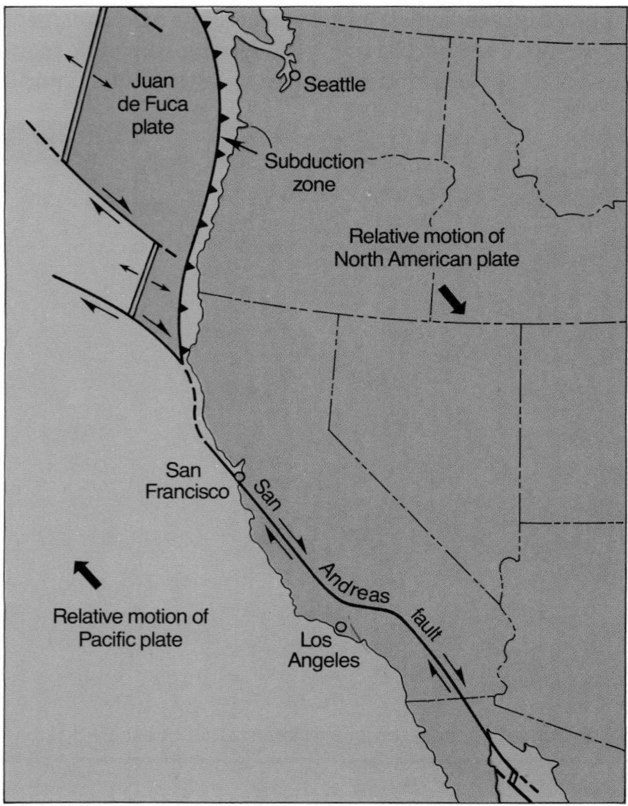


**Figure 1–22** The San Andreas fault is a transform-fault plate boundary between the Pacific plate and North American plate.

The importance of plate tectonic theory is that it allows geologists to explain many seemingly unrelated events. For example, the apparently unrelated Paleozoic mountain-building episodes in North America and Europe were actually part of a larger tectonic event involving the closing of the Iapetus Ocean and the formation of the supercontinent Pangaea (see Chapter 11). Plate tectonic theory also allows geologists to correlate periods of rapid seafloor spreading along divergent plate boundaries to times when the oceans flooded large parts of the continents (see Chapter 3). In addition, this theory helps explain some of the geographic distribution patterns of fossil and living organisms (see Chapter 5).

A revolutionary concept when it was proposed in the 1960s, plate tectonic theory has proven to have significant and far-reaching consequences. This is particularly true in regard to reconstructing and interpreting Earth history.

### The Rock Cycle and Plate Tectonics

The preceding discussion of the rock cycle was presented without reference to a mechanism responsible for recycling rock materials. Plate tectonic theory provides such a recycling mechanism (Fig. 1–23).

It is easy to visualize weathering and erosion producing sediments and transporting them by wind, water, or ice from the continents to the oceans where they accumulate, become lithified, and move along with the underlying oceanic crust. When oceanic and continental crust converge, some of the sediment and rock on the ocean bottom may be folded, faulted or metamorphosed, and incorporated into a mountain range along the continental margin. Thus, these rocks become part of the continent once again and subject to the same processes of weathering and erosion. As convergence continues and the oceanic plate is subducted under the continental plate, some of the overlying sediment and rock that were not deformed along the continental margin are carried down into the asthenosphere, where they are metamorphosed by increasing heat and pressure. Eventually, some of this metamorphic rock is carried deep enough so that it melts and produces magma. This magma then migrates upward, where it either slowly cools and crystallizes, producing coarse-grained intrusive igneous rocks, or erupts at the surface, cools rapidly, and produces fine-grained extrusive igneous rocks that are immediately attacked by the processes of weathering. Thus, the rock cycle begins anew.

## FOSSILS AND EVOLUTION

As we stated in the "Introduction", plate tectonics, organic evolution, and geologic time are three central and interrelated concepts fundamental to an understanding of Earth history. As just demonstrated, plate tectonics provides us with a model for understanding the internal workings of the Earth and how they affect the Earth's surface. Now we will consider the theory of **organic evolution**. It is not a new theory, having been seriously considered long before Charles Darwin published *On The Origin of Species by Means of Natural Selection* in 1859. However, the publication of Darwin's book marked the beginning of modern evolutionary biology.

The central claim of organic evolution is that all living organisms are the descendants of different life-forms that existed in the past. The evidence for evolution includes the way organisms are classified, similarity of

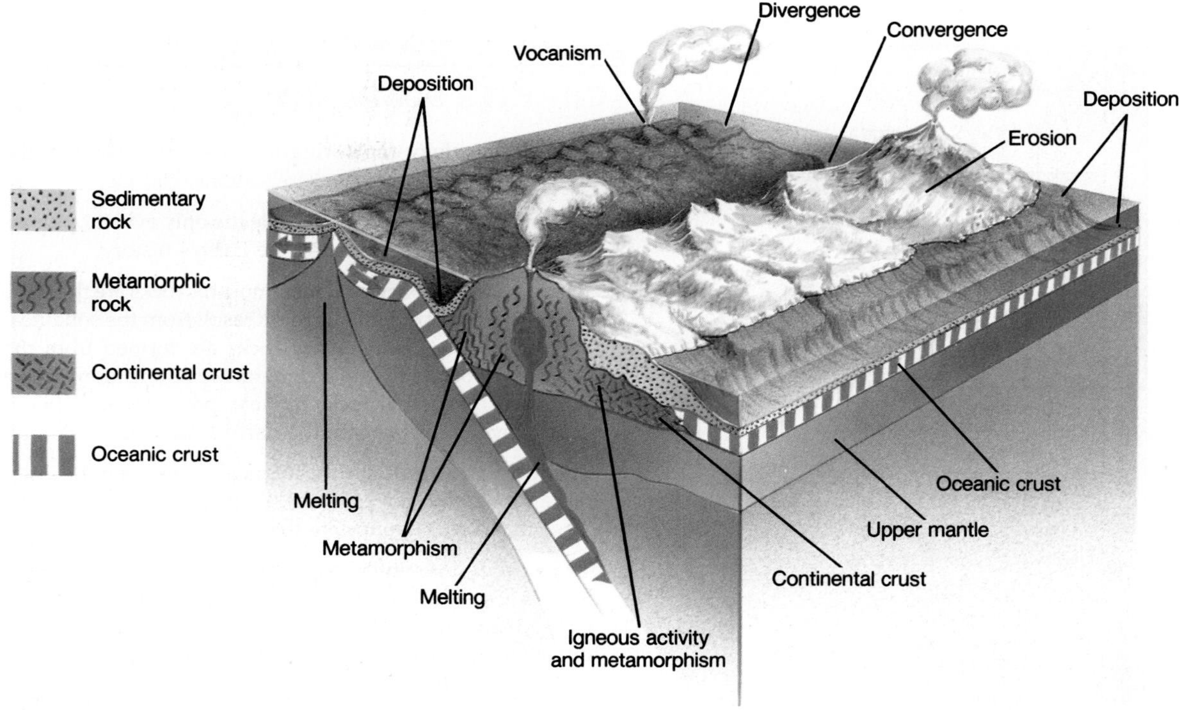


**Figure 1–23** Plate tectonics and the rock cycle. The cross section shows how the igneous, sedimentary, and metamorphic components of the rock cycle are recycled through both the continental and oceanic regions.

form and function at both the microscopic and macroscopic level, and most importantly, the fossil record. Just as the rock record allows geologists to interpret physical events and conditions in the geologic past, the fossil record provides us with a record of life in the past, and evidence that organisms have evolved through time.

**Fossils**, which are the remains or traces of once-living organisms, not only provide evidence that evolution has occurred, but also demonstrate that the Earth has a history extending beyond that recorded by humans. The succession of fossils in the rock record provides geologists with the means for age dating rocks and allowed construction of a relative geologic time scale in the 1800s.

## THE GEOLOGIC TIME SCALE

The concept of vast amounts of geologic time is central to our understanding of the evolution of the Earth and its biota. The discovery of radioactivity in 1895 provided geologists with the means to determine absolute ages of rock units in years. Prior to this, a geologic calendar was developed using the principles of relative-age dating. These principles allow geologists to place events in a sequential chronology without knowing how many years ago a particular event occurred. These principles are still used today for reconstructing the geologic history of an area and will be discussed in Chapter 2.

The development of the **geologic time scale** (Fig. 1–24) in the nineteenth century was based primarily on the succession of fossil assemblages found in the rock record. It was discovered after many years of collecting fossils from numerous locations that fossil species succeeded one another in a definite order, and that any time period in the geologic past could be recognized by its fossil content. Once this was established, geologists could identify rocks of the same age in widely separated areas on the basis of their fossil content. The construction of the geologic time scale resulted from the work of many geologists who pieced together information from numerous rock exposures and reconstructed a chronology based on changes in the Earth's biota through time.

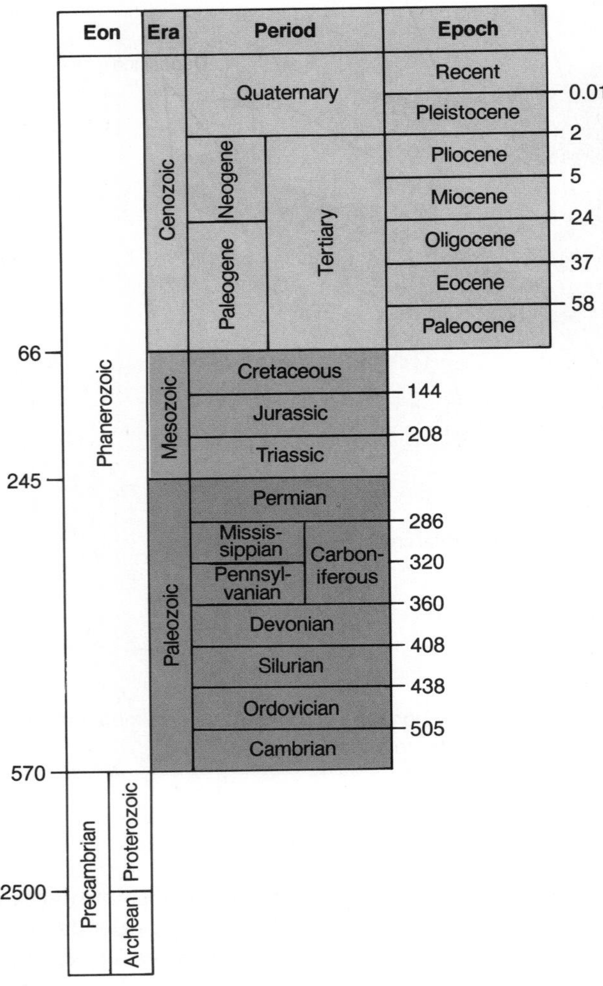

Eon
Era
Period
Epoch
Recent
Quaternary
Pleistocene
0.01
2
Pliocene
Neogene
5
Miocene
Cenozoic
24
Tertiary
Oligocene
37
Paleogene
Eocene
58
Paleocene
66
Cretaceous
144
Jurassic
Phanerozoic
Mesozoic
208
Triassic
245
Permian
286
Missis-
sippian
Carbon-
iferous
320
Pennsyl-
vanian
360
Paleozoic
Devonian
408
Silurian
438
Ordovician
505
Cambrian
570
Proterozoic
Precambrian
2500
Archean


**Figure 1–24** The geologic time scale. Numbers at the sides of the columns are ages in millions of years before present.

The division of the geologic time scale into eras, periods, and epochs was primarily based on changes in the fossil record; therefore, it was originally a relative time scale. However, with the development of various radiometric dating techniques, geologists have been able to assign absolute age dates in years to the subdivisions of the geologic time scale and thus determine how many years ago the various eras, periods, and epochs began.

We will discuss the concept of geologic time and its significance and importance as one of the three foundations of historical geology in the next chapter.

## CHAPTER SUMMARY

1. Historical geology deals with the origin and history of the Earth's continents, atmosphere, oceans, and life.
2. The rock and fossil record provide the only evidence for interpreting and reconstructing the Earth's history.
3. Igneous, sedimentary, and metamorphic rocks are the three major rock groups. Igneous rocks result from the solidification of magma. Sedimentary rocks are formed from the lithification of any type of sediment. Metamorphic rocks are produced from other rocks by heat, pressure, and chemically active fluids beneath the Earth's surface.
4. The rock cycle illustrates the interaction between the internal and external forces of the Earth and shows how the three major rock groups are interrelated.
5. The Earth is differentiated into layers. The outermost layer is the crust, which is divided into continental and oceanic crust. Below the crust is the upper mantle. The crust and upper mantle overlie the asthenosphere, a zone that behaves plastically. The asthenosphere is underlain by the lower mantle. The Earth's core is divided into an outer liquid portion and an inner solid portion.
6. The Earth's crust and upper mantle comprise the lithosphere, which is broken into a series of plates that diverge, converge, and slide past one another.
7. Plate tectonic theory provides a unifying explanation for many seemingly unrelated geologic phenomena. It also provides a mechanism for recycling rock material.
8. The central claim of the theory of organic evolution is that all living organisms evolved (descended with modification) from organisms that existed in the past.
9. Geologic time is subdivided into eras, periods, and epochs. The geologic time scale combines relative ages based on fossils with more recently determined absolute ages in years for its subdivisions.

## IMPORTANT TERMS

asthenosphere
convergent plate boundary
core
crust
divergent plate boundary
fossil
geology
geologic time scale
igneous rock
lithosphere
transform-fault plate boundary
mantle
metamorphic rock
mineral
organic evolution
plate tectonic theory
rock
rock cycle
sedimentary rock
subduction zone
theory

## REVIEW QUESTIONS

1. What are the three major rock groups?
2. Explain what a theory is and is not.
3. What are the major layers of the Earth? What is the average composition of each layer?
4. What are the three types of plate boundaries?
5. Why are the concepts of plate tectonics, organic evolution, and geologic time so important to historical geology?
6. Explain how plate tectonics and the rock cycle are related.

## ADDITIONAL READINGS

Plummer, C. C. and D. McGeary. 1988. *Physical Geology*, 4th ed. Dubuque, Iowa: Wm C. Brown Co.

Press, F. and R. Siever. 1986. *Earth*, 4th ed. New York: W. H. Freeman and Co.

Skinner, B. J. and S. C. Porter. 1987. *Physical Geology*. New York: John Wiley and Sons.

# 2

## CHAPTER OUTLINE



## PROLOGUE

*In some respects the meaning of time is defined by the methods used to measure it. In geology, we are accustomed to thinking in terms of deep time, that is, incredibly long intervals of time. Geologists talk of events occurring thousands, millions, and even billions of years ago. At the other extreme, computer engineers measure time in nanoseconds, that is, in billionths of a second. Between the two extremes most humans think of time in the more familiar terms of minutes, hours, days, and years, which relate to marked periodicities such as the change from day to night and the change of seasons.*

*Various devices and mechanisms have been used over the centuries to measure time more precisely. Many prehistoric monuments are oriented to detect the summer solstice. Sun dials were used to divide the day into measurable units. As civilizations advanced, mechanical devices were invented to measure time, the earliest being the waterclock, or clepsydra, first used by the ancient Egyptians and further developed by the Greeks and Romans. The pendulum clock was invented in the seventeenth century and provided the most accurate time-keeping for the next two and one-half centuries.*

*Major advances in accurately measuring time came with the discovery that the oscillatory motions of a vibrating solid can be used to measure time. The property of* piezoelectricity *(which literally means "pressure" electricity) in crystals like quartz is what enables them to be such accurate time-keepers. When pressure is applied to a quartz crystal, an electric current is generated. If an electric current is applied to a quartz crystal the crystal expands and compresses extremely rapidly and regularly (about 100,000 times per second). It is these vibrations from the electricity supplied by a watch's battery that provides the accuracy in a quartz watch.*

*The first clock driven by a quartz crystal was developed in 1928. Today quartz clocks and watches are commonplace. Even inexpensive quartz timepieces are extremely accurate, and precision-manufactured quartz clocks used in observatories do not gain or lose more than one second every ten years.*

# Geologic Time: Concepts and Principles

*Greater accuracy than can be obtained by a quartz timepiece is provided by atomic clocks, which now provide the basis for defining the second. This accuracy arises from the magnetic interaction of the nucleus of an atom with its electrons. Even more accurate clocks than atomic clocks are being developed based on lasers and related quantum devices.*

*It might seem that there would be no need for such precise subdivisions of time as have already been achieved, much less those that are currently being developed. Yet there are many applications where such accuracy is needed. Precision clocks are needed to measure the periods of* pulsars, *which are stars that emit their radiation in short pulses. One pulsar is so stable that it may be suitable as a standard of time over long periods.*

*Accurate measurements of time also permit measurements of things that were once thought to be constant. For example, the rotation of the Earth is now known to vary from winter to summer and from year to year. Some of the variation is regular and some is unpredictable. The reason behind such variation is still not known.*

*Time is a fascinating subject on which numerous essays and books have been written. And while we can comprehend concepts like milliseconds and understand how a quartz watch works, deep time, or geologic time, is still very difficult for most people to comprehend. As we will see in this chapter, though, geology could not have advanced as a science until the foundation of geologic time was firmly established and accepted.*

## INTRODUCTION

Time is what sets geology apart from most of the other sciences, and an appreciation of the immensity of geologic time is fundamental to an understanding of both the physical and biologic history of our planet. "So vast is the span of time recorded in the history of the Earth that it is generally distinguished from the more modest kinds of time by being called 'geologic time.'"[1]

Most people have difficulty comprehending geologic time since they tend to view time from the perspective of their own existence. Ancient history is what occurred hundreds or perhaps thousands of years ago, and yet when geologists talk in terms of ancient geologic history, they mean events that happened millions or even billions of years ago.

Geologists use two different frames of reference when speaking of geologic time. **Relative dating** involves placing geologic events in a chronologic order as determined from their position in the rock record. Relative dating will not tell us how long ago a particular event occurred, only that one event preceded another. The various principles used to determine relative dating, such as superposition and cross-cutting relationships, were discovered hundreds of years ago and since then have been used to develop the relative geologic time scale. These principles are still widely used and will be discussed later in this chapter.

**Absolute dating** results in specific dates for rock units expressed in years before the present. Radiometric dating is the most common method of obtaining absolute age dates. It is based on the natural decay of various radioactive elements that occur in trace amounts in some rocks. Not until the discovery of radioactivity near the end of the last century could ages in years before the present be accurately applied to the relative geologic time scale. Today the geologic time scale is really a dual

[1]Knopf, A. 1949. *Time and It's Mysteries.* New York University Press, Ser. 3, p. 33.

**Table 2–1** The geologic time scale. Note that the Precambrian is an informal term that comprises the Archean and Proterozoic Eons.

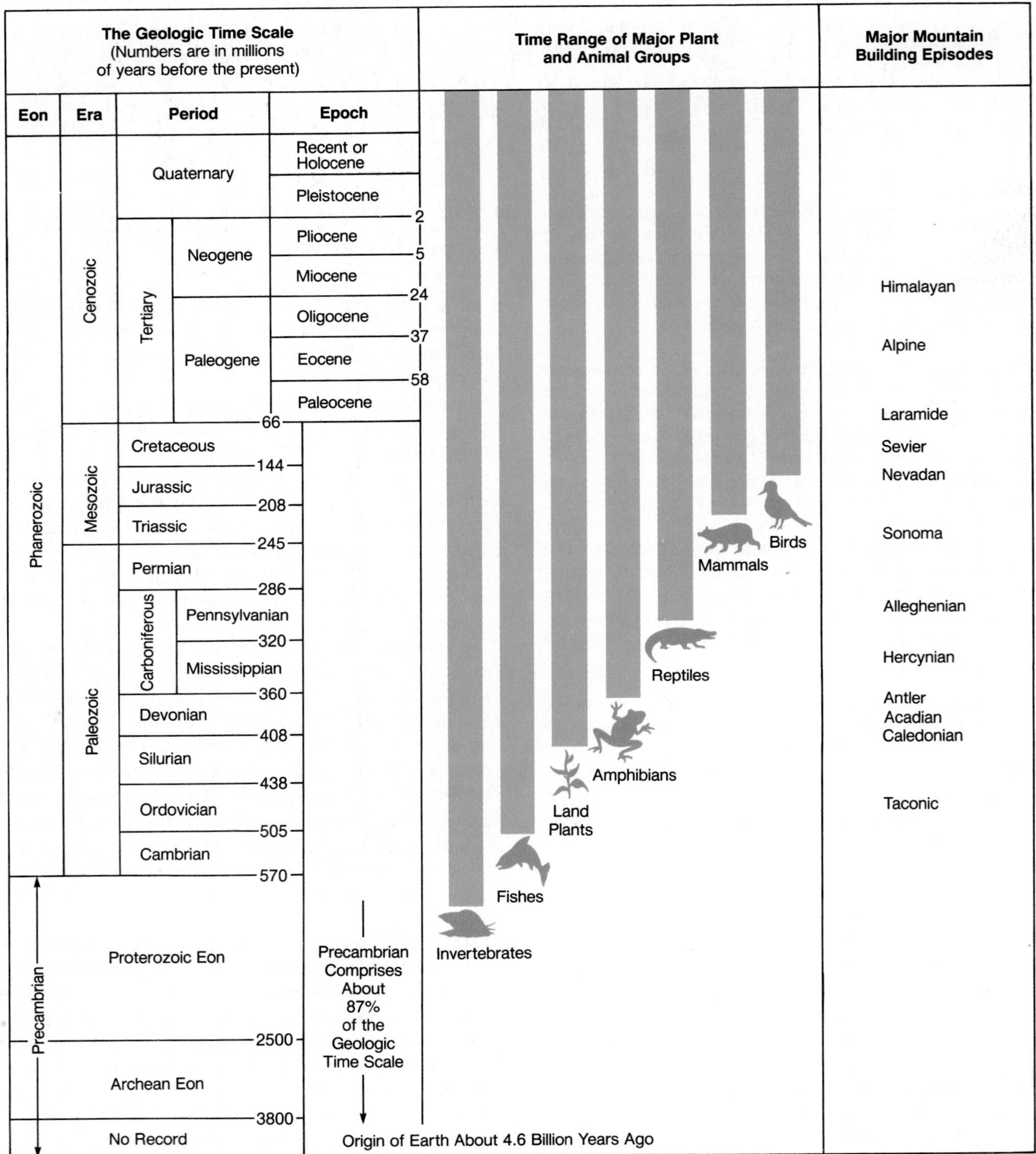

scale: a relative scale based on rock sequences fitted to an absolute scale based on radiometric dates expressed as years before present (Table 2–1).

This chapter has two objectives. The first is to establish the concept of geologic time as one of the three cornerstones of historical geology; the other two will be discussed in Chapters 5 and 6. One of the major contributions of historical geology is the realization of the immensity of time. The second objective is to introduce the major underlying principles of historical geology that allow the reconstruction of geologic history. Without an appreciation of the vastness of time and a firm understanding of basic geologic principles, the history of the Earth becomes nothing more than a recitation of seemingly unrelated facts.

## EARLY DEVELOPMENT OF THE CONCEPT OF GEOLOGIC TIME

The concept of geologic time and the way it is measured have changed over the years. For example, early Christian theologians were mostly responsible for formulating the idea that time is linear rather than circular. St. Augustine of Hippo (354–430 A.D.) stated that the Crucifixion was a unique event from which all other events could be measured, thus establishing the idea of the B.C. and A.D. time scale. This prompted many religious scholars and clerics to try to establish the date of creation by analyzing historical records and the genealogies found in Scripture.

A famous and influential Christian scholar, James Ussher (1581–1665), Archbishop of Armagh, Ireland, is generally credited as being the first to calculate the age of the Earth based on genealogies described in Genesis. Ussher stated in 1650 that the Earth was created on October 22, 4004 B.C. This date was later reproduced in many editions of the Bible and incorporated into the dogma of the Christian church. For nearly a century thereafter, it was considered heresy to assume that the Earth and all its features were more than about 6,000 years old. Thus, the idea of a very young Earth provided the basis for most chronologies of Earth history prior to the eighteenth century.

During the eighteenth and nineteenth centuries several attempts were made to determine the age of the Earth based on scientific evidence rather than on revelation. For example, the French zoologist Georges Louis de Buffon (1707–1788), assumed the Earth formed as a molten ball and gradually cooled to its present condition. He heated iron balls of various diameters to their melting point, measured how long it took for the iron balls to cool to the surrounding temperature, and extrapolated his results to account for the cooling rate of the Earth. From these experiments he determined that the Earth required at least 96,000 years to cool to its present temperature. Later he revised this estimate downward to 75,000 years based on experiments using mixtures of metallic and nonmetallic substances.

This age contrasted sharply with the much younger age based on Scripture. However, during Buffon's time it was very risky to publish views that seemed to contradict Church doctrine, so to escape censure, Buffon acknowledged his theory as pure philosophical speculation; his close connection with the French Court provided him with additional protection.

Other equally ingenious attempts at calculating the age of the Earth were tried. For example, geologists reasoned that if rates of deposition could be determined for various sedimentary rocks, they might be able to calculate the time required to deposit a given thickness of rock. They could then extrapolate how old the Earth was from the total thickness of sedimentary rock in the Earth's crust. However, even for the same type of rock, rates of deposition vary. Furthermore, it is impossible to estimate how much rock has been removed by erosion, or how much a rock sequence has been reduced by compaction. As a result of these uncertainties, estimates ranged widely—from less than a million years to over a billion years.

Another attempt at determining the age of the Earth involved calculating the age of the oceans. If the ocean basins were filled very soon after the origin of the planet, then they would be only slightly younger than the Earth itself. The best known calculations for the oceans' age were made by the Irish geologist John Joly in 1899. He reasoned that the Earth's ocean waters were originally fresh and their present salinity was the result of dissolved salt being carried into the ocean basins by rivers. By measuring the present amount of salt in the world's rivers, and knowing the volume of ocean water and its salinity, Joly calculated it would have taken about 90 million years for the oceans to reach their present salinity level. This age was too young by a factor of 50, mainly because there was no way to account for recycled salt, continental salt deposits, or salt incorporated into clay minerals deposited on the sea floors.

Although each of these historical methods yielded ages considerably younger than we now know the Earth to be, such attempts did change the way naturalists perceived the age of the Earth and represented a significant milestone in our understanding of Earth history.

## FUNDAMENTAL GEOLOGIC PRINCIPLES

The seventeenth century was an important time in the development of geology as a science because of the widely circulated writings of the Danish anatomist Nicolas Steno (1638-1686). Steno observed the present-day processes of sediment transport and deposition during stream flooding near Florence, Italy. These observations allowed him to determine the manner in which sedimentary rock layers formed and served as the basis for three fundamental principles of geology. While these principles may now seem self-evident, their discovery was an important scientific achievement and absolutely essential for interpreting geologic history.

### Steno's Principles

Steno noted that flooding streams spread out across their floodplains and deposited layers of sediment that buried floodplain-dwelling organisms. Subsequent flooding events produced new layers of sediments that were deposited or superposed over previous deposits. When lithified, these sediment layers became sedimentary rock. Thus, in a vertical succession of sedimentary rock layers, the oldest layer is at the bottom and the youngest layer is at the top. This **principle of superposition** is the basis for relative age determinations of strata and their contained fossils.

Since sedimentary particles settle from water under the influence of gravity, Steno reasoned that sediment layers were deposited essentially horizontally (the **principle of original horizontality**). Therefore, a sequence of sedimentary rock layers that is steeply inclined from the horizontal must have been tilted after deposition and lithification.

Steno's third principle, the **principle of lateral continuity**, states that when sediment layers are deposited, they extend laterally in all directions until they thin and pinch out, or terminate, against the edge of the depositional basin. The Grand Canyon in Arizona beautifully illustrates Steno's three principles (Fig. 2–1).

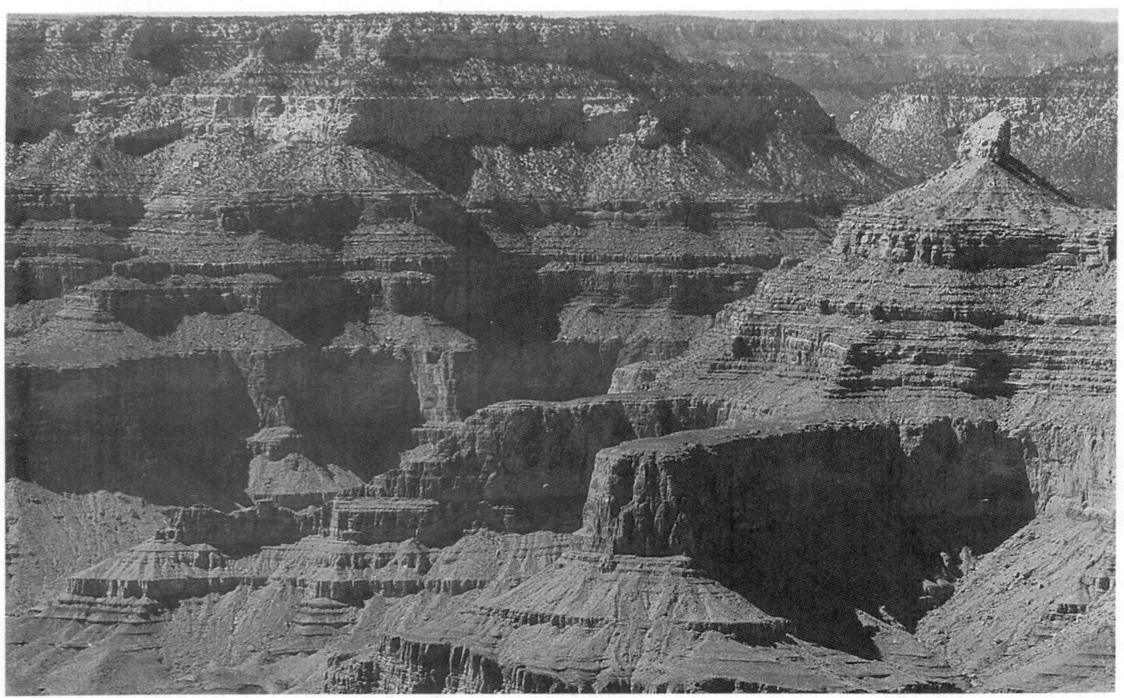

**Figure 2–1** The Grand Canyon of Arizona illustrates three of the four fundamental principles of historical geology. The sedimentary rocks of the Grand Canyon were originally deposited horizontally in a variety of marine and terrestrial environments (principle of original horizontality). The oldest rocks are therefore at the bottom of the canyon and the youngest rocks are at the top, forming the rim (principle of superposition). The exposed rock layers extend laterally for some distance (principle of lateral continuity).

## Perspective 2-1

# Leonardo da Vinci—Geologist

Leonardo da Vinci (1452-1519) is known to most people as the painter of such masterpieces as *The Last Supper* and the *Mona Lisa*. Many people are also aware that da Vinci was a great engineer, scientist, and inventor (Fig. 1). What may not be as well known are his contributions to geology. His powers of observation, imagination, and sound reasoning led him far beyond the beliefs of his time. His writings abound with references to geology, and his sketches show a sound fundamental knowledge of the geology of landscapes.

Among his contributions was the recognition that fossils were the remains of once-living organisms and that they indicated changes in the distribution of land and sea. Da Vinci correctly reasoned that fossil shells found in the hills many miles from the sea could not have traveled that distance in the 40 days and nights of the Deluge. Furthermore, they could not have been deposited there by the rising waters of the Noachian Flood since they do not occur with any other debris that surely would have been carried by a flood.

Much of da Vinci's professional engineering had to do with canals and rivers, and from these studies he realized that most valleys were eroded by the rivers occupying them. He also recognized that the material carried by streams to the sea was eventually cemented into sedimentary rock and later uplifted to form mountains.

From his study of river processes, da Vinci came to the conclusion that the Earth must be older than 6,000 years. In fact, based on observations of how slowly river deposits form, he concluded that the deposits of the Po River required 200,000 years to accumulate, and that amount of time was only a small portion of all geologic time.

In terms of his understanding of geologic concepts and principles, Leonardo da Vinci was indeed a Renaissance man.

**Figure 1** Leonardo da Vinci, inventor, geologist, and artist.

## ESTABLISHMENT OF GEOLOGY AS A SCIENCE—THE TRIUMPH OF UNIFORMITARIANISM OVER NEPTUNISM AND CATASTROPHISM

Steno's principles were significant contributions to early geologic thought, but the prevailing concepts of Earth history continued to be those that could be easily reconciled with a literal interpretation of Scripture. Two of these ideas, neptunism and catastrophism, were particularly appealing and accepted by many naturalists. In the final analysis, however, another concept, uniformitarianism, became the underlying philosophy of geology because it provided a better explanation for observed geologic phenomena than either neptunism or catastrophism.

### Neptunism and Catastrophism

**Neptunism** was a concept proposed in 1787 by the German professor of mineralogy Abraham Gottlob Werner (1749–1817). Although Werner was an excellent mineralogist, he is most remembered for his incorrect interpretation of Earth history. He believed that all rocks, including granite and basalt, were precipitated in an orderly sequence from a primeval, worldwide ocean (Table 2–2). The oldest, or Primitive rocks, were all unfossiliferous igneous and metamorphic rocks that supposedly formed entirely by precipitation from seawater. Since they are found in the cores of mountain ranges, Werner reasoned they must have been the earliest rocks precipitated from the sea (Fig. 2–2). As the ocean waters subsided, the Transition rocks were deposited. These rocks contain fossils and include the first detrital and chemical rocks. Werner believed that the fossils in the Transition rocks marked the time the Earth became suitable for habitation. The next rocks in Werner's sequence, the Secondary rocks, included a variety of fossiliferous detrital and chemical rocks, as well as basalt layers. The youngest, or Alluvial rocks, consisted of unconsolidated sediments. A fifth category of rocks included volcanics; these rocks, such as pumice and lava, were still being produced by volcanoes but were not considered important by Werner.

**Table 2–2** Standard geologic column around 1800. Werner's subdivision of the rocks of the Earth's crust in which he believed that all rocks, including granite and basalt, were precipitated from a universal ocean.

| | |
|---|---|
| Alluvial rocks | Unconsolidated sediments. |
| Secondary rocks | Various fossiliferous, detrital, and chemical rocks such as sandstones, limestones, coal as well as basalts. |
| Transition rocks | Chemical and detrital rocks including fossiliferous rocks. |
| Primitive rocks | Oldest rocks of the Earth (igneous and metamorphic). |

Werner's subdivision of the Earth's crust by supposed relative age attracted a large following in the late 1700s and became almost universally accepted as the standard geologic column. Two factors account for this. First, Werner's charismatic personality, enthusiasm for geology, and captivating lectures popularized the concept. Equally important was the fact that neptunism included a worldwide ocean that could easily be reconciled with the biblical deluge.

However, in spite of Werner's personality and arguments, his neptunian theory failed to explain what happened to the tremendous amount of water that once covered the Earth. An even greater problem was Werner's insistence that all igneous rocks were precipitated from seawater. To Werner, all volcanoes were recent and had no importance in Earth history. He believed that volcanic eruptions were the result of combustion of buried coal seams and therefore that volcanoes could not have occurred until after the deposition of coals, which were categorized among his Secondary rocks. It was this failure to correctly recognize an igneous origin for basalt that led to the downfall of neptunism.

From the late eighteenth century to the mid-nineteenth century the concept of **catastrophism**, proposed by the French zoologist Baron Georges Cuvier (1769–1832), dominated European geologic thinking. Cuvier explained the physical and biologic history of the Earth as resulting from a series of sudden widespread catastrophes. Each catastrophe accounted for significant and rapid changes in the Earth, exterminating existing life in the affected area. Following a catastrophe, new organisms were either created or migrated in from elsewhere.

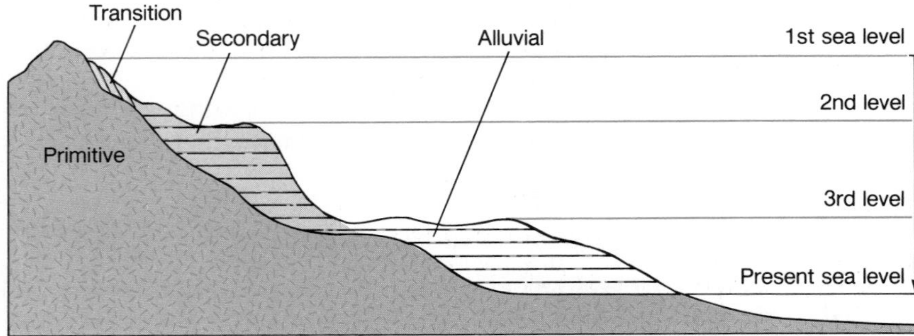


**Figure 2–2** According to neptunists, all strata were deposited from a retreating worldwide ocean. As shown here, superposition seems to be defied since older strata overlie younger strata. A. G. Werner, a prominent neptunist, believed that older strata dip away from the mountains and lie beneath younger strata.

**Figure 2–3** James Hutton, who originated the principle of uniformitarianism and through his writings profoundly influenced the course of geologic thinking.

**Figure 2–4** The principle of cross-cutting relationships. a. A dark colored dike has been intruded into older light colored granite, north shore of Lake Superior, Ontario, Canada. b. A fault, shown by the zone of rubble directly above the geologist, cutting through strata in the Bighorn Mountains, Wyoming.

According to Cuvier's concept, six major catastrophes had occurred in the past. These conveniently corresponded to the six days of biblical creation. Furthermore, the last catastrophe was taken to be the biblical deluge, so catastrophism had wide appeal, especially among theologians. While not all catastrophists accepted a 6,000-year age for the Earth, catastrophism as expounded by Cuvier was consistent with such a young age.

Eventually, both neptunism and catastrophism were abandoned as untenable hypotheses because their basic assumptions could not be supported by field evidence. The simplistic sequence of rocks predicted by neptunism (Table 2–2) was simply contradicted by field observations from many different areas. Moreover, basalt was shown to be of igneous origin, and subsequent discoveries of volcanic rocks interbedded with secondary and primitive deposits proved that volcanic activity had occurred throughout Earth history. As more and more field observations from widely separated areas were made, naturalists began to realize that far more than six catastrophes were needed to account for Earth history. With the demise of neptunism and catastrophism, the principle of uniformitarianism, advocated by James Hutton and Charles Lyell, became the guiding philosophy of geology.

## Uniformitarianism

James Hutton (1726–1797) is considered by many to be the father of historical geology (Fig. 2–3). His detailed studies and observations of rock exposures and present geologic processes served as the basis for two important geologic principles. The **principle of cross-cutting relationships** holds that an igneous intrusion or fault must be younger than the rocks it intrudes or cuts (Fig. 2–4). This principle is very important in relative dating of geologic events and in reconstructing Earth history.

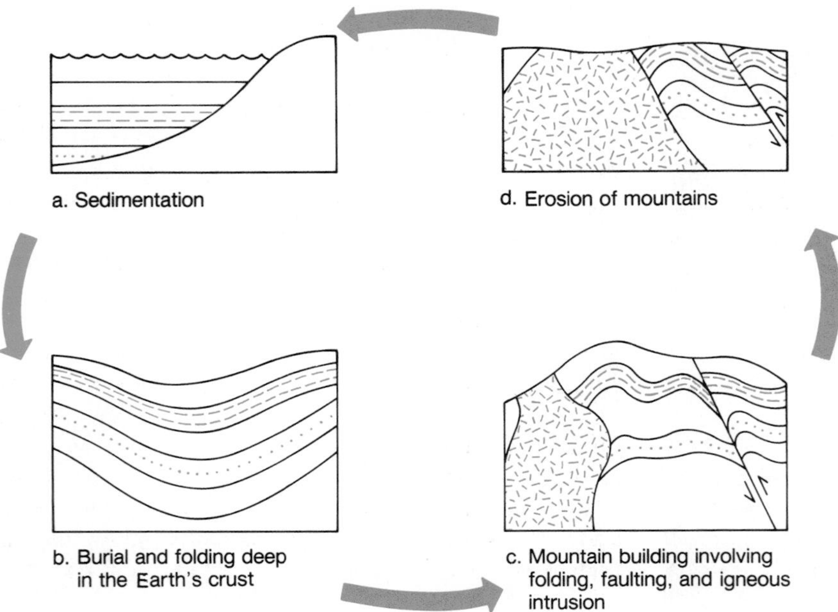


**Figure 2–5** The geologic cycle, as deducted by James Hutton, in which sediments are deposited in the sea (a), deeply buried in the Earth's crust (b), pushed up by thermal expansion from the Earth's hot interior (c), and then eroded by such processes as wind, wave and stream action (d).

**Figure 2–6** East West cross section of Northern Granite, Isle of Arran, Strathclyde, Scotland. Here granite of Tertiary age (forming the central part of the mountain) has intruded and pushed up the surrounding strata of Precambrian and younger age. The granite and surrounding strata have in turn been cut by dikes. This is one of several examples Hutton offered as proof that granite is an intrusive igneous rock. It also illustrates his concept of the geologic cycle in which rocks are pushed up by thermal expansion from the hot interior of the Earth, and the principle of cross-cutting relationships. (From James Hutton's *Theory of the Earth: the lost drawings.*)

Hutton observed the processes of wave action, erosion by running water, and sediment transport, and concluded that given enough time these processes could account for the geologic features in his native Scotland. He believed that "the past history of our globe must be explained by what can be seen to be happening now." This assumption that present-day processes have operated throughout geologic time was the basis for the **principle of uniformitarianism.**

While Hutton developed a comprehensive theory of uniformitarian geology, it was Charles Lyell (1797–1875) who became the principal advocate and interpreter of uniformitarianism. The term itself was coined by William Whewell in 1832.

Hutton viewed Earth history as being cyclical. He believed that erosional processes wore down the continents, the eroded sediment was deposited in the sea, and uplift of the seafloor created new continents, thus completing a cycle (Fig. 2–5). His idea of the mechanism for uplift was thermal expansion from the Earth's hot interior. Hutton's field observations, and experiments performed by his contemporaries involving the melting of basalt samples, convinced him that igneous rocks were the result of cooling magmas. This interpretation of the origin of igneous rocks, called *plutonism,* eventually displaced the neptunian view that igneous rocks precipitated from seawater (Fig. 2–6).

Hutton also recognized the importance of unconformities in his cyclical view of Earth history. At Siccar Point, Scotland, he observed steeply inclined rocks that had been eroded and covered by flat-lying younger rocks (Fig. 2–7). It was clear to him that severe upheavals had tilted the lower rocks and formed mountains. These were then worn away and covered by younger, flat-lying rocks. The erosional surface meant there was a gap in the rock record, and the rocks above and below this surface provided evidence that both mountain building and erosion had occurred. Although Hutton did not use

**Figure 2–7** Angular unconformity at Siccar Point, Scotland. It was here that the significance of an unconformity was first realized by James Hutton in 1788. (Photo courtesy of Edward A. (Sandy) Hay, De Anza College)

**Figure 2–8** Portrait of Sir Charles Lyell, author of *Principles of Geology,* in which uniformitarianism was established as the guiding philosophy of geology.

the word "unconformity," he was the first to understand and explain the significance of such gaps in the rock record.

Hutton was also instrumental in establishing the concept that geologic processes had vast amounts of time in which to operate. Since Hutton relied on known processes to account for Earth history, he concluded that the Earth must be very old. However, he estimated neither how old the Earth was nor how long it took to complete a cycle of erosion, deposition, and uplift. He merely allowed that "we find no vestige of a beginning, and no prospect of an end," which was in keeping with a cyclical view of Earth history.

Unfortunately, Hutton was not a particularly good writer, so his ideas were not widely disseminated or accepted. In fact, neptunism and catastrophism continued as the dominant geologic concepts well into the 1800s. However, in 1830 Charles Lyell (Fig. 2–8) published a landmark book, *Principles of Geology,* in which he championed Hutton's concept of uniformitarianism.

Lyell clearly recognized that small, imperceptible changes brought about by present-day processes could, over long periods of time, have tremendous cumulative effects. Not only did Lyell effectively reintroduce and establish the concept of unlimited geologic time, but he also discredited catastrophism as a viable explanation of geologic phenomena and firmly established uniformitarianism as the guiding philosophy of geology. The recognition of virtually limitless time was also instrumental in the acceptance of Darwin's theory of evolution (Chapter 5).

Perhaps because uniformitarianism is such a general concept, it has been interpreted by scientists in a number of different ways. Lyell's concept of uniformitarianism embodied the idea of a steady-state Earth in which present-day processes have operated at the same rate in the past as they do today, and the Earth has remained the same overall through time. For example, according to Lyell, the frequency of earthquakes and volcanic eruptions for any given period of time in the past was the same as it is today. If the climate in one part of the world got hot, it would have to get cold somewhere else so that overall the climate could remain the same. By such reasoning, Lyell claimed that conditions for the Earth as a whole had remained essentially constant and unchanging through time. However, if this were truly the case, the Earth would be a perpetual motion machine, an impossibility according to the laws of physics.

## Modern View of Uniformitarianism

Geologists today assume that the principles, or laws, of nature are constant but the rates and intensities of change have varied through time. For example, volcanic activity was more intense in North America during the Miocene Epoch than it is today, while glaciation has been more prevalent in the last 3 million years than in the previous 300 million years. Since the rates and intensities of various processes have varied through time, some geologists prefer to use the term *actualism* rather than uniformitarianism in order to remove the idea of "uniformity" from the concept. Most geologists though, still use the term uniformitarianism, since it indicates that even though the rates and intensities of change have varied in the past, the laws of nature have remained the same.

Uniformitarianism is a powerful concept that allows us through analogy and inductive reasoning to use present-day processes as the basis for interpreting the past and predicting potential future events. It does not eliminate the occurrence of occasional, sudden, short-term events such as volcanic eruptions, earthquakes, floods, or even meteorite impacts as forces that shape our modern world. In fact, some geologists view the history of the Earth as a series of such short-term, or punctuated, events; and this view is certainly in keeping with the modern principle of uniformitarianism. The Earth is in a state of dynamic change and has been since it formed. While the rates of change may have varied in the past, the natural laws governing the processes have not.

## LORD KELVIN AND A CRISIS IN GEOLOGY

Lord Kelvin (1824–1907), an English physicist, claimed in a paper written in 1866 to have destroyed the uniformitarian foundation on which Huttonian-Lyellian geology was based. Kelvin did not accept Lyell's strict uniformitarianism, in which chemical reactions in the Earth's interior were supposed to continually produce heat, allowing for a steady-state Earth. Kelvin rejected this idea as perpetual motion—impossible according to the known laws of physics—and accepted instead the assumption that the Earth was originally molten.

Kelvin knew from deep mines in Europe that the Earth's temperature increases with depth, and he reasoned that the Earth is losing heat from its interior. By knowing the melting temperature of the Earth's rocks, the size of the Earth, and the rate of heat loss, Kelvin was able to calculate back to the time when the Earth was entirely molten. From these calculations, he concluded that the Earth could not be older than 400 million years or younger than 20 million years. This wide discrepancy in age reflected uncertainties in average temperature increases with depth and the various melting points of the Earth's constituent materials.

After finally establishing that the Earth was very old, and showing how present-day processes can be extrapolated over long periods of time to explain geologic features, geologists were in a quandary. They had to either accept Kelvin's dates and squeeze events into a shorter time frame, or they had to abandon the concept of seemingly limitless time that was the underpinning of Huttonian-Lyellian geology and one of the foundations of Darwinian evolution. Some geologists objected to such a young age for the Earth, but their objections seem to have been based more on faith rather than hard facts. Kelvin's quantitative measurements and arguments seemed flawless and unassailable to many scientists.

Kelvin's reasoning and calculations were sound, but his basic premises were false, thereby invalidating his conclusions. Kelvin was unaware that the Earth has an internal heat source, radioactivity, that has allowed it to maintain a fairly constant temperature through time.[2] Kelvin's 40-year campaign for a young Earth ended with the discovery of radioactivity near the end of the nineteenth century. His "unassailable calculations" were no longer valid and his proof for a geologically young Earth collapsed. Kelvin's theory, just like neptunism, catastrophism, and a worldwide flood, became an interesting footnote in the history of geology.

While the discovery of radioactivity destroyed Kelvin's arguments, it provided geologists with a clock that could measure the Earth's age and validate what geologists had been saying—that the Earth was indeed very old! Less than ten years after the discovery that radium generated heat, radiometric calculations were providing ages of billions of years for some of Earth's oldest rocks.

## ABSOLUTE GEOLOGIC TIME

Our preceding discussions have largely concerned the concept of vast amounts of geologic time and the formulation of principles used to determine relative ages. Ironically, the very process that invalidated Lord Kelvin's calculations of absolute dates, serves as the basis for the same calculations. Some of the 92 naturally occurring elements are radioactive and spontaneously decay to other elements, releasing energy in the process. The discovery in 1903 by Pierre and Marie Curie that radioactive decay produces heat as a by-product meant that geologists no longer needed to assume the Earth had cooled from a molten state. The geologic importance of radioactive decay was soon recognized, and in 1907 Bertram B. Boltwood, a chemist-physicist at Yale University, proposed that lead is an end product of the radioactive decay of uranium, and calculated the Earth's age as somewhere between 400 million and 2.2 billion years.

Geologists now had a mechanism for explaining the internal heat of the Earth that did not rely on residual cooling from a molten origin. With these discoveries, geologists and paleontologists had the long time periods they needed for a Huttonian-Lyellian-Darwinian view of the Earth and could ignore Lord Kelvin's discredited calculations. Furthermore, geologists now had a powerful tool to accurately date geologic events.

[2]Actually, the Earth's temperature has decreased through time, but at a rate considerably less than the one that would lend any credence to Kelvin's calculations (see p. 192, Chapter 8).

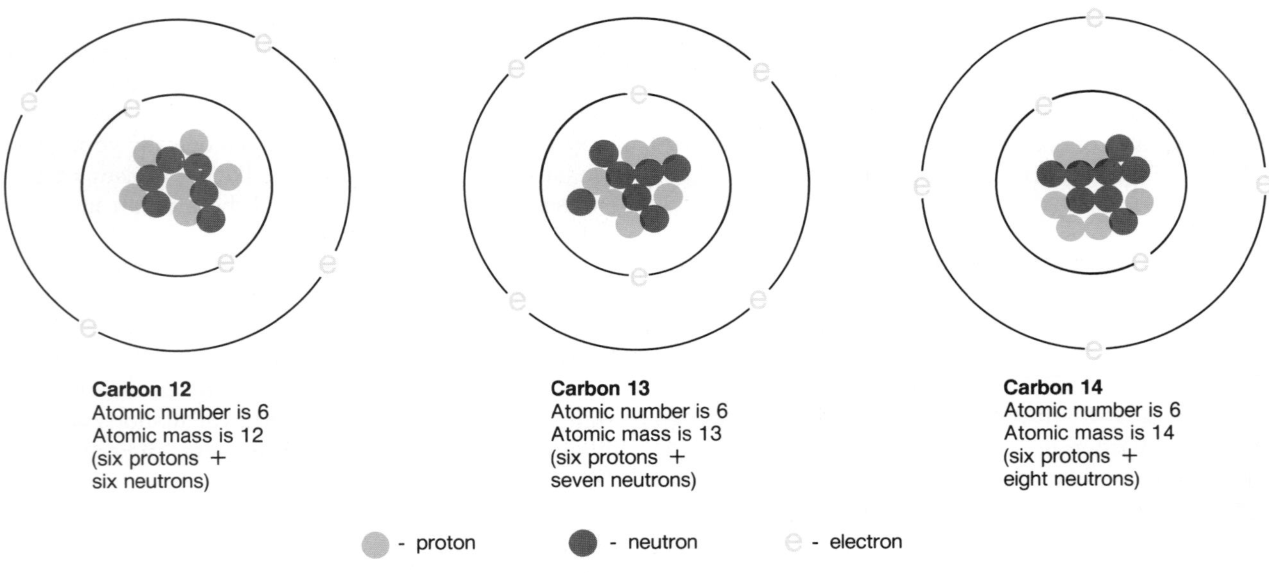


**Figure 2–9** The atomic structure, atomic number, atomic mass, and isotopes of carbon. Carbon has an atomic number of 6 and an atomic mass of 12, 13, or 14, depending on the number of neutrons. Variable forms of the same element are called isotopes.

## Atoms and Isotopes

The nucleus of an atom is composed of protons (positively charged particles) and neutrons (neutral particles). The number of protons defines its atomic number and determines its properties and characteristics. Therefore, each change in the number of protons (and hence the atomic number) forms a new element with a different atomic structure and thus different physical and chemical properties (Fig. 2–9).

The combined number of protons and neutrons in an atom closely approximates its atomic mass. Carbon, for example, has an atomic number of 6 but any of three different atomic masses, 12, 13, or 14, depending on the number of neutrons present. These variable forms of the same element are called **isotopes** (Fig. 2–9). Most isotopes, such as carbon 12 and 13 for example, are stable, but others, such as carbon 14, are unstable. It is the unstable isotopes that are radioactive, and their decay rate is what geologists measure to determine absolute ages.

## Radioactive Decay and Half-Lives

**Radioactive decay** is the process whereby an unstable atomic nucleus is spontaneously transformed into another atomic nucleus. Three types of radioactive decay, all of which result in a change of atomic structure, are recognized; alpha decay, beta decay, and electron capture decay (Fig. 2–10). **Alpha decay** is the emission of two protons and two neutrons from the nucleus, resulting in a loss of two atomic numbers and four atomic mass units. **Beta decay** is the emission of a fast-moving electron from a neutron in the nucleus, changing that neutron to a proton and consequently increasing the atomic number by one, with no resultant atomic mass change. **Electron capture decay** results when a proton captures an electron from an orbital shell and thereby converts to a neutron, resulting in a loss of one atomic number and no change in atomic mass.

Some elements undergo only one decay step in conversion from an unstable form to a more stable form. For example, rubidium 87 decays to strontium 87 by a single beta emission, while potassium 40 decays to argon 40 by a single electron capture decay step. Other radioactive elements undergo several decay steps. Uranium 235 decays to lead 207 by seven alpha steps and six beta steps, while uranium 238 decays to lead 206 by eight alpha and six beta steps (Fig. 2–11).

When discussing decay rates, it is convenient to refer to them in terms of half-lives. The **half-life** of a radioactive element is the time it takes for one-half of the atoms of the original unstable **parent element** to decay

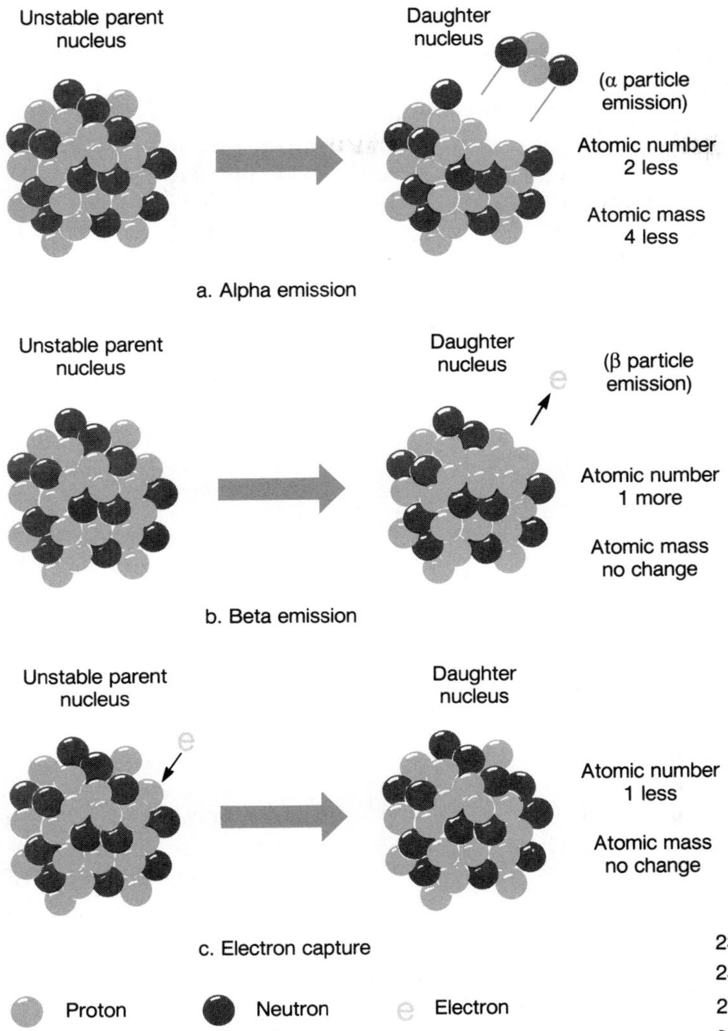

Unstable parent nucleus

Daughter nucleus

(α particle emission)

Atomic number 2 less

Atomic mass 4 less

a. Alpha emission

Unstable parent nucleus

Daughter nucleus

(β particle emission)

Atomic number 1 more

Atomic mass no change

b. Beta emission

Unstable parent nucleus

Daughter nucleus

Atomic number 1 less

Atomic mass no change

c. Electron capture

Proton Neutron Electron

**Figure 2–10** Alpha decay, in which an unstable parent nucleus emits two protons and two neutrons (a). Beta decay, in which an electron is emitted from the nucleus (b). Electron capture decay, in which a proton captures an electron, thereby converting to a neutron (c).

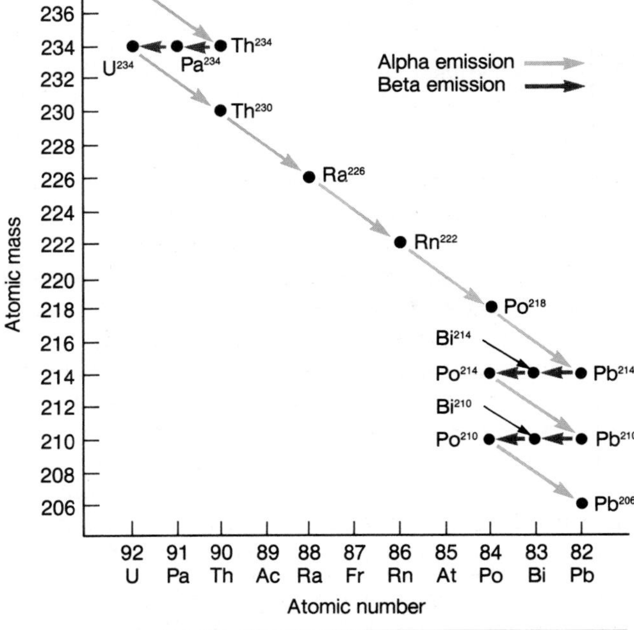


**Figure 2–11** Radioactive decay series for uranium 238 to lead 206. Radioactive uranium 238 decays to its stable end product, lead 206, by eight alpha and six beta decay emissions. A number of different isotopes are produced as intermediate steps in the decay series.

to atoms of a new, more stable **daughter element.** The half-life of a given radioactive element is constant, regardless of external conditions, and can be precisely measured in the laboratory. Half-lives of various radioactive elements range from less than a billionth of a second to 49 billion years.

Radioactive decay occurs at a geometric rate rather than a linear rate. Therefore, a graph of the decay rate produces a curve rather than a straight line (Fig. 2–12). For example, an element with *1,000,000* parent atoms after one half-life will have *500,000* parent atoms and 500,000 daughter atoms; after two half-lives, it will have *250,000* parent atoms (one-half of the previous parent atoms, which is equivalent to one-fourth of the original parent atoms) and 750,000 daughter atoms; after three half-lives, it will have *125,000* parent atoms (one-half of the previous parent atoms, or one-eighth of the original parent atoms) and 875,000 daughter atoms, and so on, until the number of parent atoms remaining is so few that they cannot be accurately measured by present-day instruments.

By measuring the parent-daughter ratio and knowing the half-life of the parent (which has been determined in the laboratory), geologists can calculate the age of a sample containing the radioactive element. The parent-daughter ratio is usually determined by a mass spectrometer, an analytical instrument that separates and measures the proportions of minute particles according to the differences in their mass. Samples of elements are vaporized in a vacuum chamber and bombarded by a stream of electrons. This bombardment knocks electrons off the atoms, leaving them positively charged. The positively charged ions are deflected as they pass between plates bearing electrical charges. Since the amount of deflection depends on the charge-to-mass ratio, in general, the heavier the ion, the less it will be deflected.

## Potential Sources of Error

The most accurate radiometric dates are those obtained from igneous rocks. As a magma cools, radioactive parent atoms are separated from previously formed daughter atoms by the crystallization process. What is being measured is the time of crystallization of the mineral containing the radioactive element, not the time of formation of the radioactive element. However, to obtain accurate radiometric dates, geologists must be sure that they are dealing with a closed system, meaning that neither parent or daughter atoms have been added or removed from the system since crystallization and that the ratio between them results only from radioactive decay. Otherwise, an inaccurate date will result. If daughter atoms have leaked out of the mineral being analyzed, the calculated age will be too young; if parent atoms have been removed, the calculated age will be too great.

Leakage may occur if the rock is heated or subjected to intense pressure such as occurs during metamorphism. If this happens, some of the parent or daughter

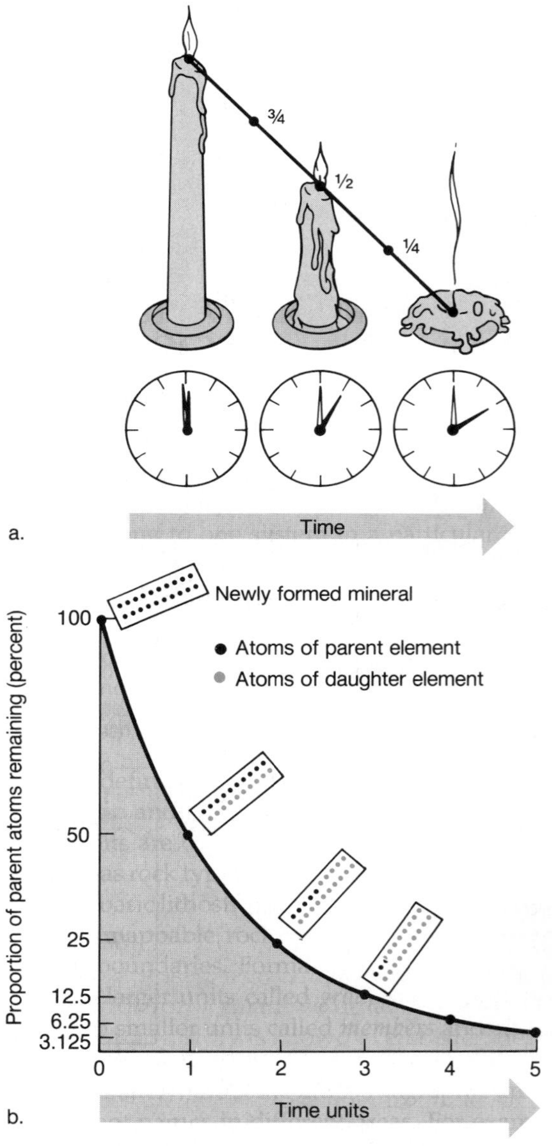


**Figure 2–12** Uniform, straight-line depletion, (a) characteristic of many familiar processes. Geometric radioactive decay curve, (b) in which each time unit represents one half-life, and each half-life is the time it takes for one-half of the parent element to decay to the daughter element.

may be driven from the mineral being analyzed, resulting in an inaccurate age determination. If the daughter product were selectively removed, then one would be measuring the time since metamorphism, and not the time since crystallization of the mineral. (Fig. 2–13). For example, in dating rocks using the potassium-argon technique, one must be very careful because argon, the measured stable daughter element, is an inert gas that is easily driven out of a mineral when the rock is metamorphosed.

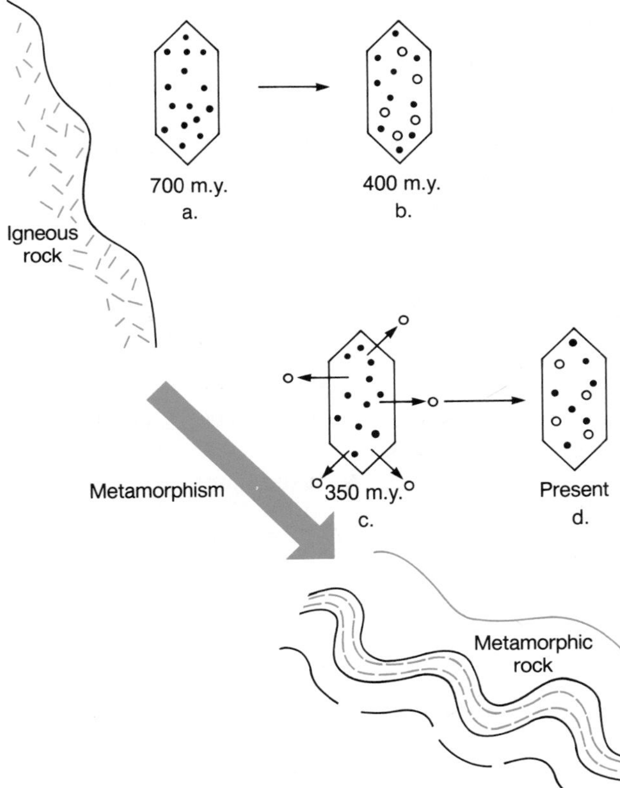


**Figure 2–13** The effect of metamorphism in driving out daughter atoms from a feldspar mineral that crystallized 700 million years ago. The feldspar crystal is shown immediately after crystallization (a), then after 400 million years (b), when some of the parent atoms (dots) had decayed to daughter atoms (circles). Metamorphism at 350 million years ago (c) drives the daughter atoms out of the feldspar into the surrounding rock. Today (d) dating of the feldspar would reveal the time of metamorphism, while dating of the rock would yield the time of original crystallization, 700 million years ago. This assumes the rock has remained a closed chemical system with respect to the daughter atoms.

There is also a source of error inherent in measuring the minute amounts of the different elements and isotopes used for age dating. Mass spectrometers are $\pm 0.2$ to 2.0 percent accurate in their measurements. Thus, for a rock 10 million years old, a possible error of 20,000 to 200,000 years exists. For a 1-billion-year- old rock, however, this is a 2-million to 20-million-year difference. Therefore, when a radiometric age is given, it usually includes a plus or minus factor in number of years appended to the age to indicate the limits of error in the dating technique.

## Long-Lived Radioactive Isotope Pairs

The most reliable dates are those obtained by at least two different radioactive decay series in the same rock. Naturally occurring uranium consists of both uranium 235 and uranium 238 isotopes in the ratio of 138 to 1. Through various decay steps, uranium 235 decays to lead 207, while uranium 238 decays to lead 206. The age of a rock can be calculated using both of these decay series. If the minerals containing them have remained closed systems, then the ages should be in close agreement, and if so, they are said to be concordant, reflecting the time of crystallization of the magma. If the ages do not closely agree, then they are said to be discordant.

Table 2–3 shows five different long-lived parent-daughter isotope pairs used in radiometric dating. Long-lived pairs have half-lives of millions or billions of years. All of these were present when the Earth formed and are still present in measurable quantities. Other, shorter-lived radioactive isotope pairs have decayed to the point that only small quantities near the limit of detection remain.

The most commonly used isotope pairs are the uranium-lead and thorium-lead series, which are used principally to date ancient igneous intrusives, lunar samples, and some meteorites. The rubidium-strontium pair is also used for very old samples and has been effective in dating the oldest rocks on Earth as well as meteorites. The potassium-argon method is typically used for dating fine-grained volcanic rocks from which individual crystals cannot be separated, hence the whole rock is analyzed. Other long-lived radioactive isotope pairs exist, but they are rather rare and used only in special situations.

## Dating by Fission Tracks

When a uranium isotope emits an alpha decay particle, the heavy, rapidly moving alpha particle damages the crystal structure. The damage appears as small linear

## Perspective 2-2

# Fossils and the History of the Earth's Rotation

Astronomers generally agree that the period of the Earth's revolution around the sun has been constant through time, but its rotation around its axis has been gradually slowing down. Based on modern and ancient astronomical data, it appears that during the past 3,000 years, the length of the day has been increasing at a constant rate of approximately 2.5 seconds per 100,000 years. If this is true, then it follows that the number of days in the year has been decreasing through time (Fig. 1). How can such a hypothesis be tested?

Many living organisms—for example, corals and clams—show daily growth increments, or rings, much like the growth-rings of a tree. These daily growth-rings can be grouped into larger yearly growth-bands. For modern organisms, there are approximately 365 growth-rings per year. If the Earth's rotation has changed in the past, that change should be reflected in the number of daily growth-rings per yearly cycle.

In 1963 Professor John Wells of Cornell University published "Coral Growth and Geochronometry." In this pioneering study, he counted the number of daily growth-rings deposited per year by living and fossil corals. He found that modern corals add approximately 365 daily growth-rings per year, while Pennsylvanian corals added about 390 growth-rings per year and Devonian corals added around 400 growth-rings per year (Fig. 2). These results indicate that the number of days per year have been decreasing since at least the Devonian Period and seemingly confirm the astronomical data. Geologists studying the growth increments of other fossil organisms have generally confirmed Wells's conclusions.

The importance of Wells's study is that it revealed the only known way to directly measure the rate of the Earth's rotation in the geologic past. Paleontological data can thus be used to verify the theoretical calculations based on astronomical and geophysical assumptions. Furthermore, a time scale based on growth-rings found in fossils provides a means for absolute dating of sedimentary rocks.

**Table 2-3** Five of the principal long-lived radioactive isotope pairs used in radiometric dating.

| Isotopes: Parent | Isotopes: Daughter | Half-life of parent (years) | Effective dating range (years) | Minerals and rocks that can be dated |
|---|---|---|---|---|
| Uranium 238 | Lead 206 | 4.5 billion | 10 million to 4.6 billion | Zircon<br>Uraninite |
| Uranium 235 | Lead 207 | 710 million | | |
| Thorium 232 | Lead 208 | 14 billion | | |
| Rubidium 87 | Strontium 87 | 47 billion | 10 million to 4.6 billion | Muscovite<br>Biotite<br>Potassium-feldspar<br>Whole metamorphic or igneous rock |
| Potassium 40 | Argon 40 | 1.3 billion | 100,000 to 4.6 billion | Glauconite, Muscovite, Biotite, Hornblende, Whole volcanic rock |

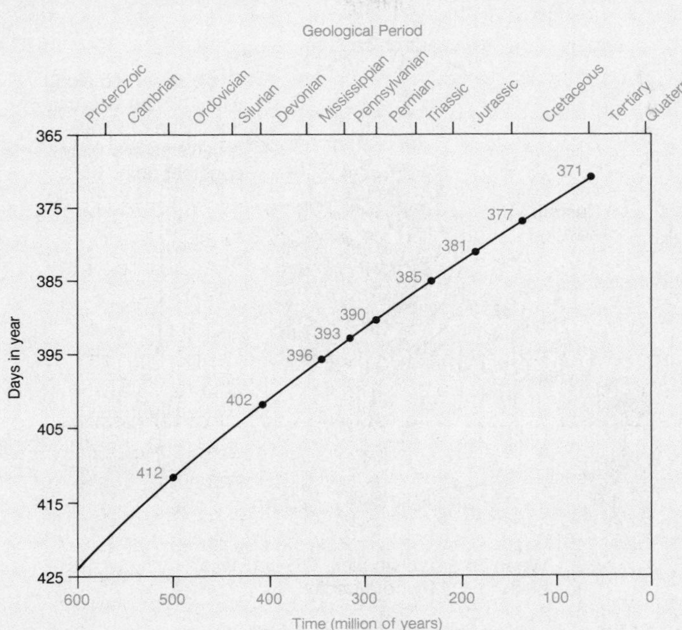


**Figure 1** Length of the year for various geologic periods based on calculations indicating that the Earth's rotation has been slowing at an average rate of 2.5 seconds per 100,000 years. Counts of the bands on corals and other invertebrates have agreed with these estimates.

**Figure 2** Devonian solitary coral showing growth-rings. The fine lines are presumed to be daily growth-rings. Counts of these growth-rings indicate there were approximately 400 days in the year during the Devonian Period. The honeycomb structure present near the base of the specimen is a colonial coral.

tracks that are visible only after etching the mineral with hydrofluoric acid. The age of the sample is determined on the basis of the number of fission tracks present and the amount of uranium the sample contains. The older the sample, the greater the number of tracks (Fig. 2–14).

**Fission track dating** is of particular interest to geologists because the technique can be used to date samples ranging from only a few hundred to billions of years in age. It is most useful for dating samples between about 40,000 and 1 million years ago, a period for which other dating techniques are not particularly suitable. As with all radiometric dating techniques, problems can occur. If the rocks have been subjected to high temperatures, the damaged crystal structures are repaired by annealing and consequently the tracks disappear. In such instances, the calculated age will be younger than the actual age.

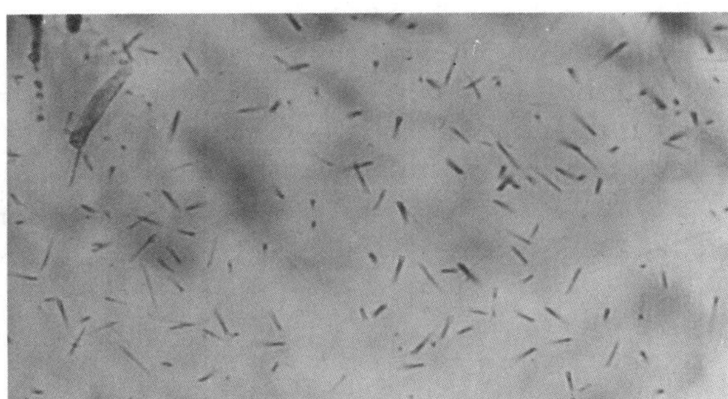

**Figure 2–14** Fossil fission tracks in a crystal of apatite. Each track, which is about 16 μm in length, is the result of the radioactive decay of an atom of uranium. The apatite crystal, which has been etched with hydrofluoric acid, comes from one of the dikes at Shiprock, New Mexico, and indicates a calculated age of 27 million years. (Photo courtesy of Charles W. Naeser, U.S. Geological Survey)

## Radiocarbon and Tree-Ring Dating Methods

Carbon is an important element in nature—one of the basic elements found in all forms of life. It has three isotopes, carbon 12 and 13, which are stable, and carbon 14, which is radioactive and has a half-life of 5,730 years plus or minus 30 years. The **carbon 14 dating technique** is based on the ratio of carbon 14 to carbon 12 and can be used only on once-living organisms.

The short half-life of carbon 14 makes this dating technique practical only for specimens no older than about 60,000 years. Consequently, the carbon 14 dating method is especially useful in archaeology and has greatly aided in unraveling the events of the Late Pleistocene Epoch.

Carbon 14 is continuously created in the upper atmosphere through the bombardment of nitrogen 14 by cosmic radiation (neutrons). The newly created carbon 14 is quickly assimilated into the carbon cycle, and along with carbon 12 and 13, is absorbed in a nearly constant ratio in all living organisms (Fig. 2–15).

Since carbon 14 is continually created in the upper atmosphere and distributed throughout the carbon cycle, the ratio of carbon 14 to carbon 12 remains constant in all living things. However, carbon 14 is not replenished when an organism dies, and hence the ratio of carbon 14 to carbon 12 decreases as the carbon 14 decays.

While the ratio of carbon 14 to carbon 12 is remarkably constant in both the atmosphere and living organisms today, there is good evidence that the production of carbon 14, and thus the ratio of carbon 14 to carbon 12, has varied somewhat over the past several thousand years. This was determined by comparing ages established by carbon 14 dating of wood samples against those established by counting annual tree-rings in the same samples. As a result, corrections in carbon 14 ages have been made to reflect such variations in the past.

**Tree-ring dating** is another useful method for dating geologically recent events. The age of a tree can be determined by counting the growth rings in the lower part of the stem. Each ring represents one year's growth, and the pattern of wide and narrow rings can be compared among trees to establish the exact year in which the rings were formed. The procedure of matching ring patterns from numerous trees and wood fragments in a given area is referred to as crossdating. By correlating distinctive tree-ring sequences from living to nearby dead trees, a time scale can be constructed that extends back to about 14,000 years ago (Fig. 2–16). By matching ring patterns to the composite ring scale, wood samples whose ages are not known can be accurately dated. Furthermore, tree-ring dating can be used to establish the year in which an event took place, provided the event involved damaging or killing the tree (Fig. 2-17).

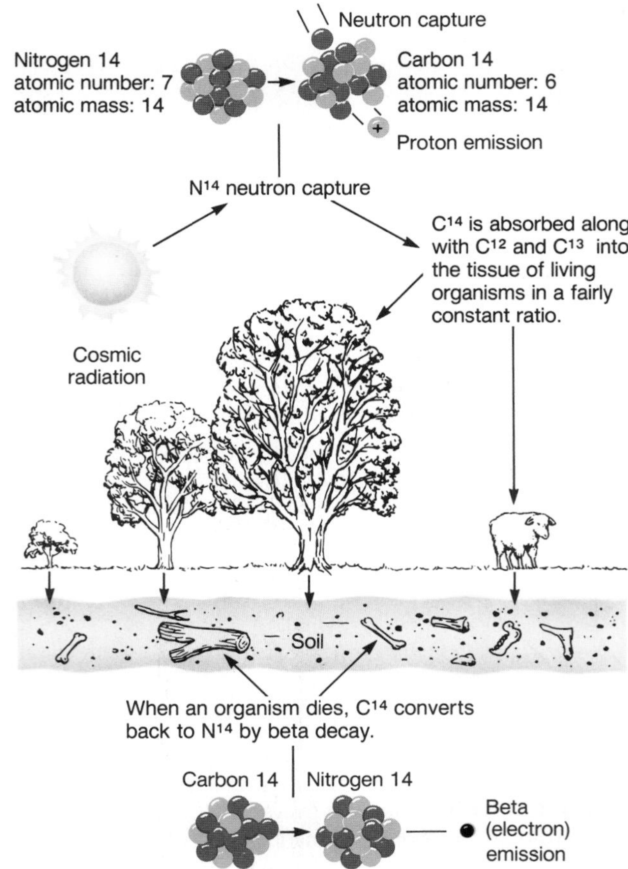


**Figure 2–15** The carbon cycle showing the formation, dispersal, and decay of carbon 14.

## CHAPTER SUMMARY

1. Early Christian theologians were responsible for formulating the idea that time is linear and that the Earth was very young. Archbishop Ussher calculated an age of about 6,000 years based on a literal interpretation of Scripture.

2. During the eighteenth and nineteenth centuries, attempts were made to determine the age of the Earth based on scientific evidence rather than revelation. While

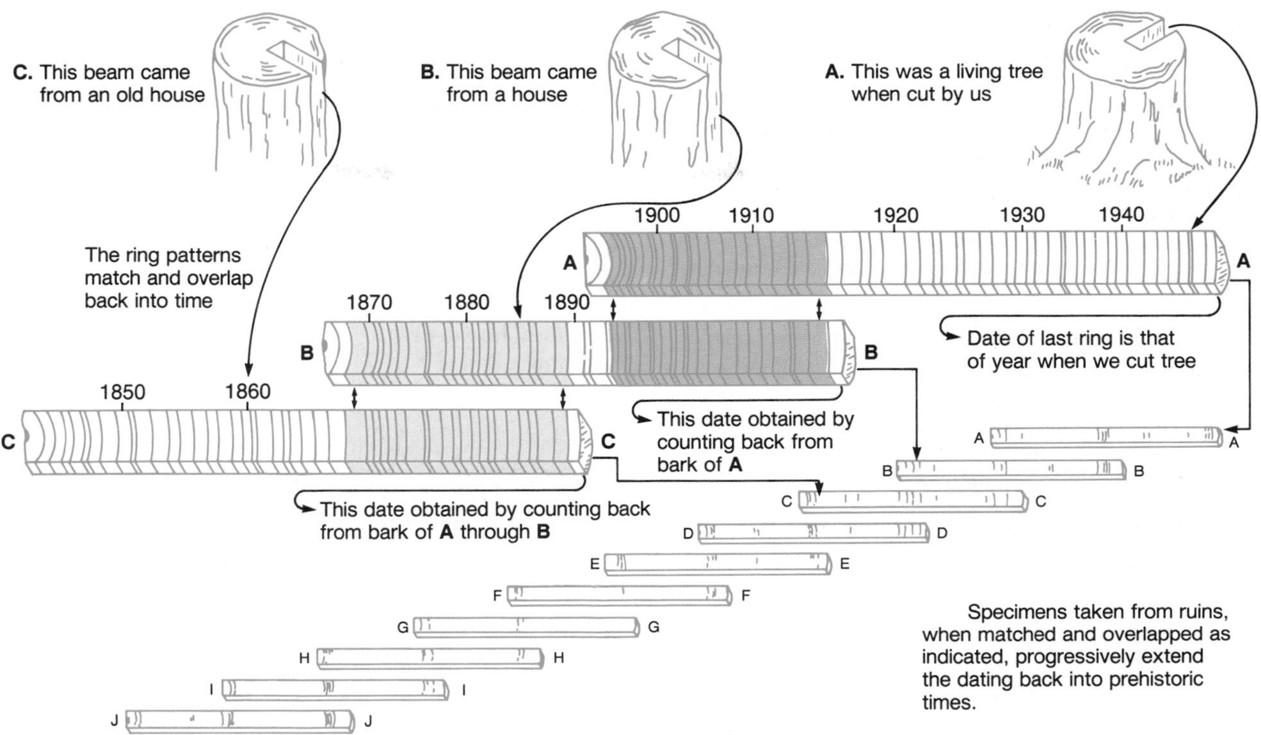
C. This beam came from an old house
B. This beam came from a house
A. This was a living tree when cut by us
The ring patterns match and overlap back into time
1900
1910
1920
1930
1940
1870
1880
1890
1850
1860
A
B
C
Date of last ring is that of year when we cut tree
This date obtained by counting back from bark of A
This date obtained by counting back from bark of A through B
A
B
C
D
E
F
G
H
I
J
Specimens taken from ruins, when matched and overlapped as indicated, progressively extend the dating back into prehistoric times.

**Figure 2–16** In the crossdating method, tree-ring patterns from different woods are matched against each other to establish a ring-width chronology backward in time.

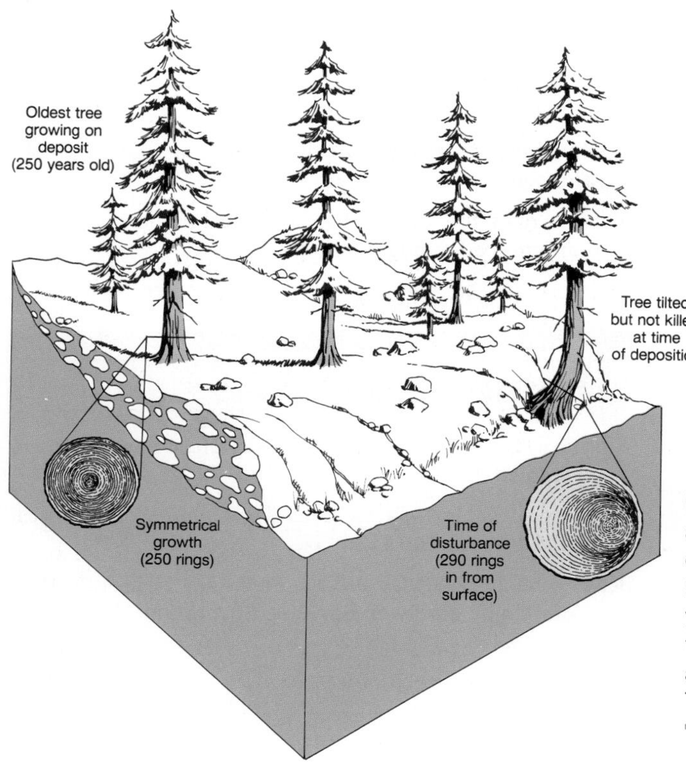
Oldest tree growing on deposit (250 years old)
Tree tilted, but not killed at time of deposition
Symmetrical growth (250 rings)
Time of disturbance (290 rings in from surface)

**Figure 2–17** Using tree-ring ages to determine the time of a landslide at the base of a slope. Each tree-ring represents one year's growth. The oldest tree growing on the landslide has 250 growth-rings indicating the landslide is at least 250 years old. The tree that is tilted but was not killed at the time of the landslide shows a change from symmetrical to asymmetrical ring growth at 290 rings in from the surface. Therefore, the landslide occurred 290 years ago.

some attempts were quite ingenious, they all yielded ages that were much too young.

3. Considering the religious, political, and social climate during the seventeenth, eighteenth, and early nineteenth centuries, it is easy to see how concepts such as neptunism, catastrophism, and a very young Earth were eagerly embraced. However, as geologic data accumulated, it soon became apparent that these concepts were not supported by evidence, and that the Earth must be older than 6,000 years.
4. James Hutton believed that present-day processes operating over long periods of time could explain all the geologic features in his native Scotland. He also viewed Earth history as cyclical and believed the Earth to be very old. Hutton was instrumental in establishing the basis for the principle of uniformitarianism.
5. Uniformitarianism as articulated by Charles Lyell soon displaced neptunism and catastrophism as the guiding doctrine of geology. The principle of uniformitarianism holds that the laws of nature have been constant through time and that the same processes operating today have operated in the past.
6. In addition to uniformitarianism, the principles of superposition, original horizontality, lateral continuity, and cross-cutting relationships are basic for determining relative geologic ages and for interpreting the geologic history of the Earth.
7. Radioactivity was discovered in the late nineteenth century, and soon thereafter radiometric dating techniques allowed geologists to determine absolute ages for geologic events.
8. Absolute age dates for rock samples are usually obtained by determining how many half-lives of the radioactive parent element have elapsed since the sample originally crystallized. A half-life is the time it takes for one-half of the radioactive parent element to decay to a daughter element.
9. The most accurate radiometric dates are obtained from igneous rocks. The five common long-lived radioactive isotope pairs are uranium 238–lead 206, uranium 235–lead 207, thorium 232–lead 208, rubidium 87–strontium 87, and potassium 40–argon 40. The most reliable dates are those obtained by at least two different radioactive decay series in the same rock.
10. Carbon 14 dating can be done only on organic matter such as wood, bones, and shells and is effective back to about 60,000 years ago. Unlike the long-lived isotopic pairs, the carbon 14 method determines age by the ratio of radioactive carbon 14 to stable carbon 12.

## IMPORTANT TERMS

absolute dating
alpha decay
beta decay
carbon 14 dating
catastrophism
cross-cutting relationships
daughter element
electron capture decay
fission track dating
half-life
isotope
lateral continuity
neptunism
original horizontality
parent element
radioactive decay
relative dating
superposition
tree-ring dating
uniformitarianism

## REVIEW QUESTIONS

1. What is the difference between relative and absolute dating of geologic events?
2. What are the four principles used in relative age dating, and why are they so important in historical geology?
3. What contributions to the development of geology did the following men make: Georges Louis de Buffon, James Hutton, Lord Kelvin, Charles Lyell, Nicolas Steno?
4. Compare the modern view of uniformitarianism to Lyell's view.
5. If you wanted to calculate the absolute age of an intrusive body, what information would you need?
6. What are the potential sources of error in radiometric dating?
7. How can geologists be sure the absolute age dates they obtain from igneous rocks are accurate?
8. How does the carbon 14 dating technique differ from uranium-lead dating methods?

## ADDITIONAL READINGS

Albritton, C. C., Jr. 1980. *The Abyss of Time.* San Francisco: Freeman, Cooper & Co.

Albritton, C. C., Jr. 1984. Geologic Time. *Journal of Geological Education,* 32, no. 1: 29–37.

Berry, W. B. N. 1987. *Growth of a Prehistoric Time Scale.* Palo Alto, Calif.: Blackwell Scientific Publications, Inc.

Eicher, D. L. 1976. *Geologic Time.* 2nd ed. Englewood Cliffs, N. J.: Prentice-Hall, Inc.

Faul, H. 1978. A History of Geologic Time. *American Scientist,* 66, no. 2: 159–165.

Gould, S. J. 1987. *Time's Arrow, Time's Cycle.* Cambridge, Mass.: Harvard University Press.

Harland, W. B., A. V. Cox, P. G. Llewellyn, C. A. G. Pickton, A. G. Smith, and R. Walter. 1982. *A Geologic Time Scale.* Cambridge, England: Cambridge University Press.

Laudan, R. 1987. *From Mineralogy to Geology: The Foundations of a Science, 1650–1830.* Chicago, Ill.: The University of Chicago Press.

Shea, J. H. 1982. Twelve Fallacies of Uniformitarianism. *Geology,* 10, no. 9: 455–460.

Wetherill, G. W. 1982. Dating Very Old Objects. *Natural History,* 91, no. 9: 14–20.

# 3

## CHAPTER OUTLINE



## PROLOGUE

*Most people today know that sand deposited on a beach may become sandstone, and that shells and bones in rocks are the remains of once-living animals. These concepts are well entrenched in modern thinking. During most of human existence, however, people had little or no idea of how rocks formed or what fossils were.*

*Humans have used rocks and fossils since prehistoric times for tools, weapons, and ornaments. Fossils, for example, were a trade item in the ancient world, and even in the Middle Ages some fossils were believed to have curative powers (Fig. 3–1). The sixth-century* B.C. *Greek philosopher Xenophanes was apparently the first to recognize that fossil seashells and bones represent the remains of once-living organisms, while the Chinese came to the same conclusion by the first century* B.C. *Through most of historic time in the West, however, fossils were believed to be inorganic in origin. Explanations for their origin included spontaneous generation within rocks, growth of seeds in rocks, and images produced in rocks by some kind of molding force. Fossils were even believed to be objects placed in the rocks by the Creator to test our faith or by Satan to sow seeds of doubt.*

**Figure 3–1** Fossil teeth of rays, a type of fish, were called toadstones in the Middle Ages. It was believed that the stones came from the heads of living toads (left), and that they were a powerful cure for several ailments (right).

# Rocks, Fossils, and Time

*One of the problems involved in understanding fossils was that no one knew how rocks formed. Some of the ancient Greeks and Egyptians had glimpses of insight into rock-forming processes, and Leonardo da Vinci (1452–1519) realized that sand transported in rivers could become sandstone (Perspective 2-1). Nevertheless, it was not until Nicolas Steno (Chapter 2) elucidated his principles in 1669 that a framework for interpreting rocks existed. In short, before Steno's work, no one knew anything about Earth history.*

*By the late 1700s and early 1800s, scientists had gained considerable insight into Earth history. However, old ideas are difficult to displace, and the idea of a young Earth shaped by catastrophes (perhaps a single catastrophe) persisted. One of the most extreme views of Earth history was proposed in 1857 by Philip Henry Gosse, a British naturalist. According to Gosse, the Earth was created a few thousands of years ago with the appearance of a lengthy history. In other words, rocks that appeared to have been deposited by rivers or that appeared to have formed from lava flows were created with that appearance. The events denoted by their appearance had not really happened. Fossils were not remains of organisms that once lived on Earth, they were simply another created aspect of rocks.*

*Gosse's proposal was immediately rejected by most people, who reasoned that if Gosse were correct, the historical record preserved in rocks was the product of a deceptive Creator. In addition to the philosophical implications of Gosse's proposal, it is important to understand why it cannot be considered scientific:*

> *[W]e have no way to find out whether it is wrong—or, for that matter, right. It is the classical example of an utterly untestable notion, for the world will look exactly the same . . . whether fossils and strata [were created] or products of an extended history.*[1]

*Even before Gosse proposed his idea, scientists had concluded that the Earth had a long history, and that that history could be read from the only available record—the geologic record.*

[1]Gould, S.J. 1984. Adam's Navel. *Natural History*, 93, no. 6: 10,14.

## INTRODUCTION

The Earth is a dynamic planet. Volcanic eruptions, earthquakes, floods, and shoreline erosion are manifestations of some of the internal and surface processes that continuously modify it. The Earth has changed, or evolved, throughout its history, but our only record of such change is the evidence preserved in rocks, the **geologic record.** All rock types are important in analysis of the geologic record; igneous rocks record volcanic activity related to rifting and plate convergence, and studies of metamorphic rocks reveal something of the dynamic processes that occur within the Earth's crust.

Sedimentary rocks have a special place in historical geology. Sediments or sedimentary rocks cover most of the seafloor and about two-thirds of the continents, and they preserve evidence of the surficial processes responsible for their origin. Furthermore, many contain the remains or traces of prehistoric life, or what we call *fossils.*

Our main objective in this chapter is to explore the principles and concepts geologists use to make sense of the geologic record. Just as the written historical record must be analyzed and interpreted, the geologic record requires analyses if we are to make inferences about geologic history—that is, if we are to read Earth history from rocks.

## STRATIGRAPHY

**Stratigraphy** is that branch of geology concerned with the composition, origin, and areal and age relationships of layered or stratified rocks. Stratification may occur in any of the major rock groups; a succession of lava flows or ash falls will be layered, metamorphosed layered rocks also show this property, and sedimentary rocks are invariably stratified (Fig. 3–2).

a.

b.

c.

**Figure 3–2** Stratification. a. These igneous rocks in the San Luis Valley, Colorado are stratified; the lowest exposed unit is a volcanic mudflow overlain by an ash bed (white) and a lava flow. b. Stratification in metamorphic rocks, the Siamo Slate of Michigan. c. Stratified sedimentary rocks, the Triassic Chugwater Formation, Wyoming.

Our main concern here is with stratification in sedimentary rocks, but some of the concepts of stratigraphy apply to any sequence of layered rocks. Where sedimentary rocks are observed in surface exposures (Fig. 3–2), the vertical relationships among the strata are easily determined. Lateral relationships, equally important in stratigraphic studies, usually must be determined from a number of separate exposures.

## Vertical Stratigraphic Relationships

The bounding surface separating one layer of strata from another is called a **bedding plane** and may be gradational or sharp. Gradational bedding planes are those where one rock type gradually changes upward into another rock type.They represent gradually changing conditions of sedimentation. Sharp bedding planes are surfaces where the rocks above and below differ in age or are distinctly different in composition (Fig. 3–3). Abrupt changes in composition indicate rapid changes in sedimentation or perhaps a period of nondeposition or erosion followed by renewed deposition. Regardless of the nature of bedding planes, it is essential that the correct vertical sequence of strata at the time of deposition be known.

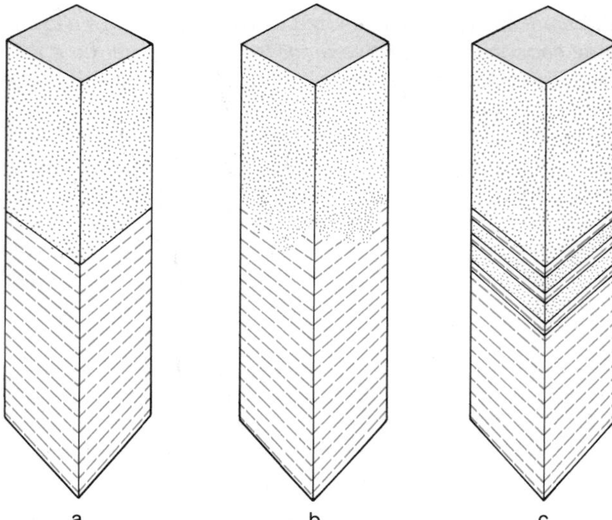


**Figure 3–3** Vertical stratigraphic relationships. a. Shale and sandstone separated by a sharp bedding plane. Such abrupt changes in composition indicate rapid changes in sedimentation or a time during which no deposition occurred. b and c. Two examples of gradual upward changes from one rock type into another.

a.

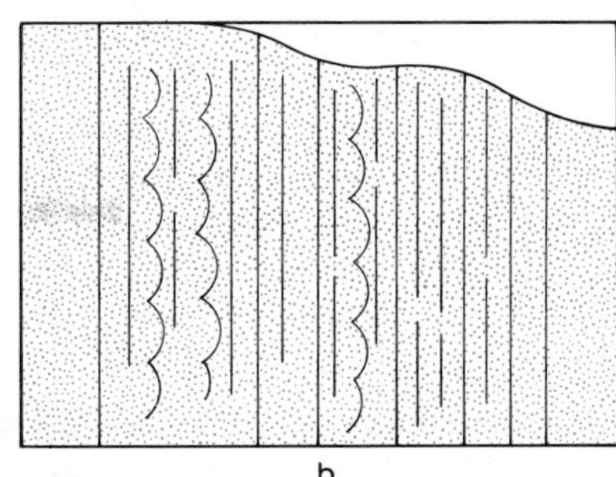

b.

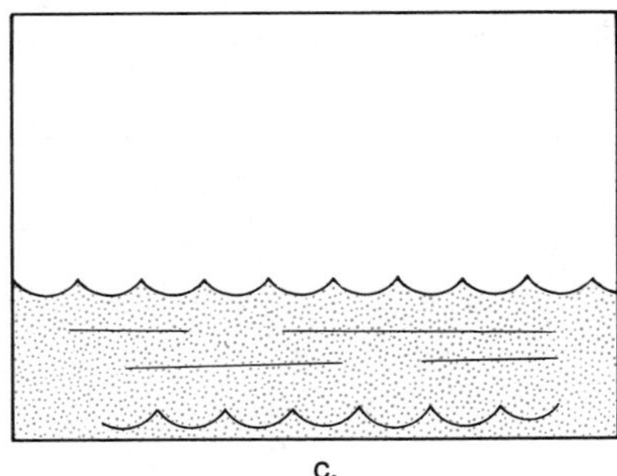

c.

**Figure 3–4** a. Precambrian Mesnard Quartzite, Michigan. These beds were deposited horizontally, but after they were lithified, they were deformed, and now the layers are oriented vertically. b. Diagramatic view of the internal structure of the Mesnard Quartzite beds. Note that the wave-formed ripple marks' crests point to the left, toward the youngest bed. c. Orientation of the ripple-marked beds when they were deposited.

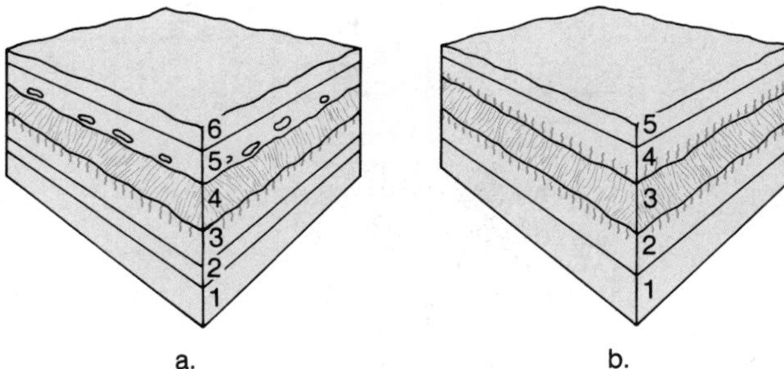


**Figure 3–5** Relative ages of lava flows, sills, and associated sedimentary rocks may be difficult to determine. a. A buried lava flow, bed 4, baked the underlying bed, and bed 5 contains inclusions of the lava flow. The lava flow is younger than bed 3 and older than beds 5 and 6. b. The rock units above and below the sill (3) have been baked, indicating the sill is younger than beds 2 and 4, but its age relative to bed 5 cannot be determined.

### Superposition

Nicholas Steno (Chapter 2) realized that correct relative ages of strata in a horizontal sequence could be determined by their position in the sequence (Fig. 2–1). If strata have been deformed, however, relative age determinations are more difficult. For example, the strata in Figure 3–4 have been tilted from their original horizontal orientation and now stand nearly vertical. Can you tell which bed is youngest?

Geologists commonly solve relative age problems such as the one in Figure 3–4 by looking for sedimentary structures, structures that formed in sediments at the time of deposition or shortly thereafter. In Figure 3–4 the strata contain wave-formed ripple marks, the crests of which point in the direction of the youngest bed, or the top of the sequence. Several other sedimentary structures and fossil organisms in their original living positions may also be used to determine which bed is youngest in a sequence. These will be discussed in Chapter 4.

In addition to deformation, the association of sedimentary and igneous rocks may cause problems in relative dating. Buried lava flows and intrusive igneous bodies called sills both look much the same in a sequence of strata (Fig. 3–5). However, a buried lava flow is older than the rocks above it, while a sill is younger than the bed immediately above it. Similar problems arise where sedimentary rocks are associated with large, irregularly shaped igneous bodies such as batholiths (Fig. 3–6).

To resolve relative age problems such as these, geologists look closely at the sedimentary rocks in contact with the igneous rocks to see if they show signs of baking by heat. Any sedimentary rock showing such effects is older than the igneous rock with which it is in contact. In Figure 3–5, for example, a sill produces a zone of baking in the rocks at its upper and lower margins, while a lava flow bakes only those rocks below.

Another way to determine relative ages is by use of the **principle of inclusions.** This principle holds that inclusions, or fragments in a rock unit, are older than the rock unit itself. If the batholith shown in Figure 3–6 were intruded into the sandstone, it might contain sandstone inclusions, and the sandstone unit might show the effects of baking. Accordingly, one would conclude that the sandstone is older than the batholith. If, on the other hand, the sandstone contained granite rock fragments, it would mean the batholith was a source rock for the sandstone and must be older than the sandstone.

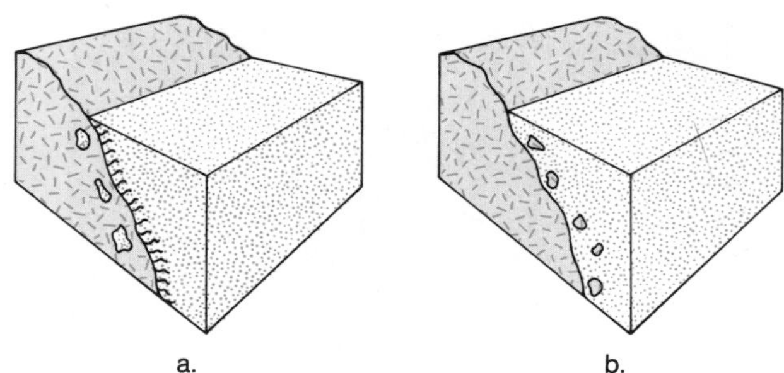


**Figure 3–6** a. The batholith is younger than the sandstone because the sandstone has been baked at its contact with the granite, and the granite contains sandstone inclusions. b. Granite inclusions in the sandstone indicate that the batholith was the source of the sandstone and therefore is older.

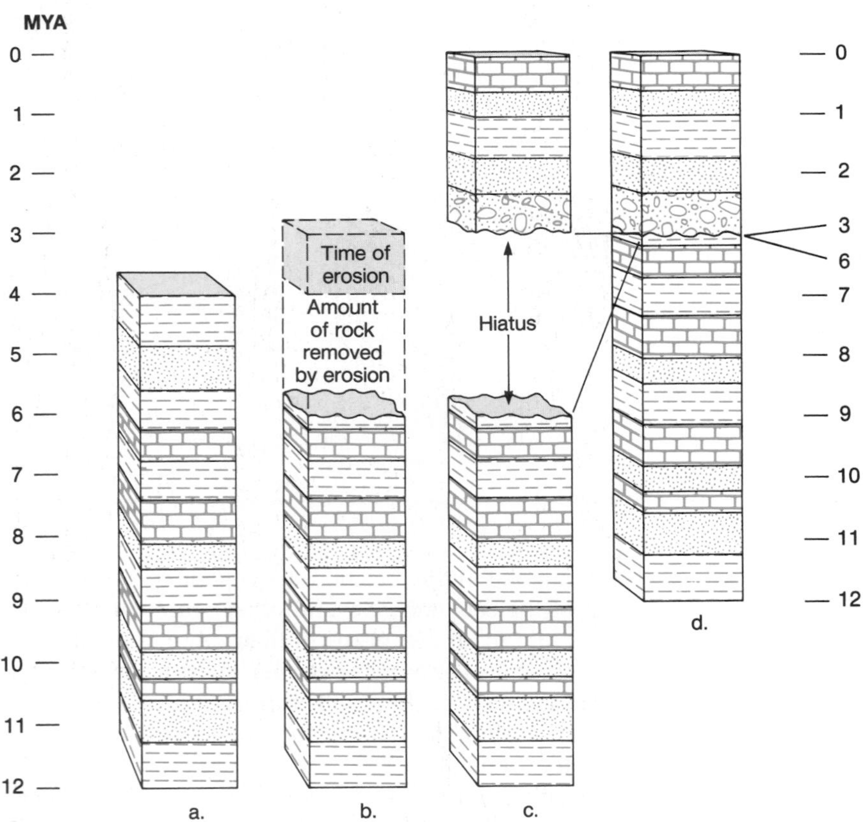


**Figure 3–7** Simplified diagram showing the development of an unconformity and a hiatus. a. Deposition began 12 million years ago (MYA) and continued more or less uninterrupted until 4 MYA. b. A 1 million-year episode of erosion occurred, and during that time strata representing 2 million years of geologic time were eroded. c. A hiatus of 3 million years exists between the older strata and the strata that formed during a renewed episode of deposition that began 3 MYA. d. The actual stratigraphic record. The unconformity is the surface separating the strata and represents a major break in our record of geologic time.

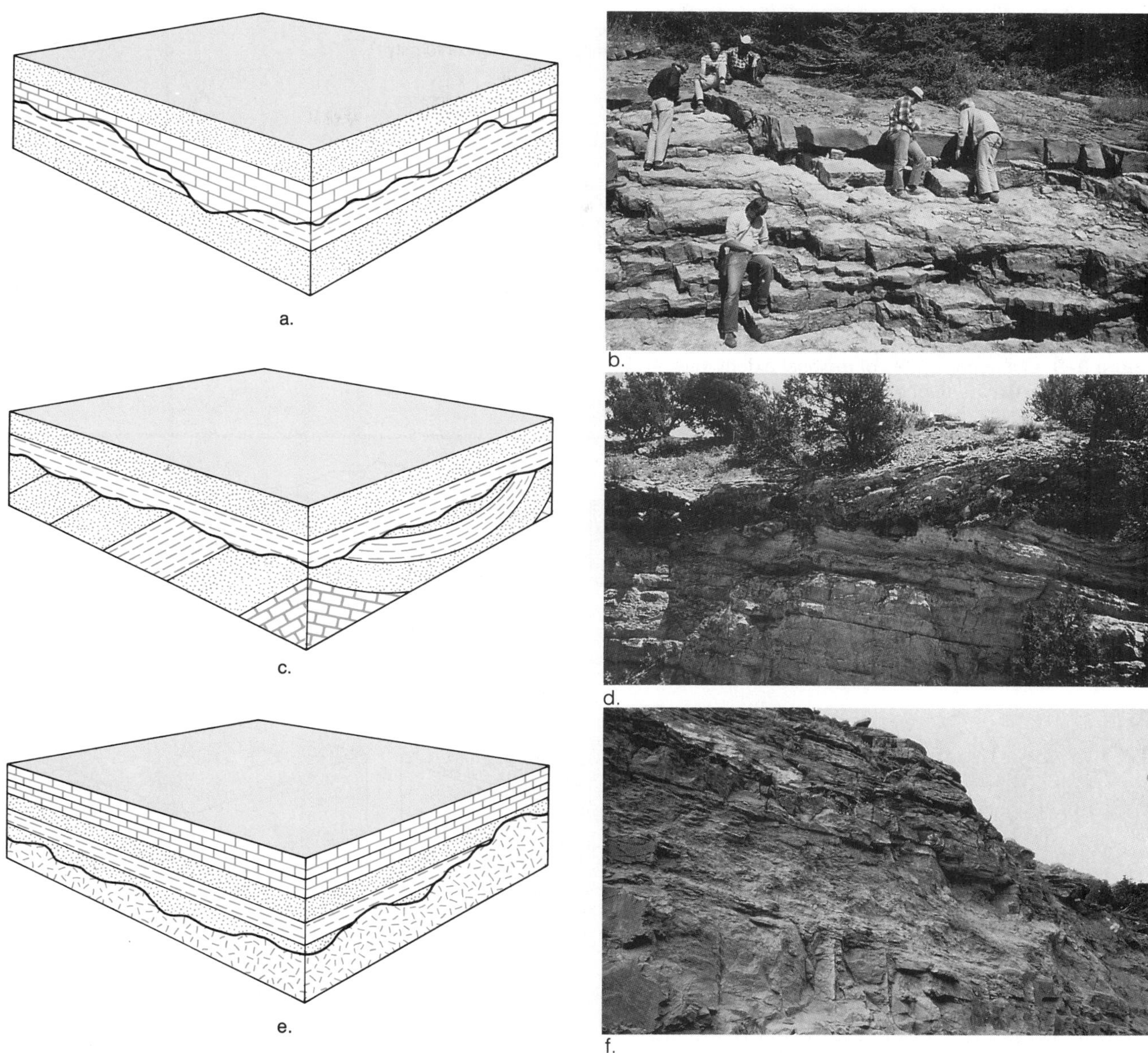


**Figure 3–8** Unconformities. a. Block diagram showing a disconformity. b. Disconformity between Mississippian and Jurassic strata in Montana. The geologist at the upper left is sitting on Jurassic strata and his right foot is resting upon Mississippian rocks. c. Angular unconformity. d. Angular unconformity separating Pennsylvanian-age strata from Tertiary gravel in Colorado. e. Nonconformity. f. Nonconformity between Precambrian granite and the overlying Cambrian-age Deadwood Formation, Wyoming.

## Unconformities

Our discussion thus far has been concerned with **conformable** sequences of strata, sequences in which no depositional breaks of any consequence occur. A sharp bedding plane (Fig. 3–3) separating strata may represent a depositional break of minutes, hours, years, or even tens of years but is inconsequential when considered in the context of geologic time. Surfaces of discontinuity representing significant amounts of geologic time are called **unconformities,** and any interval of geologic time not represented by strata in a particular area is a **hiatus** (Fig. 3–7). Thus an unconformity is a surface of nondeposition or erosion that separates younger strata from older rocks. As such, it represents a break in our record of geologic time.

Three major types of unconformities are recognized (Fig. 3–8). A **disconformity** is a surface of erosion or nondeposition between younger and older beds that are

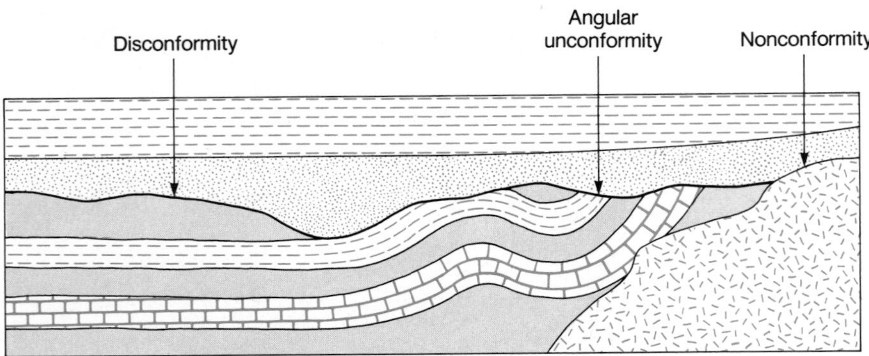


**Figure 3–9** An unconformity of regional extent may change from one type to another as shown in this cross section.

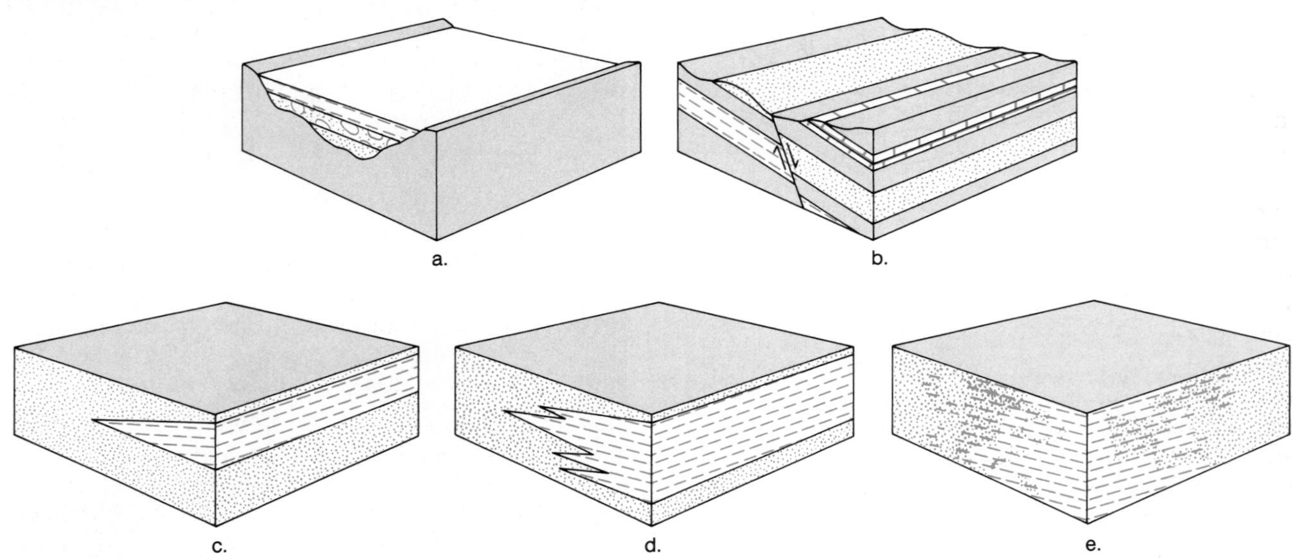


**Figure 3–10** a. Lateral termination of rock units at the edge of a depositional basin. b. Faulting and erosion causing lateral termination. Notice that the beds above the sandstone do not appear on the left side of the fault. c. Lateral termination by pinchout. d. Intertonguing. e. Lateral gradation.

parallel with one another. An **angular unconformity** is an erosional surface on tilted or folded strata over which younger strata have been deposited. Both younger and older strata may dip, but if their dip angles are different (generally the older strata dip more steeply) an angular unconformity is present. A **nonconformity** is a type of unconformity in which an erosion surface cut into metamorphic or intrusive rocks is covered by sedimentary rocks. Unconformities of regional extent may change laterally from one type to another and do not everywhere encompass equivalent amounts of time (Fig. 3–9).

## Lateral Relationships—Facies

In 1669 Nicholas Steno formulated the principle of lateral continuity (Chapter 2). He recognized that sediment layers extend outward in all directions until they terminate. They may terminate where eroded, truncated by faults, or abruptly where they abut the edge of a depositional basin (Fig. 3–10). Lateral termination of sedimentary rocks may also occur when a rock unit becomes progressively thinner until it pinches out. *Intertonguing* occurs when a unit splits laterally into thinner units, each of which pinches out. And finally, a

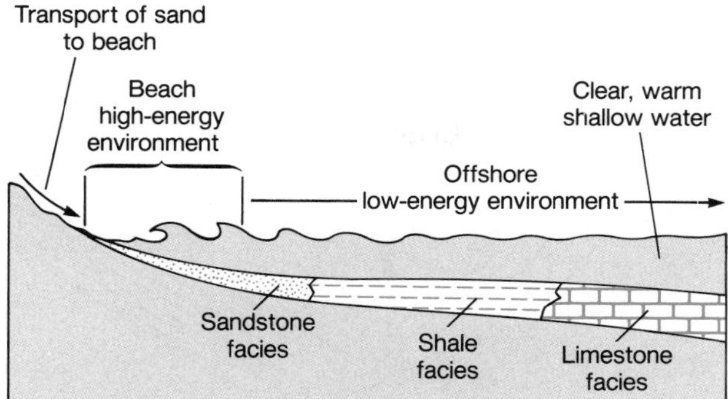


**Figure 3–11** Simultaneous deposition in different environments located adjacent to one another. Each distinct depositional unit is recognized as a sedimentary facies.

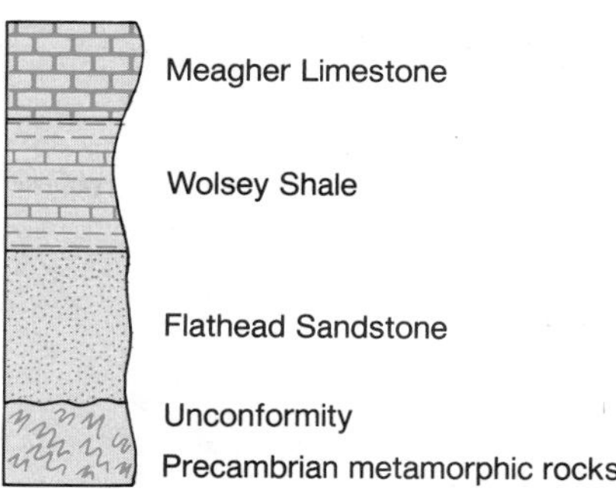


**Figure 3–12** Generalized stratigraphic column for approximately 100 m of Cambrian strata in southwestern Montana. The succession of sandstone overlain by shale and limestone is interpreted as the product of a marine transgression. Similar vertical sequences of Cambrian strata are known from many other places including Wyoming, Colorado, South Dakota, and the Grand Canyon, Arizona. These strata were deposited during a major marine transgression that covered much of western North America (see Chapter 10).

unit may change in composition or texture, or both, by lateral gradation until it is no longer recognizable (Fig. 3–10).

Intertonging and lateral gradation result from different depositional processes operating in adjacent environments. For example, sand may be deposited in the high-energy nearshore part of the continental shelf, while mud and carbonate sediments accumulate simultaneously in the laterally adjacent low-energy, offshore area (Fig. 3–11). Deposition in each of these environments produces a body of sediment, each of which is characterized by a distinctive set of physical, chemical, and biological attributes. Such distinctive bodies of sediment or sedimentary rock are called **sedimentary facies.**

Any aspect of a sedimentary rock unit that makes it recognizably different from laterally adjacent rocks of the same age, or approximately the same age, can be used to establish a sedimentary facies. In Figure 3–11 we recognize three sedimentary facies, a sandstone facies, a shale facies, and a limestone facies. Armanz Gressly in 1838 was the first to use the term *facies.* Gressly carefully traced sedimentary rock units in the Jura Mountains of Switzerland and noticed lateral changes, from shale to limestone, for example. He concluded that such lateral changes resulted from deposition in different depositional environments that had existed simultaneously.

## Transgressions and Regressions

Many rock units in the interiors of continents show clear evidence of having been deposited in marine depositional environments. The strata in Figure 3–12, for example, consist of a sandstone facies that was deposited in a nearshore marine environment overlain by shale and limestone facies that were deposited in offshore environments. To account for such vertical sequences of facies, geologists recognize times during which sea level has risen or fallen with respect to the continents. When sea level rises with respect to a continent, the shoreline moves inland, giving rise to a **marine transgression** (Fig. 3–13). As the shoreline moves inland, the depositional environments located parallel to the shoreline do likewise. Remember that each laterally adjacent environment in Figure 3–13 is the depositional site of a different sedimentary facies. As a result of a marine transgression, the facies that formed in the offshore environments are superposed over the facies that was deposited in the nearshore environment.

Another important aspect of marine transgressions is that individual rock units become younger in a landward direction. The sandstone, shale, and limestone facies (Fig. 3–13) were deposited continuously as the shoreline moved landward, but none of the facies was deposited simultaneously over the entire area it now covers. In other words, the facies are **time transgressive,** meaning that their ages vary from place to place.

The opposite of a marine transgression is a **marine regression.** If sea level falls with respect to a continent, the shoreline and environments that parallel the shoreline move in a seaward direction (Fig. 3–13). The vertical sequence produced by a marine regression has facies of the nearshore environment superposed over facies of

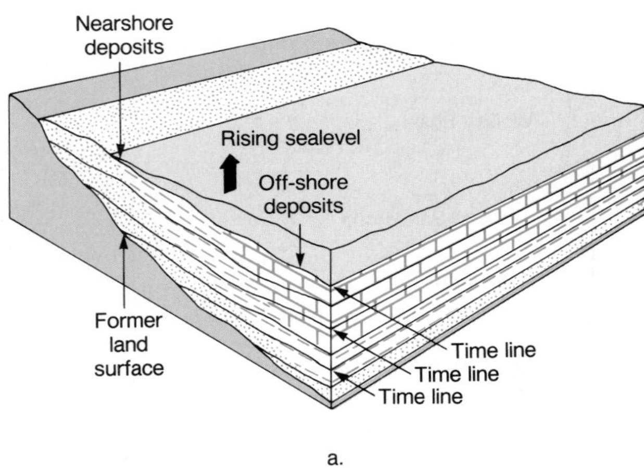


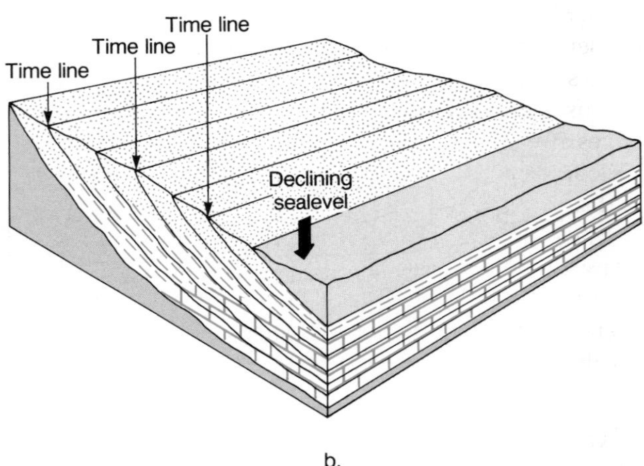


**Figure 3–13** a. Marine transgression. If sea level rises, the shoreline moves inland, as do the depositional environments that parallel the shoreline. The effect of a marine transgression is to superpose offshore facies upon nearshore facies, and rock units become younger in a landward direction. b. Marine regression. As sea level falls, the shoreline and environments move seaward. Nearshore facies are superposed over offshore facies, and rock units become younger in a seaward direction.

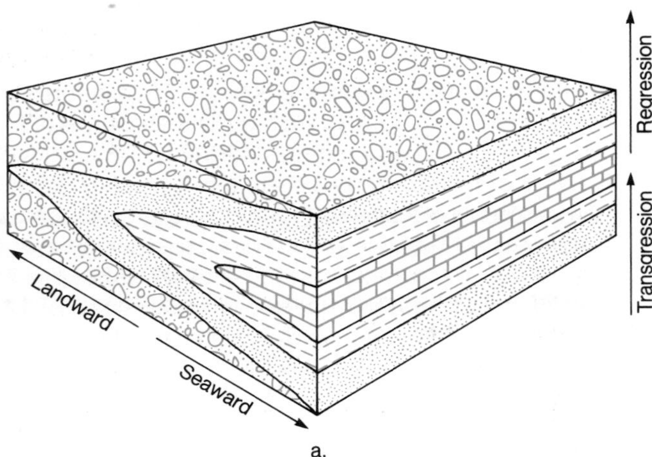


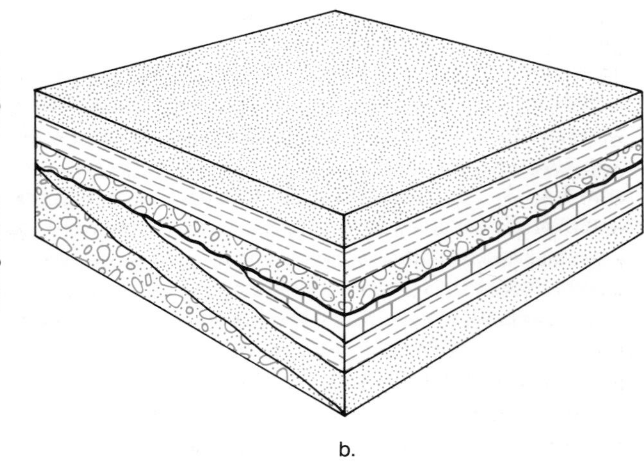


**Figure 3–14** a. This vertical sequence of facies was produced by a transgression followed by a regression. All facies are related, and their lateral relationships can be determined by the application of Walther's Law. b. This vertical sequence of facies contains an unconformity. Walther's Law can be applied successfully to the facies below the unconformity as shown. However, the facies above the unconformity are unrelated to those below.

offshore environments. In addition, individual rock units become younger in a seaward direction.

The significance of lateral and vertical facies relationships was first clearly recognized by Johannes Walther (1860–1937). When Walther traced rock units laterally, he reasoned, as Gressly had, that each sedimentary facies he encountered had been deposited in laterally adjacent but different environments. Furthermore, he realized that the same facies that occurred laterally were also superposed upon one another in a vertical sequence. Walther's observations have since been formulated into **Walther's Law**, which holds that the same facies following one another in a conformable vertical sequence will also replace one another laterally.

The value of Walther's Law is well illustrated by its application to vertical facies relationships resulting from marine transgression or regression. In practice, one commonly cannot trace individual rock units far enough laterally to demonstrate facies changes. It is much easier to observe vertical facies relationships, and from such vertical sequences one can work out the lateral relationships by using Walther's Law. However, Walther's Law applies only to vertical successions of facies that contain no unconformities. The presence of an unconformity indicates that the facies immediately above and below the surface of unconformity are unrelated, and Walther's Law is not applicable (Fig. 3–14). In areas of complex deformation, vertical facies relationships are difficult to determine, but once the correct sequence is known, Walther's Law may be applied.

## Extent, Rates, and Causes of Transgression and Regression

The seas have clearly occupied large parts of all continents at several times in the past. In North America, for example, six major episodes of transgression followed by regression occurred since the beginning of the Paleozoic Era. In fact, the sequences of strata deposited during these transgressive-regressive episodes will serve as the stratigraphic framework for our discussions of Paleozoic and Mesozoic geologic history (Chapters 10, 11, and 13).

The maximum extent of a transgression or maximum withdrawal of the sea during a regression can be established with some certainty, but rates and causes of such events are problematic. Rates of shoreline movement probably occur at centimeters per year, but these rates are deceptive because they are determined by dividing distance by time. For example, if a shoreline moved inland 1,000 km in 20 million years, the rate of transgression would be 5 cm per year. However, major transgressions and regressions are not simply events during which the shoreline moves steadily inland or seaward. As a matter of fact, both are characterized by numerous reversals in the overall transgressive or regressive trend (Fig. 3–15).

Most geologists agree that uplift or subsidence of the continents, together with seafloor spreading and the seawater contained in glaciers, is sufficient to cause a transgression or regression. If a continent is uplifted, it rises with respect to sea level, the shoreline moves seaward, and a regression ensues. The opposite type of movement, subsidence, results in a transgression (Fig. 3–16).

Seafloor spreading can cause transgressions and regressions by changing the volumes of the ocean basins. During relatively rapid episodes of extensive seafloor spreading, such as in the Cretaceous, the mid-oceanic ridges expand and displace seawater onto the continents

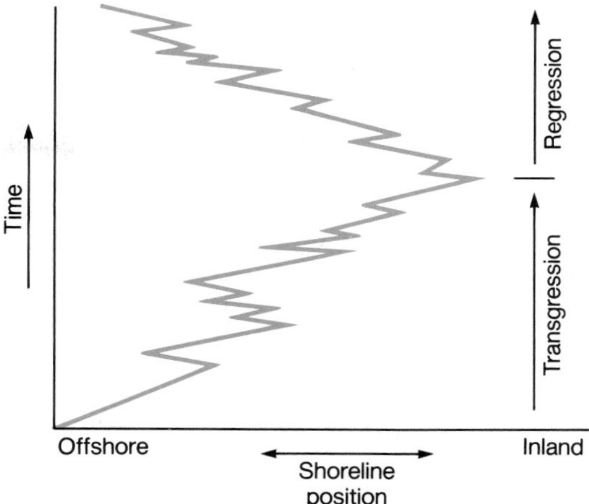


**Figure 3–15** Relative position of the shoreline during a transgressive-regressive event. Notice that while the overall trend may be one of transgression or regression, periodic reversals in the trends occur.

(Fig. 13–9). At other times, when seafloor spreading is comparatively slow, the ridges subside, the volume of the ocean basins increases, and the seas retreat from the continents. Worldwide changes in sea level related to additions or removal of water from the oceans may also cause marine transgressions and regressions. During a widespread glacial episode, large quantities of seawater are tied up on land as glacial ice, and, consequently, sea level falls. When glaciers melt, the water returns to the seas, and sea level rises (Fig. 3–16). A number of Pennsylvanian transgressions and regressions can be attributed to sea level changes caused by glaciation (Chapter 11).

# FOSSILS

In Chapter 2 and preceding sections of this chapter several principles, including superposition and cross–cutting relationships, were discussed. These principles are essential for interpretations of geologic history but can be applied only on a local scale. For example, Figure 3–17 shows two vertical sequences of strata in different areas. The relative ages of the rocks are easily determined within either area; but with the data given, the age of the rocks in one area cannot be determined relative to those in the other area. There is, however, a solution to this problem, a solution that involves the use of fossils.

**Fossils** are the remains or traces of prehistoric life that have been preserved in rocks of the Earth's crust. In

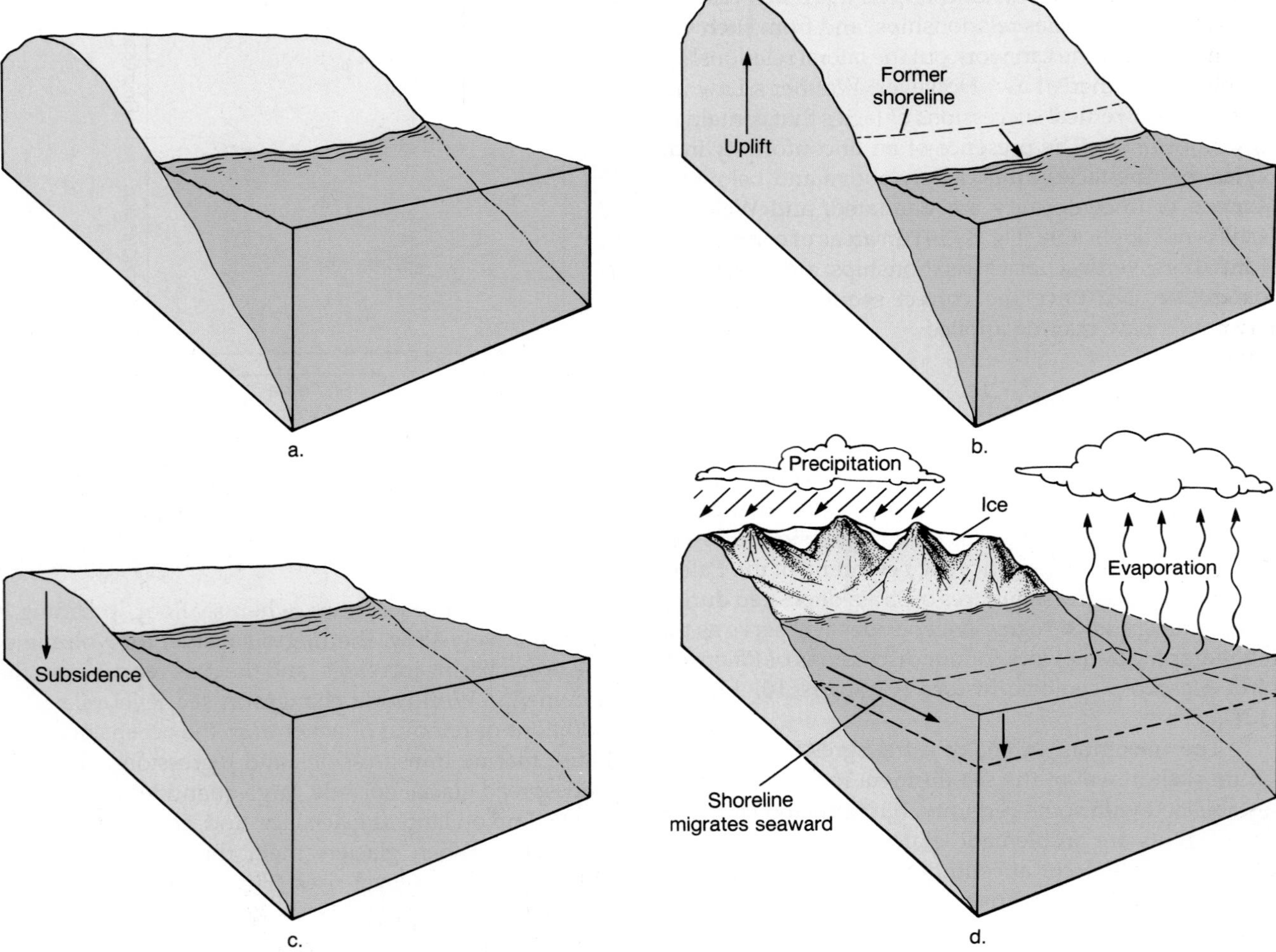


**Figure 3–16** Causes of transgressions and regressions. a. Reference diagram showing position of shoreline. b. Uplift of a continent causes the shorline to migrate seaward resulting in a marine regression. c. A marine transgression occurs when a continent subsides. d. When a large volume of sea water is tied up on land as glacial ice, sea level is lowered, and a marine regression occurs. When glaciers melt, however, the water returns to the sea, sea level rises, and a transgression occurs.

addition to their use in determining relative ages of strata, fossils are important in determining environments of deposition (Chapter 4), and they provide some of the evidence for evolution (Chapter 5).

During most of historic time most people believed that fossils were inorganic in origin (see Prologue). Such perceptive observers as Leonardo da Vinci in 1508, Robert Hooke in 1665, and Nicolaus Steno in 1667 recognized the true nature of fossils (Fig. 3–18), but their views were largely ignored. By the late eighteenth century, however, the evidence had become overwhelming that fossils had indeed once been living organisms. It was also apparent by the nineteenth century that many fossils represented organisms no longer living.

## Fossilization

Our definition of fossils includes the phrase "remains or traces." Remains, commonly called **body fossils** (Fig. 3–19), are usually the hard skeletal elements such as bones, teeth, and shells, rarely the soft parts of organisms. The exceptional cases of preservation by freezing and mummification provide us with partial or nearly complete body fossils with flesh, hair, and stomach contents. **Trace fossils** include tracks, trails, burrows, borings, nests, or any other indication of the activities of organisms (Fig. 3–20). Particularly interesting trace fossils are nests built by duckbilled dinosaurs and the so-called Devil's corkscrew (Fig. 3–20), which is the burrow of an extinct beaver. Fossilized feces, called **coprolites,** provide evidence of activity of organisms (Fig. 3–21) and may provide important information on the diet and size of the animal that produced them.

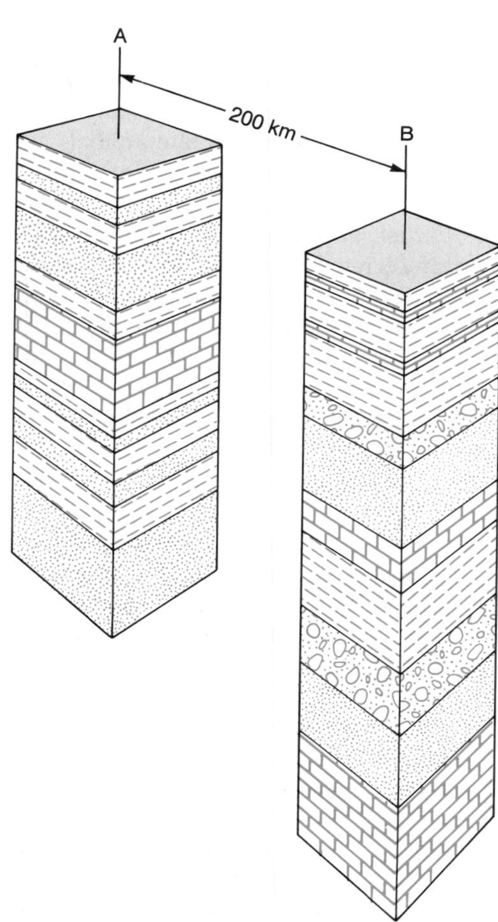


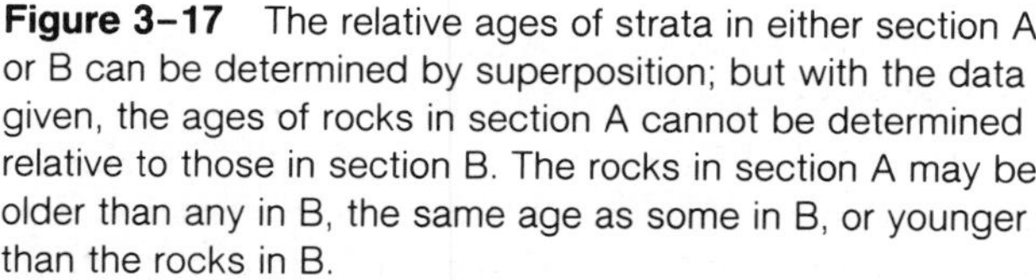

**Figure 3–17** The relative ages of strata in either section A or B can be determined by superposition; but with the data given, the ages of rocks in section A cannot be determined relative to those in section B. The rocks in section A may be older than any in B, the same age as some in B, or younger than the rocks in B.

**Figure 3–18** *Glossopterae* or tonguestones, as they were called during the Middle Ages, were thought to fall from the sky and kill people. Nicholas Steno determined that tonguestones were actually fossil shark teeth. This illustration of a shark was published in 1667 by Steno after he dissected a shark that had washed ashore in Italy.

a.

b.

**Figure 3–19** Body fossils are the actual remains of organism. a. Shells of brachiopods in Ordovician limestone. b. Dinosaur bones being excavated from the Late Cretaceous Two Medicine Formation near Choteau, Montana.

Fossils in general are quite common. The remains of various microorganisms are by far the most abundant and in many ways the most useful. Shells of marine animals are also common and are easily collected in many areas (Figure 3–19). Despite the abundance of fossils, however, they represent very few of the organisms that lived at any one time, since any potential fossil must escape the ravages of destructive processes such as waves or running water, scavengers, exposure to the atmosphere, and bacterial action. Hard skeletal elements are more resistant to destructive processes than soft parts, but even skeletal elements must be buried in some protective medium such as mud or sand if they are to be preserved; and even if buried, skeletal elements may be dissolved by groundwater or destroyed by alteration of the host rock.

Considering all the ways in which potential fossils may be destroyed, it comes as no surprise that the fossil record is biased toward those organisms with hard parts and those organisms that lived in areas of active sedimentation. Accordingly, most fossils are preserved in sediments or sedimentary rocks. Volcanic ash falls may also contain fossils (Perspective 3–1) but fossils are rare to nonexistent in other types of igneous rocks and in metamorphic rocks. Despite its biases, the fossil record is our only record of prehistoric life. Indeed, it is only through fossils that we have any knowledge of such extinct life forms as trilobites and dinosaurs.

## Types of Preservation

Fossils may be preserved as **unaltered remains,** in which case they retain their original structure and composition. The pollen and spores of plants have a tough outer cuticle that is chemically resistant to change (Fig. 3–22). The bones and teeth of vertebrate animals are composed of the mineral apatite, a variety of calcium phosphate, and are commonly unaltered in composition and structure. Partial or complete insects may be trapped in the resin secreted by some plants, especially by coniferous trees. When this resin is buried, it hardens to become amber and preserves in exquisite detail the remains contained therein (Fig. 3–22).

Unusual types of preservation of unaltered remains include preservation in tar, mummification, and freezing. The La Brea Tar Pits of southern California contain numerous tar-impregnated bones of Pleistocene animals that otherwise retain their original composition and structure (Fig. 3–23). Mummification involves air drying and shriveling of soft parts such as muscles and tendons before burial. Although very rare, partial mummies of dinosaurs show the surface pattern of the skin (Fig. 3–24), and some contain preserved stomach contents.

Perhaps the most interesting example of preservation of unaltered remains is by freezing. Only a few types of animals have been found frozen, such as mammoths, woolly rhinoceroses, musk oxen, and a few others, and all are from Late Pleistocene-age deposits. The frozen mammoths of Siberia are no doubt the best known example of this mode of preservation (Fig. 3–24).

**Altered remains** are those fossils that have been changed structurally, chemically, or both. Quite commonly, bones, teeth, and shells have mineral matter added to their pores and cavities after burial. This process, called *permineralization,* increases the preservation potential of fossils by increasing their durability.

a.

b.

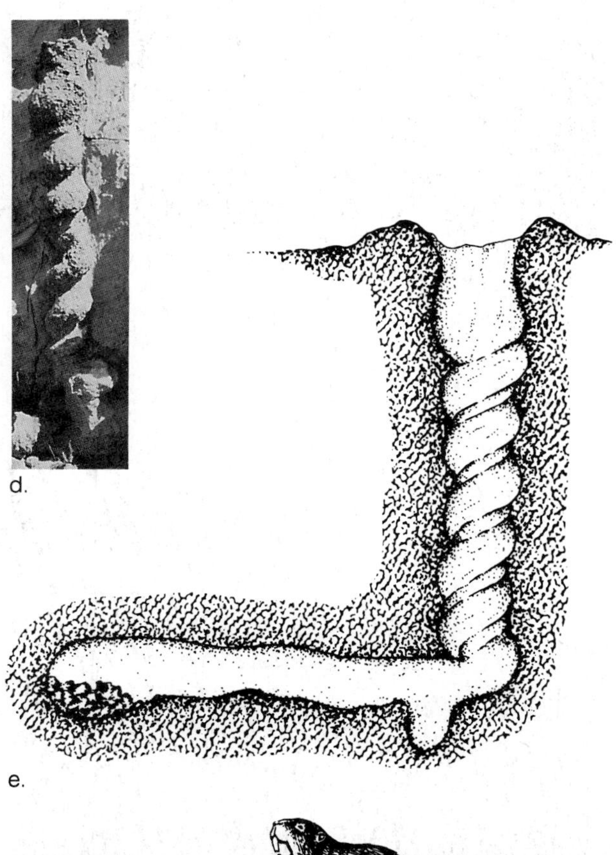

d.

e.

c.

f.

**Figure 3–20** Trace fossils. a. Vertical burrows in the Pliocene Etchegoin Formation, California. b. Crawling trace in the Lower Pennsylvanian Fentress Formation, Tennessee (Photo courtesy of Molly Fritz Miller, Vanderbilt University). c. Dinosaur tracks in Cretaceous age rocks in the bed of the Paulexy River, Texas. d. In Nebraska these spiral burrows, which may reach 2.5 m underground are called "Devil's corkscrew." (Photo courtesy of L. D. Martin and D. K. Bennett). e. Reconstruction of a complete burrow. f. Restoration of the small Miocene beaver, *Paleocastor,* that made the burrows.

**Figure 3–21** Fossilized fish feces (coprolite) about 320 million years old. Specimen is about 12 cm long. (Photo courtesy of Gordon Wood, Amoco Production Co.)

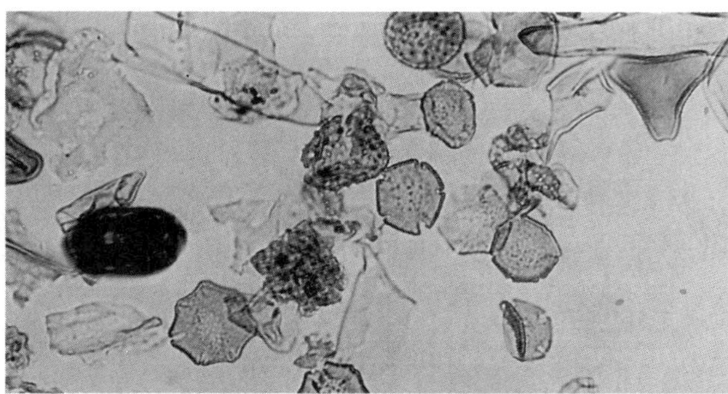

a.

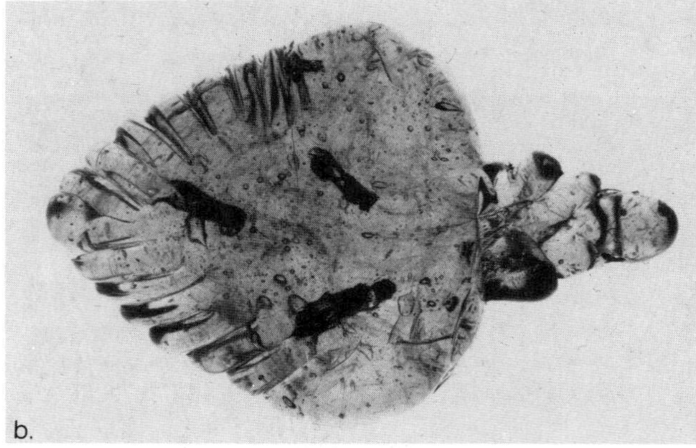

b.

**Figure 3–22** a. Late Eocene pollen from the Eurla Basin, Australia. b. Insects in amber. (Photo courtesy of R. V. Dietrich, Central Michigan University)

a.

b.

**Figure 3–23** Pleistocene fossil mammals in tar pits at Rancho La Brea, in Los Angeles, California. a. A mural by Charles R. Knight showing late Pleistocene mammals of Rancho La Brea, California. A giant ground sloth is trapped in a tar pit, while giant vultures, saber-toothed cats, and dire wolves wait nearby. A herd of mammoths is shown in the background. b. Excavation of mammoth bones from excavation Pit 9, Rancho La Brea in about 1913.

a.

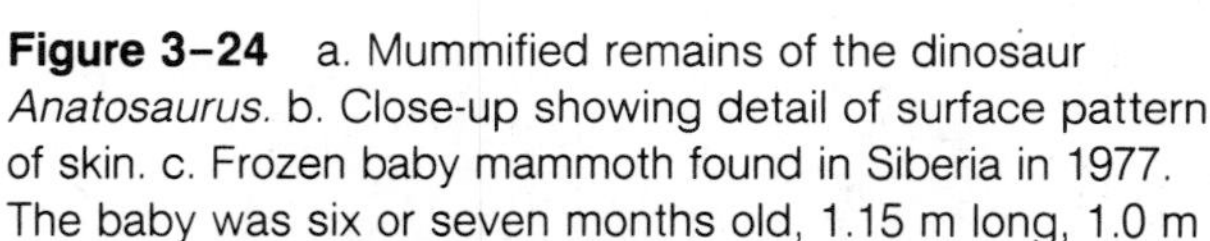

b.

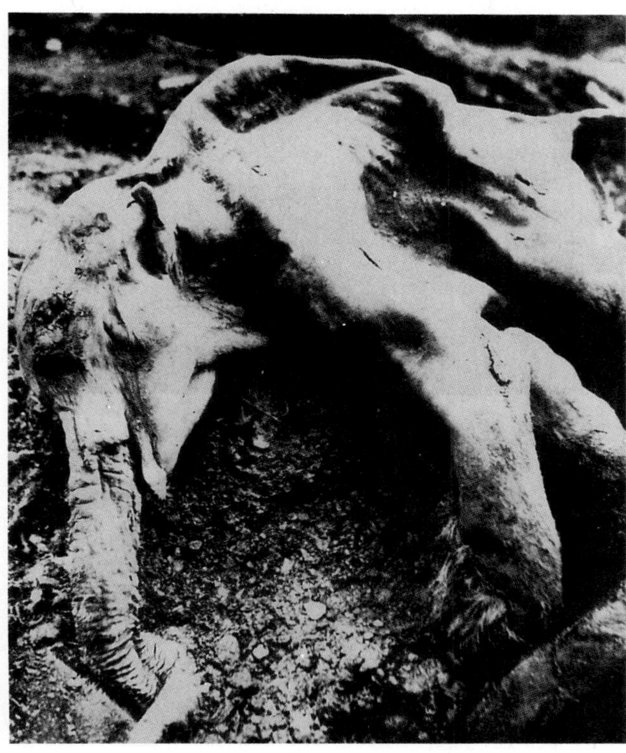

c.

**Figure 3–24** a. Mummified remains of the dinosaur *Anatosaurus.* b. Close-up showing detail of surface pattern of skin. c. Frozen baby mammoth found in Siberia in 1977. The baby was six or seven months old, 1.15 m long, 1.0 m tall, and when alive had a hairy coat. Most of the hair has fallen out except around the feet. The carcass is about 40,000 years old.

The shells of many marine invertebrates are composed of an unstable form of calcium carbonate called aragonite. When buried, these shells commonly recrystallize in the more stable form of calcium carbonate called calcite. A few plants and animals have shells or skeletal elements composed of opal, a variety of silicon dioxide ($SiO_2$). These also commonly recrystallize. While recrystallization involves no change in chemical composition or outward appearance, the microscopic crystal structure of the shell is altered.

We know that brachiopods, clams, snails, and many other invertebrates have skeletons of calcium carbonate, yet in some strata we find these shells composed of silicon dioxide or pyrite ($FeS_2$). These shells have been altered by a process called *replacement,* during which the original skeletal material is replaced by a compound of a different composition. Insect skeletons replaced by silicon dioxide is a particularly interesting example of this type of preservation (Fig. 3–25).

Wood may be preserved by the complete replacement of woody tissue by silicon dioxide in the form of chert or opal and, more rarely, by calcite. Wood preserved in this manner is often called *petrified,* a term that means to become stone. Replacement particularly when it is by silicon dioxide, may be so exact that even details of cell structure are finely preserved (Fig. 3–25). In many cases the form of the woody tissue actually remains unaltered, and the pore spaces have simply been filled in by silicon dioxide.

The altered remains of plants may be preserved by a process called *carbonization,* in which the volatile elements of organic material, such as a leaf, vaporize, leaving a thin carbon film (Fig. 3–26). Fish, graptolites, and soft-bodied animals such as worms may also be carbonized. Exceptional preservation in Eocene deposits of Germany show the carbonized body outlines of mammals.

In addition to unaltered and altered remains, fossils may be preserved as **molds** and **casts.** Molds and casts of

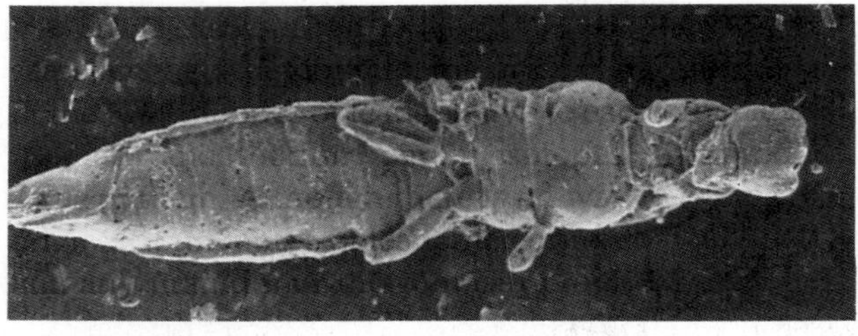

a.

b.

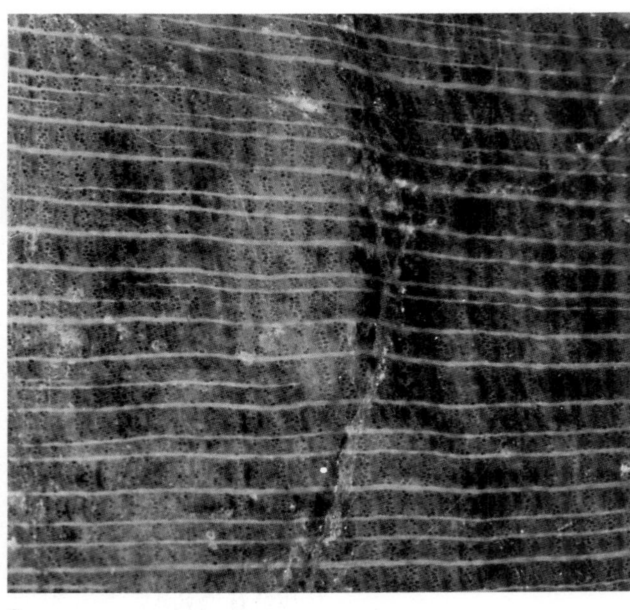

c.

**Figure 3-25** a. Insects replaced by silicon dioxide in the Miocene Button Bed of the Barstow Formation, southern California. b. Petrified wood of Miocene age in Ginkgo Petrified Forest, Washington. Replacement is by silicon dioxide. c. Section of petrified wood showing preserved cell structure.

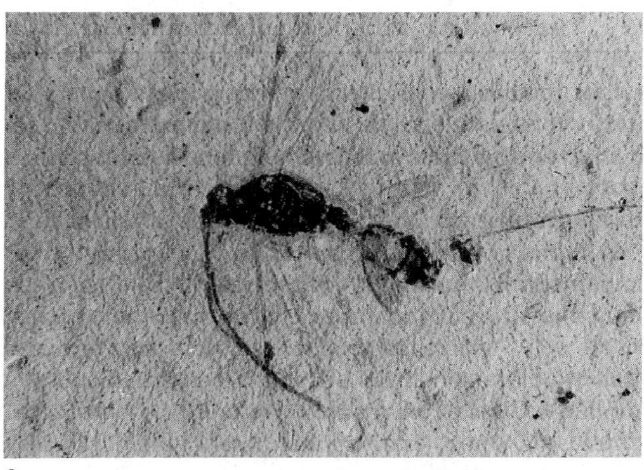

a.

b.

**Figure 3-26** Altered remains preserved by carbonization. a. Insects from the Oligocene Renova Formation, Montana. b. Leaves from the Pennsylvanian Mazon Creek flora, Illinois.

shelled marine organisms such as clams and brachiopods are quite common (Fig. 3–27). Molds are formed when the remains of a buried organism are dissolved, thereby leaving a cavity with the external shape of the organism (an external mold). If the mold is later filled with sediment or mineral matter, an external cast is formed. Internal molds are cavities showing the inner features of shells, and internal casts are formed when such cavities are filled with sediment or mineral matter (Fig. 3–27).

## Fossils and Uniformitarianism

Uniformitarianism in its simplest form can be expressed as "the present is the key to the past." While this is an

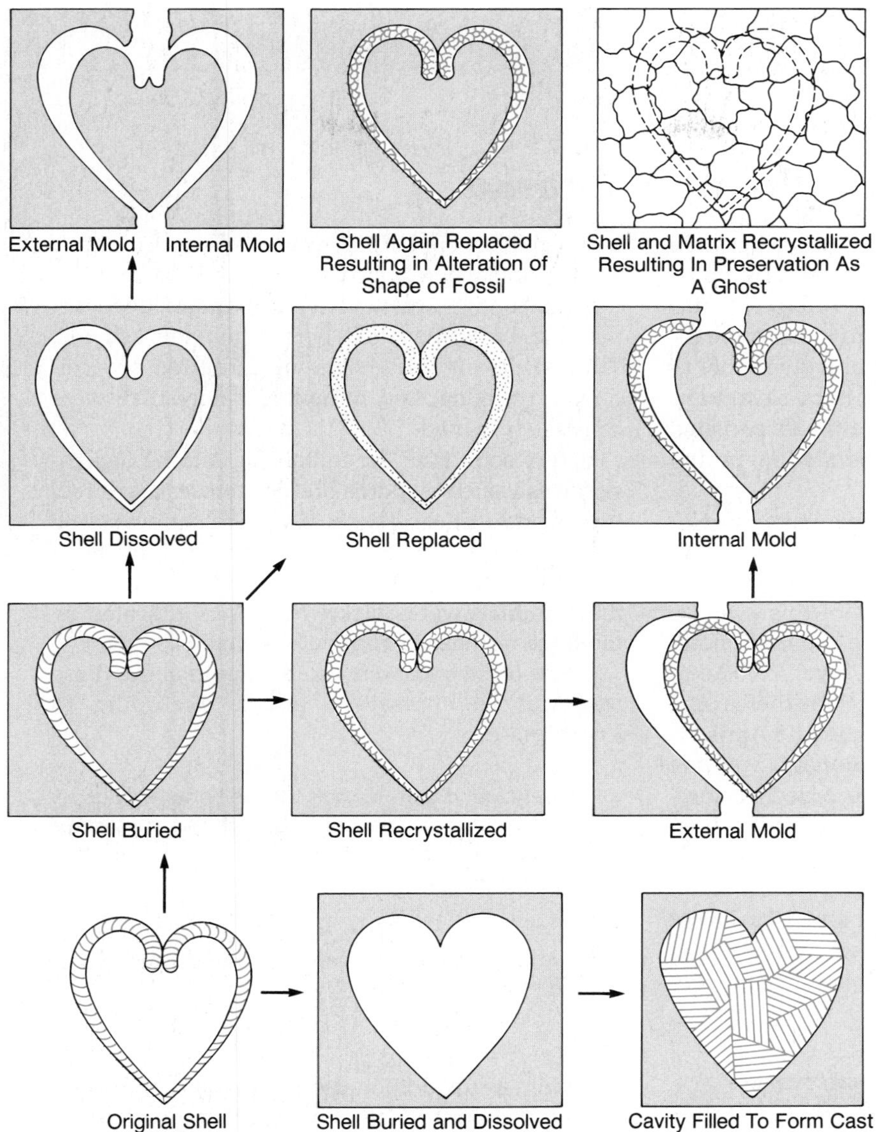


**Figure 3–27** Diagram showing how a shell may be preserved by replacement, recrystallization, and by the formation of molds and casts. The arrows show possible sequences of preservation.

oversimplification, it is true that observations of the modern processes leading to the burial of organic remains give us a better understanding of how fossils were preserved.

Fossilization begins with the death of an organism and its burial in sediment or volcanic ash. Only rarely is the cause of death of a fossil organism apparent, but the conditions of burial commonly can be inferred by comparison with modern processes. For example, Charles Darwin visited Uruguay during the 1830s and learned that millions of horses and cattle had died during a prolonged drought. When the drought broke, the rivers flooded and buried thousands of these animals in floodplain deposits. Similarly, many of the fossil mammals in the White River Badlands of South Dakota were buried in floodplain deposits (Fig. 3–28).

## Perspective 3-1

# A Miocene Catastrophe in Nebraska

Ten million years ago, in what is now northeastern Nebraska, a vast grassland was inhabited by short-legged, aquatic rhinoceroses, camels, horses, saber-toothed deer, land turtles, and many other animals. This was a temperate savannah habitat with life as varied and abundant as it is now on the savannahs of East Africa. But many of these animals perished when a vast cloud of volcanic ash rolled in, probably from the southwest.

Michael Voorhies, of the University of Nebraska State Museum, and his crews have recovered the remains of hundreds of victims of this catastrophe (Fig. 1). The magnitude of this event is difficult to imagine, since the source of the ash cloud may have been in New Mexico, over 1,000 km away. We know from historic eruptions that an ash cloud can travel this far. For example, the April 11, 1815, eruption of Tambora in Indonesia was the deadliest volcanic eruption in recorded history. About 88,000 people perished, an ash layer 22 cm thick accumulated 400 km to the west, and some ash fell more than 1,500 km from the volcano.

The Nebraska ash fall was probably 10 to 20 cm thick, although it was redistributed by wind and collected to 1 m deep in a depression where the fossils were recovered. This depression was the site of a water hole where animals congregated. Many animals, including three-toed horses, camels, deer, turtles and birds, perished here when the ash fell. Their skeletons show signs of partial decomposition, scavenging, and trampling before they were completely buried.

Very soon after the initial ash fall, the depression was visited by herds of rhinoceroses and a few horses and camels. These, too, were suffocated by ash, but in this case the ash may simply have been blown into the depression by the wind. In any case, these animals were quickly buried, as indicated by the large number of complete skeletons.

One of the most remarkable things about these fossils is the preserved detail. According to Voorhies,

> Rarely found parts such as tongue bones, cartilages, tendons, and tiny bones in the middle ear all survive in exquisite detail and in their correct positions.[1]

[1]1981, Ancient Ashfall Creates a Pompeii of Prehistoric Animals, *National Geographic,* 159, no. 1: 69.

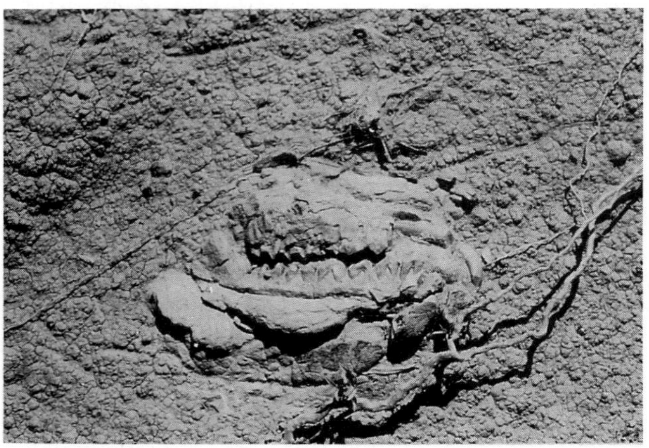

**Figure 3–28** Fossils in sediments of White River Badlands, South Dakota.

Numerous additional examples exist of the burial of organic remains. Small animals are trapped and die in the sticky residue of southern California oil seeps (Fig. 3–29). A single storm on the continental shelf may cause the burial of shallow marine invertebrates. Moroever, modern lake sediments contain leaves, insects, and occasionally fish. Following the 1980 eruption of Mt. St. Helens, Washington, mudflows transported and buried logs and stumps and partially buried trees in growth position. The same or very similar processes occurred numerous times during the Eocene in what is now Yellowstone National Park, Wyoming (Fig. 3–29). Events such as the burial of Pompeii A.D. 79 by volcanic ash remind us that humans too are potential fossils.

It is important to note from these examples that the burial of organisms by natural processes is a common occurrence. The conditions of burial range from slow sedimentation in lakes to rapid burial by ash falls and

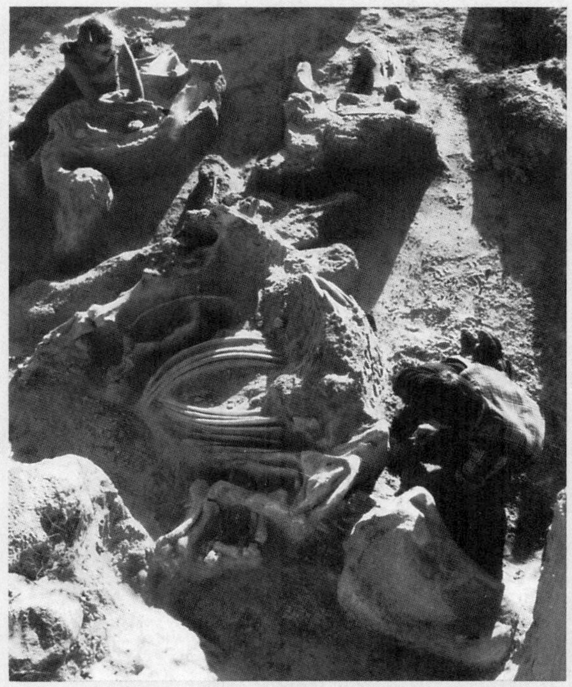

Additionally, the association of babies, juveniles, young adults, and mature adults gives us a better understanding of herd structure. The rhinoceros herds, for example, were probably made up of a single bull and several cows with their calves.

Only rarely are paleontologists fortunate enough to find and recover so many, well-preserved, associated vertebrate animals. A Late Miocene catastrophe turns out to be our good fortune since it provides us with a unique glimpse of what life was like in Nebraska 10 million years ago.

**Figure 1** Paleontologists excavating rhinoceros (foreground) and horse (background) skeletons from volcanic ash near Orchard, Nebraska. (Photo courtesy of Nebraska State Museum)

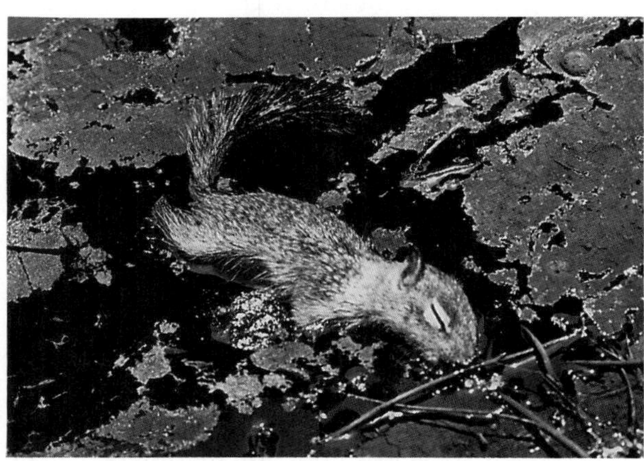

a.

b.

c.

**Figure 3–29** a. Recent victim, a ground squirrel, trapped in the sticky residue of a southern California oil seep. b. and c. Trees buried in growth position. b. This stump was partly buried by a 1980 mudflow in Smith Creek, Washington, following the eruption of Mt. St. Helens. c. Fossilized stump in the Eocene Lamar River Formation, Yellowstone National Park, Wyoming. (b and c courtesy of Amy Karowe, University of Oslo, Norway)

## Perspective 3-2

# The Irish Elk of Ballybetagh Bog, Ireland

Anthony B. Barnosky reports that at least 118 skulls of the giant deer *Megaloceros giganteus*, commonly called the Irish elk, have been recovered from Ballybetagh bog near Dublin, Ireland (Fig. 1). All of the skulls came from antlered males, a fact traditionally explained by the males' large, heavy antlers making them more susceptible to becoming mired in mud or drowning. Barnosky reasoned that if true, the miring-drowning hypothesis should be supported by several lines of evidence deduced from the hypothesis. These deductions and the field observations are tabulated below:

| *Deductions of hypothesis* | *Observations* |
|---|---|
| **1.** Overrepresentation of males. | All of the skulls are from antlered males. |
| **2.** Males with relatively large antlers. | The antlers are relatively small compared to more than 150 specimens from other areas. |
| **3.** Many skeletons should be articulated, or at least the feet and legs should be vertically oriented. | No observations consistent with deduction. Many bones broken as if trampled, and gnawed by rodents and carnivores. Bones appear to have lain on surface for several years. |
| **4.** Sticky clay deposits more than 1 m thick. | Clay deposit varies from 6 to 65 cm thick. |
| **5.** Bedding of bog deposits disturbed by trampling. | No deformation exists; clay layers are undisturbed. |
| **6.** Water deeper than height of an Irish elk, about 2.2 m. | In most places water depth was less than 2 m; bones are abundant only at edges of ancient bog. |

mudflows, but all are consistent with the concept of uniformitarianism. We have every reason to believe that organisms in the fossil record were buried under comparable circumstances.

There were no human observers through most of geologic time, so we have no way to check directly on our interpretations for the burial of fossil organisms. This does not mean, however, that we can only speculate on the nature of prehistoric events. We can make observations today that lend credence to a hypothesis or force us to reject it. For example, a hypothesis explaining the preservation of Irish elk in Pleistocene bog deposits in Ireland makes testable predictions (Perspective 3–2).

## FOSSILS AND TIME

The utility of fossils in relative dating and geologic mapping was clearly demonstrated in England and France in the early nineteenth century. William Smith (1769–1839), an English civil engineer involved in surveying and building canals in southern England, independently recognized the principle of superposition, reasoning that the fossils lowest in a sequence of strata are oldest and those higher in the sequence are younger. He made observations in many natural exposures, mines, and quarries and discovered that the relative sequence of fossils, and particularly groups or assemblages of fossils, is consistent from area to area. In short, he discovered a method whereby the relative ages of strata in widely separated areas regardless of their composition could be determined by their fossil content (Fig. 3–30).

By recognizing the relationship between strata and fossils, Smith was able to predict the rock units that would be encountered in digging a canal. This knowledge also helped him determine the best route for a canal and the best areas for bridge foundations. His observations proved to be of economic significance, and his services were in great demand.

The discoveries of Smith, and those of other geologists such as Alexandre Brongniart in France, served as

The overrepresentation of males (deduction 1) is the only observation consistent with the miring-drowning hypothesis. Accordingly, the hypothesis must be rejected for lack of supporting evidence. Barnosky proposes a winterkill hypothesis which he claims is consistent with all six deductions listed above. [The winterkill hypothesis is]

> derived from observations of modern [deer], that males visited Ballybetagh bog more often than females did during winters, when unfit animals died and decomposed near the water's edge, in some cases on the ice, and were scavenged and trampled. The fragmented, weathered bones would have fallen into the lake as they were washed, pushed, or kicked downslope from the adjacent shore or as they dropped through the melting ice.

**Figure 1** Restoration of the Irish elk *Megaloceros giganteus*, which lived in Europe and Asia during the Pleistocene. Large males had an antler spread of more than 3 m.

In addition to the six points listed earlier, the winterkill hypothesis predicts that the fossils should show evidence of: (1) seasonal deaths, (2) overrepresentation of juveniles and older adults, (3) poor physical condition, and (4) suitable modern analogs. All predictions are borne out by evidence except that very few juveniles were found. But this is explained by the fact that juveniles, with their smaller and less durable bones, have a lower preservation potential and are typically underrepresented in all fossil accumulations.

Source: Branosky, Anthony B. 1985. Taphonomy and Herd Structure of the Extinct Irish Elk, *Megaloceros giganteus, Science,* 228: 340–344.

the basis for what is now called the **principle of fossil succession.** According to this principle, fossil assemblages succeed one another through time in a regular and determinable order. The validity and successful use of this principle depend on three factors: (1) life has varied through time, (2) fossil assemblages are recognizably different from one another, and (3) the relative ages of the fossil assemblages can be determined. Observations of fossils in older versus younger strata clearly demonstrate that life-forms have changed. And since this is true, fossil assemblages (point 2) are recognizably different. And finally, superposition can be used to demonstrate the relative ages of the fossil assemblages.

## THE RELATIVE TIME SCALE

Recall from Chapter 2 that many eighteenth-century geologists, such as the Neptunists, believed that the relative ages of rocks could be determined by their composition (Fig. 2–2). This same thinking prevailed into the nineteenth century, but eventually it became apparent that composition is not a sufficient criterion for determining relative ages, and geologists in England and Europe began using stratigraphic position (superposition) and fossil content (fossil succession) for their subdivisions. Their efforts, largely between 1830 and 1842, resulted in the recognition of rock bodies called **systems** and the construction of a geologic column that is the basis for the relative time scale (Fig. 3–31). A short discussion of how some of the systems were recognized and defined will be helpful in understanding how the geologic column and relative time scale became established.

In the 1830s, Adam Sedgwick described and named the Cambrian System for rocks exposed in northern Wales, while Sir Roderick Impey Murchison, working in southern Wales, named the Silurian System. Unfortunately, Sedgwick's strata contain few fossils, and he paid little attention to the few present. His Cambrian System based on lithology could not be recognized be-

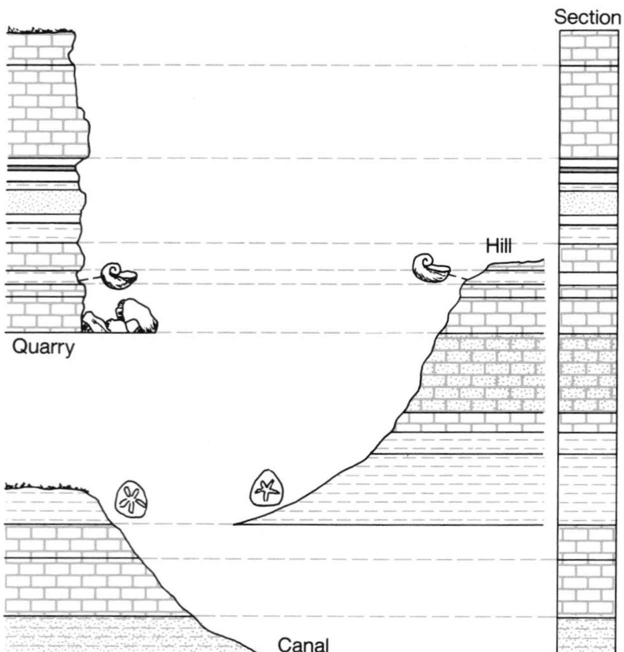


**Figure 3–30** Generalized diagram showing how William Smith used fossils to identify strata of the same age in different areas. In this example, the composite section on the right shows the relative ages of all strata in this area.

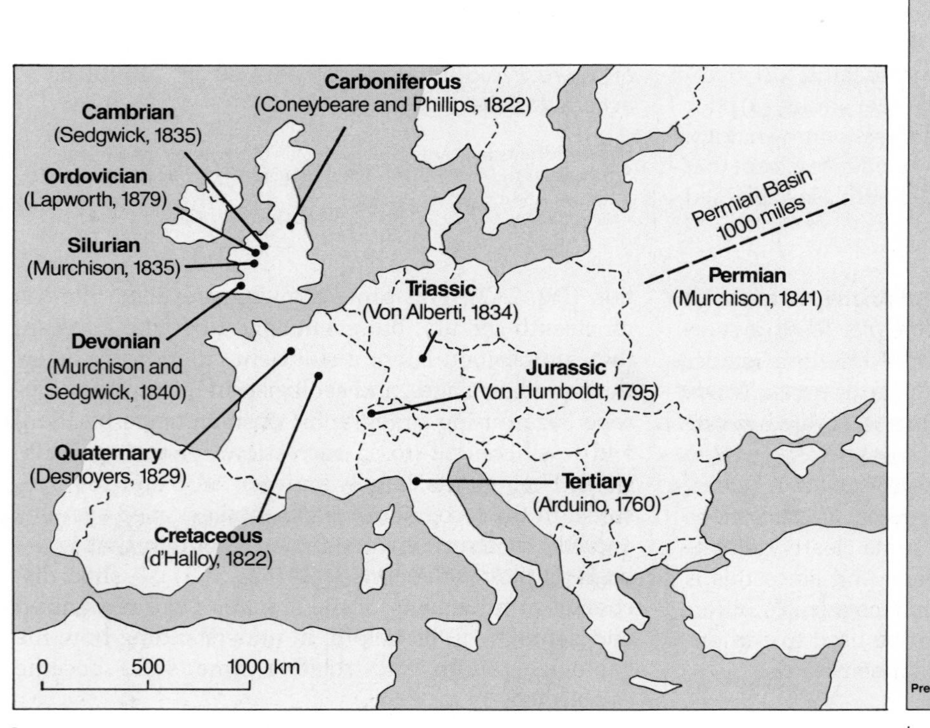


a.

| Relative durations of major geologic intervals | Era | Period | | Epoch |
|---|---|---|---|---|
| Cenozoic | Cenozoic | Quaternary | | Recent |
| Mesozoic | | | | Pleistocene |
| Paleozoic | | Tertiary | | Pliocene |
| | | | | Miocene |
| | | | | Oligocene |
| | | | | Eocene |
| | | | | Paleocene |
| | Mesozoic | Cretaceous | | |
| | | Jurassic | | |
| | | Triassic | | |
| | Paleozoic | Permian | | |
| | | Carboniferous | Pennsylvanian | |
| | | | Mississippian | |
| | | Devonian | | |
| | | Silurian | | |
| | | Ordovician | | |
| | | Cambrian | | |
| Precambrian | Precambrian | | | |

b.

**Figure 3–31** Developement of the geologic column and relative time scale. a. Geologists in Great Britain and Europe defined the geologic systems at the locations shown. b. A geologic column was constructed by arranging the systems in their correct relative sequence, so the geologic column is in effect a relative time scale. Note that the Carboniferous, which is recognized in Europe, is represented by two systems in North America, the Mississippian and Pennsylvanian.

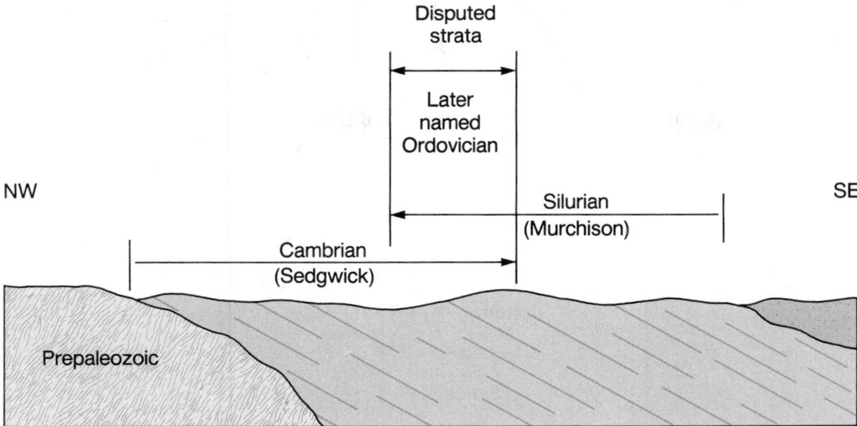


**Figure 3–32** Simplified cross section showing the disputed interval of strata that Sedgwick and Murchison each claimed belonged to their respective systems. The dispute was resolved in 1879 when Charles Lapworth assigned the disputed strata to a new system, the Ordovician.

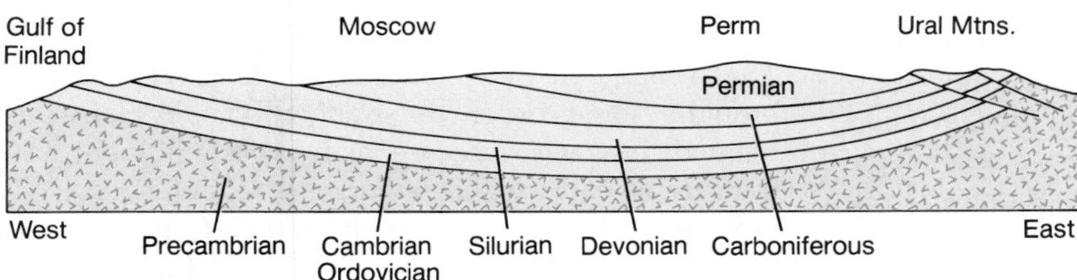


**Figure 3–33** Generalized east-west cross section showing the stratigraphic relationships of Permian and older strata in Russia. During his visit to Russia in 1840–1841, Murchison recognized Silurian, Devonian, and Carboniferous strata by their fossils. He assigned the fossiliferous strata overlying the Carboniferous to a Permian System.

yond the area where it was described. Murchison, on the other hand, carefully described fossils characteristic of his Silurian System, so strata of Silurian age could be recognized elsewhere.

In 1835 Sedgwick and Murchison jointly published the results of their studies. Stratigraphic position clearly demonstrated that Silurian strata were younger than Cambrian strata, but it was also apparent that Sedgewick's and Murchison's systems partially overlapped (Fig. 3–32). Each man claimed that the strata in the overlapping interval belonged to his system. This boundary dispute resulted in a feud that ended their long-standing friendship and was not resolved until 1879, when Charles Lapworth suggested that the disputed strata be assigned to a new system, the Ordovician. In the final analysis, three systems were named, their stratigraphic positions were known, and distinctive fossils of each system were recognized.

Before Sedgewick's and Murchison's feud, they also named the Devonian system for rocks exposed near Devonshire, England (Fig. 3–31). Rocks defined as Devonian contained fossils that differed from those in underlying Silurian strata and were overlain by rocks of the Carboniferous System,[2] which had been described earlier. In 1840–41, Murchison visited western Russia, where he identified strata as Silurian, Devonian, and Carboniferous by their fossils. Overlying the Carboniferous strata were fossil-bearing rocks that he assigned to a Permian System (Figs. 3–31, 3–33).

We need not discuss the specifics of where and when the other systems were defined, except to note that superposition and fossil content were the criteria

[2] In North America two systems, the Mississippian and the Pennsylvanian, are recognized and correspond to the Lower and Upper Carboniferous, respectively.

**Table 3–1** Classification of stratigraphic units.

| UNITS DEFINED BY CONTENT | | UNITS EXPRESSING OR RELATED TO GEOLOGIC TIME | |
|---|---|---|---|
| *LITHOSTRATIGRAPHIC UNITS* | *BIOSTRATIGRAPHIC UNITS* | *TIME-STRATIGRAPHIC UNITS* | *TIME UNITS* |
| Supergroup | Biozones | Eonothem | Eon |
| Group | | Erathem | Era |
| Formation | | System | Period |
| Member | | Series | Epoch |
| Bed | | Stage | Age |

used in their recognition. The important point is that geologists were piecing together a composite geologic column (Fig. 3–31). This geologic column is, in effect, a relative time scale because the systems are arranged in their correct chronological order. Thus, we can refer to the Devonian System when speaking of the stratigraphic position of rocks, and we can also use Devonian as a term to designate a particular interval of geologic time, the Devonian Period. We will have more to say about systems and periods in the following section.

One additional aspect of the geologic column should be mentioned. Notice in Figure 3–31 that all rocks beneath Cambrian strata are called Precambrian. Long ago geologists realized that these strata contain few fossils (none were positively identified until the 1950s) and that they could not be effectively subdivided into systems. Terminology for Precambrian rocks and time is discussed in Chapters 8 and 9.

**Figure 3–34** The Chattanooga Shale, shown here in Tennessee, straddles the boundary between the Devonian and Mississippian systems. (Photo courtesy of Molly Fritz Miller, Vanderbilt University).

## MODERN STRATIGRAPHIC TERMINOLOGY

The recognition of systems and a relative time scale brought some order to stratigraphy. Problems remained, however, because many rock units are time transgressive (discussed earlier in this chapter), so a rock unit may belong to one system in a particular area but also be included in another system elsewhere, or it may simply straddle the boundary between systems (Fig. 3–34). In order to deal with both rocks and time, modern stratigraphic terminology includes two fundamentally different kinds of units: those defined by their content and those expressing or related to geologic time (Table 3–1).

Units defined by their content include **lithostratigraphic** and **biostratigraphic units.** Lithostratigraphic[3] units are defined by physical attributes of the rocks, such as rock type, with no consideration of time of origin. The basic lithostratigraphic unit is the **formation,** which is a mappable rock unit with distinctive upper and lower boundaries. Formations may be lumped together into larger units called *groups* or *supergroups* or divided into smaller units called *members* and *beds* (Fig. 3–35 and Table 3–1).

Unfortunately, the same lithostratigraphic unit may have different names in different areas. For example, a widespread Cambrian sandstone is called the Flathead Formation in Montana and Wyoming, but in South Dakota it is called the Deadwood Formation. Situations such as this arise for two reasons. One is historic; formations were named in different areas when geologists had

[3] *Lith-* and *litho-* are prefixes meaning stone or stonelike.

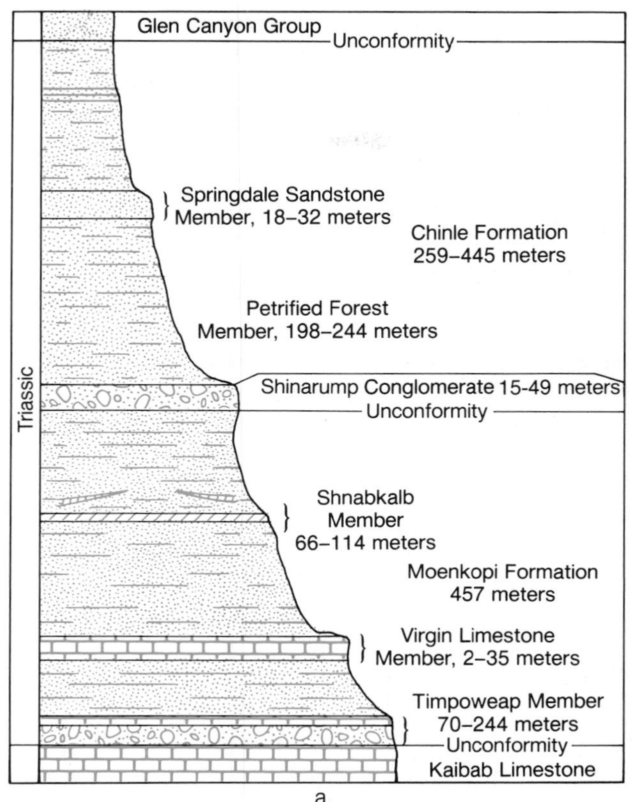


a.

b.

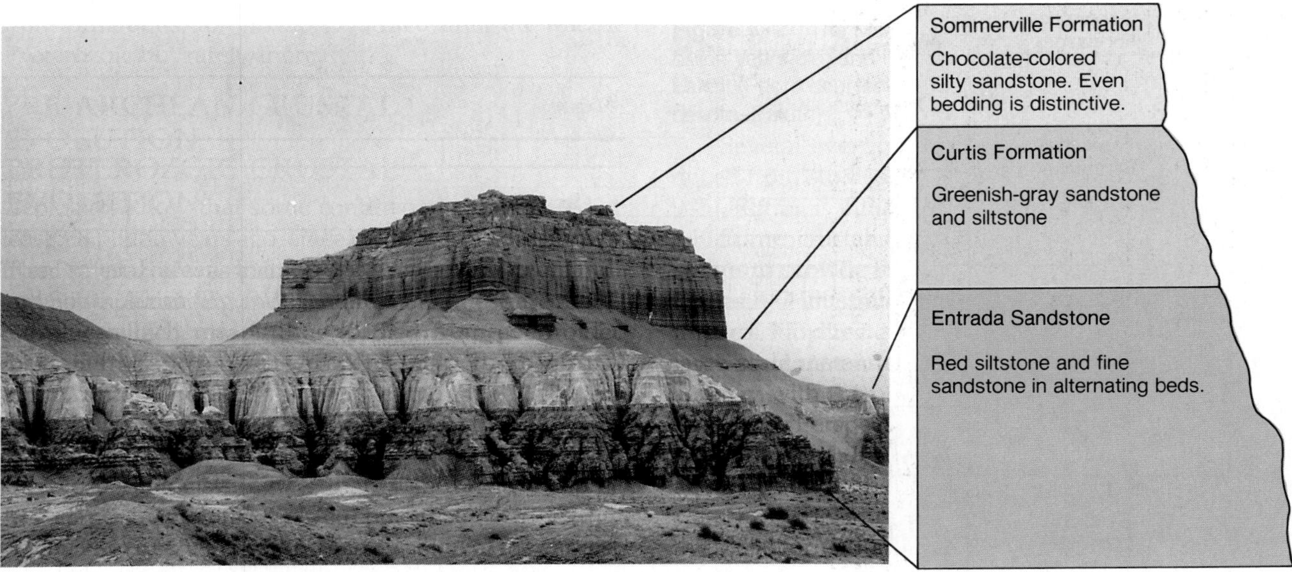


c.

**Figure 3–35** Lithostratigraphic units. a. Illustration of the formations and members in the Zion National Park area of Utah. b. The Madison Group in Montana consists of two formations, the Lodgepole Formation and the overlying Mission Canyon Formation. The Mission Canyon Formation is the rock unit exposed on the skyline. The underlying Lodgepole Formation is the rock covered on the slopes below. c. Mesozoic formations exposed in Goblin Valley State Park, Utah.

no knowledge of their total geographic distribution, and the original formation names were retained. The second reason is because of complex stratigraphic relationships (Figure 3–46 shows one way in which lithostratigraphic terminology can be applied in areas of stratigraphic complexity).

Fossil content is the only criterion used to define biostratigraphic units, which are bodies of strata containing recognizably distinct fossils. Biostratigraphic unit boundaries do not necessarily correspond to lithostratigraphic boundaries (Fig. 3–36). The fundamental biostratigraphic unit is the **biozone.** Several types of biozones are recognized, three of which are discussed in the section on correlation.

The category of units expressing or related to geologic time includes **time-stratigraphic units** and **time units** (Table 3–1). Time-stratigraphic units are units of rock that were deposited during a specific interval of time. The **system** (Fig. 3–31) is the fundamental time-stratigraphic unit. It is based on rocks in a particular area, the stratotype, and is recognized beyond the stratotype area primarily on the basis of fossil content.

Time units are simply units designating specific intervals of geologic time. For example, the Cambrian Period is defined as the time during which strata of the Cambrian System were deposited. The basic time unit is the **period,** but smaller units (epoch and age) are also recognized. The time units period, epoch, and age corre-

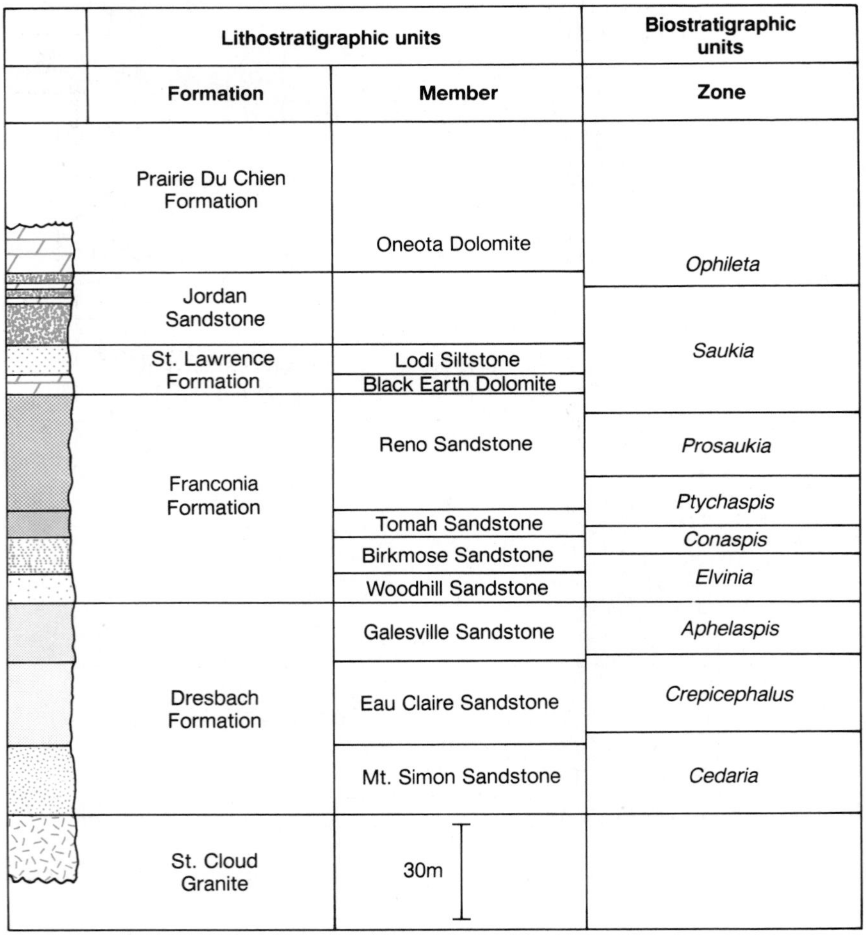

| Lithostratigraphic units | | Biostratigraphic units |
|---|---|---|
| **Formation** | **Member** | **Zone** |
| Prairie Du Chien Formation | Oneota Dolomite | *Ophileta* |
| Jordan Sandstone | | *Saukia* |
| St. Lawrence Formation | Lodi Siltstone<br>Black Earth Dolomite | |
| Franconia Formation | Reno Sandstone | *Prosaukia* |
| | | *Ptychaspis* |
| | Tomah Sandstone | *Conaspis* |
| | Birkmose Sandstone | |
| | Woodhill Sandstone | *Elvinia* |
| Dresbach Formation | Galesville Sandstone | *Aphelaspis* |
| | Eau Claire Sandstone | *Crepicephalus* |
| | Mt. Simon Sandstone | *Cedaria* |
| St. Cloud Granite | 30m | |

**Figure 3–36** Relationships of biostratigraphic units to lithostratigraphic units in southeastern Minnesota. Notice that the biozone boundaries do not necessarily correspond to lithostratigraphic boundaries.

spond to the time-stratigraphic units system, series, and stage, respectively (Table 3–1). Time units of higher rank than period also exist. Eras include several periods, while eons include two or more eras. Corresponding time-stratigraphic terms for era and eon are rarely used (Table 3–1).

Time-stratigraphic units and time units and their relationships are particularly confusing to students. Time-stratigraphic units are defined by their position in a sequence, so terms such as lower, medial, and upper may be applied (Fig. 3–37). Thus, we may refer to Upper Jurassic strata, meaning that these strata occur in the upper part of the Jurassic System. In contrast, time units are simply subdivisions of geologic time and may be designated as early, middle, and late; the Late Jurassic Period was the time during which the Upper Jurassic System was deposited (Fig. 3–37).

## CORRELATION

If we could not demonstrate the equivalency of stratigraphic units in different areas, stratigraphic studies

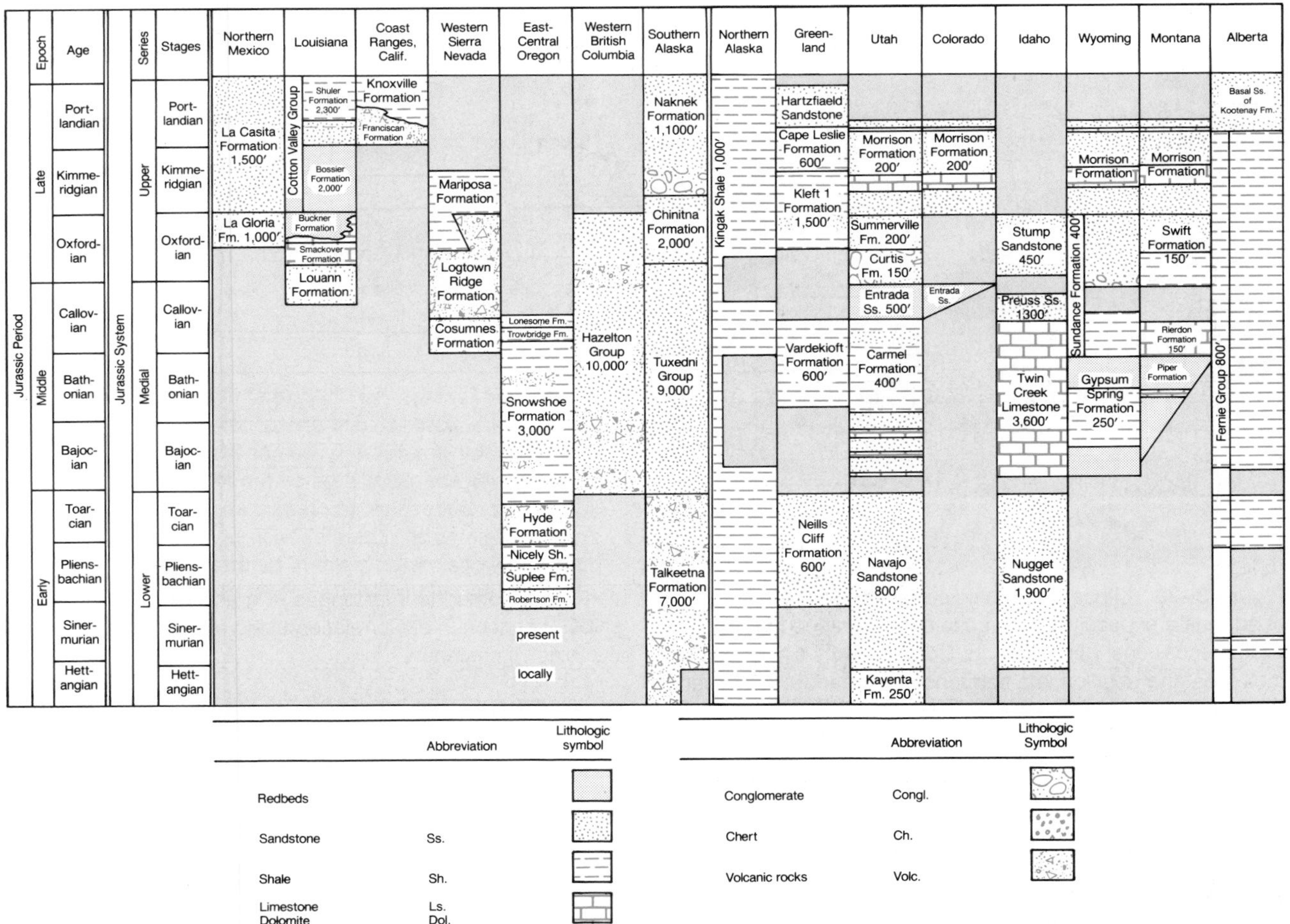


**Figure 3–37** Correlation of Jurassic strata of North America. Notice that the same-aged strata in different areas are not necessarily of the same rock types. Also notice the use of time-stratigraphic terms lower, medial, and upper, and their corresponding time terms, early, middle, and late. The Nugget Sandstone of Idaho, for example, occurs in the Lower Jurassic System, and it was deposited during the Early Jurassic Period.

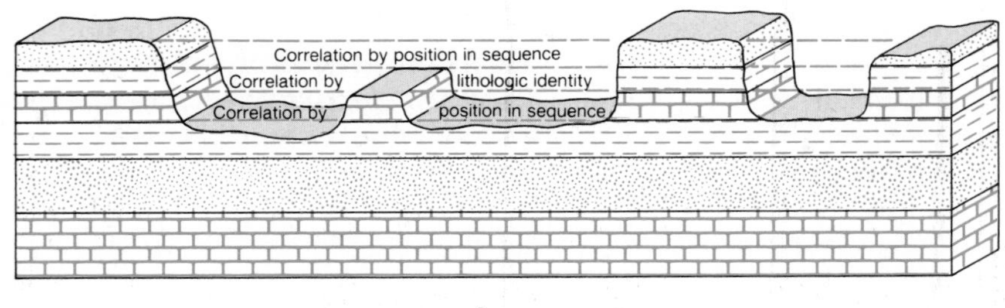

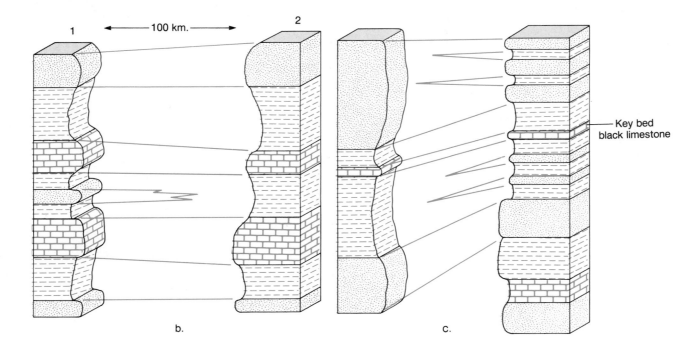


**Figure 3–38** Lithostratigraphic correlation. a. In the areas of adequate exposures, rock units can be traced laterally even if occasional gaps exist. b. Correlation by similarities in rock type and position in a sequence. The sandstone in section 1 is assumed to intertongue or grade laterally into the shale at section 2. c. Correlation using a key bed, a distinctive black limestone.

would be limited in their scope. **Correlation** is the process of demonstrating equivalency. Correlation of lithostratigraphic units is the recognition of units of similar lithology and stratigraphic position over broad geographic areas. If exposures are adequate, units may simply be traced laterally, even if occasional gaps exist (Fig. 3–38). Other criteria used to correlate are lithologic similarity, position in a sequence, and *key beds,* which are beds sufficiently distinctive to allow identification in different areas (Fig. 3–38).

Drilling operations have provided a wealth of subsurface data. When drilling for oil or natural gas, one common practice is to recover cores or rock chips called *well cuttings* from the drill hole. In addition, geophysical instruments may be lowered down the drill hole to record such rock properties as electrical resistivity, density,

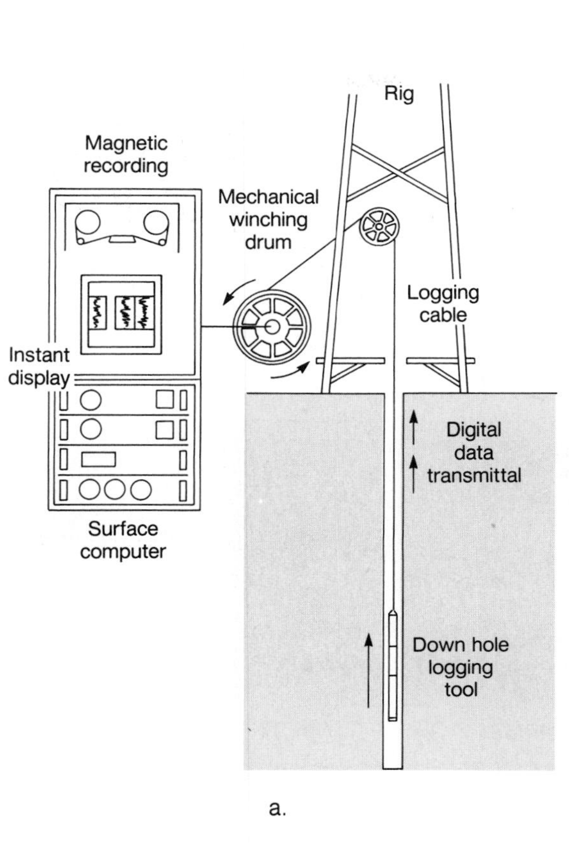


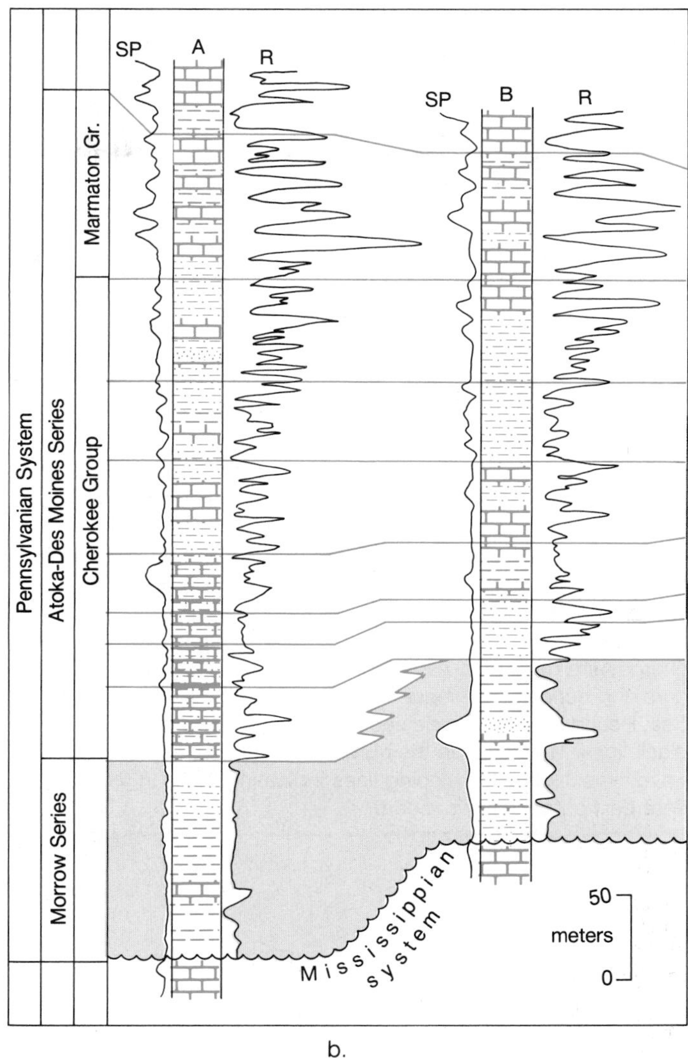


**Figure 3–39** a. Schematic diagram showing how well logs are made. A logging tool is lowered down the drill hole. As the logging tool is withdrawn from the drill hole data is transmitted to the surface where it is recorded and printed out as a well log. b. Electric logs and lithologic correlations of rocks in two wells in Colorado. The curves labeled SP are plots of self potential with depth, and the curves labeled R are plots of resistivity with depth.

and radioactivity, thus yielding a lithologic record, a *well log,* of the rocks penetrated. Cores, well cuttings, and well logs are all extremely useful in making subsurface lithostratigraphic correlations (Fig. 3–39).

Subsurface rock units may also be detected and traced by the study of seismic profiles. Energy pulses, such as those from explosions, travel through rocks at a velocity determined by rock density, and some of this energy is reflected from various horizons back to the surface, where it is recorded (Fig. 3–40). Seismic stratigraphy is particularly useful in areas such as the continental shelves where other stratigraphic techniques have limited use.

Lithostratigraphic correlation demonstrates the geographic extent of rock units, but it does not imply time equivalence. To demonstrate that strata in different areas

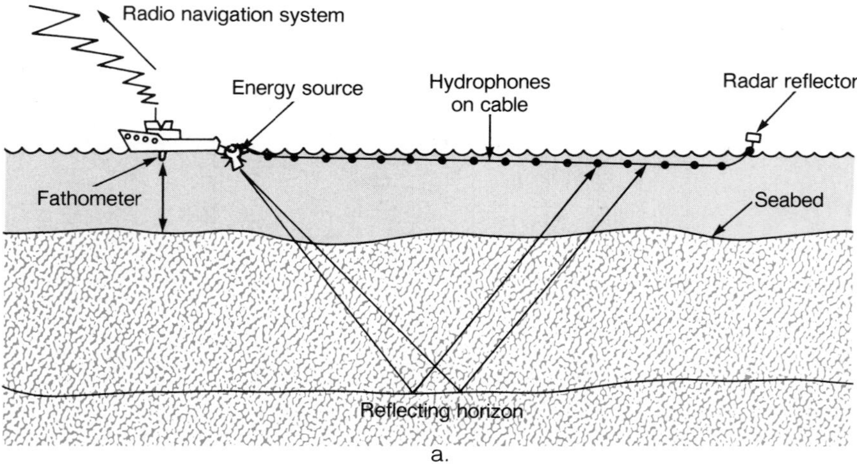


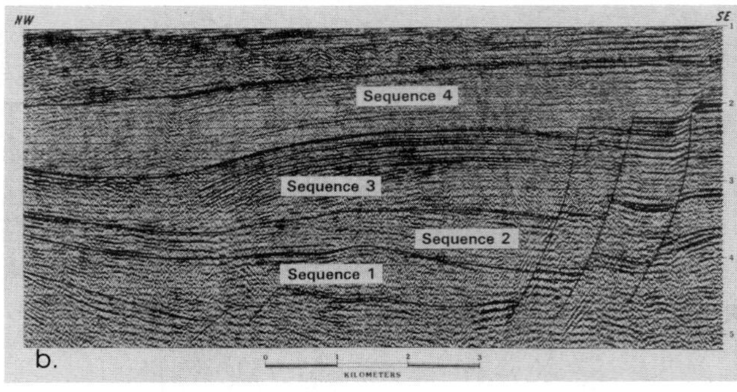


**Figure 3–40** a. Diagram showing the use of seismic reflections to detect buried rock units at sea. Sound waves are generated at the energy source. Some of the energy of these waves is reflected from various horizons back to the surface where it is detected by hydrophones. Buried rock units can also be detected on land, but here explosive charges are detonated as an energy source. b. Seismic record and depositional sequences defined in the Beaufort Sea. Boundaries of seismic sequences are shown by solid black lines. The scale on the right shows seismic wave travel time. Notice the sloping lines indicating faults in the right part of the seismic record.

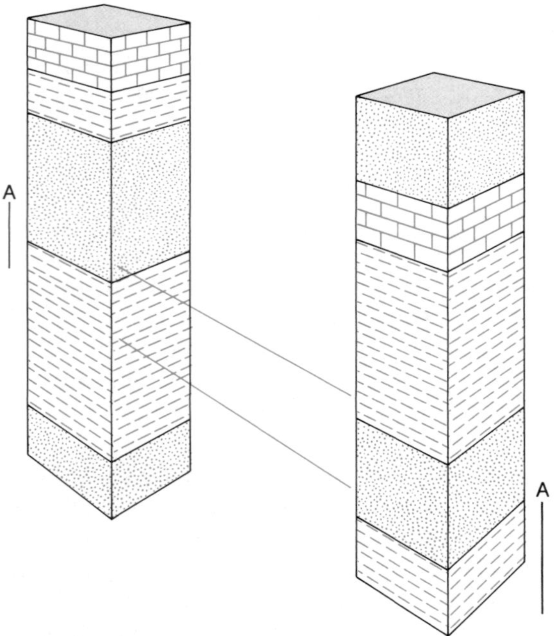


**Figure 3–41** Correlation between two sections based on the range zone of fossil A.

are the same age involves time-stratigraphic correlation. Since most rock units are time transgressive, we usually cannot use rock type in this type of correlation. In most cases, biozones are used, although several other methods are useful as well.

One type of biozone, **the range zone,** is defined by the total geologic range of a particular fossil group (a species or a group of related species, called a *genus,* for example) (Fig. 3–41). Particularly useful fossils are those that are easily identified, geographically widespread, and that have rather short geologic ranges. The brachiopod genus *Lingula* is easily identified and widespread, but its geologic range of Ordovician to Recent makes it of little use in biostratigraphy. In contrast, the trilobite *Paradoxides* and the brachiopod *Atrypa* meet all of the criteria mentioned, so are good **guide fossils** (Fig. 3–42). **Concurrent range zones** are established by plotting the overlapping ranges of fossils that have different geologic ranges. The first and last occurrences of fossils are used to establish zone boundaries (Fig. 3–43).

Correlation of range zones or concurrent range zones generally yields correlation lines that are considered time equivalent. In other words, the strata encom-

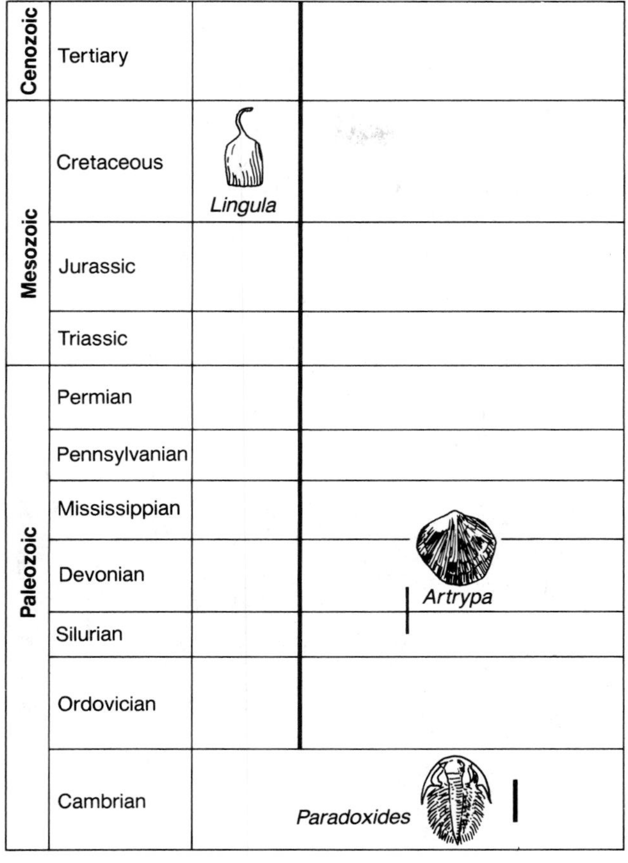


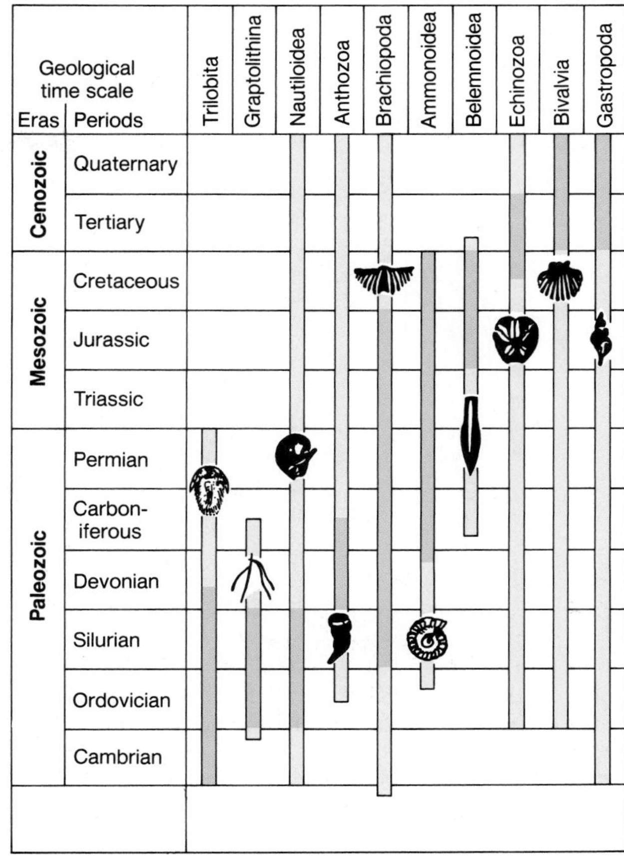


**Figure 3–42** a. Comparisons of the geologic ranges of three marine invertebrates. *Lingula* is of little use in biostratigraphy because of its long geologic range. *Atrypa* and *Paradoxides* are good guide fossils because they are widespread, easily identified, and have short geologic ranges. b. The major groups of marine invertebrates used in biostratigraphic zonation. The bars show the total ranges, while the shaded bars show the times during which many species of each group are important as guide fossils.

passed by the correlation lines so established are thought to be the same age. However, geologists are aware that zones are not of precisely the same age everywhere, because no fossil organism appeared and disappeared simultaneously over its entire geographic range. Even so, correlation of range zones can yield an error no greater than the range of the fossil used (Fig. 3–44), and the use of concurrent range zones yields even more precise correlations.

Time equivalence can also be based on some physical events of short duration. Ash falls (Fig. 3–45), for example, may cover large areas, and they occur instantaneously when considered in the context of geologic time. Furthermore, ash falls are not restricted to a particular depositional environment, so they may be continuous from continental into marine depositional environments.

In our preceding examples of correlation using biozones and ash falls, time-equivalent units of strata were identified. Such time-equivalent units are time-stratigraphic units (systems, series, and stages), and they define time units (periods, epochs, and ages) (Table 3–1). To conclude this section, we give a hypothetical example that may help students better understand lithostratigraphic units and the relationships of lithostratigraphic units to time (Fig. 3-46).

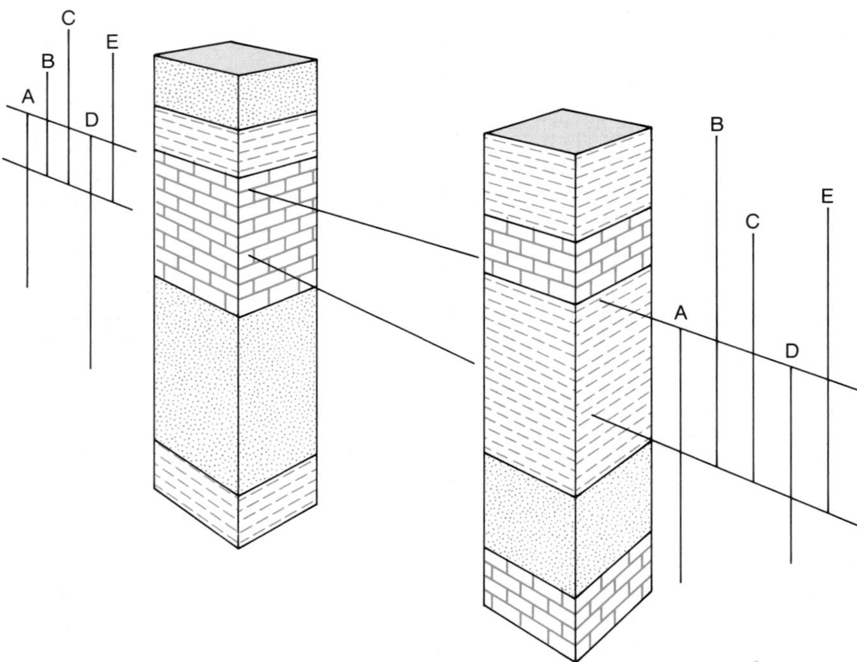


**Figure 3–43** Correlation of two sections by using concurrent range zones, which are established by the overlapping ranges of fossils A through E.

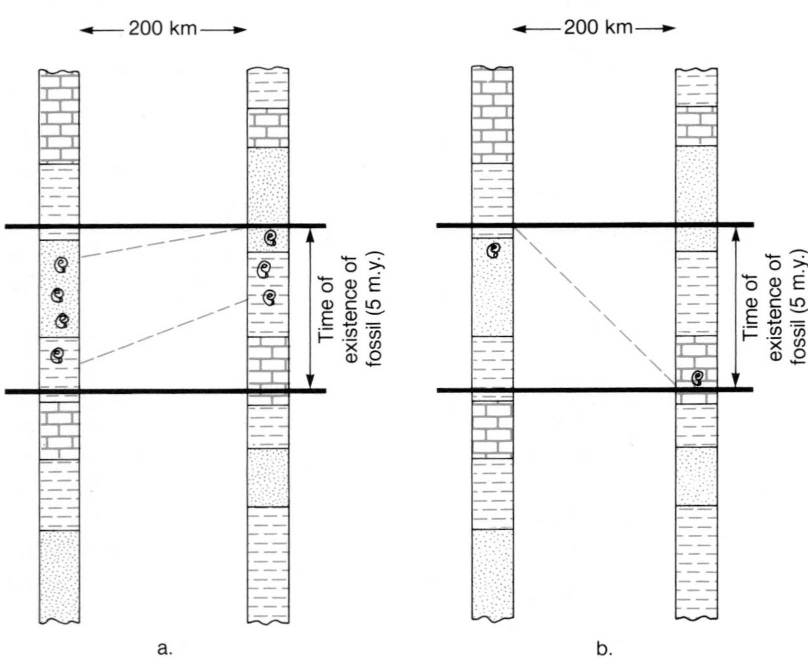


**Figure 3–44** Simplified diagrams showing possible errors in time stratigraphic correlations based on the range zone of a fossil that existed for five million years. a. In these two stratigraphic sections the fossil is found at the positions shown. The biostratigraphic correlation lines (dashed) obviously encompass strata only partly of the same age in the two sections. b. The maximum possible error would occur if the last and first occurrance of the fossil were assumed to represent time equivalent points.

## QUANTIFYING THE RELATIVE TIME SCALE

Several of the correlation techniques discussed here demonstrate time equivalence between sections, but the ages determined are relative only. In the 1840s, Murchison used fossils to demonstrate that strata in Russia, Germany, and England belong to the Permian System and are therefore of the same age. Furthermore, the relative age of Permian rocks was known with respect to Carboniferous and Triassic strata, but no one knew when the Permian began or how long it lasted.

A number of minerals in sedimentary rocks can be dated radiometrically, but the ages tell only the age of the source rock that supplied the minerals to the deposit. All this tells us is that the sedimentary rock and its fossils are younger than the age determined. Glauconite is a mineral that forms in some sediments shortly after deposition and can be dated by the potassium-argon method. Unfortunately, glauconite easily loses argon, so absolute

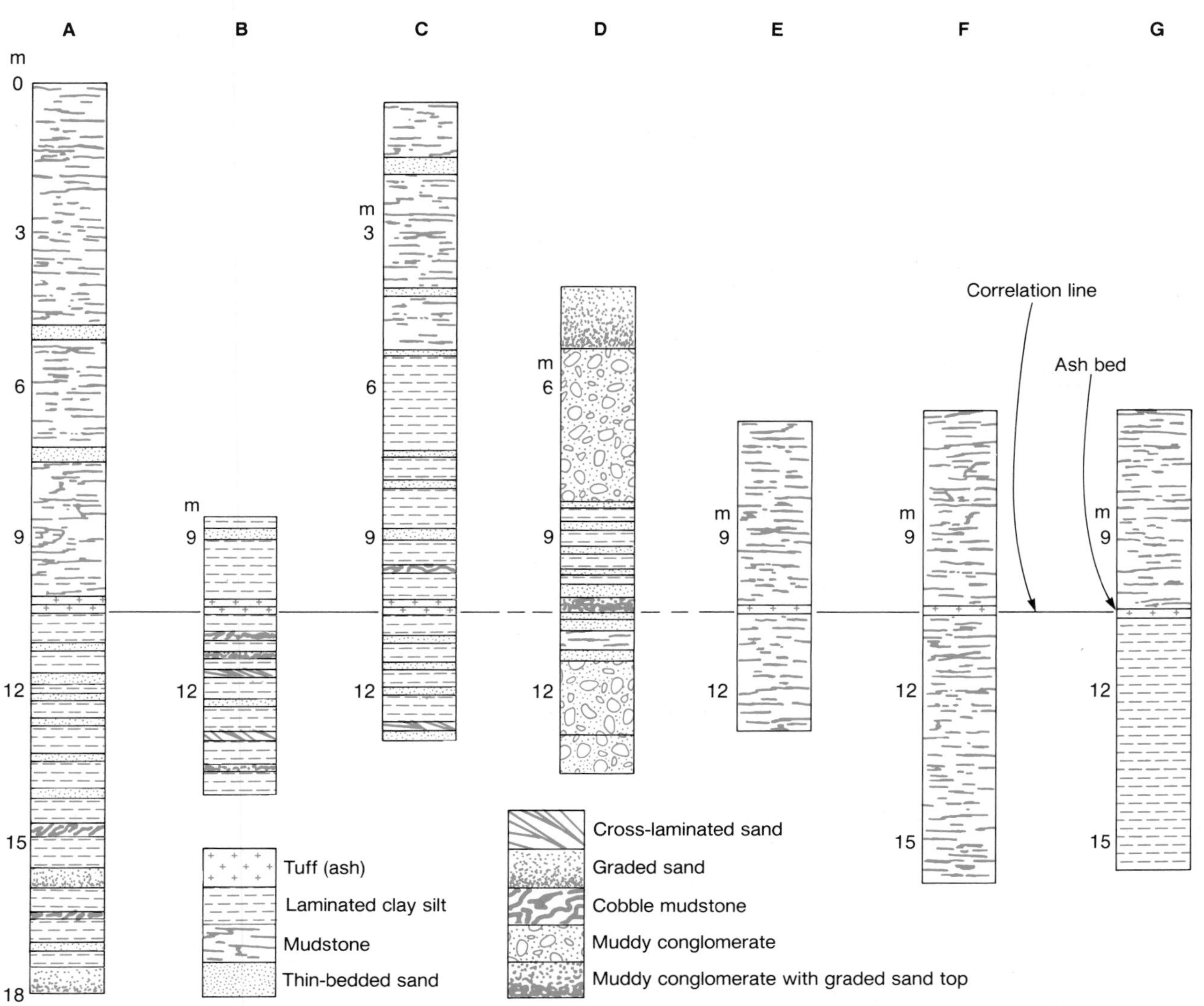


**Figure 3–45** Time-stratigraphic correlation by using an ash bed, the Late Pliocene Bailey Ash in the Ventura Basin, California.

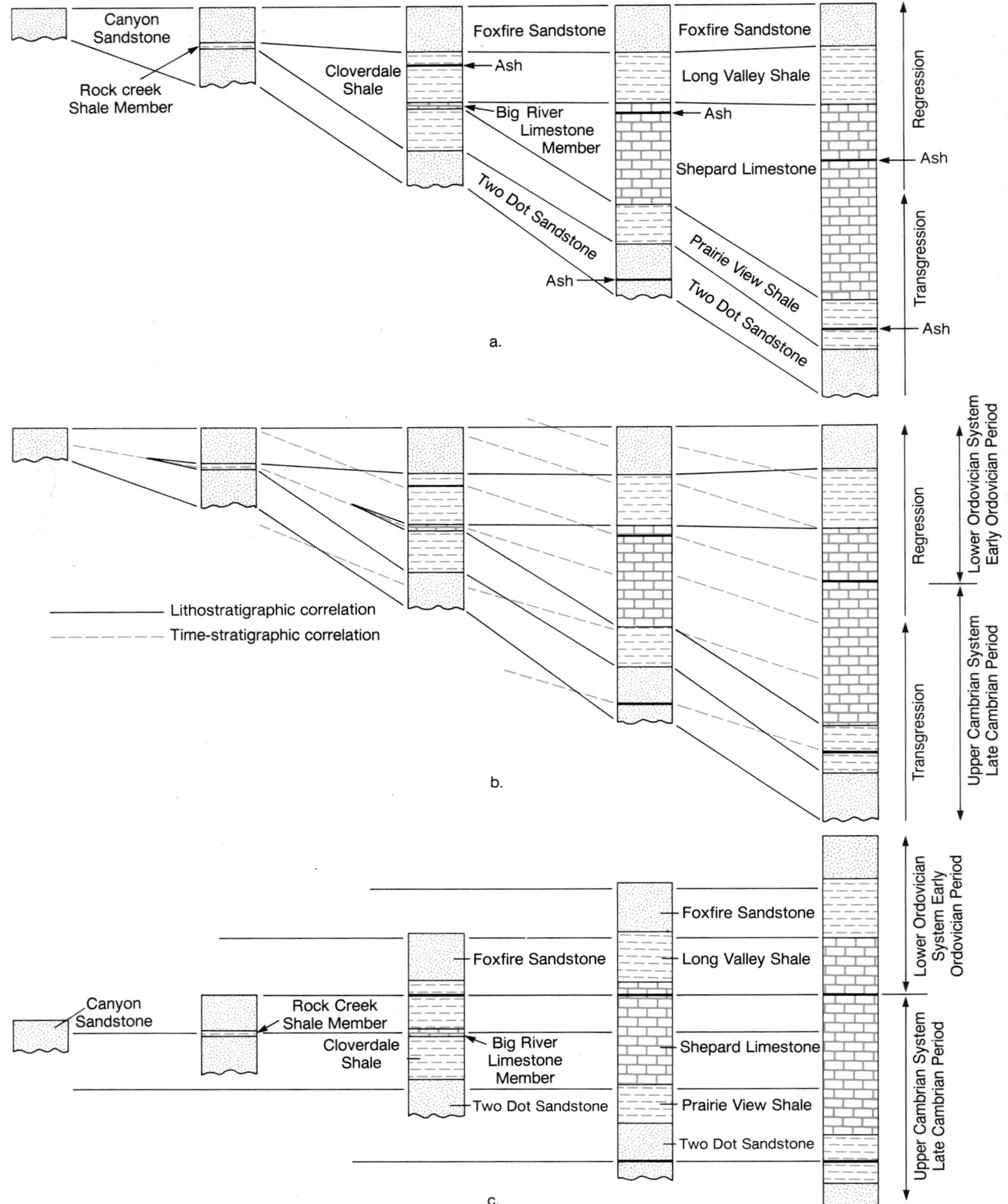

Canyon Sandstone
Rock creek Shale Member
Cloverdale Shale
Ash
Big River Limestone Member
Foxfire Sandstone
Long Valley Shale
Shepard Limestone
Two Dot Sandstone
Prairie View Shale
Regression
Transgression
a.
Lithostratigraphic correlation
Time-stratigraphic correlation
Lower Ordovician System Early Ordovician Period
Upper Cambrian System Late Cambrian Period
b.
Rock Creek Shale Member
c.

**Figure 3–46** In this idealized example, sedimentary rocks and interbedded ash falls (black) formed during a major transgressive-regressive event. a. Lithostratigraphic correlations and application of lithostratigraphic terminology. The stratigraphic relationships make it necessary to give rock units different formation names in different areas. b. In this diagram both lithostratigraphic and time-stratigraphic correlations are shown. Time-stratigraphic units are defined by correlations of biozones and ash falls. Note that time lines cut across lithostratigraphic correlation lines. c. The same stratigraphic sections in a and b are arranged here so that time lines are horizontal, illustrating the relationships of lithostratigraphic units to time-stratigraphic units and time units.

ages determined by this method must be considered minimum ages. In most cases, absolute ages of sedimentary rocks must be determined indirectly by dating associated igneous rocks and metamorphic rocks.

## Dating Associated Igneous and Metamorphic Rocks

An intrusive igneous body is younger than the rock it intrudes (principle of cross-cutting relationships). Therefore, an accurately dated intrusive igneous body provides a minimum age for the sedimentary rock it intrudes and a maximum age for the sedimentary rocks unconformably overlying it (Fig. 3–47). Age dates of regionally metamorphosed rocks give a maximum age for any overlying sedimentary rocks (Fig. 3–47).

One of the best ways to get good radiometric dates in a sedimentary sequence is by using interbedded volcanic rocks. An ash fall or lava flow provides an excellent marker bed that is a time-equivalent surface, providing a minimum age for the sedimentary rocks below and a maximum age for the rocks above. Ash falls are particularly useful because they may fall over both marine and nonmarine sedimentary environments and can provide a connection between these different environments. Multiple ash falls, lava flows, or a combination of both in a stratigraphic sequence are particularly useful in determining absolute ages of sedimentary rocks and their contained fossils (Fig. 3–48).

Thousands of absolute ages are now known for sedimentary rocks of known relative ages, and these absolute dates have been added to the relative time scale (Fig. 3–31). In this way, geologists have been able to determine the absolute ages of the various systems and to determine their durations.

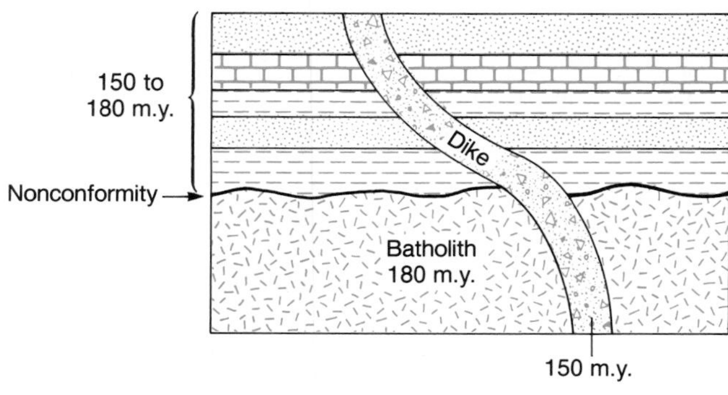


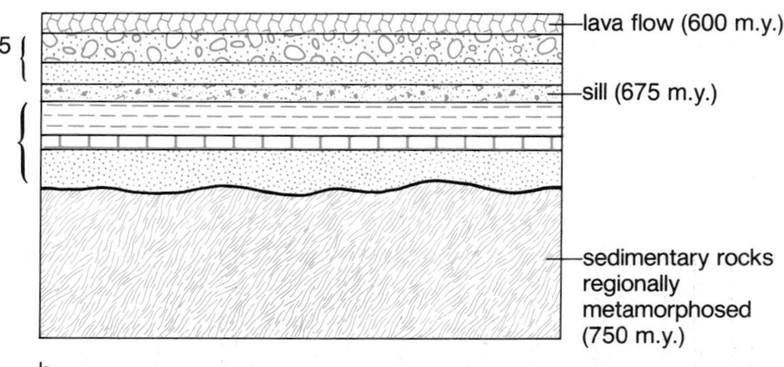

**Figure 3–47** Absolute ages of sedimentary rocks determined by dating associated igneous and metamorphic rocks. In both a and b, sedimentary rocks are bracketed by rock bodies for which absolute ages have been determined.

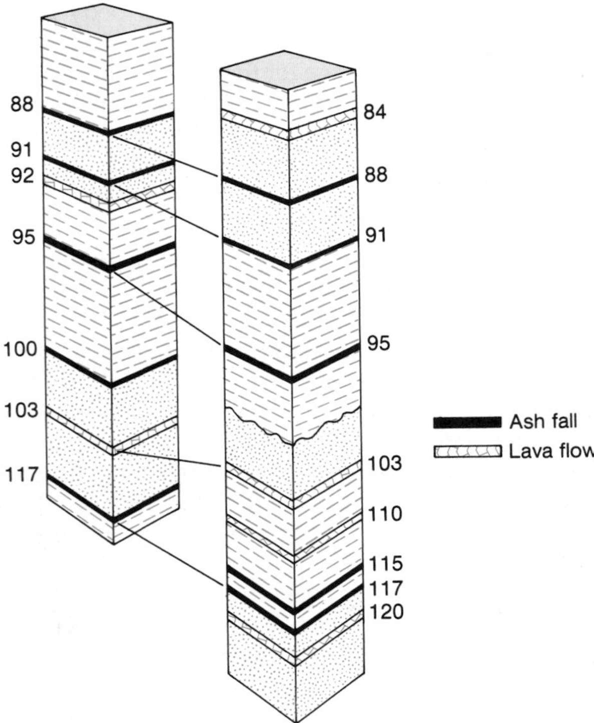


**Figure 3–48** Idealized example of determining the absolute ages of fossiliferous Cretaceous strata. Ash falls and lava flows in the two sections can be correlated to establish time-stratigraphic units, and the absolute ages of these units can be determined. Ages shown are in millions of years.

## CHAPTER SUMMARY

1. The relative ages of rock units in any sequence of strata can be determined by superposition even if the strata are complexly deformed.
2. Surfaces of discontinuity that encompassed significant amounts of geologic time are common in the geologic record. Such surfaces are called unconformities and result from times of nondeposition or erosion.
3. Lateral changes in sedimentary rock units resulting from deposition in adjacent but different environments are called sedimentary facies. The same facies that replace one another laterally may be superposed in a vertical sequence as a result of marine transgression or regression.
4. In a conformable sequence of facies, Walther's Law can be applied to determine what the lateral facies relationships were.
5. The causes of marine transgressions and regressions include uplift and subsidence of the continents, rates of seafloor spreading, and the amount of seawater contained in glaciers.
6. Most fossils are preserved in sedimentary rocks as unaltered remains, altered remains, as molds and casts, or as traces of organic activity.
7. Although fossils of some organisms are quite common, very few of the organisms that lived at any one time were actually preserved as fossils. Furthermore, the fossil record is biased toward those organisms with hard skeletal elements and those organisms that lived in areas of active sedimentation.
8. Fossils are useful in stratigraphy because fossil assemblages succeed one another through time in a predictable sequence. William Smith's work with fossils in Great Britain is the basis for what is now called the principle of fossil succession.
9. The geologic column is a composite sequence of rock bodies called systems arranged in chronological order. Superposition and fossil succession were used to determine the relative ages of the systems.
10. Modern stratigraphic terminology includes two fundamentally different kinds of units: units based on content and units related to geologic time.
11. Correlation is the stratigraphic practice of demonstrating equivalency of units in different areas. Demonstrating time equivalence is most commonly done by correlating strata with similar fossils.
12. Most absolute ages of sedimentary rocks and their contained fossils are obtained indirectly by dating associated metamorphic or igneous rocks.

## IMPORTANT TERMS

altered remains
angular unconformity
bedding plane
biostratigraphic unit
biozone
body fossil
cast
concurrent range zone
conformable
coprolite
correlation
disconformity
formation
fossil
geologic record
guide fossil
hiatus
lithostratigraphic unit
marine regression
marine transgression
mold
nonconformity
period
principle of inclusions
principle of fossil succession
range zone
sedimentary facies
stratigraphy
system
time-stratigraphic unit
time transgressive
time unit
trace fossil
unaltered remains
unconformity
Walther's Law

## REVIEW QUESTIONS

1. Explain how a geologist would determine the relative ages of a granite batholith and an overlying sandstone formation.
2. Describe a sequence of events that could account for an angular unconformity.
3. A conformable vertical sequence of facies consists of sandstone followed upward by shale and limestone. All facies contain marine fossils. Diagram the lateral facies relationships, and show which facies was deposited nearest the shoreline.
4. Explain how geologists use the concept of uniformitarianism to understand the conditions leading to the preservation of fossils.
5. Compare the processes of permineralization, replacement, and recrystallization of fossils.
6. The principles of superposition and fossil succession were used to establish the geologic column. Define both principles, and explain how they were used to determine the relative ages of the systems.
7. What is a concurrent range zone? How can such zones be used to demonstrate time equivalency of strata in widely separated areas?
8. How do lithostratigraphic units differ from time stratigraphic units?
9. In a vertical sequence of strata one observes the following; lava flow (absolute age 255 million years) followed upward by a Permian sandstone, a Triassic shale, and an ash fall (absolute age 240 million years). What is the approximate absolute age of the Permian-Triassic boundary?

## ADDITIONAL READING

Boggs, S., Jr., 1987. *Principles of Sedimentology and Stratigraphy.* Columbus, Ohio: Merrill Publishing Company.

Fritz, W.T. and Moore, N.J. 1988. *Basics of Physical Stratigraphy and Sedimentology,* New York: John Wiley & Sons, Inc.

Harbaugh, J.W., 1974. *Stratigraphy and the Geologic Time Scale.* Dubuque, Iowa: Wm. C. Brown Co.

Moody, R. 1986. *Fossils.* New York: Macmillan Publishing Co.

Simpson, G.G. 1983. *Fossils and the History of Life.* New York: Scientific American Books.

# 4

## CHAPTER OUTLINE



## PROLOGUE

*Today the Mediterranean Sea is located in an arid region where the rate of evaporation exceeds the rate at which water is added to the sea by rainfall runoff. If it were not for the connection between the Mediterranean and the Atlantic Ocean at the Strait of Gibraltar (Fig. 4–1), the Mediterranean would eventually dry up and become a vast desert basin lying far below sea level. Some geologists, particularly Kenneth Hsü of the Swiss Federal Institute of Technology, think this is precisely what happened to the Mediterranean during the Late Miocene.*

*Evaporite deposits up to 2 km thick have been discovered beneath the floor of the modern Mediterranean Sea. Sedimentological studies of these Late Miocene evaporites suggest that some were deposited in shallow-water environments rather than in a deep-ocean basin, as the Mediterranean is now. Observations such as this have led to the hypothesis that the Mediterranean Sea periodically lost its connection with the Atlantic Ocean during the Late Miocene. With its supply of water cut off, the Mediterranean Sea evaporated to near dry-*

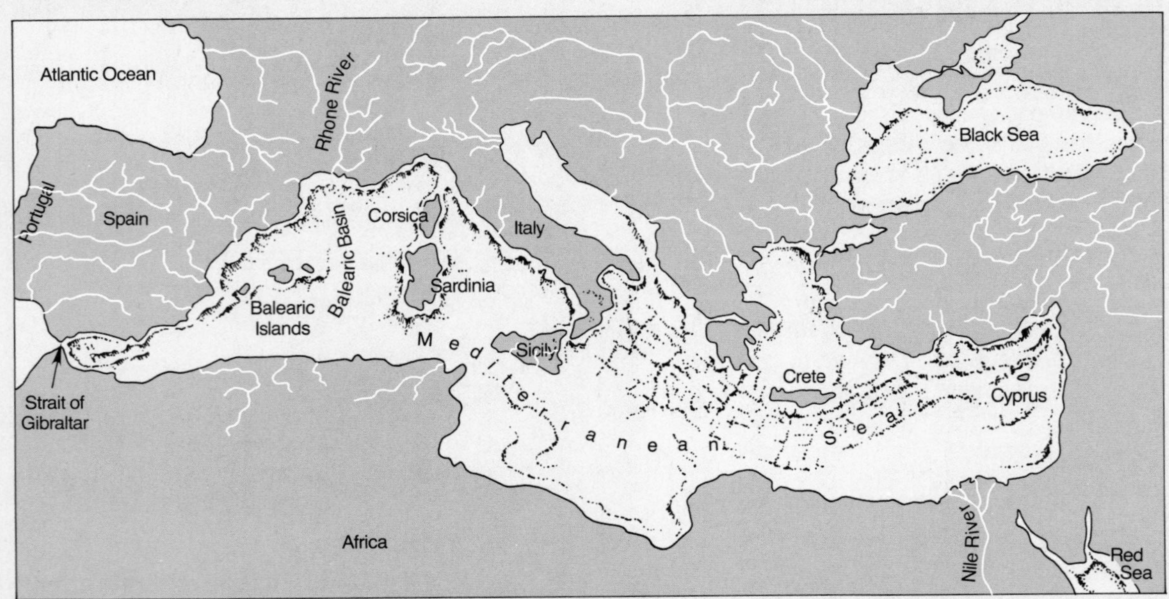


**Figure 4–1** Panoramic view showing the submarine topography of the modern Mediterranean Basin. About 6 million years ago, the Mediterranean Basin was probably a vast desert lying 3,000 m below sea level. The Balearic Basin was a salt lake where evaporite minerals were deposited. At the end of the Miocene, an oceanic connection was reestablished at the Strait of Gibraltar, and the basin rapidly refilled.

# Origin and Interpretation of Sedimentary Rocks

*ness in as little as 1,000 years and became a vast desert basin lying 3,000 m below sea level (Fig. 4–2).*

*Apparently, periods of isolation of the Mediterranean Basin, with evaporation, alternated with periods when an oceanic connection was reestablished at the Strait of Gibraltar. During those times when an oceanic connection existed, gravels and silts were deposited around the basin margin, and deep-sea sedimentation occurred near the basin center. However, when the oceanic connection was lost, deposition occurred in a body of water that became progressively shallower. Vast amounts of carbonates were deposited, followed by evaporites, especially gypsum and halite, as the waters became more and more saline.*

*A desert basin lying 3,000 m below sea level may seem preposterous, but there is additional supporting evidence for this view of Mediterranean history. The modern Mediterranean is the principle base level for rivers flowing from Africa and southern Europe, yet beneath these rivers are buried river valleys deeply incised into bedrock far below present-day sea level. For example, a buried channel has been discovered by drilling more than 900 m into the Rhone Delta in southern Europe. Several river-cut submarine canyons have been traced out into the Mediterranean Basin itself where they now lie more than 2,000 m below sea level.*

*At the Aswan High Dam, 465 km upstream on the Nile, Russian geologists discovered a buried river valley incised more than 200 m below present sea level (Fig. 4–3). It seems that this valley was cut when the water level in the Mediterranean was much lower, later to be filled with Pliocene marine sediments when sea level rose again and buried the canyon.*

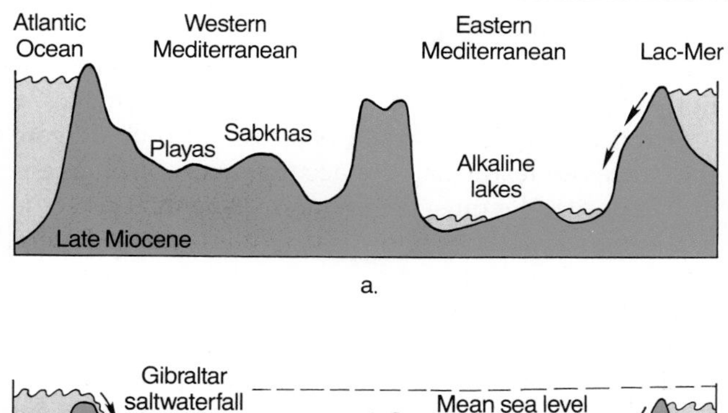


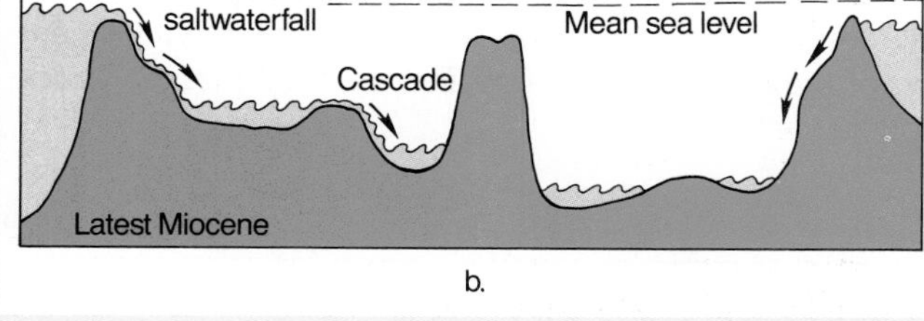


**Figure 4–2** Proposed model to explain the deposition of evaporites in the Mediterranean Basin. a. Isolation of the Mediterranean and evaporation to near dryness. b. Refilling stage. To account for the thick evaporite deposits present beneath the floor of the modern Mediterranean, this cycle of isolation, evaporation, and refilling must have occurred many times. The vertical scale is greatly exaggerated here.

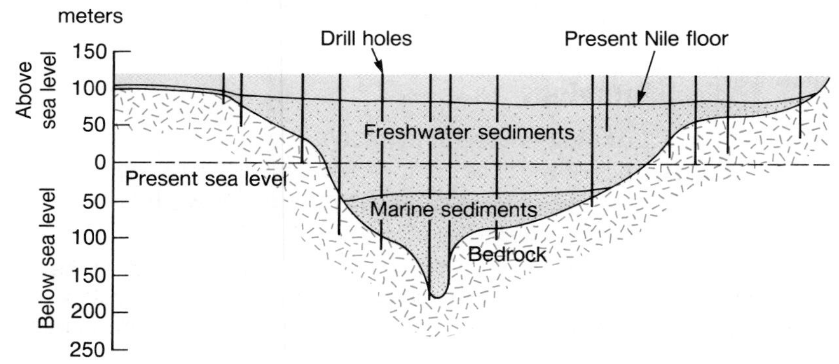


**Figure 4–3** Boreholes drilled before the construction of the Aswan High Dam on the upper Nile in Egypt revealed a deep gorge beneath the present floor of the Nile. This gorge extends more than 200 m below sea level, so its incision must have occurred when the Mediterranean was dry or at least much lower than it is now. When sea level rose in the Pliocene, the gorge was flooded and partly filled by marine sediments.

## INTRODUCTION

Recall from Chapter 1 that sedimentary rocks are composed of the solid particles called *detritus* and dissolved mineral matter derived from preexisting rocks. Such detritus and dissolved material may be transported and deposited as sediment; sand on a beach is detrital sediment, whereas chemically precipitated mineral matter is chemical sediment. Any sediment can be acted upon by the processes of lithification, (compaction and cementation) and thus be converted into sedimentary rock.

In this chapter our main objective is to investigate the processes leading to the accumulation of sedimentary deposits. We must begin our investigation with the end products of deposition and lithification, sedimentary rocks. The sedimentary rocks we observe in the geologic record acquired their various properties, in part, as a result of the physical, chemical, and biological processes that operated in the original depositional environment. Thus, our task is to determine what that environment was by examining the sedimentary rock properties.

## FEATURES OF SEDIMENTARY ROCKS

The first step in any investigation of sedimentary rocks is observation and data gathering (Fig. 4-4). Accordingly, geologists visit the areas where sedimentary rocks are exposed, carefully examine the strata, describe and measure the thickness of the strata, trace units laterally, and collect samples for further analysis in the lab. While in the field, geologists commonly make some preliminary interpretations. For example, the color of sedimentary rocks is a useful environmental parameter: red-colored rocks may indicate deposition in a continental environment, whereas greenish rocks are more typical of marine environments. There are, however, many exceptions, so color must be interpreted with caution. Geologists may also recognize types of sedimentary particles or fossils that indicate deposition in a particular environment.

After field studies have been completed, the data can be more fully analyzed in the lab. Such analyses may include microscopic and chemical examination of rock samples, identification of fossils, statistical determination of ancient current directions, and construction of diagrammatic representations of vertical and lateral facies relationships. And finally, when all available data have been analyzed, an environmental interpretation is made (Fig. 4–4).

**PARAMETERS OF ENVIRONMENTAL SIGNIFICANCE**

1. *Geometry:* Three dimensional shape of rock bodies.
2. *Lithology:*
   a. Detrital—Grain size as energy level index, texture-process controlled rather than environment controlled.
   b. Carbonates—Sedimentary facies with characteristic grain types.
3. *Fossils:*
   a. Macrofossils as rock builders (reef, for example).
   b. Macro- and microfossils as environmental indices.
   c. Trace fossils—diagnostic assemblages in specific environments.
4. *Sedimentary structures:* Associations of sedimentary structures such as ripple marks and cross-beds useful in environmental interpretations.
5. *Paleocurrents:* Ancient current directions determined from sedimentary structures.

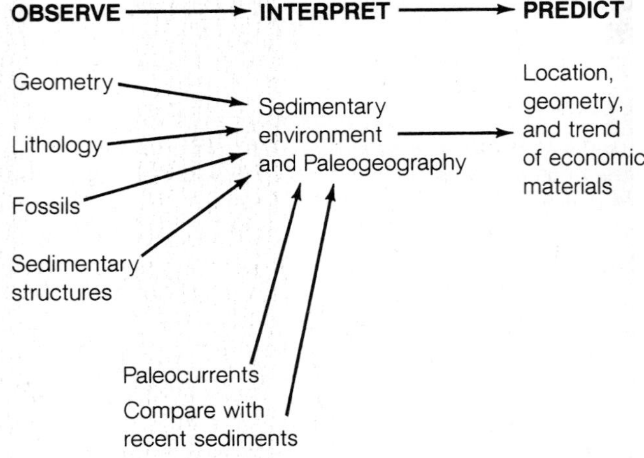


**Figure 4–4** Methodology for environmental analysis.

### Lithology

**Lithology** is a term that refers to the physical characteristics of rocks as composition and texture. Sedimentary rocks may have a **clastic texture** meaning they are composed of clasts (the broken particles of preexisting rocks) (Fig. 4–5). All detrital sedimentary rocks and some chemical rocks are composed of clasts (Table 1–3). Conglomerates and sandstones, for example, contain gravel-

**Figure 4–5** Outcrop in Montana showing the clastic texture of sandstone and conglomerate.

and sand-sized clasts (Fig. 4–5), respectively, and clastic limestones may be composed of broken shell fragments. Many chemical sedimentary rocks are composed of an interlocking mosaic of mineral crystals and are said to have a **crystalline texture**.

Other important textural features include grain size and sorting. **Grain size** is simply a measure of a sedimentary particle's size; gravel, sand, silt, and clay are size designations for detrital sedimentary particles (Table 1–3). High-energy transport agents such as running water and waves are needed to transport large particles; thus, detrital gravel and sand tend to be deposited in stream channels or on beaches. Silt and clay, on the other hand, can be transported by weak currents and accumulate under low-energy conditions such as in lakes and lagoons (Fig. 4–6) Clastic limestones are typical of

a.

b.

c.

d.

**Figure 4–6** Sedimentary textures. a. Channel deposit from a Miocene alluvial fan in Montana. This deposit is coarse-grained (particle size ranges up to 15 cm) and poorly sorted. b. Modern sand dune in Michigan. This deposit is composed of sand grains that average about .25 mm in diameter. This sand is well sorted since a narrow range of grain sizes is represented. c. This Florida beach deposit is composed of sand- and gravel-sized shell fragments. If lithified, this sediment would be a clastic limestone. d. Silt- and clay-sized particles settle from suspension in low-energy environments such as this small pond.

beaches and offshore bars where wave energy is intense (Fig. 4-6) and may be composed of sand- and gravel-sized shell fragments.

**Sorting** refers to the variation in grain sizes. Sedimentary rocks can be characterized as well sorted if the range of grain sizes is not great and as poorly sorted if the range is great (Fig. 4–6). Sorting results from processes that selectively transport and deposit particles by size. Windblown dunes are composed of well-sorted sand, because wind cannot effectively transport gravel, and it blows silt and clay beyond the areas of sand accumulation. Glaciers, however, are unselective, because their transport power allows them to move many sizes of sediment, and their deposits are typically poorly sorted.

Well over 100 different minerals have been identified in detrital sedimentary rocks, but only quartz, feldspars, and clays are very common. Unfortunately, the composition of detrital rocks reveals little about transport or depositional history. Composition is, however, very important in determining the source area or areas that yielded the sedimentary particles (Fig. 4–7).

Environmental inferences can be made from chemical sedimentary rock composition. Limestones, for example, are composed of the mineral calcite ($CaCO_3$), a mineral that most commonly indicates a warm, shallow marine environment; some limestones form in lakes, however. Evaporites such as gypsum and rock salt indicate depositional environments in which evaporation rates were high—conditions that prevail in some lake and marine environments such as the Great Salt Lake, Utah, and the Persian Gulf area.

**Figure 4–8** The chalk exposed in these Cretaceous outcrops in southern Denmark is composed of the microscopic shells and plates of marine organisms. (Photo courtesy of R.V. Dietrich, Central Michigan University.)

## Fossils

Fossils or fossil fragments are one of the most common sedimentary grains in limestone (Figs. 4–6, 4–8). Much of the sediment on the deep-sea floor is composed of

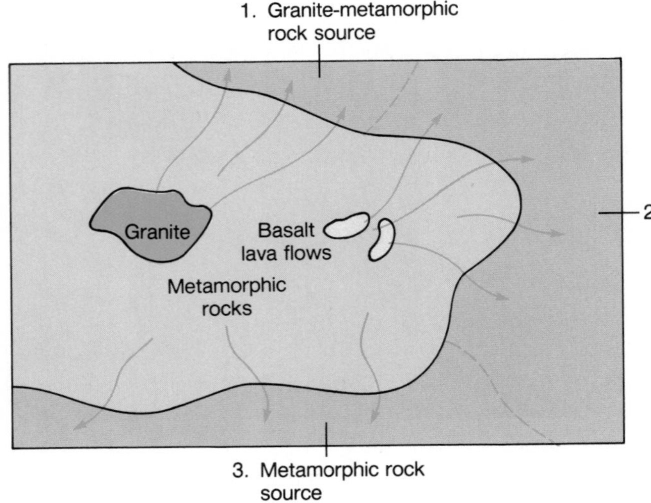


**Figure 4–7** Hypothetical geologic map showing the source areas for a sandstone. Compositional analysis of the sandstone in this example indicates a sediment dispersal pattern from the source areas as shown by the arrows. For example, in area 2 the sandstone contains minerals and rock fragments indicating that the sand was derived from basalt and metamorphic source rocks.

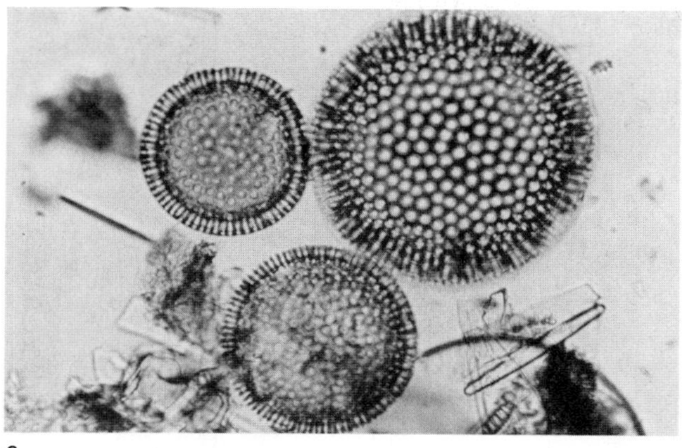

a.

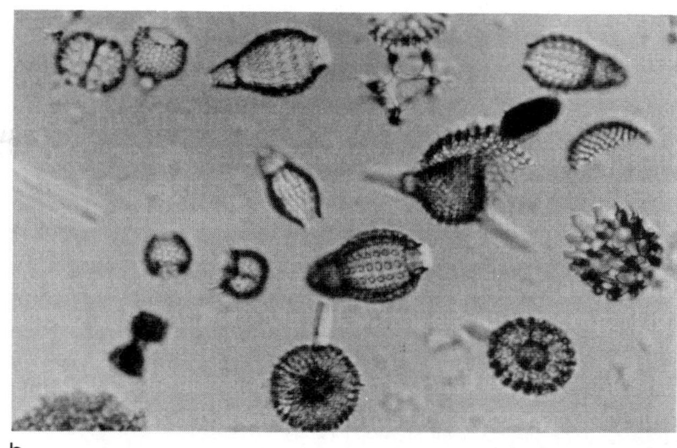

b.

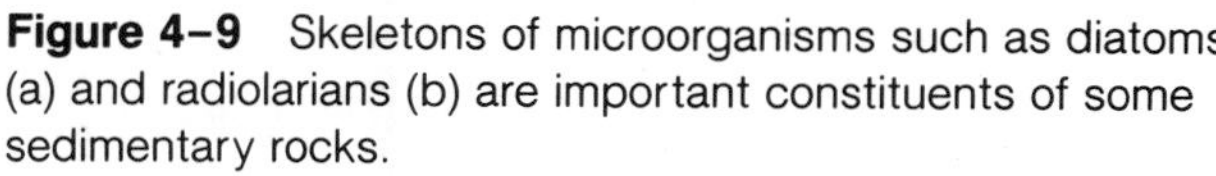

**Figure 4–9** Skeletons of microorganisms such as diatoms (a) and radiolarians (b) are important constituents of some sedimentary rocks.

microfossils, and the structural framework of reefs is made up of the skeletons of corals and other organisms. Diatomite, a rock composed of the shells of microscopic plants called *diatoms,* is mined and used for abrasives and as a filtering agent in gas purification (Fig. 4–9).

Two factors must be considered when using fossils in environmental analyses. The first is whether the fossil organisms lived in the location where they were buried or whether they were transported there. The second important factor is what kind of habitat the organisms originally occupied. Studies of a fossil's structure and its living descendants, if there are any, are helpful in this endeavor. For example, clams with heavily constructed shells typically live in shallow, turbulent water. In contrast, organisms dwelling in low-energy environments commonly have thin, fragile shells (Fig. 4–10) Also, organisms that carry on photosynthesis are restricted to the zone of sunlight penetration in the seas, which is generally less than 200 m. Thus, stromatolites, which are structures produced by photosynthesizing algae and bacteria, indicate shallow water depths (Fig. 4–11). The amount of suspended sediment is also a limiting factor on the distribution of organisms. Many corals, for example, live in shallow, clear seawater, because suspended sediment clogs their respiratory and food–gathering organs, and some have photosynthesizing algae living in their tissues.

Microfossils are particularly useful for environmental studies because the remains of numerous individual organisms can be recovered from small rock samples. In oil-drilling operations, for example, small rock chips called *well cuttings* are brought to the surface. Such samples rarely contain entire macrofossils but may contain numerous microfossils that aid in age and environmental interpretations. Trace fossils (Fig. 3–20) are not transported from the place where they formed, and certain traces are known to be characteristic of particular environments (Fig. 4–12).

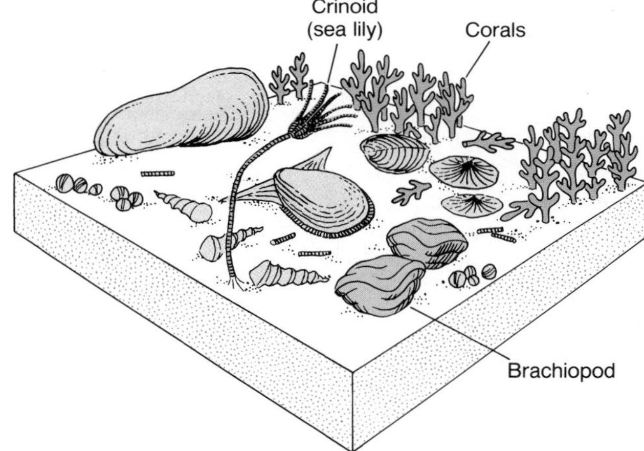


**Figure 4–10** The association of fossils shown here includes brachiopods, corals, and crinoids and indicates a shallow marine environment.

## Sedimentary Structures and Paleocurrents

Sedimentary rocks commonly contain structures that formed in sediment at the time of deposition or shortly

a.

b.

**Figure 4–11** Stromatolites. These structures are produced by the activities of photosynthesizing bacteria. a. Cross sectional view of Late Proterozoic stromatolites near Marquette, Michigan. b. Surface view of Late Proterozoic stromatolites in Glacier National Park, Montana.

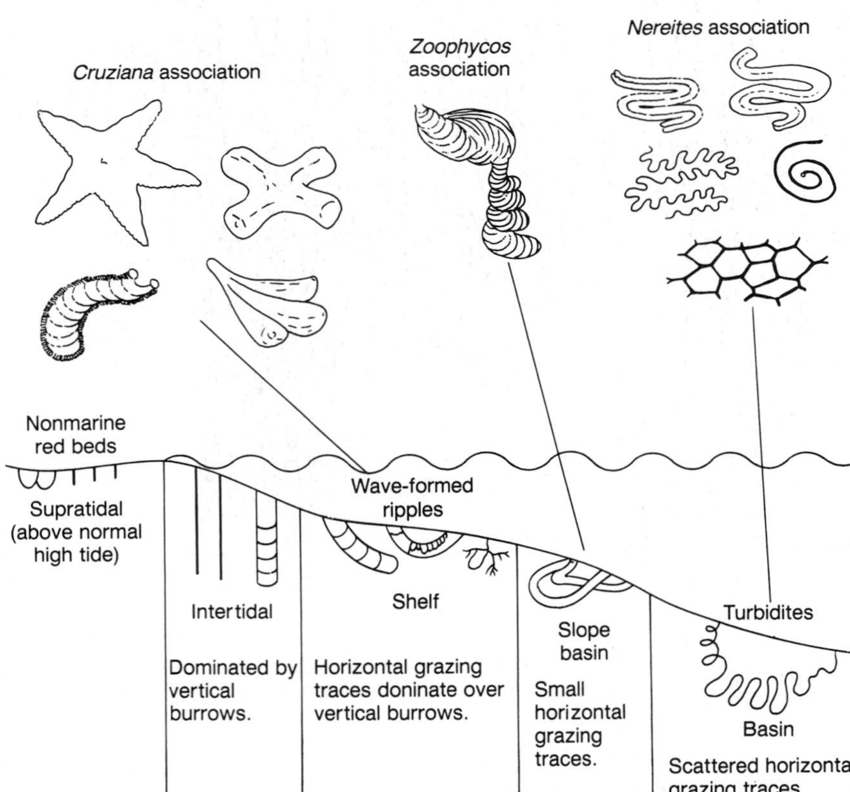


**Figure 4–12** Common associations of trace fossils and their environmental significance, especially with regard to water depth.

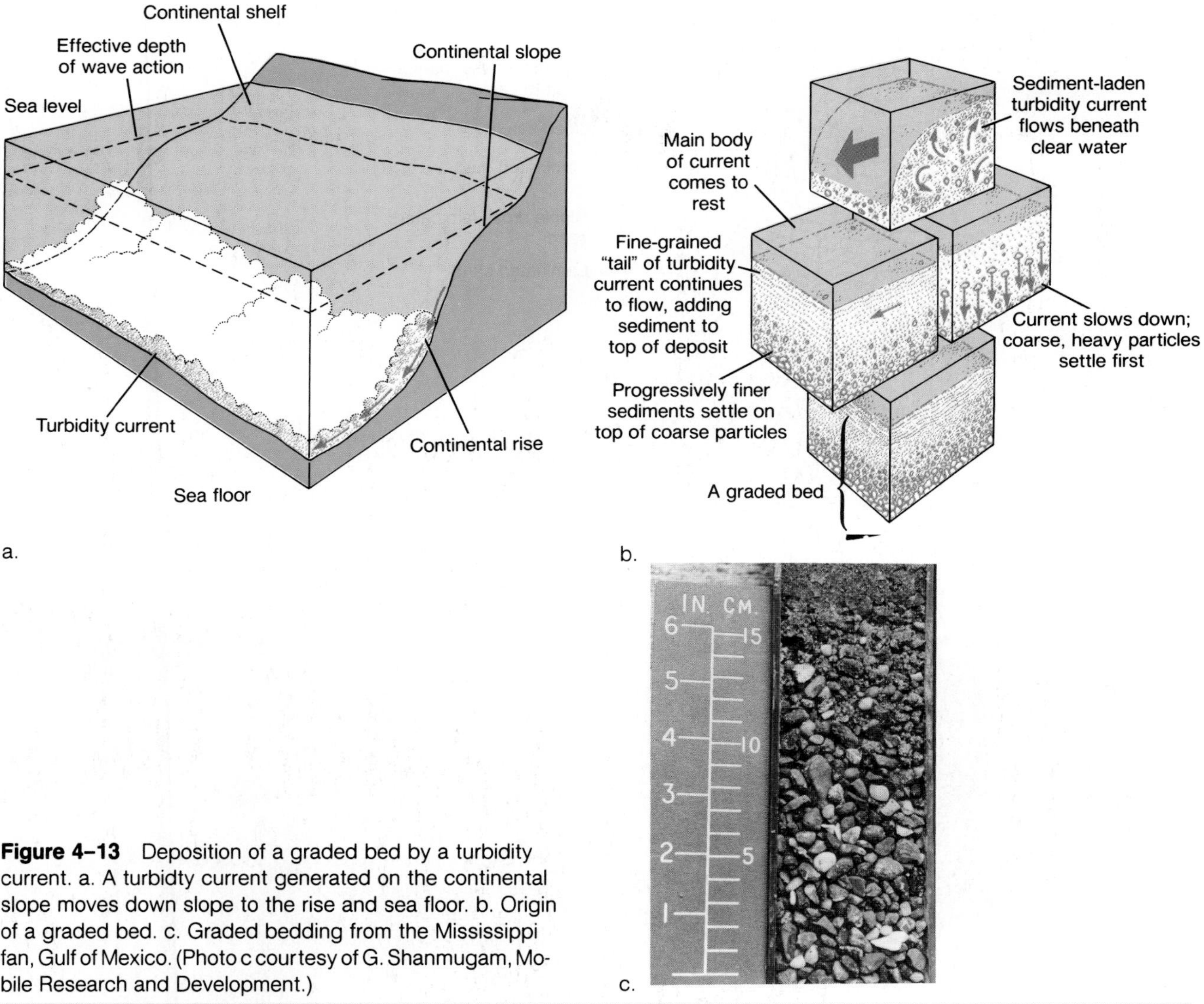


**Figure 4–13** Deposition of a graded bed by a turbidity current. a. A turbidty current generated on the continental slope moves down slope to the rise and sea floor. b. Origin of a graded bed. c. Graded bedding from the Mississippi fan, Gulf of Mexico. (Photo c courtesy of G. Shanmugam, Mobile Research and Development.)

thereafter. Such **sedimentary structures** are the manifestations of physical and biological processes that operated in depositional environments and thus provide information on what the depositional processes were. The processes whereby many sedimentary structures form are well known because they can be observed forming in modern depositional environments, and many can be formed experimentally.

Sedimentary rocks have a layered appearance called **bedding** or **stratification.** Individual layers are referred to as **laminae** if they are less than 1 cm thick and **beds** if they are thicker. Some beds show a decrease in grain size from bottom to top, a feature known as **graded bedding** (Fig. 4–13). Graded bedding is common in turbidity current deposits. *Turbidity currents* are underwater flows of sediment-water mixtures that have greater densities than sediment-free water. They move down slopes to the seafloor or lake bottoms, where their velocity diminishes and the heaviest particles settle out first, followed by progressively smaller particles (Fig. 4–13). Turbidity currents also commonly produce scour marks in the muddy sediments over which they flow. When the turbidity current loses velocity, the scour marks fill with sand, thus producing *flute casts* in the sand layer at the base of the turbidity current deposit (Fig. 4–14).

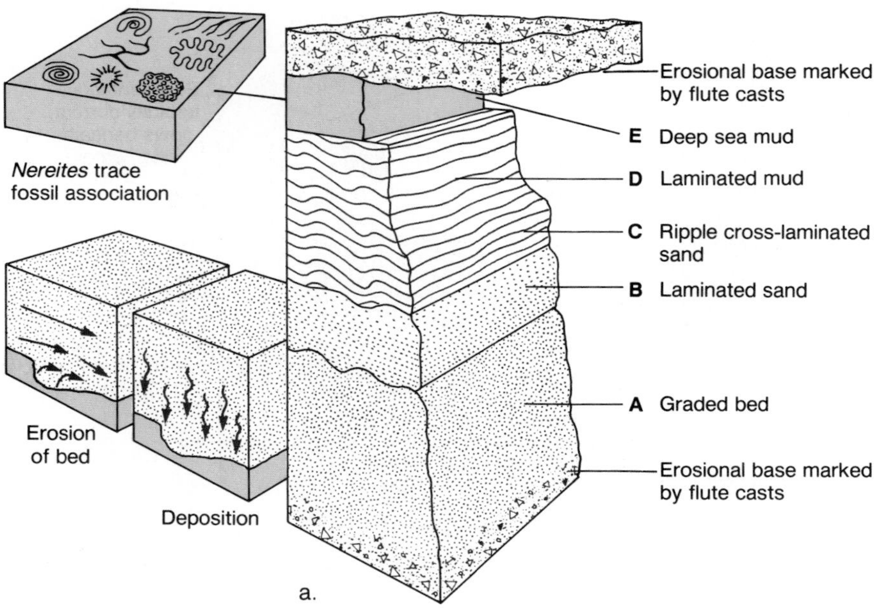

b.

c.

**Figure 4–14** Associations of sedimentary structures are useful for environmental interpretations, especially when considered along with other rock properties such as textures and fossils. This example shows an idealized vertical sequence deposited by a turbidity current. The flute casts are produced by erosion, but as a turbidity current slows down, units A through D are deposited in succession. The deep-sea mud (unit E) forms as particles settle from seawater. It commonly contains the *Nereites* trace-fossil association. b. This rock specimen from the Cambrian Levis Formation of Quebec, Canada, shows the A, B, and C units; the D and E units were present but broke off this specimen. (Photo courtesy of Roger G. Walker, McMaster University, Hamilton, Ontario, Canada.) c. Sedimentary rocks formed by turbidity currents are called turbidites, such as these 1.9 billion-year-old rocks of the Pethei Group along the shores of the Great Slave Lake, Northwest Territories, Canada. (Photo courtesy of Paul Hoffman, Geological Survey of Canada.)

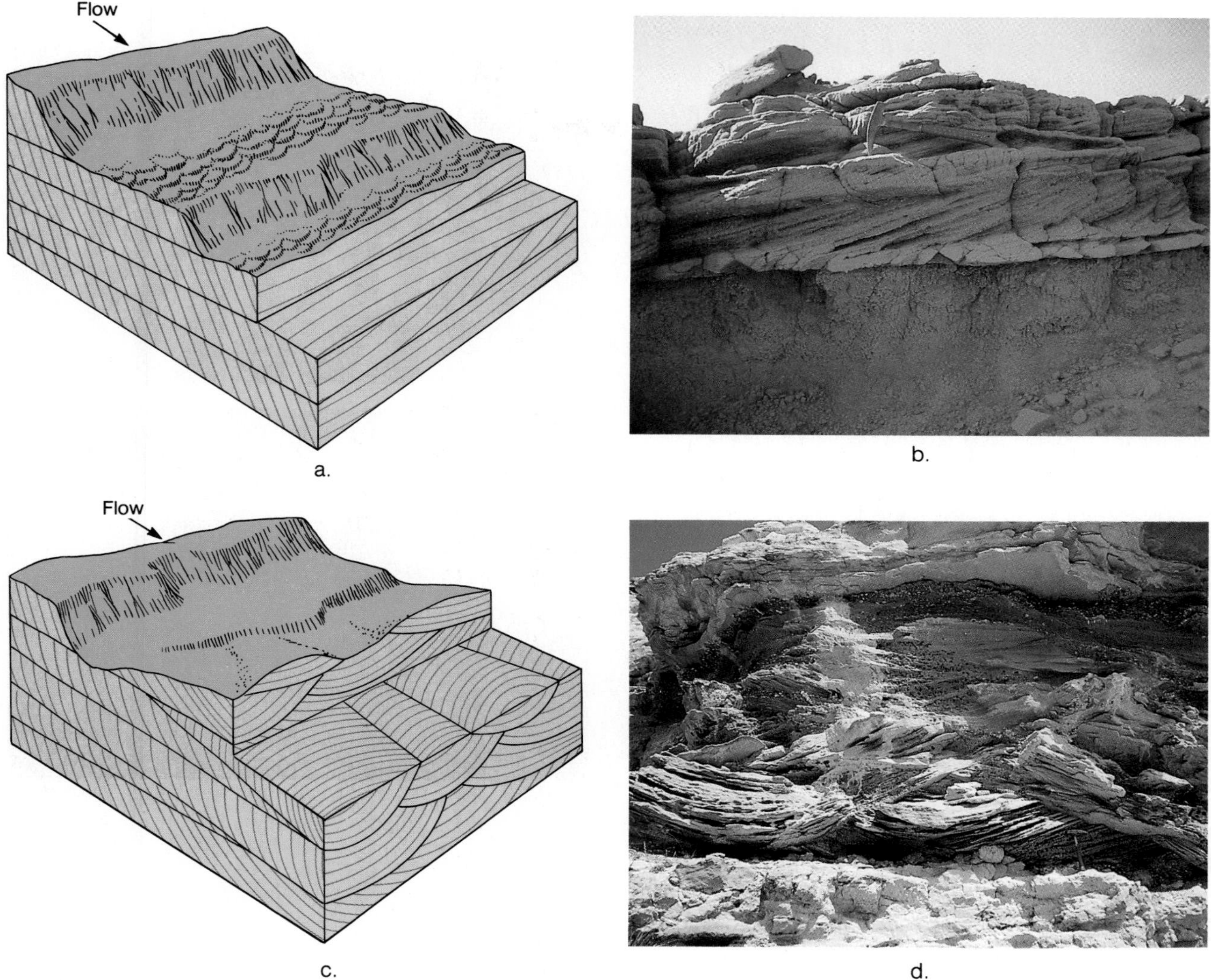


**Figure 4–15** Two types of cross-bedding. a. Tabular cross-bedding formed by migration of sand waves. b. Tabular cross-bedding in the Upper Cretaceous Two-Medicine Formation, Montana. c. Trough cross-bedding formed by migrating dunes. d. Trough cross-bedding in the Pliocene Six-Mile Creek Formation, Montana. This view shows the cross-beds as they are illustrated on the right side of diagram c.

**Cross-bedding** is formed when layers are deposited at an angle to the surface upon which they are accumulating. Invariably, cross-beds result from transport and deposition by wind or water currents, and the cross-beds dip in the direction of flow (Fig. 4–15). Since their orientation depends on direction of flow, cross-beds are good indicators of ancient current directions, or **paleocurrents.**

Bedding planes may be marked by such sedimentary structures as **ripple marks** and **desiccation cracks** (Fig 4–16). Current ripples form in response to water or wind currents that move in one direction. The asymmetrical profiles of current ripples (Fig. 4-16) allow geologists to determine paleocurrent directions. Wave-formed ripples result from the to and fro motion of waves and tend to be symmetrical in profile (Fig. 4–16). Crests of wave-formed ripples are oriented more or less parallel to shorelines. When clay-rich sediments dry, they shrink and crack into polygonal forms bounded by desiccation cracks (Fig. 4–16). These desiccation cracks indicate alternating periods of wetting and drying such as occur along lake margins or on river floodplains.

**Figure 4–16** a. through c. Sedimentary structures from present depositional environments. a. Current ripples in a stream channel. b. Wave-formed ripple marks. c. Desiccation cracks. d. Bioturbation in siltstone caused by burrowing organisms.

**Biogenic sedimentary structures** are produced by organisms and include tracks, trails, and burrows; such organically produced marks are also called *trace fossils* (Chapter 3). We also pointed out in Figure 4–12 that trace fossils are important in environmental analyses. Extensive burrowing by organisms, a process called **bioturbation,** may alter the physical and chemical properties of sediments and modify or destroy other sedimentary structures (Fig. 4–16).

It is important to realize that no single sedimentary structure is unique to a particular depositional environment. Current ripples, for example, are common in stream channels but may also be found in tidal channels, on the seafloor, or near lake margins. Associations of sedimentary structures are particularly useful in environmental analyses, especially when considered with other aspects of a rock unit such as composition, textures, and fossils (Fig. 4–14).

## Geometry

The three-dimensional shape, or the geometry, of a rock body may be helpful in environmental analyses, but geometry must be interpreted with caution. The same geometry may be produced in different sedimentary environments. Moreover, rock-body geometry may be modified by sediment compaction, erosion, and deformation. Nevertheless, geometry can be useful when considered in conjunction with other properties of a rock body.

Some of the most extensive sedimentary rock units in the stratigraphic record are those deposited during marine transgressions and regressions (Figs. 3–13). Such rock units may cover tens or hundreds of thousands of square kilometers, but they are not very thick when compared to their other dimensions of length and width. Rock units with these dimensions are said to have a *blanket,* or *sheet, geometry.*

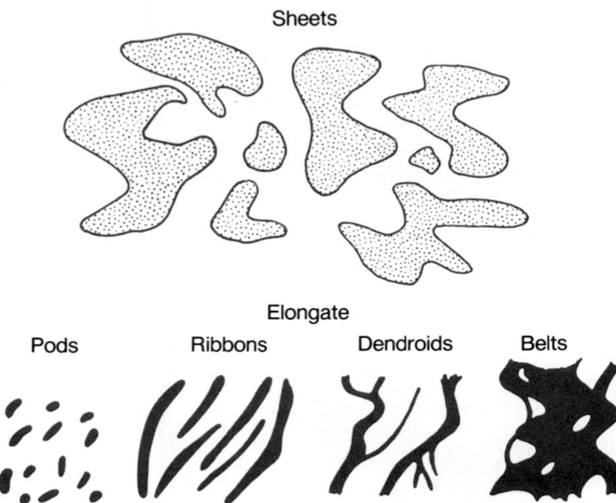


**Figure 4–17** Diagram showing rock-body geometry.

Some sand bodies have an *elongate,* or *shoestring, geometry* (Fig. 4–17) especially those deposited in stream channels or barriers islands. Delta deposits tend to be lens-shaped in cross profile or long profile but lobate when observed from above. Buried reefs are irregular in shape, but many are elongated.

## DEPOSITIONAL ENVIRONMENTS

Any area in which sediment accumulates is called a **depositional environment.** No completely satisfactory classification of depositional environments exists; however, geologists generally recognize three major depositional settings: continental, transitional and marine, each of which contains several specific environments (Fig. 4–18). As a starting point in environmental interpretations, geologists rely upon the concept of uniformitarianism. We assume, for example, that current ripple marks and cross-beds formed in the past just as they do in modern depositional environments. Thus, the study of modern depositional processes and environments is fundamental to our understanding of ancient processes and environments.

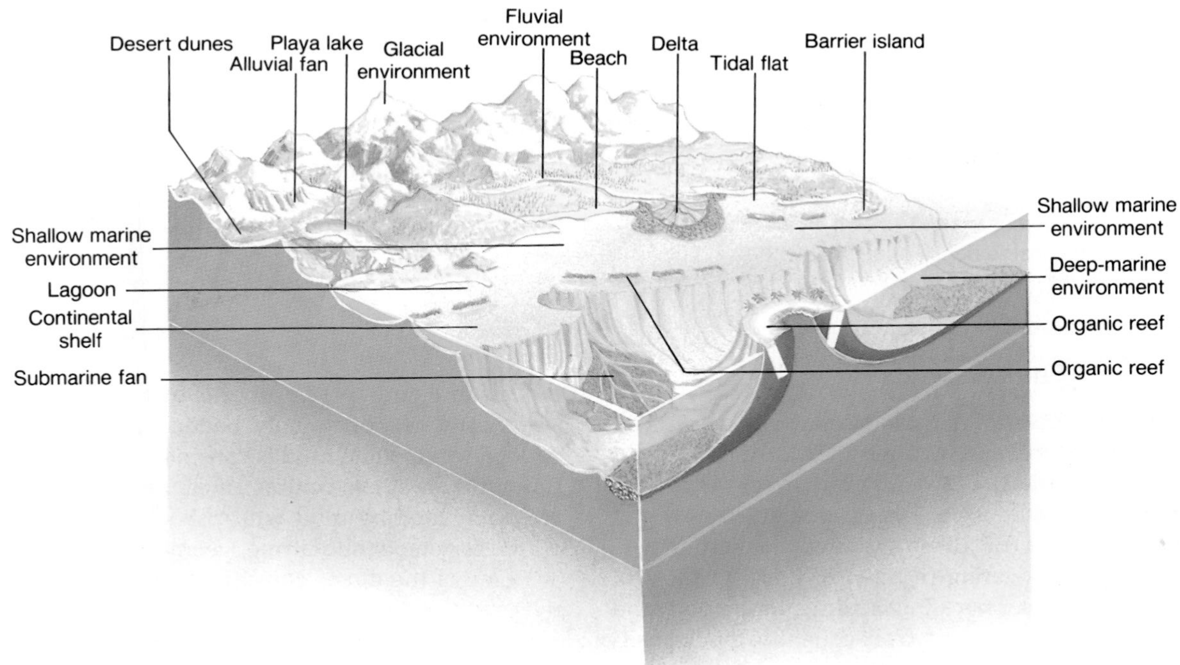


**Figure 4–18** Major depositional environments are shown in this generalized diagram. The environments located along the marine shoreline are transitional from continental to marine. The shallow marine environment corresponds to the continental shelf and can be the site of either detrital or carbonate deposition.

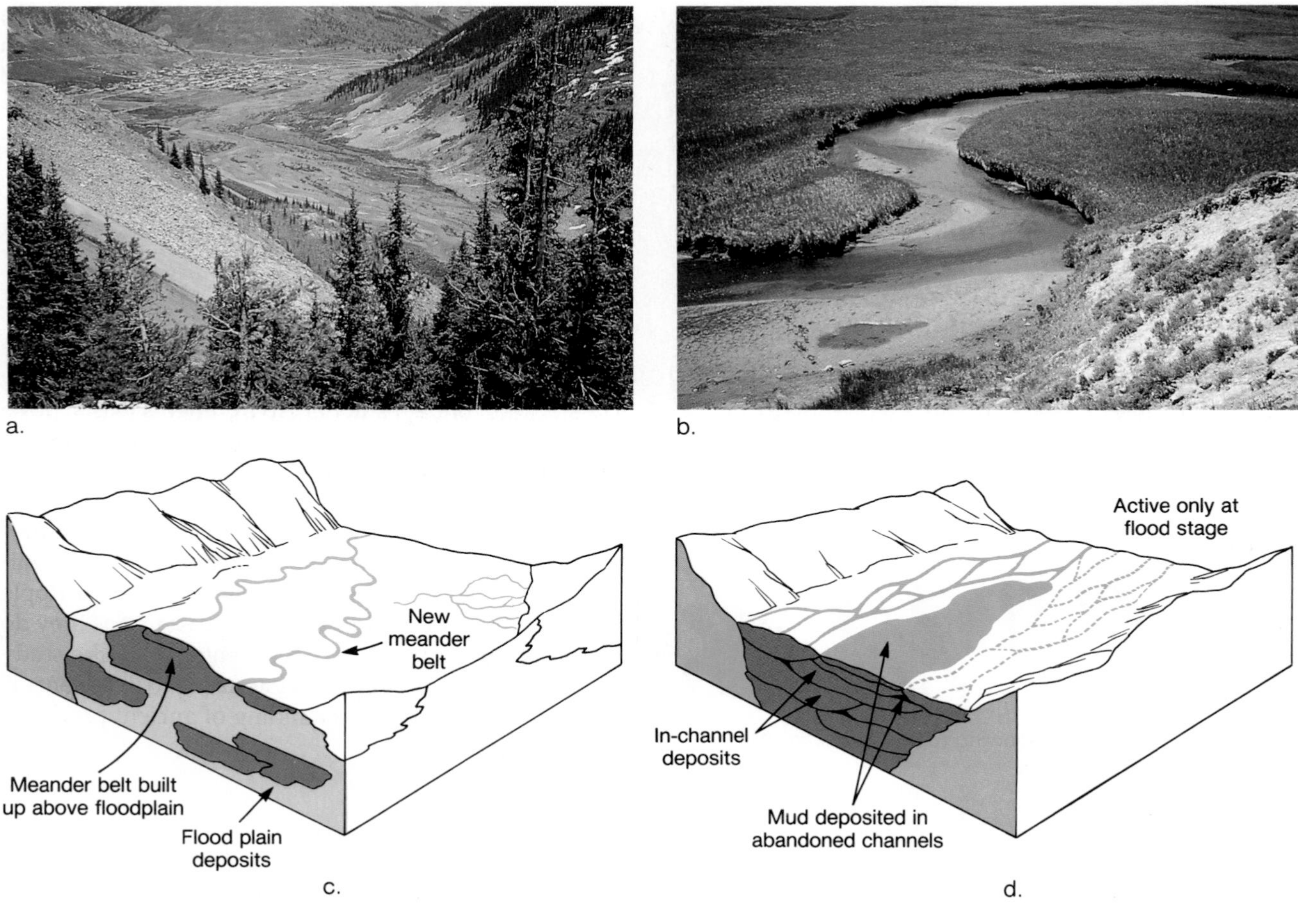


**Figure 4–19** a. The Animas River in Colorado is braided. Notice that the river is divided into numerous channels by sand and gravel bars. b. Otter Creek in Yellowstone National Park, Wyoming is a meandering stream. Point bars are well developed on the inside banks of each meander. c. Facies model for a meandering river system. Meandering river deposits are shoestring sands surrounded by mudstones. d. Facies model for a braided river system. Braided rivers deposit mostly sand and gravel with subordinate mud.

## Continental Environments

Major environments of deposition on the continents include river (fluvial) systems, deserts, and glacial environments (Table 4–1). **Fluvial** is a term referring to river activity and to the deposits that accumulate in river systems. Two types of river systems are recognized: braided and meandering (Fig.4–19). Both do most of their geologic work of erosion, sediment transport, and deposition when they flood. Consequently, the deposits of rivers do not record the continuous day-to-day activities of rivers, but rather record those periodic, large-scale events of sedimentation associated with flooding. The deposits of braided and meandering rivers, however, differ considerably, as shown in the idealized facies models in Figure 4–19. Deposits of braided rivers consist mostly of horizontally bedded gravel and cross-bedded sand, while mud is conspicuous by its near absence (Fig. 4–20). In contrast, meandering river deposits are dominated by mud with subordinate sand bodies with elongate, or shoestring, geometries (Fig. 4–19).

One of the most distinctive aspects of meandering river sand bodies is the **point bar** sequence. When meandering rivers flood, they erode the steep, outer banks of meanders and simultaneously deposit sediment on the inner, gently sloping banks as point bars. Successive episodes of erosion and deposition result in an eroded surface overlain by a sequence of point bars, each of which consists of a cross-bedded sand body (Fig. 4–21).

**Figure 4–20** Gravel bar in the Smith River of northern California.

In addition to deposition of point bars, flooding meandering rivers transport silt and clay into the floodplain, where they settle out to form layers of mud.

Windblown desert dunes are recognized largely on the basis of textures, sedimentary structures, and facies relationships. All dune sands are typically well sorted and cross-bedded, on a scale measured in meters or tens of meters (Fig. 4–22). Large-scale cross-bedding also occurs in dunelike sands of the continental shelf, but such sands contain marine fossils and are associated with other marine rocks. Desert dunes, in contrast, may contain the remains or traces of land animals.

The association of windblown dunes with other deposits typical of deserts is most helpful. For example, the margins of desert basins commonly are the sites where **alluvial fans** accumulate, and the desert floor may have

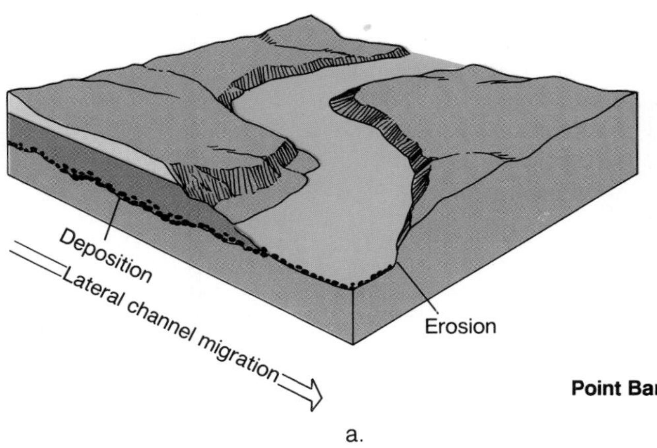


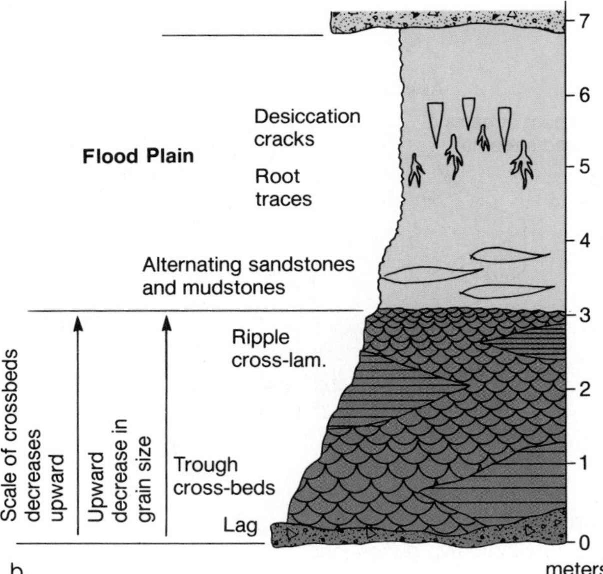

**Figure 4–21** a. Origin of a point bar in a meandering river. b. Idealized point bar sequence overlain by floodplain deposits. c. Point bar in the Cretaceous Belly River Formation, Alberta, Canada. Note sharp base to sand body, and cross-bedding (by notebook) in lower part. Upper part of sand body (toward the left) is ripple cross-laminated and overlain by fine-grained deposits. (Photo courtesy of Roger G. Walker, Dept. of Geology, McMaster University, Hamilton, Ontario, Canada.)

**Figure 4–22** Large-scale cross-bedding in Jurassic dune sands of the Navajo Formation, Zion National Park, Utah.

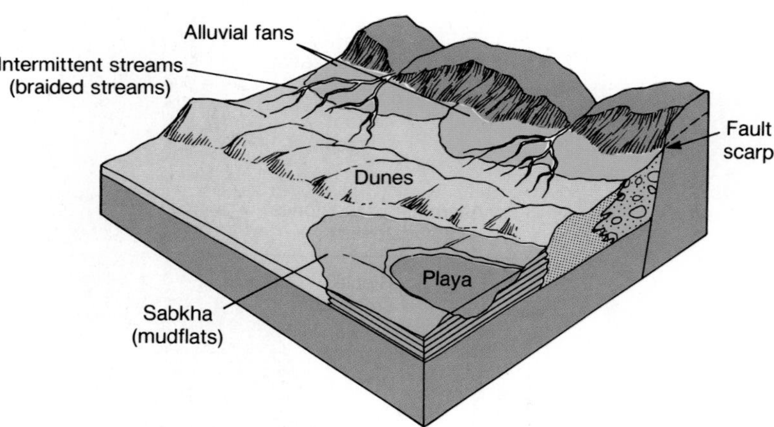


**Figure 4–23** Desert basin showing the association of alluvial fans, windblown dunes, and playa lakes.

temporary lakes called **playa lakes** in which mud and evaporites accumulate (Fig. 4–23).

Glaciers today are comparatively restricted in distribution. Many small glaciers exist in high mountains, but the only extensive ice sheets of continental proportions are in Greenland and Antarctica. Widespread glaciation occurred at several times in the past, however—in the Late Precambrian (Chapter 9), the Ordovician (Chapter 10) and Pennsylvanian (Chapter 11), and most recently during the Pleistocene (Chapter 17).

Glaciers are very effective agents of erosion, transport, and deposition, Many of the surficial deposits in the northern tier of states and in parts of Canada were deposited by Pleistocene glaciers or in associated subenvironments (Fig. 4–24). All sediments deposited in glacial environments are collectively called **drift,** but we must distinguish between two types of drift. **Till** is unsorted, unstratified drift deposited directly by glacial ice, mostly as ridgelike deposits called *moraines* (Fig. 4–24). A second type of drift, called **outwash,** is deposited

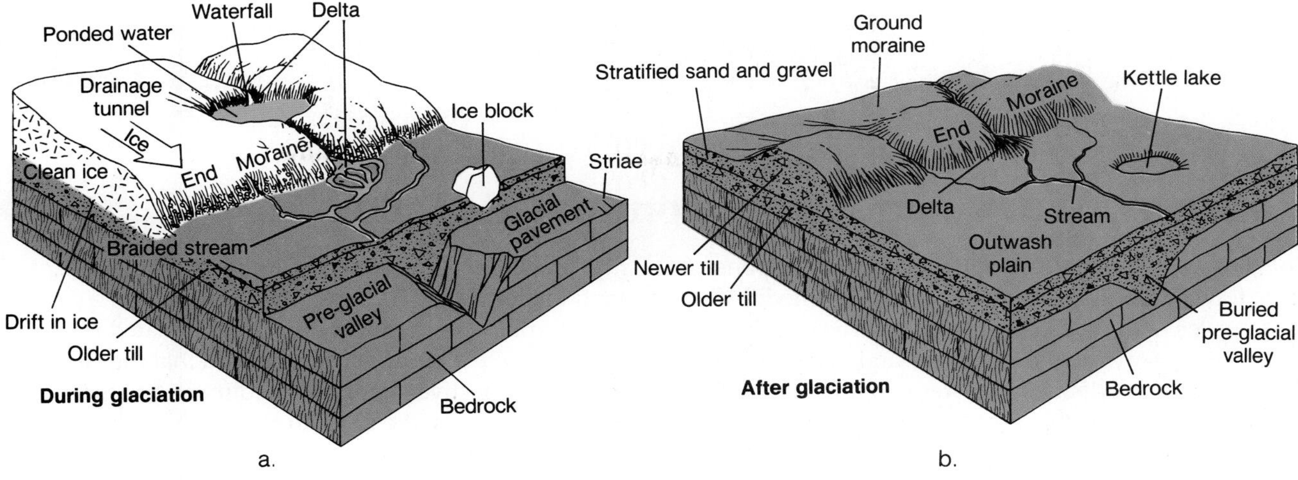

**Figure 4–24** a. and b. Diagrams showing the origin of glacial drift. c. Ice-rafted dropstone in Late Proterozoic varved sediments of the Gowganda Formation, Canada.

by fluvial processes, mostly in braided streams, that issue from melting glaciers (Fig. 4–24). Such streams are heavily loaded with sediment, much of which is deposited as sheets of sand and gravel. Since the terminus of a glacier may advance or retreat, it is not uncommon for till and outwash to be interbedded.

Deposits of glacial lakes deserve special mention. Such lakes typically accumulate finely laminated muds consisting of alternating light and dark laminae; each light-dark couplet is called a **varve** (Fig. 4–24). Each varve represents an annual episode of deposition: the light layers form in the spring and summer and consist of silt and clay; dark layers form in the winter when fine-grained clay and organic matter settle from suspension as the lake freezes over. Another distinctive feature of varved deposits is *dropstones,* which are pieces of gravel dropped from floating ice (Fig. 4–24).

## Transitional Environments

Such environments as deltas, barrier islands, and tidal flats (Fig. 4–18, Table 4–1) are transitional from continental to marine. The deposits of deltas, for example, accumulate where a fluvial system enters the sea, but the deposits are modified by such marine processes as waves and tides (Perspective 4–1).

The process whereby **deltas** form is rather simple: when a stream or river enters a standing body of water, a lake or the sea, its velocity decreases and deposition occurs. Deltas in lakes are common, but marine deltas

## Perspective 4–1

# River-, Wave-, and Tide-Dominated Deltas

As we discussed earlier, deltas form in response to fluvial processes, but they may be modified by waves and tides, so geologists recognize river-dominated, wave-dominated, and tide-dominated deltas. The Mississippi River delta, for example, is dominated by river processes; the sediment supply rate is high, and it lies on a shallow shelf where it is protected from waves and tides. Distributary channels prograde far out to sea as a series of fingerlike sand bodies (Fig. 1).

Distributary channels of river-dominated deltas eventually prograde so far seaward that they are no longer effective avenues of sediment transport. When this occurs, the river abandons that channel and establishes a new one elsewhere (Fig. 1). The new distributary channel then progrades until it,

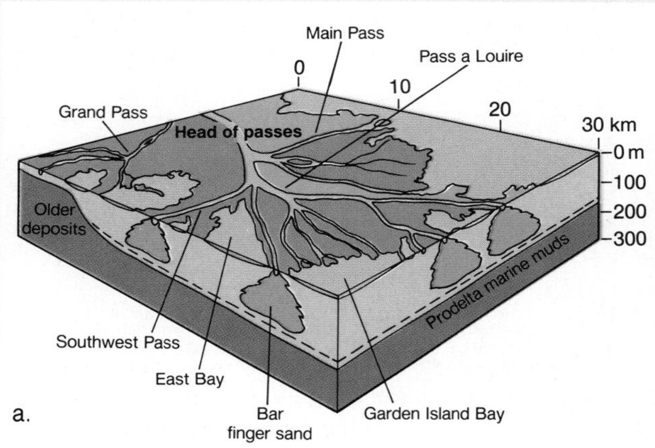


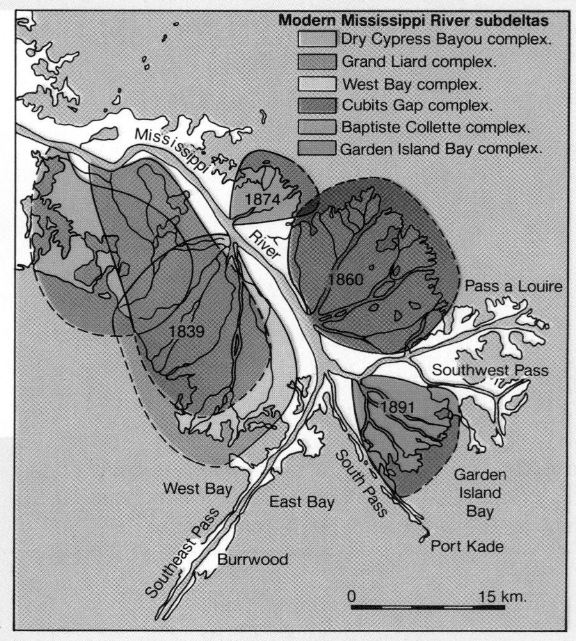


**Figure 1** a. Block diagram illustrating progradation of distributary channel sand bodies (bar finger sands). b. The modern lobe of the Mississippi River delta consists of several subdeltas that formed when the main distributary channel was abandoned. The area shown in this view corresponds to the Balize lobe in Figure 2.

are even more common in the geologic record, much larger, and far more important economically (Perspective 4–1). Marine deltas form where a fluvial system supplies sediments to a shoreline faster than marine processes can redistribute the sediments along the shoreline or carry it out to sea. As a result of deltaic deposition, the shoreline builds out, or **progrades**, into the sea.

Prograding deltas deposit a characteristic vertical sequence in which prodelta deposits are overlain successively by delta-front deposits and delta-plain deposits (Fig. 4–25). Such vertical sequences result from the lateral migration of environments, illustrating Walther's Law (Chapter 3). The flow velocity of a river entering the sea diminishes rapidly, but the finest sediments are carried some distance beyond the river mouth, where they settle from suspension and form prodelta muds. Nearer the river mouth, sand and silt are deposited as gently inclined layers on the delta front. The delta plain is traversed by a network of distributary channels in which fluvial deposits of silt and sand accumulate (Fig. 4–26). Between distributary channels are low marshy areas where fine-grained sediment and organic matter are deposited.

It should be clear from the preceding discussion that deltas are composites of marine and continental depos-

too, is abandoned. In short, river-dominated deltas build a major deltaic lobe seaward by the progradation and abandonment of distributary channels. The active lobe of the modern Mississippi River delta records a series of such events (Fig. 1).

In addition to abandoning individual distributary channels, river-dominated deltas periodically abandon entire deltaic lobes. The Mississippi River is currently depositing sediment on a single lobe, but the delta is composed of several lobes that formed during the last few thousand years (Fig. 2). As a matter of fact, if it were not for the efforts of the Army Corps of Engineers, the currently active lobe of the Mississippi River delta would probably have been abandoned by now.

Deltas that are strongly modified by marine processes are common; the Nile River delta of Egypt is wave-dominated, and the Ganges-Brahmaputra delta of Bangledesh is tide-dominated (Fig. 3). Waves continually modify the seaward margin of the Nile delta; therefore, progradation of fingerlike lobes does not occur. Rather, the river-deposited sediment is reworked and redistributed along the delta margin as a series of barrier sand bodies that grade seaward into silt and clay. The Nile Delta's triangular shape resembles the Greek capital letter

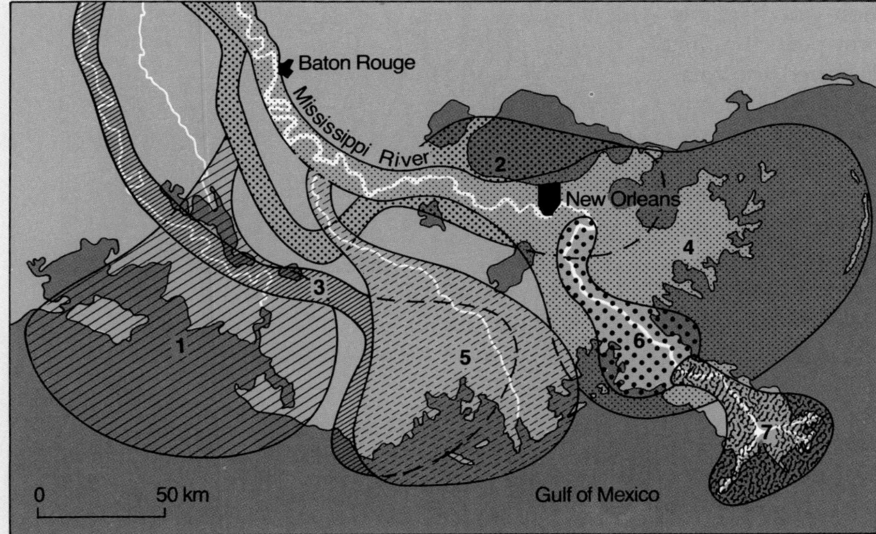


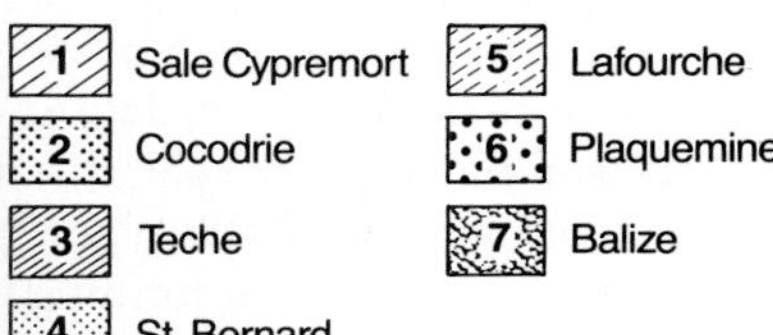


**Figure 2** Major deltaic lobes deposited by the Mississippi River during the last 7,000 years. The oldest lobe, the Sale Cypremort lobe, was deposited for about 1200 years over 150 km west of the modern lobe. Deposition of the modern Balize lobe began 600 to 800 years ago.

its. The facies of the prodelta and delta-front environments are marine and commonly contain marine fossils. However, the delta-front deposits may also contain pieces of waterlogged wood and other land-plant remains. The distributary channel deposits of the delta plain show clear evidence of deposition by fluvial processes, and any fossils they contain will be of land plants and animals.

Shorelines where marine processes effectively redistribute sediments as fast as they are supplied are characterized by linear beaches and barrier islands paralleling the shoreline (Fig. 4–18). Beaches are simply narrow sand bodies, constructed and modified by waves and nearshore currents, grading seaward into interbedded silt and sand. On their landward sides beaches are commonly bordered by windblown sand dunes.

Elongate sand bodies separated from the mainland by a lagoon are called **barrier islands** (Fig 4–27). These are high-energy environments where windblown dune and beach deposits accumulate; and during storms, waves may overtop barrier islands and deposit washover lobes in the lagoon (Fig. 4–27). Sand dunes consist of well-sorted, cross-bedded sand. Quartz is the dominant mineral in dune sands, but sand-sized fragments of marine shells are commonly present, too. In

## Perspective 4–1 (continued)

*delta* (Δ), which is why, in about 490 B.C., the Greek philosopher Herodotus coined the term.

Deltas on shorelines affected by strong tidal currents are shaped into elongate tidal sand bodies that parallel the direction of tidal flow (Fig. 3). When such deltas prograde, they produce a vertical sequence of prodelta deposits overlain by sand bodies that look much like those deposited in braided river channels.

Coal, oil, and gas are important resources found in deltaic deposits. Coal can form in several depositional environments, one of which is the marshes between distributary channels (Fig. 1). Such marshes are dominated by non-woody plants whose remains accumulate to form peat, the first stage in the origin of coal. If peat is buried, the volatile components of the plants are lost, and mostly carbon remains, thus forming coal.

Deltaic progradation is one way in which potential reservoirs for oil and gas form in marine basins. Distributary sand bodies, because of their porosity and permeability, together with their association with organic-rich marine sediments, commonly contain oil and gas. Much of the oil and gas production of the Gulf Coast of Texas comes from subsurface deltaic deposits. The Niger River delta of Africa and the Mississippi River delta are also known to contain reserves of oil and gas.

a.

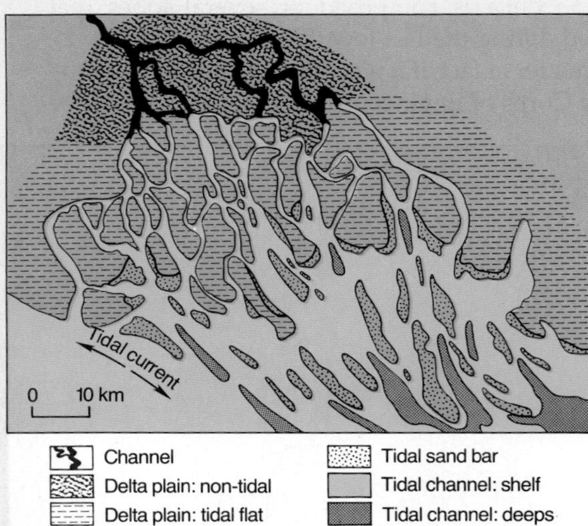


b.

**Figure 3** Wave- and tide-dominated deltas. a. The Nile River delta of Egypt is wave-dominated. Sand supplied by distributary channels is redistributed along the front of the delta as a series of barrier sand bodies. b. The Ganges-Brahmaputra delta is tide-dominated.

fact, the presence of marine fossils and facies relationships are the criteria used to distinguish these dunes from desert dunes. Beach sands are typically coarser grained than dune sands, contain fragmented marine fossils, and grade seaward into sediments of the shoreface environment (Fig. 4–27).

Because of their proximity to potential source rocks, barrier islands are good reservoirs for hydrocarbons. Accordingly, geologists have thoroughly studied such modern barrier islands as those along the Gulf Coast of Texas and the east coast of the United States (Fig. 4–28), so that their search for ancient hydrocarbon-bearing deposits is likely to be more fruitful.

### Marine Environments

Marine environments (Fig. 4–18, Table 4–1) are those that lie seaward of the transitional environments. They include such depositional settings as the continental shelf, slope, rise, and deep-sea floor. Much of the detritus eroded from continents is deposited in marine environments, but other types of sediment may accumulate as well. For example, limestones are common in warm, shallow seas, and evaporites may be deposited in arid regions.

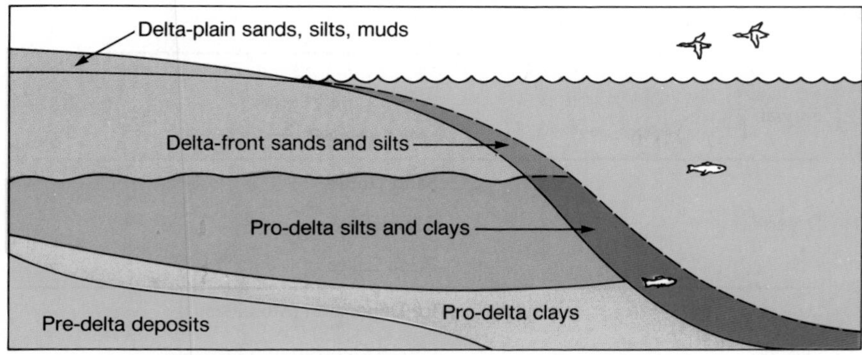


**Figure 4–25** Generalized cross section of a prograding delta. Deposition occurs in a seaward direction as indicated by the bold arrows. As a result of progradation, prodelta deposits are overlain by delta-front and delta-plain deposits.

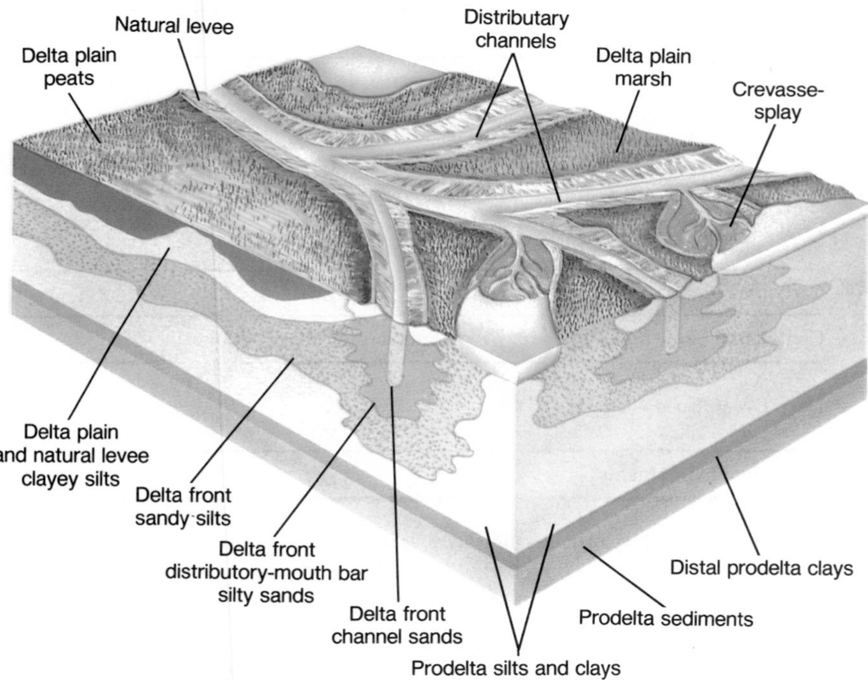


**Figure 4–26** Schematic view of a delta showing facies deposited during progradation. Notice the branching network of distributary channels and the delta plain marsh. Progradation of distributary channel produces a vertical succession of facies like that in Figure 4–25.

## Detrital Marine Environments

The continental shelf is a gently seaward-sloping surface bounded at its outer edge by a shelf-slope break. Beyond this break lie the continental slope and continental rise, and finally the deep-ocean basins (Fig. 4–29). Continental shelves are subdivided into a shallow, high energy inner shelf, and a deeper, low-energy outer shelf (Fig. 4–30). Waves and tides more or less continuously stir up and sort inner-shelf sediments until mostly sand is left. This sand is commonly shaped into large, cross-bedded dunes and sand waves. The outer shelf is a low-energy area largely unaffected by tides

**Table 4–1** Depositional environments discussed in the text.

| | | |
|---|---|---|
| **Continental Environments** | Fluvial | Braided River<br>Meandering River |
| | Desert | Sand Dunes<br>Alluvial Fans<br>Playa Lakes |
| | Glacial | Ice Deposition (Moraines)<br>Fluvial Deposition (Outwash)<br>Glacial Lakes |
| **Transitional Environments** | Delta | River-Dominated<br>Wave-Dominated<br>Tide-Dominated |
| | Beach | |
| | Barrier Island | Beach<br>Sand Dune<br>Lagoon |
| **Marine Environments** | Continental Shelf | Detrital Deposition<br>Carbonate Deposition |
| | Carbonate Platform | |
| | Continental Slope and Rise | Submarine Fans (Turbidites) |
| | Deep Ocean Basin | Oozes<br>Brown Clay |
| | Evaporite Environments* | |

* Evaporites may be deposited in a variety of environments including playa lakes, saline lakes, lagoons and tidal flats in arid regions, and in marine environments.

and waves, except during major storms when surface waves may stir up sediments of the entire shelf. Mud is the most common sediment on the outer shelf, but layers of silt and sand are present too, especially near the boundary between the inner and outer shelf.

Much of the sediment derived from the continents crosses the continental shelf in submarine canyons and eventually comes to rest on the continental slope and rise as a series of overlapping **submarine fans** (Fig. 4–31). Sedimentologically, the edge of the continental shelf is a dividing line between two major realms in the oceans. Landward from the shelf edge, sediments are at least periodically affected by waves and tides, but in a seaward direction these processes are no longer of any importance in sediment transport and deposition. Once sediment has passed the shelf edge, it is transported and deposited by gravity processes such as slumps, submarine debris flows, and particularly by turbidity currents (Fig. 4–13).

Beyond the continental slope and rise system, the seafloor is covered mostly by fine-grained deposits. Local exceptions occur near oceanic islands where coarse-grained detrital sediments accumulate, and melting icebergs may carry sand and gravel far from continents. Otherwise no mechanisms exist that can transport significant quantities of coarse-grained detritus from continents into the deep-ocean basins.

There are, however, sources of fine-grained sediments. Windblown dust and ash from oceanic islands or continental volcanoes falls far out to sea; meteorite dust

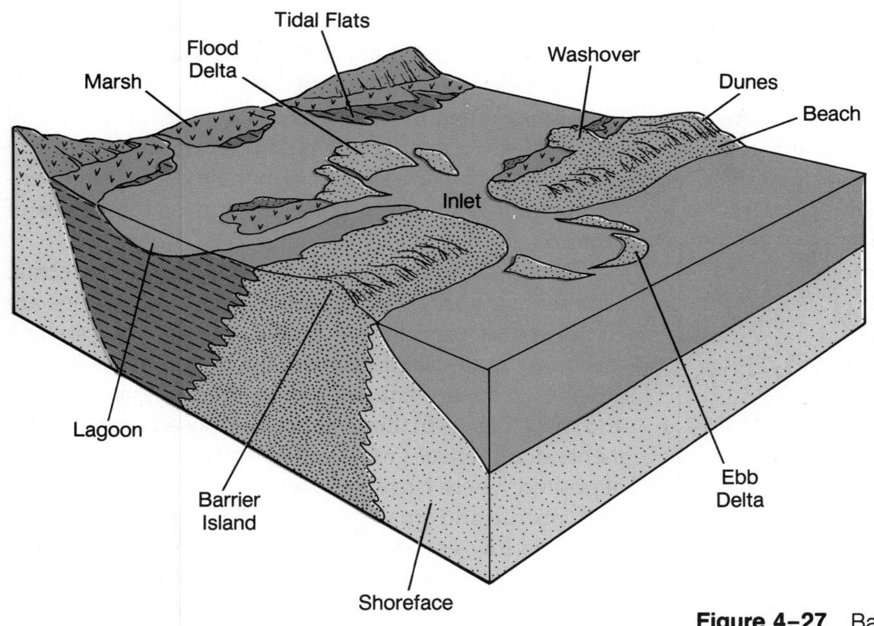


**Figure 4–27** Barrier island and associated environments.

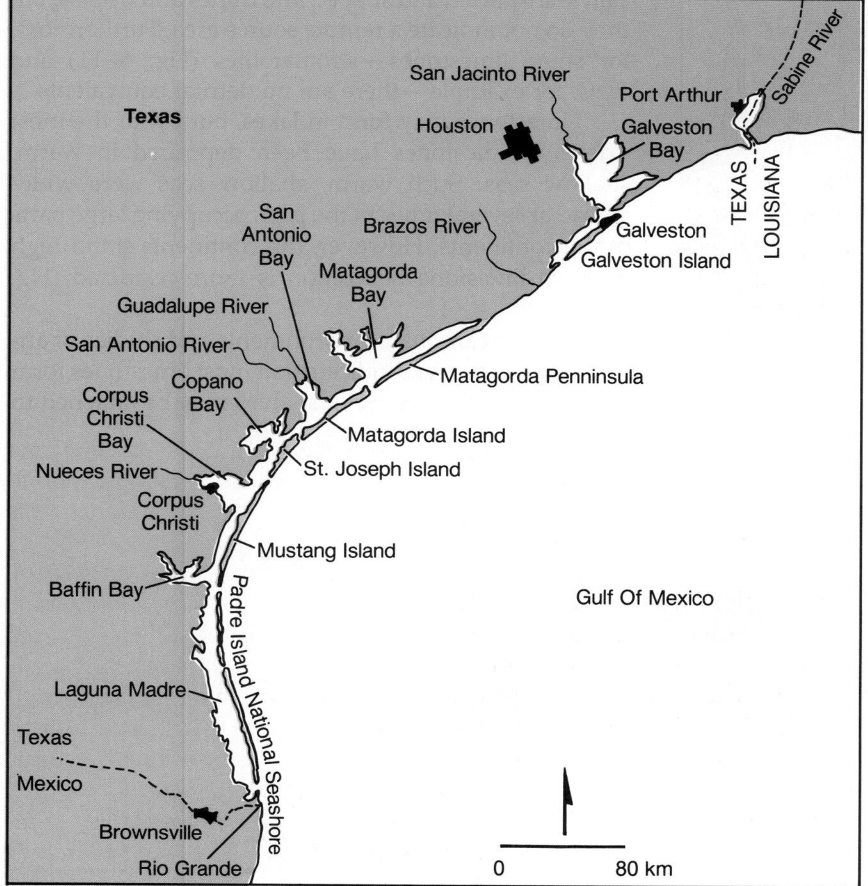


**Figure 4–28** Barrier islands of the Texas Gulf Coast.

comes from space; and the organically productive surface waters of the oceans contribute the shells of floating microorganisms. Figure 4–32 shows that the seafloor is covered mostly by three types of sediment—brown clay and two types of oozes that are composed of the shells of single-celled plants and animals.

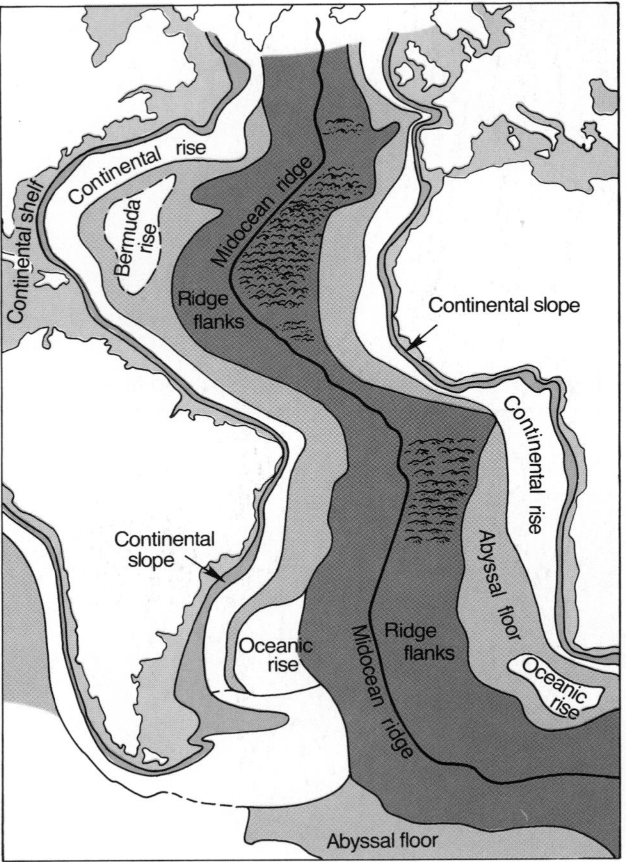


**Figure 4–29** Continental shelves, slopes, and rises in the Atlantic Ocean basin.

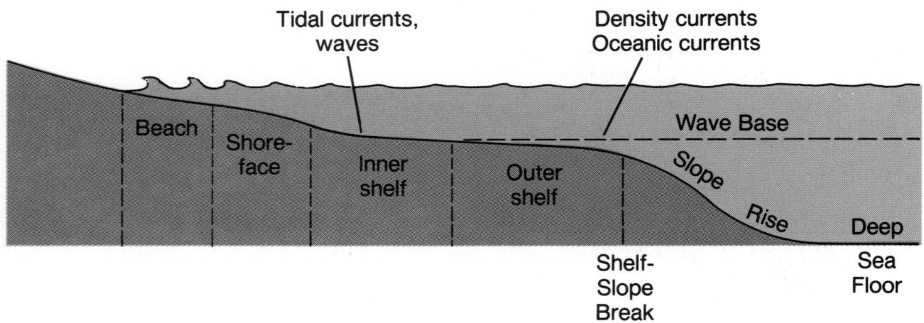


**Figure 4–30** Environments of the continental shelf.

### Carbonate Depositional Environments

Limestone and dolomite are the only common carbonate rocks. Most dolomite, however, was originally limestone. When limestone is dolomitized, the mineral calcite, $CaCO_3$, changes to dolomite, $CaMg(CO_3)_2$. Accordingly, our discussion of carbonate depositional environments is mostly a consideration of the origin of limestones.

In some respects limestones are similar to detrital sedimentary rocks. For example, many limestones are composed of particles, or grains, and microcrystalline carbonate mud called **micrite.** Such grains and micrite are the carbonate equivalents to detrital gravel, sand, and mud (Fig. 4–6). Limestones composed of shell fragments or spherical grains called **oolites** (Fig. 4–33) are deposited where currents or waves are intense. In contrast, micrite and small grains such as fecal pellets accumulate in low-energy environments such as lagoons.

Despite textural similarities, limestones and detrital rocks also have some fundamental differences. Composition is one obvious difference; but more importantly, carbonate grains and micrite originate within the basin of deposition. Such carbonate constituents may be locally transported and shaped into dunes and ripples, but they do not indicate a remote source area. Furthermore, for some limestones—stromatolites (Fig. 4–11) and reefs, for example—there are no detrital equivalents.

Limestones may form in lakes, but by far the most extensive limestones have been deposited in warm, shallow seas. Such warm, shallow seas were widespread at several times in the past, occupying large parts of the continents. However, the continents stand high now, so limestone deposition is more restricted (Fig. 4–34).

Modern carbonate environments and studies of ancient carbonate rocks indicate that most limestones form on carbonate shelves. Such shelves may be attached to

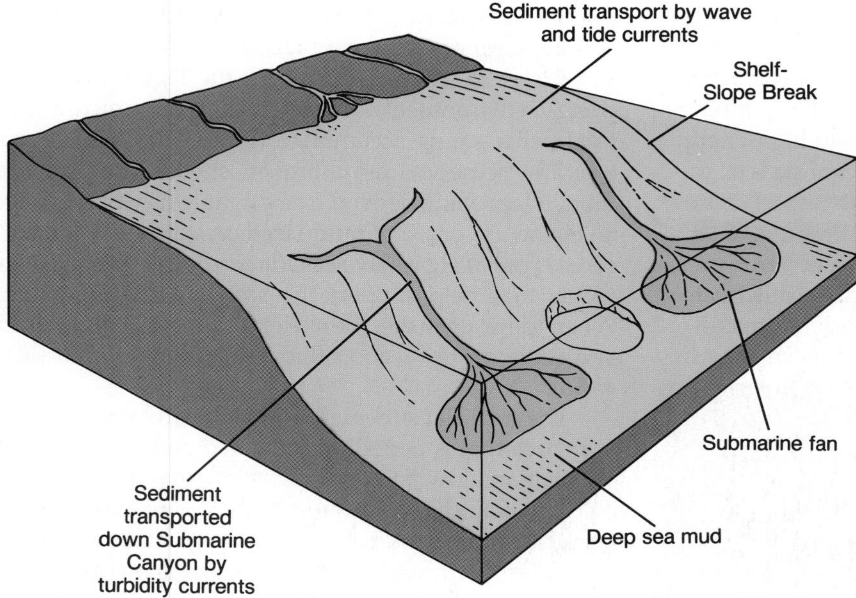


**Figure 4–31** One of the main avenues of transport of sediment across the shelf is in submarine canyons. Turbidity currents carry sediments beyond the shelf into deeper waters of the slope and rise where they accumulate as submarine fans.

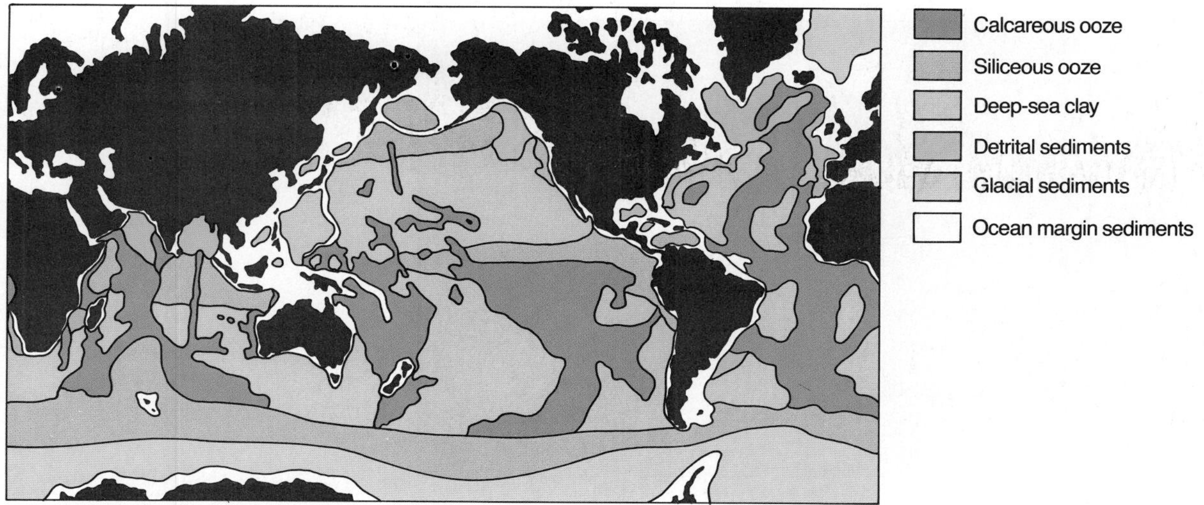


**Figure 4–32** Distribution of sediments on the seafloor.

a continent or may lie across the tops of offshore banks such as the Great Bahama Bank (Fig. 4–35). In either case deposition occurs in shallow-water areas where there is little or no influx of detritus, especially mud.

Carbonate barriers (Fig. 4–35) form in high-energy environments and may be reefs or banks of skeletal carbonate sand or oolites. **Reefs** have a structural framework of coral skeletons, mollusks, sponges, and many other organisms and form wave-resistant structures. Within lagoons small, isolated reef masses called *patch reefs* may occur (Fig. 4–35). Reef rock is massive, structureless limestone, large pieces of which may be torn loose by storm waves and deposited in the forereef and backreef areas as reef breccia (Fig. 4–35).

**Figure 4–33** Small, spherical or ovate, calcium carbonate grains, such as those in this oolitic limestone, are common constituents of some limestones. Most oolites measure 0.5 to 1.0 mm in diameter.

Except during major storms, the lagoon is a low-energy environment, where fossiliferous micrite and fecal-pellet sands accumulate (Fig. 4–35). Fossils are typically preserved as unbroken shells. Much of the mud in lagoons is derived from organisms whose skeletal elements consist of mud-sized crystals. For example, some types of algae have calcium carbonate skeletal elements that help support the soft tissues during life. When these algae die, the skeletal elements are released and settle on the lagoon floor as carbonate mud (Fig. 4–36).

The tidal-flat subenvironment of carbonate shelves is an area that is periodically flooded and exposed during high and low tides, respectively. Tidal flats are traversed by tidal channels in which the sediments can vary from carbonate sand to mud, and stromatolites may be present on the upper parts of some tidal flats (Figs. 4–11, 4–35). On the landward side of tidal flats is a supratidal marsh, an area flooded only during exceptionally high tides (Fig. 4–35). Algal mats, desiccation cracks, and laminated micrite are common features of supratidal marshes. In arid regions supratidal areas are called **sabkhas.** High evaporation rates on sabkhas cause seawater contained within the sediments to evaporate. When this occurs, evaporites, especially gypsum ($CaSO_4 \cdot 2H_2O$), form within the sediments as irregular masses called *nodules.*

Carbonate rock units covering tens of thousands of square kilometers are common in the geologic record. Many of these carbonate units were deposited during

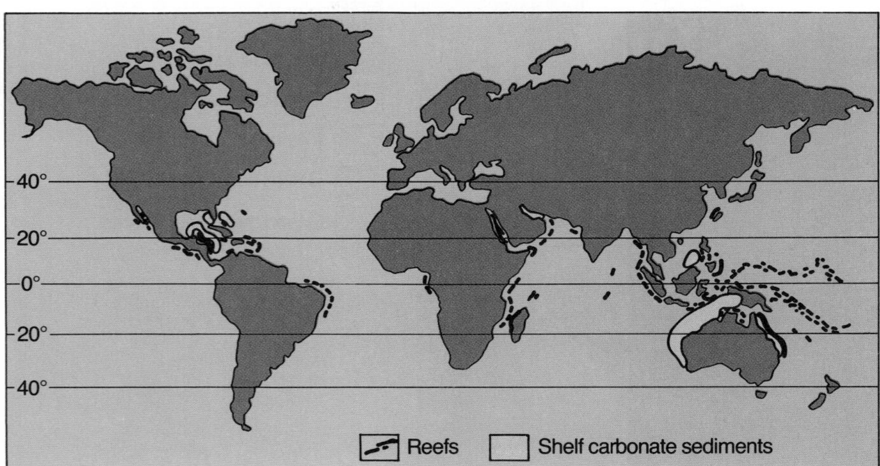


**Figure 4–34** Deposition of modern, shallow-water carbonate sediments occurs mostly in warm seas at low latitudes.

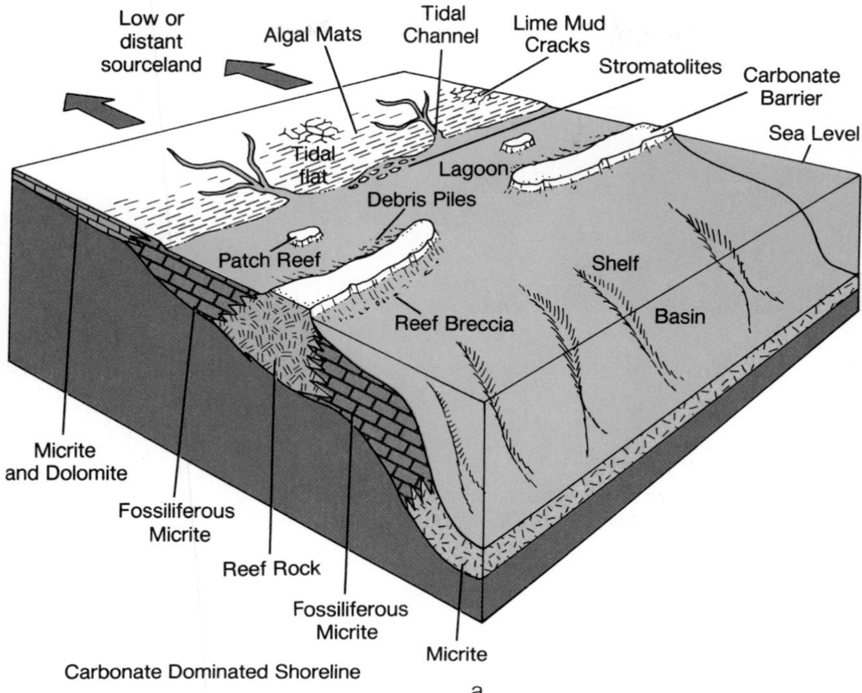


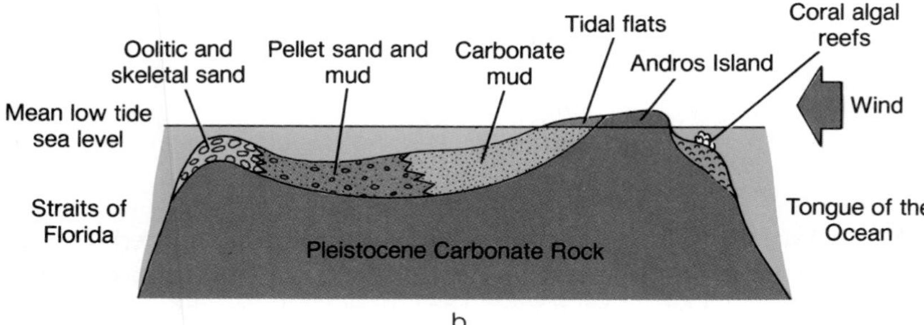


**Figure 4–35** Carbonate depositional environments. a. A shelf attached to a continent. Modern carbonate deposition is occurring on such attached shelves in southern Florida and the Persian Gulf. b. Generalized cross section of the Great Bahama Bank. The carbonate shelf lies on top of a bank that rises from the seafloor.

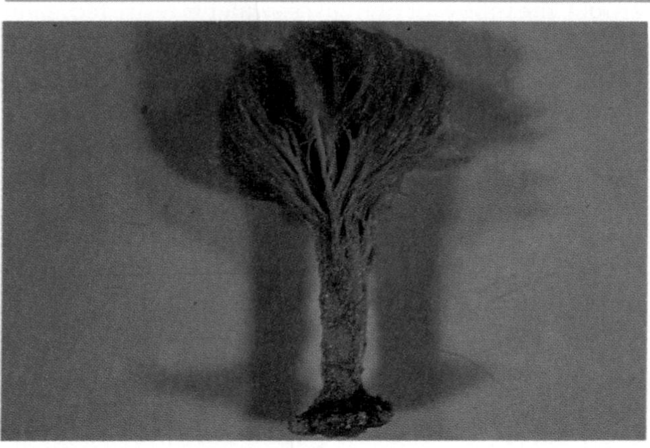

**Figure 4–36** The alga *Penicillus* contributes carbonate particles to the sediments.

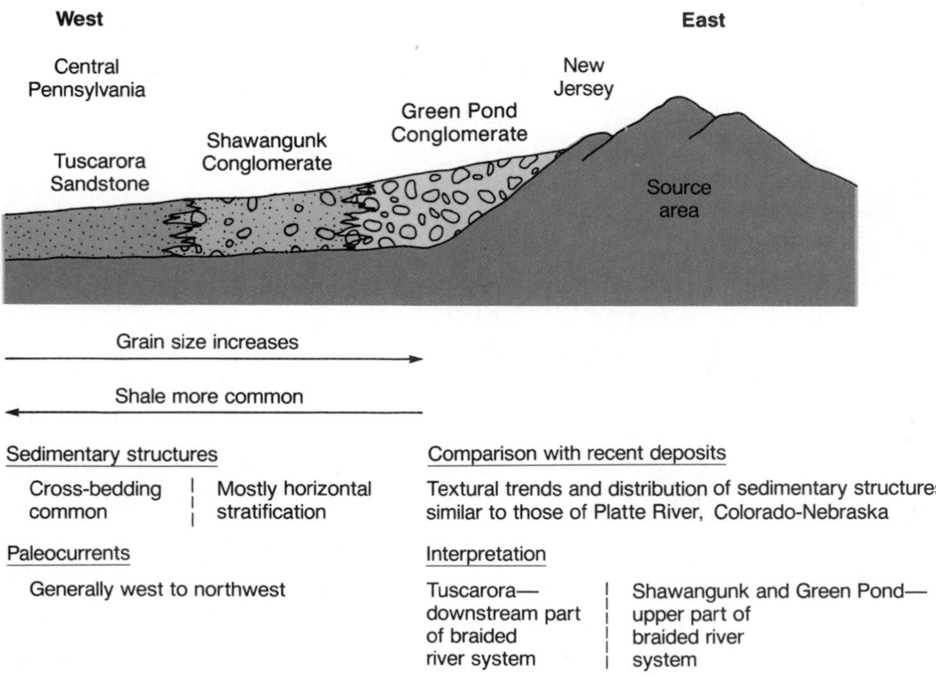


**Figure 4–37** Simplified cross section showing the probable lateral relationships and environmental interpretation for Lower Silurian strata in the eastern United States.

major transgressions or regressions and so have overall sheetlike geometries. However, individual depositional units within carbonates may be circular (patch reefs) or shoestring-shaped (oolite and skeletal sand barriers). The carbonate depositional environment is truly unique in one respect, however; it is the only major depositional environment that generates its own sediment.

### Evaporite Environments

Evaporites, mostly gypsum[1] and rock salt, are not particularly common when compared with the abundance of sandstone, shale, and limestone. Nevertheless, some evaporite deposits are of impressive dimensions, and some are locally important as resources. Evaporite rocks may form in saline lakes such as the Dead Sea, in playa lakes, and in marine environments including sabkhas (see Prologue).

Marine evaporites are the most common type in the geologic record, but little agreement exists on the specific environment in which they formed. There is agreement, however, on some of the conditions necessary for evaporite deposition. Obviously, evaporation rates would have to be high so that seawater would become increasingly saline until it reached a point at which minerals would begin to precipitate. Such conditions are best met in nearshore environments such as lagoons or embayments in arid regions. In an embayment, for example, the inflow of normal seawater is restricted, salinity increases because of evaporation, and evaporite minerals form. One such modern environment is the Gulf of Kara Bogaz, an embayment of the Black Sea.

Some marine evaporites appear to have been deposited in long, narrow seas that formed in response to continental rifting. When a continent is rifted, the crust is stretched and thinned and eventually subsides below sea level. Chapter 6 addresses this subject further. If this newly formed sea is in an arid region, evaporite deposition occurs. The Louann Salt (Chapters 13 and 15) of the Gulf Coast of the United States and several other evaporite deposits are thought to have formed in such a plate tectonic setting.

[1]Gypsum ($CaSO_4 \cdot 2H_2O$) is the common sulfate precipitated from seawater; but when deeply buried, gypsum loses its water and is converted to anhydrite ($CaSO_4$).

## ENVIRONMENTAL INTERPRETATIONS AND HISTORICAL GEOLOGY

The ultimate goal in historical geology is to determine (1) what events occurred in the past and (2), the order in which they occurred. Such determinations are based, in part, on analyses of sedimentary rocks and inferences that can be drawn from such analyses. In Chapter 3 we discussed the use of sedimentary rocks and their contained fossils in establishing a meaningful stratigraphic succession, whereas Chapter 4 has been devoted to environmental analyses. In short, we now have the information necessary to determine what specific events occurred and when. For example, Lower Silurian strata in New Jersey and Pennsylvania possess characteristics of grain size, rock types, and sedimentary structures that indicate deposition in a braided-stream environment (Fig. 4–37). Vertical facies relationships, rock types, fossils and sedimentary structures of Ordovician carbonate rocks in Arkansas are interpreted as transgressive carbonate shelf sediments (Fig. 4–38). Environments of deposition will figure importantly in the chapters on geologic history. Such interpretations are based on the kinds of data and analyses covered in this chapter. Furthermore, environmental interpretations are not of academic interest only; they allow geologists to more effectively predict the locations and geometries of rock units that contain important mineral resources or oil and gas.

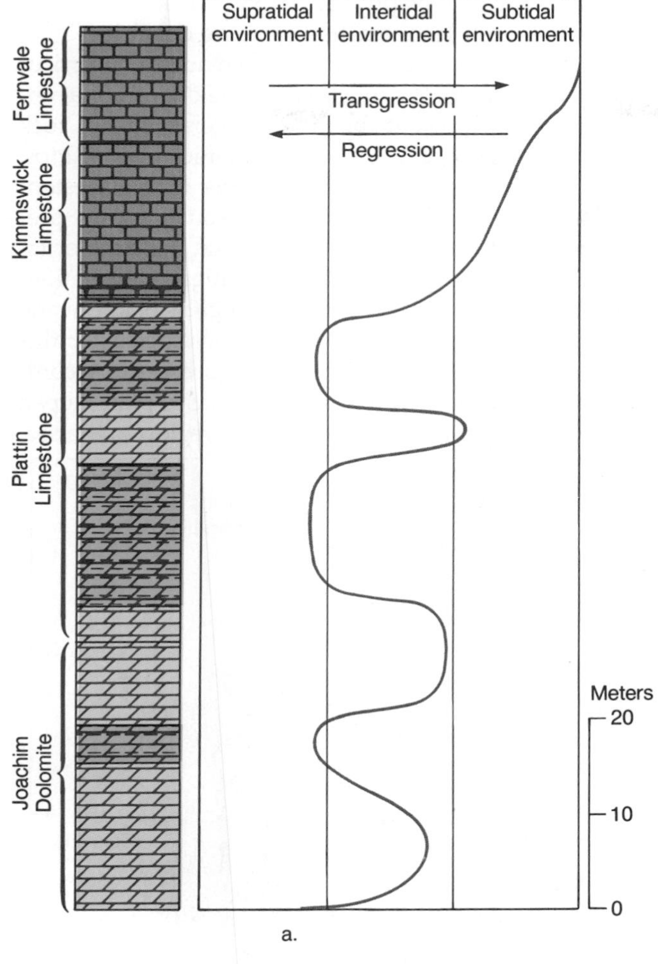


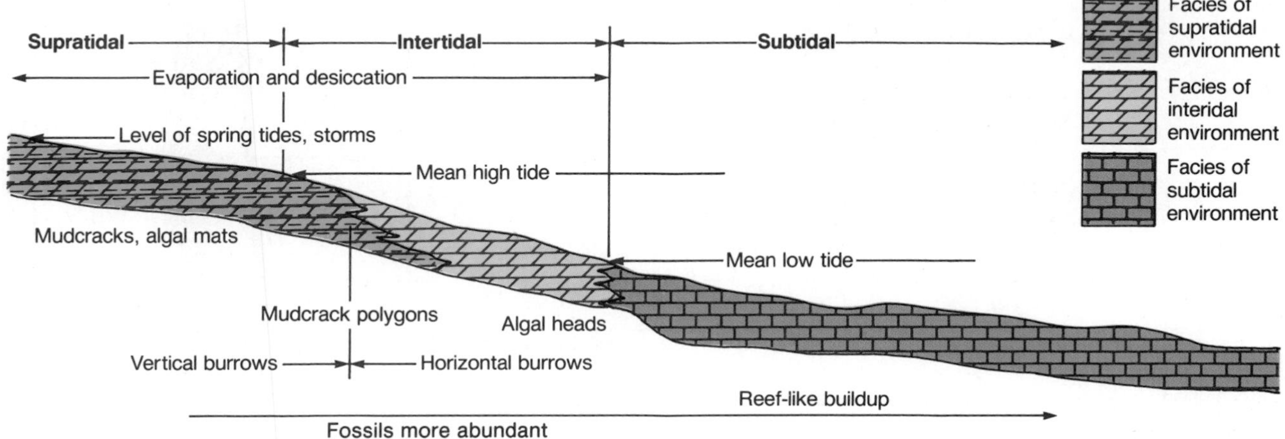


**Figure 4–38** Middle and Upper Ordovician strata in northern Arkansas. a. Vertical stratigraphic relationships and inferred environments of deposition. Notice that even though several minor transgressions and regressions occurred that the trend was transgressive. b. Simplified cross section showing lateral facies relationships.

## PALEOGEOGRAPHY AND PALEOCLIMATES

The surface patterns of the Earth have not remained unchanged through time. In the Late Paleozoic, for example, a single supercontinent, Pangaea, existed, but it began breaking up during the Triassic and has continued to do so ever since. Thus, the present geographic distribution of continents evolved through time. **Paleogeography** involves a study of the Earth's or a region's surface patterns for a particular time in the past. A series of paleogeographic maps will show the relative positions of the continents through time.

The paleogeography of several Rocky Mountain states has been reconstructed in some detail in Figure 4–39. Such reconstructions must be made for a particular time, so geologists must determine depositional environments, work out the physical continuity of rock units, and determine time-stratigraphic relationships. Figure 4–40 shows the major basins in the western United States in which Upper Cretaceous strata are preserved. Notice in the cross sections that three major facies are recognized, and that the stratigraphic relationships indicate transgressions and regressions. Several paleogeographic maps have been constructed for this region, two of which are shown in Figure 4–41. According to these reconstructions, a broad coastal plain sloped gently eastward from a mountainous region to the sea.

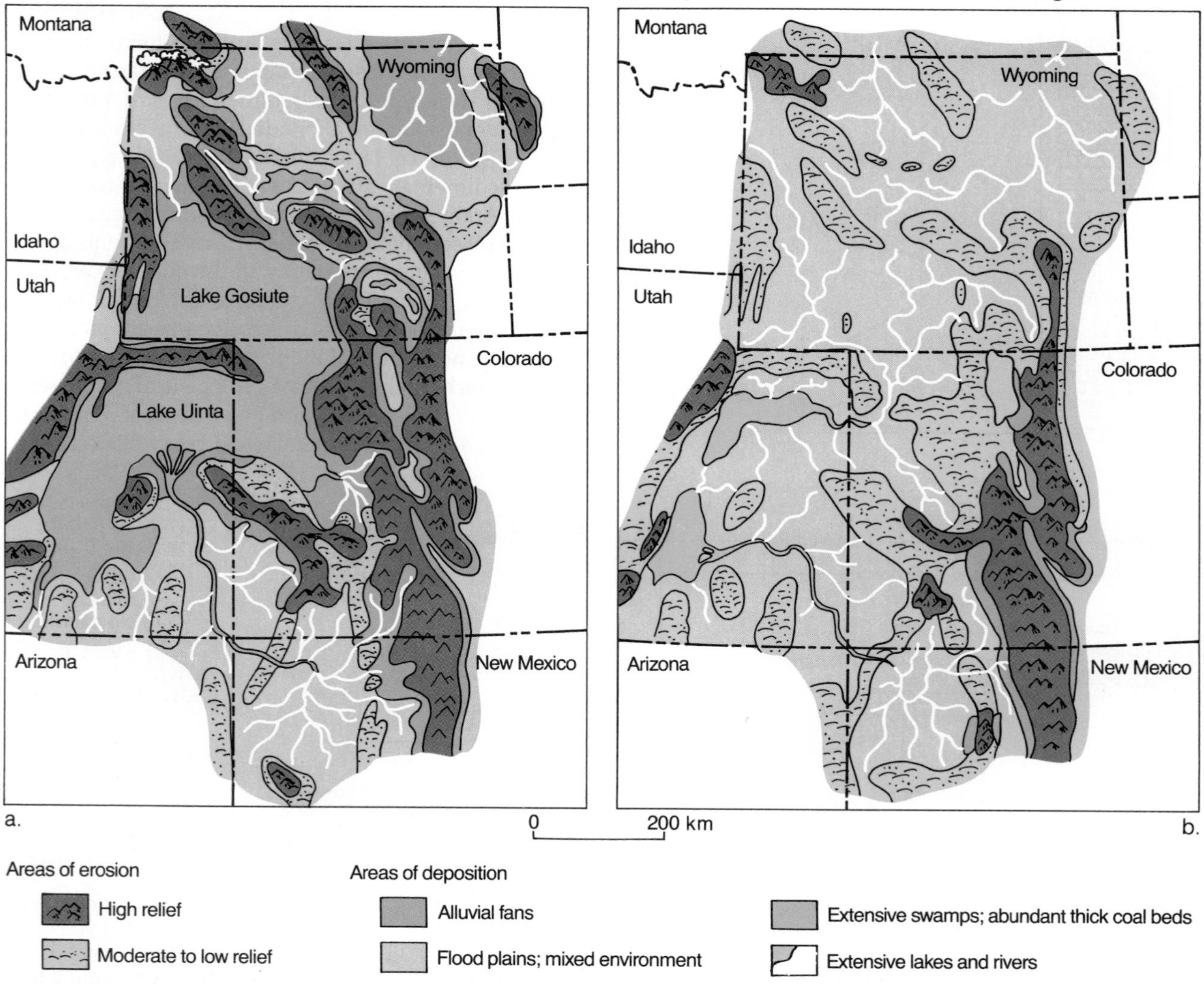

**Figure 4–39** Eocene paleogeography of some of the Rocky Mountain states. a. Paleogeography in Late Early to Middle Eocene time. b. Paleogeography in Late Eocene time.

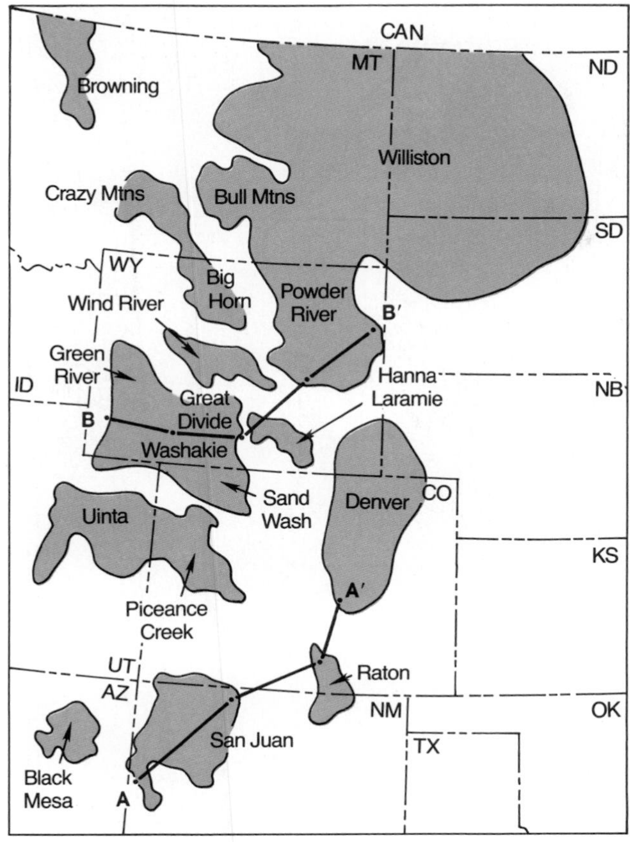


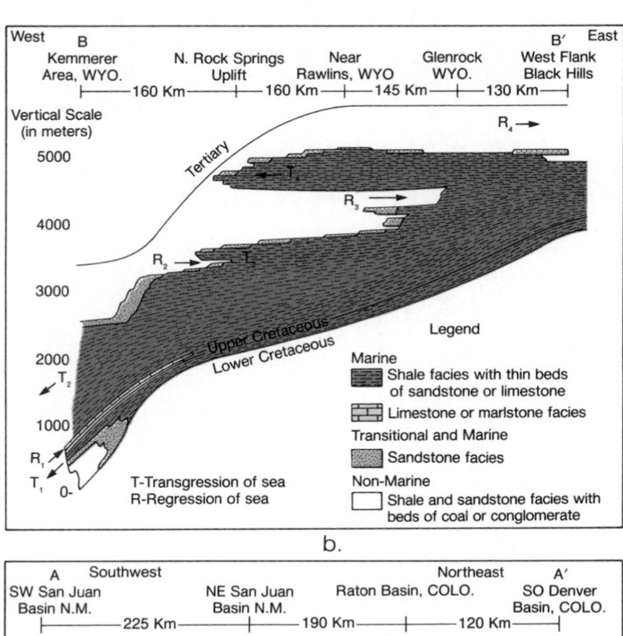


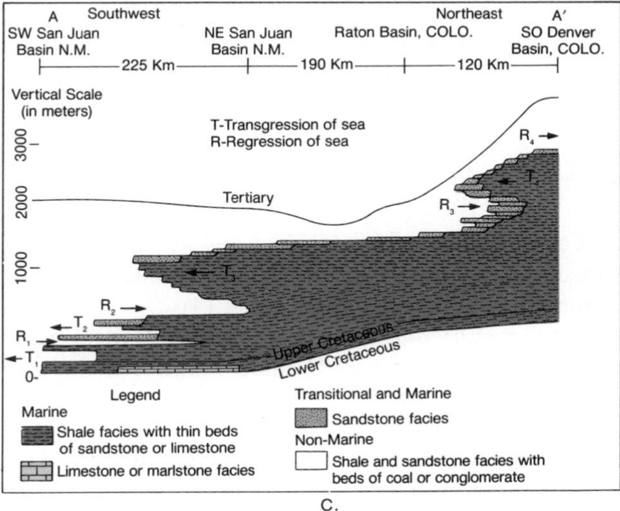


**Figure 4–40** a. Map showing the major basins in which Upper Cretaceous strata are present in the western United States. b and c. Cross sections along the lines indicated. Three major facies are recognized, and the stratigraphic sequence of facies records transgressions and regressions.

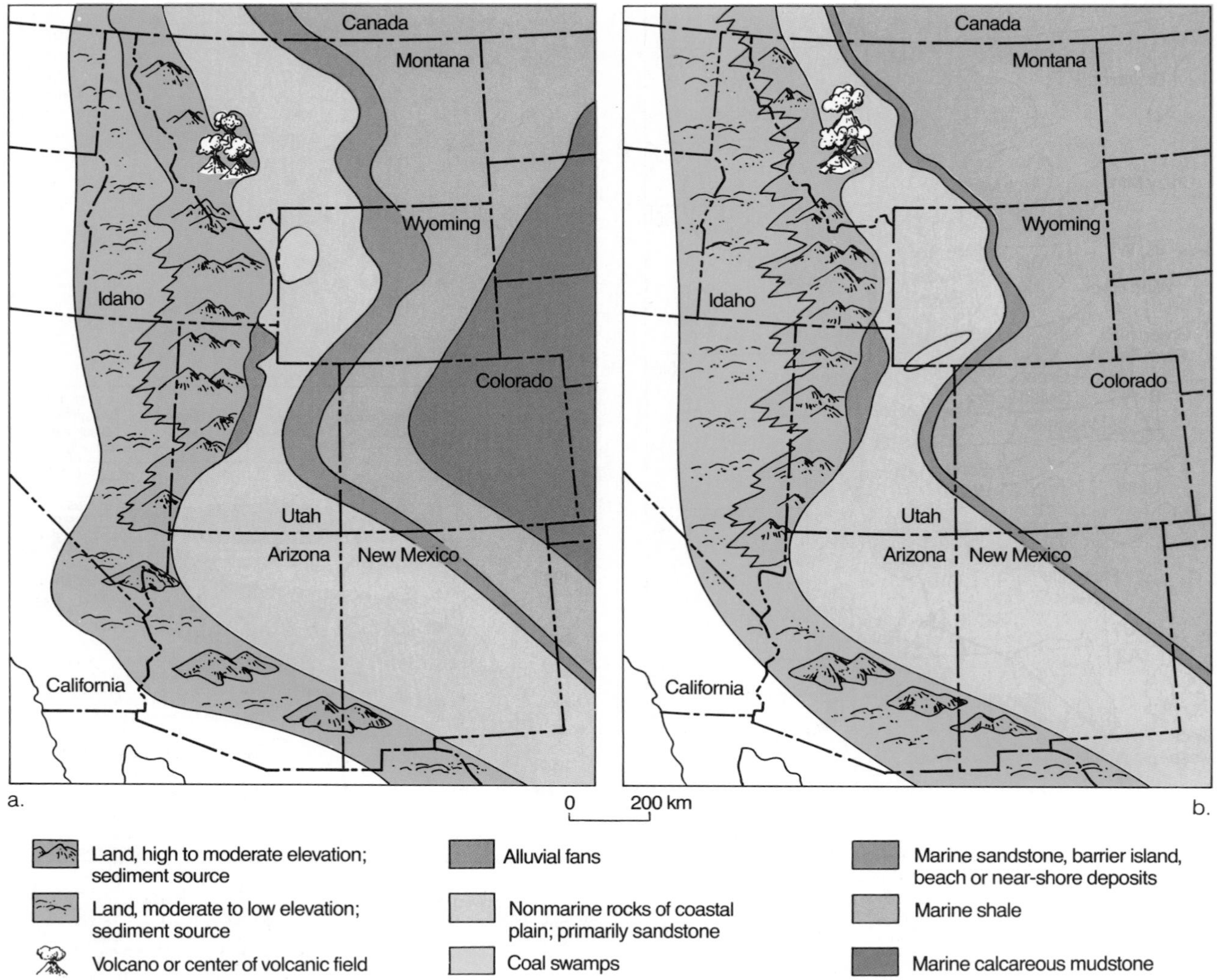

Canada
Montana
Wyoming
Idaho
Colorado
Utah
Arizona
New Mexico
California
a.
0
200 km
b.
Land, high to moderate elevation; sediment source
Land, moderate to low elevation; sediment source
Volcano or center of volcanic field
Alluvial fans
Nonmarine rocks of coastal plain; primarily sandstone
Coal swamps
Marine sandstone, barrier island, beach or near-shore deposits
Marine shale
Marine calcareous mudstone

**Figure 4–41** Paleogeographic maps prepared from the data in Figure 4–40. Both maps show Late Cretaceous paleogeography but (a) is older than (b).

This coastal plain habitat was occupied by lush vegetation, dinosaurs, flying reptiles, lizards, turtles, and early mammals (see Prologue to Chapter 14).

Sedimentary rocks and fossils also tell something of ancient climates, or **paleoclimates.** Pennsylvanian and Permian strata of the southwestern United States indicate desert conditions, whereas extensive coal beds in the east indicate warm, moist swamps (Chapter 11). The recognition of glacial deposits including tillite (lithified glacial till) and varved deposits (Fig. 4–24) in Precambrian, Ordovician, Pennsylvanian, and Pleistocene strata have allowed geologists to outline several areas of glacial climates. Deposits of evaporites indicate hot, arid climates.

## CHAPTER SUMMARY

1. In order to determine what environment sedimentary rocks were deposited in, geologists must observe and interpret the physical and biological features of those rocks.
2. Sedimentary structures and fossils are the most useful sedimentary rock features for environmental analyses, but composition, textures, and rock-body geometry are helpful as well.
3. Three major depositional settings are recognized: continental, transitional, and marine; each setting includes several specific depositional environments.

4. Fluvial systems may be braided where deposits are mostly gravel and sand with subordinate mud, or meandering where deposits are mostly mud with elongate sand bodies.
5. The association of alluvial fans, playa-lake deposits, and wind-blown dunes is typical of desert depositional environments. Deposits in areas affected by glaciation are mostly till and outwash.
6. Marine deltas form where a fluvial system supplies sediments to a shoreline. Deltas prograde seaward and produce vertical sequences of prodelta deposits overlain by delta-front and delta-plain deposits.
7. A barrier island complex includes lagoon, beach, and dune subenvironments, each characterized by a unique association of lithologies, fossils, and sedimentary structures.
8. Waves and tides are important agents on continental shelves, but beyond the shelf-slope break, gravity processes—especially turbidity currents—predominate. Detrital shelf deposits consist of inner-shelf sands and outer-shelf muds. Deposits on the continental slope-rise systems are mostly submarine fans.
9. Sediments on the deep-sea floor are mostly fine-grained because there is no mechanism that transports large amounts of coarse detritus from continents into deep seas.
10. Most carbonates are deposited in warm, shallow seas. Many limestone textural features are indicators of depositional conditions. Carbonate shelves may be attached to a landmass or lie on top of banks rising from the seafloor.
11. The best conditions for the origin of evaporites occur along marine shorelines in arid regions. Marine embayments, lagoons, and sabkhas are areas in which many modern evaporites are deposited.

## IMPORTANT TERMS

alluvial fan
barrier island
bed
bedding (stratification)
biogenic sedimentary structure
bioturbation
clastic texture
cross-bedding
crystalline texture
delta
depositional environment
desiccation crack
drift
fluvial
graded bedding
grain size
laminae (lamination)
lithology
micrite
oolite
outwash
paleoclimate
paleocurrent
paleogeography
playa lake
point bar
progradation
reef
ripple mark
sabkha
sedimentary structure
sorting
submarine fan
till
varve

## REVIEW QUESTIONS

1. Define grain size and sorting, and explain how each is used in environmental analyses.
2. Why are microfossils and trace fossils particularly useful for determining environments of deposition?
3. Illustrate three sedimentary structures that can be used to determine paleocurrent directions. Explain how each of these structures forms.
4. How do the deposits of braided and meandering rivers differ from one another?
5. What criteria would you use to distinguish desert dunes from dunes on a barrier island?
6. Illustrate and describe the vertical sequence of facies produced by a prograding, river-dominated delta.
7. How and why do inner and outer continental-shelf sediments differ?
8. In what ways are limestones and detrital sedimentary rocks similar and dissimilar?
9. Draw a generalized cross section of a carbonate shelf, and show the types of sediments you would expect on various parts of the shelf.
10. Why are deep-sea sediments mostly fine-grained? What are the sources of these sediments?

## ADDITIONAL READING

Boggs, S., Jr. 1987. *Principles of Sedimentology and Stratigraphy.* Columbus, Ohio: Merrill Publishing Company.

Collinson, J. D., and D. B. Thompson. 1982. *Sedimentary Structures.* London, England: George Allen & Unwin.

Davis, R. A., Jr. 1983. *Depositional Systems: A Genetic Approach to Sedimentary Geology.* Englewood Cliffs, N. J.: Prentice-Hall, Inc.

Fritz, W. J. and J. N. Moore. 1988. *Basics of Physical Stratigraphy and Sedimentology.* New York: John Wiley & Sons, Inc.

Hallam, A. 1981. *Facies Interpretation and the Stratigraphic Record.* San Francisco: W. H. Freeman and Company.

Reading, H. G. (ed.). 1986. *Sedimentary Environments and Facies.* Boston: Blackwell Scientific Publications.

Selley, R. C. 1978. *Ancient Sedimentary Environments.* Ithaca, N. Y.: Cornell University Press.

Walker, R. G. (ed.) 1984. Facies Models. *Geoscience Canada,* Reprint Series 1. The Geological Association of Canada.

# 5

## CHAPTER OUTLINE



## PROLOGUE

*On December 27, 1831, Charles Robert Darwin departed from Devonport, England, aboard the H.M.S.* Beagle *as an unpaid naturalist. Nearly five years later the* Beagle *returned to England, and Darwin never ventured far from home again. Nevertheless, the 40,000 mile voyage (Fig. 5-1) was the most important experience in his life, an experience that changed his view of nature and ultimately revolutionized all of science.*

*When Darwin sailed aboard the* Beagle *he was a little known recent graduate in theology from Christ's College. In fact, his father, Dr. Robert Darwin, had sent him to Christ's College as a last resort; Charles showed little aptitude for academics, except perhaps for science, and had already withdrawn from studies in medicine at Edinburgh. Although he completed his theological studies, he was rather indifferent to religion. He nevertheless fully accepted the biblical account of creation as historical fact. Darwin's belief in biblical creation initially endeared him to the* Beagle's *captain, Robert Fitzroy, but his views changed during the voyage, and his relationship with Captain Fitzroy became strained.*

*When the voyage began, Darwin was nominally a clergyman with interests in the sciences, especially botany, zoology, and geology. He suffered from prolonged attacks of seasickness, but nevertheless he kept detailed notes on his observations; he collected, catalogued, and dissected specimens; and he eventually became a professional naturalist. The* Beagle *made*

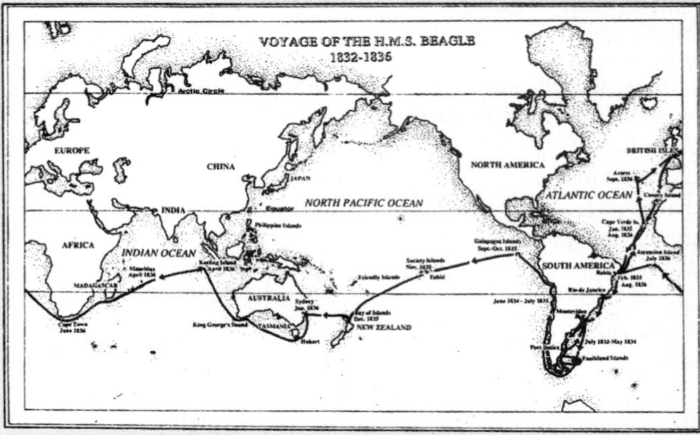


**Figure 5-1** Route of the H.M.S. *Beagle* during its five-year voyage. Important localities visited by Darwin are shown.

# Evolution

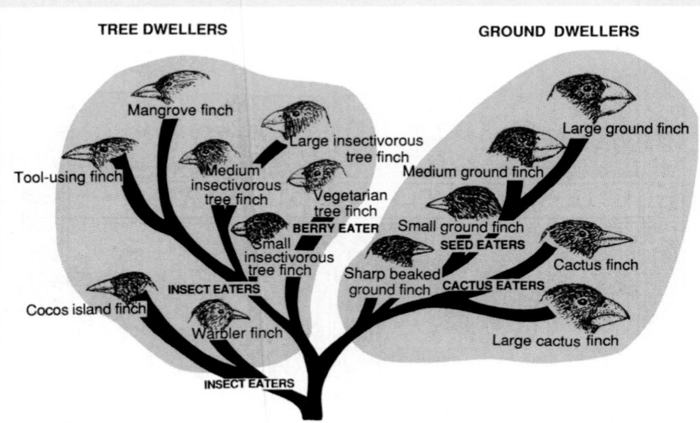


**Figure 5–2** Darwin's finches arranged to show evolutionary relationships. The six species on the right are ground-dwellers, and the others are adapted for life in trees. Notice that the shapes of the beaks of finches in both groups vary depending on diet.

*several lengthy stops in South America where Darwin explored rain forests, experienced an earthquake, and collected fossils. The fossils he collected were clearly related to the living sloth, armadillo, and llama, and implied that living species descended from fossil species. Such evidence caused him to question the concept of fixity of species, a concept held by those who accepted biblical creation.*

*Darwin was particularly fascinated by the plants and animals of oceanic islands. The Cape Verde Islands and the Galapagos Islands are comparable distances west of Africa and South America, respectively (Fig. 5–1). Each island group has its own unique plants and animals, yet these plants and animals most closely resemble those of the nearby continent. The Galapagos, for example, are populated by 13 species of finches (Fig. 5–2), but only one species exists in South America. These finches are adapted to the different habitats occupied on the mainland by various species such as parrots, flycatchers, and toucans.*

*Darwin reasoned that the Cape Verdes and Galapagos had received colonists from the nearby continents and that "such colonists would be liable to modification—the principle of inheritance still betraying their original birthplace."[1] The significance of this statement is that it revealed a change in Darwin's view of nature; he no longer accepted the idea of fixity of species. For example, he proposed that an ancestral species of finch had somehow reached the Galapagos Islands from South America and differentiated into the various types he observed (Fig. 5–2). Furthermore, he had read Charles Lyell's* Principles of Geology *during his voyage and accepted uniformitarianism and the great age of the Earth. In short, he came to view nature as dynamic rather than static.*

## INTRODUCTION

The term *evolution* comes from the Latin, meaning unrolled. Roman books were written on parchment and rolled on wooden rods, so they were unrolled, or evolved, as they were read. In modern usage evolution usually refers to change through time. If evolution is defined simply as change through time, it is a pervasive phenomenon; stars evolve, the Earth has evolved and continues to do so, languages and social systems evolve, and life evolves. Change through time is, of course, demonstrable. We know from historical records that languages have changed, and the fossil record reveals that a succession of different life-forms existed through time.

The biological concept of evolution, most appropriately called *organic evolution*, is concerned with changes in organisms that are inheritable. Evolution so defined is also observable, at least on a small scale. For example, the proportion of hereditary determinants in populations of moths in Great Britain changed in such a way that their color changed from light to dark in several generations. Large-scale evolution is not directly observable but can be inferred from fossil evidence and from studies of comparative anatomy, embryology, and biochemistry of living organisms.

[1]Darwin, C. 1958. *The Origin of Species*. New York: New American Library, 377.

As we discussed in Chapter 1, theories are scientific explanations of natural phenomena that have a large body of supporting evidence. The **theory of evolution** is such an explanation. The central claim of this theory is that all living organisms are the evolutionary descendants of life-forms that existed in the past. As a scientific theory it meets the usual criteria for theories; it is a naturalistic explanation, and it can be tested so that its validity can be assessed.

## THE EVOLUTION OF AN IDEA

Evolution as a biological concept was seriously considered by some naturalists even before Charles Darwin was born. Indeed, as early as 600 B.C., the ancient Greeks speculated on possible interrelationships among organisms; but overall there was little thought of evolution in the ancient world or during the Middle Ages. In fact, the prevailing belief well into the eighteenth century was that all important knowledge was contained in the works of Aristotle and the Bible. One did not have to observe nature to understand it, since all the answers were in these two sources, particularly in the first two chapters of Genesis. Literally interpreted, Genesis was taken as the final word on the origin and diversity of life and much of Earth history. According to this view, all species of plants and animals were perfectly fashioned by God during the days of creation, and they had remained fixed and immutable ever since (Fig. 5–3). To question divine creation in any way was considered an act of heresy.

In eighteenth-century Europe the social and intellectual climate changed, and the absolute authority of the Church in all matters declined. Ironically, those naturalists determined to find physical evidence supporting Genesis as a factual, historical account were, in fact, finding more and more evidence that could not be explained by a literal reading of Scripture. This was particularly true of the evidence for a worldwide catastrophic flood. For example, the sedimentary deposits that traditionally had been attributed to the biblical deluge showed clear evidence of a noncatastrophic origin. Many naturalists reasoned that this evidence truly reflected the conditions that prevailed when the strata were deposited, for to infer otherwise implied a deception on the part of the Creator, a thesis that they could not accept.

It should not be concluded from this short discussion that eighteenth-century naturalists abandoned the Judeo-Christian tradition. As a matter of fact, most did not. Rather, they came to accept Genesis as a symbolic account containing important spiritual messages but not a factual account of creation, a view shared by most modern theologians and biblical scholars.

**Figure 5–3** Illustration from a 16th-century Bible showing the six days of creation described in Genesis Chapter 1. According to this account God created heaven and Earth on the first day (*die primo*), and on the second day (*die secundo*) He separated the sky and the waters. He created the dry land and plants on the third day (*die tertio*). He made the sun, moon, and stars on the fourth day (*die quarto*). Birds and fishes were created on the fifth day (*die quinto*), and on the sixth day (*die sexto*) God created land-animals and humans.

These changing attitudes can be attributed in large part to the Enlightenment, an eighteenth-century philosophical movement that relied on rationalism. In science this meant that one could discover natural laws to explain observations of natural systems; all did not depend on divine revelation. The sciences, particularly biology and chemistry, began to acquire a modern look.

In this changed intellectual atmosphere the concepts of uniformitarianism and the great age of the Earth were becoming accepted. The French zoologist Georges Cuvier clearly demonstrated that many types of plants

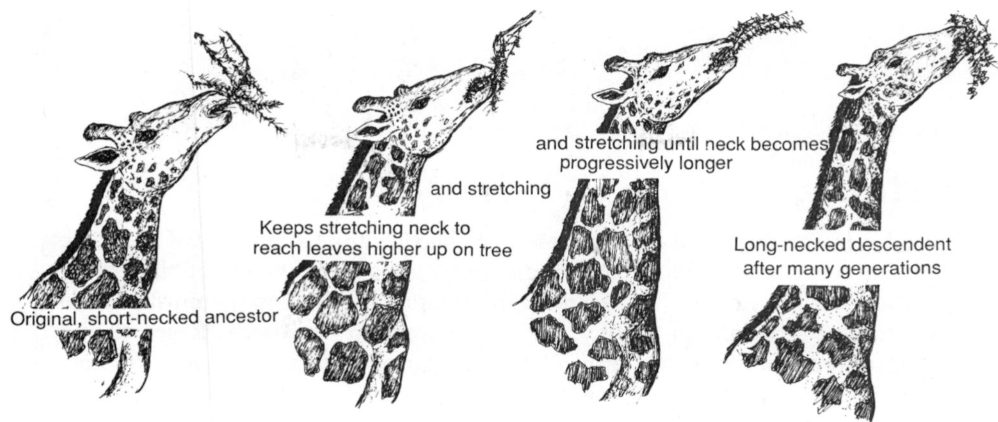


**Figure 5–4** According to Lamarck's theory of inheritance of acquired characteristics, ancestral short-necked giraffes stretched their necks to reach leaves higher up on trees and their descendants were born with longer necks.

and animals had become extinct, and that fossils differed through time. In view of the fossil evidence, as well as observations and studies of living organisms, many naturalists became convinced that species were not immutable. Change from one species to another—that is, evolution—became an acceptable idea. What was lacking, however, was an overall theoretical framework to explain evolution.

## Lamarck and the Giraffe's Neck

Physician, poet, and naturalist Erasmus Darwin, Charles Darwin's grandfather, was the first to propose a process by which gradual evolution could occur. But it was Jean Baptiste de Lamarck (1744–1829), a French botanist-geologist, who was first taken seriously by his colleagues. Pointing out that organisms are adapted to their environments, he proposed that evolution is an adaptive process whereby organisms acquire traits or characteristics during their lifetimes and pass these acquired traits on to their descendants.

According to Lamarck's theory of **inheritance of acquired characteristics**, ancestral short-necked giraffes, for example, stretched their necks to browse on leaves of trees (Fig. 5–4). The environmentally imposed characteristic of neck-stretching was thus expressed in future generations as a longer neck. In other words, the capacity for giraffes to produce offspring with longer necks was acquired through the habit of neck-stretching.

Considering the data available in the early 1800s, Lamarck formulated a reasonable theory. Little was known about how traits are passed from parent to offspring, so Lamarck's inheritance mechanism seemed logical and was accepted by many naturalists as a viable mechanism for evolution. In fact, Lamarck's concept of inheritance was not completely refuted until more modern concepts of inheritance were developed. Based on the work of Gregor Mendel in the 1860s, and developments in the early 1900s, we now know that all organisms possess hereditary determinants called *genes* that cannot be modified by any effort of the organism during its lifetime. Numerous attempts have been made to validate Lamarkian inheritance, including one in this century in the Soviet Union (Perspective 5–1), but all have failed.

## Charles Darwin's Contribution

In 1859 Charles Robert Darwin published *The Origin of Species* in which his ideas on evolution were outlined (Fig. 5–5). Most naturalists almost immediately recognized that evolution provides a unifying theory that explains an otherwise encyclopedic collection of biologic facts. Even most, but not all, churches and theologians eventually accepted the idea of evolution as a naturalistic explanation that does not conflict with the purpose and intent of Scripture.

While 1859 may mark the beginning of modern evolutionary biology, Darwin had in fact formulated his ideas more than 20 years earlier, but being aware of the furor his ideas would generate, was reluctant to publish. As a matter of fact, when the furor did erupt, Darwin, a

## Perspective 5-1

# The Lysenko Affair

One of the most bizarre episodes in the annals of science occurred in the Soviet Union between 1938 and 1965. Prior to that time the state promoted a vigorous program in the sciences, and especially in agricultural science. In fact, it appeared that the Soviets might challenge the United States for leadership in the selective breeding of plants and animals. However, beginning in 1929, Joseph Stalin initiated a new program in the sciences, a program in which scientists were expected to develop concepts compatible with Marxist-Leninist philosophy. One such scientist was Troffim Denisovich Lysenko (1898–1976).

Lysenko rejected the inheritance theories developed in the West and claimed that plants and animals could be changed in desirable ways simply by exposing them to a new environment. For example, seeds could be subjected to cold, acquire a resistance to cold temperatures, and cold-resistance would be inherited by future generations of plants. In short, Lysenko accepted the Lamarckian concept of inheritance of acquired characteristics.

In 1938 Lysenko became president of the USSR Academy of Agricultural Sciences, and soon he and his colleagues had nearly total control of all agricultural research. The Central Committee of the Soviet Union, driven by the apparent compatibility of Lamarckian inheritance and Marxist-Leninist philosophy, endorsed Lysenko's research. Research into any other inheritance mechanism was forbidden; scientists holding views contrary to those of Lysenko had to recant or face imprisonment.

Lysenko was fired as president of the USSR Academy of Sciences in 1953, but he regained some of his power that same year when he was appointed director of the Institute of Genetics, a post he held until 1965. Unfortunately for the Soviet people, agricultural research lagged far behind that of the West. Lysenko and his colleagues had not significantly increased food production, and the results of their experiments could not be repeated by anyone outside the Soviet Union.

There is an important lesson to be learned from the Lysenko affair; scientific research must be based on scientific realities, not on philosophical preference. Lamarckian inheritance was not mandated as the correct way to proceed in agricultural research because of its scientific merit, but because it seemed compatible with a belief system.

semi-invalid most of his life, never defended his theories in public; he left their defense to others.

As discussed in the Prologue, Darwin concluded during the voyage of the *Beagle* that species are not fixed. In fact, by the time Darwin returned home in 1836, he had what he thought was clear evidence that species had indeed changed through time, but he had no idea what might bring about change. However, by 1838 his observations of the selection practiced by animal and plant breeders, and a chance reading of Thomas Malthus's essay on population, gave him the elements necessary to formulate his theories.

Animal and plant breeders selected organisms for desirable traits, bred these organisms with those traits, and thereby induced a great amount of change. This practice of **artificial selection** yielded a fantastic variety of domestic pigeons, dogs, and plants (Fig. 5–6). If this much change could be produced artificially, was there perhaps a process acting on natural populations that also selected among variant types?

Darwin came to appreciate fully the power of selection after reading Malthus's book on population. Malthus argued that far more animals are born than reach maturity, yet the numbers of adult animals of a species remain rather constant. He reasoned that a high infant mortality rate resulted because of competition for resources, thus limiting the size of the population. Darwin proposed that a natural process was selecting only a few animals for survival, a process he called **natural selection**.

**Figure 5–5** Charles Darwin in 1840.

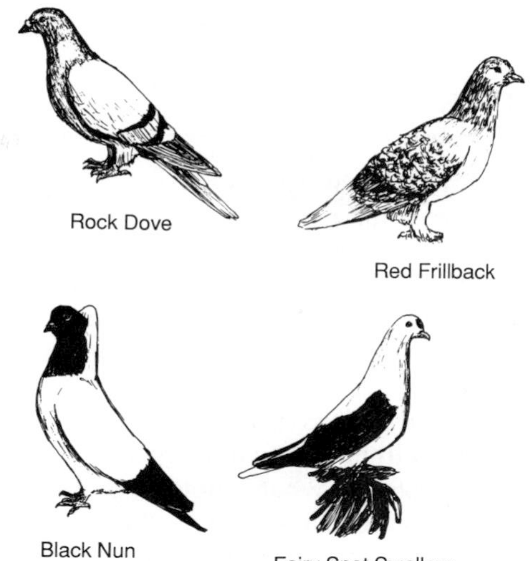


**Figure 5–6** Artificial selection has yielded about 300 domesticated breeds of pigeons, three of which are shown here. The common rock dove (upper left) is thought to be the parent stock of the domestic breeds.

## Natural Selection

For 20 years Darwin kept his ideas on evolution and natural selection largely to himself. But in 1858 he received a letter from Alfred Russell Wallace, a young naturalist working in southern Asia. Wallace had also read Malthus's essay on population and had come to precisely the same conclusion that Darwin had. Friends convinced Darwin that both his and Wallace's papers should be read simultaneously before the Linnaean Society of London. Later some of Darwin's correspondence was published, establishing his scientific priority, and indeed Wallace even insisted that Darwin be given credit in recognition of his earlier discovery and more thorough documentation of the theory.

The Darwin-Wallace theory of natural selection can be summarized as follows:

1. All populations contain heritable variations— size, speed, agility, coloration, and so forth.

2. Some variations are more favorable than others. That is, some variant types have an edge in the competition for resources and avoidance of predators.

3. Not all young survive to reproductive maturity.

4. Those with favorable variations are more likely to survive and pass on their favorable variations.

Evolution by natural selection is then largely a matter of reproductive success, for it is only those that reproduce that can pass on favorable variations. Of course, favorable variations do not guarantee survival for an individual, but within a population, those with favorable variations are more likely to survive and reproduce.

Natural selection is the process proposed by Darwin and Wallace to account for evolution. But is their statement substantially different from Lamarck's evolution by the inheritance of acquired characteristics? In other words, how would Darwin and Wallace explain the giraffe's long neck?

Suppose in an ancestral population of giraffes some individuals had longer necks than most of the others (Fig. 5–7). These long-necked animals could obviously browse higher on trees. The important point here is that some simply have longer necks and therefore enjoy a selective advantage in that they can more effectively acquire resources. Consequently, they are more likely to survive, reproduce, and pass on their favorable variations. Even in the second generation, there might be a few more giraffes with longer necks than others, which would confer on them a competitive edge. And so it would go, generation by generation, thereby increasing the proportion of giraffes with longer necks.

## MENDEL AND THE BIRTH OF GENETICS

Critics of natural selection were quick to point out that Darwin and Wallace could not account for the origin of variation or how it was maintained in populations. They reasoned that should a variant trait arise, it would simply blend with other traits in the population and be lost. This was a valid criticism at the time, one that neither Darwin nor Wallace could effectively answer. Actually, information that could have given them the answer existed, but it remained in obscurity until 1900.

A monastery located in what is now Czechoslovakia may seem to be an unlikely setting for the discovery of the rules of inheritance, and an Austrian monk may seem an unlikely candidate for "father of genetics." Nevertheless, Gregor Mendel, who later became abbot of the monastery, was doing research in the 1860s that answered some of the inheritance problems that plagued Darwin and Wallace. Unfortunately, Mendel's work was published in an obscure journal and went largely unnoticed. Mendel even sent Darwin a copy of his manuscript, but it apparently lay unread on his bookshelf.

Mendel performed a series of controlled experiments with true-breeding strains of garden peas (strains that when self-fertilized always display the same trait, such as a particular flower color). In one experiment, he transferred pollen from white-flowered plants to red-flowered plants, which produced a second generation of all red-flowered plants. But when left to self-fertilize, these plants yielded a third generation with a ratio of red-flowered plants to white-flowered plants of slightly over 3 to 1 (Fig. 5–8).

Mendel concluded from his experiments that traits such as flower color are controlled by a pair of factors, or what we now call **genes**. He also concluded that genes controlling the same trait occur in alternate forms, or **alleles**; that one allele may be dominant over another; and that offspring receive one allele of each pair from each parent. When an organism produces sex cells—pollen and ovules in plants, and sperm and eggs in animals—only one allele for a trait is present in each sex cell (Fig. 5–8). For example, if R represents the allele for red flower color, and r represents white, the offspring may inherit the combinations of alleles symbolized as RR, Rr, or rr. And since R is dominant over r, only those offspring with the rr combination will have white flowers.

The most important aspects of Mendel's work can be summarized as follows: the factors (genes) controlling traits do not blend during inheritance, but are transmitted as discrete entities; and even though particular traits may not be expressed in each generation, they are not lost. Therefore, some variation in populations is accounted for by alternate expression of genes (alleles), and variation can be maintained.

Even though Mendelian genetics explains much about heredity, we now know the situation is much more complex. For example, our discussion has relied

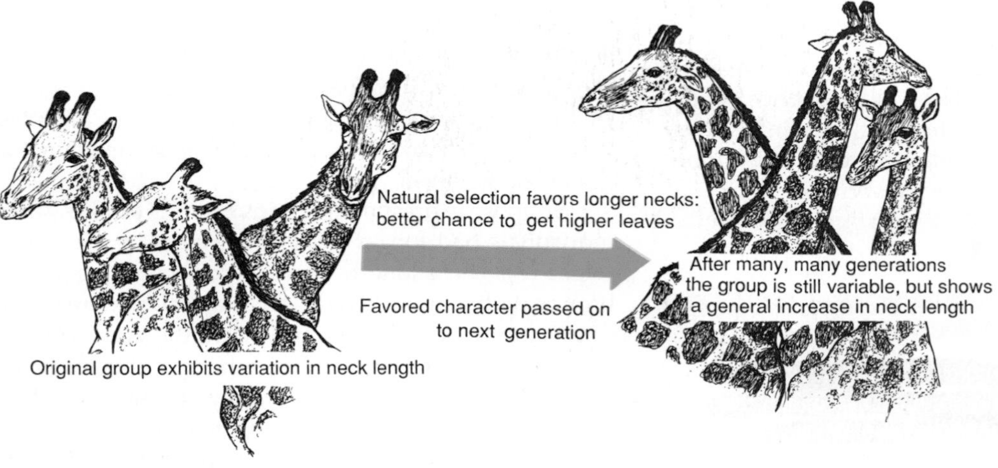
Natural selection favors longer necks:
better chance to get higher leaves
Favored character passed on
to next generation
After many, many generations
the group is still variable, but shows
a general increase in neck length
Original group exhibits variation in neck length

**Figure 5–7** According to the Darwin-Wallace theory of natural selection, the giraffe's long neck evolved as animals with favorable variations were selected for survival.

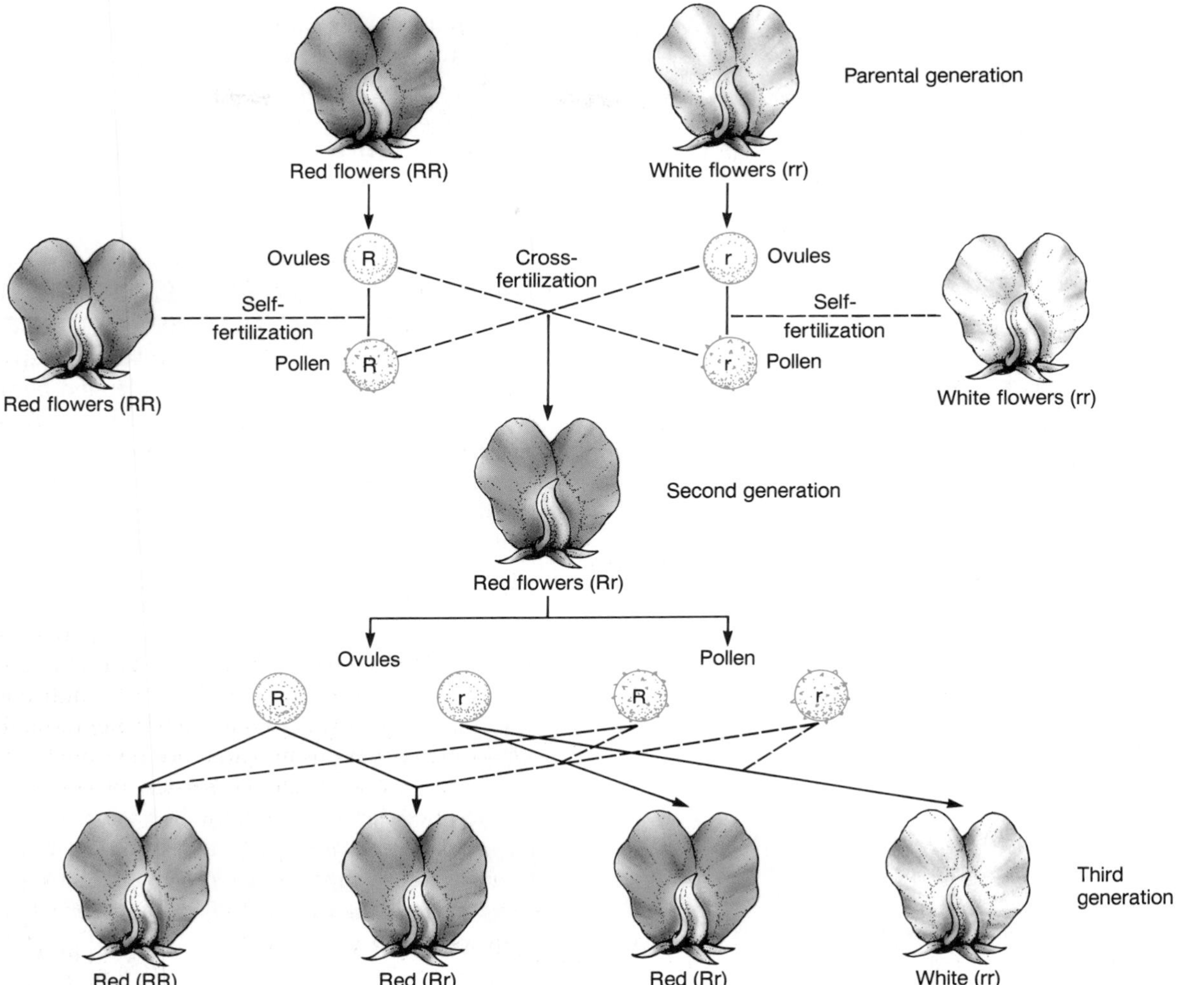


**Figure 5–8** In his experiments with flower color in garden peas, Mendel used true-breeding strains. Such plants, shown here as the parental generation, when self-fertilized always yield offspring with the same trait as the parent. However, if the parental generation is cross-fertilized, all plants in the second generation will receive the combination of alleles symbolized as Rr; these plants will have red flowers, because R is dominant over r. The second generation of plants produces ovules and pollen with the alleles shown, and when left to self-fertilize produces a third generation with a ratio of three plants with red flowers to one plant with white flowers.

upon a single gene controlling a trait, but, in fact, most traits are controlled by many genes. Nevertheless, the answers Darwin and Wallace needed were discovered by Mendel, though they went unnoticed until rediscovered by three independent researchers in 1900.

## Genes and Chromosomes

The cells of all organisms contain thread-like structures called **chromosomes.** Chromosomes are complex, double-stranded, helical molecules of **deoxyribonucleic acid (DNA)** (Fig. 5–9). Specific segments, or regions, of the DNA molecule are the basic hereditary units, the genes.

The number of chromosomes is specific for a single species but varies among species. For example, the fruit fly *Drosophila* has 8 chromosomes, humans have 46, and horses have 64. However, chromosomes occur in pairs, pairs that carry genes controlling the same characteristics. Remember that the genes on chromosome pairs may occur in different forms, alleles.

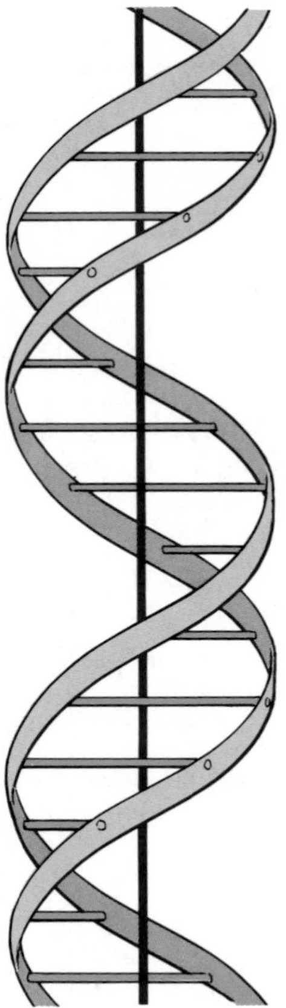

**Figure 5–9** Chromosomes are double-stranded, helical molecules of deoxyribonucleic acid (DNA) shown here diagramatically. Specific segments of chromosomes are genes.

In sexually reproducing organisms, the production of eggs and sperm results when parent cells undergo a type of cell division known as **meiosis**. The meiotic process yields cells containing only one chromosome of each pair (Fig. 5–10). Accordingly, eggs and sperm have only one-half the chromosome number of the parent cell; for example, human eggs and sperm have 23 chromosomes, one of each pair.

During reproduction, a sperm fertilizes an egg, producing a fertilized egg with the full chromosome complement for that species—46 in humans. As Mendel deduced from his garden pea experiments, one-half of the genetic makeup of the fertilized egg comes from each parent. The fertilized egg, however, develops and grows by a cell division process called **mitosis** that does not reduce the chromosome number (Fig. 5-10).

According to the chromosomal theory of inheritance, chromosomes carrying the genetic determinants, the genes, are passed from one generation to the next. However, changes called **mutations** may occur in genes, and any such change occurring in sex cells is inheritable. We shall have more to say about mutations in the section on "Sources of Variation."

## THE MODERN SYNTHESIS

As noted previously, the Darwin-Wallace concept of evolution by natural selection was challenged because it could not account for how variation arose nor how it was maintained in populations. The developing science of genetics partly answered the inheritance problems, but early in this century the geneticists believed that mutation rather than natural selection was the mechanism whereby evolution occurred.

In the 1930s and 1940s, the ideas developed by geneticists, paleontologists, population biologists, and others were brought together to form a **modern synthesis** or neo-Darwinian view of evolution. The chromosome theory of inheritance was incorporated into evolutionary thinking; mutation was seen as one source of variation in populations; Lamarckian concepts of inheritance were completely rejected; and the importance of natural selection was reaffirmed. The modern synthesis also emphasized that evolution is a gradual process, a point that has been challenged by some recent investigators, as will be discussed later.

### Sources of Variation

The raw materials for evolution by natural selection are variations within populations. Most variations can be accounted for by sexual reproduction and the reshuffling of alleles from generation to generation. Considering that each one of thousands of genes may have several alleles, and that offspring receive one-half of their genes from each parent, the potential for variation is enormous (Fig. 5–11).

New variation may be introduced into a population by a change in genes, a mutation. To understand mutations we must explore the function of chromosomes. One function of chromosomes is to direct the synthesis of molecules called proteins. During protein synthesis, information in the chromosome structure directs the formation of a protein by selecting the appropriate amino acids in a cell and arranging these amino acids into a se-

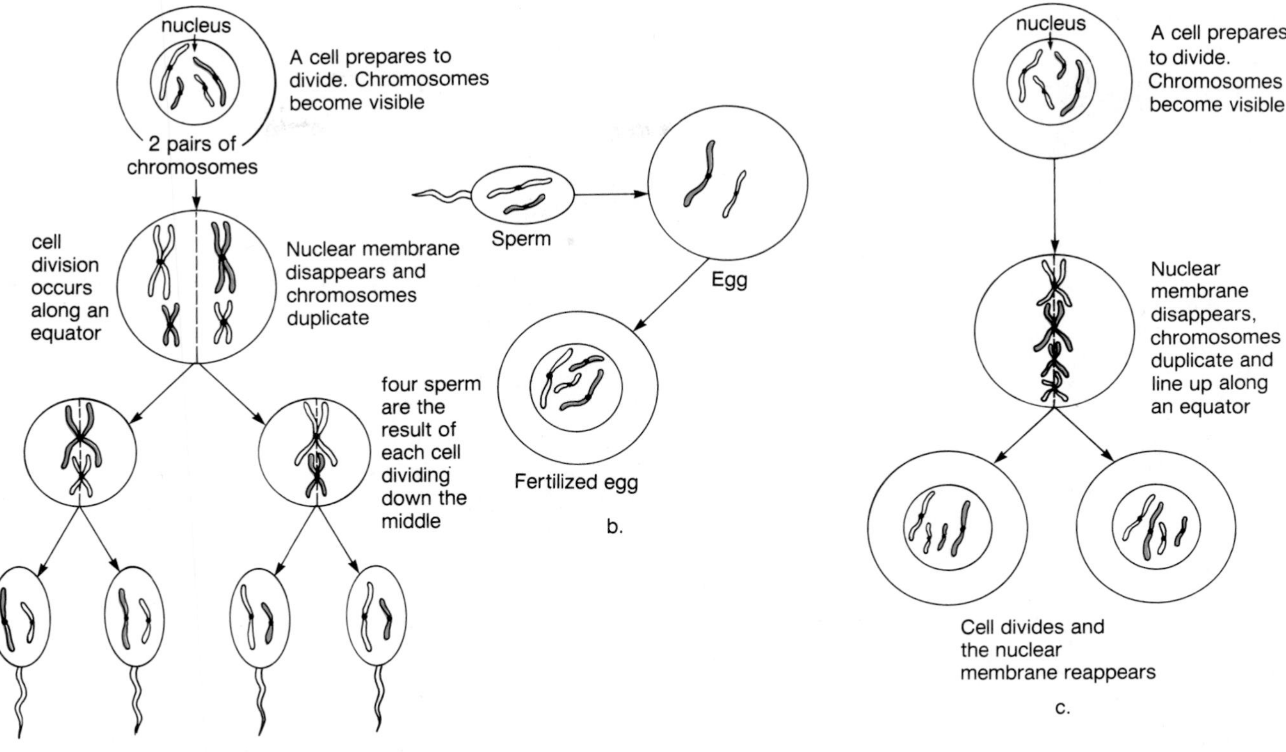


**Figure 5–10** a. Meiosis is a type of cell division that forms sex cells, each of which contains one member of each chromosome pair. The formation of sperm cells is shown; eggs form in a similar manner, but only one of the four final cells is a functional egg. b. When a sperm fertilizes an egg, the fertilized egg has the full complement of chromosomes restored. c. Mitosis results in the complete duplication of a cell. In this simplified example a cell with four chromosomes (two pairs) produces two cells, each with four chromosomes. Mitosis occurs in all body cells except sex cells. Once an egg is fertilized, the developing embryo grows by mitosis.

quence (Fig. 5–12). The proteins produced determine the characteristics of an organism. Any change in the information directing protein synthesis is a mutation.

Two additional aspects of mutations are important. First, only those mutations that occur in sex cells, eggs and sperm, are inheritable. And second, mutations are random with respect to fitness. That is, there is no evidence of a predetermination of a mutation's effect; it may be harmful (many are), neutral, or beneficial. However, the attributes of harmful versus beneficial can be considered only with respect to the environment.

If a species is well adapted to the environment in which it lives, most mutations would not be particularly useful and perhaps would be harmful. But if the environment changes, what was once a harmful mutation may become beneficial. For example, some plants have developed a tolerance for contaminated soils around mines. Plants of the same species from the normal environment do poorly or die in these contaminated soils, while contaminant-resistant plants do poorly in the normal environment. The mutations for contaminant resistance probably occurred repeatedly in the population, but their adaptive significance was not realized until contaminated soils were present.

## Speciation and the Rate of Evolution

In classifying organisms the term **species** is used for populations of similar individuals that in nature can interbreed and produce fertile offspring. Sheep and goats belong to different species because in nature they do not interbreed, yet in captivity they may hybridize and produce fertile offspring. Obviously barriers to reproduction are not complete in this case. If reproductive barri-

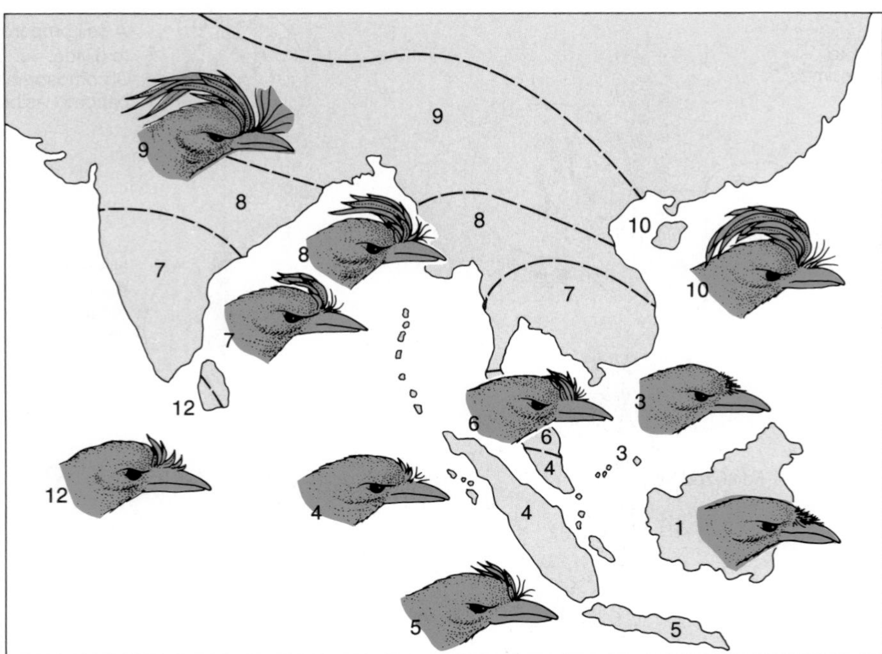


**Figure 5–11** Variation in the form of the crest in the drongo (*Dicrurus paradiseus*) of southern Asia.

Tryptophon-Glycine-Argenine-Theronine-Glycine-Lysine-Valine-Phenylalanine-

Tryptophon-Glycine-Argenine-Teronine-Glycine-Isolysine-Valine-Phenylalanine-

**Figure 5–12** A segment of a hypothetical protein consisting of amino acids linked together in a specific sequence. Chromosomes contain the information that directs the synthesis of proteins in cells. Any change in the sequence of amino acids results from a change in the genetic material, a mutation. In this example, a mutation occurred, resulting in the substitution of isolysine for lysine in the protein structure. A substitution such as this one may or may not have an effect on the function of the protein, since proteins have hundreds of sites occupied by amino acids. Only those changes that affect the function of the protein will be of any consequence.

ers are complete, different species cannot interbreed under any circumstances.

The process of speciation involves a change in the genetic makeup of the population, that is, a change in the frequency of alleles in its **gene pool**. A gene pool is simply all of the genes available in the genetic makeup of a population. There may also be a marked change in form and structure (**morphology**), but morphologic change is not necessary in the origin of a new species. In fact, a descendant species may look very much like its ancestral species (Fig. 5–13).

Darwin proposed that the origin of new species is a gradual process. That is the gradual accumulation of minor changes brings about a transition from one species to another, a process commonly called **phyletic gradualism** (Fig. 5–14). According to this concept of speciation, an ancestral species grades imperceptibly into a descendant species. Evolutionary changes in large

**Figure 5–13** a. Eight species of recent horses are recognized, although one, the quagga, became extinct in 1883. Differences in physical appearance of many of these species are slight, but they are legitimate species because in nature they do not commonly interbreed. In fact, those that do occasionally interbreed in nature or that are crossbred in captivity most often produce sterile offspring because of genetic incompatibility (notice that the chromosome numbers, in parentheses, vary among the species). The geographic range of each species is shown. b. The mule is the sterile offspring of a male ass (*Equus asinus*) and a female horse (*Equus caballus*). c. Breeding between zebras and horses produces the zebroid, which is sterile.

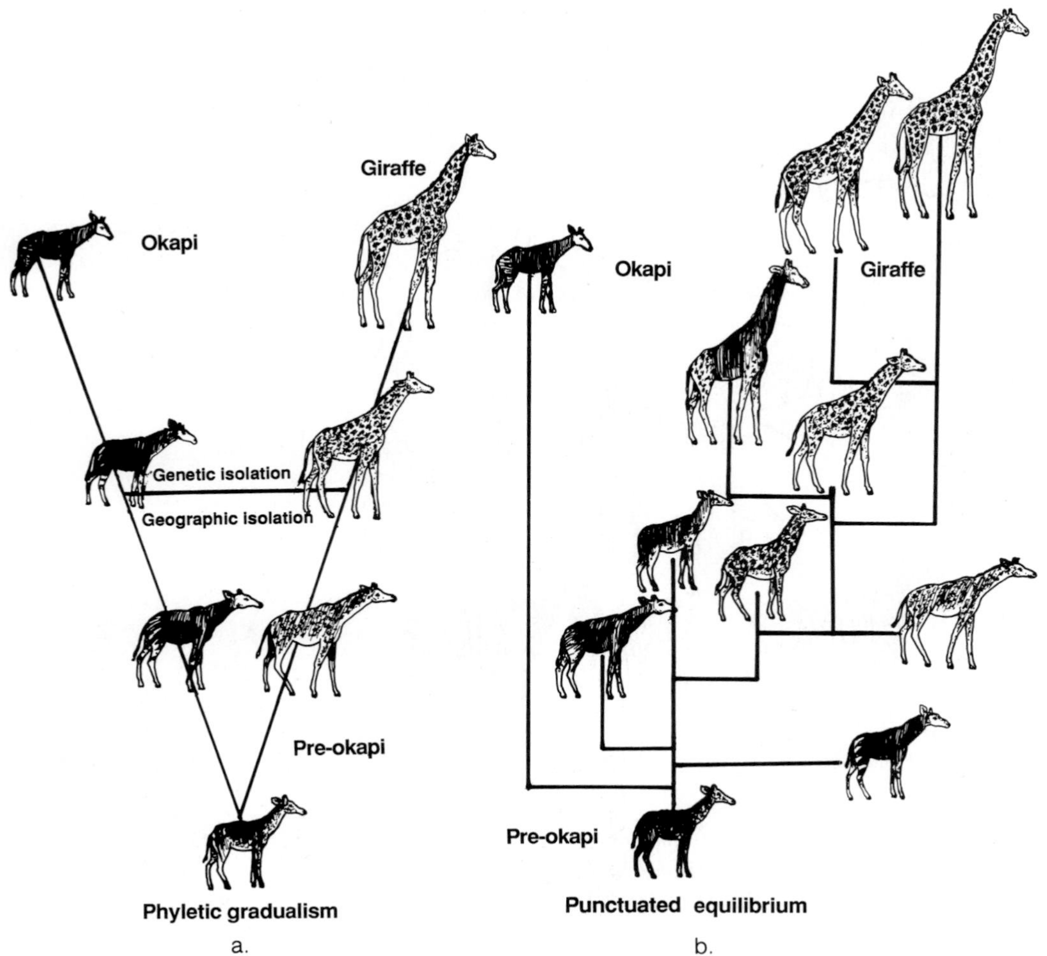


**Figure 5-14** Comparison of two models for differentiation of the modern okapi and giraffe from a common ancestor. a. In the phyletic gradualism model, slow, continuous change occurs from one species to the next. A pre-okapi species was divided into two populations by a geographic barrier, and the populations eventually became genetically isolated. b. According to the punctuated equilibrium model, change occurs rapidly, and new species evolve rapidly. However, little or no change occurs in a species during most of its existence.

populations, perhaps all the members of a species, are involved. Some evolutionists still accept this view, but it has been challenged.

Another concept of speciation, **allopatric speciation**, holds that most species arise after a small part of a population is isolated from its parent population by some kind of barrier. A rising mountain range, a river, or an invasion of part of a continent by the sea may effectively isolate parts of a once interbreeding population (Fig. 5–15). Isolation may also be accomplished by the accidental introduction of a species to a remote area. Once a small population becomes isolated, it no longer shares in the parent population's gene pool, and it is subjected to different selection pressures. Under these conditions each population evolves independently, and eventually may give rise to reproductively isolated species. Numerous examples of allopatric speciation have been well documented, especially speciation events in extremely remote areas such as oceanic islands; the finches of the Galapagos Islands are a good example (Fig. 5–2).

Most investigators agree that allopatric speciation is

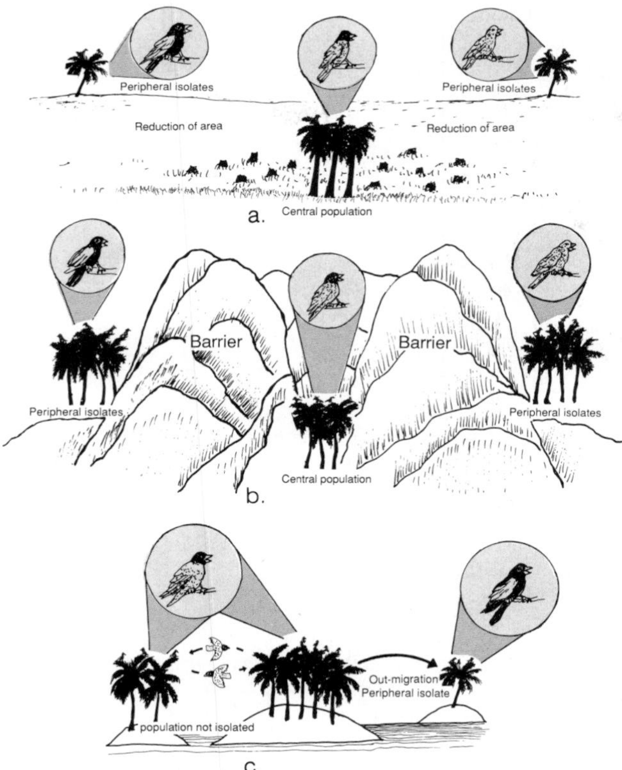


**Figure 5–15** Allopatric speciation. a. Reduction of area occupied by a species may leave small, isolated populations, *peripheral isolates*, at the periphery of the once more extensive range. In this example, members of both peripheral isolates have evolved into new species. b. Barriers have formed across parts of a central population's range, thereby isolating small populations. c. Out-migration and the origin of a peripheral isolate.

a common phenomenon, but they disagree about how rapidly new species evolve. According to one view, species evolve more or less continuously in a gradual manner. Another view, called **punctuated equilibrium**, holds that evolution is characterized by long periods of little or no change punctuated by short periods of rapid evolution. Species arise rapidly, perhaps in a few thousands of years, but once a new species evolves, it remains much the same for its several millions of years of existence (Fig. 5–14).

Proponents of punctuated equilibrium claim that the fossil record supports their hypothesis, because few examples of gradual transitions from one species to another can be found, and many species appear to have existed for millions of years without noticeable change. According to this view, species arise rapidly in small, geographically isolated populations, and the transitional forms connecting ancestral and descendant species are unlikely to be preserved in the fossil record. And, as a matter of fact, many species do appear abruptly in the fossil record with no evidence of direct ancestors.

Critics, however, point out that the abrupt appearance of fossil species may be an artifact of the fossil record. After all, the fossil record is incomplete, and most fossils are skeletal elements only, but it is known that significant evolution, certainly the origin of a new species, does not necessarily involve any significant morphological change (Fig. 5–13). Their contention is that, the fossils record those events that yield morphologically distinct organisms but do not tell us about the equally important evolutionary changes in biochemistry, physiology, and reproduction. The critics also point out that some gradual transitions from ancestral to descendent species are known (Fig. 5–16).

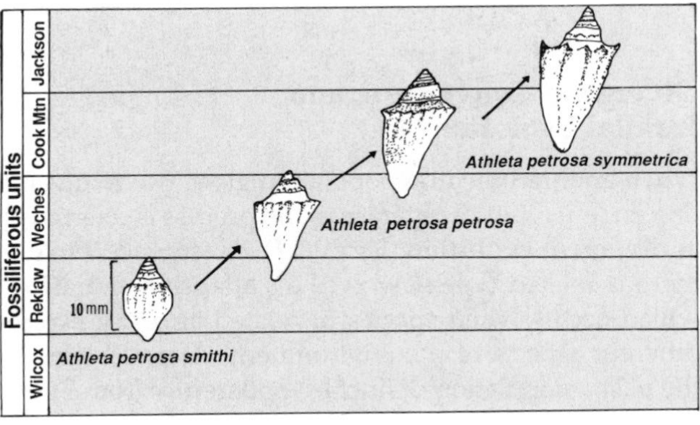


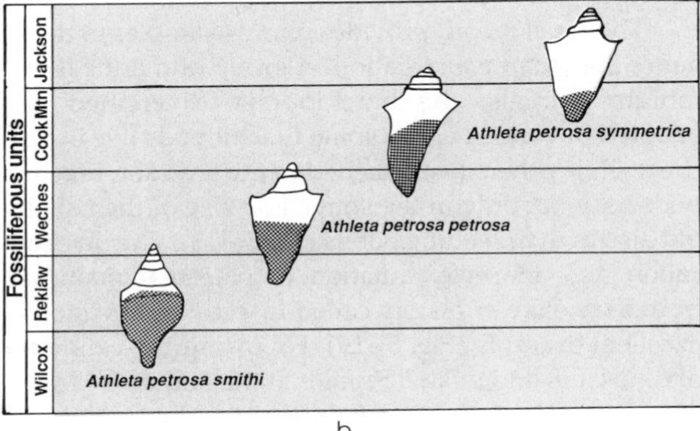


**Figure 5–16** Gradual evolution of the small snail *Athleta*. a. Changes in size, shape, and the development of spines on the shell. b. Changes in the surface ridge pattern of the shell. The time span represented is about 5 million years.

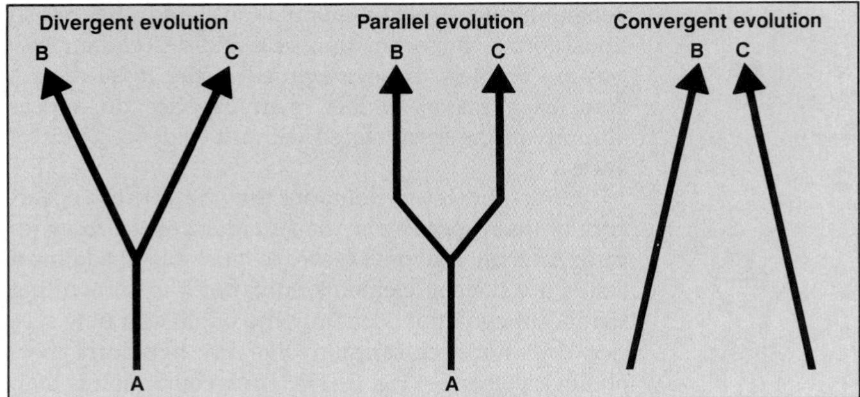


**Figure 5–17** Divergent evolution in this example results as species A diverges and gives rise to two descendant species, B and C. Parallel evolution also involves divergence from a common ancestor, A, but descendant species, B and C, then develop in a similar way as they adapt to a similar life style. Convergent evolution is much like parallel evolution except that species B and C are distantly related but resemble one another in some aspects because of similar adaptations.

## Divergent, Convergent, and Parallel Evolution

When an interbreeding population gives rise to diverse descendent types of organisms, the process is referred to as **divergent evolution** (Fig. 5–17). Divergence into numerous related types involves an **adaptive radiation**, which occurs when species of related ancestry exploit different aspects of the environment. At some time in the past a population of finches, apparently from South America, reached the Galapagos Islands. This ancestral population diversified into the adaptive types of species of finches that Darwin observed (Fig. 5–2).

The fossil record provides many examples of divergence and adaptive radiation. A group of marine invertebrate animals, the brachiopods, diversified and adapted in various ways; some brachiopods live in burrows while others live attached to the seafloor, brachiopods vary not only in the shape and size of their shells, but also in their feeding habits (Fig. 5–18). The diversification and adaptive radiation of placental mammals from shrewlike ancestors called insectivores is another excellent example (Fig. 5–19). Following the extinction of dinosaurs and related reptiles at the end of the Mesozoic Era, placental mammals rapidly diversified and eventually gave rise to such adaptive types as whales, bats, elephants, and monkeys.

While divergent evolution leads to organisms that differ markedly from their ancestors, **convergent evolution** and **parallel evolution** are processes whereby similar adaptations arise in different groups (Fig. 5–17).

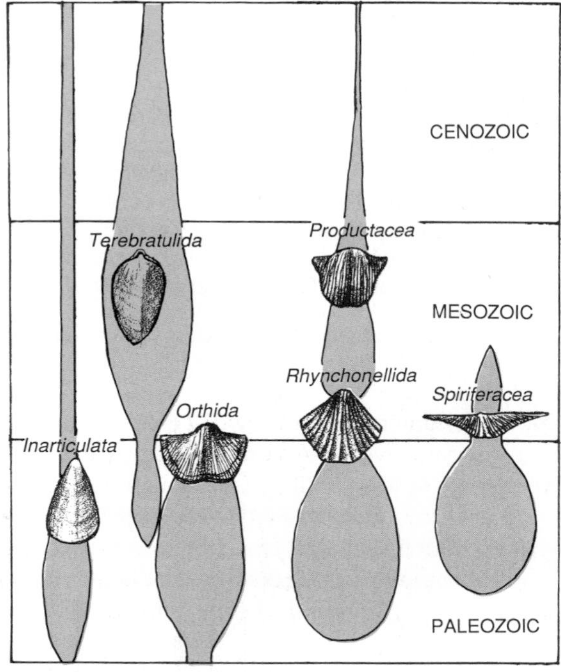


**Figure 5–18** Divergence and adaptive radiation of brachiopods. Relative abundances of the major groups are shown by the patterns, and representatives of the groups are shown. The inarticulate brachiopods adapted to burrowing into muddy sediments near seashores and have persisted with little change from the Early Paleozoic to the present. Most of the other brachiopods shown live on the seafloor and show a wider range of adaptive types.

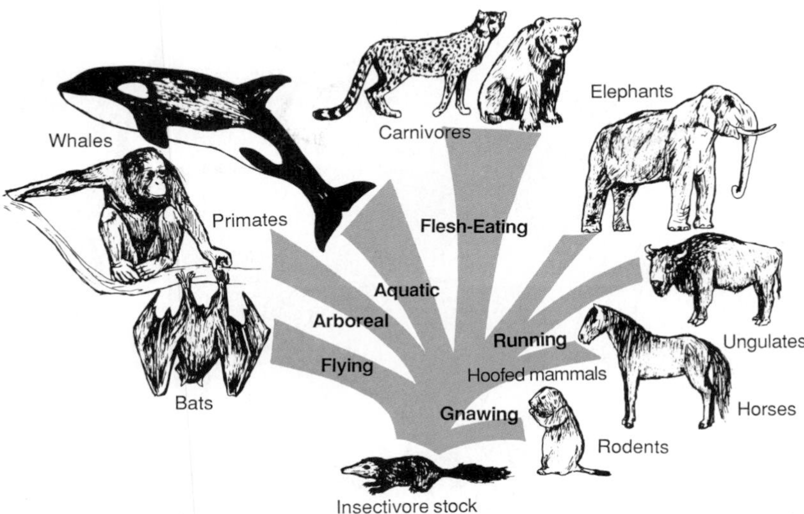


**Figure 5–19** Divergent evolution of the placental mammals. An ancestral stock, probably small, shrewlike insectivores, gave rise to diverse descendant groups, each of which adapted to different life-styles.

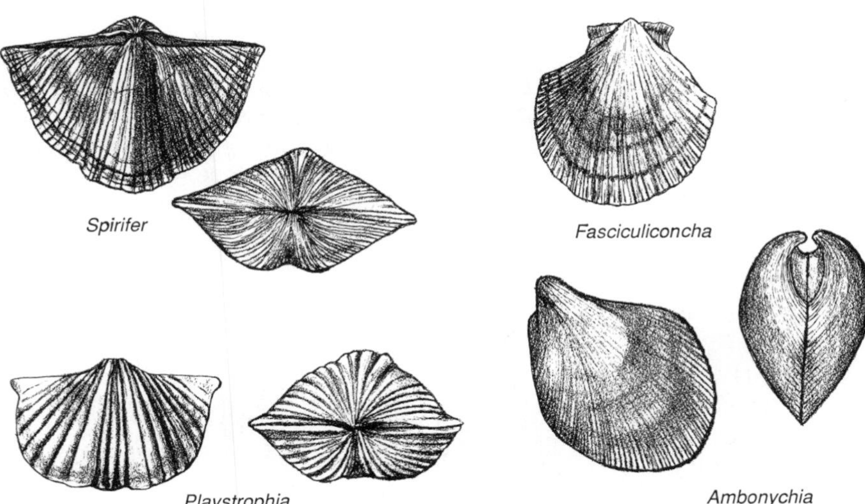


**Figure 5–20** Brachiopods (left) and clams (right) are distantly related, but both have adapted to a very similar habitat. Convergence is shown by the similarities in clam and brachiopod shells; in both groups the shell consists of two parts hinged together. However, the details of the shells of each group differ in a number of aspects.

Convergent evolution can be defined as the development of similar characteristics in distantly related organisms, and parallel evolution as the development of similar characteristics in closely related organisms. The distinction between these two types of evolution depends on the degree of relatedness between the organisms in question, which is not always easy to determine. However, both convergent and parallel evolution occur as a result of different organisms adapting to similar ways of life.

The shells of clams and brachiopods, although not identical, are similar, and they represent an adaptive response to a similar life-style (Fig. 5–20). Since clams and brachiopods are very distantly related, their similar

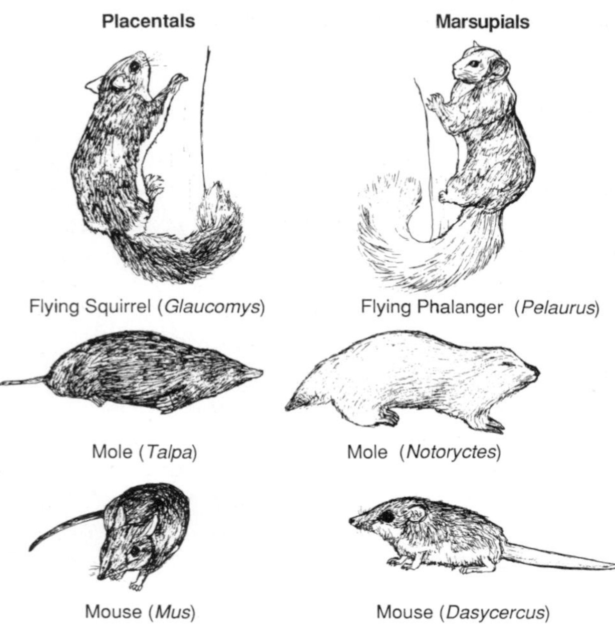


**Figure 5–21** Convergent evolution among Australian marsupial (pouched) mammals and placental mammals of the other continents. Marsupials and placentals are distantly related, but many adapted in similar ways, giving rise to superficially similar animals.

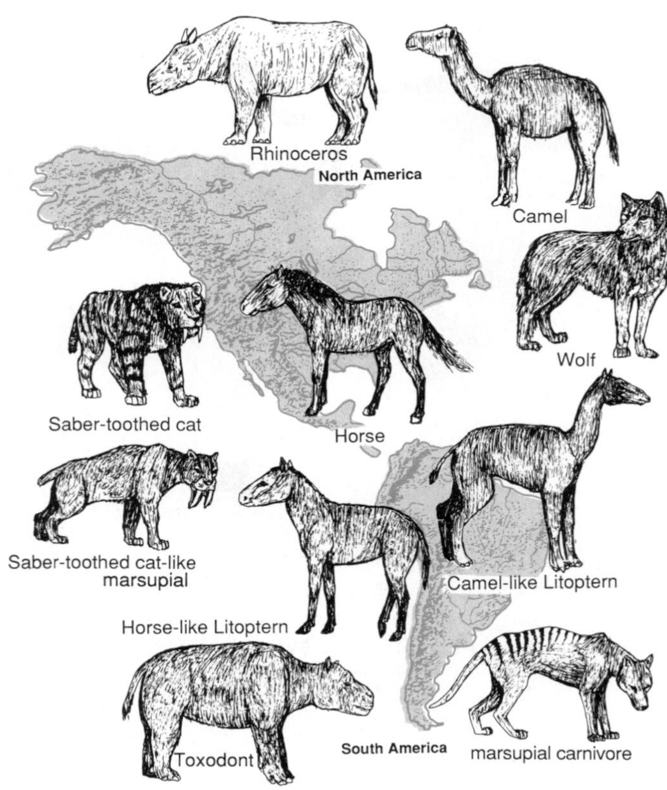


**Figure 5–22** Before the origin of the Isthmus of Panama a few million years ago, mammals of North and South America evolved independently. Similar adaptations of animals on each continent account for convergence.

**Figure 5–23** The New World porcupine, *Erithizon*, and the Old World porcupine, *Hystrix*, have several similarities in addition to quills. Both are placental mammals, but whether they are related closely enough to be an example of parallel evolution is not clear. In any case, adaptations to similar lifestyles have resulted in similarities in appearance.

shells are an example of convergent evolution. Convergent evolution is also well illustrated by the similar adaptations of the Australian marsupial (pouched) mammals and the placental mammals of the other continents (Fig. 5–21). Convergence accounts for the fact that many of the Australian marsupials superficially resemble placental mammals elsewhere.

Another impressive example of convergent evolution is shown by the Cenozoic mammals of North and South America (Fig. 5–22). During most of the Cenozoic, South America was an island continent, and its mammalian fauna, consisting of both marsupials and placentals, evolved independently. A number of mammals on each continent adapted to similar environments and developed many features in common.

Parallel evolution is also a common phenomenon, but, as noted, it is not always easy to distinguish from convergent evolution. For example, the Old World and New World porcupines (Fig. 5–23) are both placental mammals and are similar in many aspects other than having quills. Whether this is an example of parallelism or convergence is debatable, but one thing is clear; both animals solved the same adaptive problem in a similar way. There is little doubt, however, that parallel evolution accounts for the independent development of enlarged, saber-like canine teeth in two families of cats (Perspective 5–2).

## Evolutionary Trends

The evolutionary history, or *phylogeny*, of a group can be worked out in some detail if sufficient fossil material is known. Furthermore, constructing such phylogenies reveals the *evolutionary trends* that characterize a group of organisms. The phylogeny of ammonites, an extinct group related to the modern squid and octopus, is well-known; one evolutionary trend in this group was an increasingly complex shell structure (Fig. 5–24).

Titanotheres were mammals that lived in North America and eastern Asia from Early Eocene to Early Oligocene time. A good fossil record records their phylogeny and several evolutionary trends (Fig. 5–25). One trend was simply an increase in size, but titanotheres also developed large horns, and the shape of the skull changed.



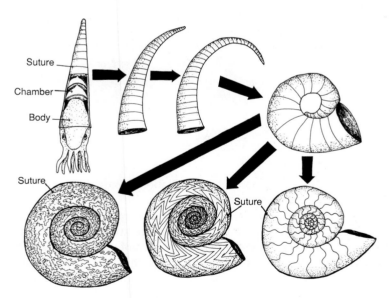

**Figure 5–24** Evolutionary trends in the ammonites. Ammonites with coiled shells evolved from straight-shelled ancestors. Notice that an ammonite's shell is internally divided into chambers, and that the animal lives only in the outermost chamber. The sutures, the lines formed where chamber walls meet the wall of the outer shell, became increasingly complex as ammonites evolved.

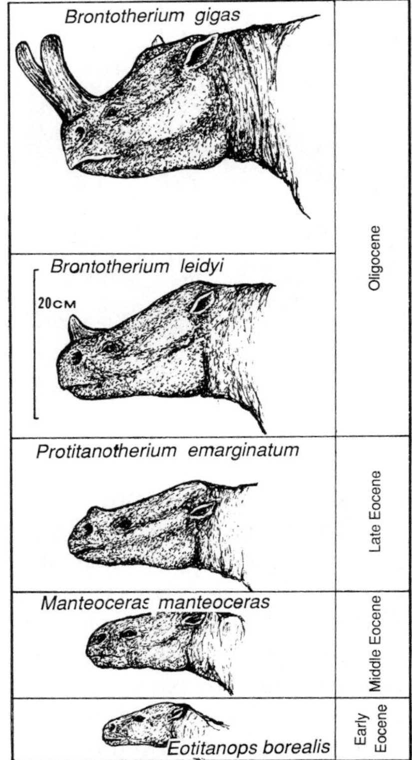


**Figure 5–25** Evolutionary trends in the titanotheres, an extinct group of mammals related to modern horses and rhinoceroses. Titanotheres evolved from small ancestors to giants about 2.4 m tall at the shoulder, and they developed large horns.

## Perspective 5–2

# Parallelism and Convergence of Saber-toothed Carnivorous Mammals

All living carnivorous mammals have well-developed canine teeth, but none have canines as large as those that developed in some fossil carnivores. Saber-toothed cats of the Pleistocene, with canines 15 cm long, are the best-known carnivores with enlarged canines. However, they were not the only mammals with this specialization. Large, saber-like canines developed independently four times during the Cenozoic: once in an extinct group of mammals called creodonts, once in marsupials, and twice in the cats (Fig. 1). Parallel evolution is the best explanation for saber-like canines developing twice in the same family of cats, while convergent evolution accounts for enlarged canines in creodonts, marsupials, and cats.

Parallelism or convergence does not end with saber-like canines, however. Skull modifications were also necessary so that the mouth could be

a.

b.

Increase in size is one of the most common evolutionary trends. However, trends are complex in that they do not always proceed at the same rate, they may be reversed, or several may occur in the same group. Horses evolved from fox-sized ancestors that lived during the Eocene. One trend was an increase in size, but it was not a steady, uniform increase, and, in fact, some horses actually show a reversal in this trend (Fig. 5–26). Horse evolution is also characterized by a reduction in the number of toes, lengthening of the legs and feet, and changes in the teeth and skull. All of these trends are well documented by fossils, but they all did not occur at the same rate (Fig. 5–26).

One can view evolutionary trends as a series of adaptations to changing environments, or adaptations that occur in response to exploitation of new habitats. The same trends occur repeatedly in the fossil record, however. Increase in size is one of the most often repeated trends, but a number of others are known as well. Ammonites, for example, nearly became extinct at the end of the Paleozoic Era, but those that survived into the Mesozoic diversified and gave rise to many adaptive types very similar to those that had existed earlier. The evolutionary significance of this phenomenon is that very similar adaptations have occurred in many groups of organisms in different places throughout the history of life.

Some organisms, however, show little evidence of any evolutionary trends for long periods of time. One good example is the brachiopod *Lingula,* whose shell

opened widely enough for effective use (Fig. 2). It is not surprising that the skulls of all saber-toothed carnivorous mammals are similarly constructed, because there are only a few ways to modify the basic mammalian skull to accommodate enlarged canines.

How saber-toothed carnivores used their canines is a debatable point. Pleistocene saber-toothed cats are commonly portrayed inflicting deep stab wounds into large prey animals such as elephants. This mode of predation seems unlikely, however, because the curvature of the canines would have prevented stabbing thrusts even though the mouth could be opened very widely. It also seems unlikely that the canines were used to pierce the top of the victim's neck or skull; saber-like canines were too weak to penetrate or break bones.

Leonard Radinsky and Sharon Emerson[1] have proposed that the saber-like canines were used to

[1]See "The Late, Great Sabertooths," *Natural History*, 91, no. 4 (1982): 50–57.

**Figure 1** Saber-toothed carnivores. a. *Smilodon*, a Pleistocene cat, is shown feeding on the carcass of a mammoth trapped in the La Brea Tar Pits of Southern California. b. In this scene the Late Miocene cat *Barbourfelis* is about to kill its prey by slashing its throat. Both *Smilodon* and *Barbourfelis* belong to the cat family, the Felidae, so the independent development of saber-like canines in each is an example of parallel evolution. c. The marsupial saber-toothed carnivore *Thylacosmilus* from the Pliocene of South America. The development of saber-like canines in marsupials and cats, which are distantly related, is an example of convergent evolution.

has not changed significantly since the Ordovician. Such organisms that show little or no change for long periods of time are commonly called **living fossils** (Fig. 5–27). *Lingula* burrows into the muddy sediments of shallow seas. Apparently, once it developed its adaptive strategy, it stayed in an unchanging environment and has changed very little, at least in the appearance of its shell. Of course, *Lingula* may have evolved physiologically or reproductively, but fossils do not reveal changes such as these.

## Extinctions

James Audubon (1785–1851) once estimated that a single flock of passenger pigeons numbered over 1 billion individuals. In 1857 a committee of the Ohio senate said the passenger pigeon needed no protection. In 1914 the last passenger pigeon died in the Cincinnati Zoo. Steller's sea cow was discovered in 1742, and 27 years later had been hunted to extinction. Rhinoceroses in the wild are now on the verge of extinction.

The preceding examples of extinction were brought about by humankind; and, indeed, humans have a degree of control over their environment unprecedented in the history of life. They can bring about changes that may not have occurred otherwise or that would have occurred at very different rates. However, extinctions are the rule in the history of life. In fact, extinction seems to be the ultimate fate of all species; probably over 99 percent of all species that ever lived are now extinct.

## Perspective 5-2 (continued)

make a well-placed, shallow slash across the throat (Fig. 1). Modern cats that prey on animals as large or larger than themselves bite the victim's throat and hold on until it suffocates. Perhaps saber-toothed carnivores also preyed on large animals, but simply inflicted a slash across the throat and then waited for the victim to die.

Perhaps the most curious aspect about saber-toothed carnivores is that none exist now. The development of saber-like canines must have been a successful adaptation, since it occurred four times, and saber-tooths of one kind or another existed from the Early Cenozoic until about 10,000 years ago. The cause of their final extinction is unknown, but hypotheses for the extinction event at the end of the Pleistocene are discussed in Chapter 17.

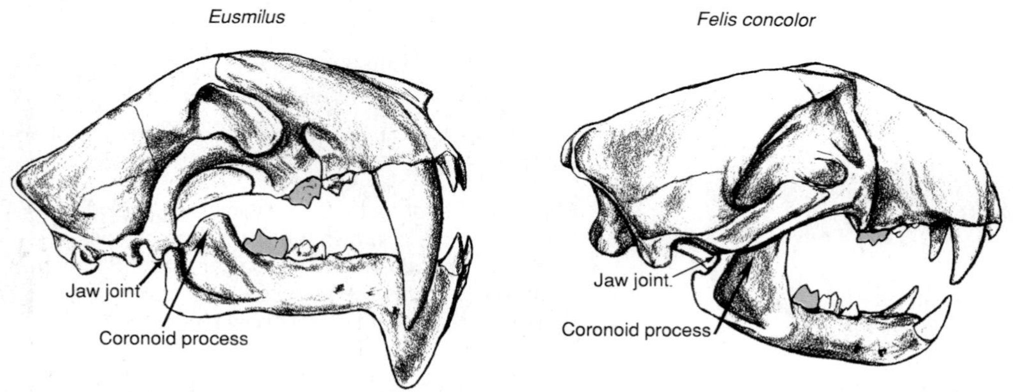


**Figure 2** Comparison of the skulls of a modern mountain lion, *Felis concolor* (right), and an Oligocene saber-toothed cat, *Eusmilus* (left). The skull of *Eusmilus* shows the basic modifications in saber-tooths that allowed the mouth to be opened widely enough for effective use of the enlarged canines. The jaw joint is lower and the coronoid process is shortened, allowing for longer muscle fibers. The shearing teeth (shaded) are positioned closer to the jaw joint, and the face is rotated upward relative to the braincase.

The term *extinct* seems so final, and, in a very real sense, it is. But we must distinguish between two types of extinction. In one case a species develops into a new species that differs so much from its ancestral group that the parent species can be considered extinct. Perhaps this is best called *pseudoextinction* since, strictly speaking, no extinction event occurred. The second case of extinction is terminal. That is, a species dies out without giving rise to anything else. Many examples of terminal extinction are known from the fossil record (Fig. 5-28).

Extinction by one method or another appears to be a continual occurrence in the history of life. But so is the origin of new species that usually quickly exploit the opportunities created by the disappearance of other species. In some cases of extinction an ecologically equivalent organism may not appear for some time. Ichthyosaurs were Mesozoic marine reptiles that lived like and superficially resembled modern porpoises and dolphins (Fig. 5- 29). Yet, whereas ichthyosaurs are unknown after the Mesozoic, porpoises and dolphins did not occupy their niche until some 30 million years later.

There have been times when extinction rates were greatly accelerated, called *mass extinctions*. The two greatest crises in the history of life were the mass extinctions at the end of the Paleozoic and Mesozoic eras. In

Equus 1 toed, springing feet
Pleistocene
Pliocene
Nannippus
Neohipparion
Hypohippus Calippus
Megahippus
1 toed, springing feet
Pliohippus
Miocene
Archaeohippus
3 toed, springing feet
Merychippus
Oligocene
Mesohippus 3 toed, padded feet
Eocene
1 meter
Hyracotherium 4 toed, padded feet

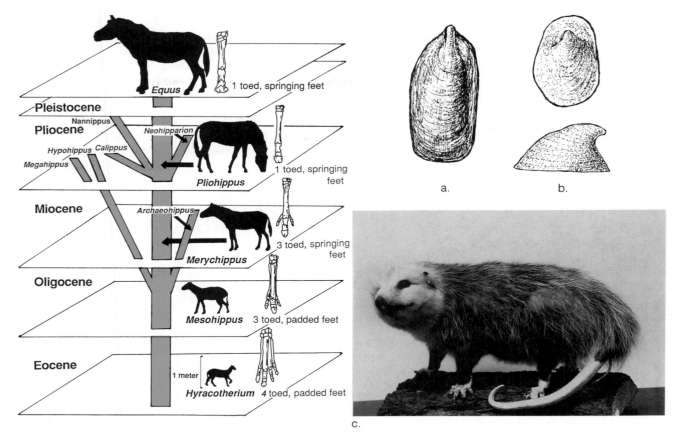

**Figure 5–26** Two of the evolutionary trends in horses. Horses increased in size during the Cenozoic but not at a uniform rate, and some horses reversed the trend and became smaller. The modern horse, *Equus*, has one functional toe on each foot, but its oldest known direct ancestor, *Hyracotherium*, had four-toed forefeet and three-toed back feet. The forefeet illustrated in this diagram show the trend in toe reduction in the forefoot. Forefeet shown are not to scale.

both cases, the diversity of life was sharply reduced. These extinction events, and some of lesser magnitude, will be discussed in greater detail in the following chapters.

## EVIDENCE FOR EVOLUTION

When Charles Darwin proposed his theory, he cited such supporting evidence as classification, embryology,

a. b.

c.

d.

**Figue 5–27** Four examples of so-called living fossils. a. The brachiopod *Lingula* looks much like its Ordovician ancestors. b. *Neopilina* is a deep-sea-dwelling mollusk that closely resembles Paleozoic mollusks. c. The American or Virginia opossum, *Didelphis marsupialis*, has not changed significantly in skeletal structure for the last 60 million years. d. *Latimeria* is a member of a group of fishes that was thought to have become extinct at the end of the Mesozoic Era. A specimen was caught in 1938 off the east coast of Africa, and since then several more have been caught.

comparative anatomy, the geographic distribution of organisms, and, to a limited extent, the fossil record (Fig. 5–30). However, Darwin had little knowledge of the mechanism of inheritance or biochemistry, and molecular biology was completely unknown during his lifetime. Studies in these areas coupled with a better understanding of the fossil record have added to the body of evidence. Today, most scientists accept that the theory of evolution is as well supported by evidence as any major theory.

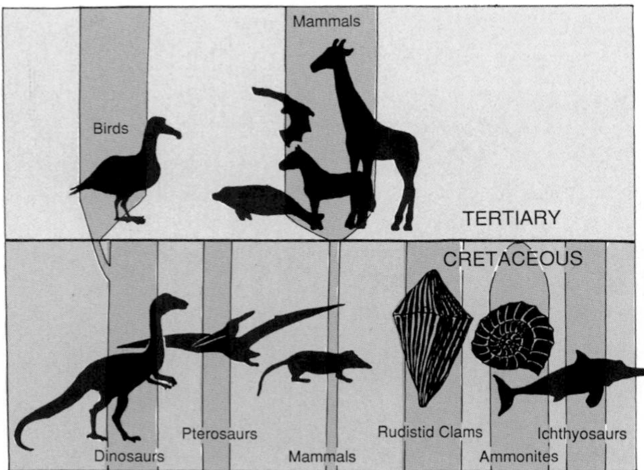


**Figure 5–28** Some of the victims of the mass extinction at the end of the Cretaceous Period. One group of small carnivorous dinosaurs probably gave rise to birds, and so live on in their evolutionary descendants; but all other dinosaurs suffered terminal extinction. Among the marine animals, rudists (a type of clam) and ammonites both died out without leaving evolutionary descendants. Extinctions are common phenomena in the history of life. When they occur, they create opportunities for other organisms. For example, mammals coexisted with dinosaurs but were small and not very diverse. However, when the dinosaurs and their relatives became extinct, mammals rapidly exploited the opportunities created by the extinction event.

## Classification—A Nested Pattern of Similarities

The Swedish naturalist Carolus Linnaeus (1707–1778) proposed a formal classification of organisms. According to the Linnaean system an organism is given a two-part name; the coyote, for example, is referred to by its genus (plural, *genera*) name and its species name, *Canis latrans*.

Table 5–1 shows the basic arrangement of the Linnaean classification although the scheme now used contains more categories. The arrangement is hierarchical. That is, as one proceeds up the list the categories become more inclusive. The coyote (*Canis latrans*) and the wolf (*Canis lupus*) are members of different species, but they share a large number of characteristics, so both belong to the same genus. The red fox (*Vulpes fulva*) also shares some, but fewer, characteristics with coyotes and wolves, so it, along with other related dog-like animals, make up the family Canidae. In turn, all canids share some characteristics with cats, bears, weasels, and raccoons, all of which belong to the order Carnivora. Likewise, all carnivores, rodents, bats, primates, and elephants have hair and mammary glands and constitute the class Mammalia (Fig. 5–31). And so it goes, up to kingdom, the most inclusive category in the classification scheme.

It should be clear from the preceding discussion that shared characteristics are the bases for inclusion in a specific category. Linnaeus clearly recognized similarities among organisms; however, he simply intended to classify species that he believed were specially created and immutable. He viewed his classification as a reflection of the Great Chain of Being, a sequence from simple to complex, with humans at the apex.

**Figure 5–29** Ichthyosaurs were fish-eating, marine reptiles that became extinct at the end of the Mesozoic Era. Following their extinction no vertebrate animal occupied their niche until the dolphins and porpoises appeared about 30 million years later. Notice that ichthyosaurs and dolphins resemble one another in several features. Both groups adapted to a similar life-style and provide an excellent example of convergent evolution.

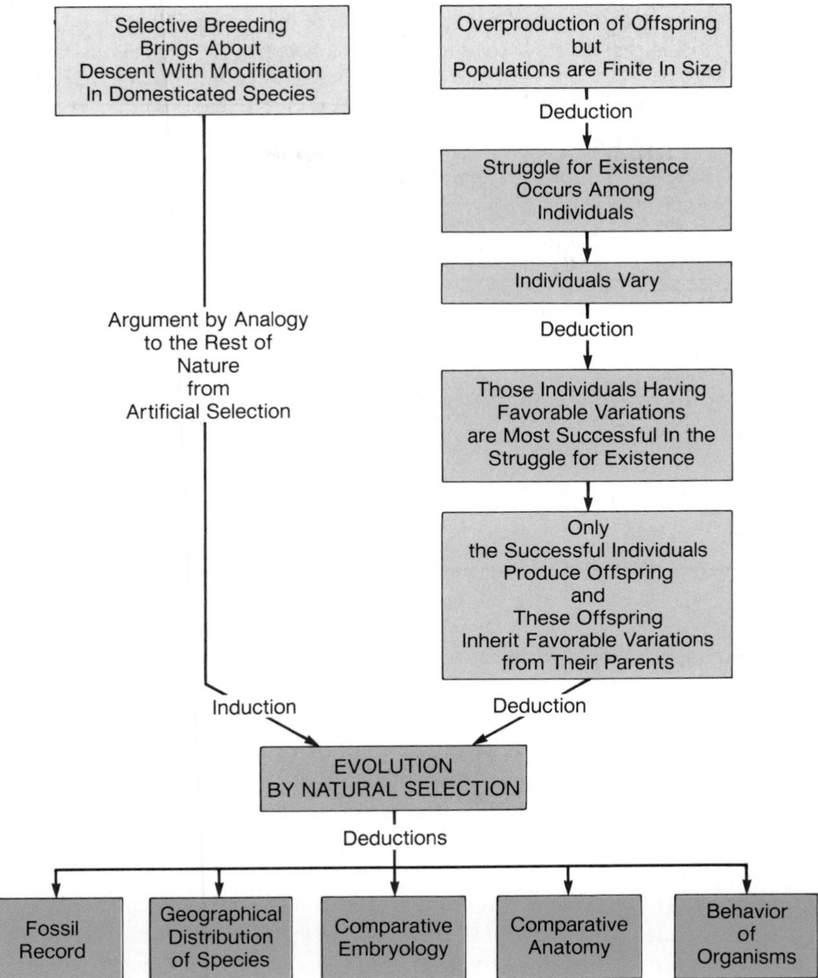


**Figure 5–30** Evolution by natural selection was the product of both inductive and deductive reasoning. Darwin used an inductive argument from artificial selection to formulate his theory of natural selection. By deduction he proposed a set of causes that produced natural selection. He further deduced the kinds of observations that would be made in nature if evolution by natural selection operated in the way he predicted.

Following the publication of *The Origin of Species,* biologists soon realized that shared characteristics among organisms constitute a strong argument for evolution. In our example, coyotes and wolves share many characteristics because they evolved from a common ancestor in the not too distant past. They also share some characteristics with bears and cats because all members of the order Carnivora had a common ancestor in the more remote past.

## Biological Evidence for Evolution

According to evolutionary theory, all life-forms are related, and all living organisms descended with modification from ancestors that lived in the past. If this statement is correct, then there should be evidence of fundamental similarities among all life-forms, and closely related organisms should be more similar to one another than to more distantly related organisms. As a

**Table 5-1** The classification scheme now in use showing the hierarchical arrangement of the categories. The boxes include those animals to which the coyote, *Canis latrans*, is most closely related at various levels in the classification scheme (see Figure 5-31).

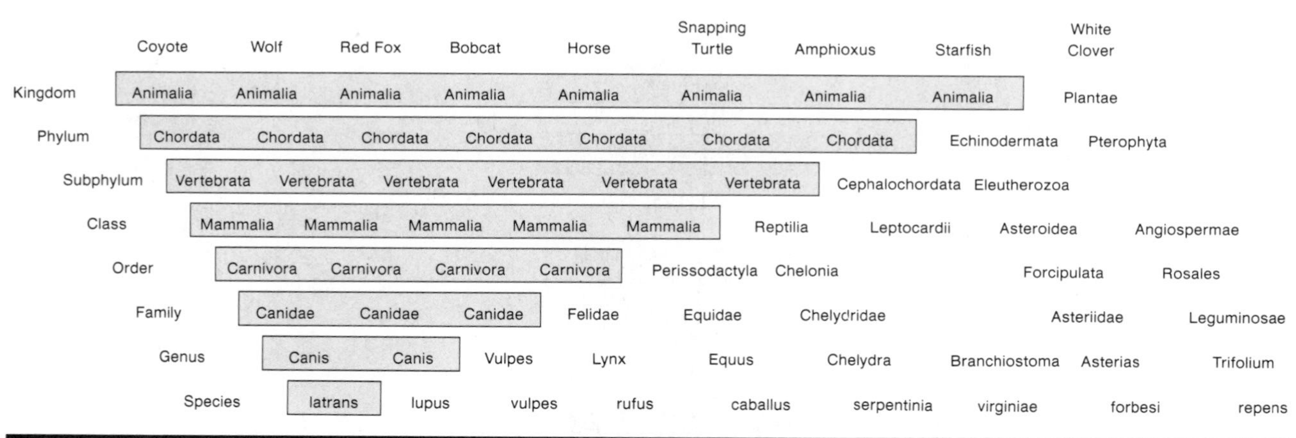

| | Coyote | Wolf | Red Fox | Bobcat | Horse | Snapping Turtle | Amphioxus | Starfish | White Clover |
|---|---|---|---|---|---|---|---|---|---|
| Kingdom | Animalia | Animalia | Animalia | Animalia | Animalia | Animalia | Animalia | Animalia | Plantae |
| Phylum | Chordata | Chordata | Chordata | Chordata | Chordata | Chordata | Chordata | Echinodermata | Pterophyta |
| Subphylum | Vertebrata | Vertebrata | Vertebrata | Vertebrata | Vertebrata | Vertebrata | Cephalochordata | Eleutherozoa | |
| Class | Mammalia | Mammalia | Mammalia | Mammalia | Mammalia | Reptilia | Leptocardii | Asteroidea | Angiospermae |
| Order | Carnivora | Carnivora | Carnivora | Carnivora | Perissodactyla | Chelonia | | Forcipulata | Rosales |
| Family | Canidae | Canidae | Canidae | Felidae | Equidae | Chelydridae | | Asteriidae | Leguminosae |
| Genus | Canis | Canis | Vulpes | Lynx | Equus | Chelydra | Branchiostoma | Asterias | Trifolium |
| Species | latrans | lupus | vulpes | rufus | caballus | serpentinia | virginiae | forbesi | repens |

matter of fact, all living things, from bacteria to whales, are composed of the same chemical elements—mostly carbon, nitrogen, hydrogen, and oxygen. Furthermore, the chromosomes of all organisms are composed of DNA, except bacteria, which have RNA; and in all cells, proteins are synthesized in essentially the same way.

Thousands of biochemical tests of numerous organisms have provided compelling evidence for evolutionary relationships. Blood proteins, for example, are similar among all mammals, but they also indicate that among the primates humans are most closely related to the great apes, followed, in order, by the Old World monkeys, the New World monkeys, and the lower primates such as lemurs. Biochemical tests indicate that all birds are related among themselves and more closely related to turtles and crocodiles than to snakes and lizards, a finding corroborated by the fossil record.

Studies of developing embryos reveal that some organisms are similar to one another until very late in embryonic development, while others are less similar (Fig. 5-32). Fish, chimpanzee, and human embryos are very similar in the earliest stages, but differences soon become apparent in fish embryos. Chimpanzees and humans, in contrast, retain remarkable embryonic similarities until very late in the developmental process. The inference is that chimpanzees and humans are more closely related to one another than they are to fish.

Evidence for evolution is also provided by comparing anatomical structures of organisms (Perspective 5-3). The forelimbs of humans, whales, dogs, and birds (Fig. 5-33) are superficially dissimilar, yet all are composed of the same bones and have basically the same arrangement of muscles, nerves, and blood vessels, all are similarly arranged with respect to other structures, and all have a similar pattern of embryonic development. Such structures are said to be **homologous**. Descent with modification from a common ancestor, that is, a common genetic heritage, is the best explanation for the existence of homologous organs.

Not all similarities among organisms are evidence of evolutionary relationships; in fact, some similarities are rather superficial. Insects are not closely related to bats and birds, but all three of these animals have wings. Such structures that serve the same function are called **analogous organs** (Fig. 5-34). The wings of bats and birds, but not of insects, are homologous, however, and provide evidence for a common ancestry. Insect wings, on the other hand, represent convergence; a similar solution to an adaptive problem, flight, but they differ from bat and bird wings in both structure and embryological development.

Another type of evidence for evolution is provided by observations of small-scale evolution in modern organisms (Fig. 5-35). We have already mentioned one example, the adaptations of plants to contaminated soils. New insecticides and pesticides must be developed continually as insects and rodents develop resistance to the existing ones. Development of antibiotic-resistant strains of bacteria constitutes a continuing problem in medicine.

The important points regarding small-scale evolution are that variation existed, and that some variant

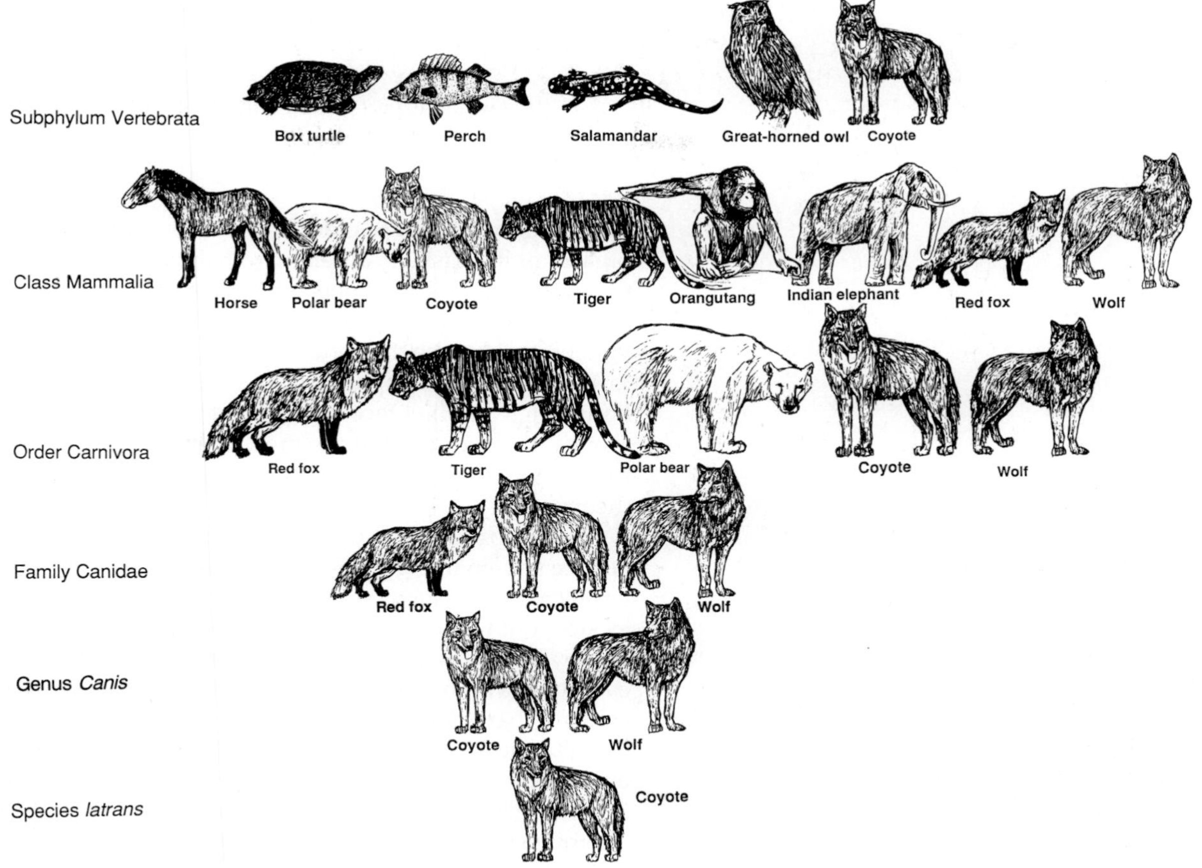


**Figure 5–31** Classification of organisms by shared characteristics. All members of the subphylum Vertebrata, including fishes, amphibians, reptiles, birds, and mammals, have a segmented vertebral column. Among the vertebrates, however, only those warm-blooded animals having hair or fur and mammary glands are mammals. Eighteen orders of mammals are recognized, including the order Carnivora shown here. All members of this order have teeth specialized for a diet of meat. The family Canidae includes only the dog-like carnivores, and the genus *Canis* includes closely related species. The coyote, *Canis latrans*, stands alone as a species.

types had a better chance to survive and reproduce. The source of variation is irrelevant; it may have already existed, or it may have been introduced by mutations. Regardless of their source, the survival and reproduction of some variant types caused the genetic makeup of the population, its gene pool, to change through time.

## Fossils and Evolution

The occurrence of fossil marine animals far from the sea, high in mountains, led many early naturalists to conclude that they were deposited during the worldwide biblical flood. As early as 1508, Leonardo da Vinci realized that the fossil distribution was not what one would expect in a rising flood. Nevertheless, the flood explanation persisted, and John Woodward (1665–1728) proposed that the fossils had separated out of the floodwater according to their density. In other words, Woodward's hypothesis predicts that the fossils should be arranged in a vertical sequence, with the densest ones at the bottom followed upward by those of decreasing density. This hypothesis was quickly rejected since field observations clearly did not support the prediction; fossils of various densities are found throughout the fossil record.

The fossil record does show a sequence, but not one based on density, size, or shape of the fossils. Rather, it shows a sequence of appearances of different life-forms

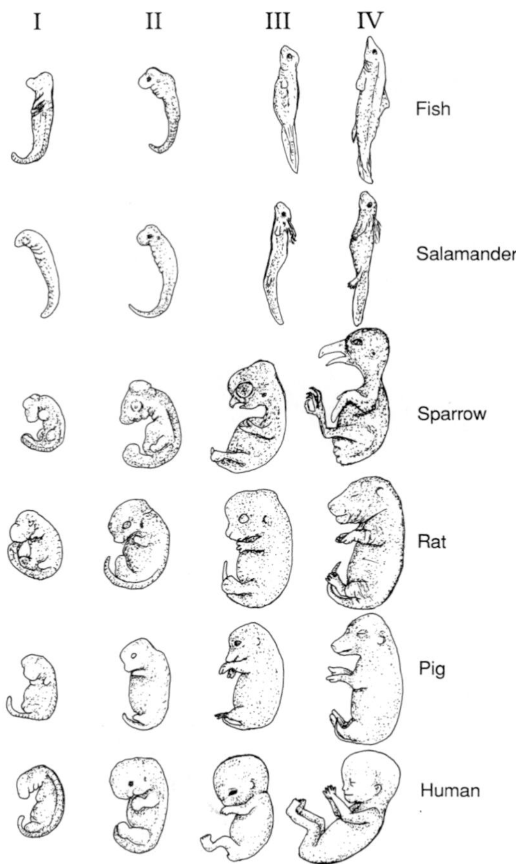


**Figure 5–32** Comparison of several embryos of vertebrate animals at four stages of development. All look much the same in the early stage, but differences soon become apparent between those distantly related. Similarities persist, however, between those more closely related.

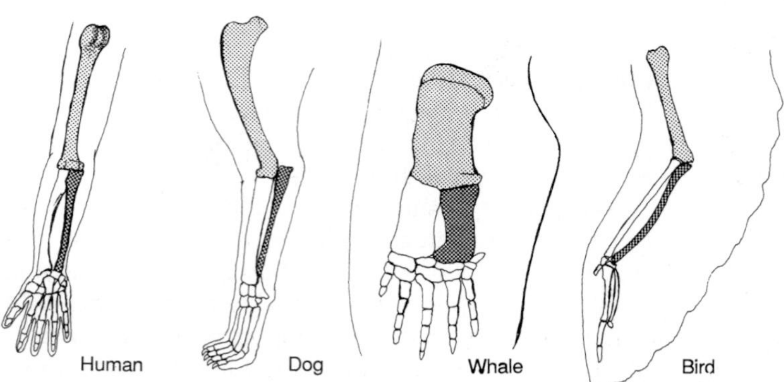


**Figure 5–33** Homologous organs such as the forelimbs of humans, dogs, whales, and birds serve different functions but are composed of the same elements and have similar embryological developments.

through time (Fig. 5–36). In fact, according to evolutionary theory,

> . . . [W]e would expect to find only the simplest organisms in the very oldest fossiliferous strata and the more complex to appear in more recent strata.[2]

The fact is that older and older fossiliferous strata contain organisms increasingly different from those living today. One-celled organisms appeared before multicelled organisms, plants before animals, and invertebrates before vertebrates. Among the vertebrates, fish appear first, followed in order by amphibians, reptiles, mammals, and birds (Fig. 5–36). Once established, many primitive groups have persisted to the present, but their order of first appearance is consistent with evolutionary theory.

If the fossil record provided no more than the sequence shown in Figure 5–36, however, it could be interpreted as a succession of independently created groups. But fossils also provide evidence for the evolutionary origins of some of these groups, and evidence for evolution within groups. Some examples of such fossil evidence have already been discussed in other contexts, the evolution of ammonites and titanotheres for example.

The evolution of horses and their living relatives, the rhinoceroses and tapirs, is well documented by fossils (Figs. 5–26, 5–37). Horses, rhinoceroses, and tapirs

[2]J. A. Moore 1984. Science as a Way of Knowing—Evolutionary Biology. *American Zoologist,* 24: 467–534.

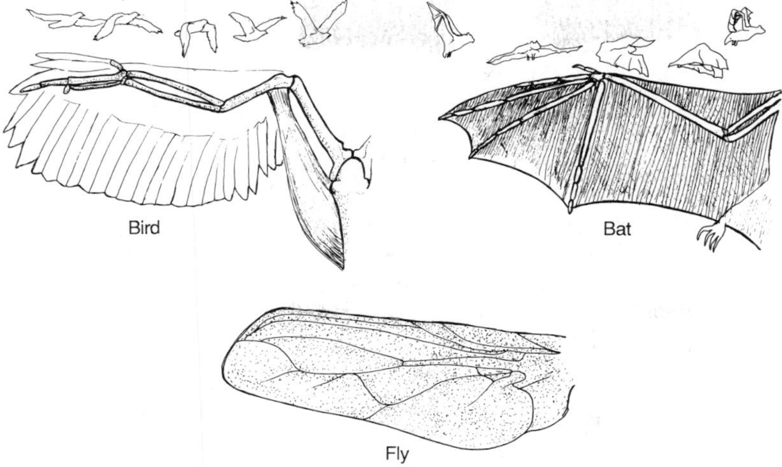


**Figure 5–34** The fly's wings serve the same function as wings of birds and bats but have a different structure and different embryological development. Organs that serve the same function are analogous, but the bird and bat wings are also homologous.

**Figure 5–35** Small-scale evolution of the peppered moths of Great Britain. In pre-industrial England most of these moths were gray and blended well with the lichens on trees. Black varieties were known, but rare, since they stood out against the light background and were easily spotted by birds. With industrialization and pollution, the trees became soot-covered and dark, and the frequency of dark-colored moths increased while that of light-colored moths decreased. However, in rural areas unaffected by pollution the light-dark frequency did not change.

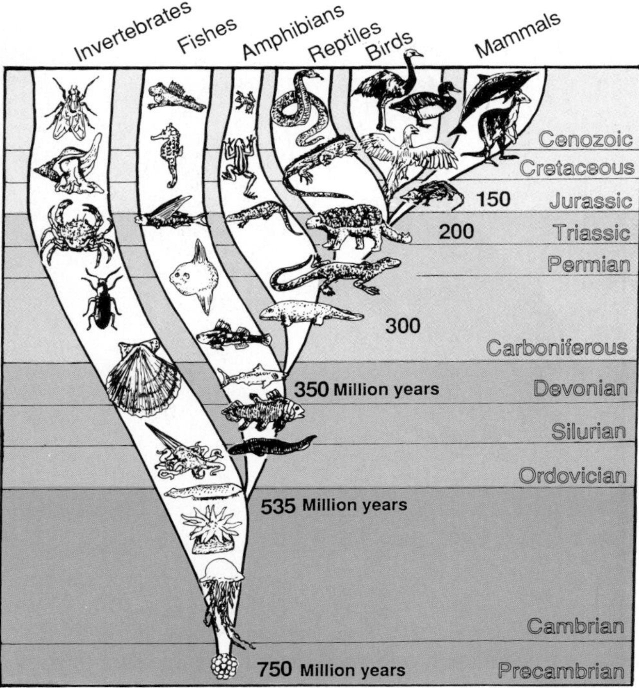


**Figure 5–36** The invertebrates and major groups of vertebrate animals made their appearances in a time sequence that is consistent with predictions of evolutionary theory. Also consistent with evolution but not shown here were the Precambrian appearances of the simplest one-celled organisms followed by more complex cells and multicelled organisms.

## Perspective 5-3

# The Evolutionary Significance of Vestigial Structures

Probably every living species has **vestigial structures**—nonfunctional or partly functional remnants of organs that were functional in their ancestors. Vestigial structures are leftovers in the evolutionary sense, structures that are probably being lost. For example, one trend in human evolution has been shortening of the jaw, which leaves too little room for the third molars, the so-called wisdom teeth. Some people never develop wisdom teeth at all, but in others, these teeth may not erupt through the gums, or may come in crooked and useless. Anyone who has suffered the pain of impacted wisdom teeth can attest to the fact that the human jaw has too little room for these teeth.

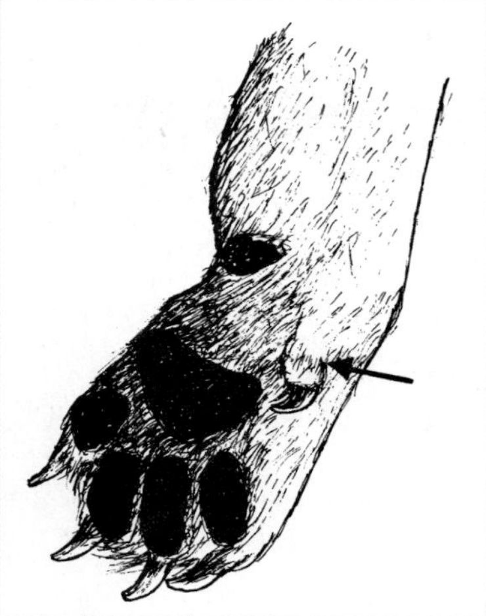

**Figure 1** Left hindfoot of a dog showing the dewclaw, which is a vestige of the big toe.

Many mammals have dewclaws, which are small, functionless, vestiges of fingers and toes. Dewclaws are particularly obvious in dogs (Fig. 1). The ancestors of dogs were rather flat-footed animals with five functional digits on each foot, but as dogs evolved they became toe-walkers, and only four digits contact the ground. Consequently, the first digit of each foot, the thumbs and big toes, lost their function and were reduced in size.

Modern horses have only one functional toe in each foot, the third, which corresponds to the middle finger and middle toe of the human hand and foot. Horses, however, retain remnants of the second and fourth toes as so-called splint-bones, one on each side of the main toe (Fig. 2). These splint-bones are not usually visible because they lie beneath the skin, but a few horses are born with extra toes:

> In rare cases, both fore and hind feet may each have two extra digits fairly developed, and all of nearly equal size, thus corresponding to the feet of the extinct [horse] *Protohippus*[1]

Whales adapted to an aquatic life-style and in the process lost the rear limbs possessed by their land-dwelling ancestors. Yet whales still have a remnant of the pelvic girdle (Fig. 3), and a few whales have been caught that have rear limbs. The

[1]Marsh, April 1982. Recent Polydactyl Hornes. *American Journal of Science*, 43: 342.

may seem to be an odd assortment of animals, but they all share several characteristics that imply they are related by evolutionary descent. If so, one would predict that as each family is traced back in the fossil record, differentiating one from the other would become increasingly difficult. In fact, the earliest members of each family are differentiated from one another by minor differences in size and characteristics of their teeth.

Fossils from the Jurassic Solnhofen Limestone of Germany show the first appearance of features we associate with birds (Fig. 5-38). Several fossil specimens showing feather impressions have been discovered, but in almost every other known physical feature these fossils are more similar to small carnivorous dinosaurs. As a matter of fact, two specimens were classified as dinosaurs until the feather impressions were noticed.

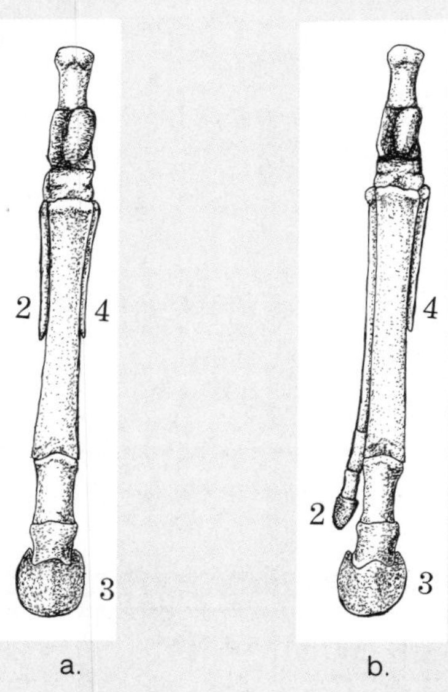

kiwi, a small flightless bird living only in New Zealand, has tiny remnants of the typical bones of a bird wing.

It seems doubtful that anyone would seriously argue that dogs were specially created with dewclaws, or that extra toes in horses really serve some function. Their presence makes sense, however, if they are the remnants of once functional organs. Charles Darwin had this to say:

> They [vestigial organs] have been partially retained by the power of inheritance, and relate to a former state of things.[2]

[2]1958. *The Origin of Species*. New York: New American Library: 419–420.

**Figure 2** a. Normal condition of a horse's back foot; the only functional toe is the third, but side splints representing toes 2 and 4 are still present. b. Modern horse with an extra toe. c. Modern horse with an extra toe on each foot.

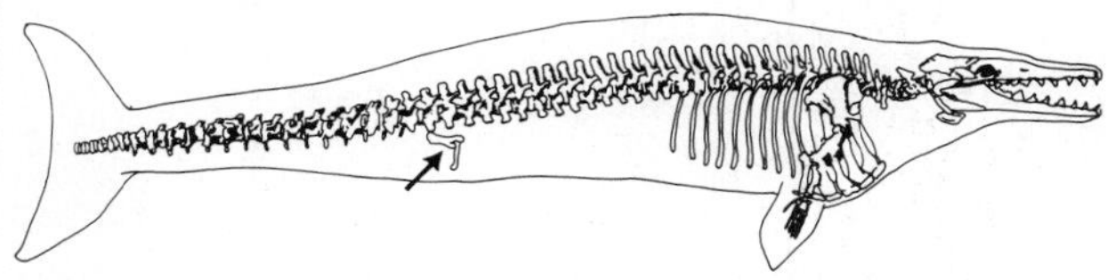

**Figure 3** Vestigial hind limbs in the fossil whale *Zeuglodon* indicate that whales evolved from animals with functional rear limbs. Modern whales retain only a remnant of the pelvis.

This early bird-like creature, *Archaeopteryx* (ancient feather), illustrates the concept of **mosaic evolution**. All organisms are mosaics of characteristics, some of which are retained from the ancestral condition, while others are more recently evolved. *Archaeopteryx*, for example, retains dinosaur-like teeth, tail, hind limb structure, and brain size but also possesses feathers and a wishbone, characteristics typical of birds.

Many people have the idea that the fossil record provides the main evidence for evolution. Fossils are unquestionably important in evolutionary studies, but one must not overlook the fact that fossils are only one of several lines of evidence that support the concept of evolution. Shared characteristics, biochemistry, comparative anatomy, and small-scale evolution of modern organisms are equally important.

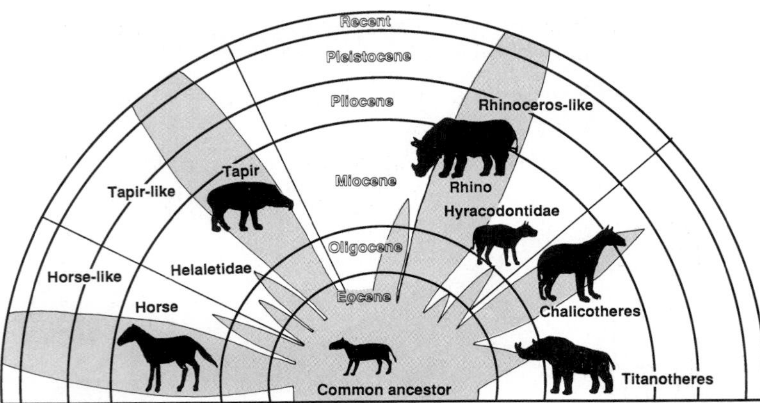


**Figure 5–37** The divergence and adaptive radiation of horses, rhinoceroses, and tapirs, and their extinct relatives, is well documented by fossils. The evolution of these animals is discussed in more detail in Chapter 16.

**Figure 5–38** *Archaeopteryx*, a Jurassic-age fossil from Germany, has feathers and a wishbone, so is classified as a bird. However, in almost every other anatomical feature *Archaeopteryx* is most similar to small carnivorous dinosaurs.

## CHAPTER SUMMARY

1. The first formal theory of evolution to be taken seriously was proposed by Jean Baptiste de Lamarck. Inheritance of acquired characteristics was his mechanism for evolution.
2. In 1859 Charles Darwin and Alfred Russell Wallace simultaneously published their views on evolution. They proposed natural selection as the mechanism for evolutionary change.
3. Darwin's observations of variation in natural populations and artificial selection, and his reading of Thomas Malthus's essay on population, helped him formulate the idea that natural processes select favorable variants for survival.
4. Gregor Mendel's breeding experiments with garden peas provided some of the answers regarding how variation is maintained and passed on. Mendel's work is the basis for modern genetics.
5. Genes are the hereditary determinants in all organisms. This genetic information is carried in the chromosomes of cells. Only those genes in the chromosomes of eggs and sperm are inheritable.
6. Most variation is maintained in populations by sexual reproduction and mutations.
7. Evolution by natural selection is a two-step process. First, there must be the production and maintenance of variation in interbreeding populations, and second, favorable variants must be selected for survival.

8. An important way in which new species evolve is by allopatric speciation. When a group is isolated from its parent population, gene flow is restricted or eliminated, and the isolated group is subjected to different selection pressure.
9. Divergent evolution involves an adaptive radiation during which an ancestral stock gives rise to diverse species. The development of similar adaptive types in different groups of organisms results from parallel and convergent evolution.
10. The evidence for evolution includes the way organisms are classified, comparative anatomy and embryology, biochemical similarities, small-scale evolution, and the fossil record.

## IMPORTANT TERMS

adaptive radiation
allele
allopatric speciation
analogous organ
artificial selection
chromosomes
convergent evolution
divergent evolution
DNA (deoxyribonucleic acid)
gene
gene pool
homologous organ
living fossil
inheritance of acquired characteristics
meiosis
mitosis
modern synthesis
morphology
mosaic evolution
mutation
natural selection
parallel evolution
phyletic gradualism
punctuated equilibrium
species
theory of evolution
vestigial structure

## REVIEW QUESTIONS

1. Compare and contrast evolution by inheritance of acquired characteristics with evolution by natural selection.
2. How does sexual reproduction maintain variation in interbreeding populations?
3. What observations led Darwin to the conclusion that natural selection is a mechanism for evolution?
4. What is a mutation? Explain what is meant by harmful, neutral, and beneficial, with respect to mutations.
5. Explain the process whereby an organism accidentally introduced to a remote island may give rise to a new species.
6. Give examples of divergent and convergent evolution.
7. How does the way in which organisms are classified provide evidence for evolution?
8. Explain how the concept of punctuated equilibrium accounts for the origin of new species. Compare this with phyletic gradualism.

## ADDITIONAL READING

Ayala, F. J. 1978. The Mechanisms of Evolution. *Scientific American*, 239, no. 3: 56–69.

Bajema, C. J. 1985. Charles Darwin and Selection as a Cause of Adaptive Evolution 1837–1859. *The American Biology Teacher*, 47: 226–232.

Bingham, R. 1982. On the Life of Mr. Darwin. *Science, 82*, no. 3: 34–39.

Dobzhansky, T., F. J. Ayala, G. L. Stebbins, and J. W. Valentine. 1977. *Evolution*. San Francisco: W. H. Freeman and Company.

Holton, N. II. 1968. *The Evidence of Evolution*. New York: American Heritage Publishing Co., Inc.

Mayr, E. 1978. Evolution. *Scientific American*, 239, no. 3: 47–55.

# 6

## CHAPTER OUTLINE



## PROLOGUE

*When Maurice Ewing first made a series of seismic measurements of the continental shelf off Norfolk, Virginia, in 1935, virtually nothing was known about the deep-ocean floor. Analysis of his seismograms indicated that the continental shelf was not a permanent feature of the continent but was composed of sediment up to 4,000 m thick that had been deposited on the ocean-floor bedrock. Since these thick sediments probably contained oil deposits, he tried to interest oil companies in supporting further studies of the continental shelf. However, he was told that oil was so easily accessible on land that there was no reason to look for it under the sea. Undiscouraged, Ewing pursued his research of the ocean floor and made many important discoveries.*

*In 1947, Ewing was commissioned by the National Geographic Society to explore the Mid-Atlantic Ridge and the sea floor around it. Little was known about this undersea area, and Ewing incorporated explosion seismic and depth sounding techniques in his studies as well as equipment for sampling seawater, determining water temperature at various depths, and for coring the ocean floor itself. His initial core samples and seismic tests produced surprising results. Instead of a thick layer of ocean sediments that had been accumulating for some 3 billion years, his data indicated a thin layer of sediments (the thickest only several hundred meters thick) representing 100 to 200 million years of sedimentation. Furthermore, after dredging across the slopes of the Mid-Atlantic Ridge, Ewing hauled up pieces of* pillow lava, *a type of lava that is formed when lava is extruded into water. Not only was the ocean floor covered by a thin layer of sediments, but the floor itself was of geologically recent volcanic origin.*

*Following two more expeditions to the Mid-Atlantic Ridge in 1948, Ewing founded the Columbia Lamont Geological Observatory in 1949, with the main mission of studying the ocean floor. One of the early discoveries was that the oceanic crust was composed of dense basalt and not sunken continental material. Furthermore, the oceanic crust was only 5 to 10 km thick, compared to continental crust that is 20 to 70 km thick.*

# Plate Tectonics—A Unifying Theory

*In the early 1950s, Ewing decided that it was time to transfer all of the available seismic profiles of the North Atlantic Ocean floor onto a topographic map. He assigned the job to Bruce Heezen, a graduate student who enlisted the help of Marie Tharp, a cartographer at the observatory. As the profiles were converted into a map, both Heezen and Tharp were surprised to see a deep canyon (or rift valley) that ran down the center of the Mid-Atlantic Ridge. While initially not believing such a large-scale rift existed, Heezen and Ewing began plotting the locations of all mid-ocean earthquakes they had data for. What emerged was a band of earthquakes running through not only the middle of the rift valleys mapped by Tharp, but through all of the world's oceans.*

*In 1959 Ewing, Heezen, and Tharp published their description of the North Atlantic ocean floor and a stunning three-dimensional physiographic map of the North Atlantic Ocean. The map revealed vast plains and conical plateaus called* sea mounts, *as well as the Mid-Atlantic Ridge, with its still mysterious rift valley. As more of the world's ocean floors were explored, this original regional map was expanded, revealing a mountain chain 25,000 km long that ran through all the world's oceans.*

*Hailed as one of the most exciting and important discoveries in earth science in the last 20 years, the recognition of a curving ridge located halfway between and parallel to the coasts of South America and Africa forced earth scientists to question the traditional theories about the Earth. As we will see in this chapter, the realization that new crust was forming along the rift valley of the Mid-Atlantic Ridge hastened the acceptance of sea-floor spreading and the theory of plate tectonics.*

## INTRODUCTION

The decade of the 1960s was a time of social and cultural as well as geological revolution. The ramifications of the newly proposed plate tectonic theory radically changed the way in which geologists viewed the Earth. No longer could it be thought of as an unchanging planet on which continents and ocean basins remained fixed through time. Instead, it was perceived as a dynamically changing planet whose surface consists of constantly moving plates.

It is not an exaggeration to say that plate tectonics has been the dominant process affecting the evolution of the Earth. The interaction of plates determines the locations of continents, ocean basins, and mountain chains, which in turn affect atmospheric and oceanic circulation patterns that ultimately determine global climates. Furthermore, the physical changes resulting from plate interactions have profoundly influenced the evolution of the Earth's biota.

Geologists now view Earth history in terms of interrelated events that are part of a global panorama of dynamic change through time. For example, the Paleozoic history of the Appalachian Mountains is no longer considered an isolated regional event but rather part of a global interaction of plates that culminated in the creation of a large supercontinent at the end of the Paleozoic.

This chapter provides an overview of plate tectonic theory as it is used to interpret the history of the Earth from the rock and fossil record. This powerful and unifying theory accounts for many seemingly unrelated geologic phenomena, allowing them to be viewed as part of a continuing story rather than as a series of isolated events.

## EARLY OPINIONS ABOUT CONTINENTAL DRIFT

The earliest maps showing the east coast of South America and the west coast of Africa probably provided people with the first evidence that continents may have once been joined together, then broken apart and moved to their present positions. It was not until 1858, however, that the Frenchman Antonio Snider-Pellegrini suggested in his book, *La Création et ses mystères devoilés*, that all the continents were assembled together during the Pennsylvanian Period and then later split apart. He based his hypothesis on the fact that similar plant fossils occur in coal measures in Europe and North America. Snider-Pellegrini envisioned the rupturing of this single continent as the result of the Great Flood. Following this first suggestion of continental drift, another Frenchman, Elisee Reclus, discussed the idea of continental movement in his book, *The Earth*, published in 1872. Reclus believed continental drift, not the Great Flood, was responsible for mountain building, volcanic island arcs, and earthquake activity.

It was also during this time that the Austrian geologist Edward Suess noted the similarity between the Late Paleozoic plant fossils of India, Australia, South Africa, and South America as well as evidence of glaciation in the rock sequences of these southern continents. The plant fossils comprise a unique flora that occurs in the coal layers just above the glacial tillites of these southern continents. This flora is very different from the contemporaneous coal swamp flora of the northern continents and is collectively known as the ***Glossopteris*** flora after its most conspicuous genus (Fig. 6–1).

In his book, *The Face of the Earth*, published in 1885, Suess coined the term *Gondwanaland* (or **Gondwana**) for the supercontinent comprising the aforementioned southern continents. Gondwana is a locality in India where abundant fossils of the *Glossopteris* flora are found in coal seams. He believed these continents were connected by land bridges over which plants and animals migrated. In this way, Suess explained the similar fossils of the continents as due to the appearance and disappearance of the land bridges connecting them.

In 1908 the American geologist Frank B. Taylor privately published a pamphlet in which he presented his own theory of continental drift. In it he envisioned the present-day continents as parts of larger polar continents that eventually broke apart and migrated toward the equator. He explained the breakup of these polar continents as the result of gigantic tidal forces. Taylor believed that these giant tides were the result of the moon becoming a satellite of the Earth about 100 million years ago (although we now know that this is incorrect), and thereby slowing the Earth's rotational rate. One of Taylor's most significant contributions was his suggestion that the submarine ridge discovered in the Atlantic Ocean by the *Challenger* expeditions of 1872–1876 might mark the site along which one ancient continent split apart to form the present-day Atlantic Ocean.

### Alfred Wegener and Continental Drift

Alfred Wegener (Fig. 6–2) is generally credited with developing the theory of **continental drift**. In his monumental book, *The Origin of Continents and Oceans*, first published in 1915, he portrayed his grand concept of

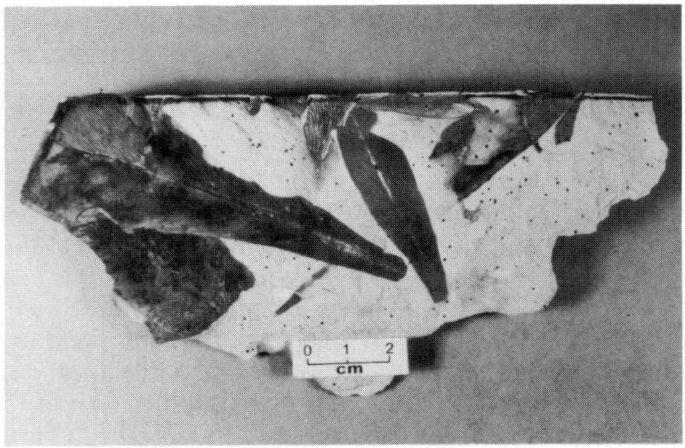


a.

b.

**Figure 6–1** Representative members of the *Glossopteris* flora. Fossils of these plants are found on the continents that make up Gondwana. *Glossopteris* leaves from the Late Permian Dunedoo Formation (a), and the Late Permian Illawarra Coal Measures, Australia (b). (Photos courtesy of Patricia G. Gensel, University of North Carolina.)

**Figure 6–2** Alfred Wegener proposed the theory of continental drift based on geological, paleontological, and climatological evidence. He is shown here waiting out the Arctic winter in an expedition hut.

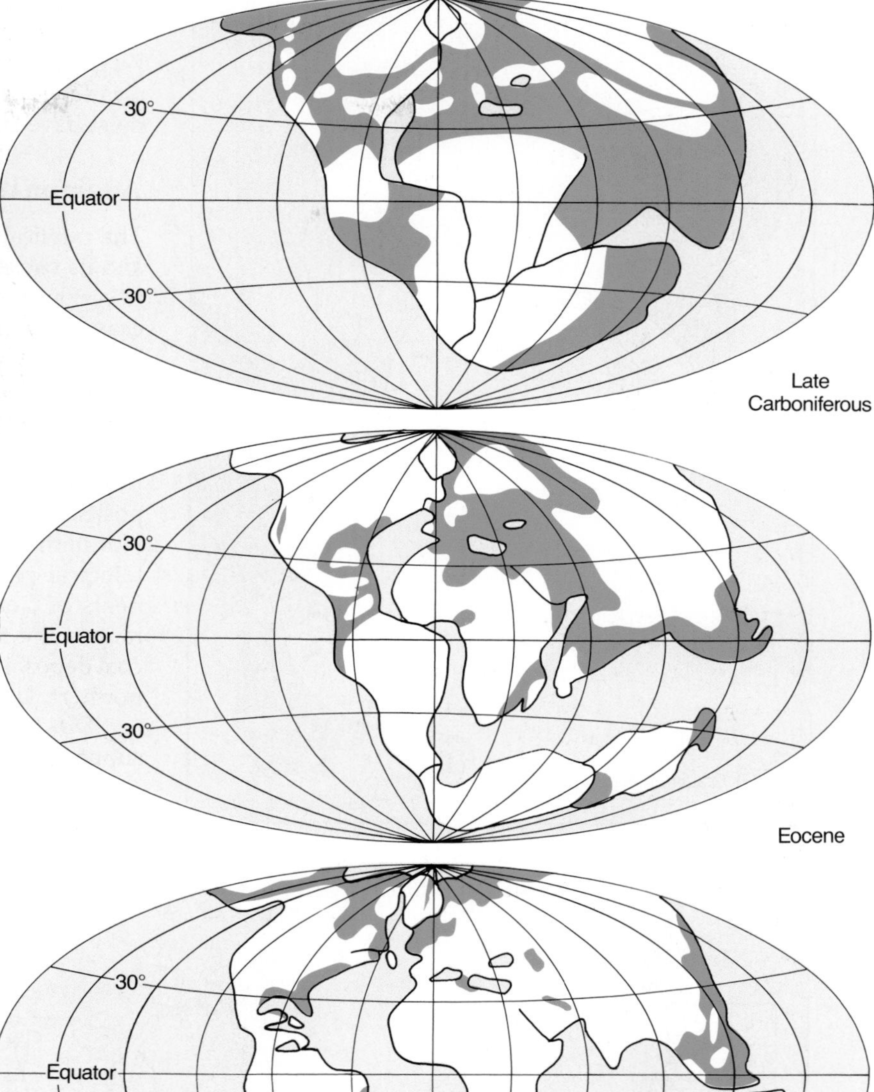


**Figure 6–3** Location of continents in the geologic past as envisioned by Wegener in his book, *The Origin of Continents and Oceans*, first published in 1915. Africa is placed in its present position as a point of reference. The outlines of the continents are for identification purposes. Brown shading represents shallow seas. (After A. Wegener, Die Eutstehung der Kontinents und Ozeano, Friedrich Vieweg und Sohn, Brunswick, Germany, 1915).

continental movement in a series of maps (Fig. 6–3). He proposed the name **Pangaea** for the supercontinent that resulted from the unification of the present continents into a single landmass. He also stated that Pangaea began breaking apart during the Cenozoic Era. We now know that this fragmentation began during the Triassic Period.

Wegener amassed a tremendous amount of geological, paleontological, and climatological evidence in support of continental drift. He noted that marine and nonmarine rock sequences of the same age are found on widely separated continents; mountain ranges and glacial deposits match up when continents are united into a single landmass; the shorelines of continents fit together, forming a large supercontinent; many of the same extinct plant and animal groups are found today on widely separated continents, indicating the continents must have been in close proximity at one time;

and the rock record indicates that the global climatic belts were not always parallel to the equator (Fig. 6–4). Wegener argued that this vast amount of evidence from a variety of sources surely indicated that the continents must have been close together at one time in the past.

## Additional Support for Continental Drift

The publication of *The Origin of Continents and Oceans* and its subsequent four editions caused some scientists to seriously consider Wegener's unorthodox views. Alexander du Toit, a South African geologist, was one of his more ardent supporters. He further developed Wegener's arguments and introduced more geologic evidence in support of continental drift. In 1937 du Toit published *Our Wandering Continents,* in which he contrasted the glacial deposits of Gondwana with coal deposits of the same age found in the continents of the Northern Hemisphere. To resolve this apparent climatological paradox, du Toit placed the southern continents of Gondwana at or near the South Pole and arranged the northern continents together such that the coal deposits were located at the equator. He named this northern landmass **Laurasia**.

Du Toit also provided additional paleontological support for continental drift by noting that fossils of the

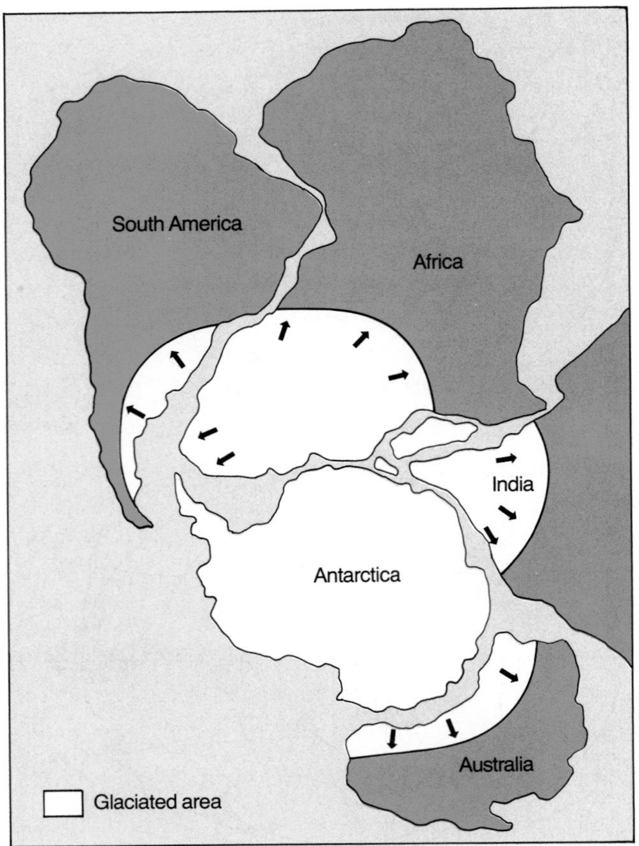


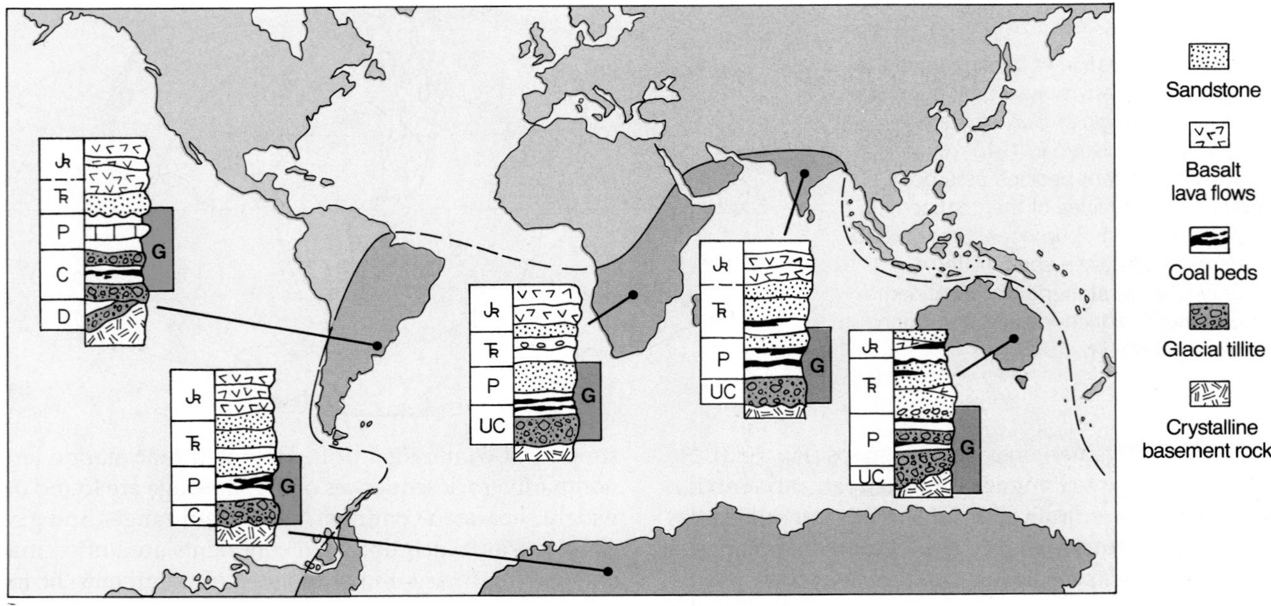


**Figure 6–4** Two of the various lines of evidence for continental drift. a. Reconstruction of Gondwana showing area of glaciation and directions of ice movement. b. Similarities of rock sequences of the same age from the Gondwana continents. The range indicated by G is that of the *Glossopteris* flora.

Permian freshwater reptile *Mesosaurus* occurred in rocks of the same age in both Brazil and South Africa (Fig. 6–5). Since the physiology of freshwater and marine animals is completely different, most paleontologists find it hard to imagine how a freshwater reptile could have swum across the Atlantic Ocean and then found a freshwater environment nearly identical to its former habitat. It is more logical to assume that *Mesosaurus* occupied a large lake that covered portions of both continents when they were united.

In spite of what seemed to be overwhelming evidence presented by Wegener and later by du Toit and others, most geologists simply refused to accept the idea that continents might have moved in the past. It was not necessarily a question of being obstinate to new ideas, but rather that the proposed mechanisms for continental drift were inadequate and unconvincing.

## PALEOMAGNETISM AND POLAR WANDERING

Interest in continental drift revived in the 1950s as a result of new evidence from paleomagnetic studies of the Earth. **Paleomagnetism** is simply the remnant magnetism in ancient rocks recording the direction of the Earth's magnetic poles at the time of the rock's formation. The Earth can be thought of as a giant dipole magnet in which the magnetic poles essentially coincide with the geographic poles (Fig. 6–6). Such an arrangement means that the strength of the magnetic field is not constant, but varies, being weakest at the equator and strongest at the poles. The Earth's magnetic field is believed to result from the rotation of the Earth around its liquid outer core.

When a magma cools, the iron-bearing minerals align themselves with the Earth's magnetic field, recording both its direction and strength. The temperature at which iron-bearing minerals gain their magnetization is called the **Curie point**. As long as the rock is not subsequently heated above the Curie point, it will preserve that remnant magnetism. Thus, an ancient lava flow provides a record of the orientation and strength of the Earth's magnetic field at the time the lava flow cooled.

As paleomagnetic research in the 1950s progressed, some unexpected results emerged. When geologists measured the paleomagnetism of recent rocks, they found it was consistent with the Earth's current mag-

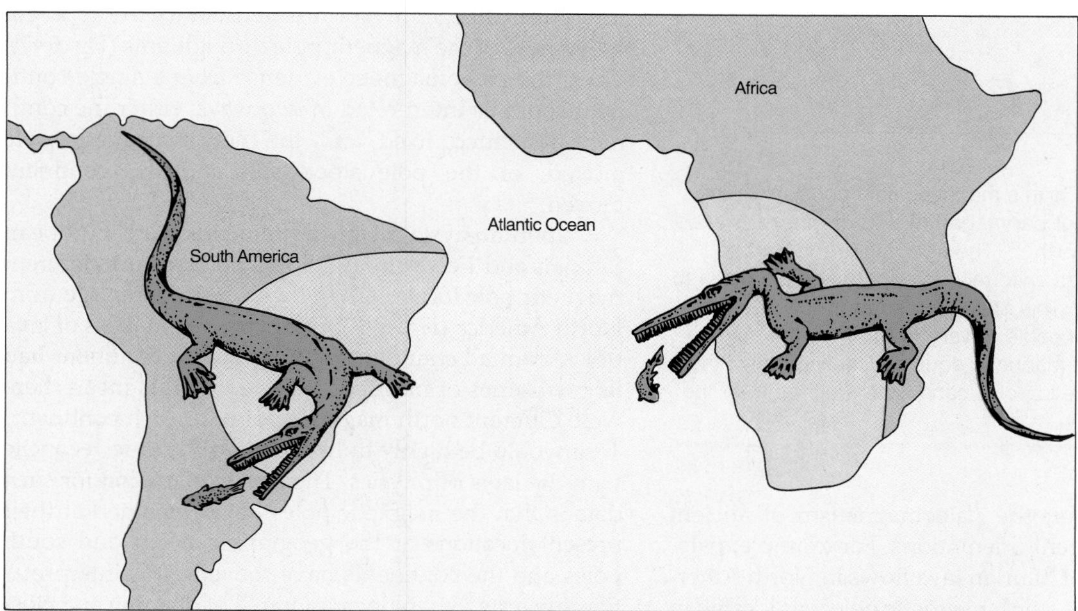


**Figure 6–5** Fossils of the Permian freshwater reptile *Mesosaurus* are found on both sides of the South Atlantic and nowhere else in the world. Such evidence supports the theory that continents have moved in the past.

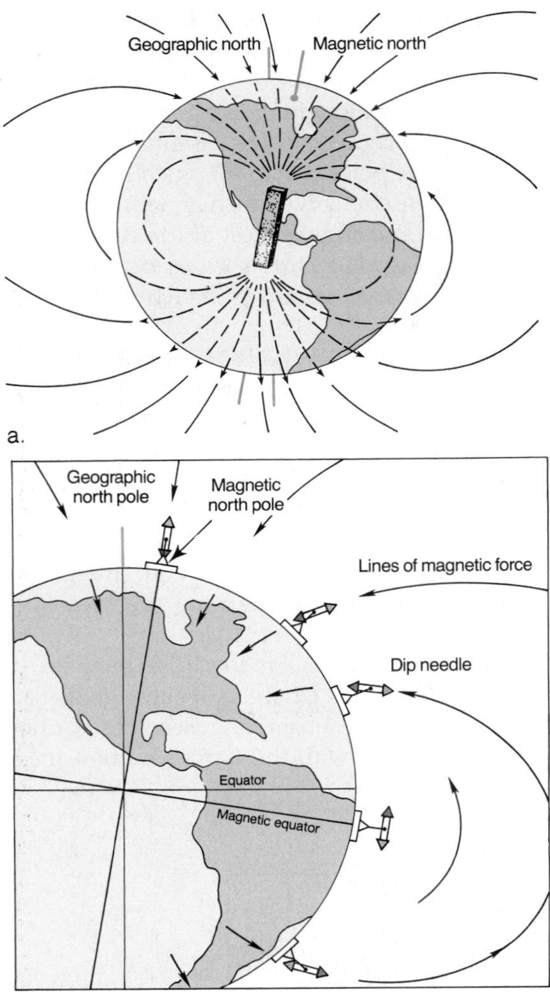


**Figure 6–6** a. The Earth's magnetic field consists of lines of force just like a giant bar magnet would produce if placed at the center of the Earth. b. The strength of the Earth's magnetic field changes uniformly from the magnetic poles to the magnetic equator. This change in strength causes the angle of dip on a dip needle to vary from 90° at the magnetic pole to 0° at the magnetic equator. Consequently, the distance to the magnetic poles can be determined from the dip angle.

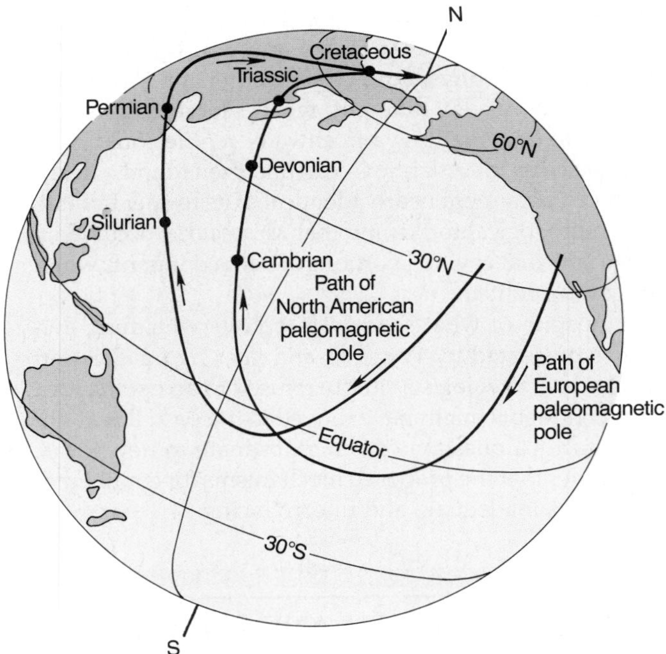


**Figure 6–7** The apparent paths of polar wandering for North America and Europe. The apparent location of the north magnetic pole is shown for different periods on each continent's polar wandering path.

netic field. However, the paleomagnetism of ancient rocks showed different orientations. For example, paleomagnetic studies of Silurian lava flows in North America indicated that the north magnetic pole was located in the western Pacific Ocean at that time, while the paleomagnetic evidence from Permian lava flows indicated a pole in Asia, and that of Cretaceous lava flows pointed to yet another location in northern Asia. When plotted on a map, the paleomagnetic readings of numerous lava flows from all ages in North America trace the apparent movement of the magnetic pole through time (Fig. 6–7). Thus, the paleomagnetic evidence from a single continent could be interpreted in two ways: either the continent remained fixed and the north magnetic pole moved, or the pole stood still and the continent moved.

Upon analysis, magnetic minerals from European Silurian and Permian lava flows pointed to a different magnetic pole location than those of the same age from North America (Fig. 6–7). Furthermore, analysis of lava flows from all continents indicated each continent had its own series of magnetic poles. Does this mean there were different north magnetic poles for each continent? That would be highly unlikely and difficult to reconcile with the laws of physics. The best explanation for such data is that the magnetic poles have remained at their present locations at the geographic north and south poles and the continents have moved. This interpretation suggests a changing panorama of opening and closing ocean basins with mountain ranges forming along the margins of colliding continents. Furthermore, when the continental margins are fitted together so that the paleomagnetic data points to only one magnetic pole, we find, just as Wegener did, that the rock sequences and glacial deposits match up, and the fossil evidence is consistent with the reconstructed paleogeography.

## MAGNETIC REVERSALS AND SEA-FLOOR SPREADING

Geologists refer to the Earth's present magnetic field as being normal, that is, with the north and south magnetic poles located roughly at the north and south geographic poles. At numerous times in the geologic past, the Earth's magnetic field has completely reversed. The existence of such **magnetic reversals** was discovered by dating and determining the orientation of the remnant magnetism in lava flows on land (Fig. 6–8). Once their existence was well established, magnetic reversals were discovered as well in ocean basalts as part of the extensive mapping of the ocean basins that took place in the 1950s. While the cause of magnetic reversals is still unknown, their occurrence in the geologic record is well documented.

In addition to the discovery of magnetic reversals, mapping of the ocean basins also revealed a 70,000 km long ridge system constituting the most extensive mountain range in the world. Perhaps the best known part of the ridge system is the Mid-Atlantic Ridge, which divides the Atlantic Ocean basin into two nearly equal parts (Fig. 6–9).

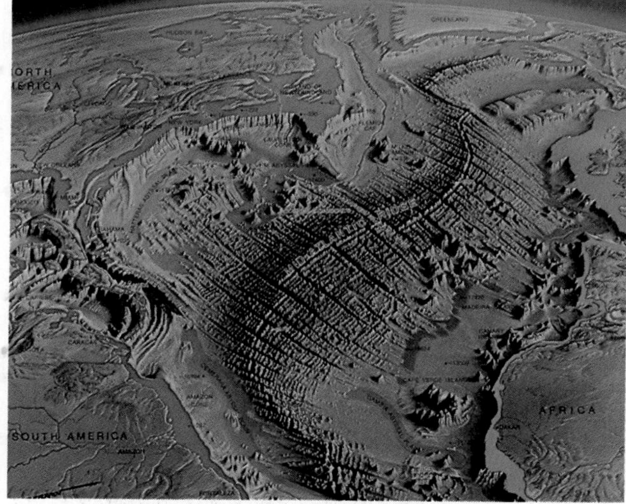


**Figure 6–9** Artistic view of what the Atlantic Ocean basin would look like if all the water were removed. The major feature is the Mid-Atlantic Ridge.

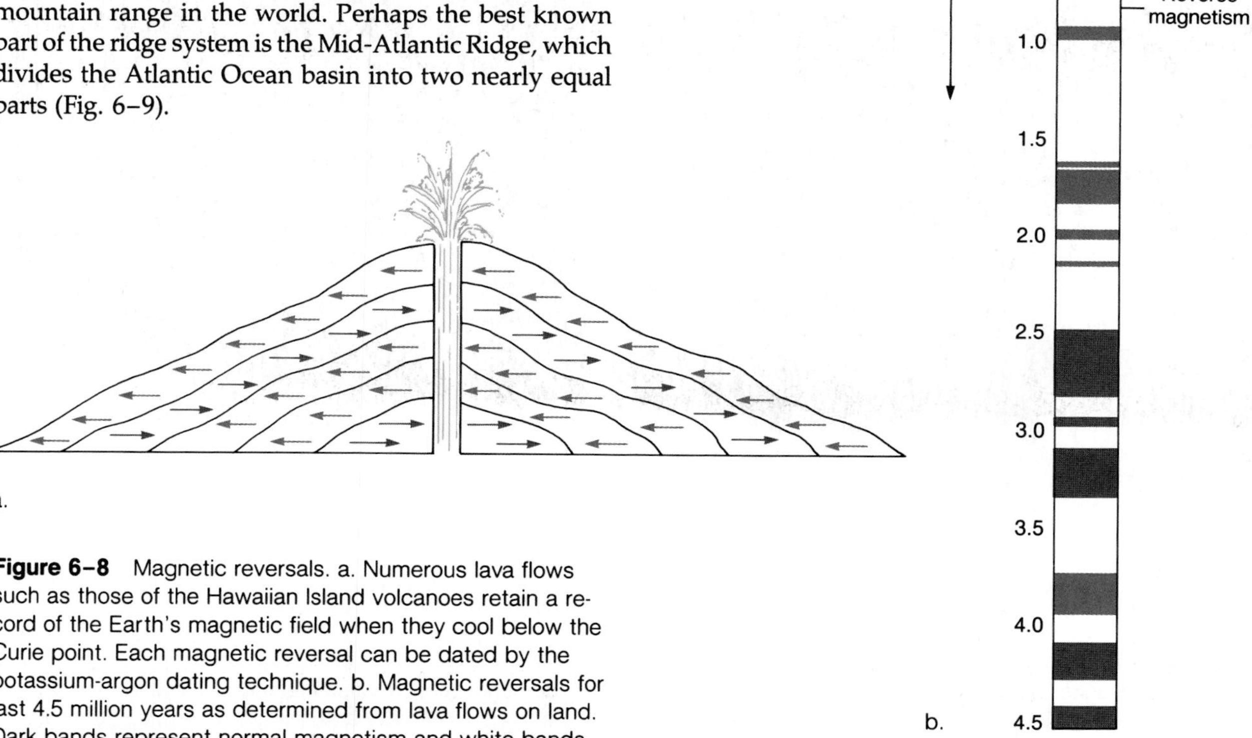


**Figure 6–8** Magnetic reversals. a. Numerous lava flows such as those of the Hawaiian Island volcanoes retain a record of the Earth's magnetic field when they cool below the Curie point. Each magnetic reversal can be dated by the potassium-argon dating technique. b. Magnetic reversals for last 4.5 million years as determined from lava flows on land. Dark bands represent normal magnetism and white bands represent reverse magnetism.

As a result of the oceanographic research conducted in the 1950s, Harry Hess of Princeton University in 1962 proposed the theory of **sea-floor spreading** to account for continental movement. Hess suggested that continents do not move across oceanic crust, but rather the continents and oceanic crust move together. He suggested that sea floors separate at oceanic ridges where new crust is created by upwelling magma. As the magma cools and forms oceanic crust, it moves laterally away from the ridges. As a mechanism to drive this system, Hess revived the idea of **thermal convection cells** in the mantle. That is, hot magma rises from the mantle and spills out at the oceanic ridges. Cold crust is subducted back into the mantle at deep-sea trenches, where it is heated and recycled as part of a thermal convection cell.

How could Hess's theory be confirmed? Magnetic surveys of the oceanic crust revealed striped **magnetic anomalies** (deviations from the average strength of the Earth's magnetic field) in the rocks that were both parallel to and symmetrical with the ridges (Fig. 6–10). Furthermore, the pattern of oceanic magnetic anomalies matched the pattern of magnetic reversals already known from studies of continental lava flows (Fig. 6–8). When magma wells up and cools along a ridge summit, it records the Earth's magnetic field at that time as either normal or reversed. As new crust forms at the summit, the previously formed crust moves away from the ridge. The fact that these magnetic stripes, representing times of normal or reversed polarity, are parallel to and symmetrical around oceanic ridges (where upwelling magma creates new oceanic crust) conclusively confirmed Hess's theory of sea-floor spreading.

One of the consequences of the sea-floor spreading theory is its confirmation that ocean basins are geologically young features whose openings and closings are partially responsible for continental movement (Fig. 6–11). Radiometric dating reveals that the oldest oce-

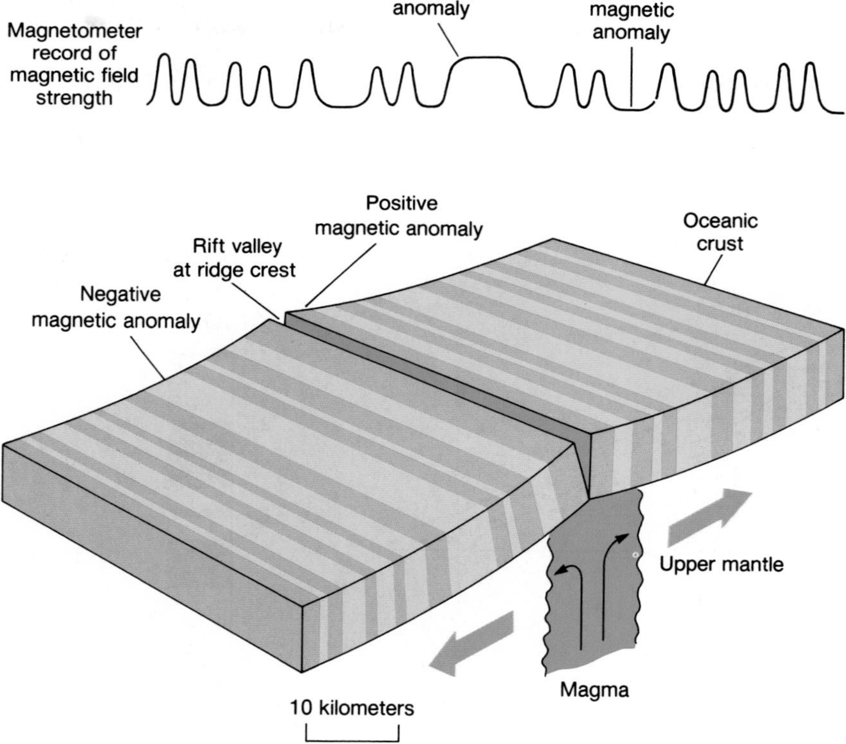


**Figure 6–10** Marine magnetic anomalies. The line shows the positive and negative magnetic anomalies as recorded by a magnetometer. In the cross section of the oceanic crust, the positive anomalies are shown as orange bars and the negative anomalies are shown as yellow bars. The magnetic anomalies are both parallel to and symmetrical with the ridge and match the pattern of magnetic reversals already known from continental lava flows.

anic crust is less than 180 million years old, while the oldest continental crust is 3.8 billion years old. Although geologists do not universally accept the idea of thermal convection cells as a driving mechanism for plate movement, most accept that plates are created at oceanic ridges and destroyed at deep-sea trenches, regardless of the driving mechanism involved.

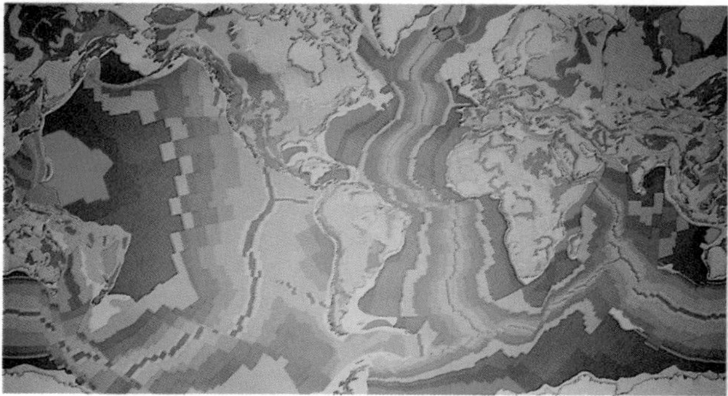

**Figure 6–11** The age of the ocean basins as determined from magnetic anomalies.

## PLATE TECTONICS AND PLATE BOUNDARIES

According to the theory of **plate tectonics**, the Earth's surface is divided into rigid plates bounded by mid-oceanic ridges, oceanic trenches, faults, and mountain belts (Fig. 6–12). Most geologists by the late 1960s accepted the theory of plate tectonics, in part because the evidence was overwhelming, but also because it is a unifying theory that can explain many seemingly unrelated geologic phenomena.

Since it is believed that plate tectonics has operated since at least the Proterozoic (Chapter 9), it is important that we understand how plates move and interact with each other and how ancient plate boundaries are recognized. After all, the movement of plates has had a profound effect on the geologic and biological history of this planet.

Geologists recognize three major types of plate boundaries: divergent, convergent, and transform faults. It is along these boundaries that new plates are generated, consumed, or slide past one another. To un-

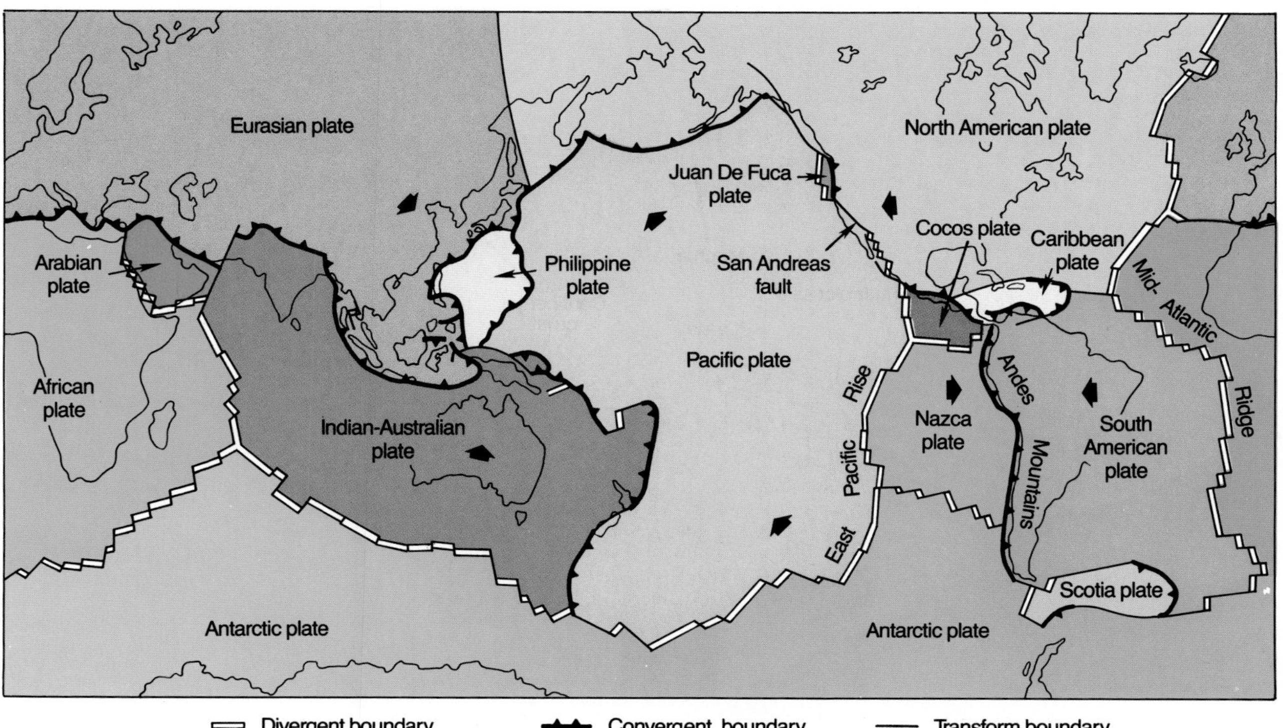


**Figure 6–12** Map of the world showing the major plates, their boundaries, and direction of movement.

derstand the implications of plate interactions as they have affected the history of the Earth, geologists must study modern plate boundaries.

## Divergent Plate Boundaries

**Divergent plate boundaries** occur where plates are separating and new crust is forming. These boundaries most commonly occur along oceanic ridges or, less commonly, beneath continents where the boundary can be recognized by rift valleys. Along a divergent plate boundary at an oceanic ridge, magma wells up and the plates move apart. As the magma cools, new strips of oceanic crust are formed and record the magnetic field of the Earth at that time (Fig. 6–10). A rugged topography, high topographic relief, normal faulting with many associated shallow-focus earthquakes, high heat flow, and basaltic pillow lavas (Fig. 6–13) are common features associated with these oceanic ridges.

**Figure 6–13** Pillow lavas forming along the Mid-Atlantic Ridge. Their distinctive bulbous shape is the result of underwater eruption.

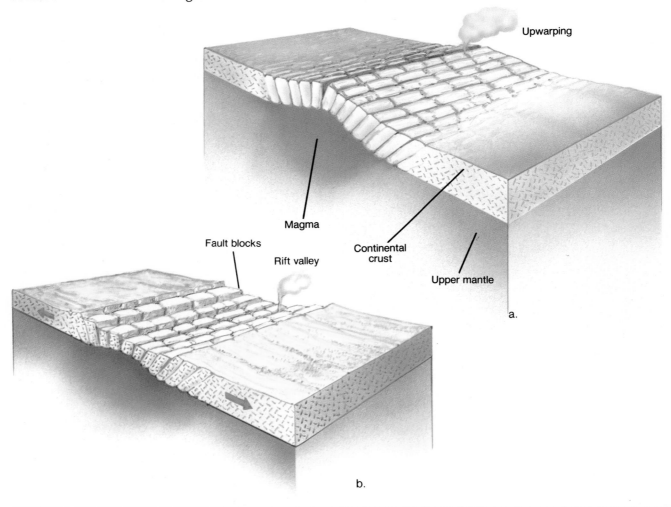

Divergent plate boundaries also occur under continents during the early stages of rifting (Fig. 6–14). As a result of magma welling up beneath it, continental crust is initially elevated, stretched, and thinned. Normal faults and a rift valley then begin to form along a central *graben* (the down-dropped fault block) producing shallow-focus earthquakes. During this stage, magma typically intrudes into fractures, forming sills and dikes, as well as flowing onto the graben floor. An example of this stage is the East African Rift (Fig. 6–15). As rifting proceeds, the continental crust eventually breaks. If magma continues to well up, the two parts of the continent will move away from each other, as is happening today beneath the Red Sea. As this newly created narrow ocean basin continues to enlarge, it will eventually become an expansive ocean basin such as the Atlantic or Pacific Ocean basins are today (Fig. 6–14d).

It should be pointed out that when rifts develop, they often form three arms that meet at a junction. Such a feature is called a **triple junction**. It is not uncommon for all three arms of a triple junction to develop into segments of plate boundaries. Sometimes one of the arms fails to develop, however. A good example of a triple junction and one that illustrates the model of continental rifting in progress is the East African Rift in the Red Sea and Gulf of Aden region (Fig. 6–15). The Red Sea and Gulf of Aden represent newly formed ocean basins, while the East African Rift is an arm in which the continental crust has not yet broken. Whether it will break or become a failed arm remains to be seen.

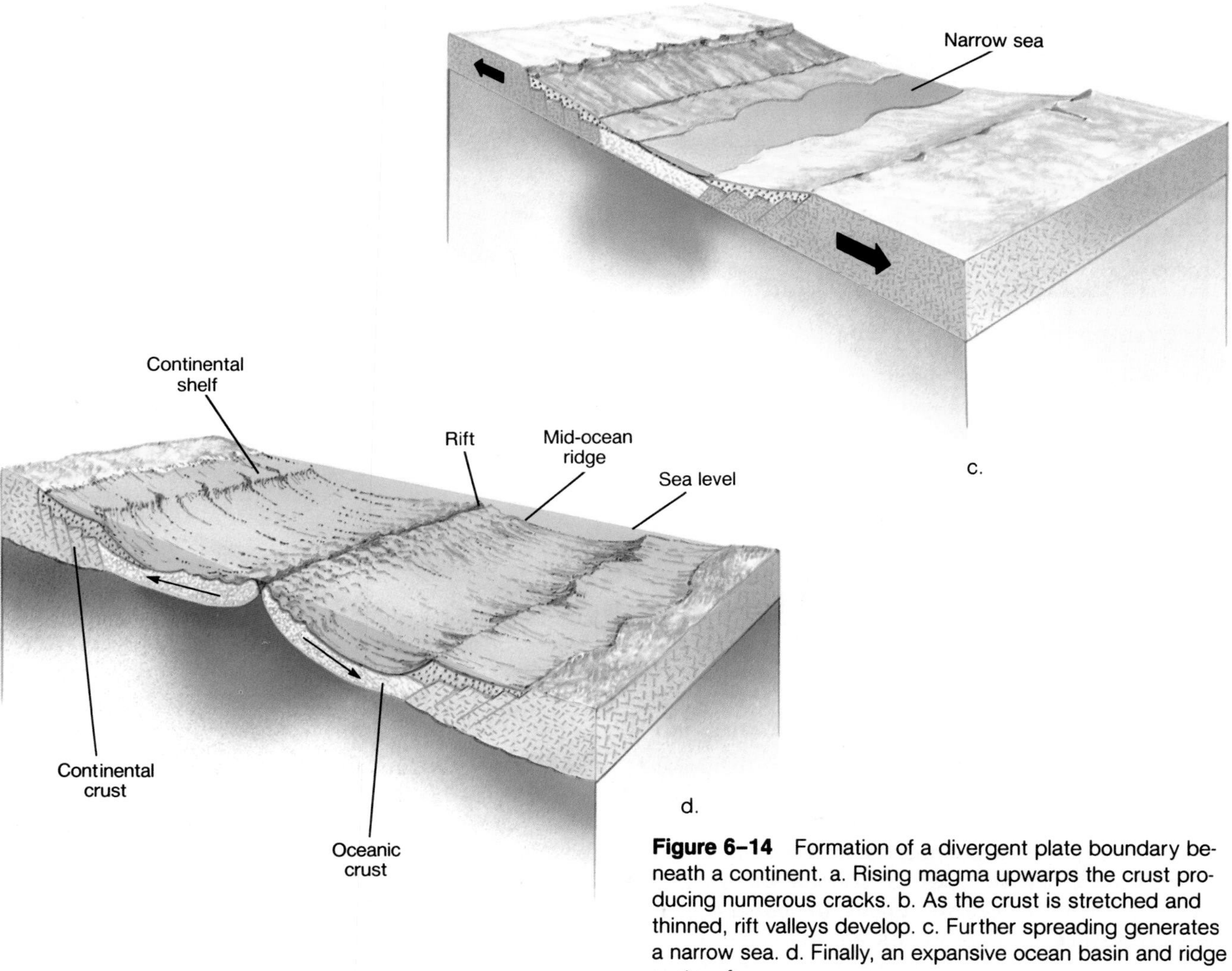


**Figure 6–14** Formation of a divergent plate boundary beneath a continent. a. Rising magma upwarps the crust producing numerous cracks. b. As the crust is stretched and thinned, rift valleys develop. c. Further spreading generates a narrow sea. d. Finally, an expansive ocean basin and ridge system form.

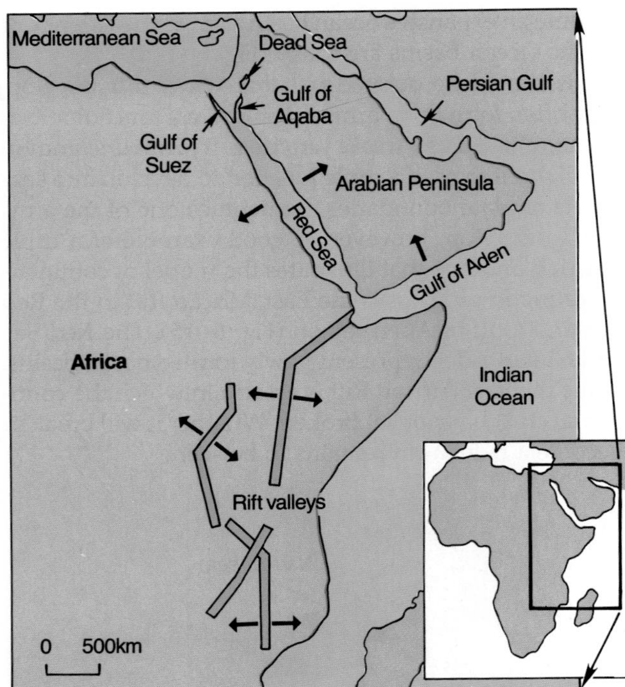


**Figure 6–15** East African Rift valley and associated features. Arrows show direction of movement of crust. The East African rift valleys, Red Sea, and Gulf of Aden form a triple junction.

## An Example of Ancient Rifting

What features in the rock record can geologists use to recognize ancient rifting? Associated with regions of continental rifting are normal faults, dikes, sills, lava flows, and thick sedimentary sequences within the rift valleys. A good example of ancient continental rifting is the Triassic fault basins of the eastern United States (Fig. 6–16). These fault basins mark the zone of rifting that occurred when North America split apart from Africa. They contain thousands of meters of continental sediment and are riddled with dikes and sills.

## Convergent Plate Boundaries

Whereas new crust is created at divergent plate boundaries, old crust is destroyed at **convergent plate boundaries.** One plate is subducted under another and eventually is resorbed in the asthenosphere. When we talk about convergent plate boundaries, we are really talking about three different types of boundaries: oceanic-oceanic, oceanic-continental, and continental-continental. The basic processes are the same for all three types of boundaries, but because of the different types of crust involved, the results are different.

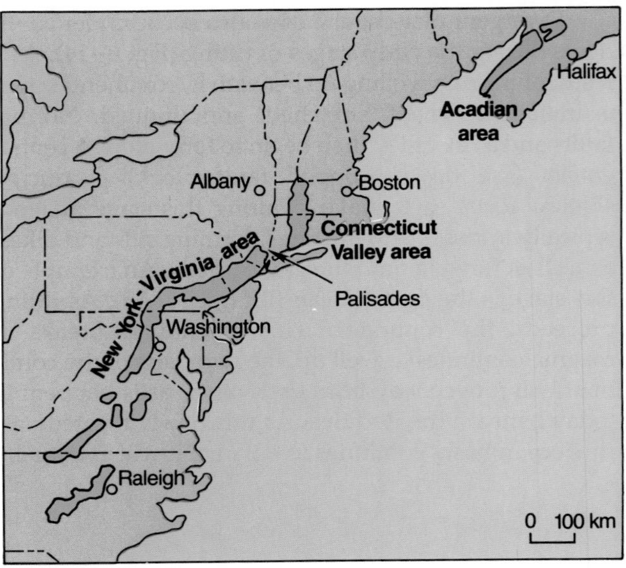


a.

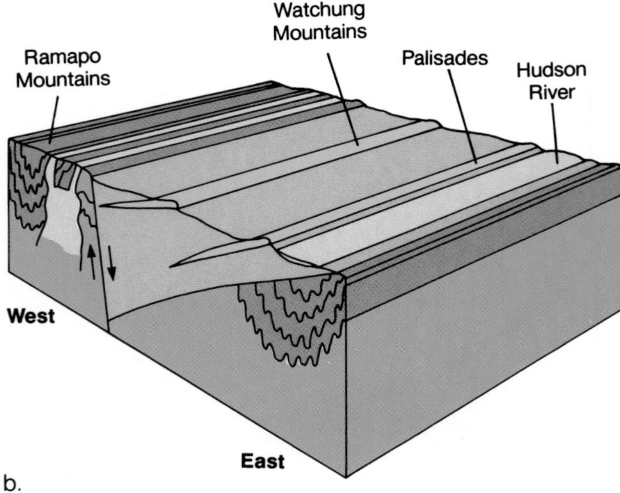


b.

**Figure 6–16** a. Location of the Triassic fault basins in eastern North America. b. Cross section of a typical Triassic fault basin, showing the thick accumulation of sediments and the two sills forming the Watchung Mountains and the Palisades.

### Oceanic-Oceanic Plate Boundaries

When two oceanic plates converge, one of them is subducted under the other along an **oceanic-oceanic plate boundary** (Fig. 6–17). The subducting plate bends downward to form the outer wall of a deep-sea trench, while a subduction complex and forearc basin form along the inner wall. A subduction complex consists of highly contorted, thrust-faulted marine sediments. The lower part of the complex is pulled down by the move-

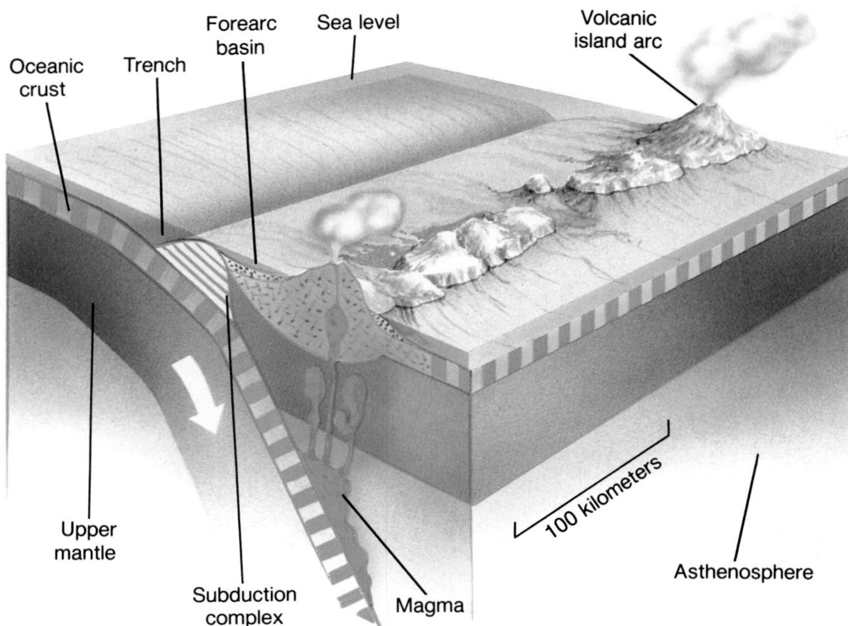


**Figure 6–17** Oceanic-oceanic plate boundary. A deep-sea trench forms where one oceanic plate is subducted beneath another. On the nonsubducted oceanic plate, a volcanic island arc forms as the result of rising magma.

ment of the subducting plate, forming wedge-shaped slices. The underthrusting of new slices added to the bottom of the stack pushes the subduction complex up, forming a forearc basin between it and the volcanic island arc. As the subducting plate descends into the mantle, it is heated. Eventually, partial melting and assimilation of the subducting plate occurs, generating a magma of andesitic composition. This magma rises to the surface of the nonsubducted plate and forms a chain of volcanic islands termed a **volcanic island arc.** It is parallel to the oceanic trench and separated from it by a distance of up to several hundred kilometers—the distance depends on the angle of dip of the subducting plate. The Aleutian Islands, Japanese Island chain, and Philippine Island region are good examples of volcanic island arcs resulting from oceanic-oceanic plate convergence.

### Oceanic-Continental Plate Boundaries

When oceanic crust is subducted under continental crust along an **oceanic-continental plate boundary**, a subduction complex and forearc basin form the inner wall of the trench, while the edge of the continent is deformed into a young mountain range (Fig. 6–18). The andesitic magma generated by subduction rises beneath the continent and either crystallizes before reaching the surface or erupts at the surface, producing a line of andesitic volcanoes. Behind the volcanoes are a series of thrust sheets forming a backarc thrust belt. This occurs because the cold crust of the continental interior is moving toward the trench (since the plates are converging) and encounters the thick, mobile core of the mountain range, resulting in a series of thrust sheets that move over each other. The weight of the backarc thrust sheets depresses the crust downward, forming a backarc basin that receives sediments from the mountain range and continental interior. An excellent example of an oceanic-continental plate boundary is the Pacific coast of South America where the oceanic Nazca plate is presently being subducted under South America. The Peru-Chile Trench marks the site of subduction, and the Andes Mountains are the resulting volcanic mountain chain on the nonsubducting plate (Fig. 6–19). This particular example demonstrates the effect plate tectonics has on our lives. For instance, earthquakes are commonly associated with subduction zones, and the western side of South America is the site of frequent and devastating earthquakes. Furthermore, the southern Andes Mountains act as an effective barrier to moist, easterly blowing

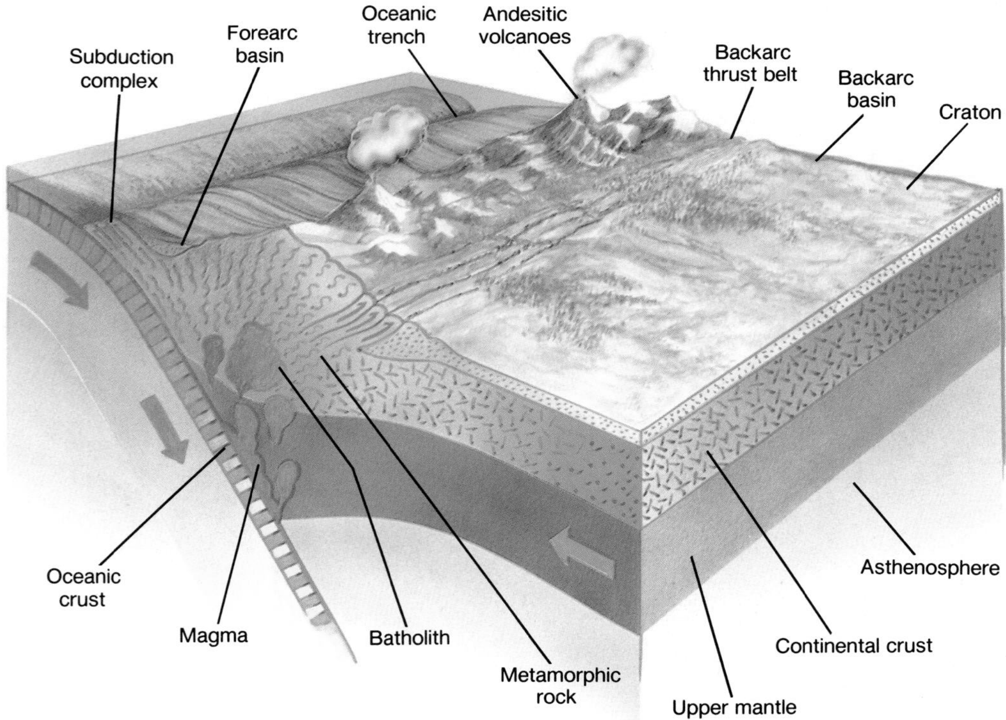


**Figure 6–18** Oceanic-continental plate boundary. The subduction of an oceanic plate beneath a continental plate produces an andesitic volcanic mountain range.

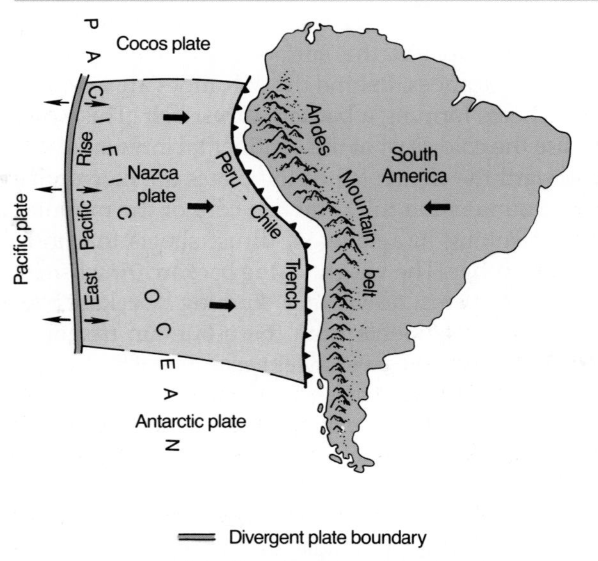


**Figure 6–19** Oceanic-continental plate boundary between the Nazca plate and South American plate producing the Andes Mountains. Subduction of the Nazca plate beneath South America causes numerous earthquakes and volcanic eruptions.

Pacific winds, resulting in a desert east of the southern Andes that is virtually uninhabitable.

### Continental-Continental Plate Boundaries

Two continents approaching each other will initially be separated by an ocean floor that is being subducted under one continent. The edge of that continent will display the features characteristic of oceanic-continental convergence. As the ocean floor continues to be subducted, the two continents will come closer together until they eventually collide. Since continental lithosphere, which consists of continental crust and the upper mantle, is less dense than oceanic lithosphere (oceanic crust and the upper mantle), it cannot sink into the asthenosphere. While one continent may slide a short distance under the other, it cannot be pulled or pushed down into a subduction zone (Fig. 6–20).

When two continents collide, they are welded together along a zone marking the former site of subduction. At this **continental-continental plate boundary**, an interior mountain belt is formed consisting of thrusted and folded sediments, igneous intrusions, metamorphic rocks, and fragments of oceanic crust. In addi-

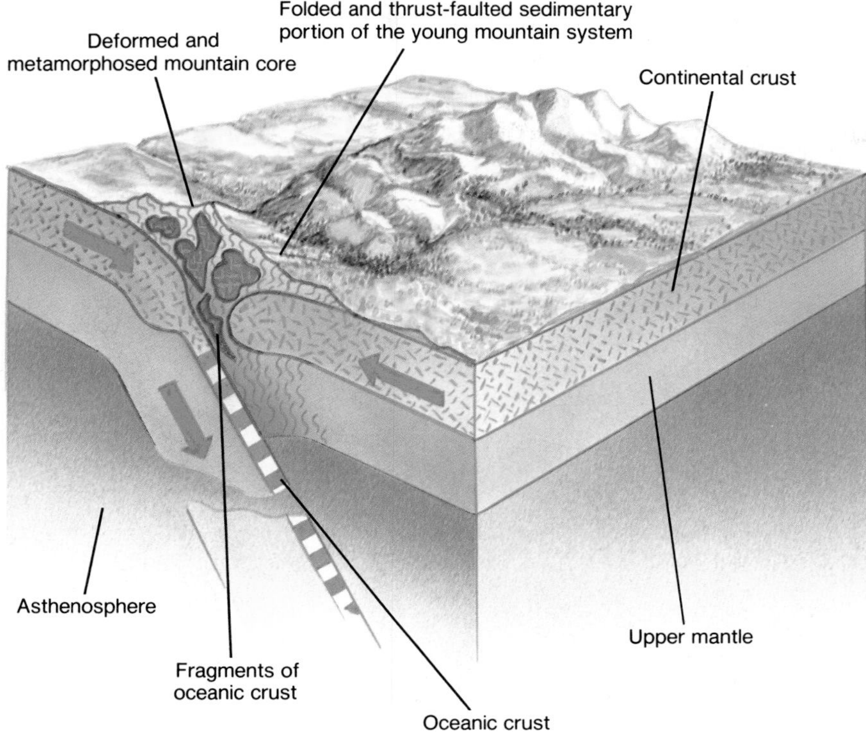


**Figure 6–20** Continental-continental plate boundary. When two continental plates collide, neither one is subducted and a mountain range is formed in the interior of a new, larger continent.

tion, the entire region is subjected to numerous earthquakes. The Himalaya Mountains in central Asia are the result of a continental-continental collision between India and Asia that began about 55 million years ago and continues today.

## Ancient Convergent Plate Boundaries

How can former subduction zones be recognized in the rock record? One clue is provided by igneous rocks. The magma that erupts on the Earth's surface, forming island arc volcanoes and continental volcanoes, is of andesitic composition. Another clue can be found in the zone of intensely deformed rocks that occur between the deep-sea trench where subduction is taking place and the area of igneous activity (Fig. 6–21). Here, sediments and submarine rocks are folded, faulted, and metamorphosed into a chaotic mixture of rocks termed a *melange*.

Within the melange one frequently finds a distinctive assemblage of deep-sea sediments, including graywackes (poorly sorted sandstones containing abundant feldspar and rock fragments, usually in a clay-rich matrix), black shales, and cherts, as well as oceanic crustal rocks such as submarine lavas, altered peridotite, and gabbro. Together these rocks constitute an **ophiolite suite** (Fig. 6–21), which represents slices of oceanic lithosphere that were scraped off the upper portion of the descending plate and squeezed against the edge of the continent. Ophiolites are key features in recognizing plate convergence along a subduction zone (see Perspective 9–1 in Chapter 9).

Elongate belts of folded and thrust-faulted marine sediments, andesites, and ophiolite suites are found in the Appalachians, Alps, Himalayas, and Andes mountains. The combination of such features is good evidence that these mountain ranges resulted from deformation along convergent plate boundaries.

## Transform-Fault Plate Boundaries

The third type of major plate boundary is a **transform fault**. These form when either a new divergent or new convergent margin breaks the lithosphere. The faults can be recognized by zones of intensely shattered rock,

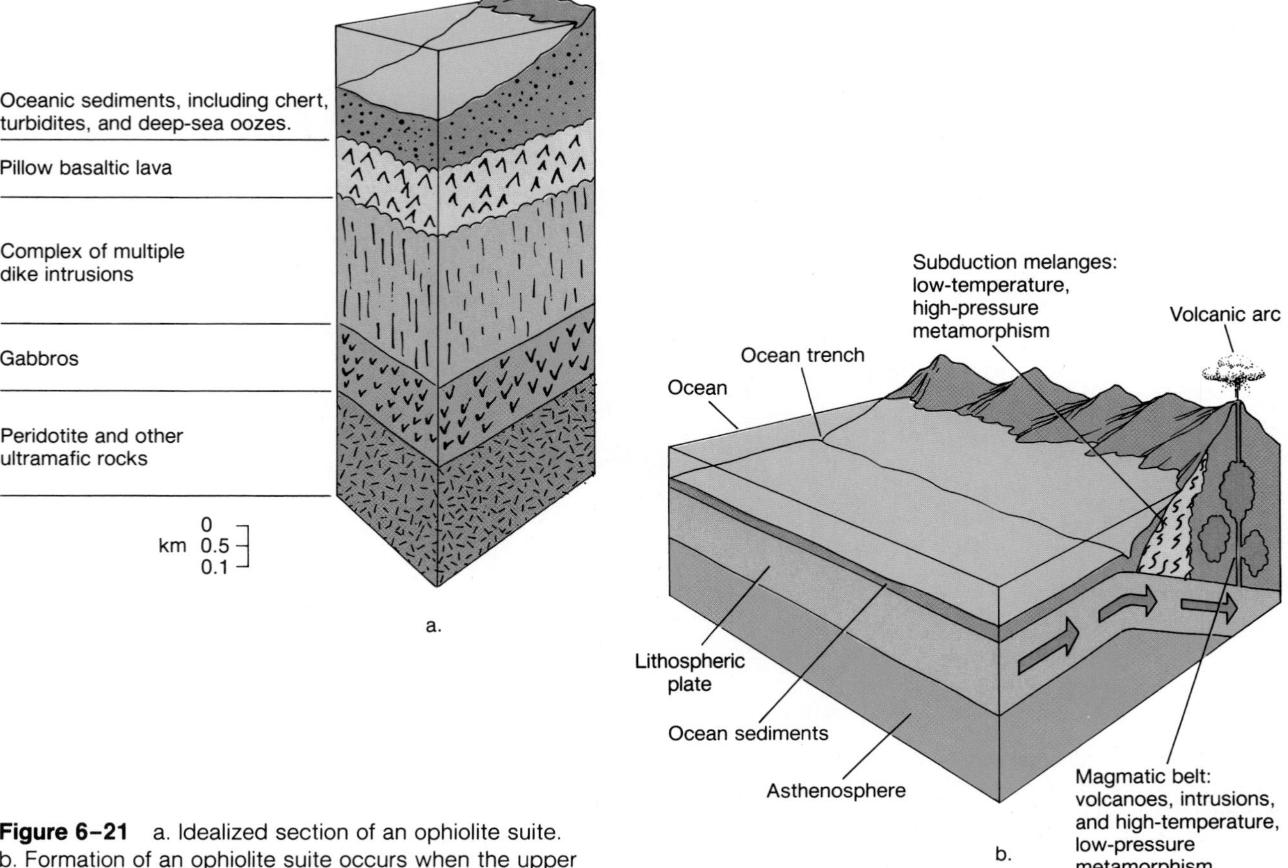


**Figure 6–21** a. Idealized section of an ophiolite suite. b. Formation of an ophiolite suite occurs when the upper part of a descending oceanic plate is squeezed against a continental margin.

the result of two plates sliding past each other. The majority of transform faults cut oceanic crust and are marked by elongate zones of narrow ridges and valleys on the ocean floor. When transform faults cut continental crust, their expression is usually subdued, and they tend to be marked by parallel faults in zones as wide as 100 km. Probably the best-known transform fault is the San Andreas fault in California, which separates the Pacific plate from the North American plate (Fig. 6–22). The many earthquakes that affect California are the result of movement along this fault.

Unfortunately, transform faults generally do not leave any characteristic or diagnostic features except for the obvious displacement of the rocks with which they are associated. This displacement is commonly large, on the order of tens to hundreds of kilometers. Such large displacements in ancient rocks can sometimes be related to transform-fault systems.

## MANTLE PLUMES AND HOT SPOTS

Before leaving the topic of plate boundaries, we should briefly mention an intraplate feature found beneath both oceanic and continental plates. **Hot spots** are locations where plumes of magma rise through the crust and manifest themselves as volcanoes (Fig. 6–23). These plumes generally remain stationary in the mantle while the plates move over them. As a result, the hot spots leave a trail of extinct, progressively older volcanoes called **aseismic ridges**. Dating the volcanoes enables geologists to determine the direction and rate of movement of the plates that pass over the hot spots (Fig. 6–23). Some examples of hot spots and aseismic ridges are the Hawaiian Islands, Emperor Seamounts, and Yellowstone National Park in Wyoming.

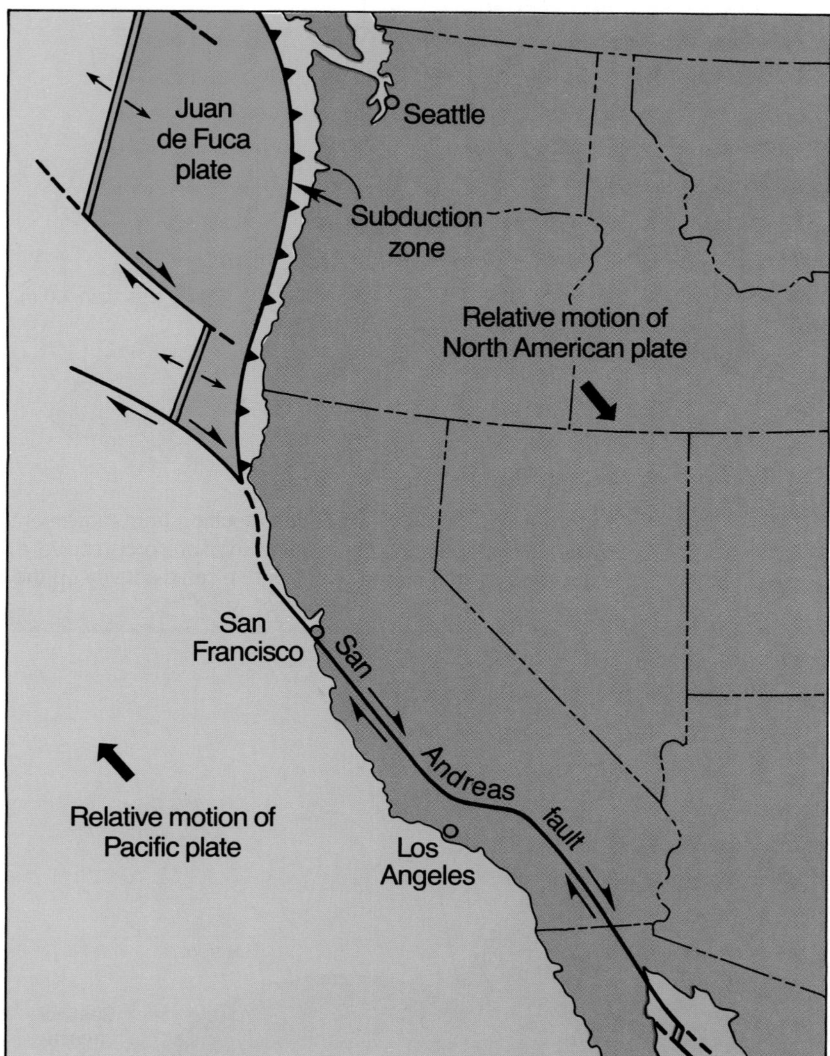


**Figure 6–22** Transform-fault plate boundary. The San Andreas fault is a fault boundary between the Pacific plate and North American plate.

## THE DRIVING MECHANISM OF PLATE TECTONICS

A major obstacle to the acceptance of continental drift was the lack of a mechanism to explain continental movement. When it was shown that continents and ocean floors moved together and oceanic lithosphere was created at oceanic ridges by rising magma, most geologists were willing to accept some type of a convection cell system to explain the movements of plates.

Today, plate motions can be described in great detail (Perspective 6–1), yet the question of what exactly drives the plates is still being debated. Most theories involve some type of convection system, in which mantle rock is heated and flows like a very viscous liquid. Several types of convection cell systems within the mantle have been proposed (Fig. 6–24). These include convection cells in which all movement is confined to the

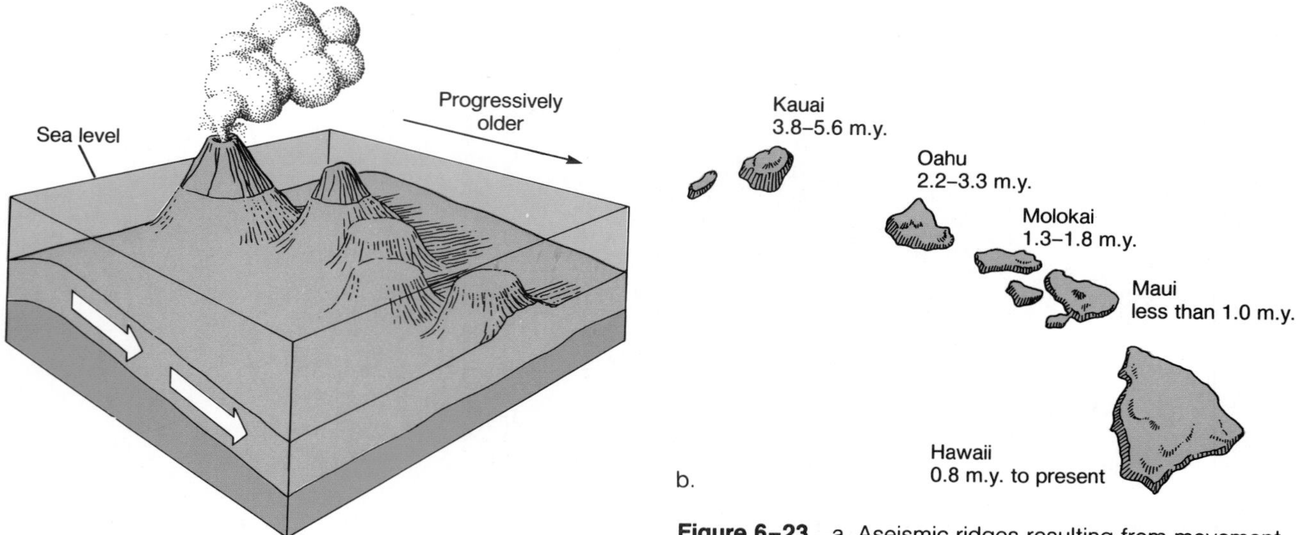


**Figure 6–23** a. Aseismic ridges resulting from movement over a hot spot. b. The Hawaiian Islands are progressively older away from the area of current volcanic activity on the island of Hawaii. Ages in millions of years (m.y.).



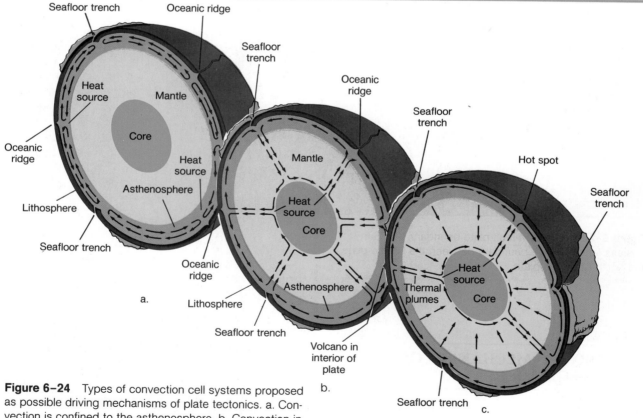

**Figure 6–24** Types of convection cell systems proposed as possible driving mechanisms of plate tectonics. a. Convection is confined to the asthenosphere. b. Convection involves the entire mantle. c. Thermal plumes rise from the mantle-core boundary and cause local hot spots at the Earth's surface.

## Perspective 6–1

# How Fast are the Plates Moving?

How fast are the plates moving, and can we prove that plates are moving at all? Recently, the results from space-based measuring devices have confirmed continental movement within the limits of error for such measuring techniques. Large radio antennas located near Fairbanks, Alaska, and on the islands of Hawaii and Japan were used to record the arrival times of radio signals from the same *quasar* (the most distant energy-emitting objects in the universe). These quasars act as stationary reference points, and the millisecond difference in the arrival time of the same radio signal at different receiving stations yields the distance between the receiving sites. By taking many such measurements over a three-year period, researchers have been able to determine that Hawaii is moving toward Japan at the rate of 8.3 cm (with a potential error of plus or minus 8 mm) per year if Fairbanks is considered stationary.

The technique used to make such precise measurements of distance between two known locations on different plates is known as Very Long Baseline Interferometry (VLBI). Using the VLBI technique, researchers can accurately measure the distance between two distant points with a very small potential source of error. The difference in the distance between the same two points over a given period of time yields the rate at which one point is moving relative to the other. The problem in the past has been that when the potential source of error in the measurement is added to the observed rate of movement, it has always been slightly more than the movement calculated from sea-floor spreading rates. Now, however, no one doubts that the observed rapid plate movement in the Pacific far exceeds any possible source of error. Furthermore, the observed rates for the Pacific are only a few mm per year less than the predicted geologic rates, and the plates on which the radio antennas are located are moving in much the way plate tectonics predicted.

In the Atlantic, movement is much slower. While the methods of VLBI could theoretically measure the distance between Europe and North America with a precision of 3 mm if the Earth had no atmosphere, the effects of the atmosphere on measurements limit the precision across the Atlantic to only 2 to 3 cm. This is slightly more than the predicted rate of movement of 1.7 to 3.0 cm per year, based on the age of the oceanic crust in relation to its distance from the Mid-Atlantic Ridge. Nevertheless, geologists now believe there is an 80 to 90 percent probability that they are observing the actual separation of North America and Europe.

asthenosphere and lithosphere, convection cells involving the entire mantle, and thermal plumes that rise from the mantle-core boundary, causing local hot spots at the Earth's surface. Whatever the precise mechanism involved, it most certainly is a highly complex type of convective system.

## PLATE TECTONICS AND MOUNTAIN BUILDING

An **orogeny** is an episode of intense rock deformation or mountain building. The process of mountain building, termed *orogenesis*, while still not completely understood, is known to be related to plate movements.

Prior to the advent of plate tectonic theory, the classic interpretation of an orogeny involved the disruption of a **geosyncline** (Perspective 6–2) by forces of unknown origin. Today however, geologists believe that orogenies are the consequence of compressive forces related to plate movement. As one plate is subducted under another, sedimentary and volcanic rocks are folded and faulted along the plate margin while the more deeply buried rocks are subjected to regional metamorphism. Magma generated within the mantle either rises to the surface to erupt as andesitic volcanoes or cools and crystallizes beneath the Earth's surface, forming intrusive igneous bodies. Typically, most orogenies occur along either oceanic-continental or continental-continental plate boundaries.

## Perspective 6–2

# Geosynclines and Plate Tectonics

Prior to the emergence of plate tectonics in the 1960s, the theory of mountain building was dominated by the concept of the geosyncline, a term still widely used. The concept of a geosyncline was introduced in 1859 by the famous American geologist James Hall of the New York State Geological Survey. Later, James D. Dana, a contemporary of Hall's, proposed the term *geosyncline*. A geosyncline is defined as an elongated downwarp of the Earth's crust in which thousands of meters of volcanic and sedimentary rocks accumulated. At some point these rocks are folded, faulted, and uplifted to create mountain belts.

Both Hall and Dana used the Appalachian Mountains as the basis for their concept of a geosyncline. They noted that in addition to volcanic rocks, the Appalachians also contained thousands of meters of shallow-water, marine sediments. To account for this, Hall envisioned subsidence of the crust due to the weight of the accumulating sediments. In this way, earlier deposited sediments would be pushed into the crust, while the current sediments would continue to be deposited in shallow water. This would explain how such thick accumulations of shallow-water sediments were possible.

The concept of the geosyncline was debated and modified for more than a century. In the 1940s, Marshall Kay introduced the term *miogeosyncline* for the shallow-water nearshore deposits of clean sands and carbonates, and the parallel *eugeosyncline* for the deeper-water black shales, cherts, and volcanics (Figure 1).

The concept of geosynclines developed during the time when most geologists did not believe continents moved laterally. It provided a model to account for the geologic features associated with mountain belts, such as thick accumulations of shallow-water sediments juxtaposed with deep-water lithologies. However, there were not many modern examples of geosynclines to be found, although the Gulf of Mexico was often cited as an example. The problem was that while the shelf area bordering the southern United States was viewed as a miogeosyncline, the area beyond the shelf was not typical of a eugeosyncline because volcanics were not common. Another problem was explaining how the thick sedimentary deposits of the Gulf of Mexico could be uplifted and deformed to form a mountain belt.

The emergence of plate tectonic theory provided geologists with an entirely new framework for studying mountain building. What was once called the miogeosyncline is actually the continental shelf, where thick deposits of sediment accumulate, and the eugeosyncline corresponds to the deep-water environment at the base of the continental slope (Fig. 1). The associated volcanics and ophiolites seen in mountain belts are the result of subduction taking place along an oceanic-continental plate boundary.

The rise and fall of the geosynclinal theory is a good example of how science works. For about 100 years, the geosynclinal concept appeared to explain much of the geologic data and was thus accepted. When new observations and information led to the formulation of plate tectonic theory, geologists recognized that it better explained the data, and they abandoned the geosynclinal concept.

**Figure 1** a. Old concept of miogeosyncline and eugeosyncline that were supposed to have existed in eastern North America during the Ordovician. b. Modern interpretation of miogeosyncline and eugeosyncline as settings along a marine depositional slope. (Vertical exaggeration 135:1).

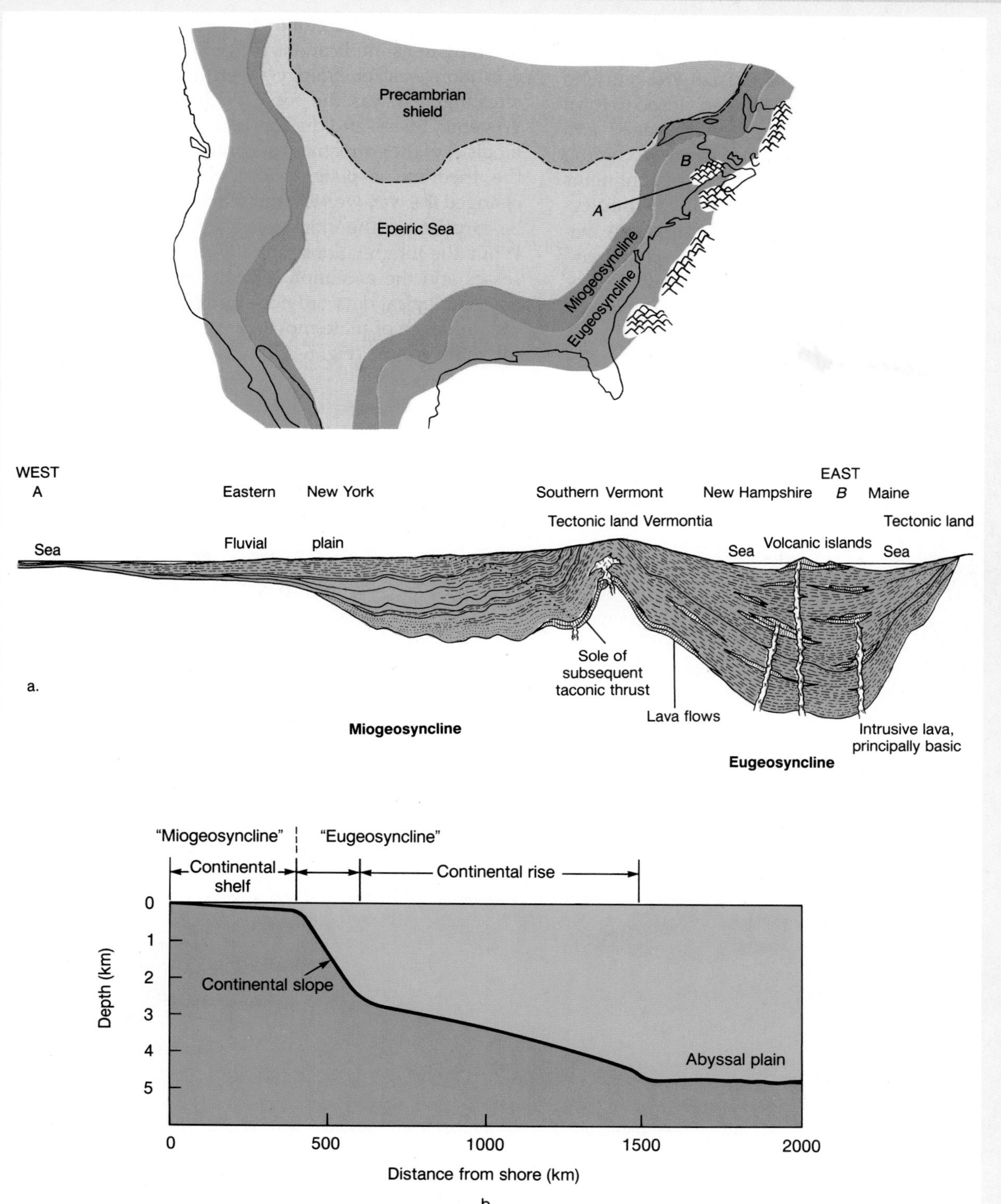

Precambrian shield
Epeiric Sea
Miogeosyncline
Eugeosyncline
A
B
WEST
A
Eastern
New York
Southern Vermont
New Hampshire
EAST
B
Maine
Tectonic land Vermontia
Tectonic land
Sea
Fluvial
plain
Sea
Volcanic islands
Sea
Sole of subsequent taconic thrust
Lava flows
Intrusive lava, principally basic
Miogeosyncline
Eugeosyncline
a.
"Miogeosyncline"
"Eugeosyncline"
Continental shelf
Continental rise
Continental slope
Abyssal plain
Depth (km)
0
1
2
3
4
5
0
500
1000
1500
2000
Distance from shore (km)
b.

As we discussed earlier in this chapter, ophiolite suites are evidence of ancient convergent plate boundaries. The slivers of ophiolites found in the interiors of such mountain ranges as the Alps, Himalayas, and Urals mark the sites of former ocean basins. The relationship between mountain building and the opening and closing of ocean basins is called the **Wilson Cycle** in honor of the Canadian geologist J. Tuzo Wilson, who first suggested that an ancient ocean had closed to form the Appalachian Mountains and then reopened and widened to form the present Atlantic Ocean. According to some geologists, much of the geology of continents can be described in terms of a succession of Wilson cycles! We shall see in Chapter 13, however, that new evidence concerning the movement of microplates and their accretion at the margin of continents must also be considered when dealing with the tectonic history of a continent.

## PLATE TECTONICS AND EVOLUTION

The theory of plate tectonics is as revolutionary and far-reaching in its implications for geology as the theory of evolution was for biology when it was proposed. Interestingly, it was the fossil evidence that convinced Wegener, Suess, and du Toit, as well as many other geologists, of the correctness of continental drift. Together, the theories of plate tectonics and evolution have changed the way we view our planet, and we should not be surprised at the intimate association between them. While the interrelationship between plate tectonic processes and the evolution of life is incredibly complex, paleontological data provides undisputable evidence of the influence of plate movement on the distribution of the Earth's biota (Fig. 6–25).

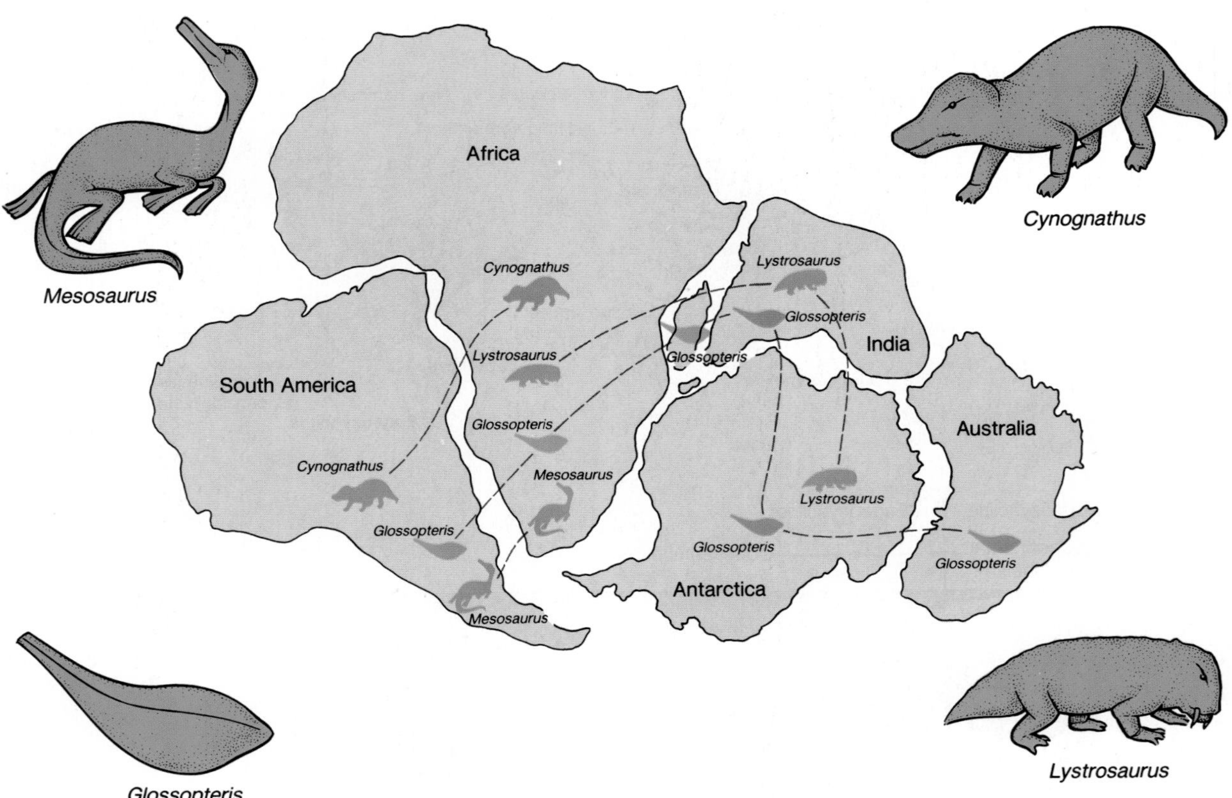


**Figure 6–25** Some of the plants and animals whose fossils are found today on the widely separated continents of South America, Africa, India, Australia, and Antarctica. These continents were joined together to form Gondwana. *Mesosaurus* is a freshwater reptile whose fossils are found in Brazil and Africa; *Glossopteris* is a Permian plant found on all the continents shown; *Lystrosaurus* is an Early Triassic reptile whose fossils are found in Africa, India, and Antarctica; *Cynognathus* is another Early Triassic reptile whose fossils are found in South America and Africa.

The two principal factors influencing the modern distribution of organisms are climate and geographic barriers, both of which are to a large degree controlled by plate movement. For example, movement along divergent plate boundaries forms ocean basins that limit the migration of land plants and animals as well as shallow-water marine invertebrates. Mountain building occurs at convergent plate boundaries and forms barriers to migration as well as influencing climate patterns. Continental-continental plate collisions affect evolution by bringing together different faunas and floras that intermingle and compete.

The movement of plates also affects climate by influencing atmospheric and oceanic circulation patterns. Changes in climate have had a profound effect on the distribution and evolution of organisms. For example, the climatic deterioration of the Late Cenozoic greatly altered the fauna and flora of the northern continents. However, the climate of the Mesozoic Era was, in general, more equable than today, and this is reflected in the wide distribution of the fauna and flora (See Chapter 14).

When a geographic barrier separates a once uniform fauna, species may undergo divergence. If conditions on opposite sides of the barrier are different, then each species of the fauna must adapt to the new conditions, migrate, or become extinct. Adaptation by the various species in the fauna to the new environment may involve enough change over time that new species arise. An excellent example of divergence in the fossil record is exhibited by the marine invertebrates found on opposite sides of the Isthmus of Panama. Prior to the rise of this land connection between North and South America, the shallow seas of the area were inhabited by a homogeneous population of bottom-dwelling invertebrates. After the rise of the Isthmus of Panama, around 5 million years ago, the original population was separated, and new species evolved on opposite sides of the isthmus in response to the changing environment (Fig. 6–26).

The creation of the Isthmus of Panama also influenced the evolution of the mammalian faunas of North and South America (See Chapter 16). During most of the Cenozoic Era, South America was an island continent, and its mammalian fauna evolved in isolation from faunas of the rest of the world. When the Isthmus of Panama formed, most of the indigenous South American mammalian fauna was replaced by migrants from North America. Surprisingly, only a few South American mammal groups migrated northward (Fig. 16–45 in Chapter 16).

Other examples of the role that plate tectonics plays in the distribution and the evolution of the Earth's biota will be covered in succeeding chapters.

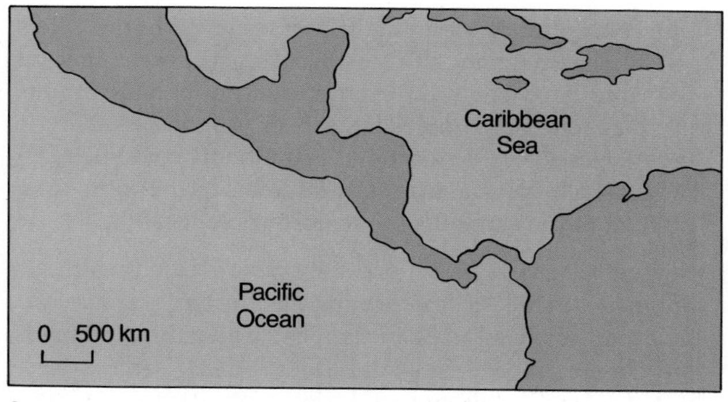


a.

Pacific species

Caribbean species

b.

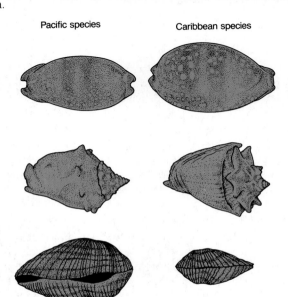

**Figure 6–26** a. The Isthmus of Panama forms a barrier that divides a once uniform fauna. b. Divergence of molluscan species after the formation of the Isthmus of Panama. Each pair belongs to the same genus but is a different species.

## CHAPTER SUMMARY

1. The concept of continental movement is not new. The earliest maps showing the similarity between the east coast of South America and the west coast of Africa provided people with the first evidence that continents may once have been united and subsequently drifted away from each other.

2. Alfred Wegener is generally credited with developing the concept of continental drift. He gathered information from many sources to show that the continents were once connected as one supercontinent he called Pangaea. Unfortunately, Wegener could not explain how the continents moved, and most geologists ignored his ideas.

3. The concept of continental drift was revived in the 1950s when paleomagnetic studies of rocks of different ages on the same continent indicated there were apparently multiple magnetic north poles instead of just one as there is today. However, the location of continents could be rearranged such that the magnetic minerals in the rocks then pointed to the same magnetic north pole location.
4. Magnetic surveys of the oceanic crust revealed magnetic anomalies in the rocks indicating that the Earth's magnetic field had reversed itself in the past. Since the anomalies are parallel and symmetrical around the oceanic ridges, this means that new oceanic lithosphere was forming and the sea floor has been spreading.
5. Radiometric dating reveals the oldest oceanic crust is less than 180 million years old, while the oldest continental crust is 3.8 billion years old. Clearly, the ocean basins are recent geologic features.
6. Plate tectonic theory became widely accepted in the 1960s because of the overwhelming evidence supporting it. Most geologists believe that some type of convective cell mechanism probably drives the plates.
7. There are three types of plate boundaries: divergent boundaries, where plates move away from each other; convergent boundaries, where one plate is subducted under another plate; and transform-fault boundaries, where two plates slide past each other.
8. Ancient plate boundaries can be recognized by their associated rock assemblages and geologic structures. For divergent boundaries these may include rift valleys with thick sedimentary sequences and numerous dikes and sills. For convergent boundaries, ophiolite suites and andesitic rocks are two characteristic features. Transform faults generally do not leave any characteristic or diagnostic features in the rock record.
9. Plate tectonics provides geologists with a powerful theory for explaining such things as mountain building, global climatic changes, and distributional patterns of the world's biota.

## IMPORTANT TERMS

aseismic ridge
continental-continental plate boundary
continental drift
convergent plate boundary
Curie point
divergent plate boundary
geosyncline
*Glossopteris* flora
Gondwana
hot spot
Laurasia
magnetic anomalies
magnetic reversals
oceanic-continental plate boundary
oceanic-oceanic plate boundary
ophiolite suite
orogeny (orogenesis)
paleomagnetism
Pangaea
plate tectonics
sea-floor spreading
thermal convection cell
transform-fault plate boundary
triple junction
volcanic island arc
Wilson cycle

## REVIEW QUESTIONS

1. What is the significance of polar wandering in relation to continental drift?
2. Give a summary of the geologic activities characterizing the three different types of plate boundaries.
3. How would you recognize ancient convergent plate boundaries?
4. What are hot spots, and how can they be used to determine the direction and rate of movement of plates?
5. How does plate tectonic theory of mountain building differ from the geosynclinal theory?
6. What is the apparent driving mechanism for plate movement?
7. Why is plate tectonics such an elegant unifying theory?
8. What features would an astronaut look for on the moon or another planet to find out if plate tectonics is currently active or if it was active in the past?

## ADDITIONAL READINGS

Ben-Avraham, Z. 1981. The Movement of Continents. *American Scientist*, 69, no. 3: 291–299.

Bonatti, E. 1987. The Rifting of Continents. *Scientific American*, 256, no. 3: 96–103.

Dewey, J. F. 1972. Plate Tectonics. *Scientific American*, 226, no. 5: 56–72.

Gass, I. G. 1982. Ophiolites. *Scientific American*, 247, no. 2: 122–131.

Hallam, A. 1972. Continental Drift and the Fossil Record. *Scientific American*, 227, no. 5: 56–69.

Jordon, T. H., and J. B. Minster. 1988. Measuring Crustal Deformation in the American West. *Scientific American*, 259, no. 2: 48–59.

Molnar, P. 1986. The Structure of Mountain Ranges. *Scientific American*, 255, no. 1: 70–79.

Nance, R. D., T. R. Worsley, and J. B. Moody. 1988. The Supercontinent Cycle. *Scientific American*, 259, no. 1: 72–79.

Toksoz, M. 1975. The Subduction of the Lithosphere. *Scientific American*, 233, no. 5: 88–101.

Vink, G. E., W. J. Morgan and P. R. Vogt. 1985. The Earth's Hot Spots. *Scientific American*, 252, no. 4: 50–57.

# 7

## CHAPTER OUTLINE

PROLOGUE
INTRODUCTION
THE ORIGIN AND EARLY HISTORY OF THE UNIVERSE
THE CHANGING COMPOSITION OF THE UNIVERSE
THE ORIGIN AND HISTORY OF THE SOLAR SYSTEM
General Characteristics of the Solar System
Evolutionary or Catastrophic Origin of the Solar System?
*Evolutionary Theories*
*Catastrophic Theories*
Current Theory on the Origin and History of the Solar System
METEORITES—EXTRATERRESTRIAL VISITORS
Stones, Irons, and Stony-Irons

PERSPECTIVE 7-1: METEORITE SHOWERS AND THEIR IMPACT ON PLANETS

METEORITES AND THE AGE OF THE EARTH
ORIGIN AND DIFFERENTIATION OF THE EARLY EARTH
Homogeneous Accretion
Inhomogeneous Accretion
Conclusions on Homogeneous versus Inhomogeneous Accretion

PERSPECTIVE 7-2: THE ORIGIN AND EARLY HISTORY OF THE MOON

CHAPTER SUMMARY

## PROLOGUE

*Sometimes the most important scientific discoveries are made quite by accident or while searching for something else. Such was the case for Arno Penzias and Robert Wilson of Bell Telephone Laboratories. What began as a seemingly straightforward research project in the early 1960s resulted in the 1978 Nobel Prize for physics for these two scientists. Penzias and Wilson were originally interested in mapping the radio waves coming from directions away from the main plane of our galaxy. They needed a particularly sensitive radio telescope to pick up the extremely faint signals emanating from distant parts of the universe. Such a radio telescope was available at the Bell Telephone Laboratories at Holmdel, New Jersey.*

*Penzias and Wilson began their research by calibrating the radio telescope to a frequency at which they did not expect to find much galactic radio waves. After the calibration, they would then turn to a still lower frequency to map the distant radio waves they were interested in. However, the calibration frequency was far too "noisy," or in the words of Penzias, "The antenna was considerably hotter than expected." This was hardly a promising beginning to a discovery that ranks as one of the most important in astronomy in this century. They concluded that something must be wrong with the telescope. One possibility was the two pigeons nesting on it. However, even after the pigeons were dislodged, the noise still persisted.*

*Early in 1965 Penzias and Wilson reached a point in their observations where they could dismantle the telescope without harming the results of their research. After painstakingly cleaning everything and carefully reassembling it, they found the bothersome noise still there. "We frankly did not know what to do," lamented Arno Penzias. If the radio noise was not in the telescope, it must be coming from space. If this was the case, then it was spread uniformly throughout the universe. It was always there, day and night, and in every direction the scientists aimed their telescope. The radio noise corresponded to a constant temperature of 3° C above absolute zero. Absolute zero equals −273° C, the temperature at which all atomic motion stops. The problem was that there was no astronomical explanation for a uniform "noise" permeating the entire universe.*

*At about the same time Penzias and Wilson were puzzling over their findings, a colleague of theirs recalled reading a paper by Jim Peebles of Princeton University in which he predicted that*

# The Origin of the Universe, Solar System, and Planet Earth

*the universe should have a background temperature of about 10° C above absolute zero. If our present expanding universe began with a "Big Bang," then there should be a relic of that initial high temperature pervading all of space. Peebles's calculations pointed to a very low background temperature because of the cooling that has occurred since the formation of the universe.*

*Based on these calculations, Peebles and another Princeton physicist, Robert Dicke, had just started searching for the background radiation with a radio telescope at Princeton when Penzias contacted Dicke about Peebles's paper. The two groups published their findings later in 1965 as companion letters in* The Astrophysical Journal, *with Penzias and Wilson referring readers to the "theory" paper of Dicke and Peebles for interpretation of their observations.*

*After checking Penzias and Wilson's results using other telescopes and at different frequencies, astronomers and physicists embraced their findings enthusiastically. What was initially interpreted as some type of problem with the radio telescope turned out to be the fading afterglow of the explosion in which the universe began.*

The Andromeda Galaxy, 2.2 million light years away, is the closest major galaxy to our Milky Way. Its full diameter is nearly 150,000 light years and it contains some 300,000,000,000 stars.

## INTRODUCTION

Most people are interested in the question of when, where, and how the universe began as well as the equally intriguing aspect of our place in it. The origin of the universe is not only of interest to scientists, but also to theologians and philosophers. How does one comprehend infinity from a finite viewpoint or conceptualize the creation of matter and energy under conditions where the known laws of physics do not apply? Robert Browning wrote, "Ah, but a man's reach should exceed his grasp." In attempting to understand the origin of the universe, our reach must certainly exceed our grasp!

Most scientists today believe the universe originated approximately 15 to 20 billion years ago in what is popularly called the "**Big Bang**." At the moment of the Big Bang, in a region billions of times smaller than a proton, both space and time were set at zero. Therefore there is no "before the Big Bang," only what occurred after it. The reason for this is that space and time are unalterably linked to form a space-time continuum as demonstrated by Einstein's theory of relativity. Without space there can be no time.

How do we know the Big Bang took place between 15 and 20 billion years ago? Why couldn't the universe have always existed as we know it today? To answer those questions, we turn our optical and radio telescopes skyward and observe two fundamental phenomena. The first is that the universe is expanding; the second is that there is a background radiation permeating it.

That the universe is expanding was first recognized by Edwin Hubble in 1929. By measuring the optical spectra of distant galaxies, Hubble noted that the velocity at which a galaxy moves away from the Earth increases proportionally to its distance from Earth. He observed that the spectral lines (wavelengths of light) of the galaxies are shifted toward the red end of the spectrum; that is, there is a shift toward longer wavelengths. Such a redshift would be produced by galaxies receding from each other at tremendous speeds (Fig. 7–1). This is an example of the **Doppler effect**, which is the change in the frequency of a sound, light, or other wave caused by movement of its source relative to the observer (Fig. 7–2).

One way to envision how velocity increases with increasing distance is by reference to the commonly used analogy of a rising loaf of raisin bread in which the raisins are uniformly distributed throughout the loaf (Fig. 7–3). Suppose that before the dough begins to rise, the raisins are 1 cm apart. After one hour the dough has risen to the point where the raisins are now 2 cm apart. Any raisin is now 2 cm from its nearest neighbor and 4 cm away from the next one, 6 cm from the next one, and so on. From the perspective of any raisin, its nearest neighbor has moved away from it at a speed of 1 cm per hour (it originally was 1 cm away and is now 2 cm away), the next raisin over moved away at a speed of 2 cm/hr (it was 2 cm away, and is now 4 cm away), while the next one moved away at 3 cm/hr (it was originally 3 cm away and is now 6 cm away from the reference raisin), and so on. If the dough continues to rise until the distance between any two raisins is 4 cm apart, then we would find all the distances between raisin pairs have quadrupled what they were to start with, and the speeds needed to accomplish this are directly proportional to the distances between any two raisins. The farther away a given raisin is to begin with, the farther it must move to maintain the regular spacing during the expansion, and hence the greater the velocity must be. In the same way that raisins move apart in a rising loaf of bread, galaxies are receding from each other at a rate proportional

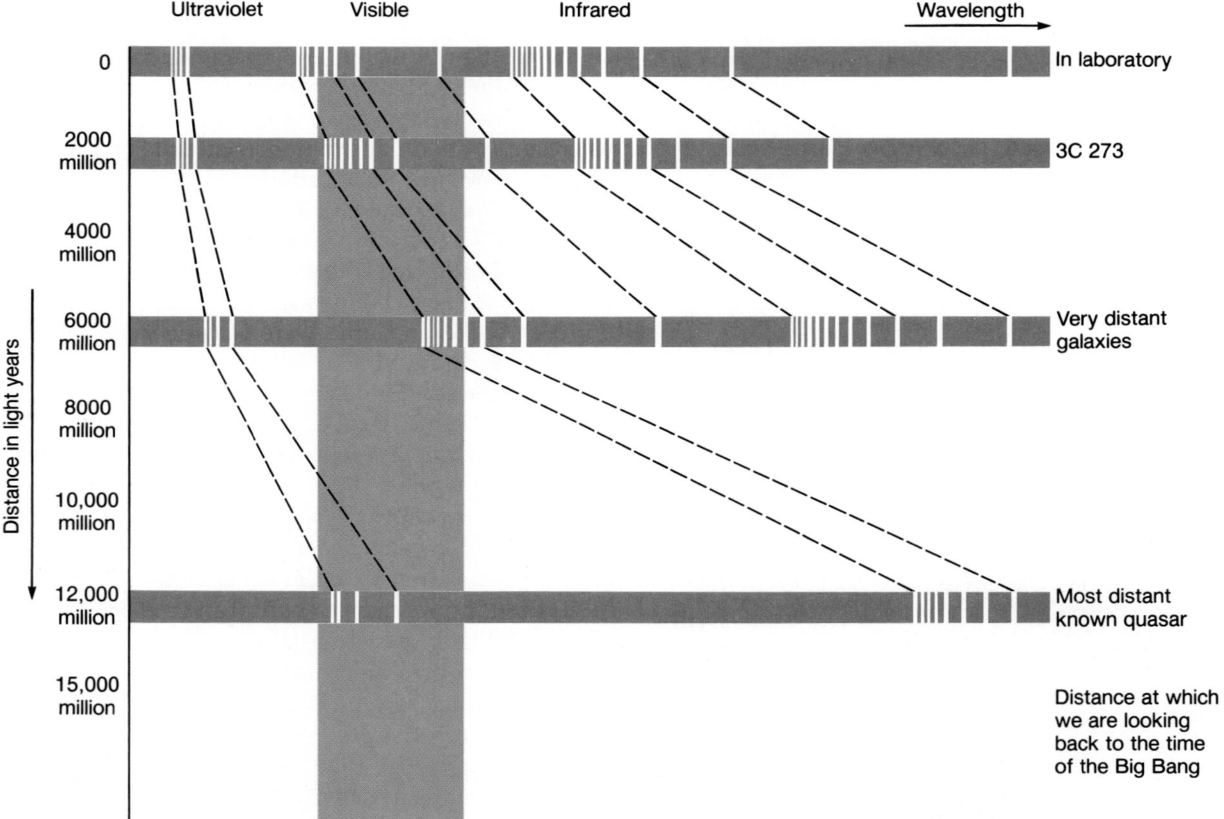


**Figure 7–1** Spectra showing the redshift of the galaxies. Just as the Doppler effect makes a rapidly receding train whistle drop in pitch, the expansion of the universe stretches light waves from receding objects toward the longer-wavelength end of the spectrum. This light is said to be *redshifted*, and the amount of the redshift is proportional to the object's distance from us. The spectral lines of hydrogen in this diagram are stretched in proportion to their original wavelengths. The ratio of wavelength increase to the original wavelength is the same, however, for all lines in the spectrum of any object.

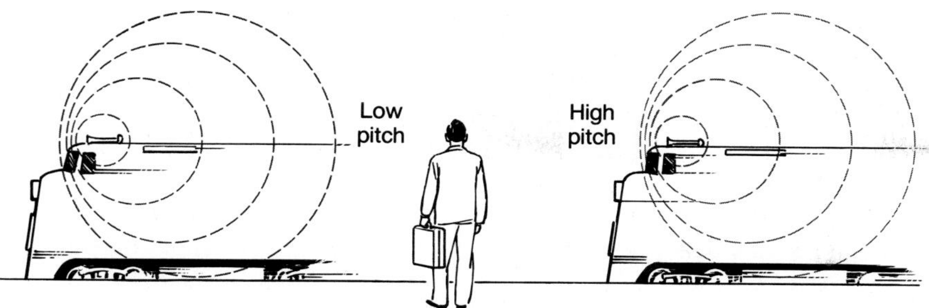


**Figure 7–2** The Doppler effect as applied to sound waves. The sound waves of an approaching whistle are slightly compressed so that the individual hears a shorter-wavelength, higher-pitched sound. As the whistle passes and recedes from the individual, the sound waves are slightly spread out, and a longer-wavelength, lower-pitched sound is heard.

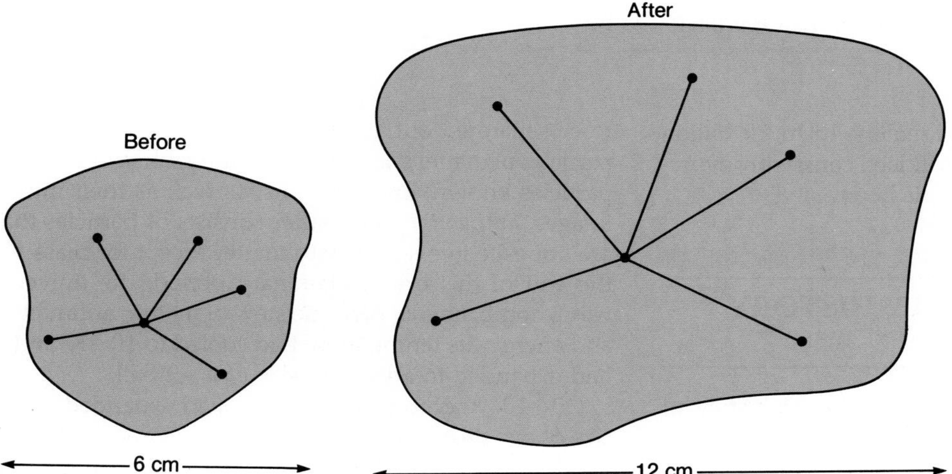


**Figure 7–3** The motion of raisins in a rising loaf of raisin bread illustrates the relationship that exists between distance and speed and is analogous to an expanding universe. In this diagram, adjacent raisins are located 2 cm apart before the loaf rises. After one hour, any raisin is now 4 cm away from its nearest neighbor and 8 cm away from the next raisin over, and so on. Therefore, from the perspective of any raisin, its nearest neighbor has moved away from it at a speed of 2 cm per hour, and the next raisin over has moved away from it at a speed of 4 cm per hour. In the same way raisins move apart in a rising loaf of bread, galaxies are receding from each other at a rate proportional to the distance between them.

to the distance between them, which is exactly what astronomers see when they observe the universe (Fig. 7–4).

A corollary to Hubble's expanding universe is that astronomers can use this expansion rate to calculate how long ago the galaxies were all together in a single point. Such calculations yield an estimated age for the universe of 15 to 20 billion years.

The second important observation providing evidence of the Big Bang was made in 1965 by Arno Penzias and Robert Wilson of Bell Telephone Laboratories when they discovered that there is a pervasive background radiation of 3 K, that is, 3° above absolute zero (zero degree Kelvin, (0 K), equals −273° C) everywhere in the universe. This background radiation is believed to be the fading afterglow of the Big Bang.

From these two observations (that the universe is expanding and that it has a 3 K background radiation), it is concluded that the universe, matter, and energy as

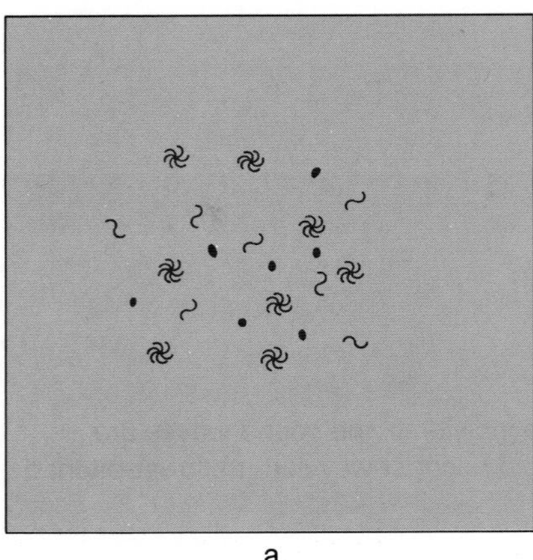


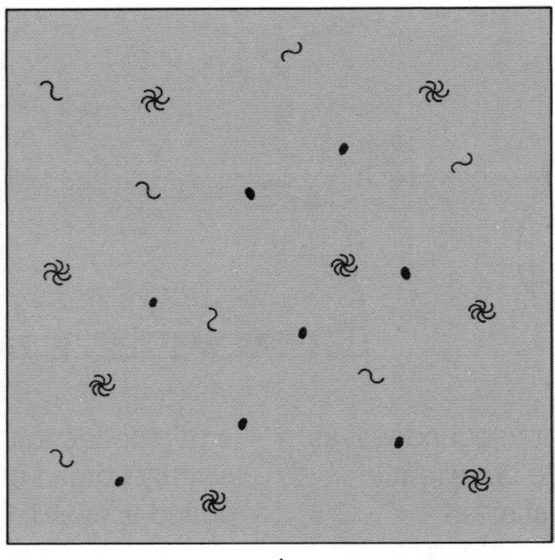


**Figure 7–4** The expanding universe. a. The position of a number of galaxies at a given point in time. b. The same location at a later time. The galaxies have moved away from each other at a rate proportional to the distance between them.

we know it originated approximately 15 to 20 billion years ago. The universe then is indeed constantly evolving rather than remaining a static body.

## THE ORIGIN AND EARLY HISTORY OF THE UNIVERSE

Presently, astrophysicists can reconstruct the history of the universe back to $10^{-43}$ seconds following the Big Bang (Table 7–1). What was its history before $10^{-43}$ seconds? No one really knows, since it is impossible at this stage in our knowledge to deal with the infinitely high densities and temperatures that must have been present. Matter as we know it today could not have existed under those conditions, and the universe consisted of pure energy. Many physicists suspect that at the extreme temperatures prevailing before $10^{-43}$ seconds, the four basic forces—**gravity, electromagnetic force, strong nuclear force, and weak nuclear force** (Table 7–2)—were united into a single unified force. These forces are defined in the table.

By $10^{-43}$ seconds after the Big Bang, gravity separated from the other basic forces, which remained united, and physics as we know it began. At a temperature estimated at $10^{32}$ K, the expanding universe at this time is believed to have been only $10^{-28}$ centimeters in diameter.

Between $10^{-35}$ and $10^{-32}$ seconds after the Big Bang, a major inflationary period took place. The strong force separated out, and energy began to congeal into particles of matter such as quarks (the most fundamental particles known) and electrons, as well as their mirror images, *antimatter*. Antimatter consists of particles that are opposite in every way to matter except for mass. At the end of this brief inflationary period, the universe was a homogeneous, opaque stew of matter, antimatter, and energy. Its temperature had cooled to $10^{27}$ K, and it had expanded to about the size of a softball.

By $10^{-6}$ seconds the universe had expanded to the size of our solar system and had cooled enough ($10^{13}$ K) so that quarks could bind into protons and neutrons. During that brief time after its formation, the universe was not as symmetrical as it should have been. As it cooled to $10^{13}$ K, matter and antimatter collided and annihilated each other. However, because the universe was not symmetrical there was a "slight" excess of matter left over that would become our present universe of galaxies, stars, and planets. If there had not been this asymmetry, the universe would instead be an ever-expanding and slowly cooling emptiness.

By the time the universe was 1 second old, the electromagnetic and weak nuclear forces separated. Three minutes after the Big Bang, the temperature had cooled to $10^{9}$ K, and at this temperature protons and neutrons fused to form the nuclei of hydrogen and helium atoms. At around 100,000 years the temperature had cooled to 3,000 K, and electrons then combined with the previously formed nuclei to create complete atoms of hydrogen and helium. At that time, photons (the energetic particles of light) separated from matter, and the uni-

**Table 7–1** Summary of the early history of the universe.

| | |
|---|---|
| Big Bang | The origin of the universe. |
| $10^{-43}$ seconds | Physics starts here, and gravity separates from the other basic forces. |
| $10^{-35}$ to $10^{-32}$ seconds | A major inflationary period takes place. The strong force separates, and energy begins to congeal into quarks and electrons and their mirror images, antimatter. |
| $10^{-6}$ seconds | Quarks bind into protons and neutrons. Matter and antimatter collide with a slight excess of matter left over that comprises the matter in the universe today. |
| 1 second | Electromagnetic and weak nuclear forces separate. |
| 3 minutes | Protons and neutrons fuse into atomic nuclei. |
| $10^5$ years | Electrons join with nuclei to make atoms. Photons separate from matter, and the universe bursts forth with light. |
| $10^5$ to $10^9$ years | The universe becomes clumpy. |

**Table 7–2** The forces that bind and other quarks of nature.

Four forces appear to be responsible for all interactions of matter. While scientists do not yet know the mechanism that makes the four forces work, these forces are responsible for all interactions of matter.

1. **Gravity** is the attraction of one body toward another.
2. The **electromagnetic force** combines electricity and magnetism into the same force and binds atoms into molecules. It also transmits radiation across the various spectra at wavelengths ranging from gamma rays (shortest) to radio waves (longest) through massless particles called *photons.*
3. The **strong nuclear force** binds protons and neutrons together in the nucleus of an atom.
4. The **weak nuclear force** is responsible for the breakdown of an atom's nucleus, producing radioactive decay.

Atoms are composed of protons, neutrons, and electrons. Protons and neutrons are composed of even smaller and more elementary particles called *quarks.*

verse became transparent as light burst forth for the first time. This liberation of photons is what we observe today as the 3 K background radiation permeating the universe.

Sometime between 100,000 and 1 to 2 billion years after the Big Bang, the universe started to become clumpy. For reasons still not understood, matter began gathering into clouds of different sizes that eventually collapsed to form clusters of galaxies and stars. These galaxies tended to form like beads on a string into superclusters, the largest celestial objects known.

## THE CHANGING COMPOSITION OF THE UNIVERSE

As the universe continued expanding and cooling, its chemical makeup changed. Today it is 98 percent hydrogen and helium by weight, while early in its history, it was 100 percent hydrogen and helium. How did the heavier elements form? The formation of heavier elements from lighter elements results from fusion reactions in which atomic nuclei combine to form more massive nuclei. Such reactions convert hydrogen to helium and occur in the cores of stars. Stars more massive than our sun may undergo many nuclear-reaction steps during their history, in which hydrogen is initially converted to helium, then to carbon, and from carbon to even heavier elements (Fig. 7–5). When these stars die, often explosively, the heavier elements are returned to interstellar space and are available for inclusion in new stars. When new stars form, they will have a small component of these heavier elements. In this way the chemical composition of the galaxies, which are made up of billions of stars, is gradually enhanced in heavier elements.

The chemical composition of the Milky Way Galaxy has thus been changing in the period between the Big Bang and the formation of our solar system. Today, nearly 2 percent of the total mass of the Milky Way Galaxy is in the form of elements heavier than helium.

## THE ORIGIN AND HISTORY OF THE SOLAR SYSTEM

Having looked at the origin and history of the universe, we can now examine how our own solar system formed. It is interesting that in many ways astronomers know more about the birth, life, and death of distant stars and

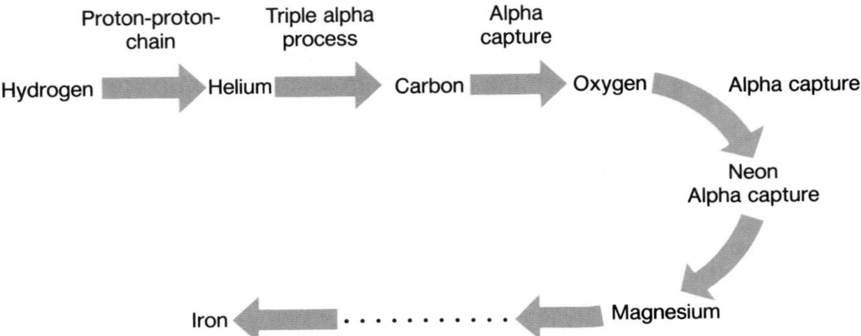


**Figure 7–5** Stars undergo nuclear-reaction steps in which light elements are converted to heavier elements. The first step, the conversion of hydrogen to helium, occurs in all stars. The subsequent conversion to heavier elements depends on the mass of the star. The conversion of helium into carbon involves the triple alpha process. A helium nuclei is called an *alpha particle*, and in the triple alpha process, three helium nuclei combine to form one carbon nucleus. An *alpha capture* involves the fusion of an alpha particle with the nucleus of an atom, creating a heavier element.

galaxies than they do about the history of our own solar system. This is because stars are generally too far away for us to observe how planets may be forming around them. However, the first visual evidence of a stellar disk outside our own solar system was made in April 1984 by astronomers at the Las Campañas Observatory in Chile. They observed a huge, disk-shaped cloud of matter surrounding Beta Pictoris, a star twice as large as our sun in the constellation Pictor, 50 light-years away. This disk-shaped cloud of matter, which may contain planets, provides us with our first telescopic image of what appears to be an evolving solar system (Fig. 7–6). Further study of it should provide astronomers with new insights into the early history of our own solar system.

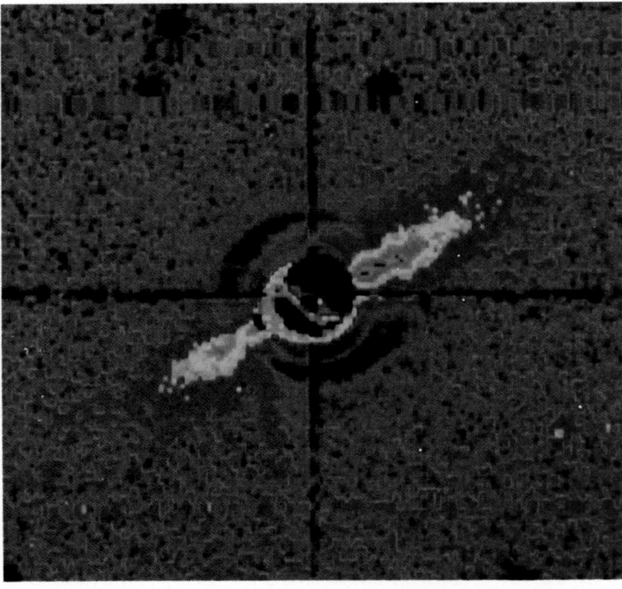

**Figure 7–6** Beta Pictoris, a star about twice the size of our sun in the constellation Pictor is seen to have a disk-shaped cloud of matter surrounding it that may have planet-sized objects in it. It is the first visual evidence of a stellar disk outside our own solar system.

## General Characteristics of the Solar System

Any theory that attempts to explain the origin and history of our solar system must take into account certain general characteristics. These can be conveniently grouped into four categories (Table 7–3).

The first category concerns planetary orbits and rotation. All the planets revolve around the sun in a counterclockwise direction when viewed from a point high in space above the Earth's North Pole; the orbits around the sun are nearly circular, and all planetary orbits lie in a common plane, called the *plane of the ecliptic*. Furthermore, the rotation of all the planets but Venus, and nearly all the planetary satellites, is counterclockwise, and the axes of rotation of the planets, except for those of Uranus and Pluto, are nearly perpendicular to the plane of the ecliptic (Fig. 7– 7).

The second general category concerns the chemical and physical properties of the planets. It is convenient to divide the planets into two groups. The four inner rocky, or terrestrial, planets—Mercury, Venus, Earth, and Mars—are all small and have high mean densities (Table 7–4), indicating they are composed of rock and metallic elements. The four outer Jovian planets— Jupiter, Saturn, Uranus, and Neptune—are all large and have low

**Table 7–3** General characteristics of the solar system.

1. **PLANETARY ORBITS AND ROTATION**
   - Planetary and satellite orbits lie in a common plane.
   - Nearly all of the planetary and satellite orbital and spin motions are in the same direction.
   - The rotation axes of nearly all the planets and satellites are roughly perpendicular to the plane of the ecliptic.
2. **CHEMICAL AND PHYSICAL PROPERTIES OF THE PLANETS**
   - The terrestrial planets are small, have a high density (4.0 to 5.5 g/cm$^3$), and are composed of rock and metallic elements.
   - The Jovian planets are large, have a low density (0.7 to 1.6 g/cm$^3$), and are composed of gases and frozen compounds.
3. **THE SLOW ROTATION OF THE SUN**
4. **INTERPLANETARY MATERIAL**
   - The existence and location of the asteroid belt.
   - The distribution of interplanetary dust.

**Table 7–4** Densities of the planets in the solar system.

| Planet | Density |
|---|---|
| Mercury | 5.4 g/cm$^3$ |
| Venus | 5.3 g/cm$^3$ |
| Earth | 5.5 g/cm$^3$ |
| Mars | 4.0 g/cm$^3$ |
| Jupiter | 1.3 g/cm$^3$ |
| Saturn | 0.7 g/cm$^3$ |
| Uranus | 1.2 g/cm$^3$ |
| Neptune | 1.6 g/cm$^3$ |
| Pluto | 0.9 g/cm$^3$ |

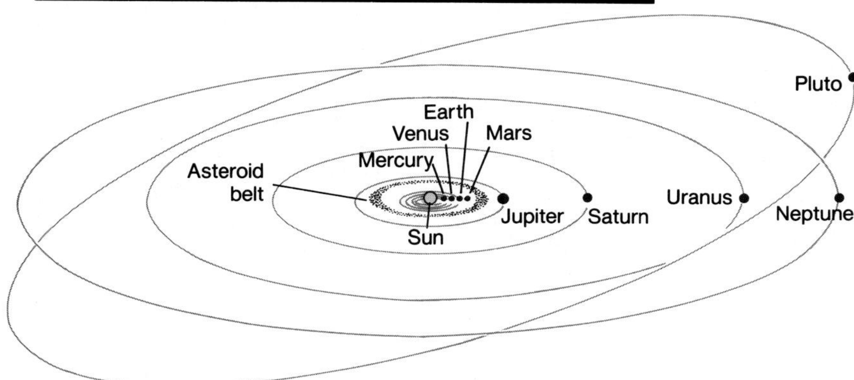


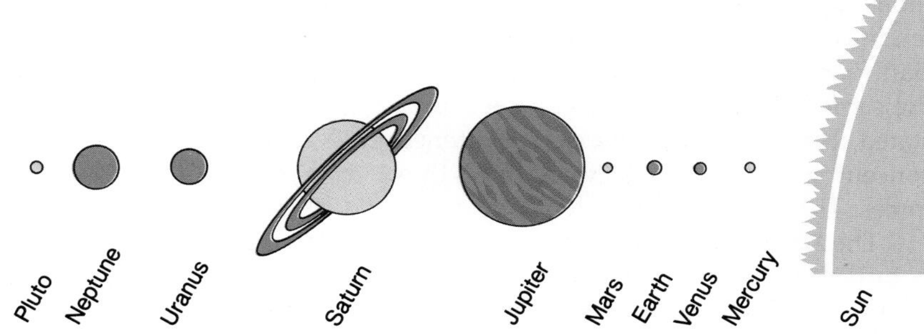


**Figure 7–7** Diagramatic representation of our solar system showing the relative size of the planets and their orbits around the sun.

mean densities, indicating they are composed of lightweight gases such as hydrogen and helium, as well as frozen compounds such as ammonia and methane. The outermost planet, Pluto, is small and has a low mean density of 0.9 $g/cm^3$.

The slow rotation of the sun is the third characteristic of interest here, and it constituted a major problem for many of the early theories of the origin of the solar system. The nature and distribution of the various types of interplanetary objects such as the asteroid belt, comets, and interplanetary dust comprise the fourth category that must be explained in any theory of the origin of the solar system.

## Evolutionary or Catastrophic Origin of the Solar System?

The various theories proposed to explain the origin and history of the solar system can be divided into two groups: evolutionary and catastrophic. *Evolutionary theories* state that the formation of the solar system occurred as part of the normal sequence of events that produced the sun. *Catastrophic theories* involve the formation of the sun, followed by a singular cataclysmic event that disrupted the sun and formed the planets.

### Evolutionary Theories

The first serious evolutionary theory on the origin of the solar system was proposed in 1644 by the French scientist and philosopher René Descartes. He proposed the solar system formed from some gigantic whirlpool within a universal fluid. Within this vortex, smaller eddies formed the planets and their satellites. While never specifying the nature of the cosmic material from which the sun and planets formed, the theory did account for the fact that all the orbital motions are in the same direction.

In 1755 the German philosopher Immanuel Kant elaborated on Descartes' idea by using Newton's laws of motion to show that a rotating cloud of gas would flatten into a disk as it contracted. The French mathematician Pierre Simon de Laplace independently proposed the same theory as Kant with some modifications. He stated that as the spinning cloud flattened into a disk, concentric rings of material would form due to rotational forces. These rings would then condense into planets.

The Kant and Laplace theories came to be called the **nebular theory** of the origin of the solar system (Fig. 7–8). This theory had great appeal because it explained many of the orbital and rotational characteristics of the planets and satellites in the solar system. It also explained how one could get a disk from a collapsing ball of interstellar material.

There was, however, one major flaw in the theory. According to the laws of physics, the angular momentum of such a system must remain constant unless some outside force acts on it. This means that if a spinning object shrinks in size, it must increase its rotation rate to compensate for its smaller size, thereby maintaining constant angular momentum (Fig. 7–9). Thus, the sun, which formed at the center of a collapsing cloud of interstellar material, should have a rapid rotation rate. Instead, it has a rather leisurely 25-day rotation period. Because of this apparent contradiction to the laws of

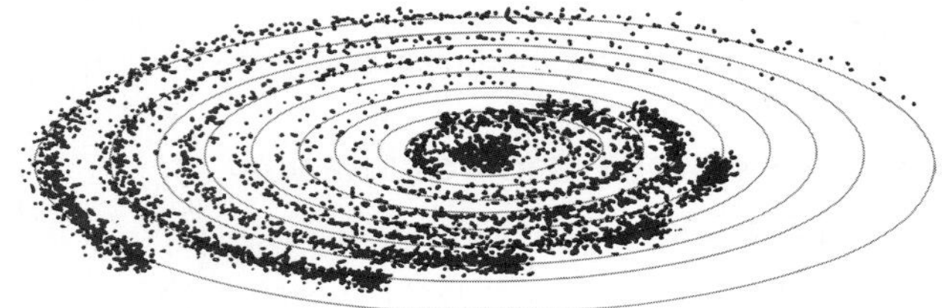

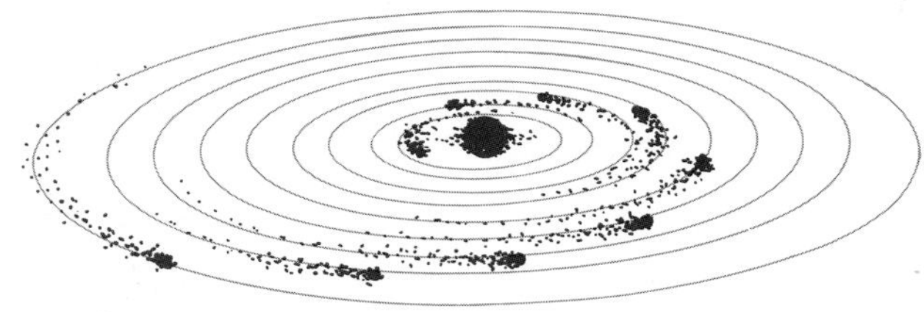

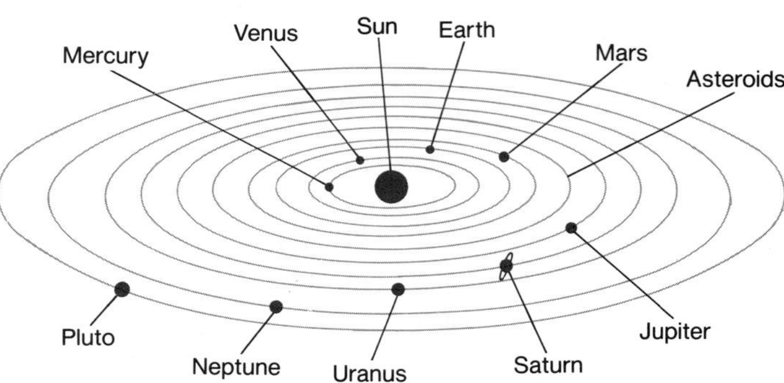


**Figure 7–8** The Kant-Laplace nebular theory for the origin of the solar system involves formation of the sun and planets from an interstellar cloud that flattened to a rotating disk in which detached rings condensed into the planets.

physics, the nebular theory could not be completely accepted until a way was found to explain the sun's slow rotation.

### Catastrophic Theories

Even before Kant proposed his theory, the famous French scientist, Georges Louis de Buffon, suggested the first catastrophic theory for the origin of the solar system in 1745. He proposed that a comet passed so close to the sun that it pulled out gaseous material and dust, which then condensed to form planets. Buffon's idea was largely ignored until the beginning of this century, when the problem of the sun's slow rotation forced scientists to consider alternatives to the nebular theory.

Probably the best known catastrophic theory is the **encounter theory** proposed by the English astronomer Forest R. Moulton and the American geologist Thomas C. Chamberlin in 1900 (Fig. 7–10). Their theory was essentially a modification of Buffon's original idea and envisioned another star passing very close to the sun and pulling away huge filaments of matter that accreted into larger bodies called **planetesimals** that ultimately evolved into the planets and their satellites.

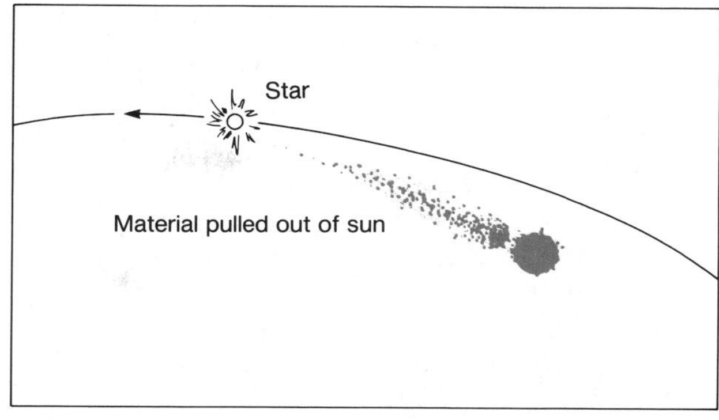


**Figure 7–10** The encounter theory for the origin of the solar system involves a star passing very close to the sun and pulling away huge filaments of matter that condensed to form planets.

**Figure 7–9** A spinning skater illustrates conservation of angular momentum. When the skater pulls in his arms, he spins at a faster rate. Similarly, when a rotating nebula contracts, its speed of rotation increases.

The problem with this theory and all similar catastrophic theories is that near collisions of stars are extremely rare events. Furthermore, calculations showed that even if a near encounter did occur, the material pulled out from the sun would be so hot that it would expand and dissipate into space rather than condense into planetary bodies.

## Current Theory of Origin and History of Solar System

With the failure of catastrophic theories to explain the origin of the solar system, research shifted back to evolutionary models to explain the problem of a slowly rotating sun. Work by the German physicists C. F. von Weiszacker and Gerard P. Kuiper, and the discovery of solar winds, resulted in the current **solar nebula theory** for the origin of the solar system (Fig. 7–11).

Our solar system was formed about 4.6 billion years ago when interstellar material in a spiral arm of the Milky Way Galaxy condensed and began collapsing. As this cloud gradually collapsed under the influence of gravity, it began to flatten and rotate counterclockwise, with about 90 percent of its mass concentrated in the central part of the cloud. As rotation and concentration continued, an embryonic sun, surrounded by a turbulent, rotating cloud of material called a *solar nebula*, formed. The turbulence in this solar nebula resulted in localized eddies forming where condensation of gas and solid particles took place.

As planetesimals formed in these eddies, they all rotated in the same direction around the sun and in the

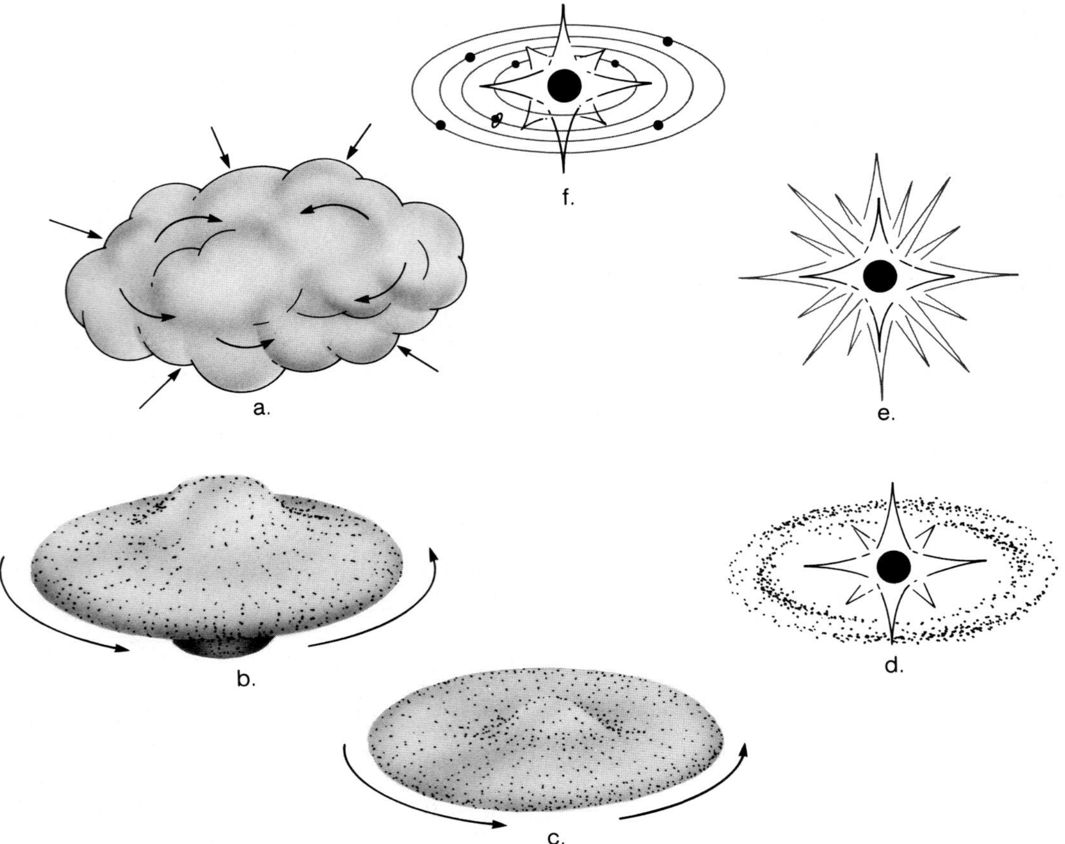


**Figure 7–11** The current solar nebula theory for the origin of the solar system involves (a) a huge nebula condensing under its own gravitational attraction, then (b) contracting, rotating, and (c) flattening into a disk, with (d) the sun forming in the center and eddies gathering up material to form planets. As the sun contracts and begins to shine visibly (e) intense solar radiation blows away unaccreted gas and dust until finally, (f) the sun begins burning hydrogen and the planets, complete their formation.

same direction around their own axes. Something happened, however—perhaps an unusually large collision—that caused Venus to rotate around its axis in the opposite direction. A collision could also explain why Uranus and Pluto do not rotate nearly perpendicular to the plane of the ecliptic. During this early accretionary phase in the solar system's history, collisions between bodies were common, as indicated by the cratering that many planets and satellites exhibit (Perspective 7–1).

The composition of the various planets can be explained by the fact that every element and compound has a temperature and pressure at which it will condense out of the gaseous phase. In the hot inner portions of the solar nebula, refractory elements, which condense at high temperatures, began to condense into solid particles. It was still too hot in this inner region for volatile gases to condense, so they remained in the gaseous state. In the outer regions of the solar nebula, however, these gases began condensing to form ices.

As condensation took place, gaseous, liquid, and solid particles began accreting into ever larger masses. The masses became planetesimals and continued to accrete into true planetary bodies with their composition reflecting their distance from the sun. The inner terrestrial planets are composed of rock and metallic elements that condense at high temperatures. The outer Jovian planets are composed of hydrogen, helium, ammonia, and methane, all of which condense at low temperatures.

While the planets were accreting, material that had been pulled into the center of the nebula also condensed, collapsed, and was heated to several million degrees by gravitational compression. The result was the birth of a star, our sun.

Early in the sun's history, it emitted a tremendous blast of energy that blew away the solar system's unaccreted gases and dust into interstellar space. Such a blast is a normal phase in the evolution of a star and explains why the solar system is so free of extraneous debris. Also during its early history, the sun's magnetic field interacted with the ionized gases of the solar nebula, slowing down its rotation through a magnetic braking process (Fig. 7–12). The discovery that the sun's magnetic field exerted a force on the surrounding nebular gas solved the problem of why the sun has such a slow rotation.

The final feature of the solar system that must be explained by this theory is the asteroid belt. Asteroids probably formed as coalescing matter in a localized eddy between what eventually became Mars and Jupiter. Due to the tremendous gravitational field of Jupiter, however, this matter was prevented from forming a planet.

The solar nebula theory for the formation of our solar system accounts for the similarities of orbits and rotation of the planets and their satellites, the differences in composition of the inner and outer planets, the slow rotation of the sun, as well as the asteroid belt. While some details still need to be worked out, the scenario for the origin and history of the solar system as explained by the solar nebula theory is probably essentially correct.

## METEORITES—EXTRATERRESTRIAL VISITORS

**Meteorites** are thought to be original pieces of material created during the formation of the solar system. During the first 700 million years after the solar system formed, planets and their satellites were frequently bombarded by meteorites. Since then, meteorite activity has greatly diminished. Presently, only about 500 meteorites larger than a baseball reach the Earth's surface each year. Those reaching the Earth's surface have different compositions, reflecting different origins. Most of the meteorites found on Earth are probably fragments resulting from collisions of asteroids.

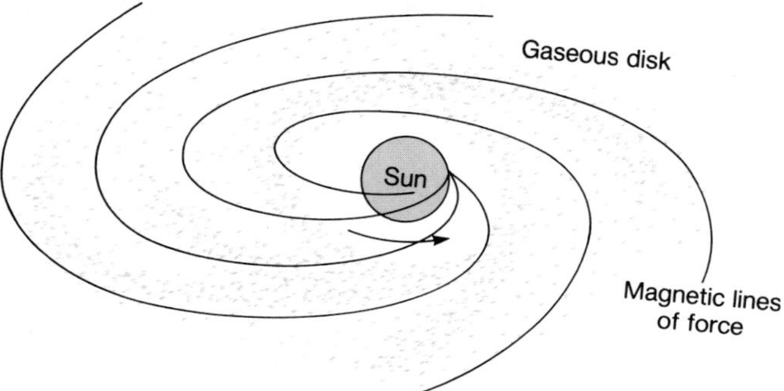


**Figure 7–12** The slow rotation of the sun can be explained by the fact that during its early history, its magnetic force lines interacted with the ionized gases of the solar nebula to slow down its rotation through a magnetic braking process.

### Stones, Irons, and Stony-Irons

Based on their proportions of metals and silicates, meteorites are classified into three broad groups: stones, irons, and stony-irons (Fig. 7–13).

The largest meteorite group is that of **stones**, which comprise about 93 percent of all meteorites. Stony meteorites are not all the same and can themselves be divided into three different types.

**Ordinary chondrites** are the most abundant type, composed of high-temperature ferromagnesian silicate minerals such as olivine and certain pyroxenes. They are 4.6 billion years old and represent material that existed when the solar system was forming. Most chondrites contain **chondrules**, which are small mineral bodies that formed by rapid cooling when the first solid material of the meteorite was condensing.

**Carbonaceous chondrites** have the same general composition as ordinary chondrites, but contain about 5 percent organic compounds, including inorganically produced amino acids. While these organic molecules are not biogenic, they may represent chemical precursors of organic evolution. One of the interesting things about carbonaceous chondrites is that they have a high volatile content, indicating they were never exposed to much heat. This suggests they are samples of early planetary material that formed when the sun did and have not undergone melting or mineral fractionation since then.

Ordinary and carbonaceous chondrites probably represent parts of small asteroids. They are believed to be the most primitive of meteorites, because they have not undergone any of the processing that takes place in the interior of large asteroids, and for this reason are thought to be representative of the original material of the solar system.

**Acondrites**, the third type of stony meteorite, do not contain chondrules. Their composition is similar to that of terrestrial basalts, and their sharp, angular texture may be the result of collisions between larger asteroids.

The second major group of meteorites, the **irons**, account for about 6 percent of all meteorites. Iron meteorites are intergrowths of two varieties of iron and nickel alloys. The large crystal size and chemical composition

## Perspective 7–1

# Meteorite Showers and Their Impact on Planets

A remarkable episode in Earth history occurred soon after the Earth originated and yet left very few traces of its occurrence. When geologists look at the other inner planets and satellites in the solar system they see evidence of a period when meteorite impacts and cratering were far more frequent. The time of this great meteorite shower can be deduced by analyzing the frequency of cratering on the moon and dating the lunar rocks from different areas. When we look at the moon we see that it is heavily cratered. Many of the large craters are floored by basalt flows, which are younger than the craters and are themselves only lightly to moderately cratered.

Radiometric dates of samples brought back from the moon indicate that most of the large craters as well as the highlands are between 3.9 and 4.6 billion years old. The basalts flooring these craters are slightly younger, ranging from 3.2 to 3.8 billion years. The relatively light cratering of these floor basalts indicates that the high intensity of meteorite showering, estimated to be more than 1,000 times the present rate, subsided sometime after 4 billion

a.

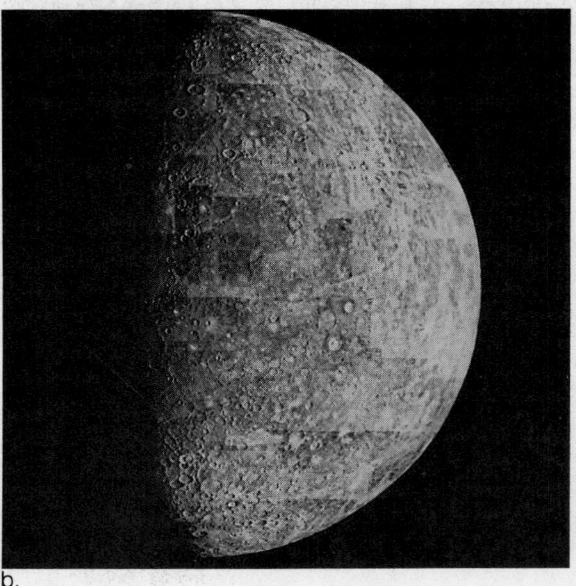

b.

**Figure 1** Cratering of the moon (a) and Mercury (b) by meteorites. Notice the difference between the light-colored cratered highlands and the dark-colored, less cratered maria of the moon.

indicate these meteorites cooled very slowly, as slowly perhaps as 1° C per million years. Such a slow rate of cooling would be possible only in objects as large as asteroids where the hot iron-nickel interior could be insulated from the cold of space. Collisions between such slowly cooling asteroids produced the iron meteorites we find today.

The rarest of meteorites are the **stony-irons,** which make up less than 1 percent of all meteorites. Stony-irons are composed of nearly equal amounts of iron and nickel and silicate minerals. They are generally considered to represent fragments from the interface between the silicate and metallic portions of a large differentiated asteroid.

Meteorites are important to us because their age—about 4.6 billion years—and composition provide us with information about the origin and history of the solar system. Furthermore, their frequency of collision with Earth and the other planets has affected the topography of the planets and, while there is no unanimity on this point, perhaps the evolution of life on Earth (Perspective 7–1).

years ago. Analysis of the density of craters on other planets also shows a period of intense meteorite shower activity between 4.6 and 4.0 billion years ago with only relatively minor meteorite activity since then (Fig. 1).

This period of heavy bombardment took place when the solar system was clearing itself of the many pieces of material that had not yet accreted into planetary bodies or satellites. The reason we see little evidence of this impacting activity on Earth is that weathering, erosion, and recycling of crustal rocks have destroyed almost all such evidence.

Since that period of intense meteorite bombardment around 4.0 billion years ago, meteorite impacts on Earth, while not as intense, have still been numerous. Most of these meteorites come from the asteroid belt, where there are uncounted millions of pieces of rock and small planetoids left over from the origin of the solar system. When these asteroids smash into each other, they send fragments of material into space where they may collide with a planet or moon. The gravitational pull of Jupiter's huge mass also alters the orbits of some asteroids, increasing their chances for collision with planets.

Just as two highways intersect at a crossroad, so, too, do the orbits of Earth and these asteroids, making an eventual collision inevitable. Astronomers have identified at least 40 and estimate there may be as many as 1,000 asteroids larger than a kilometer in diameter whose orbits cross the Earth's. Calculations indicate that we can expect several collisions every million years with asteroids of this size. One of the most famous results of an Earth-asteroid collision is Meteor Crater, Arizona, which was formed between 25,000 and 50,000 years ago and left a crater 1 km in diameter (Fig. 2).

While collisions with Earth are rare, they do happen and can have devastating results. Geologists can identify 116 impact sites on Earth, of which 13 are definitely associated with meteorites, while

**Figure 2** Meteor Crater, Arizona, is the result of an Earth-asteroid collision that occurred between 25,000 and 50,000 years ago. It produced a crater 1 km in diameter.

## METEORITES AND THE AGE OF THE EARTH

The oldest terrestrial rocks so far discovered indicate the Earth is older than 3.8 billion years. To determine how much older, geologists must examine two types of less direct evidence.

The first comes from the radiometric dating of meteorites and lunar samples. We assume that meteorites and lunar rocks have not undergone the extensive recycling and alteration that have affected the rocks of the Earth's crust. Their ages indicate when they formed and, by extrapolation, when the Earth and other planets also formed.

Uranium-lead and rubidium-strontium dating methods show that almost all meteorites are between 4.5 and 4.7 billion years old. Furthermore, these same independent dating techniques indicate the oldest rocks and soils from the moon are also around 4.6 billion years old (Perspective 7–2). Since the solar nebula theory states

## Perspective 7–1 (continued)

the other 103 are probably impact structures, based on chemical and mineralogical evidence that is distinctive to meteorite impacts (Fig. 3).

A collision with a meteorite about 10 km in diameter is believed to have occurred 65 million years ago. Many scientists think this collision resulted in the extinction of the dinosaurs and all large reptilian groups (see Chapter 14). If such a collision occurred, it would have generated a tremendous amount of dust. This dust would have blocked out the sun, causing a cessation in photosynthesis and lowering global temperatures, resulting in massive extinctions.

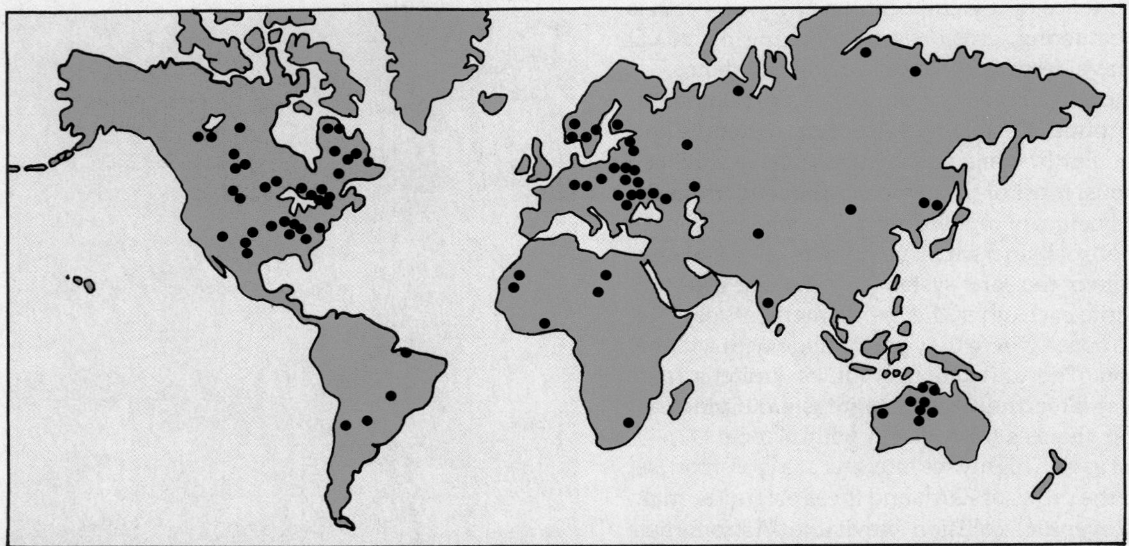

**Figure 3** Map showing the location of major craters thought to be the result of collisions with massive objects. It should be noted that the distribution is largely a result of our knowledge of these areas and does not purport to be a real distribution of craters.

that the sun, planets, and their satellites all formed at the same time from a contracting cloud of stellar material, we would expect the Earth to be the same age as the moon and meteorites.

The second line of evidence for the overall age of the Earth is based on the present-day abundances of the various isotopes of lead occurring in the Earth's crust. Natural lead is a mixture of four stable isotopes: lead 204, 206, 207, and 208. Three of these result from radioactive decay (uranium 235 to lead 207, uranium 238 to lead 206, and thorium 232 to lead 208). The fourth isotope, lead 204, is not the result of radioactive decay. All of the lead 204 that is present on the Earth today originated when the Earth was initially forming. However, only a portion of the other leads (206, 207, and 208) found in the Earth today originated when the Earth formed, and the rest has been added due to the radioactive decay of uranium 238, uranium 235, and thorium 232, respectively.

**Figure 7–13** a. Relative proportion of the different types of meteorites. b. Polished slab of an ordinary chondrite from Pinto Mountains. c. Polished slab of a carbonaceous chondrite from Allende, Mexico. d. Acondrite from Palo Blanco Creek. e. Polished slab of an iron meteorite from Bogou, Upper Volta, Brazil. f. Polished slab of a stony-iron from Thiel Mountain, Antarctica. The white minerals are iron-nickel and the grey minerals are olivine. (Photos c, e, and f courtesy of Brian Mason, Smithsonian Institution, photos b and d courtesy of Ken Nichols, University of New Mexico)

From extensive sampling of the Earth's crust, researchers have determined the present-day abundances of the four isotopes of lead relative to each other and the uranium and thorium parent isotopes that produced three of them. Although there is no way to estimate the original abundances of lead, meteorites that do not contain any uranium or thorium can provide a reasonable approximation of the Earth's original abundances of lead (Fig. 7–14).

Comparing the amounts of primordial lead 208, 207, 206, and 204 to their respective present values, geologists can determine how much lead has been added by radioactive decay since the Earth was formed. They can then calculate, using the half-life of each parent, how long it took to create the difference in the amount of present-day lead and primordial lead for each of the uranium-lead isotope pairs. These calculations yield an age of about 4.6 billion years, which is consistent with the ages determined from meteorites and lunar rocks. This is, of course, what we would expect, based on the solar nebular theory for the origin of the solar system.

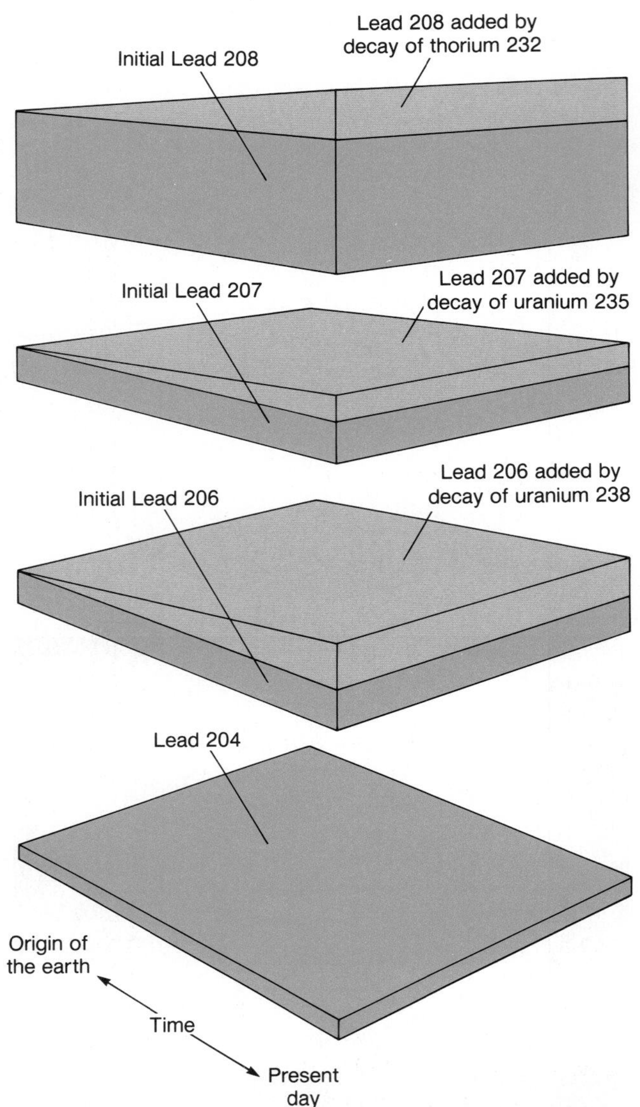


**Figure 7–14** Determination of the age of the Earth based on its initial and present-day abundances of lead isotopes. The initial abundance is estimated from meteorites, while the present abundance is calculated from the average of numerous rock measurements. The difference between the two represents the addition from the decay of radioactive uranium and thorium since the Earth originated.

## ORIGIN AND DIFFERENTIATION OF THE EARLY EARTH

As matter was accreting in the various turbulent eddies that swirled around the protosun, one such eddy eventually gathered enough material to form a planet which we call the Earth.

The Earth, as noted in Chapter 1, consists of a series of concentric layers of different composition and densities (Fig. 7–15). These density layers are a fundamental feature of the Earth, and presumably this differentiation occurred very early in the Earth's history.

Geologists know that the Earth is 4.6 billion years old. However, the oldest known rocks are 3.8-billion-year-old highly deformed gneisses from Greenland. These gneisses, just like the younger crustal rocks, are composed of relatively light silicate minerals. It appears that a crust, a heavier silicate mantle, and an iron-nickel core were already present by 3.8 billion years ago, or 800 million years after the Earth was formed. Two general theories have been proposed to explain the core-mantle-crust density layering of the Earth (Fig. 7–16), and both theories are consistent with the solar nebula theory of the origin of the solar system.

### Homogeneous Accretion

The **homogeneous accretion** model begins with an early solid Earth of generally uniform composition and density throughout. The iron and nickel of the present core were distributed fairly evenly throughout a larger mass of lighter silicate minerals.

The early protoearth is visualized as being rather cool, so that the elements and nebular rock fragments accreting to it were solids, rather than gases or liquids. In order for the iron and nickel to concentrate in the core,

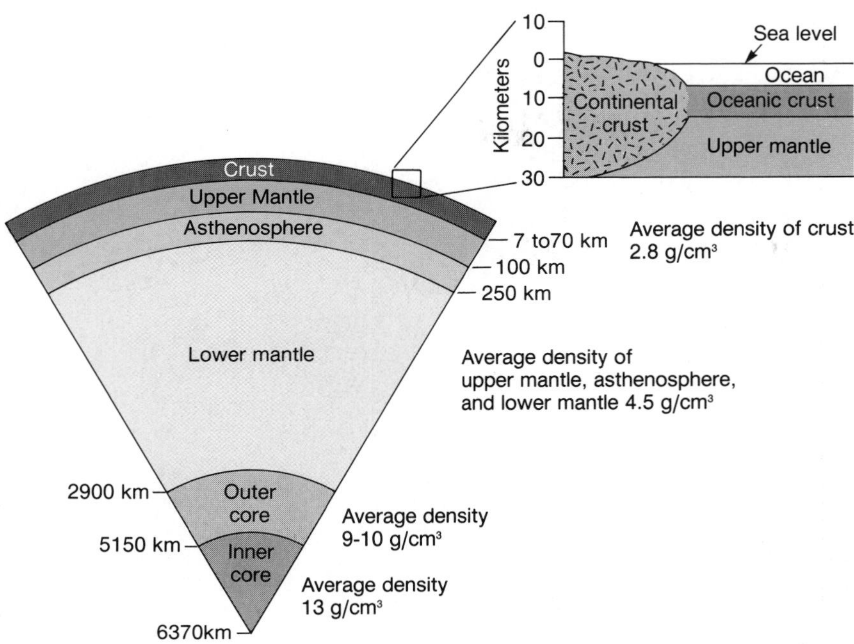


**Figure 7–15** Cross section of the Earth showing the various layers and their average density. The crust is divided into a continental and oceanic portion. Continental crust is 20 to 70 km thick; oceanic crust is about 7 km thick.

the interior of the early Earth must have been hot enough for them to melt and sink through the surrounding lighter silicate minerals.

Iron and nickel melt at lower temperatures than silicates, and because they are denser than silicates, would settle to the center of the Earth. Meantime the silicates would soften and slowly flow upward, beginning the differentiation of the mantle from the core. The source of the heat would come primarily from the decay of short-lived radioactive isotopes, which were much more abundant during the early history of the Earth than they are today. Additional heat would also have been generated from gravitational compression and from the energy of meteorites impacting on the Earth.

Calculations indicate that with a uniform distribution of elements in a solid protoearth, enough heat could be generated to begin melting iron and nickel at depths of about 650 km. Melting would have to begin at shallow depths because the temperature at which melting begins increases with pressure, and so the melting point of any material becomes greater toward the Earth's center.

During this early history of the Earth, many of the easily melted lighter elements such as calcium, potassium, and sodium would move upward through the mantle and concentrate near the Earth's surface, where they would begin to form the crust. The heavy elements such as iron and nickel would sink to the center to form the core.

The major problem with this model is that it supposes the Earth underwent a period of radioactive heating and internal reorganization. Also, current research indicates that gravitational accretion energy, even when combined with radiogenic heat, may be insufficient to produce the extensive early melting called for from a cold, planetary body.

## Inhomogeneous Accretion

To overcome the problems presented by the homogeneous accretion model, the **inhomogeneous accretion** model has been proposed. It states that the Earth's core, mantle, and crust could have condensed sequentially from the hot nebular gases forming the protoearth.

Even though iron and nickel melt at lower temperatures than most silicates, they condense from the gaseous state at a slightly higher temperature than most silicates. Calculations show that in a cooling cloud of hot nebular gases, iron and nickel would condense first and,

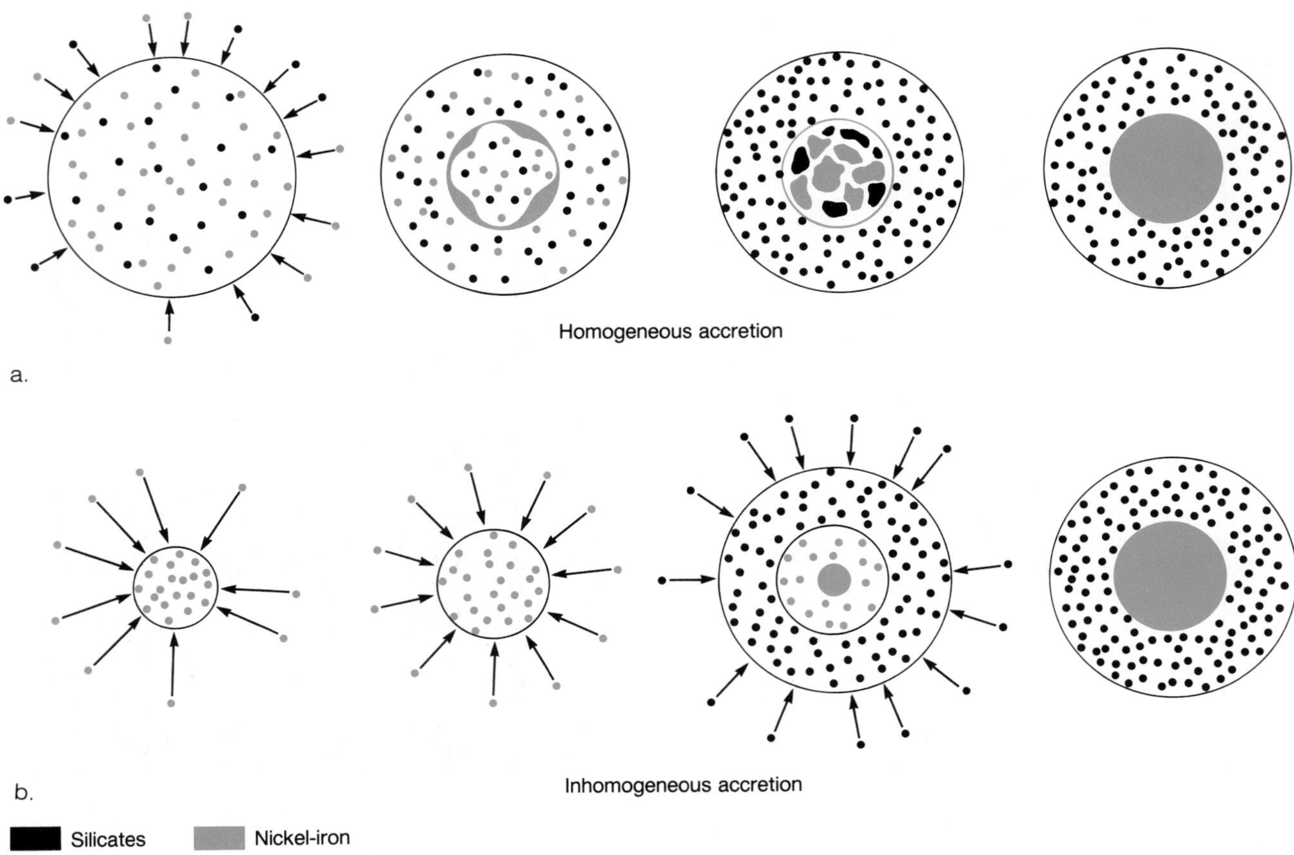


**Figure 7–16** Two principal theories are proposed to explain the core-mantle-crust density layering of the Earth. In the homogeneous accretion model (a) the Earth was initially a uniform mixture of silicates and iron-nickel. Subsequent radioactive heating led to a concentration of iron and nickel in the core and silicate minerals in the mantle and crust. In the inhomogeneous accretion model (b) the iron and nickel of the core formed first, followed by accumulation of silicate minerals in the mantle and crust.

if they accreted, form the core of the protoearth. As the cloud further cooled, iron and magnesium silicates would then condense to form the mantle, while the lightest and most volatile elements that form the crust would condense last.

## Conclusions on Homogeneous versus Inhomogeneous Accretion

Presently there is no unequivocal evidence favoring one model over the other. Both are based on the currently accepted solar nebula origin model of the solar system. Regardless of which model eventually proves correct, geologists do know that the differentiation of the Earth occurred very early in its history. And once the Earth was differentiated, geologists also know from physical, chemical, and petrological evidence, the Earth's surface underwent a pre-Archean period of intense volcanism. This volcanism was the result of convection cells in the upper mantle and radiogenic heating of the crust. During this time, the Earth's crust differentiated into oceanic and continental regions that developed around upwelling and downwelling segments of upper-mantle convection cells. These oceanic and continental regions, respectively, resembled the greenstone belts and the high-grade granite-gneiss terrains of the Archean Eon, which are discussed in the next chapter.

## Perspective 7-2

# The Origin and Early History of the Moon

We probably know more about the Earth's moon than any other planet except for the Earth. However, even though we have studied the moon for centuries through telescopes and have even sampled it directly, it still remains one of the most enigmatic of celestial bodies. It is one-fourth the diameter of the Earth, has a low density (3.3 $g/cm^3$) relative to the terrestrial planets, and an unusual chemistry in that it is bone-dry, having been heavily depleted in most volatile elements. The moon orbits the Earth and rotates on its own axis at the same rate, so we always see the same side. Furthermore, the Earth-moon system is unique among the terrestrial planets. Neither Mercury or Venus have a moon, and the two moons of Mars, Phobos and Deimos, are probably captured asteroids.

The surface of the moon can be divided into two major divisions, the topographically lower, dark-colored *marias*, or "seas," and the light-colored, higher *highlands* (Fig. 1). The highlands are the oldest parts of the Moon and are heavily cratered, providing striking proof of the massive meteorite bombardment that occurred in our solar system more than 4 billion years ago (See Perspective 7–1).

**Figure 1** Photograph of the Moon showing its major features. The dark areas are the marias, which are covered by basaltic lava flows, and the light-colored, heavily cratered areas are the highlands.

Study of the several hundred kilograms of material returned by the Apollo missions indicates that four general rock types dominate the lunar surface. The greater part of the marias are covered by *basaltic rocks*, which differ slightly in their composition from those on Earth. The second lunar rock type is *gabbro*, a coarse-grained mineralogic equivalent of basalt. The third rock type is a *breccia* composed of fragments of other rocks, including *anorthosite*, a coarse-grained igneous rock composed primarily of the feldspar mineral anorthite.

In addition to the four main rock types, the lunar surface is covered with a *regolith* or "soil," estimated to be 3 m to 4 m thick. This regolith is composed of fragments of the various aforementioned igneous rocks, glass spherules, and meteorite material. It is believed to be the result of debris formed by meteorite impact.

The interior structure of the moon is quite different from that of the Earth, reflecting a different evolutionary history (Fig. 2). The highland crust is thick (65 to 100 km) and comprises about 12 percent of the moon's volume. It was formed about 4.4 billion years ago immediately following the moon's accretion. The highlands are composed principally of anorthosite, which is primarily responsible for its white appearance.

A thin covering (1 to 2 km thick) of basaltic lava forms the marias. They cover about 17 percent of the lunar surface, mostly on the side facing the Earth. These maria lavas come from partial melting of a thick underlying mantle of varying silicate composition. Moonquakes occur at a depth of about 1,000 km, and below this depth, seismic S waves apparently are not transmitted. Since S waves do not travel through liquid, their lack of transmission implies the innermost mantle may be partially molten. There is increasing evidence that the moon has a small (300 km to 500 km in radius) metallic core comprising 2 to 5 percent of its volume.

## Perspective 7–2 (continued)

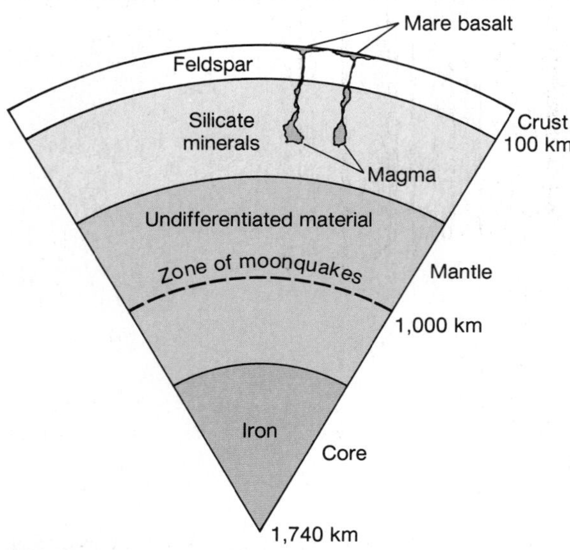


**Figure 2** The internal structure of the moon is different from that of the Earth. A white feldspathic crust overlies a mineralogically differentiated upper mantle. The upper mantle is the source for the maria basalt lavas. The rest of the mantle consists of undifferentiated material. Moonquakes occur at a depth of 1,000 km. Because seismic S waves are apparently not transmitted below this depth, it is believed the innermost mantle is liquid. Beneath this layer is a small metallic core.

While the origin and earliest history of the moon is still unclear, the basic stages in its history are well understood. It formed some 4.6 billion years ago and shortly thereafter was partially or wholly melted. As this silicate melt started cooling, the feldspar anorthite began crystallizing. Due to the low density of the anorthite crystals and the anhydrous nature of the silicate melt, the thick anorthosite highland crust formed. The remaining silicate melt cooled and crystallized to produce the zoned mantle, while the heavier metallic elements formed the small metallic core.

The formation of the lunar mantle was completed by about 4.4 to 4.3 billion years ago. The maria basalts, derived from partial melting of the upper mantle, were extruded during the great lava floods between 3.8 and 3.2 billion years ago.

Numerous models have been proposed for the origin of the moon, including capture from an independent orbit, formation with the Earth as part of an integrated two-planet system, breaking off from the Earth during accretion, and formation resulting from impact with the Earth by a Mars-sized planetesimal. These various models are not mutually exclusive, and elements of some occur in others. At this time, geologists cannot agree on a single model, as each one has some inherent problems. However, the model that seems to account best for the moon's particular composition and structure involves an impact with the Earth by a Mars-sized planetesimal (Fig. 3).

In this model, a giant planetesimal crashed into the Earth about 4.6 to 4.4 billion years ago and ejected a hot disk of material that formed the moon. The material that was ejected was mostly in the liquid and vapor phase and came primarily from the mantle of the colliding planetesimal. As it cooled, the various lunar layers crystallized out in the order we have discussed.

Such a model accounts for the unusual chemistry and internal structure of the moon and warrants further study, particularly in light of the current theories of Earth-meteorite impact events.

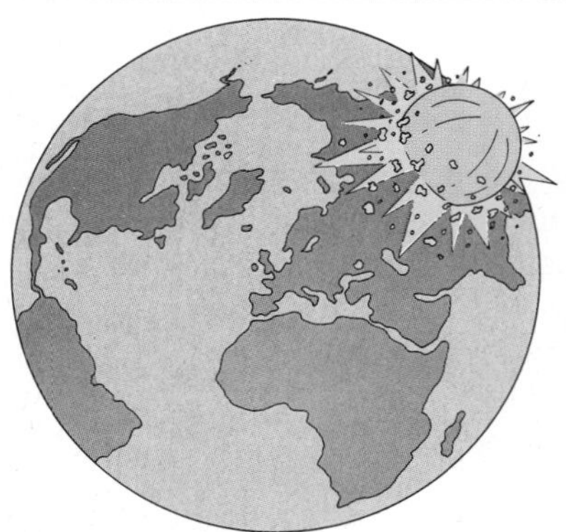

**Figure 3** According to one hypothesis for the origin of the moon, a Mars-sized planetesimal crashed into the Earth 4.6 to 4.4 billion years ago, and ejected a hot disk of material that formed the moon.

## CHAPTER SUMMARY

1. The origin and history of the universe are among the most compelling and yet frustrating questions science can address. The reason is that they must be reconstructed from the limited human perspective.
2. Astronomers and physicists can calculate how old the universe is, and what the basic laws governing it are. They can reconstruct its history from the present back to $10^{-43}$ seconds after its origin. They cannot, however, explain what happened in that first $10^{-43}$ seconds when none of the physical and chemical laws that we know existed.
3. The universe began with a Big Bang approximately 15 to 20 billion years ago. Astronomers have deduced this from the fact that celestial objects can be seen moving away in what appears to be an ever-expanding universe. Furthermore, the universe has a background radiation of 3 K, which represents the cooling remnant of that explosion.
4. During its first 2 billion years the universe expanded greatly; stars and galaxies formed that combined into such larger structures as clusters and superclusters. During this time the universe became clumpy and enriched in the heavier elements as stars underwent nuclear reactions.
5. In an arm of the Milky Way Galaxy, our own solar system formed from a rotating interstellar cloud of matter about 4.6 billion years ago. As this cloud condensed, it eventually collapsed under the influence of gravity and flattened into a counterclockwise rotating disk. Within this rotating disk, the sun, planets, and then satellites accreted from the turbulent eddies of nebular gases and solids.
6. Meteorites provide vital information about the age and composition of the solar system. The three major groups of meteorites are stones, irons, and stony-irons. Each has a different composition, reflecting a different origin.
7. Temperature as a function of distance from the sun played a major role in the type of planets that evolved. The inner terrestrial planets are composed of rock and metallic elements that condense at high temperatures. The outer Jovian planets plus Pluto, are composed of hydrogen, helium, ammonia, and methane, all of which condense at lower temperatures.
8. During the early period of the solar system's history the sun emitted a tremendous blast of energy that swept the solar system free of unaccreted gas and dust. It was during this first 500 to 600 million years of existence that most of the cratering of the planets and satellites took place.
9. The Earth formed from one of the swirling eddies of nebular material 4.6 billion years ago, and by 3.8 billion years ago was differentiated into its present-day structure. It either accreted as a solid body which then underwent differentiation during a period of heating, or the core-mantle-crust condensed in sequence from a cooling cloud of nebular gas.

## IMPORTANT TERMS

acondrite
Big Bang
carbonaceous chondrite
chondrule
Doppler effect
electromagnetic force
encounter theory
gravity
homogeneous accretion
inhomogeneous accretion
iron meteorite (irons)
meteorite
nebular theory
ordinary chondrite
planetesimal
solar nebula theory
stony meteorite (stones)
stony-iron meteorite (stony-irons)
strong nuclear force
weak nuclear force

## REVIEW QUESTIONS

1. What two fundamental phenomena indicate that the Big Bang occurred 15 to 20 billion years ago?
2. If the oldest terrestrial rocks are 3.8 billion years old, why do geologists think the Earth is 4.6 billion years old?
3. Why is the Doppler effect important to the concept of an expanding universe?
4. Why is an evolutionary theory for the origin of the solar system more appealing than a catastrophic one?
5. How does the solar nebula theory account for the four major characteristics of the solar system?
6. What are the three major groups of meteorites, and how do they aid geologists in determining the age and composition of the solar system?
7. What other important roles have meteorites played in the history of the solar system?
8. What are the strengths and weaknesses of the homogeneous and inhomogeneous accretion models for the origin and differentiation of the early Earth?

## ADDITIONAL READINGS

Barrow, J. D., and J. Silk. 1980. The Structure of the Early Universe. *Scientific American*, 242, no. 4: 118–129.

Shu, F. H. 1982. *The Physical Universe*. San Francisco: W. H. Freeman and Company.

Snow, T. P. 1985. *The Dynamic Universe, An Introduction to Astronomy*, 2d ed. St. Paul, Minn.: West Publishing Co.

Taylor, S. R. 1987. The Origin of the Moon. *American Scientist*, 75, no. 5: 468–477.

Weisskopf, V. F. 1983. The Origin of the Universe. *American Scientist*, 71, no. 5: 473–480.

Wetherill, G. W. 1981. The Formation of the Earth from Planetesimals. *Scientific American*, 244, no. 6: 162–175.

# 8

## CHAPTER OUTLINE



## PROLOGUE

*Imagine a barren, lifeless, waterless, hot planet with a poisonous atmosphere. Volcanoes erupt nearly continuously, meteorites and comets flash through the atmosphere, and cosmic radiation is intense. The planet's crust is thin and unstable, composed entirely of dark-colored igneous rock. Storms form in the turbulent atmosphere, and lightning discharges are common, but no rain falls; and since there is no oxygen in the atmosphere, nothing burns. Rivers and pools of molten rock emit a continuous reddish glow.*

*Such a description sounds like something from science fiction, but it is probably reasonably accurate for the Earth shortly after it formed. We emphasize "probably" because no record exists for the earliest chapter of Earth history, the interval from 4.6 to 3.8 billion years ago. We can only speculate about what the Earth was like during this time, based on our knowledge of how planets form and about other Earth-like planets.*

*When the Earth formed, it had a tremendous reservoir of primordal heat, heat generated by colliding particles as the Earth accreted, by compression, and by the decay of short-lived radioactive elements. Many geologists believe that the early Earth was so hot that it was partly or perhaps almost entirely molten. No one knows what the Earth's surface temperature was during its earliest history, but it was almost certainly too hot for liquid water to exist or for any known organism to survive. Volcanism must have been ubiquitous and nearly continuous. Molten rock later solidified to form a thin, discontinuous, dark-colored crust, only to be disrupted by upwelling magmas.*

*Assuming that visitors to the early Earth could tolerate the high temperatures, they would also have to contend with other inhospitable factors. The atmosphere would be unbreathable by any of today's inhabitants. It probably contained mostly carbon dioxide, methane, ammonia, and water vapor, but little or no oxygen. No ozone layer existed in the upper atmosphere, so our hypothetical visitors would receive a lethal dose of ultraviolet radiation, unless protected, and would be threatened constantly by comet and meteorite impacts. The view of the moon would have been spectacular because it was much closer to the Earth. However, its gravitational attraction would have*

# Precambrian History—The Archean Eon

*caused massive Earth tides. And finally, our visitors would experience a much shorter day because the Earth rotated on its axis in as little as 10 hours.*

*Eventually, much of the Earth's primordial heat was dissipated into space and its surface cooled. As the Earth cooled, water vapor began to condense, rain fell, and surface water began to accumulate. The bombardment by comets and meteorites slowed. By 3.8 billion years ago a few small areas of continental crust existed. The atmosphere still lacked oxygen and an ozone layer, but by at least 3.5 billion years ago life appeared. Some inconclusive evidence indicates that the most primitive life-forms existed even earlier.*

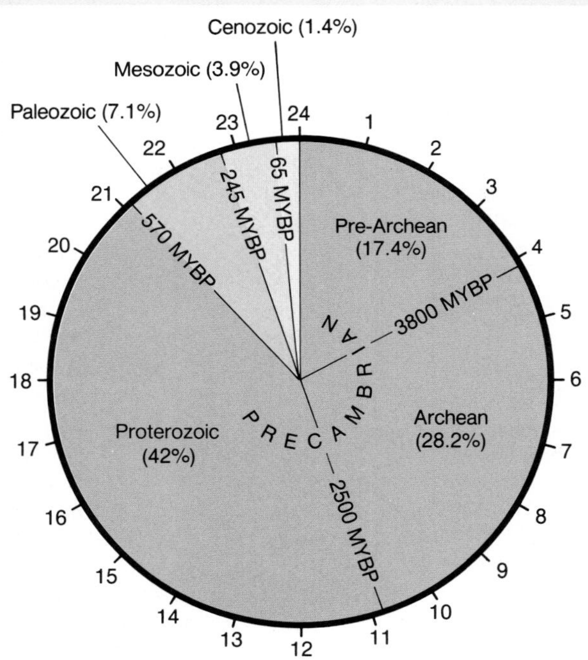


**Figure 8–1** Geologic time represented on a 24-hour clock. Precambrian time (shaded) includes more than 21 hours on this clock, or more than 87 percent of all geologic time.

## INTRODUCTION

**Precambrian** is a widely used informal term that refers to both rocks and time. All crustal rocks lying beneath strata of the Cambrian System are called Precambrian. As a geochronologic term, Precambrian includes all geologic time from the Earth's origin 4.6 billion years ago to the beginning of the Phanerozoic Eon 570 million years ago (Fig. 1–24). If all geologic time were represented by a 24-hour day, slightly more than 21 hours of it would be Precambrian (Fig. 8–1). Unfortunately for geologists, not all of this vast interval of time is recorded; rocks older than 3.8 billion years have not yet been discovered on Earth.

Establishing formal, widely recognized subdivisions of the Precambrian is a difficult task. Precambrian rocks are exposed on all continents, but many have been complexly deformed and altered by metamorphism, and much of Precambrian Earth history is recorded by nonstratified rocks. Thus, the principle of superposition cannot be applied, particularly in older Precambrian terranes, making relative-age determinations difficult. In addition, most correlations must be based on radiometric age dates, since these rocks contain few fossils of biostratigraphic significance. Considering the difficulties inherent in working with these ancient rocks, it is not surprising that Precambrian terminology has developed locally (Table 8–1).

In an effort to standardize usage, the North American Commission on Stratigraphic Nomenclature in 1982 approved a proposal that recognizes two Precambrian eons, the **Archean Eon** and **Proterozoic Eon** (Table 8–1). This usage has gained wide acceptance in North America and is followed in this book.

Notice in Table 8–1 that the Archean Eon began 3.8 billion years ago, a date that corresponds with the oldest known crustal rocks. The pre-Archean is an informal term for that part of Earth history unrecorded by rocks, fully 17 percent of all geologic time (Fig. 8–1). The

**Table 8–1** Comparison of classification schemes for Precambrian rocks and time. The scheme proposed by Harrison and Peterman (1982) is followed in this book.

Time, in millions of years before present: 0, 500, 1000, 1500, 2000, 2500, 3000, 3500, 4000, 4500

| Canada | Australia | United States | |
|---|---|---|---|
| Stockwell, 1964 | Dunn, Plumb, Roberts, 1966 | U.S. Geological Survey, 1971 | Harrison, Peterman, 1982 |
| Phanerozoic | Phanerozoic | Phanerozoic | Phanerozoic |
| Proterozoic: Hadrynian | Proterozoic: Adelaidian | Precambrian Z | Proterozoic: Late Proterozoic |
| | | 800 | |
| 880 | | | |
| | | | 900 |
| Helikian | | Precambrian Y | Middle Proterozoic |
| | 1400 | | |
| | Carpentarian | | |
| | | 1600 | 1600 |
| 1640 | | | |
| Aphebian | | | |
| | 1800 | | |
| | Nullaginian | Precambrian X | Early Proterozoic |
| | 2300 | | |
| 2390 | | | |
| | | 2500 | 2500 |
| Archean | Archean | Precambrian W | Archean: Late Archean |
| | | | 3000 |
| | | | Middle Archean |
| | | | 3400 |
| | | | Early Archean |
| | | | (3800?) |
| | | | (pre-Archean) |

Archean-Proterozoic boundary at 2.5 billion years ago is marked by major unconformities in many areas. Furthermore, a change in the style of crustal evolution occurred at about that time, but the change from the Archean to the Proterozoic style did not occur simultaneously on all continents. The end of the Proterozoic 570 million years ago was the time when organisms with preservable hard parts appeared in abundance.

The Precambrian subdivisions are geochronologic, based on radiometric-age dates rather than time-stratigraphic ages. This departs from normal practice in which geologic systems based on stratotypes (Chapter 3) are the basic time-stratigraphic units. An example will help clarify this point. The Cambrian Period is a geochronologic term, but it corresponds to the Cambrian System, a time-stratigraphic unit based on a body of rock with a stratotype in Wales (Chapter 3). By contrast, Precambrian terminology is strictly geochronologic. There are no stratotypes for the subdivisions of the Precambrian.

An alternative scheme for designations of Precambrian time was adopted in 1971 by the United States Geological Survey (USGS) (Table 8–1). Simple letter designations are used instead of names, and these, too, are based on age dates rather than stratotypes. Although this usage is not followed in this book, students will encounter it on recent USGS maps and in some books and articles.

**Figure 8–2** The Morton Gneiss of Minnesota is at least 3.6 billion years old, and is one of the oldest known rock units on Earth. (Photo courtesy of James A. Grant, University of Minnesota, Duluth).

## PRE-ARCHEAN CRUSTAL EVOLUTION

Geologists know that some continental crust existed at least 3.8 billion years ago, since rocks this old are known from several areas, including Minnesota, Greenland, and South Africa (Fig. 8–2), and many of these are metamorphic, which means they formed from even older rocks. Furthermore, Archean sedimentary rocks in Australia contain detrital minerals dated at 4.2 billion years, indicating that source rocks at least that old were present.

Most geologists agree that some kind of pre-Archean crust probably existed, but since no rocks this old are known, the origin and composition of the earliest crust must be inferred from geochemical considerations. Many investigators think there was an early episode of partial melting, magma formation, and the rise of magmas that formed a crust. Whether this crust was worldwide or more restricted is debatable, as is the cause of partial melting.

Some crustal evolution models call for one-time events to provide the heat necessary for partial melting. Figure 8–3 illustrates such a model based on meteorite impacts. No direct evidence exists for an early episode of large-scale meteorite bombardment of the Earth, but by analogy with the other three terrestrial planets and the moon (see Perspective 7–1), it seems likely that such an event occurred. If so, it probably ended 3.9 billion years ago as it did on the moon.

The model illustrated in Figure 8–3 may account for the initial formation of crust, but subsequent crustal evolution would have been largely unrelated to the formative process. By contrast, Kent Condie of the New Mexico Institute of Mining and Technology has proposed a model that relies on internally generated radiogenic heat and decreasing heat production within the Earth through time (Figs. 8–4, 8–5). Thus, the earliest crust

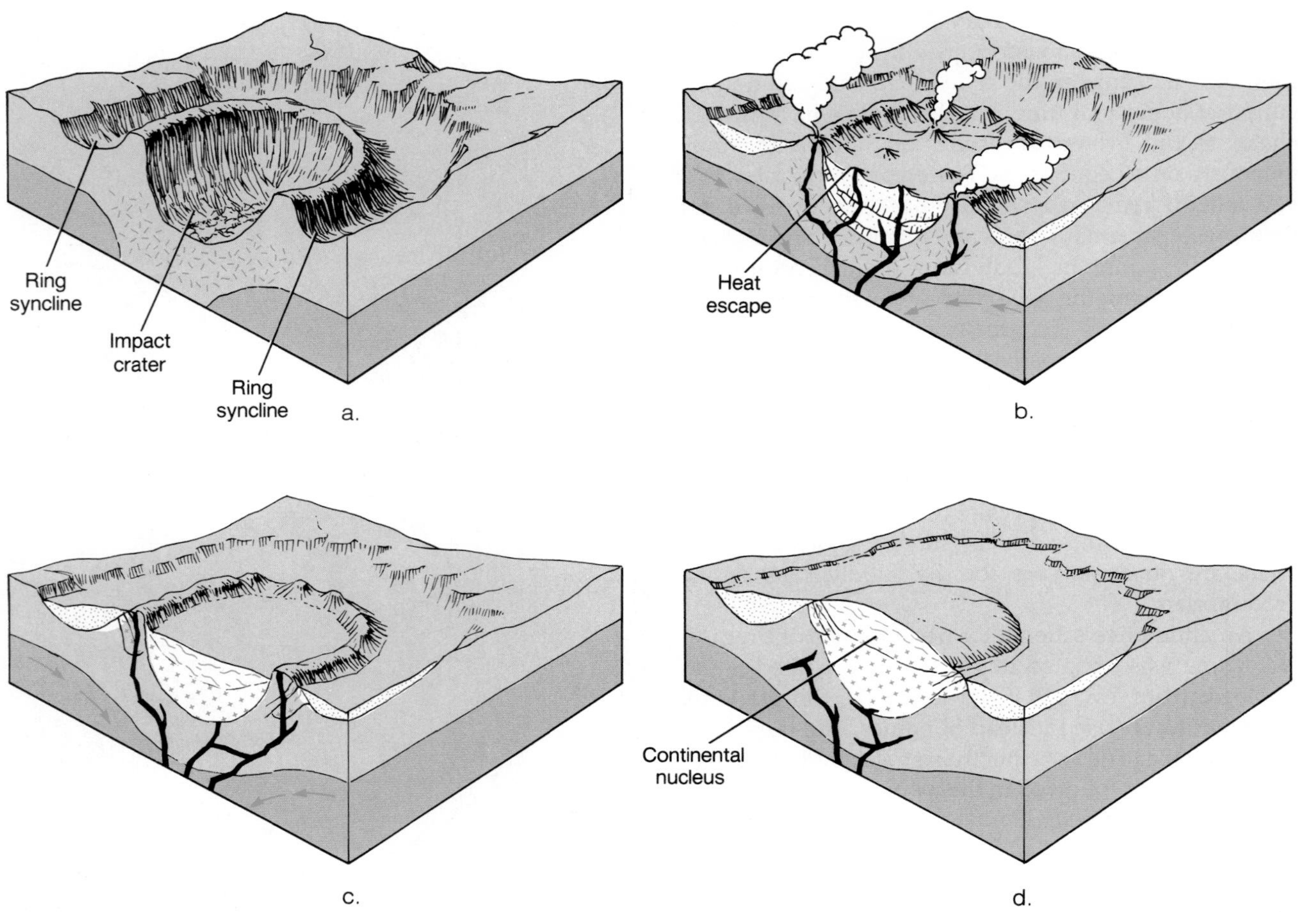


**Figure 8–3** Meteorite impact model for the origin of a continental nucleus. a. Immediately after a meteorite impact. b. Volcanism and erosion of the crater rim fill the crater with low-density volcanics and sediments. Release of heat at the surface causes cooling and downflow in the mantle. c. Downflow continues and causes compression. d. Continental nucleus rises isostatically when heat is lost and downflow ceases.

and its continued evolution are related to the same process.

An important aspect of Condie's Model is the evolution of sialic continental crust. Recall from Chapter 1 that sialic crust contains considerable silicon, oxygen, and aluminum. The earliest crust, however, was probably thin, unstable, and composed of ultramafic igneous rock. Such rock is relatively low in silica ($SiO_2$) content compared with other igneous rocks. According to Condie's model, this early ultramafic crust was disrupted by upwelling basaltic magmas at ridges and was consumed at subduction zones (Fig. 8–4). Ultramafic crust would therefore have been destroyed, since its density was great enough to make recycling by subduction very likely. Apparently, only more sialic crust, because of its lower density, is immune to destruction by subduction.

With a decrease in the Earth's radiogenic heat production, a second stage of crustal evolution began (Fig. 8–4). Partial melting of earlier formed basaltic crust resulted in the formation of andesitic island arcs, and partial melting of lower crustal andesites yielded granitic magmas that were emplaced in the earlier-formed crust. Plate motions accompanied by subduction and collisions of island arcs formed several sialic continental nuclei by the Early Archean.

We hasten to point out that the model described is by no means accepted by all geologists, nor is crustal evolution restricted to the pre-Archean. Episodes of continental growth during the Archean occurred more rapidly, however.

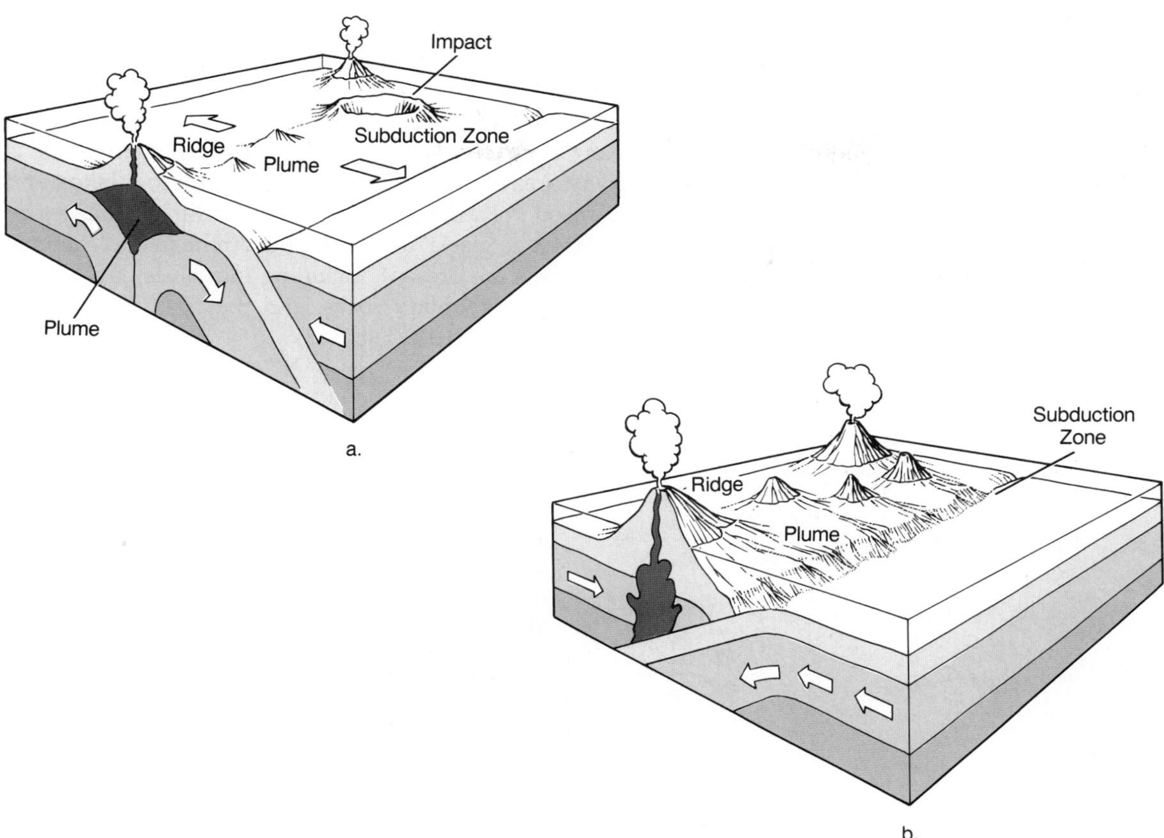


**Figure 8–4** Model for origin of pre-Archean crust. The earliest crust may have been composed of ultramafic rock but was disrupted by rising basaltic magmas. a. Basaltic crust is generated at ridges underlain by mantle plumes. Because of its high density, basaltic crust is consumed at subduction zones and is recycled. b. Andesitic island arcs form at convergent plate margins. Sialic crust grows by collisions of island arcs and intrusions of granitic magmas.

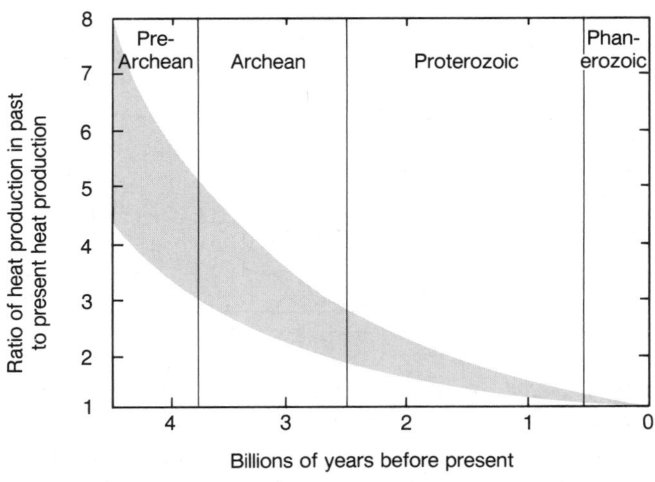


**Figure 8–5** Ratio of radiogenic heat production in the past to present heat production. The shaded band encloses the ratios according to different models, all of which show an exponential decay of radioactive elements through time. Estimates of heat production for 4 billion years ago range from three to six times the present heat production, whereas heat production at the beginning of the Phanerozoic Eon was only slightly greater than it is now.

## SHIELDS AND CRATONS

Each continent is characterized by a vast area of exposed ancient rocks called a **Precambrian shield**. Continuing outward from the shields are broad platforms of buried Precambrian rocks that underlie much of the continents. The shields and buried platforms are collectively called **cratons** (Fig. 8–6). We can think of the cratons, which are composed of both Archean and Proterozoic rocks, as the ancient nuclei of the continents.

In North America the **Canadian Shield** includes most of northeastern Canada, a large part of Greenland, parts of the Lake Superior region in Minnesota, Wisconsin, and Michigan, and the Adirondack Mountains of New York (Fig. 8–7). In general, the Canadian Shield is a vast area of subdued topography, numerous lakes, and exposed Precambrian rocks, thinly covered in places by Pleistocene glacial deposits. Both Archean and Proterozoic rocks are present, including intrusives, lava flows, various sedimentary rocks, and metamorphic equivalents of all of these (Fig. 8–8).

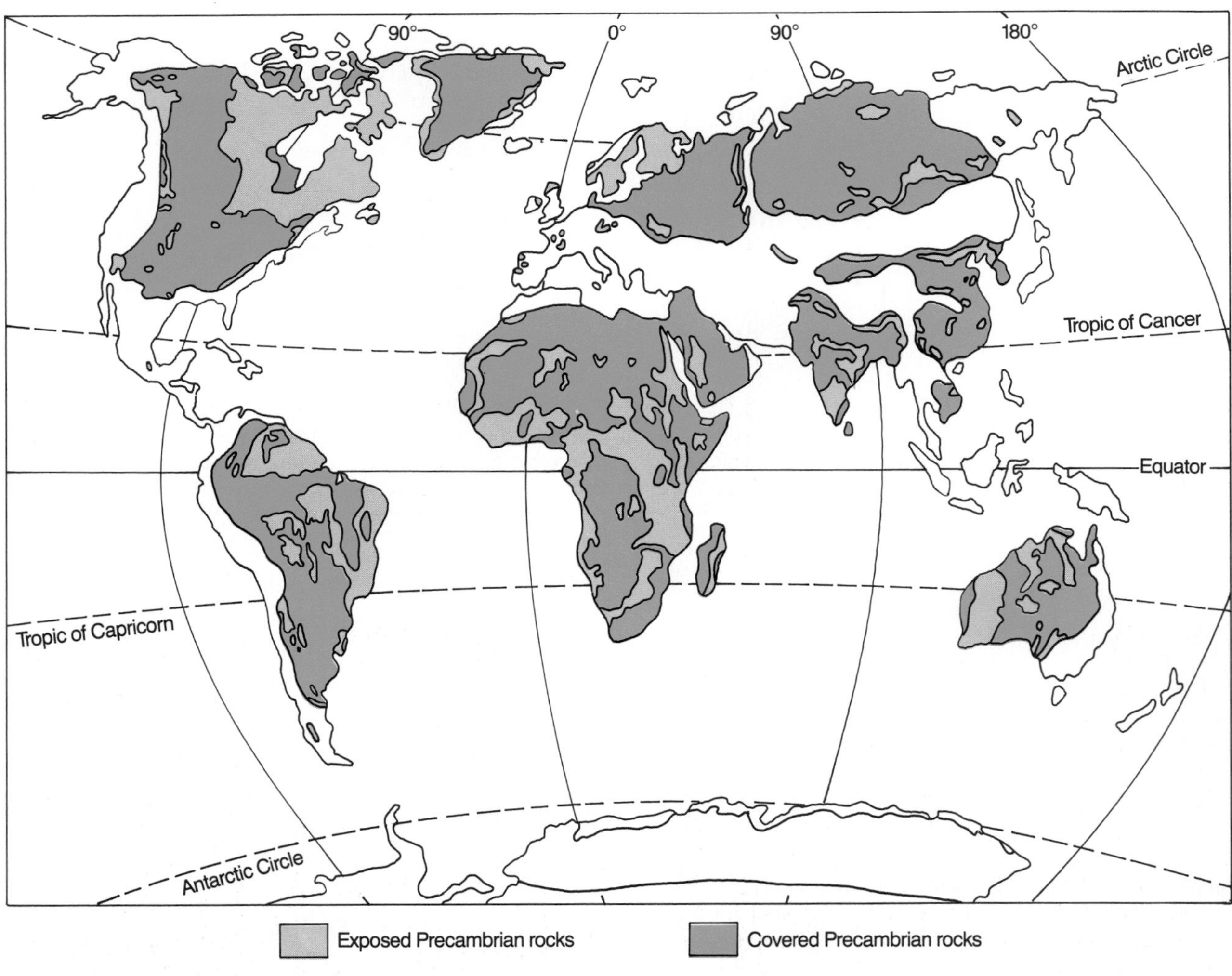


**Figure 8–6** Precambrian cratons of the world. The areas of exposed Precambrian rocks are the shields, while the buried Precambrian rocks are the platforms. Shields and platforms collectively make up the cratons.

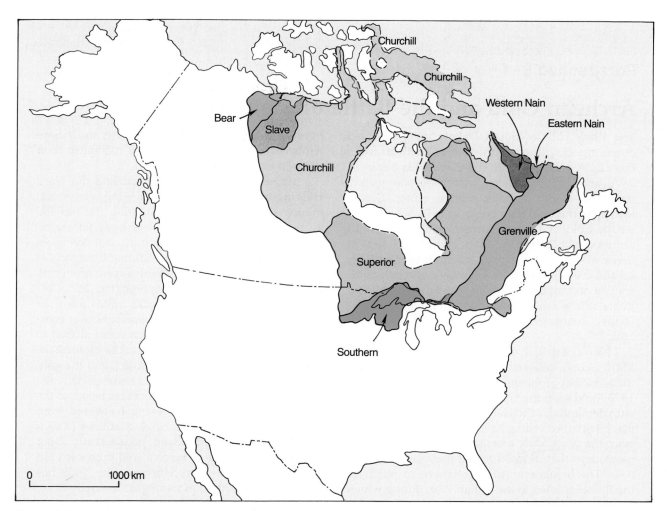

**Figure 8–7** Provinces of the Canadian Shield.

a.

b.

**Figure 8–8** Rocks of the Canadian Shield. a. Outcrop of gneiss near Wawa, Ontario, Canada. b. Contact between basalt (dark) and granite (light) along the banks of the Chippewa River, Ontario, Canada.

## Perspective 8-1

# Archean Gold and the Battle of the Little Bighorn

The Black Hills of South Dakota are an eroded domal uplift with a central core of Archean rocks (Fig. 1). A gold rush into the Black Hills began in 1876, and what had been a wilderness was quickly transformed into a major mining center. Events leading to this gold rush began in 1874, when Lieutenant Colonel George Armstrong Custer led an army expedition into the Black Hills, the Holy Wilderness of the Sioux Indians. For three months Custer and the 1,000 soldiers, engineers, and gold miners that accompanied him explored the region, and in his official report Custer said that "gold in satisfactory quantities can be obtained in the Black Hills."[1]

News of gold in the Black Hills spread rapidly. Many people believed that a gold rush would cure the economic problems that resulted in the "Panic of 1873," and soon the U.S. government was besieged with demands by thousands of voters to acquire the Black Hills. According to the treaty of 1868, however, the Black Hills were to forever belong to the Sioux, and they refused to sell their Holy Wilderness. The breakdown of negotiations to purchase the Black Hills led to the Indian War, during which Custer and some 260 of his men were annihilated in June, 1876, at the Battle of the Little Bighorn in Montana. Despite this stunning victory, the Sioux could not sustain a war against the U.S. Army, and in September, 1876, they were forced to relinquish their claim to the Black Hills.

Miners and settlers began entering the Black Hills in early 1876, and by the following year several mining towns were thriving. During the next 50 years more than $230,000,000 worth of gold was recovered. Indeed, large-scale mining still continues today. In fact, the Homestake Mining Company at Lead, South Dakota is one of the largest producers of gold in the Western Hemisphere (Fig. 2).

Most of the Black Hills' gold comes from the Homestake Formation, an Archean rock unit composed of iron- and carbonate-rich rocks altered to schist by regional metamorphism. The Homestake Formation has been badly deformed, and the gold ores are concentrated along fold axes, especially the axes of synclines. Mining of these ores began at the surface, but now they are being recovered from depths as great as 2.5 km. Huge quantities of rock must be mined and processed, because only about one-third ounce of gold is recovered from each ton of rock. Nevertheless, production since 1953 has been about 600,000 ounces of gold annually.

[1]Quoted in T. H. Watkins. 1971. *Gold and Silver in the West.* New York: Bonanza Books, p. 109.

**Figure 1** Diagramatic view of the Black Hills in western South Dakota. This broad domal uplift has a central core of Archean rocks, but some Proterozoic rocks are present as well. Most of the gold in the Archean rocks has been mined in the northern Black Hills near Lead and Deadwood.

Beyond the Canadian Shield, exposures of Precambrian rocks are limited to areas of uplift and erosion, as in the Appalachians, the Rocky Mountains, and the southwestern United States (Fig. 8–9) (Perspective 8–1). Geophysical evidence and deep drilling, however, demonstrate that Precambrian rocks underlie most of North America (Fig. 8–6).

The geologic history of the Canadian Shield is complex and not fully understood. Nevertheless, we can recognize several provinces within the shield, each of which is delineated on the basis of radiometric ages and structural trends (Fig. 8–7). These provinces and other buried provinces beyond the shield are the subunits that constitute the North American craton. Each may have been independent minicontinents that were later assembled into the larger cratonic unit. The amalgamation of these small cratons occurred along deformation belts in the Early Proterozoic and will be considered more fully in Chapter 9.

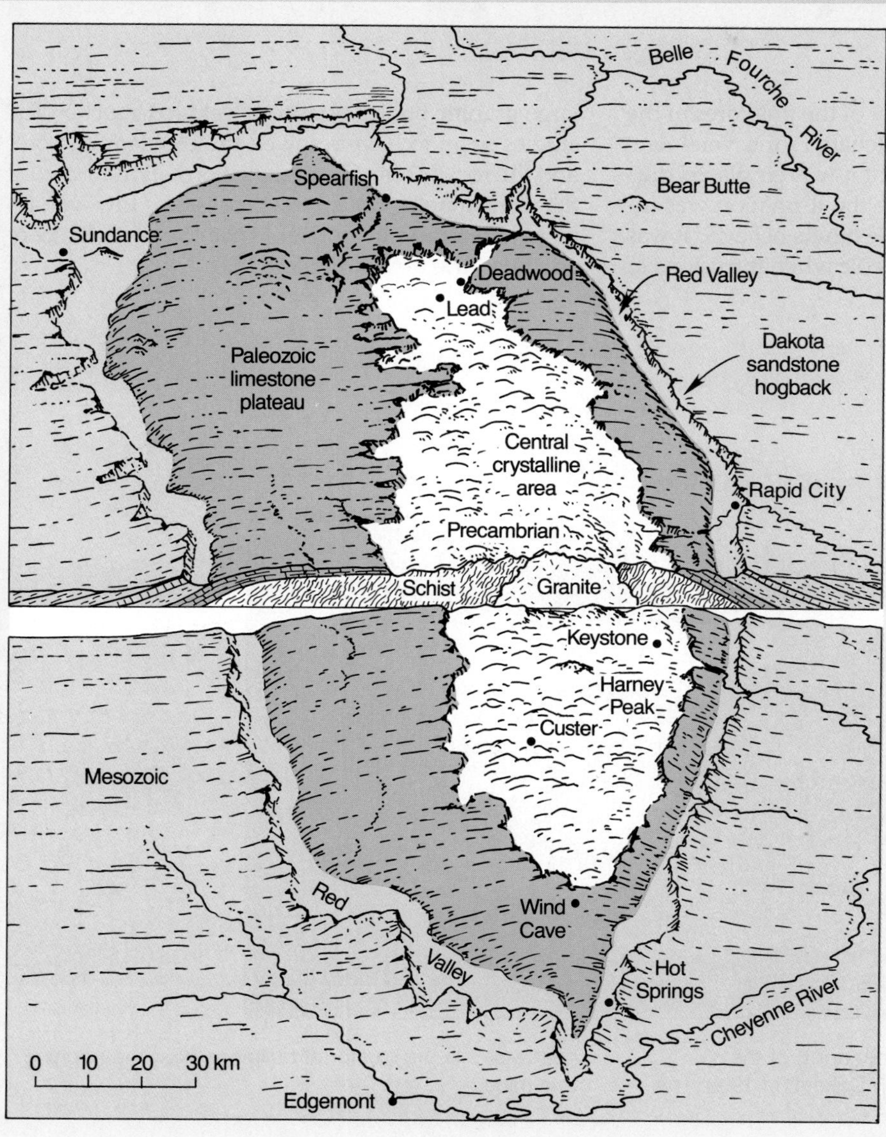


## ARCHEAN ROCKS

Areas underlain by Archean rocks (Fig. 8–10) are characterized by two main types of rock bodies; **greenstone belts** and **granite-gneiss complexes**. By far the most abundant rocks are granites and gneisses. Figure 8–11 is a geologic map of the Godthab region of western Greenland. Clearly gneisses and granites predominate, but **supracrustal sequences** are also present. *Supracrustal* refers to rocks that were deposited upon a basement of preexisting crust. The Isua supracrustal sequence (Fig. 8–11), consisting of metamorphosed lava flows, schists, quartzites and banded iron formations (Fig. 8–12), has been dated at nearly 3.8 billion years. Since sedimentation was occurring at this early date, it is safe to assume that even older source rocks must have been present.

## Perspective 8–1 (continued)

The time of emplacement of the gold ores in the Homestake Formation was debated for several decades. Some geologists thought it was emplaced during the Tertiary Period when the formation was intruded by rhyolite dikes, while others believed it was emplaced during the Proterozoic when intrusions of large granitic bodies occurred. Detailed geochemical studies seem to have resolved the problem; the ores and the rocks of the Homestake Formation formed at the same time by hot spring processes. However, regional metamorphism later in the Precambrian concentrated the ores along fold axes.

**Figure 2** The headworks (upper right) of the Homestake Mine at Lead, South Dakota, in 1900. The headworks is the cluster of buildings near the opening to a mine.

Supracrustal rocks are much more common in younger Archean terranes, but these are usually called *greenstone belts*. Their relationships to other rocks, however, are much like those shown in Figure 8–11. That is, the supracrustal sequences (or greenstone belts) occur as linear bodies in much more extensive granite-gneiss terranes.

### Greenstone Belts

The oldest large, well-preserved greenstone belts are those of South Africa, which date from 3.6 billion years ago. In North America, greenstone belts are most common in the Superior and Slave provinces of the Canadian Shield (Fig. 8–13), and most formed between 2.7 and 2.5 billion years ago.

An idealized greenstone belt consists of three major rock units; the lower and middle units are dominated by volcanic rocks, and the upper unit is sedimentary (Fig. 8–14). However, older ($>$ 2.8 billion years) and younger belts do differ somewhat. Older belts have an ultramafic lower unit succeeded upward by a basaltic unit, whereas the sequence in younger belts is from basaltic upward to

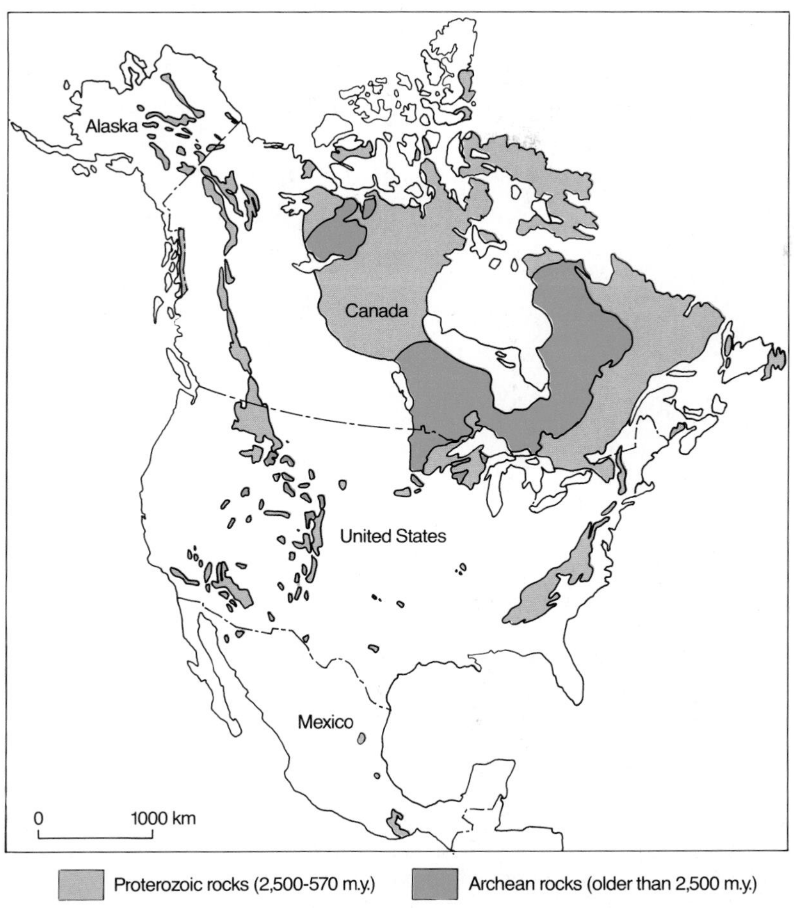


**Figure 8–9** Exposures of Precambrian rocks in North America. Note that exposures of Archean rocks are rare except in the Canadian Shield.

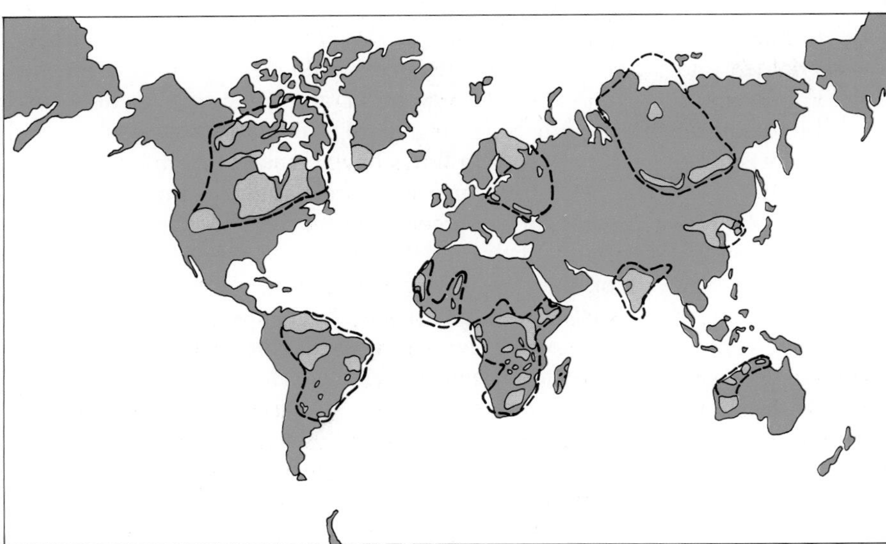

**Figure 8–10** Known distribution of Archean rocks. The dashed lines outline areas probably underlain by Archean rocks.

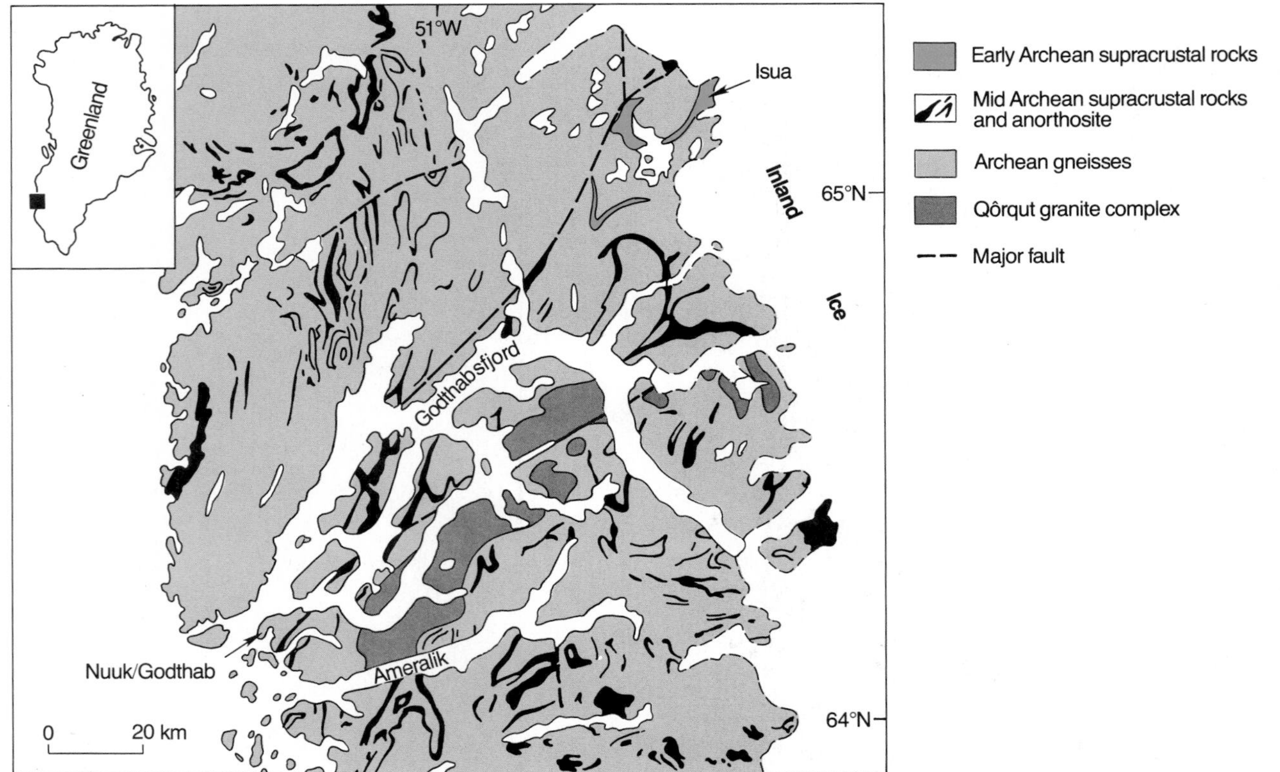


**Figure 8–11** Geologic map of the Godthab region, western Greenland. The Isua supracrustal rocks in the northeast part of the map are some of the oldest known rocks on Earth. Note that the supracrustal sequences or greenstone belts occur mostly as linear bodies in much more extensive areas of gneiss and granite.

an andesite-rhyolite unit. Most greenstone belts have a synclinal structure and are intruded by granitic magmas (Fig. 8–15), and many are complexly folded and cut by thrust faults. The volcanic rocks of greenstone belts are typically greenish due to the abundance of the mineral chlorite, which formed during low-grade metamorphism.

Much of the volcanism responsible for greenstone belt igneous rocks was subaqueous as indicated by the common occurrence of pillow structures (Fig. 8–16). Shallow water and subaerial eruptions are indicated by pyroclastics, and in some areas large volcanic centers built up above sea level. Perhaps the most interesting igneous rocks in greenstone belts are ultramafic lava flows; such flows are rare in rocks younger than Archean. In order to erupt, an ultramafic magma requires near-surface magma temperatures over 1,600° C; the highest recorded surface magma temperature for modern Hawaiian basalt lava flows is 1,350° C. Early in Earth history, however, there was considerably more radiogenic heat; thus the mantle was hotter, perhaps 300° C hotter; and ultramafic magmas could be erupted onto the surface. Since the amount of radiogenic heat has decreased through time (Fig. 8–5), the Earth has cooled; and ultramafic lava flows have ceased to form.

Sedimentary rocks are a minor component in the lower parts of greenstone belts but become increasingly abundant toward the top (Fig. 8–14, 8–15). The most common sedimentary rocks are successions of **graywacke** and **argillite** (Fig. 8–17). Graywacke is a variety of sandstone containing abundant clay, and those in the greenstone belts are rich in volcanic rock fragments. Argillites are simply slightly metamorphosed mudrocks such as shale. Small-scale graded bedding and cross bedding indicate that the graywacke-argillite successions were deposited by turbidity currents (Fig. 4–13).

Turbidity-current deposition in deep, tectonically active basins apparently accounted for many of the

a.

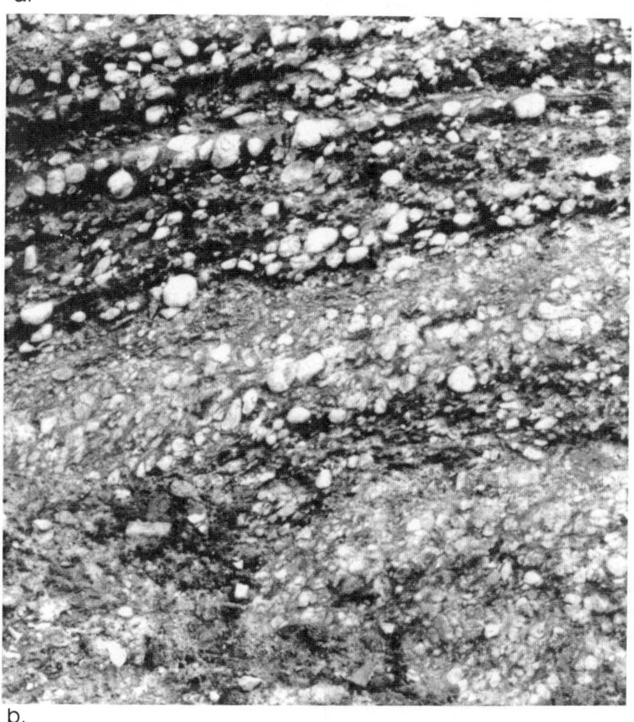

b.

**Figure 8–12** Isua supracrustal rocks, western Greenland. a. Banded iron formation showing magnetite (dark) and quartz-rich layers (light). b. Conglomerate with quartz pebbles up to 3 cm in diameter (Photos by A. P. Nutman courtesy of Stewart Watt, Geological Survey of Greenland)

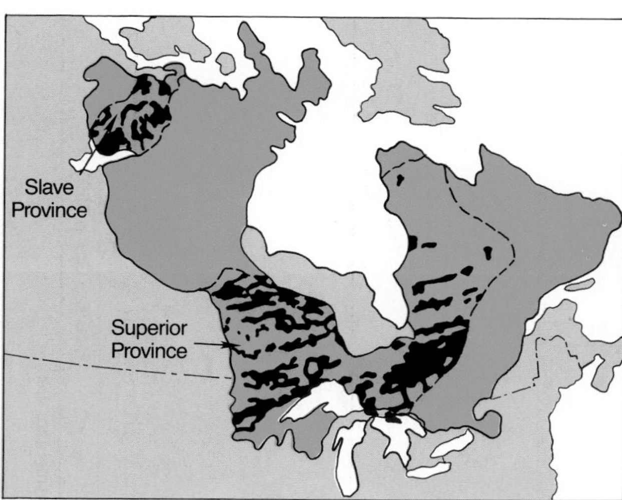


**Figure 8–13** Greenstone belts (shown in black) of the Canadian Shield are mostly in the Superior and Slave provinces.

graywacke-argillite associations (Fig. 8–17). The Fig Tree Group of South Africa (Fig. 8–14) seems to have been deposited in this way. The overlying quartz sandstones and shales of the Moodies Group (Fig. 8–14), however, show clear evidence of shallow-water deposition. These rocks were originally deposited in delta, tidal-flat, barrier-island, and shallow marine-shelf environments (Fig. 8–18).

Other sedimentary rocks occurring in greenstone belts include conglomerate, chert, carbonates, and banded iron formations. Some conglomerates are associated with graywackes and probably represent submarine slumps. Neither chert nor carbonates are very abundant; marbles 1 km thick in South Africa, however, indicate that carbonate deposition was important in some areas. Banded iron formations occur but are much more common in Proterozoic terranes, so will be discussed in the next chapter.

## Greenstone Belt Evolution

Geologists disagree on the evolutionary development of greenstone belts, but most currently popular models rely on Archean plate movements. Figure 8–19 shows a model in which greenstone belts develop in **backarc marginal basins** that subsequently close. Thus, there is an early stage of extension when the backarc marginal basin opens, accompanied by volcanism and sedimentation, followed by an episode of compression. During this

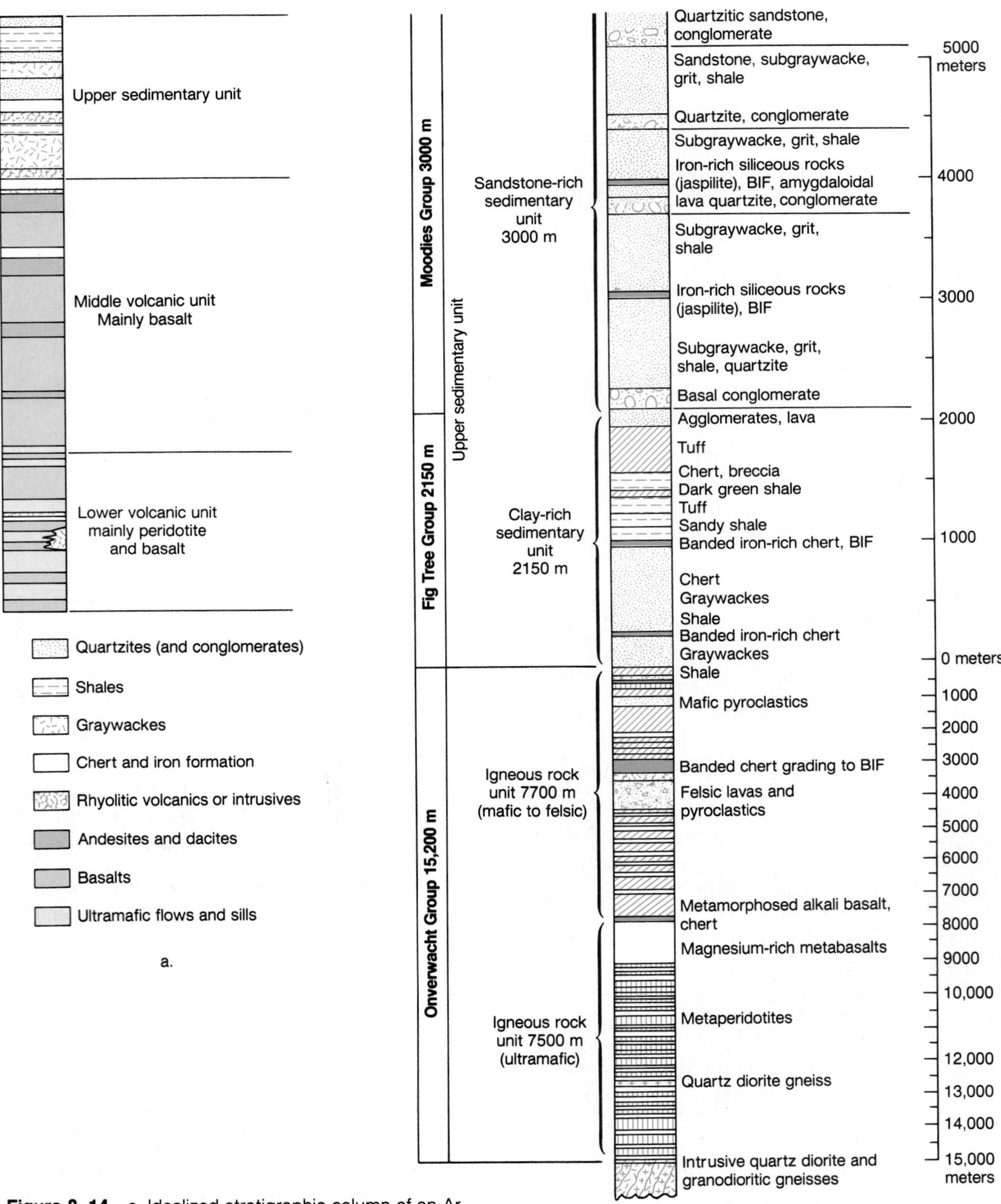


**Figure 8–14** a. Idealized stratigraphic column of an Archean greenstone belt. b. Detailed reconstruction of the Barberton greenstone belt of South Africa. Notice that the scales for the upper and lower parts of the column differ.

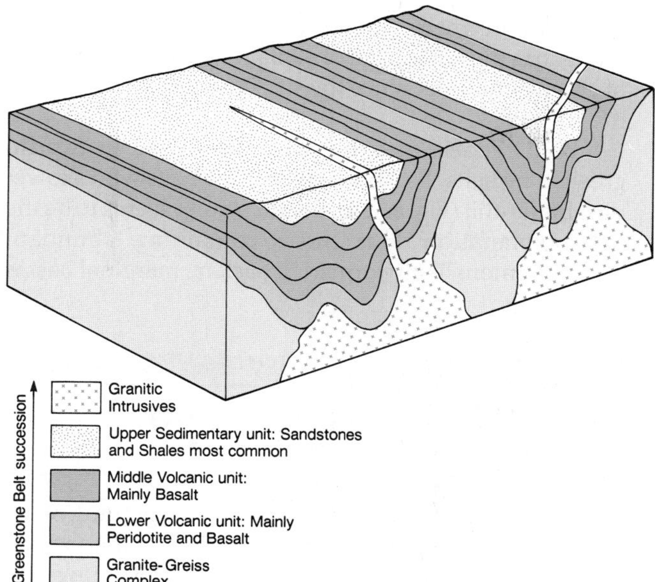


**Figure 8–15** Two adjacent greenstone belts showing their synclinal structure. Older greenstone belts, those more than 2.8 billion years old, have an ultramafic lower unit succeeded upward by a basaltic unit as shown here. In younger greenstone belts the succession is a basaltic lower unit overlain by an andesite-rhyolite unit. The upper unit in both older and younger greenstone belts consists of sedimentary rocks.

a.

b.

**Figure 8–16** Pillow structures in greenstone belt volcanic rocks. a. Ely Greenstone, Ely, Minnesota. b. Ishpeming Greenstone Belt, Marquette, Michigan. (Photo (a) courtesy of Richard W. Ojakangas, University of Minnesota, Duluth.)

compressional stage, the greenstone belt assumes its synclinal form (Fig. 8–15) and is metamorphosed and intruded by granitic magmas.

An alternate model proposes that some greenstone belts formed in **intracontinental rifts**. Several variations of the rift model exist; Figure 8–20 is one proposed to account for the origin of the Barberton greenstone belt (Fig. 8–14). This model assumes a preexisting sialic crust and requires an ascending mantle plume. As the plume rises and spreads, it generates tensional forces that cause intracontinental rifting. The plume also serves as the source of the lower and middle volcanic units. Erosion of the rift flanks accounts for the upper sedimentary units, the Fig Tree and Moodies groups in Figure 8–14. And finally, there is an episode of subsidence, deformation, low-grade metamorphism, and plutonism.

The rift model has certain appealing aspects. For one, the ultramafic volcanics can be more easily accounted for by a mantle plume rather than a backarc marginal basin setting, because subduction-related

Figure 8–17 Outcrop of Archean graywacke in Sweden. (Photo courtesy of R. V. Dietrich, Central Michigan University)

volcanics are more commonly andesitic. Also, the rift model has the advantage of explaining the variable sizes of greenstone belts, since the size is related to how much the rift opens.

Both models may be reasonably used to explain greenstone belts. For those greenstone belts with a lower ultramafic unit (Fig. 8–14), the rift setting seems to be the best explanation. But those containing abundant andesites more likely formed in backarc marginal basins (Fig. 8–19).

## DEVELOPMENT OF ARCHEAN CRATONS

We have already mentioned that by the beginning of the Archean several sialic continental nuclei or cratons had formed. They may have been rather small, however, since rocks older than 3.0 billion years are of limited geographic extent, especially compared with those 3.0 to 2.5

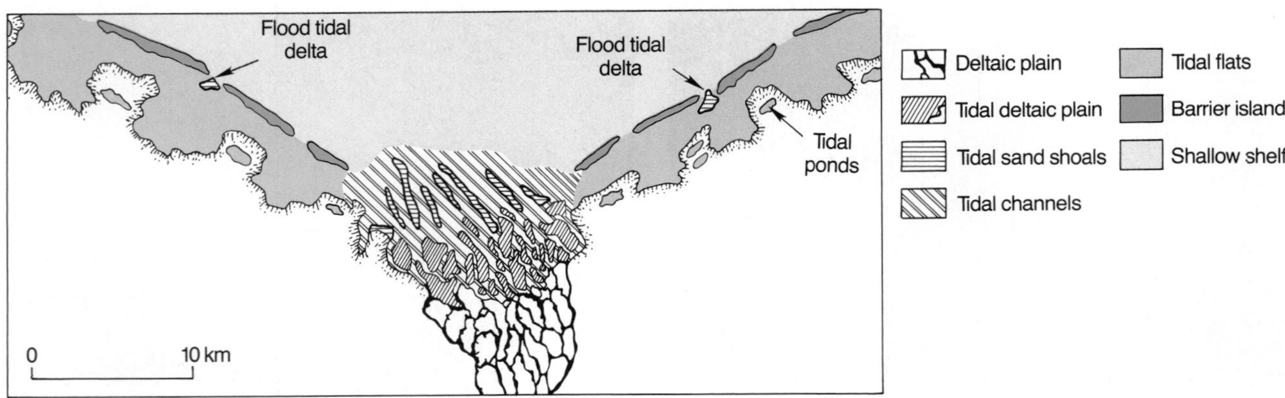


a.

b.

Figure 8–18 Moodies Group, South Africa. a. Reconstruction of depositional environments. b. Cross-bedded sandstone with crossbeds dipping in opposite directions indicating deposition in a tidal channel. (Photo b courtesy of Kenneth A. Eriksson, Virginia PolyTechnic Institute and State University.)

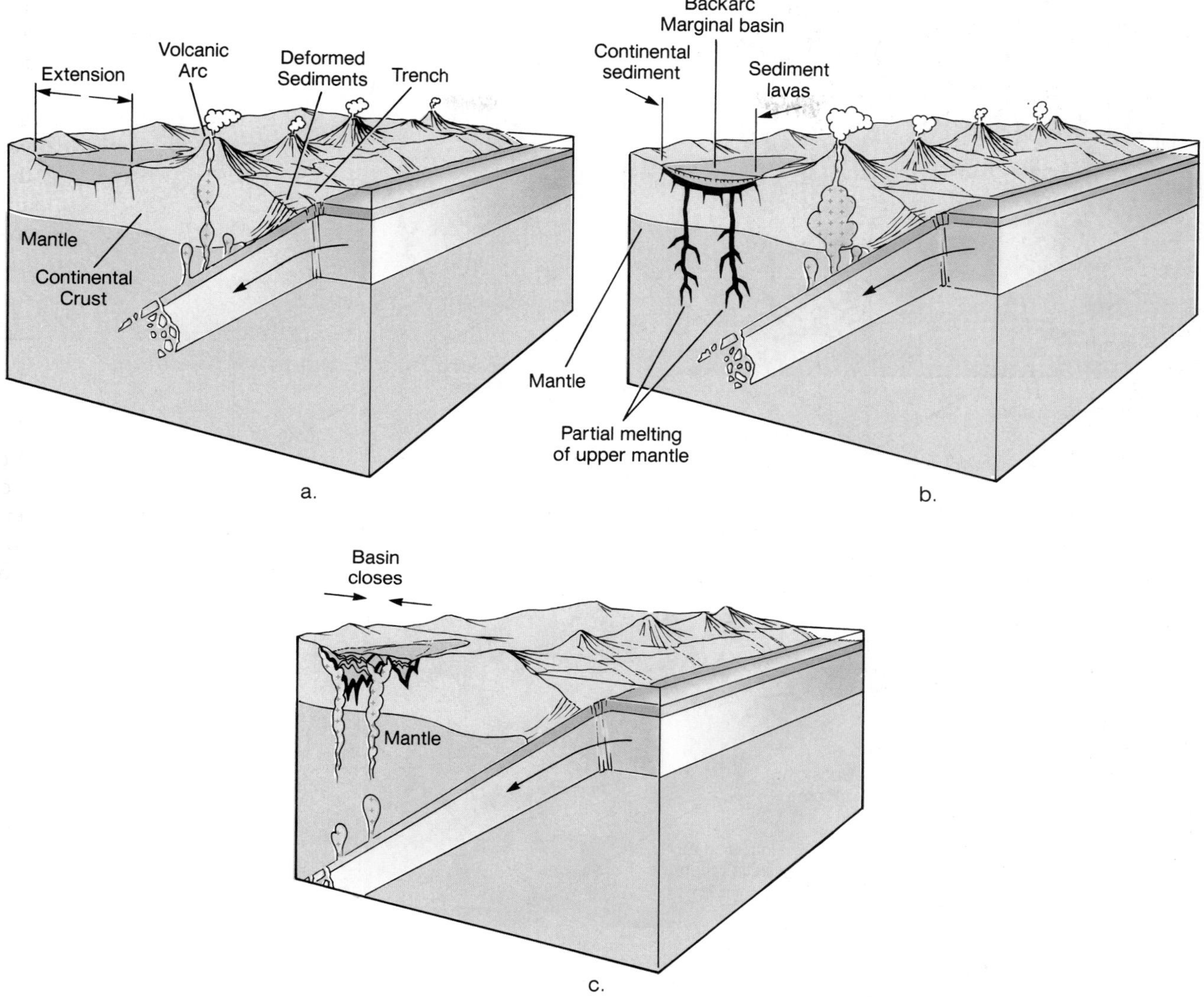


**Figure 8–19** Formation of a greenstone belt in a backarc marginal basin. a. Rifting behind a volcanic island arc forms a backarc marginal basin. Partial melting of subducted oceanic crust supplies andesitic and dioritic magmas to the island arc. b. Basaltic lavas and sediments derived from the continent and island arc fill the backarc marginal basin. c. Closure of the backarc marginal basin causes compression and deformation. The greenstone belt is deformed into a synclinal structure and is intruded by granitic magmas.

billion years old; (this latter interval seems to have been a time of rapid crustal evolution). Figure 8–21 shows a plate tectonic model for Archean crustal evolution that incorporates sialic plutonism, greenstone belt formation in marginal basins, and collisions of minicontinents. It accounts for the origin of both greenstone belts and granite-gneiss terranes.

The southern Superior province of Canada (Fig. 8–7) appears to have evolved in a manner broadly consistent with this model. In this region several east-west trending subprovinces consisting of greenstone-granite terranes alternate with granite-gneiss terranes (Fig. 8–22). These subprovinces are bordered in large part by faults, and radiometric-age dates show that they become younger toward the south. The growth of the southern Superior province may be accounted for by the sequential accretion of island arcs. The greenstones formed in the backarc portions of the island arcs, while the granite-gneiss complexes formed in the forearc portion (Fig. 8–22).

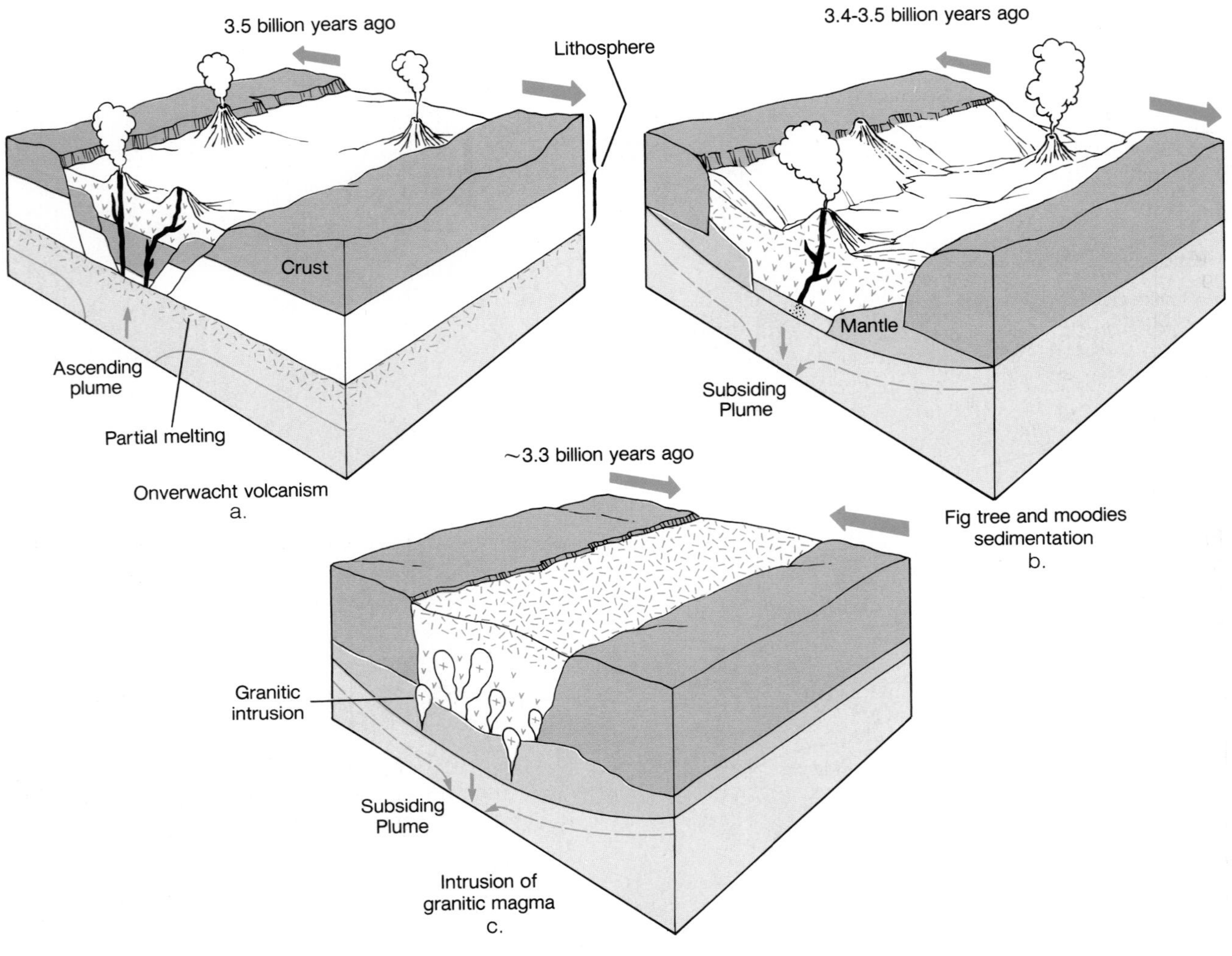


**Figure 8–20** Formation of the Barberton greenstone belt of South Africa in an intracontinental rift. a. An ascending mantle plume causes rifting and Onverwacht volcanism. b. As the plume subsides, erosion of the rift flanks accounts for deposition of the Fig Tree and Moodies groups. c. Closure of the rift causes compression and deformation. Granitic magma intrudes the greenstone belt.

The events leading to the origin of the southern Superior province are part of a more extensive orogenic episode that occurred near the end of the Archean. This episode, the **Kenoran orogeny**, was responsible for the formation of the Superior and Slave provinces, as well as some Archean terranes within the Churchill and Nain provinces (Fig. 8–7). It also affected Archean rocks in Wyoming and Montana and accounts for the Archean terranes of the Minnesota River Valley (Fig. 8–23).

Deformation during the Kenoran orogeny was the last major Archean event in North America. Several sizable cratons had formed, but each was an independent unit. Much of the continental crust of the other continents had also formed by the end of the Archean. The areas enclosed by dashed lines in Figure 8–10 are underlain by Archean rocks, but the Archean continents did not resemble these areas. For example, the older parts of the Canadian Shield seem to have been independent units, or microcontinents, that were assembled in the Early Proterozoic to form a craton like that outlined in Figure 8–10.

Table 8–2 is a summary of the Archean crust-forming events discussed in the preceding sections.

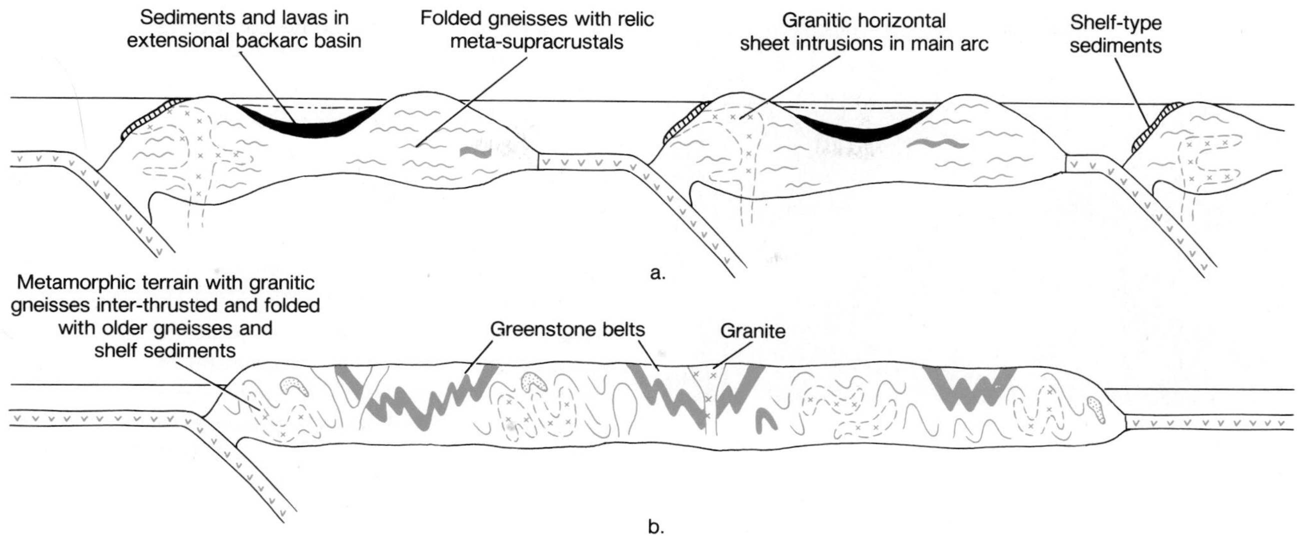


**Figure 8–21** A plate tectonic model for the growth of continents in the Archean. a. Lateral movements of Early Archean minicontinental plates with shelf sediments, mantle-derived granitic magmas in arcs, and volcanics in backarc marginal basis. b. Amalgamation of minicontinents gives rise to an extensive continental plate consisting of greenstone belts and granite-gneiss terranes by the end of the Archean.

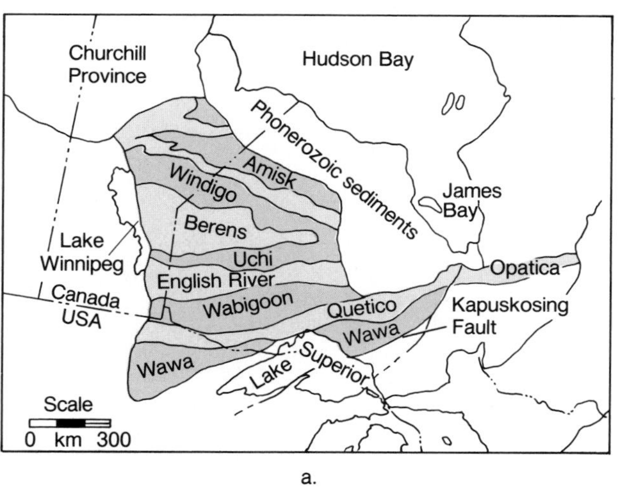


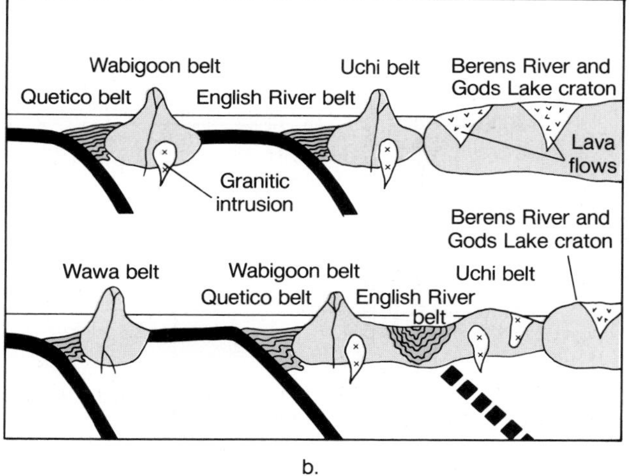


**Figure 8–22** Origin of the southern Superior province. a. Geologic map showing greenstone belts (darker areas and granite gneiss subprovinces (lighter areas). b. Plate tectonic model for development of the southern Superior province. The figure represents a north-south section, and the upper diagram is an earlier stage of the lower diagram.

## ARCHEAN PLATE TECTONICS

Undoubtedly, the present tectonic regime of opening and closing oceans has been a primary agent in Earth evolution for at least the last 2 billion years. In fact, this regime probably became established in the Early Proterozoic. Many geologists are becoming convinced that some sort of plate tectonics was operating in the Archean as well, but they disagree about the details.

Some Archean terranes appear to record plate movements in their apparent polar wandering curves and in deformation belts between presumed colliding cratons

**Figure 8–23** Archean rocks from the Beartooth uplift of Wyoming and Montana.

and island arcs. But ophiolite complexes, which mark younger convergent plate margins, are rare, although Late Archean ophiolites have recently been reported from several areas.

Apparently the Earth's radiogenic heat production has diminished through time (Fig. 8–5). Thus, in the Archean, when more heat was available, seafloor spreading and plate motions probably occurred faster, and magma was generated more rapidly. Nevertheless, Archean plates seem to have behaved differently from those in the Proterozoic. For example, sedimentary sequences typical of passive continental margins are uncommon in the Archean but quite common in the Proterozoic. Their near absence in the Archean suggests that continents with adjacent shelves and slopes were either not present or only poorly developed.

**Table 8–2** Chronologic summary of events important in the Archean development of cratons. Ages in thousands of millions of years.

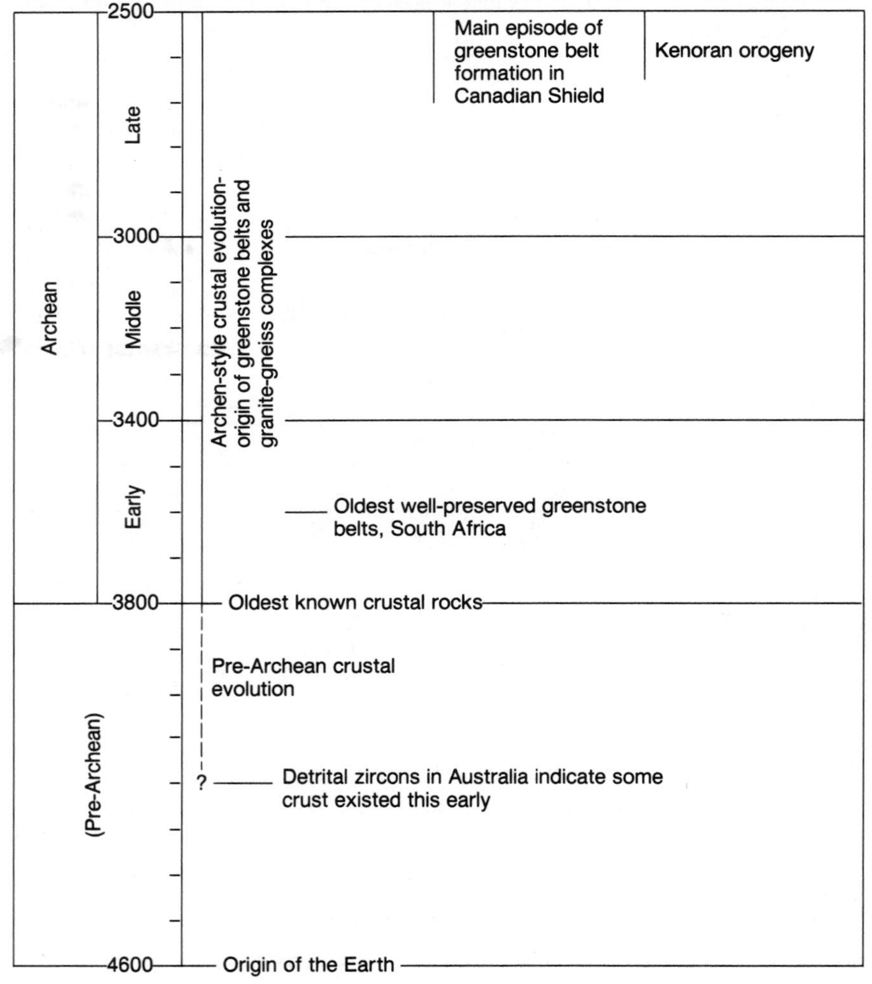

Another factor that favors some kind of Archean plate tectonics is the 3.0- to 2.5-billion-year-old episode of rapid crustal growth. Like continents today, Archean continents probably grew by accretion at convergent plate margins. They probably grew more rapidly since plate motions were faster, thus accounting for an accelerated rate of accretion at convergent margins.

Today most geologists would agree that plate tectonics comparable to that pictured in Figure 8–21 was operative in the Archean. Its details differed, however, from the modern style of plate tectonics, which began in the Proterozoic when large, stable cratons were present.

## THE ATMOSPHERE AND OCEANS

During its earliest history the Earth was a very inhospitable place. If we could somehow go back and visit it then, we would witness a barren, waterless surface and numerous meteorite impacts. We would be subjected to intense ultraviolet radiation and would be unable to breathe the atmosphere. Today, our atmosphere is rich in nitrogen and oxygen and contains important trace amounts of carbon dioxide, water vapor, and other gases (Table 8–3). In the upper atmosphere, ozone ($O_3$) blocks most of the sun's ultraviolet radiation.

Since the most abundant elements in the universe are hydrogen and helium, the Earth's earliest atmosphere was probably composed of these gases. If so, gases of such low molecular weight would have escaped into space, since Earth's gravity is insufficient to retain them. Before the Earth had a differentiated core, it lacked a magnetic field and *magnetosphere*, the area around the Earth within which the magnetic field is confined. The absence of a magnetosphere insured that a strong solar wind, an outflow of ions from the sun, would sweep away any gases that might otherwise have formed an atmosphere. Once the magnetosphere was established, internally derived volcanic gases began to accumulate. The derivation of atmospheric gases from within the Earth is a process called **outgassing** (Fig. 8–24).

Gases emitted by modern volcanoes are mostly water vapor with lesser amounts of carbon dioxide, sulfur dioxide, carbon monoxide, sulfur, chlorine, nitrogen, and hydrogen. Archean volcanoes probably emitted these same gases. The gases formed an early atmosphere but one notably deficient in free oxygen; and without free oxygen, there could have been no ozone layer. Another probable attribute of the early atmosphere was its ammonia ($NH_3$) and methane content ($CH_4$), both of which were produced by volcanic gases reacting chemically in the atmosphere.

An oxygen-deficient atmosphere appears to have persisted throughout the Archean as indicated by detrital deposits containing pyrite ($FeS_2$) and uraninite ($UO_2$). Both of these minerals are quickly oxidized in the presence of free oxygen. Iron in the oxidized state became quite common in the Proterozoic, indicating that at least some free oxygen was present by that time (Chapter 9).

Two processes, both of which began in the Archean, can account for the introduction of free oxygen into the atmosphere. The first was **photochemical dis-**

**Table 8–3** Composition of the atmosphere near the Earth's surface.

| PERMANENT GASES | | | VARIABLE GASES | | |
|---|---|---|---|---|---|
| Gas | Symbol | Percent (by volume) | Gas | Symbol | Percent (by volume) |
| Nitrogen | $N_2$ | 78.08 | Water vapor | $H_20$ | 0 to 4 |
| Oxygen | $O_2$ | 20.95 | Carbon dioxide | $CO_2$ | 0.034 |
| Argon | Ar | 0.93 | Ozone | $O_3$ | 0.000004[a] |
| Neon | Ne | 0.0018 | Carbon monoxide | CO | 0.00002[a] |
| Helium | He | 0.0005 | Sulfur dioxide | $SO_2$ | 0.000001[a] |
| Methane | $CH_4$ | 0.0001 | Nitrogen dioxide | $NO_2$ | 0.000001[a] |
| Hydrogen | $H_2$ | 0.00005 | Particles (dust, soot, etc.) | | 0.00001[a] |
| Xenon | Xe | 0.000009 | | | |

[a]Average value in polluted air.

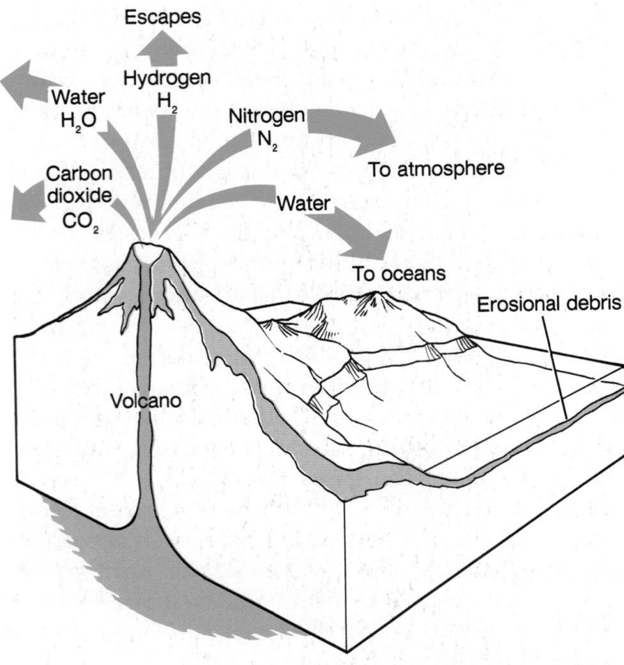


**Figure 8–24** Outgassing supplied gases to form an early atmosphere composed of the gases shown. Chemical reactions in the atmosphere also yielded methane ($CH_4$) and ammonia ($NH_3$).

**sociation** of water vapor, a process that involved the breakup of water molecules by ultraviolet radiation in the upper atmosphere (Fig. 8–25). This process eventually may have supplied up to 2 percent of present-day oxygen levels. At 2 percent free oxygen, ozone ($O_3$) will form, creating a barrier against incoming ultraviolet radiation, thus limiting the formation of more free oxygen by photochemical dissociation. Even more important were the activities of photosynthesizing organisms. During **photosynthesis** carbon dioxide and water combine into organic molecules, and oxygen is released as a waste product (Fig. 8–25). Even so, the atmosphere at the end of the Archean may have contained no more than 1 percent of its present oxygen level.

Recall that the major gas emitted by volcanoes is water vapor (Fig. 8–24), so the early atmosphere was also rich in this compound. Once the Earth cooled sufficiently, water vapor condensed and surface waters began to accumulate. Good evidence exists for oceans in the Early Archean, although their volumes and extent are unknown. We can envision an early hot Earth with considerable volcanic activity and a rapid accumulation of surface waters. Is the volume of oceanic waters still increasing? Perhaps it is; but if so, it is at a much lower rate, since the amount of heat to generate magmas has decreased (Fig. 8–5), and isotopic studies of modern volcanic emissions indicate that much of the water vapor emitted is recycled surface water.

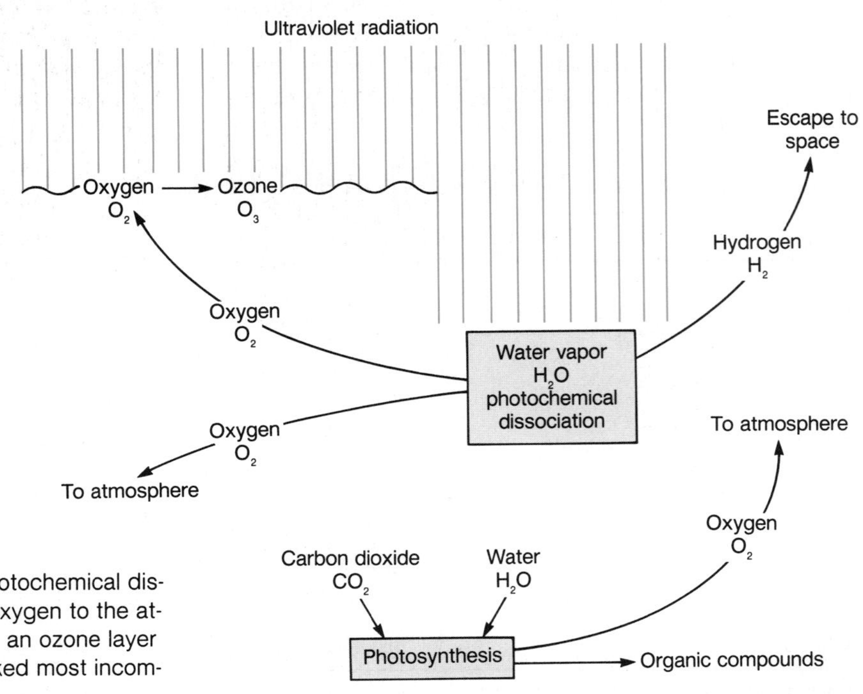


**Figure 8–25** Diagram illustrating how photochemical dissociation and photosynthesis added free oxygen to the atmosphere. Once free oxygen was present, an ozone layer formed in the upper atmosphere and blocked most incoming ultraviolet radiation.

## ARCHEAN LIFE

The fossil record reveals that life existed on Earth as much as 3.5 billion years ago. Compared to the present, however, the Archean seems to have been biologically impoverished. Today the Earth's biosphere consists of millions of species of animals, plants, and other organisms, all of which are thought to have evolved from one or a few primordial types. In Chapter 5 we considered the evolutionary processes whereby the diversification of life occurred, but here we are concerned with how life originated in the first place.

First, we must be very clear about what life is; that is, what is living and what is nonliving? Minimally, a living organism must reproduce and practice some kind of metabolism. Reproduction insures the long-term survival of a group of organisms as a species; metabolism insures short-term survival of an individual organism as a chemical system.

Using this reproduction-metabolism criterion, it would seem a simple matter to decide whether something is alive. Yet, this is not always the case. Viruses (Fig. 8–26), for example, behave like living organisms when in an appropriate host cell, but outside a host cell they neither reproduce nor manufacture organic molecules. They are composed of a bit of genetic material, either DNA or RNA, enclosed in a protein capsule. So are they living or nonliving? Most biologists think not, but viruses do illustrate that our definition of life is not as clear-cut as it would seem to be.

Microspheres are simple organic molecules that form spontaneously and show greater organizational complexity than inorganic objects such as rocks. In fact, microspheres can even grow and divide, but in ways that are more like random chemical processes than like organisms. Thus, like viruses, they are not considered living.

Having concluded that viruses and microspheres are not living, one may wonder what bearing they have on the origin of life. The answer is a two-part one. First, they demonstrate that no matter how we define life, there is no absolute criterion for deciding whether or not some things, such as viruses, are living. And second, if life originated on Earth by natural processes, it must have passed through a prebiotic stage, perhaps similar to microspheres.

As early as 1924 the great Russian biochemist, A. I. Oparin, postulated that life originated when the Earth had an atmosphere containing little or no free oxygen. With no oxygen to destroy organic molecules, and no ozone layer to block ultraviolet radiation, life could indeed have come into existence from nonliving matter.

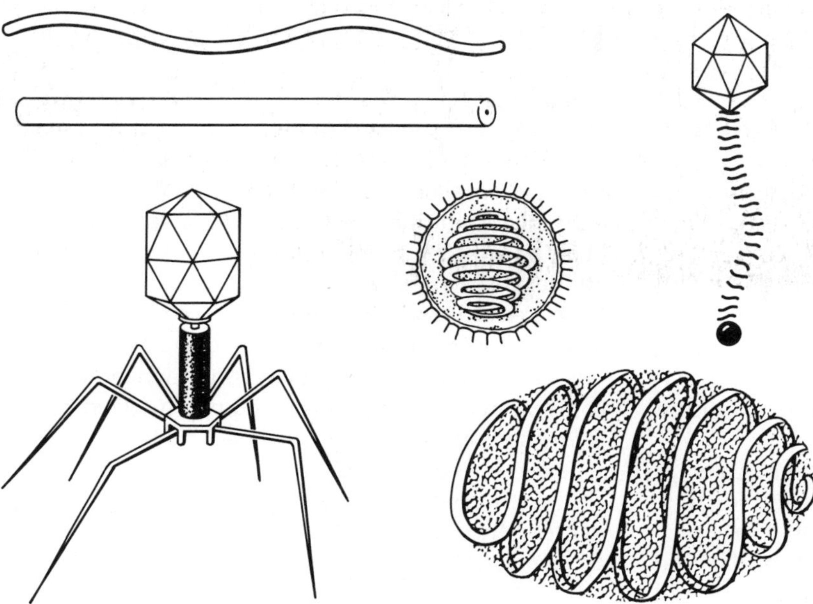

**Figure 8–26** Many diverse forms of viruses are known; the comparative sizes and shapes of six viruses are shown. Viruses show some of the features of organisms, but most biologists think that they are not living.

## The Origin of Life

All investigators agree that two requirements were necessary for the origin of life. First, there must have been a source of the appropriate elements from which organic molecules could have been synthesized. And second, there must have been an energy source to promote chemical reactions that synthesized organic molecules. All organisms are composed mostly of carbon, hydrogen, nitrogen, and oxygen, all of which were present in the early atmosphere in the form of carbon dioxide ($CO_2$), water vapor ($H_2O$), methane ($CH_4$), and ammonia ($NH_3$). It is postulated that the elements necessary for life, C, H, N, and O, combined to form simple organic molecules called **monomers**. The energy sources that promoted these reactions were probably ultraviolet radiation and lightning. Typical monomers characteristic of organisms are amino acids.

Monomers are the basic building blocks of more complex organic molecules, but is it plausible that they originated in the manner postulated? Experimental evidence suggests so. In the 1950s, Stanley Miller synthesized several amino acids by circulating gases approximating the composition of the early atmosphere in a closed glass vessel (Fig. 8–27). This mixture of gases was subjected to an electric spark (simulating lightning), and in a few days the mixture became cloudy. Analysis showed that several amino acids typical of organisms had formed. In more recent experiments all 20 amino acids common in organisms have been successfully synthesized.

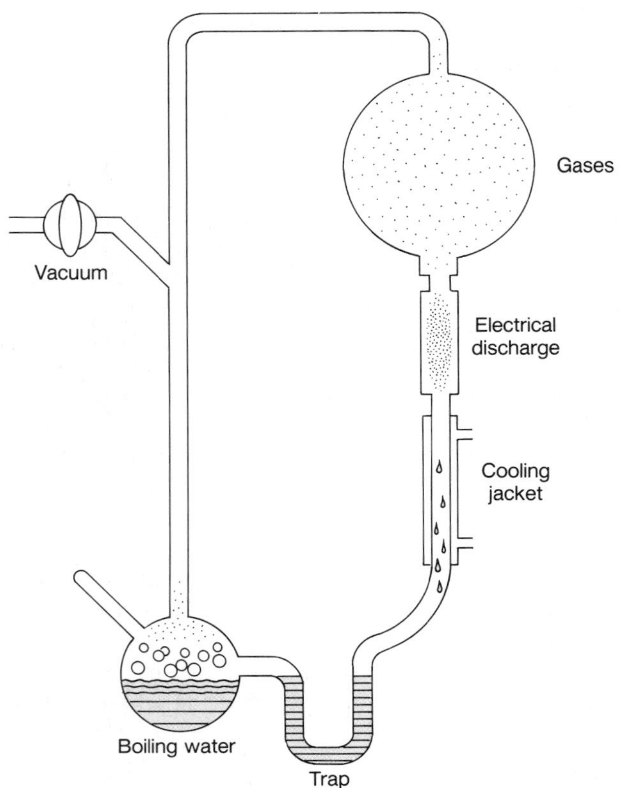


**Figure 8–27** Experimental apparatus used by Stanley Miller. Several amino acids characteristic of organisms were artificially synthesized during Miller's experiments.

Making monomers in a test tube is one thing, but since the molecules of organisms are **polymers** such as proteins and nucleic acids, which consist of linked monomers, the next question is, how did the process of polymerization occur? This is more difficult to answer, especially if polymerization occurred in an aqueous solution, which usually results in depolymerization. However, Sidney Fox of the University of Florida has synthesized small molecules he calls *proteinoids*, some consisting of more than 200 amino acid units (Fig. 8–28). He dehydrated concentrated amino acids and found that when heated they spontaneously polymerized to form proteinoids.

At this stage we can refer to these molecules as *protobionts*, which are intermediate between inorganic chemical compounds and living organisms. These protobionts, however, would have been diluted and would have ceased to exist if some kind of outer membrane had not developed. In other words, they had to be self-contained as modern cells are. Fox's experiments demonstrated that proteinoids will spontaneously aggregate into microspheres (Fig. 8–28), which, bounded by a cell-like membrane, grow and divide much as bacteria do.

Fox's experimental results are interesting, but how can they be related to what may have taken place in the early history of life? Monomers likely formed continuously and in great abundance, accumulated in the oceans, and formed what the British biochemist J. B. S. Haldane characterized as a hot, dilute soup. According to Fox, the amino acids in this hot dilute soup may have washed up onto a beach or perhaps cinder cones, where they were concentrated by evaporation and polymerized by heat. The polymers were then washed back into the sea, where they reacted further.

Not much is known about the next step in the origin of life—the development of a reproductive mechanism. Fox's microspheres divide, but in a nonbiological fashion, and they are not living. In fact, it seems that we face a paradox at this point. Nucleic acids, either as ribonucleic acid (RNA) or deoxyribonucleic acid (DNA), are necessary for reproduction. The problem is that nucleic

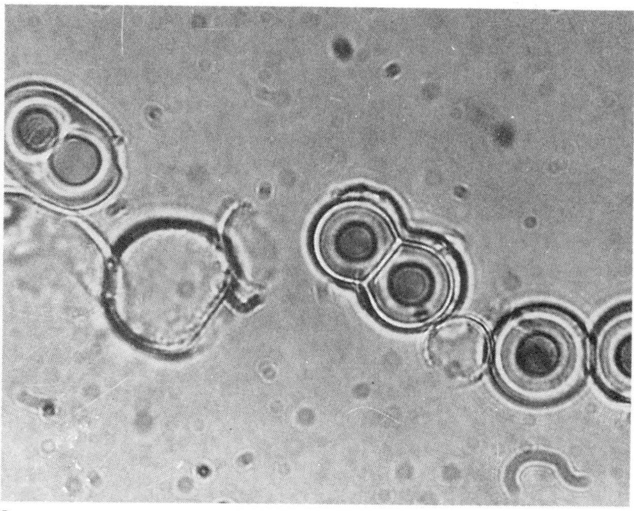

a.

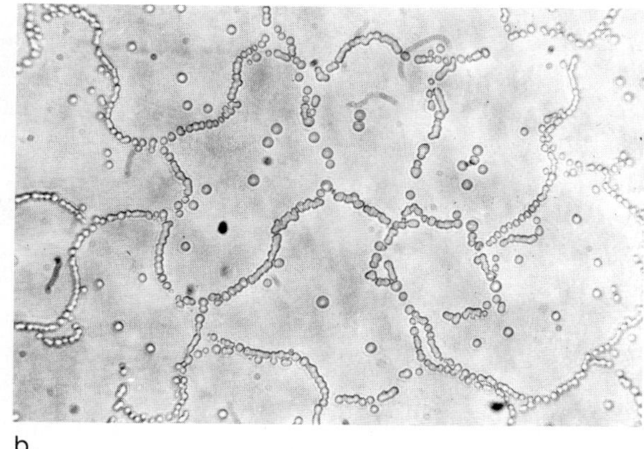

b.

**Figure 8–28** a. Bacterium-like proteinoid. b. Proteinoid microspheres. (Photos courtesy of Sidney W. Fox, Institute for Molecular Evolution, University of Miami)

acids cannot replicate without enzymes, and enzymes cannot be made without nucleic acids. Or so it seemed until recently.

Recent experimental evidence has demonstrated that "RNA has the capacity to replicate itself in the absence of protein enzymes."[1] In view of this evidence, it seems that the first replicating system may have been an RNA molecule. Just how an RNA molecule was naturally synthesized, however, remains a mystery.

It should be apparent from our discussion that much remains to be learned about the origin of life by natural processes. Considerable disagreement exists as to the steps in abiotic synthesis and the significance of some experimental results. A common theme among many investigators, however, is that the earliest organic molecules were synthesized from atmospheric gases. But even this has been questioned by some who propose that life originated in hydrothermal vent systems on the seafloor (Perspective 8–2).

## The Earliest Organisms

Prior to the mid-1950s we had very little knowledge of Precambrian life. Investigators had long assumed that the fossils so abundant in Cambrian strata must have had a long earlier history, but no such record was known. Some enigmatic Precambrian fossils were known, but these were mostly dismissed as inorganic in origin. In fact, many geologists considered Precambrian rocks to be unfossiliferous.

In the early 1900s Charles Walcott described layered mound-like structures from the Early Proterozoic Gunflint Iron Formation of Ontario (Fig. 8–29). These structures are now called **stromatolites**. Walcott proposed that they represented reefs constructed by algae, but paleontologists did not demonstrate that stromatolites are the products of organic activity until 1954. Studies of modern stromatolites show that these structures originate by the entrapment of sediment grains on sticky mats of photosynthesizing cyanobacteria or blue-green algae (Fig. 8–30). Morphologically distinct types of stromatolites may form (Fig. 8–30), all of which are known from Precambrian rocks, and some contain microfossils.

Currently, the oldest known undisputed stromatolites are from 3.0-billion-year-old rocks in South Africa. Even older probable stromatolites have recently been discovered in the 3.3- to 3.5- billion-year-old Warrawoona Group near North Pole, Australia (Fig. 8–31). Indirect evidence for even more ancient life comes from the Isua supracrustal rocks of western Greenland (Fig. 8–12). These 3.8-billion-year-old rocks contain small carbon spheres that may be of biological origin, but the evidence is not conclusive.

The oldest known fossils are of photosynthesizing organisms, but photosynthesis is a complex metabolic process, and it seems reasonable that it was preceded by

[1]Lewin, R. 1986. *Science*, 231: 545.

## Perspective 8-2

# Submarine Hydrothermal Vents and the Origin of Life

Near spreading centers, seawater seeps down into the oceanic crust through cracks and fissures, is heated by magma, then rises to form hydrothermal vents on the sea floor (Fig. 1). Over 20 years ago geologists reasoned that a logical implication of plate tectonic theory was the presence of such vents, but none were observed until 1977. The first vents were sighted from the submersible *Alvin* at a depth of about 2,500 m on the Galapagos Rift in the eastern Pacific Ocean basin. Since then, such vents have been observed in several other areas.

Submarine hydrothermal vents are interesting for several reasons. First, the rising hot-water solutions contain large quantities of dissolved mineral matter. When the hot water cools, precipitation of iron, copper, and zinc sulfides and other minerals occurs near the vents. Economic geologists are interested in this phenomenon because it may help them understand the origin of some ore deposits.

We are mainly concerned, however, with the biological implications of these vents. Communities of vent organisms (Fig. 1) are the only major communities known not dependent on sunlight as a source of energy. Chemosynthetic bacteria that oxidize sulfur compounds from the hot vent waters lie at the base of the food chain. Thus, the ultimate source of energy for the vent communities is radiogenic heat.

Apparently, hydrothermal vents have been present since the Early Archean. Their presence then is indicated by massive sulfide deposits from the Archean that appear to have been produced by hydrothermal activity. In fact, if the early Earth really did have more radiogenic heat and more magmatism and faster spreading rates, hydrothermal vents should have been more common than they are now.

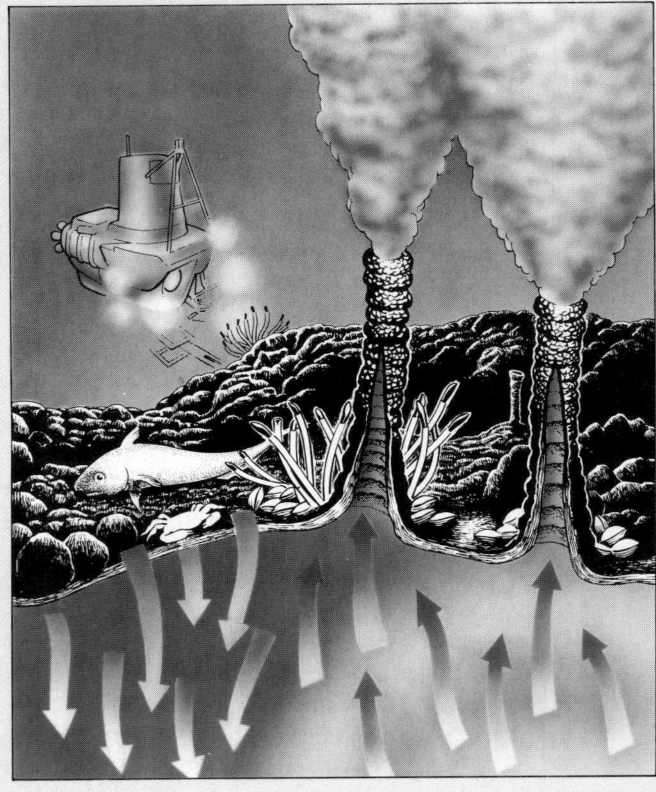

**Figure 1** The submersible *Alvin* sheds light on hydrothermal vents at the Galapagos Rift, a branch of the East Pacific Rise. Seawater seeps down through the oceanic crust and becomes heated. The heated seawater rises and builds chimneys on the sea-floor. The plume of heated water discharged through the chimney is saturated with dissolved minerals. Chemosynthetic bacteria oxidize sulfur compounds from the hot vent waters and form the base of a food chain for tubeworms, giant clams, anemones, crabs, and several types of fish.

an even simpler process. In other words, nonphotosynthesizing organisms must have been present before cyanobacteria appeared.

We have no fossils of these earliest organisms, but they probably resembled tiny bacteria. Since the early atmosphere contained little or no free oxygen they must have been **anaerobic**, meaning they needed no oxygen. And very likely they were completely dependent on an external source of nutrients. We refer to such organisms as **heterotrophic** as opposed to **autotrophic** organisms that make their own nutrients, as in photosynthesis. All were unicellular and lacked a cell nucleus. Cells of this type are called **prokaryotic cells**.

We can characterize these earliest organisms as anaerobic, heterotrophic prokaryotes. Their nutrient source was probably adenosine triphosphate (ATP), which was

Recently, investigators at the Oregon State University School of Oceanography have proposed that submarine hydrothermal vents were the sites at which life originated. They indicate that the more traditional hypotheses for the origin of life rely upon

1. atmospheric gases as the source of the elements for organic molecules,
2. the synthesis of monomers like amino acids,
3. the formation of polymers, and
4. some kind of primitive organism.

Recall from our earlier discussion that both monomers and polymers have been synthesized experimentally. The Oregon State investigators claim that

> The physical and chemical conditions of those experiments are found in a natural environment which has been present in the oceans since they first formed, in hot springs associated with submarine volcanism.[1]

The elements and energy source necessary for the synthesis of organic molecules were present in the hydrothermal vents (Fig. 2). In fact, amino acids have been detected in modern hydrothermal vent solutions. Polymerization may have occurred on the surfaces of clay minerals, a process that has been proposed by other investigators. And finally, protocells would have been discharged with vent fluids and deposited on the seafloor.

One of the most intriguing aspects of this hypothesis is that the process may be continuing at present! This does not mean that life is continually originating, since any polymers or protocells would be quickly devoured by existing vent community organisms. Nevertheless, this hypothesis can be tested, at least in part, in a natural environment.

[1] J. B. Corliss, et al. 1981. *Oceanologica Acta*, 4: 61.

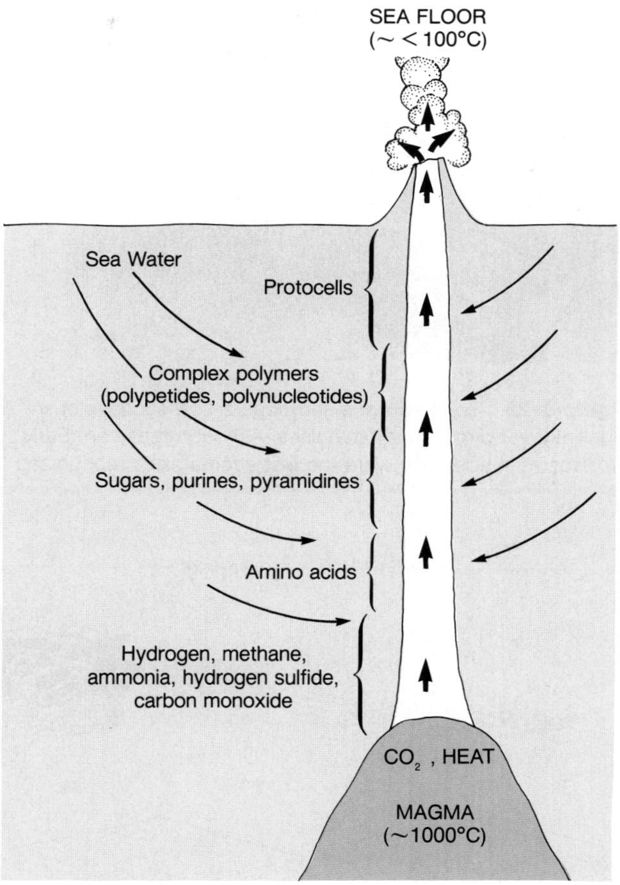

**Figure 2** Proposed sequence of events leading to the origin of protocells in a hydrothermal vent system. According to this hypothesis, seawater heated by magma rises through a cylindrical fracture and cools as it rises. Increasingly complex structures may have formed in the vent system, eventually giving rise to protocells and chemosynthetic bacteria.

used to drive the energy-requiring reactions in cells. ATP can be synthesized from simple gases and phosphate, so it was probably available in the early Earth environment. The earliest life-forms may have simply acquired their ATP from their surroundings. This situation could not have persisted for long, though, because as increasingly more cells competed for the same resources, the supply must have diminished. The first organisms to develop a more sophisticated metabolism probably used **fermentation** to meet their energy needs. Fermentation is an anaerobic process of splitting molecules such as sugars to release carbon dioxide, alcohol, and energy. In fact, most living prokaryotes ferment.

Other than the origin of life itself, the most significant biological event of the Archean was the development of the autotrophic process of photosynthesis (Fig.

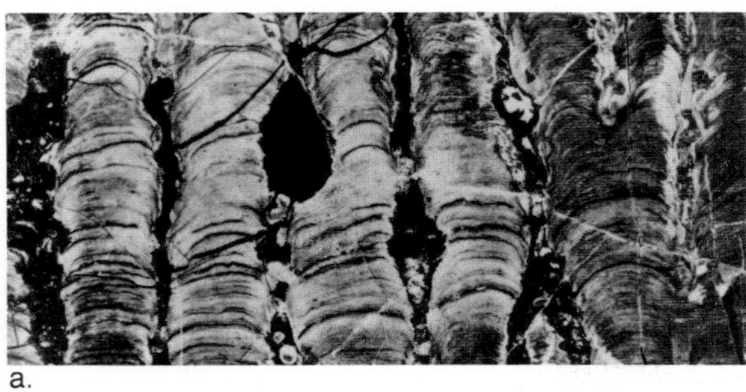
a.

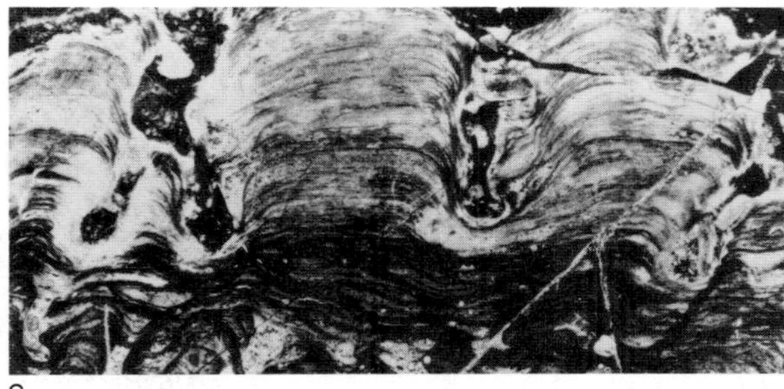
c.

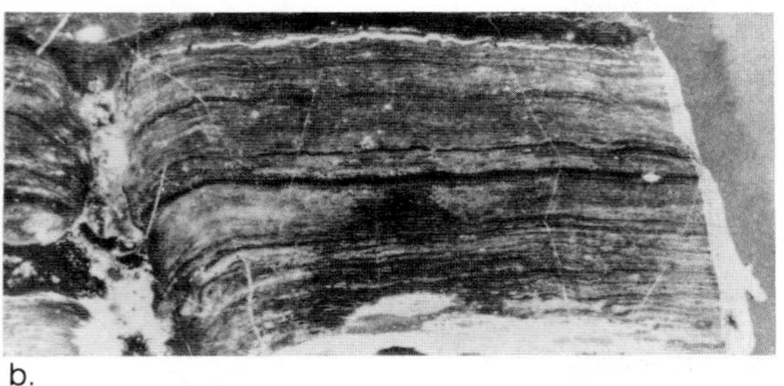
b.

**Figure 8–29** Polished slabs showing vertical sections of Gunflint Iron Formation stromatolites. Although these are Early Proterozoic in age, they were the first stromatolites recognized as the products of organic activity. Compare with Figure 8–30. (Photos courtesy of Stanley M. Awramik, University of California, Santa Barbara)

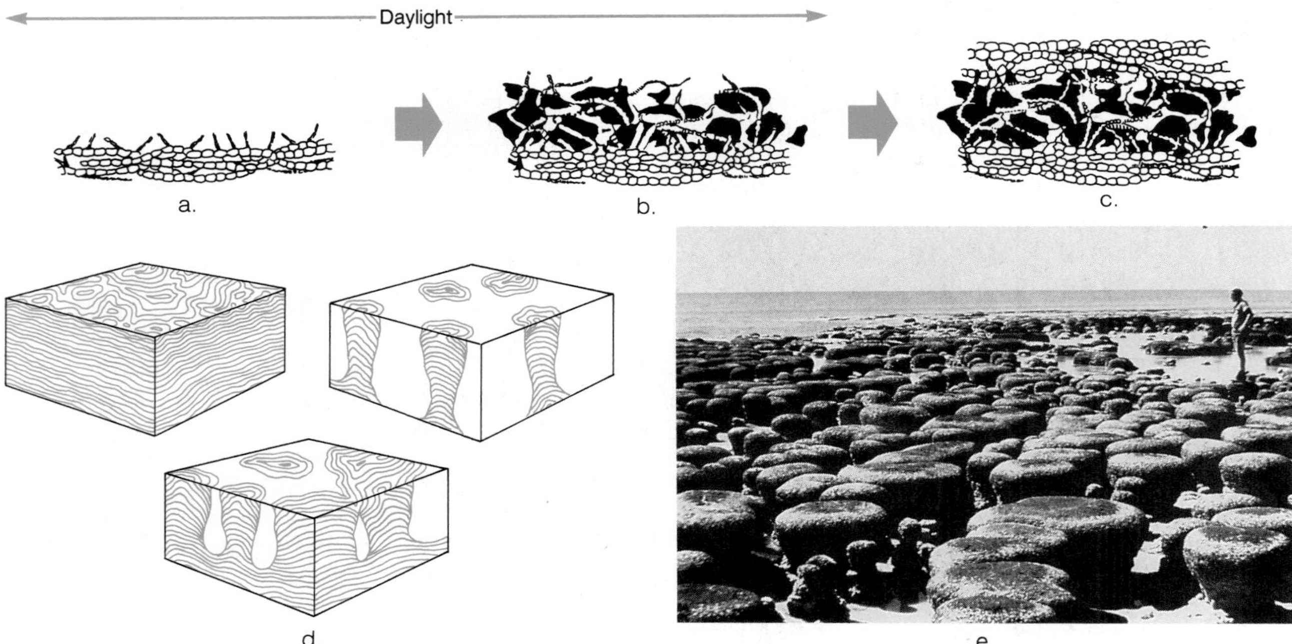


**Figure 8–30** a through c. Diagrams illustrating entrapment of sediment by a mat of cyanobacteria; dark masses are sediment grains. a. Uncovered mat at beginning of daylight period. b. Sediment trapping during daylight. c. Regrowth and sediment binding during darkness. This process is repeated many times and yields the layered structure of stromatolites. d. Morphological types of stromatolites include irregular mats, columns, and columns linked by mats. e. Recent stromatolites, Shark Bay, Australia. (Photo courtesy of Phillip E. Playford, Geological Survey of Western Australia)

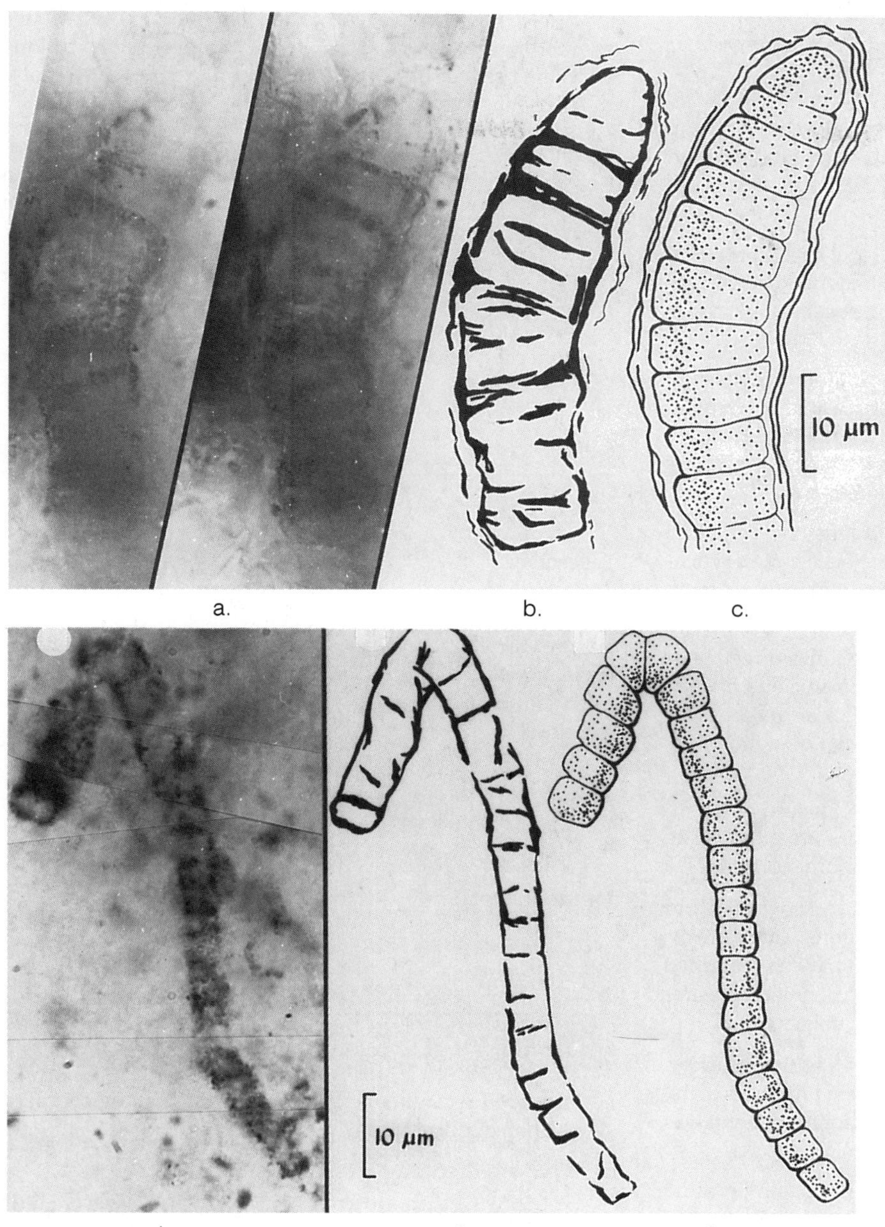


**Figure 8–31** Microfossils from Archean stromatolites. a. Filamentous prokaryote from stromatolites of the 2.8 billion year old Fortescue Group, Western Australia. b and c. Schematic reconstructions of a. d. Filamentous fossil prokaryote from the 3.3–3.5 billion year old Warrawoona Group, Western Australia. e and f. Schematic reconstructions of d. (Photos courtesy of J. William Schopf, University of California, Los Angeles)

8–25) at least 3 billion years ago. These cells were still anaerobic prokaryotes, but as autotrophs they were no longer completely dependent on an external source of preformed organic molecules as a source of nutrients. The Archean microfossils shown in Figure 8–31 are anaerobic, autotrophic prokaryotes. They belong to the kingdom Monera, which is represented in the modern world by bacteria and cyanobacteria or blue-green algae (Appendix B).

## CHAPTER SUMMARY

1. Precambrian time is divided into two eons, the Archean and Proterozoic. Pre-Archean is used for that time in Earth history from 4.6 to 3.8 billion years ago, a time not recorded by rocks.
2. The earliest pre-Archean crust may have been composed of ultramafic rocks. None of this crust has been preserved, but by the beginning of the Archean, several sialic continental nuclei existed.
3. Each continent has an ancient, stable craton composed in part of Archean rocks. The exposed part of a craton is called a Precambrian shield. The Canadian Shield of North America is made up of several provinces, which are delineated by age dates and structural trends.
4. Archean rocks are predominantly greenstone belts and granite-gneiss complexes. Greenstone belts have a synclinal structure and occur as linear bodies within much more extensive granite-gneiss terranes.
5. Typical greenstone belts can be divided into three major rock sequences. The lower units are volcanic and may be either ultramafics grading up into mafics or mafics overlain by felsic rocks. The upper unit consists of sedimentary rocks.
6. Most Archean sedimentary rocks are preserved in greenstone belts, and the most common rocks are graywacke-argillite assemblages deposited by turbidity currents.
7. The evolution of greenstone belts is controversial, but most models rely on Archean plate motions. Those greenstone belts with a lower ultramafic unit may have formed in intracontinental rifts, while those rich in andesites probably formed in backarc marginal basins.
8. During the Late Archean Kenoran orogeny the Superior, Slave, and parts of the Churchill and Nain provinces of the Canadian Shield were formed. The Kenoran orogeny was the last major Archean event in North America.
9. A modern style of plate tectonics did not begin until the Early Proterozoic, but many geologists think some type of Archean plate tectonics occurred. These plates, however, may have moved more rapidly because the Earth possessed more radiogenic heat.
10. The atmosphere and surface waters were derived from internally generated volcanic gases by a process called outgassing. The atmosphere so formed was deficient in free oxygen.
11. Most models for the origin of life require a nonoxidizing atmosphere. Atmospheric gases contained the elements necessary for simple organic molecules. Energy sources to synthesize organic molecules were probably ultraviolet radiation and lightning.
12. The first naturally synthesized organic molecules were monomers such as amino acids. Monomers formed long chains giving rise to more complex molecules called polymers, such as proteins.
13. The method whereby a reproductive mechanism developed is unknown. Some investigators think RNA molecules may have been the first molecules capable of self-replication.
14. The Archean fossil record is very poor. A few localities contain unicellular, prokaryotic bacteria of the kingdom Monera. Stromatolites formed by photosynthesizing bacteria may date from 3.5 billion years ago.

## IMPORTANT TERMS

anaerobic
Archean Eon
argillite
autotrophic
backarc marginal basin
Canadian Shield
craton
fermentation
graywacke
granite-gneiss complex
greenstone belt
heterotrophic
intracontinental rift
Kenoran orogeny
monomer
outgassing
photochemical dissociation
photosynthesis
polymer
Precambrian
Precambrian shield
prokaryotic cell
Proterozoic Eon
stromatolite
supracrustal sequence

## REVIEW QUESTIONS

1. Why are Precambrian rocks so difficult to study compared to Phanerozoic rocks?
2. What are Precambrian shields and cratons?
3. Describe the vertical succession of rock types in a typical greenstone belt.
4. Explain how greenstone belts may have formed in backarc marginal basins and in intracontinental rifts.
5. What sequence of events led to the origin of the southern Superior province of the Canadian shield?
6. How does a greater amount of radiogenic heat account for a different style of plate tectonics in the Archean?
7. Explain how photosynthesis and photochemical dissociation of water vapor supplied oxygen to the Archean atmosphere.

8. Summarize the experimental evidence suggesting that monomers could have formed by natural processes in the Archean.
9. Describe the earliest life-forms as thoroughly as possible.

## ADDITIONAL READING

Condie, K. C. 1982. *Plate Tectonics and Crustal Evolution*. New York: Pergamon Press.

Dickerson, R. E. 1978. Chemical Evolution and the Origin of Life. *Scientific American*, 239, no. 3: 70–85.

Margulis, L. 1982. *Early Life*. New York: Van Nostrand Reinhold Co.

McCall, G. J. H. (ed.). 1977. *The Archean*. Stroudsburg, Penn: Dowden, Hutchinson & Ross, Inc.

Moorbath, S. 1977. The Oldest Rocks and the Growth of Continents. *Scientific American*, 236, no. 1: 92–104.

Windley, B. F. 1984. *The Evolving Continents*. New York: John Wiley & Sons.

# 9

## CHAPTER OUTLINE

PROLOGUE
INTRODUCTION
PROTEROZOIC CRUSTAL EVOLUTION
Hudsonian Orogeny
Colorado-Central Plains and Southwestern Provinces
Granite-Rhyolite Provinces
Grenville-Llano Orogeny
Midcontinent Rift
PROTEROZOIC SUPERCONTINENTS
PROTEROZOIC ROCKS
Greenstone Belts
Quartzite-Carbonate-Shale Assemblages

PERSPECTIVE 9–1: OPHIOLITES—EVIDENCE FOR ANCIENT CONVERGENT PLATE MARGINS

Glaciation and Glacial Deposits
THE EVOLVING ATMOSPHERE
Banded Iron Formations
Red Beds
PROTEROZOIC LIFE
A New Type of Cell Appears
Multicellular Organisms

PERSPECTIVE 9–2: SYMBIOSIS AND THE ORIGIN OF EUKARYOTES

*Multicellular Algae*
*The First Animals*
CHAPTER SUMMARY

## PROLOGUE

*In 1947 the Australian geologist R. C. Sprigg discovered impressions of soft-bodied animals in rocks of the Ediacara Hills of South Australia. Additional discoveries by geologists and amateur paleontologists turned up what appeared to be impressions of algae, jellyfish, segmented worms, worm tracks, and animals that bear no resemblance to living organisms. These discoveries have partly resolved one of the great mysteries in the history of life.*

*Before the discovery of these Ediacaran fossils, geologists were perplexed by the apparent absence of animal fossils in strata older than the Cambrian. The shelly faunas of the Cambrian Period seemed to have appeared abruptly in the fossil record. In fact, the evidence for life before the Cambrian was so sparse that all pre-Paleozoic time was once referred to as the* Azoic, *meaning without life. Geologists assumed that the Cambrian shelly faunas must have been preceded by more primitive animals, but since none were known, a worldwide unconformity, the* Lipalian interval, *was proposed to account for the absence of such fossil-bearing rocks.*

*The rock unit in which the Ediacara Hills fossils were discovered, the Pound Quartzite, was initially thought to be Cambrian in age. However, a joint investigation by the South Australian Museum and the University of Adelaide revealed that the fossil-bearing strata lie more than 150 m below the oldest recognized Cambrian strata. Eventually it became clear that what had been discovered was a unique assemblage of soft-bodied animals preserved as molds and casts on the undersides of sandstone layers (Fig. 9-1). Martin Glaessner of the University of Adelaide, believes that the animals lived in a nearshore, shallow marine environment.*

*Some investigators, including Glaessner, are of the opinion that at least three modern invertebrate phyla are represented; jellyfish and sea pens (phylum Coelenterata), segmented worms (phylum Annelida), and primitive members of the phylum Arthropoda, which includes modern insects, spiders, and crabs. One worm-like Ediacaran fossil,* Spriggina, *has been cited as a possible ancestor of trilobites, and another,* Tribachidium, *may be a primitive echinoderm (Fig. 9–1).*

*Disagreement exists, however, on exactly what these Ediacaran animals were and how they should be classified. Adolph*

# Precambrian History—The Proterozoic Eon

*Seilacher of Tubingen, West Germany, for example, thinks that these animals represent an early evolutionary radiation quite distinct from the ancestry of the modern invertebrate phyla. In other words, he believes that the Ediacaran animals are not members of the phyla just noted, and that they are not ancestral to the complex invertebrates that appeared in abundance during the Cambrian Period.*

*Ediacara-type faunas are now known on all continents except Antarctica. Collectively these* **Ediacaran faunas**, *as they are commonly called, range from 570 to 670 million years old. Animals were widespread during this time, but their fossils are rare, since all lacked durable skeletons.*

*The discovery of pre-Paleozoic fossil-bearing strata has prompted Preston Cloud, of the University of California at Santa Barbara, and Martin Glaessner to propose a new geologic period and system, the Ediacarian. According to this proposal, the Ediacarian Period constitutes the first period of the Phanerozoic Eon, and the stratotype for the Ediacarian System is located in South Australia (Fig. 9–2). Thus, strata of the Ediacarian System were deposited during the Ediacarian Period which began 670 million years ago, when multicelled organisms appeared in the fossil record, and ended 570 million years ago when shelly faunas of the Cambrian Period first appeared.*

*Cloud's and Glaessner's proposal is a reasonable one, and, as a matter of fact, it has been accepted by many geologists. However, many others prefer the more traditional time scale (See Chapter 1), and in this book we consider the Ediacaran fauna to be latest Proterozoic in age.*

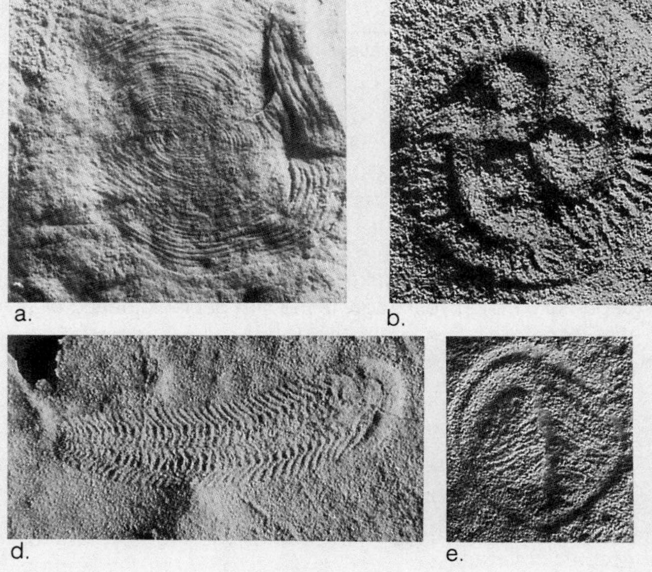

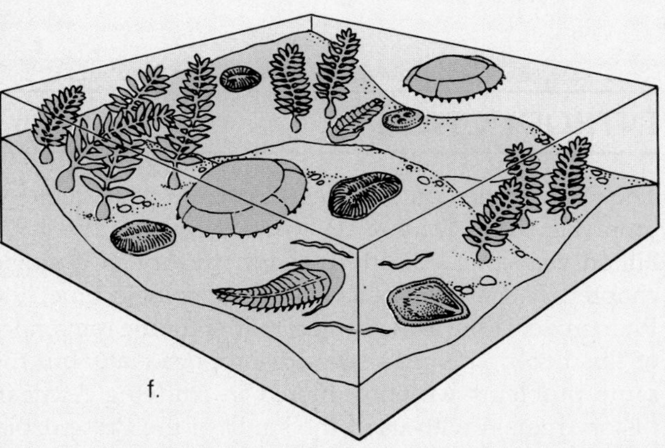

**Figure 9–1** The Ediacaran fauna of Australia. Impressions of multicelled animals: a. *Ovatoscutum concentricum.* b. *Tribrachidium heraldicum*, a possible primitive echinoderm. c. *Charniodiscus arboreus.* d. *Spriggina floundersi*, a possible ancestor of trilobites. e. *Parvancorina minchami.* f. Reconstruction of the Ediacaran environment. (Photos courtesy of Martin F. Glaessner, University of Adelaide, Adelaide, Australia.)

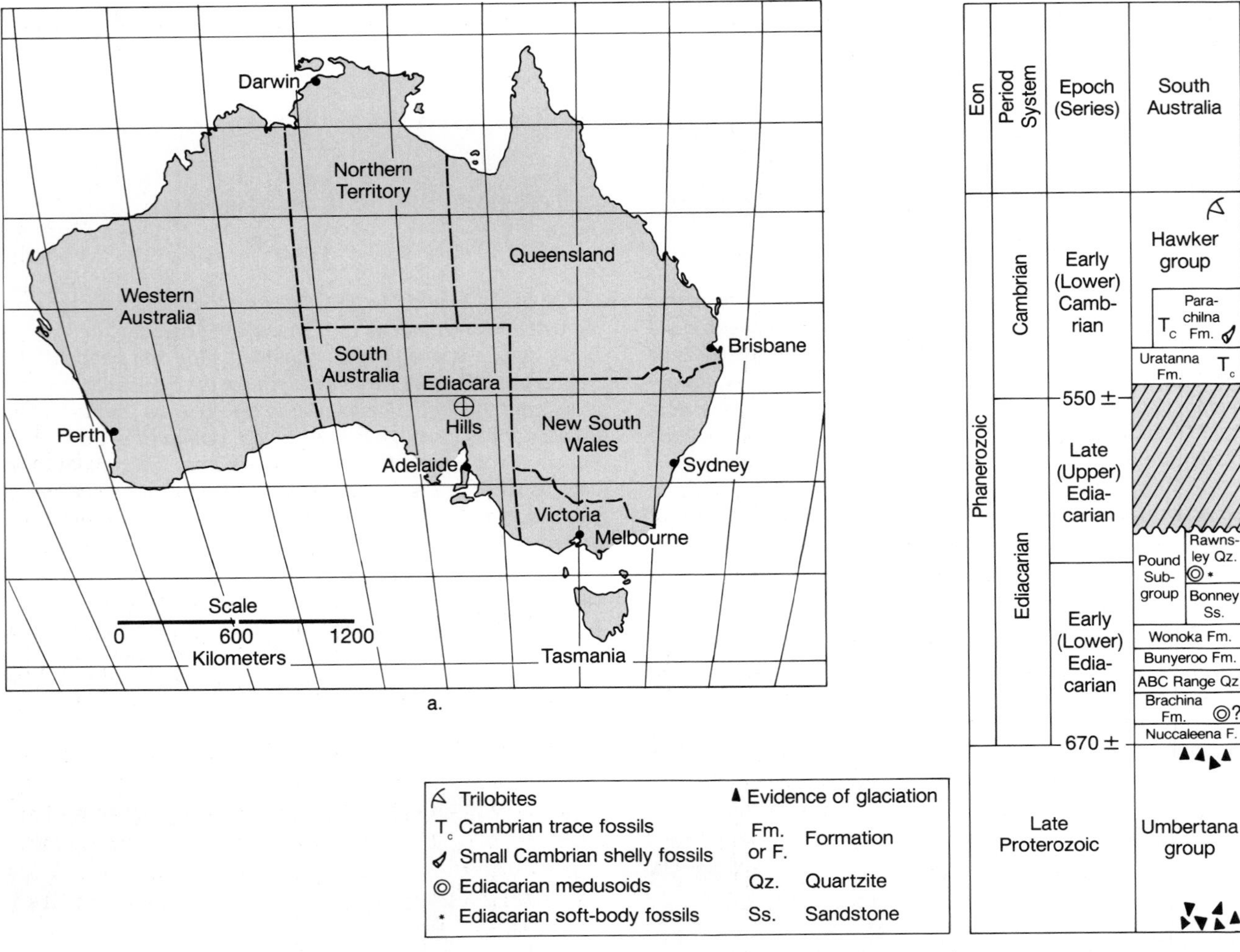


**Figure 9–2** a. Map showing the stratotype area for the Ediacarian System about 480 kilometers north of Adelaide, Australia. b. Proposed names for a new system and period. According to this proposal the Ediacarian Period, which began 670 million years ago, is the first geologic period of the Phanerozoic Eon.

## INTRODUCTION

The Proterozoic Eon includes 42 percent of all geologic time (Fig. 8–1, Table 8–1), yet we review this 1.93-billion-year-long episode of Earth history in a single chapter. A mere 12 percent of all geologic time, the Phanerozoic Eon, is the subject matter for the remainder of this book. This may seem disproportionate, but the same problems we encountered in studying Archean history remain with us in the study of the Proterozoic, although to a lesser degree. In short, we know more about the Proterozoic than we do about the Archean, but Phanerozoic history is known even better, hence its greater coverage.

How does the Proterozoic differ from the Archean? The basic difference is the style of crustal evolution. Archean crust-forming processes generated greenstone belts and granite-gneiss complexes that were shaped into cratons; and although these rock assemblages continued to form during the Proterozoic, they did so at a considerably reduced rate. Many Archean rocks have been metamorphosed, although the degree of metamorphism varies considerably, and some are completely unaltered. In contrast, many Proterozoic rocks are unmetamorphosed and little deformed, and in many areas Proterozoic rocks are separated from Archean rocks by a profound unconformity. Finally, the Proterozoic is characterized by widespread sedimentary-rock assemblages that are rare in the Archean and by a plate tectonic style that is essentially modern.

Changes definitely occurred across the Archean-Proterozoic boundary, but these changes were not synchronous. For example, Archean crustal evolution in the Kaapvaal craton of South Africa was completed nearly 3.0 billion years ago when sedimentary deposits more typical of the Proterozoic began to accumulate. In North America, this same change did not occur until 2.6 billion years ago.

The change in crust-forming processes actually occurred from 2.95 to 2.45 billion years ago, so the Archean-Proterozoic boundary at 2.5 billion years ago is rather arbitrarily placed. It marks, "the major lull that followed a nearly worldwide epoch of cratonization, an epoch unmatched in later Earth history, and the approximate beginning of major reworking of preexisting crust."[1]

We can think of the Proterozoic as a time of modernization. That is, the plate tectonic style and many of the rock assemblages are essentially like those of the Phanerozoic. There are differences, though—banded iron formations, for example, formed until the Middle Proterozoic but rarely thereafter.

## PROTEROZOIC CRUSTAL EVOLUTION

The Proterozoic was a time during which plate motions, extensive plutonism, and regional metamorphism thickened, stabilized, and increased the area of continental crust. In the last chapter we noted that the Archean cratons assembled through a series of island arc and minicontinent collisions. These provided the nuclei around which Proterozoic continental crust accreted, thereby forming much larger cratons (Fig. 9–3). One large landmass, called **Laurentia**, consisted mostly of North America and Greenland, parts of northwestern Scotland, and perhaps parts of the Baltic shield of Scandinavia. This chapter will emphasize the geologic evolution of Laurentia. Table 9–1 summarizes the crust-forming events discussed in the following sections.

### Hudsonian Orogeny

The first major episode in the Proterozoic evolution of Laurentia, the **Hudsonian orogeny**, occurred between 1.9 and 1.8 billion years ago. Several major orogens developed that resulted in suturing of Archean cratons along deformation belts (Fig. 9–3, Table 9–1). The Hudsonian orogeny was an episode of continental growth during which new crust was accreted at the margins of Archean cratons. By 1.8 billion years ago, much of what is now Greenland, central Canada, and the north-central United States formed a large craton (Fig. 9–3).

Rocks of the Trans-Hudson orogen in northern Saskatchewan record initial rifting, extrusive volcanism, sedimentation, and the formation of oceanic crust, followed by ocean basin closure (Fig. 9–4). As the ocean basin closed, a subduction zone formed resulting in island arc collisions, granitic plutonism, intense deformation, and regional metamorphism. The events leading to the formation of the Trans-Hudson orogen resulted in the suturing of the Superior and Churchill cratons (Fig. 9–3).

Another notable event was the development of the Wopmay orogen in northwestern Canada (Fig. 9–3). Rocks here also record rifting followed by closure of an ocean basin. In fact, these rocks record the oldest completely preserved Wilson cycle (from Chapter 6). Thus, the Wopmay orogen deserves more detailed discussion.

Several events in the Wopmay orogen can be deduced from studies of the Coronation Supergroup (Fig. 9–5). Rifting is indicated by the fault-bounded basins in which turbidites and volcanics of the Akaitcho Group accumulated. Continental rifting established a passive continental margin upon which sediments of the Epworth Group were deposited. These sediments are of particular interest because they form a quartzite-carbonate-shale assemblage, a suite of rocks that characterize passive margins and first became abundant and widespread in the Proterozoic.

Conformably overlying the passive-margin rocks are deep-water sediments indicating abrupt deepening of the shelf. Deformation caused by closure of the ocean basin was contemporaneous with deposition of these rocks and resulted in eastward thrusting of the Coronation Supergroup over the craton's edge.

South of the Superior craton, the Penokean orogen accounts for the origin of another large segment of continental crust (Fig. 9–3). Considerable disagreement exists

[1]K. A. Plumb and H. L. James. 1986. Subdivision of Precambrian Time: Recommendations and Suggestions by the Subcommission on Precambrian Stratigraphy, *Precambrian Research*, 32: 80.

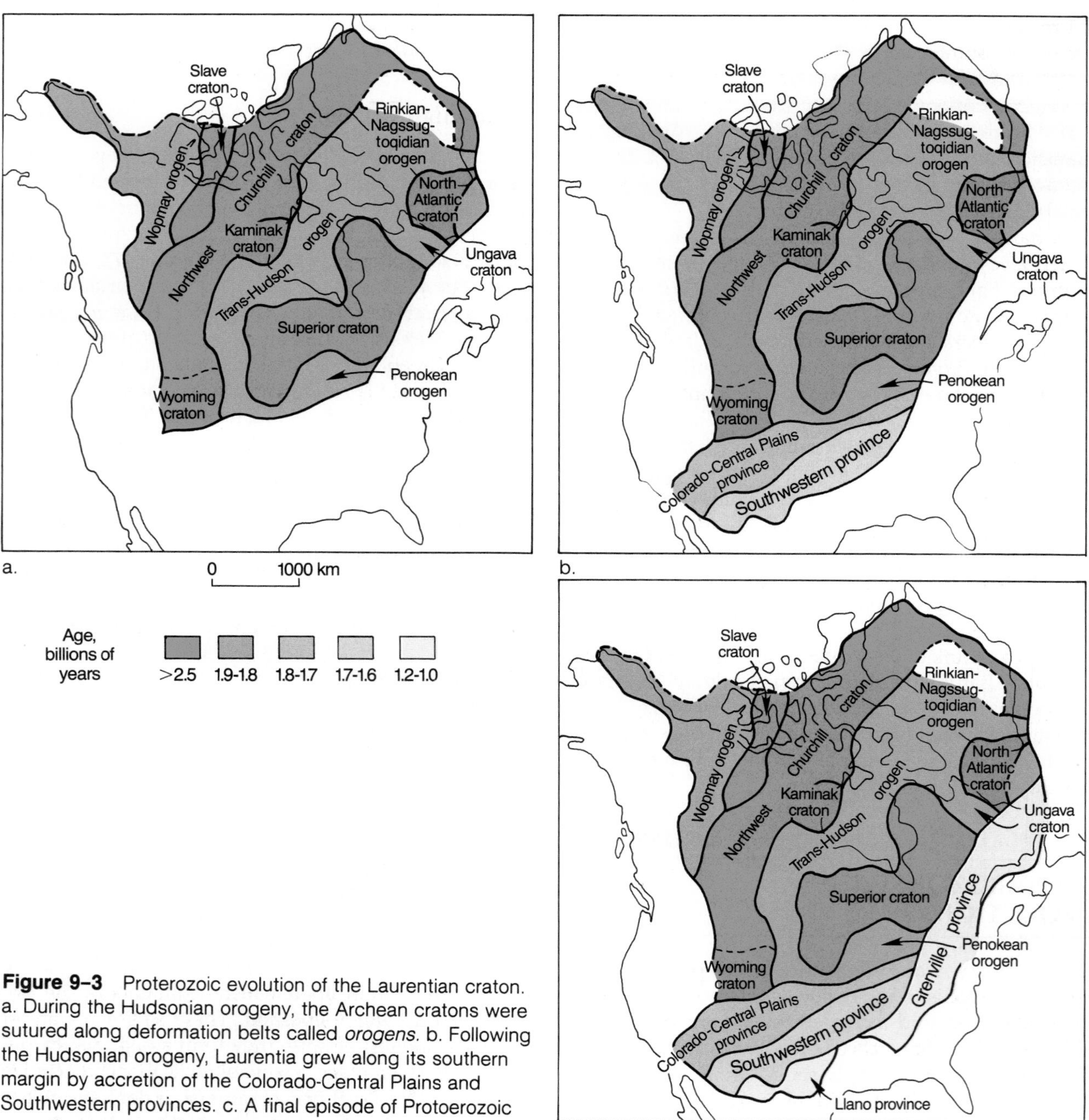


**Figure 9-3** Proterozoic evolution of the Laurentian craton. a. During the Hudsonian orogeny, the Archean cratons were sutured along deformation belts called *orogens.* b. Following the Hudsonian orogeny, Laurentia grew along its southern margin by accretion of the Colorado-Central Plains and Southwestern provinces. c. A final episode of Protoerozoic accretion occurred during the Grenville-Llano orogeny.

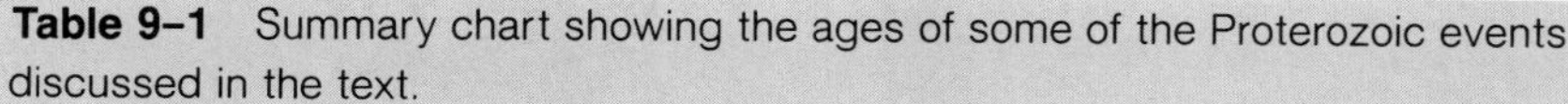

**Table 9-1** Summary chart showing the ages of some of the Proterozoic events discussed in the text.

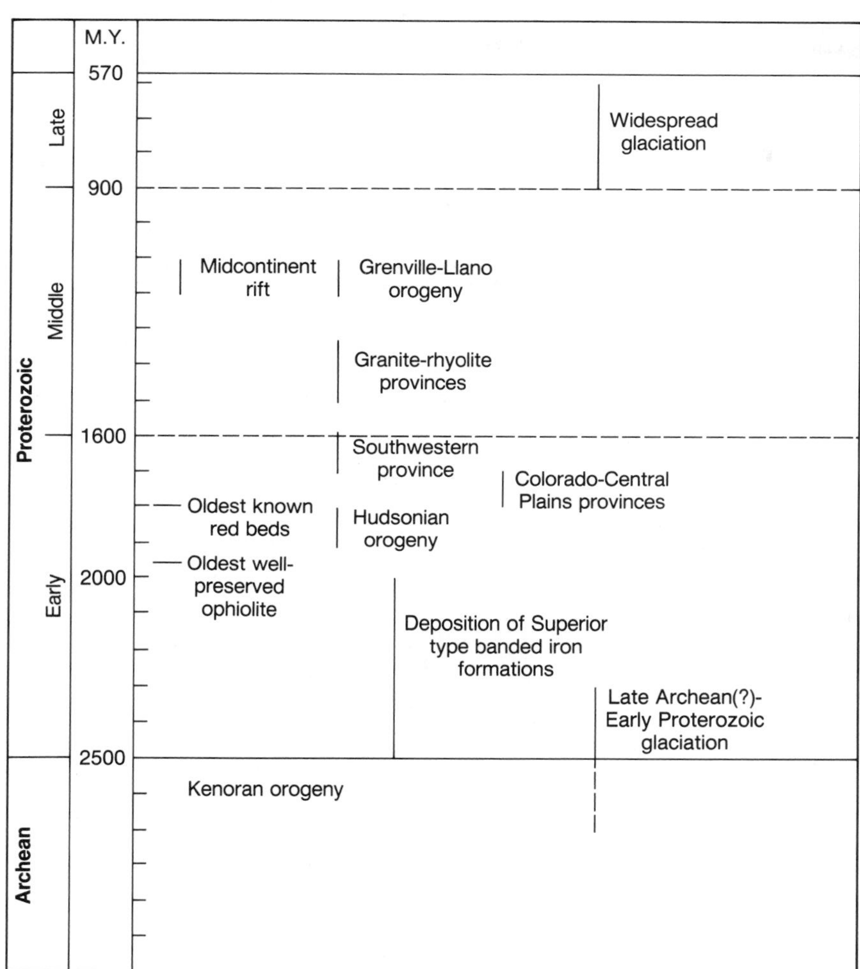


on the tectonic processes responsible for it, though. Some geologists think it formed in a manner similar to the Wopmay orogen, but other geologists appeal to intracratonic deformation for an explanation.

## Colorado-Central Plains and Southwestern Provinces

Following the Hudsonian orogeny, more episodes of accretion of continental crust resulted in the formation of the **Colorado-Central Plains** and **Southwestern provinces**. Growth continued in what is now the southwestern and central United States as successively younger belts were sutured to the post-Hudsonian craton (Figs. 9–3, 9–6).

Clearly this is a plate tectonic model containing both extensional and compressional phases. Note particularly in Figure 9–6 that several quartzite-carbonate-shale assemblages and greenstone belts are shown. The greenstone belts formed in backarc marginal basins (Fig. 8–19), while the quartzite- carbonate-shale assemblages were deposited in fluvial, tidal, and shallow marine environments on a south-facing continental margin.

Continental accretion occurred as northward-migrating island arcs collided with the continent, resulting in orogenies and the emplacement of granitic batholiths. The net effect was the accretion of more than 1,000 km of continental crust along the southern margin of Laurentia.

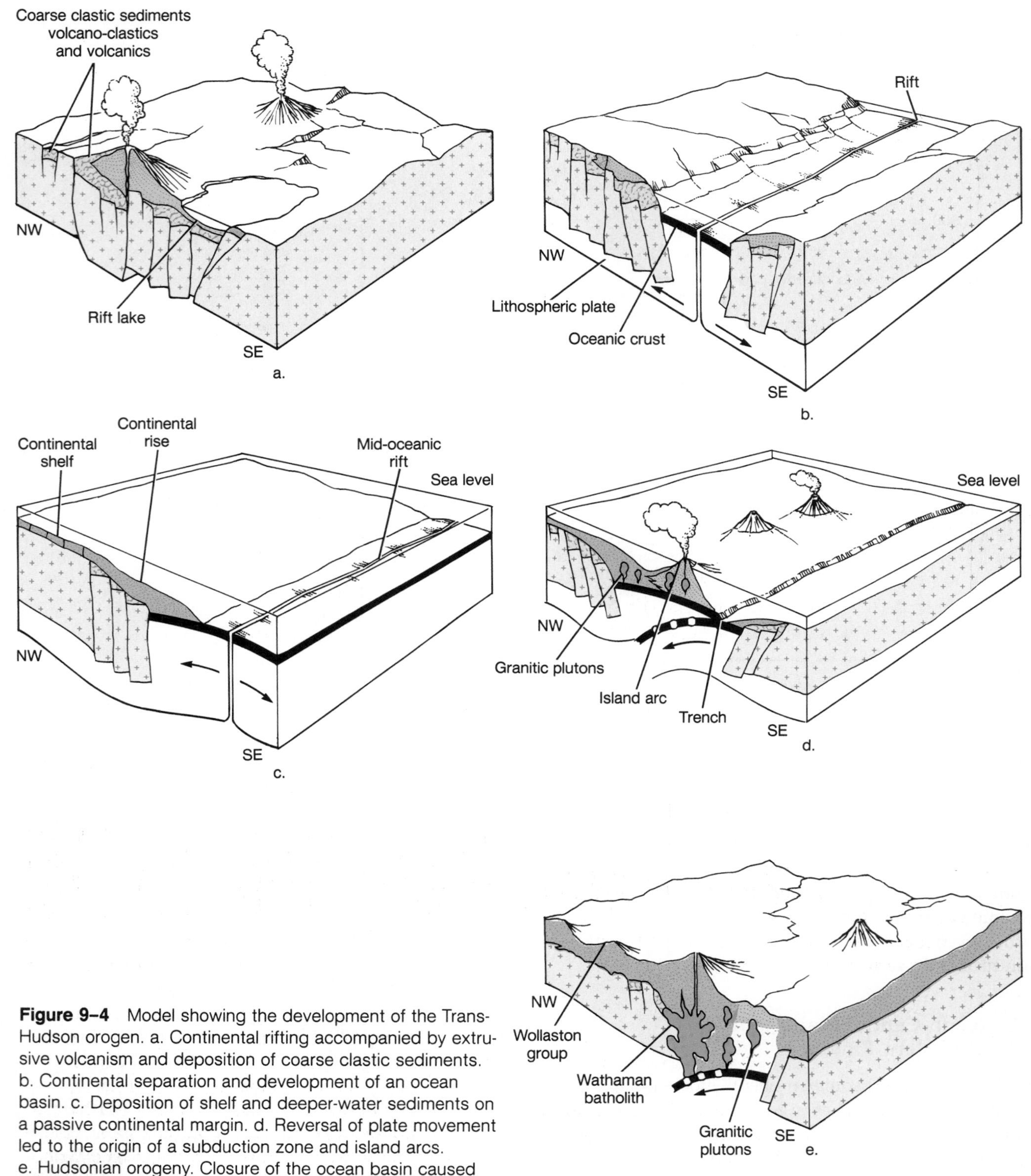


**Figure 9–4** Model showing the development of the Trans-Hudson orogen. a. Continental rifting accompanied by extrusive volcanism and deposition of coarse clastic sediments. b. Continental separation and development of an ocean basin. c. Deposition of shelf and deeper-water sediments on a passive continental margin. d. Reversal of plate movement led to the origin of a subduction zone and island arcs. e. Hudsonian orogeny. Closure of the ocean basin caused intrusive and extrusive volcanism and deformation of passive margin sediments.

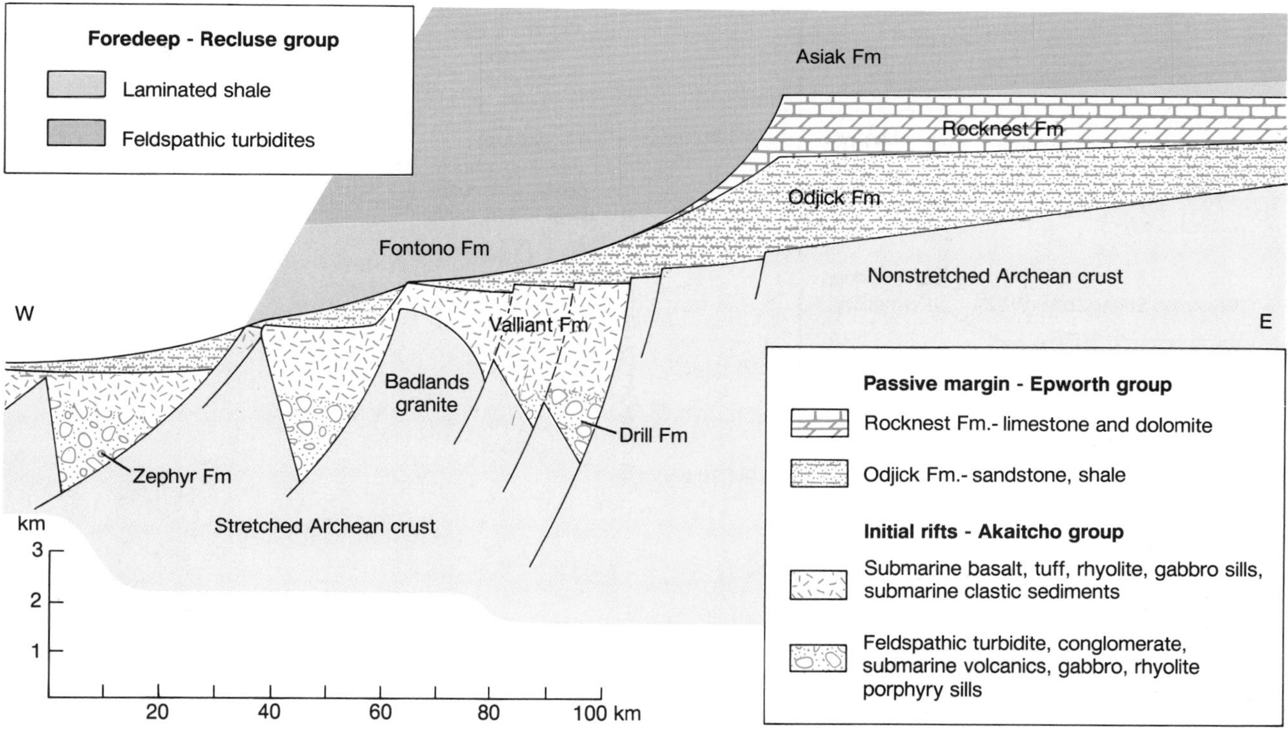


**Figure 9–5** Rocks of the Coronation Supergroup in northwestern Canada record the oldest known complete Wilson cycle. See text for discussion.

## Granite-Rhyolite Provinces

No major episode of crustal growth occurred between 1.6 and 1.2 billion years ago. Nevertheless, during the interval from 1.5 to 1.34 billion years ago, extensive volcanic activity resulted in the development of **granite-rhyolite provinces** extending from the north-central to the southwestern United States (Fig. 9–7, Table 9–1).

These rocks did not increase the area of Laurentia since they were intruded into or erupted upon already existing crust. Rocks of the granite-rhyolite provinces are buried beneath Phanerozoic strata, but "recent investigations have shown that the rocks represent the deeply eroded remains of calderas, their ash-flow fill and outflow sheets, and their subvolcanic plutons."[2]

The parallels between these rocks and those of the Late Cenozoic of the Yellowstone National Park region of Wyoming, Montana, and Idaho are striking. The latter are much more restricted in areal extent, but they, too, represent calderas and their fill, and ash flows (Fig. 9–8).

[2]Bickford, M. E. and Van Schmus, W. R. 1986. Proterozoic History of the Midcontinent Region of North America. *Geology*, 14, p. 494.

## Grenville-Llano Orogeny

Another major episode in the evolution of Laurentia occurred between 1.2 and 1.0 billion years ago. Both the **Grenville province** in the eastern United States and Canada and the **Llano province** of the Southwest formed during this interval (Fig. 9–3, Table 9–1).

Rocks of the Grenville province are extensively exposed in southeastern Canada, where they abut the Archean Superior craton (Fig. 9–3). This belt of deformed rocks extends northeast into Greenland, and continues further into Scandinavia. It also continues southwest through the area of the modern Appalachian Mountains and may be continuous with the Llano province (Fig. 9–3).

The Grenville Supergroup, consisting of sandstones, shales, and carbonates, is highly deformed, met-

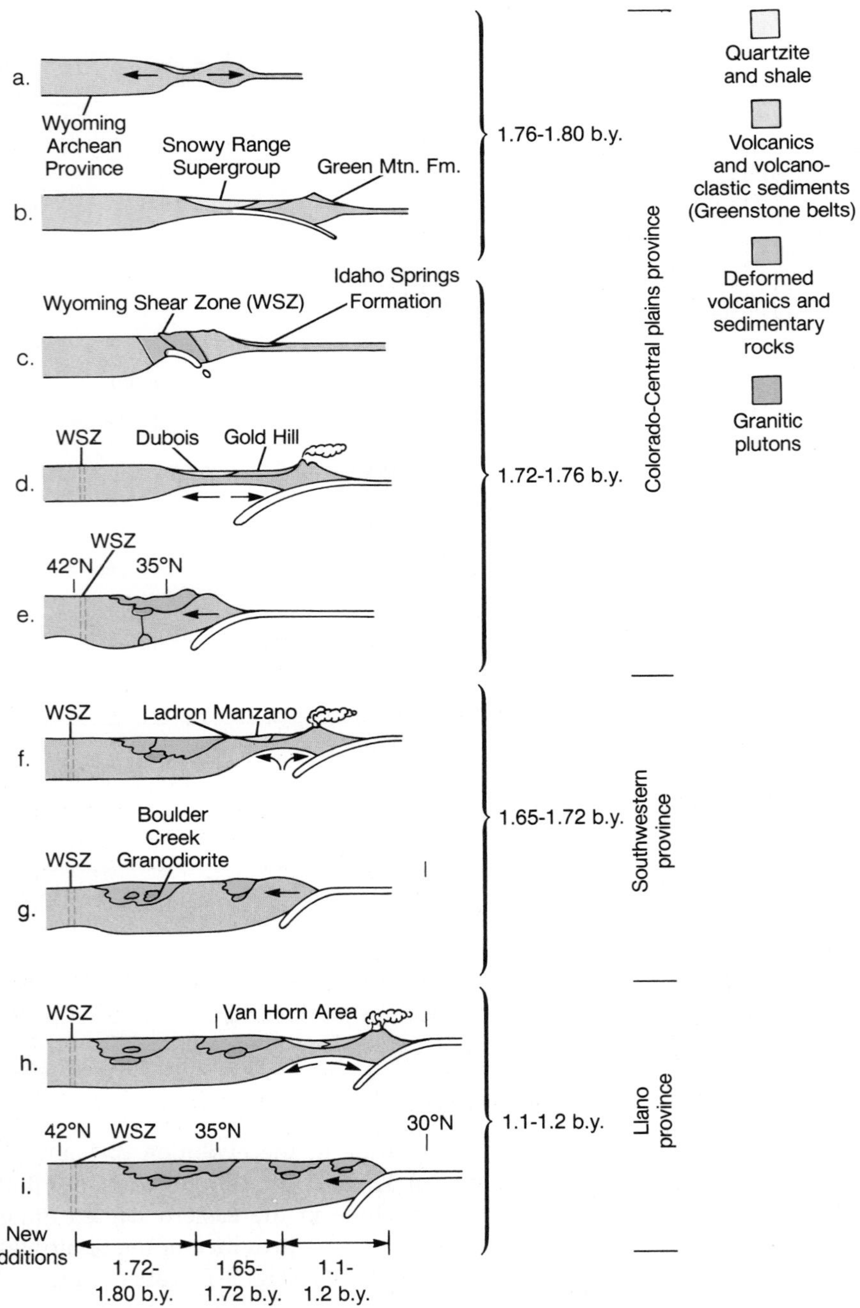


**Figure 9–6** Diagram showing accretion along the southern margin of Laurentia. a through e. Events leading to the origin of the Colorado-Central Plains province. f through g. Accretion of the Southwestern province. h through i. Accretion of the Llano province.

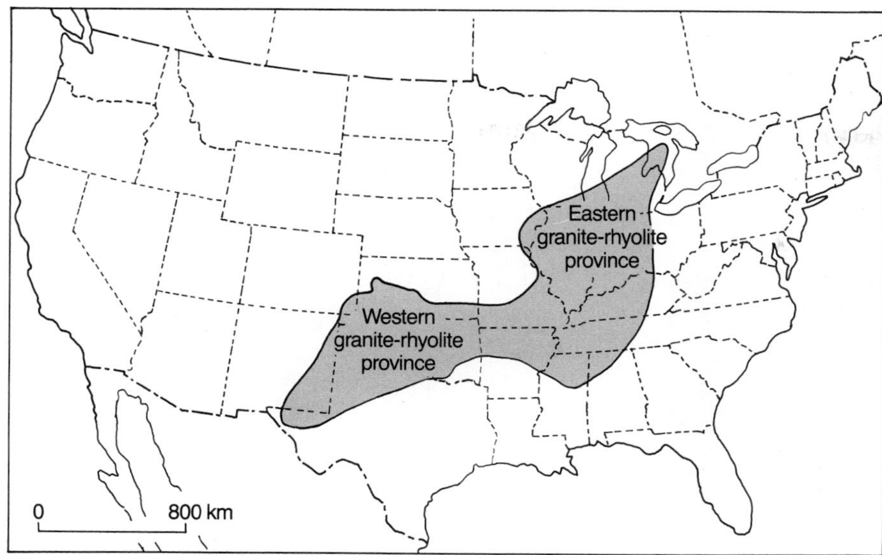


**Figure 9–7** Location of the granite-rhyolite provinces. These rocks consist of plutons and ash-flow tuffs that were intruded into or erupted onto preexisting crust.

**Figure 9–8** Roadcut near Ashton, Idaho, showing Late Cenozoic rocks similar to those of the granite-rhyolite provinces. 1. Gray rhyolite tuff. 2. Windblown dust deposit (loess). 3. White tuff. 4. Pumice. 5. Orange-pink rhyolite tuff.

amorphosed, and intruded by various igneous bodies (Fig. 9–9). Many geologists think the Grenville orogeny can be explained in terms of an opening and then a closing ocean basin. If so, the Grenville Supergroup may represent sediments deposited on a passive continental margin. Therefore, the processes responsible for the Grenville province may be similar to those that produced the Trans-Hudson and Wopmay orogens (Figs. 9–4 and 9–5). However, some geologists think that Grenville deformation occurred in an intracontinental setting or was caused by major shearing. Whatever the cause, it represents the final episode of Proterozoic continental accretion of Laurentia. In the latest Proterozoic, Laurentia assumed the shape shown in Figure 9–3.

## Midcontinent Rift

Contemporaneous with Grenville deformation was an episode of continental rifting in Laurentia. The **Midcontinent rift** is a major feature extending from the Lake Superior Basin southwest into Kansas, and a southeasterly trending branch extends into Michigan (Fig. 9–10). It cuts across Archean and Early Proterozoic structural provinces and, in the east, terminates against the Grenville province.

Most of the rift is concealed by younger strata except in the Lake Superior region where rocks that filled the rift are well exposed. The central part of the rift contains thick accumulations of basaltic lava flows that form extensive lava plateaus. For example, the Portage Lake Volcanics consist of numerous overlapping lava flows, forming a volcanic pile several kilometers thick (Fig. 9–11).

The rift formed between 1.2 and 1.1 billion years ago, but the volcanism apparently occurred during the comparatively short interval of only 20 to 30 million years. For the entire rift, the volume of volcanic rocks

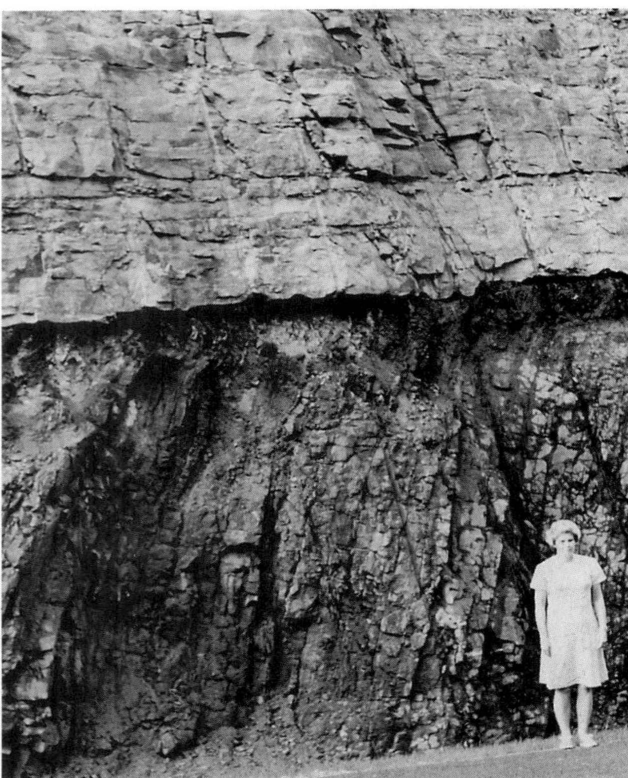

**Figure 9–9** Metamorphosed sedimentary rocks of the Grenville Supergroup near Alexandria, New York. At this locality the Grenville is unconformably overlain by the Late Cambrian Potsdam Formation. (Photo courtesy of R. V. Dietrich, Central Michigan University.)

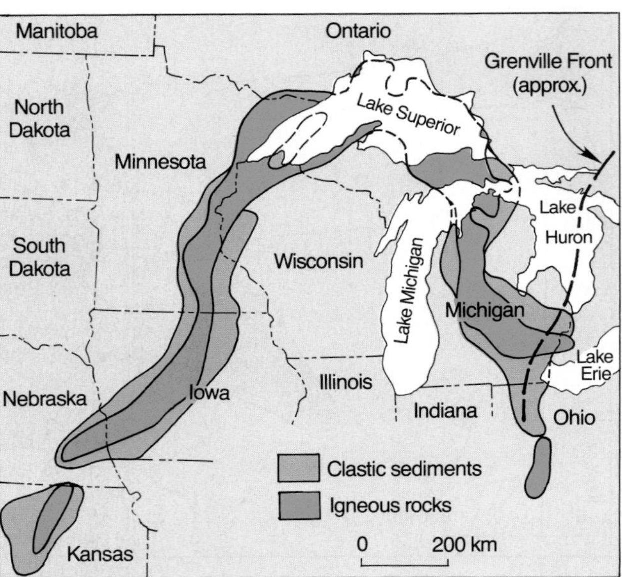


**Figure 9–10** Location of the Midcontinent rift. The volcanic and sedimentary rocks that fill the rift are exposed in the Lake Superior region, but elsewhere they are buried beneath younger rocks.

has been estimated at 300,000 to 1,000,000 km$^3$. This compares in volume but not areal extent to the great outpourings of lava during the Cenozoic (Chapter 15).

Intertonguing with and overlying the volcanics are thick sequences of sedimentary rocks. Near the rift margins, coarse-grained rocks such as the Copper Harbor Conglomerate (Fig. 9–11) formed large alluvial fans that graded into sandstones and shales of the rift axis.

## PROTEROZOIC SUPERCONTINENTS

Thus far this chapter has largely reviewed the geologic evolution of Laurentia, a large landmass composed mostly of North America and Greenland (Fig. 9–3). Paleomagnetic studies and the continuation of the Grenville province into Scandinavia suggest that the Baltic Shield may also have been part of or situated close to Laurentia during the Proterozoic (Fig. 9–12). Indeed, some geologists suggest that a collision between the Baltic and Laurentian shields may have caused the Hudsonian orogeny.

But what of Laurasia and Gondwana, the two major components of the Late Paleozoic supercontinent Pangaea? The components that ultimately formed Pangaea were evolving during the Precambrian, but few investigators agree on their size, shape, or associations. One group holds that a single Pangaea-like supercontinent persisted from the Late Archean to the end of the Proterozoic. Other investigators, pointing to uncertainties in the data, maintain that this reconstruction is not justified.

Geologists generally agree that three or more supercontinents existed in the Late Proterozoic. Their relative locations are unknown, however (Fig. 9–13). As indicated in Figure 9–13, East Gondwana consisted of Australia, India, and Antarctica, while West Gondwana included Africa and South America. The third supercontinent, Laurasia, consisted of Laurentia, Europe, and most of Asia.

During the latest Proterozoic, all Southern Hemisphere continents were strongly deformed. These events may mark the time during which East and West Gondwana were assembled into the larger supercontinent of Gondwana.

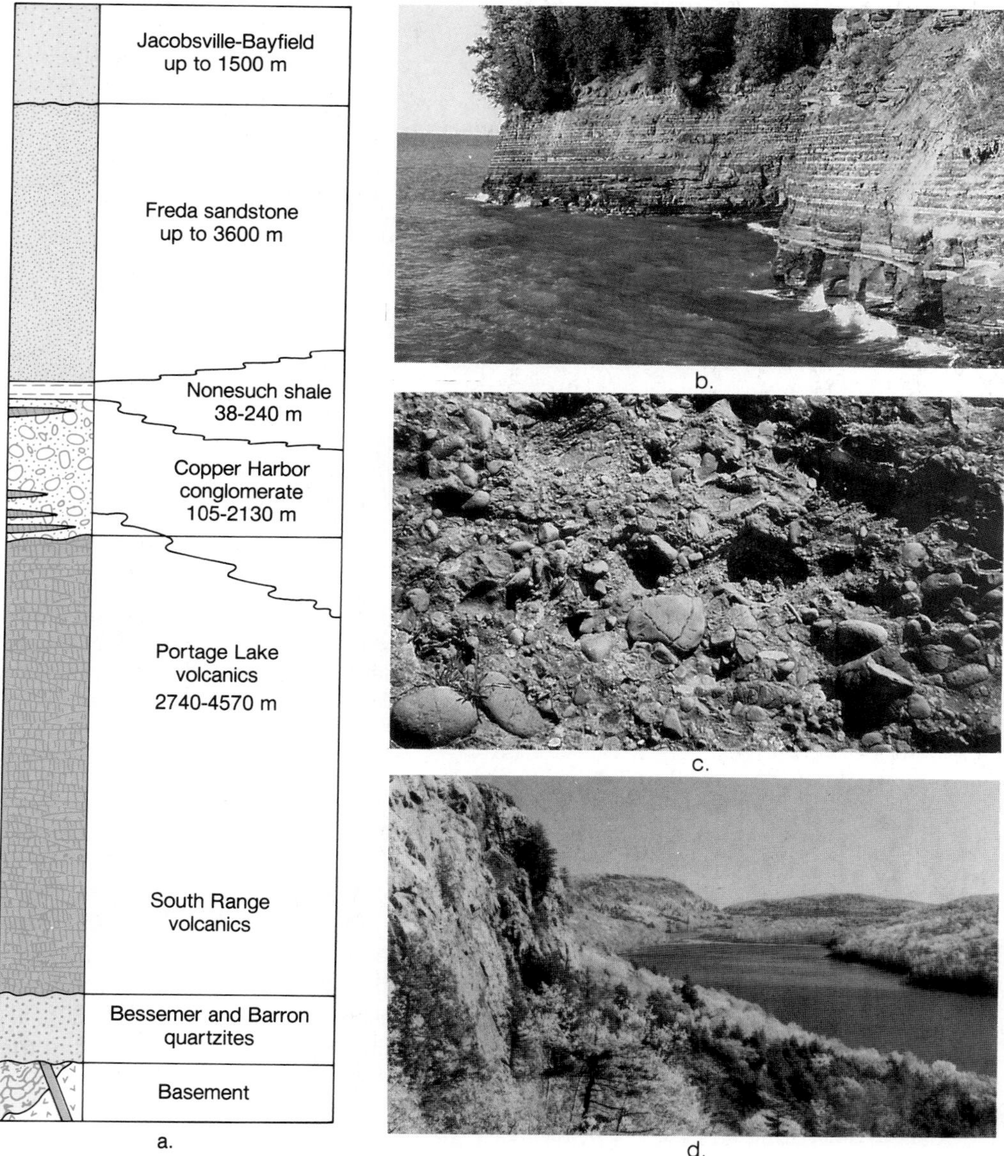


**Figure 9–11** Rocks of the Midcontinent rift exposed in the Lake Superior region. a. Section showing vertical relationships. b. Freda Sandstone. c. Copper Harbor Conglomerate. d. Portage Lake Volcanics. (Photo (b) courtesy of Albert B. Dickas, University of Wisconsin, Superior, Photo (d) by Dan Urbanski, White Pine, Silver City, Michigan.)

## PROTEROZOIC ROCKS

In the preceding section we briefly mentioned several Proterozoic rock types or rock associations. Some of these deserve special attention. Greenstone belts, for example, are more typical of the Archean, but they continued to form during the Proterozoic, even though they differed in detail. Quartzite-carbonate-shale assemblages also merit our attention since they are fully modern in appearance and first became widespread during the Proterozoic. Deposits attributed to two episodes of Proterozoic glaciation are recognized as well.

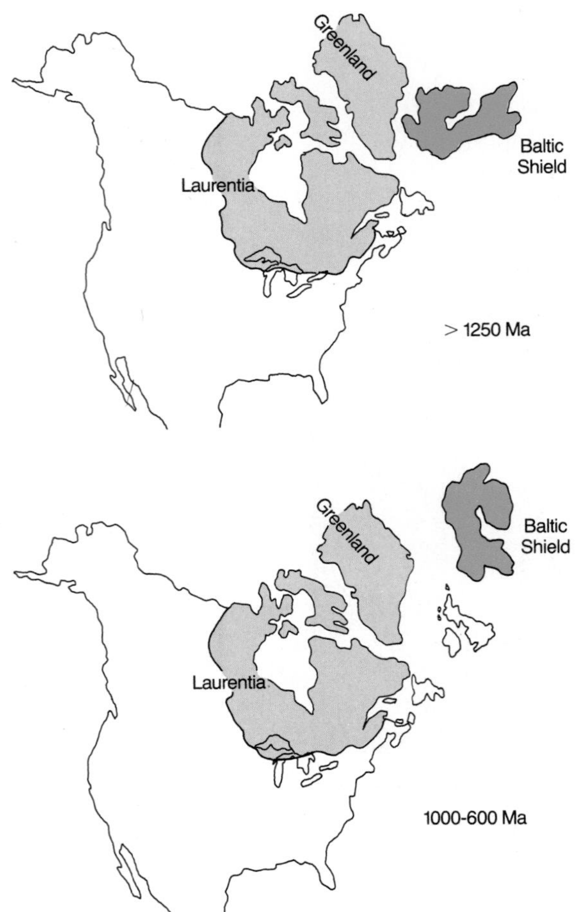


**Figure 9–12** Reconstructions of the Baltic Shield relative to Laurentia for two times during the Proterozoic.

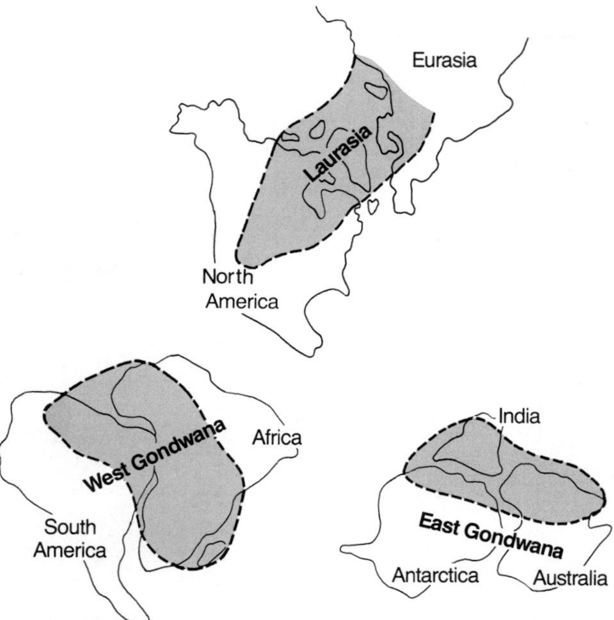


**Figure 9–13** Reconstructions of three possible Late Proterozoic supercontinents. Dashed lines indicate areas known or suspected to be underlain by Archean crustal rocks. The relative locations of Laurasia, East, and West Gondwana are unknown.

The onset of a modern style of plate tectonics probably occurred in the Early Proterozoic, and the first well-preserved **ophiolites**, which mark convergent plate margins, are of this age (Perspective 9–1). Banded iron formations and red beds, too, are important Proterozoic rock types, but these are more conveniently discussed in the section on the atmosphere.

## Greenstone Belts

Greenstone belts are characteristic of the Archean, and most seem to have formed in the interval between 3.0 and 2.7 billion years ago. They are not, however, restricted to the Archean. Proterozoic greenstone belts occur on several continents, including North America, and at least one has been reported from the Cambrian of Australia. Several formed as crustal accretion occurred south of the Wyoming province (Fig. 9–6).

Ultramafic rocks are rare in Proterozoic greenstone belts. The near absence of Proterozoic and younger ultramafics is probably accounted for by the decreasing amount of radiogenic heat produced within the Earth (Fig. 8–5). Otherwise, Proterozoic and Archean greenstone belts differ little one from the other. Both appear to have formed in similar tectonic settings, rifts and backarc marginal basins (Figs. 8–19, and 8–20), although the latter setting became more prevalent in the Late Proterozoic.

## Quartzite-Carbonate-Shale Assemblages

Fully 60 percent of known Proterozoic rock successions are composed of **quartzite-carbonate-shale assemblages**. Widespread deposition of this assemblage occurred throughout the Proterozoic in backarc basins, along rifted continental margins, and in intracratonic basins.

Early Proterozoic quartzite-carbonate-shale assemblages are widely distributed in the Great Lakes region (Fig. 9–14). Though not as apparent as quartzite and shale, conglomerates and graywackes are also present. The quartzites were derived from mature quartz sandstones, and many show wave-formed ripple marks and cross-bedding (Fig. 9–15). Thick carbonates, mostly dolomite, containing abundant stromatolites, are also present (Fig. 9–15).

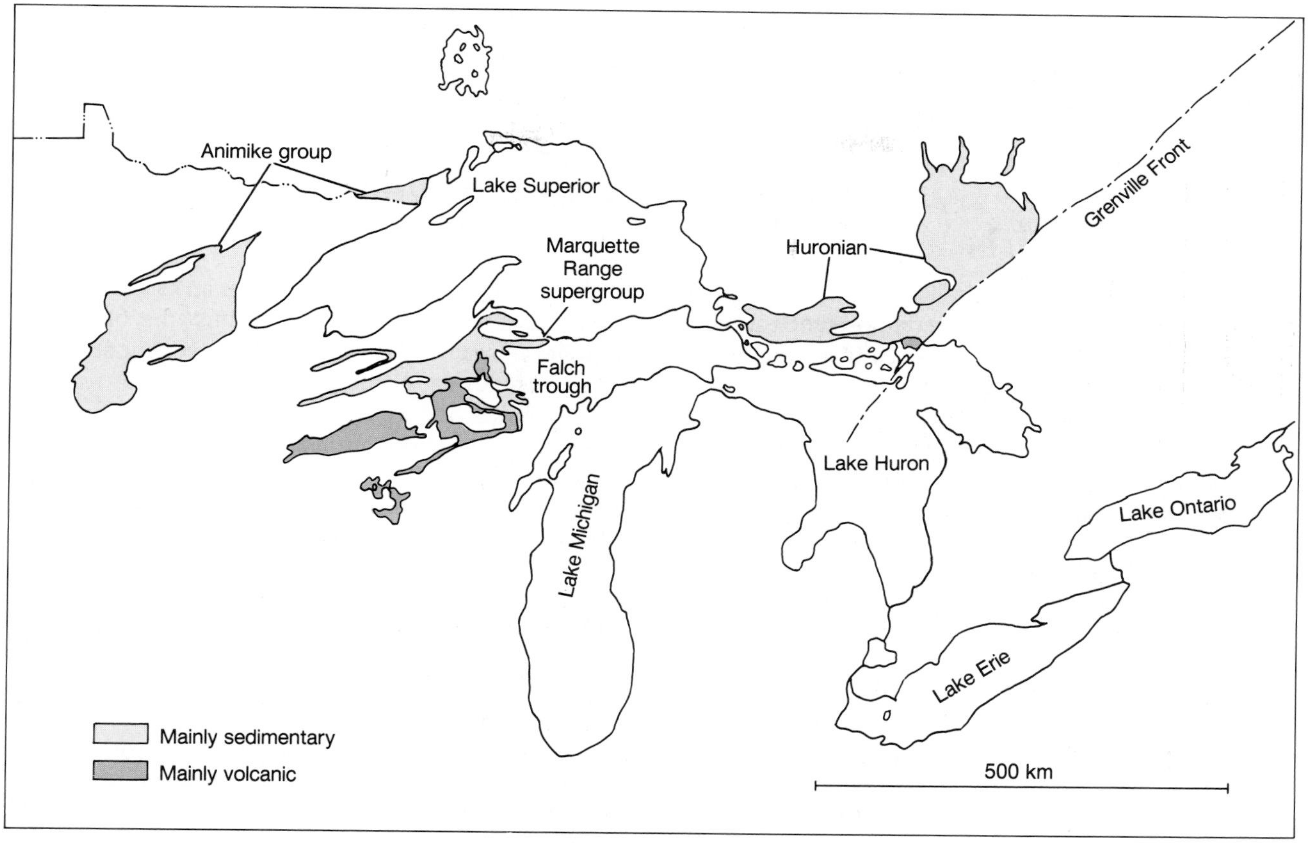

Animike group
Lake Superior
Marquette Range supergroup
Huronian
Grenville Front
Falch trough
Lake Michigan
Lake Huron
Lake Ontario
Lake Erie
Mainly sedimentary
Mainly volcanic
500 km

**Figure 9–14** Map showing the geographic distribution of Early Proterozoic rocks of the Great Lakes region.

a.

b.

**Figure 9–15** Early Proterozoic sedimentary rocks of the Great Lakes region. a. Wave-formed ripple marks in the Mesnard Quartzite. The crests of the ripples point toward the observer. b. Outcrop of the Kona Dolomite showing algal bedding.

## Perspective 9–1

# Ophiolites—Evidence for Ancient Convergent Plate Margins

The available evidence indicates that the upper mantle and oceanic crust are composed of a succession of rock types such as shown in Figure 6–21. Upper mantle rocks are thought to be peridotites, which are ultramafic igneous rocks composed mostly of iron- and magnesium-rich minerals such as olivine and pyroxenes. The oceanic crust, on the other hand, is characterized as *mafic*, meaning it is composed of fewer iron and magnesium minerals and more silica. Partial melting of the upper mantle is the probable source of the gabbros, sheeted dikes, and pillow basalts of the oceanic crust.

According to plate tectonic theory, oceanic crust is produced at the axes of oceanic ridges, then moves away from these ridges by sea-floor spreading, and is eventually destroyed at subduction zones. Continuous destruction of oceanic crust accounts for the fact that none older than 180 million years is known, except as fragments that were tectonically emplaced on continents. Such fragments, consisting of a sequence of upper mantle and oceanic crust rocks, are called *ophiolites*. Ophiolites are one of the primary features used to recognize ancient convergent plate margins.

Probably less than .001 percent of all oceanic crust that ever existed is now preserved as ophiolites. Nevertheless, these small fragments provide geologists with information on the processes whereby oceanic crust is generated at oceanic ridges, and where ancient plate collisions occurred. Furthermore, their presence indicates an essentially modern style of plate tectonics. Thus, the oldest ophiolites mark the time when a plate tectonic style comparable to that of the present began.

Ophiolites are well-known from Phanerozoic-Age rocks, and a probable 2.65-billion-year-old (Late Archean) ophiolite has recently been reported from Wyoming. The oldest complete ophiolite so far recognized, however, is the 1.96-billion-year-old (Early Proterozoic) Jormua mafic-ultramafic complex of Finland (Fig. 1). Even though the Jormua complex is highly deformed, it is similar to younger, well-documented ophiolites. Clearly, the Jourma complex represents a sequence of rocks that can be interpreted as upper-mantle, lower-oceanic-crust, and submarine eruptions represented by pillow basalts (Fig. 1), and metamorphosed deep-sea clay, oozes, and turbidites.

A. Kontinen of the Geological Survey of Finland suggests that the Jormua complex formed at a divergent plate margin during an extensional stage in the breakup of an Archean craton. A major ocean basin formed in which deposition of deep-sea sediments and turbidites occurred. During a compressional stage about 1.9 billion years ago, most of the oceanic crust was destroyed by subduction. The Jormua complex, however, was tectonically emplaced on the craton and represents a small fragment of this ancient oceanic crust.

The earliest deposits of the Great Lakes region seem to have accumulated on a stable, shallow-water shelf, perhaps a rifted continental margin. Later, however, deposition in tectonically active basins predominated, as indicated by the presence of graywackes. Some intermittent deformation took place during the deposition of Great Lakes rocks, but all were further deformed and intruded by granite batholiths during the Hudsonian orogeny. These rocks are now part of the Penokean orogen (Fig. 9–3).

In the western United States and Canada, quartzite-carbonate-shale assemblages were deposited along the continental margin. These rocks are Middle to Late Proterozoic in age and are best preserved in three large basins (Fig. 9–16). In the Belt basin of the northwestern United States and adjacent parts of Canada, a thick sequence of sedimentary rocks, the Belt-Purcell Supergroup, was deposited between 1.45 billion and 850 million years ago (Fig. 9–17). Like the deposits of the Great

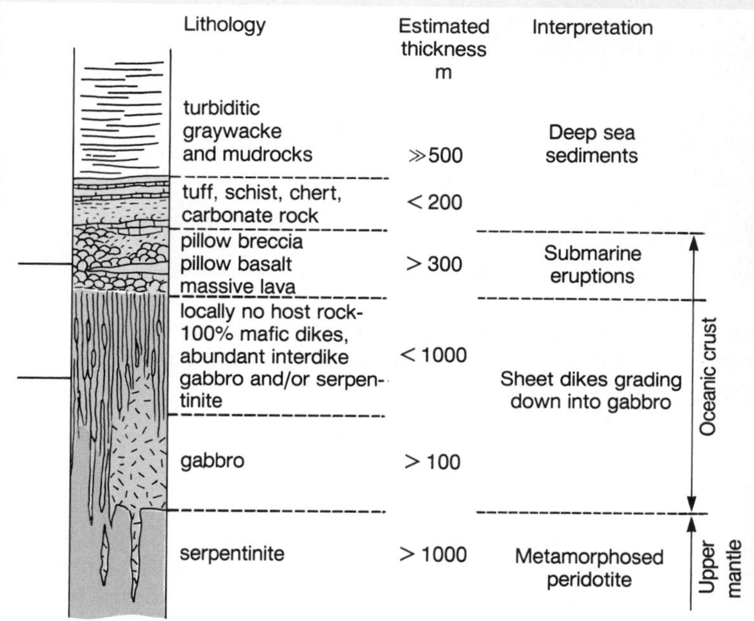

**Figure 1** Reconstruction of the highly deformed Jourma mafic-ultramafic complex of Finland. This 1.96 billion-year-old sequence of rocks is interpreted as the oldest known complete ophiolite sequence. The upper photograph shows a metamorphosed basaltic pillow lava. Length of the code plate is 12 cm. The bottom photograph shows metamorphosed gabbro screens between mafic dikes. The hammer shaft is 65 cm long. (Photos courtesy of Asko Kontinen, Geological Survey of Finland.)

Lakes region, the Belt-Purcell rocks are predominantly quartzite, shale, and thick stromatolite-bearing carbonates (Fig. 9-18). Some of the rocks are interpreted as the deposits of large alluvial fans and braided streams.

The most important aspect of these quartzite-carbonate-shale assemblages is that they represent suites of rocks similar to those of the Phanerozoic. Thus, they indicate the widespread distribution of stable cratons and depositional environments much like those of the Phanerozoic.

## Glaciation and Glacial Deposits

Glaciers deposit unsorted, unstratified sediment called *till* that when lithified is called **tillite**. Tillites or till-like deposits have been reported from more than 300 Precambrian localities. Many of these may not be glacial deposits, but some almost certainly are. In fact, two major episodes of Proterozoic glaciation have been recognized—one in the Early Proterozoic and one in the Late Proterozoic (Table 9–1).

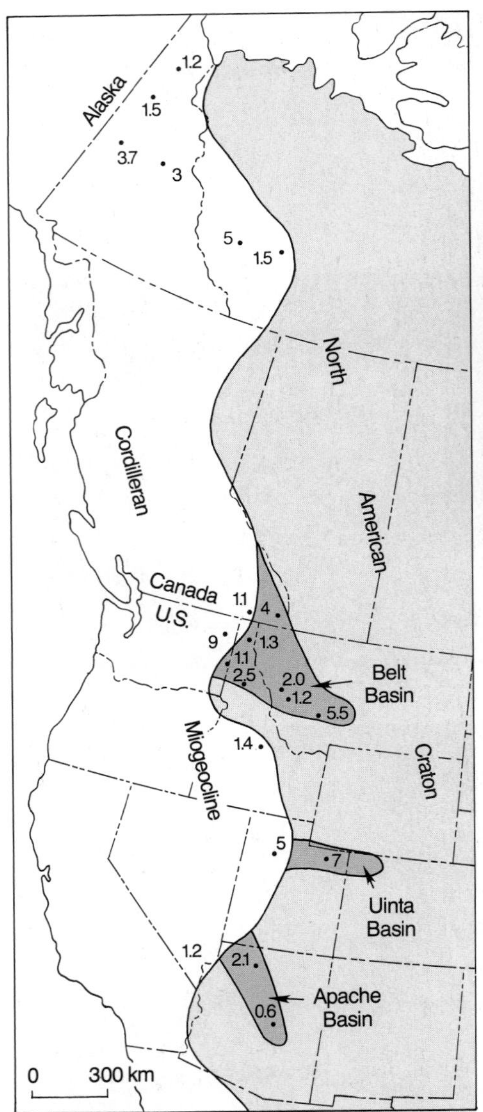


**Figure 9–16** Late Proterozoic basins of sedimentation in the western United States and Canada. Approximate thickness in kilometers of Late Proterozoic rocks shown by numbers.

**Figure 9–17** Rocks of the Late Proterozoic Belt Supergroup in Glacier National Park, Montana. The present topography resulted from erosion by glaciers during the Pleistocene and Recent epochs.

But how can we be sure there were any Proterozoic glaciers? Tillite is simply a type of conglomerate, but other processes, such as *mass wasting*, can produce very similar deposits. Mass-wasting deposits are very limited in areal extent, however, whereas true tillites cover much larger areas. In addition, a suspected tillite can be established as a glacial deposit by its association with a striated bedrock pavement (Fig. 9–19) and with finely laminated argillites (varved sediments) that contain large clasts dropped from floating ice (Fig. 9–20).

Early Proterozoic glacial deposits are known from Canada, the United States, Australia, and South Africa. Figure 9–20 shows a partial section of the Huronian Supergroup of Ontario. The tillite of the Bruce Formation may date from 2.7 billion years ago, making it Late Archean in age. Tillites of the Gowganda Formation are widespread, overlie a striated pavement, and are associated with varved deposits (Fig. 9–20).

Similar deposits thought to be the same age as the Gowganda tillite also occur in Michigan, the Medicine Bow Mountains of Wyoming, and Quebec, Canada. This distribution suggests that North America may have had an extensive Early Proterozoic ice sheet centered southwest of Hudson Bay (Fig. 9–21). Tillites of about the same age also occur on other continents; the dating of these deposits is not precise enough to determine whether there was a single, widespread glacial episode, or a number of glacial events at different times in different places.

Widespread Late Proterozoic glaciation occurred between 900 and 600 million years ago (Table 9–1). Tillites and other glacial deposits have been recognized on all continents except Antarctica. Glaciation was not continuous during this entire interval but was episodic, with four major glacial periods recognized. Figure 9–22 shows the approximate distribution of these Late Proterozoic glaciers, but we should emphasize that they are approximate, because the geographic extent of glacial

**Figure 9–18** Late Proterozoic sedimentary rocks in Glacier National Park, Montana. a. Outcrop of Snowlip Formation consisting of dolomitic mudstone and quartzite. b. Shale in the Grinnell Formation. (Photo a courtesy of Robert J. Horodyski, Tulane University.)

**Figure 9–19** Evidence for Proterozoic glaciation in Norway. The Bigganjarga tillite overlies a striated bedrock surface on sandstone of the Veidnesbotn Formation. (Photo courtesy of Anna Siedlecka, Geological Survey of Norway.)

ice is unknown. In addition, the glaciers shown in Figure 9-22 were not all present at the same time.

## THE EVOLVING ATMOSPHERE

In the last chapter we noted that many geologists are convinced that the Archean atmosphere contained little or no free oxygen. In other words, the atmosphere was not strongly oxidizing as it is now. Preston Cloud, a specialist on the early history of life, estimates that the free-oxygen content of the atmosphere increased from about 1 percent to 10 percent through the Proterozoic. It was not until about 400 million years ago that oxygen reached its present level. Most of the atmospheric oxygen was released as a waste product by photosynthesizing cyanobacteria (Fig. 8–25). As indicated by fossil stromatolites, cyanobacteria became common about 2.3 billion years ago (Fig. 9–15).

### Banded Iron Formations

**Banded iron formations (BIF)** are sedimentary rocks consisting of alternating thin layers of silica (chert) and iron minerals (Fig. 9–23). The iron in these formations is mostly iron oxide (hematite and magnetite), but iron silicate, iron carbonate, and iron sulfide also occur. Banded

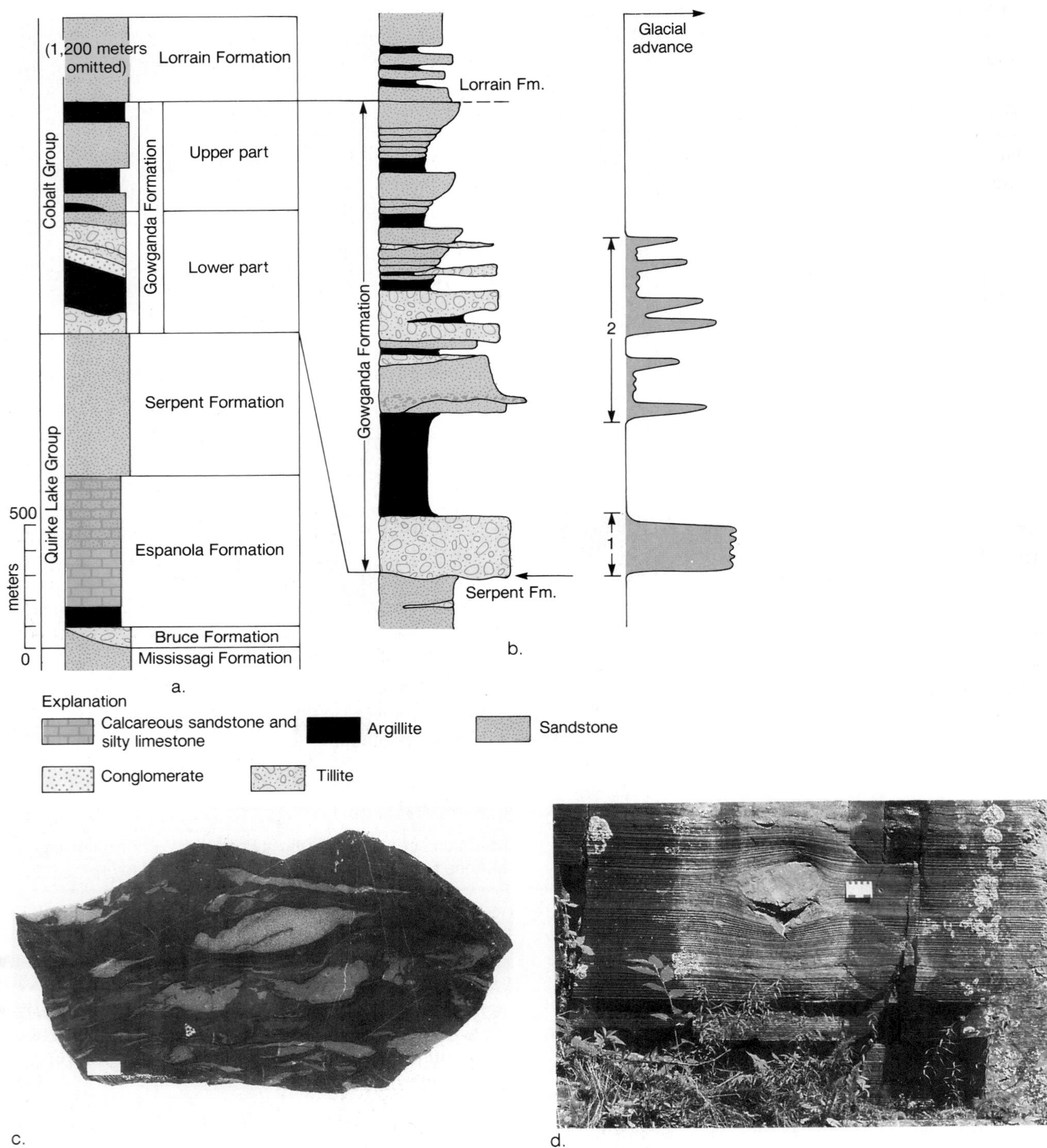


**Figure 9–20** Precambrian glacial deposits. a. Partial section of the Huronian Supergroup, Ontario, showing tillites of the Bruce and Gowganda formations. The Bruce Formation may represent Late Archean glacial deposits. b. Expanded view of the Early Proterozoic Gowganda Formation. Two major glacial advances are recognized. c. Tillite, and d. Finely laminated argillite with clast dropped from floating ice in the Gowganda Formation. (Photos courtesy of the Geological Survey of Canada.)

iron formations are present in all Precambrian cratons and account for most of the iron ore that is mined in the world today (Fig. 9–24).

Two major types of Precambrian BIF are recognized. **Algoma-type** iron formations are relatively small lenticular bodies measuring a few kilometers across and a few meters thick. Most are Archean in age, although Proterozoic and even some Phanerozoic examples are known, and most appear to have been deposited in tectonically unstable environments such as greenstone belts. **Superior-type** iron formations are the most common (Fig. 9–23). These are typically hundreds of meters thick and can be traced for hundreds of kilometers. They were mainly deposited in shallow marine waters during the Early Proterozoic, although some are also found in the Archean.

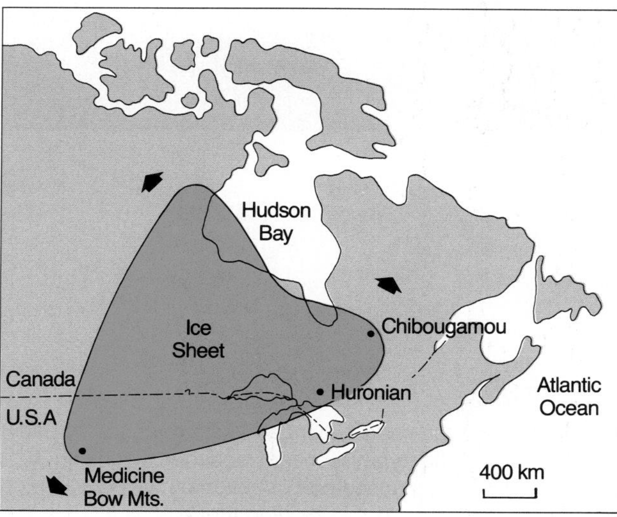


**Figure 9–21** Glacial deposits of about the same age in Ontario and Quebec, Canada, and in Michigan and Wyoming suggest that an Early Proterozoic ice sheet was centered southwest of Hudson Bay.

BIF are found throughout the geologic column, but most formed during three main episodes of deposition (Table 9–2). These three depositional periods are bounded by major orogenies representing periods of increased global plate tectonic activity. The BIF accumulated during the intervening quiescent intervals. The period from 2.5 to 2.0 billion years represents a unique time in Earth history, a time during which most of the Earth's BIF (92 percent) formed (Table 9–2). Also remarkable is the consistent layering and extent of BIF in shallow marine environments and their chemical purity; BIF are composed mostly of iron oxides and silicon dioxide as chert.

Iron is a highly reactive element. In the presence of oxygen it combines to form rust-like oxides that are not readily soluble in water. In the absence of free oxygen, however, iron is easily taken into solution in water and can accumulate in large quantities in the world's oceans. In Chapter 8 we mentioned that the Archean atmosphere was deficient in free oxygen, making it unlikely that much oxygen was dissolved in seawater. The increase in abundance of stromatolites around 2.3. billion years ago resulted in an increase in free oxygen in the oceans, since oxygen is a metabolic waste product of

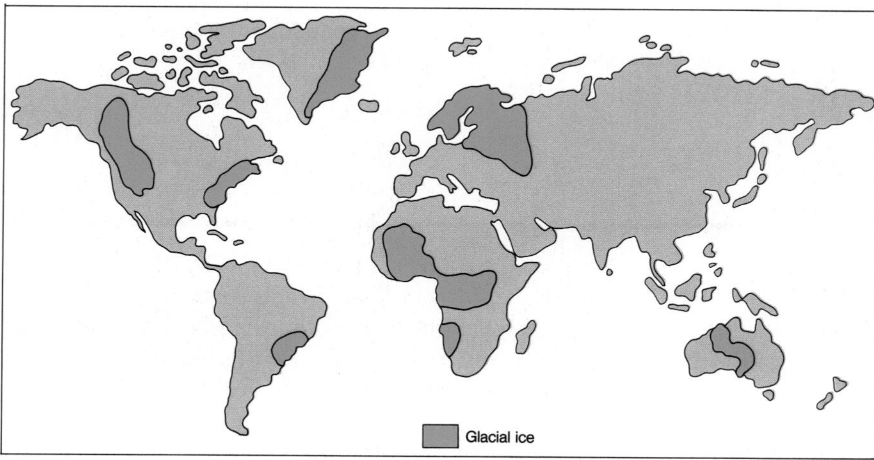


**Figure 9–22** Major Late Proterozoic glacial centers shown on a map with the continents in their present positions. Extent of ice centers is hypothetical and approximate.

a.

b.

**Figure 9–23** Precambrian banded iron formations. a. Early Proterozoic Superior-type BIF near Negaunee, Michigan. b. Close-up of the Negaunee Iron Formation on Jasper Knob, Ishpeming, Michigan. At this location the rocks are brilliantly colored alternating layers of red chert (dark) and silver iron minerals (light).

**Table 9–2** Times of deposition of banded iron formations.

| Time of Deposition (billions of years) | Percent of Total BIF Deposited | Predominant Type of BIF |
|---|---|---|
| 3.5–3.0 | 6% | Algoma type |
| 2.5–2.0 | 92% | Superior type |
| 1.0– .5 | 2% | |

photosynthesizing bacteria and cyanobacteria that make up stromatolites. Apparently this introduction of free oxygen into the world's oceans helped cause the precipitation of dissolved iron and silica, forming the BIF.

The current model favored for BIF precipitation involves a Precambrian ocean with an upper oxygenated layer overlying a large volume of oxygen-deficient water that contained reduced iron and silica. Upwelling of these deep, iron- and silica-rich waters onto newly created shallow continental shelves resulted in large-scale precipitation of iron oxides and silica as BIF (Fig. 9–25). A likely major source for this iron and silica was submarine volcanism, although weathering of the continents may have supplied iron as well. Recent research has shown that hydrothermal vents near spreading centers (Perspective 8–2) remove large quantities of iron and silica from the oceanic crust. The hydrothermally derived iron and silica combined with free oxygen released by photosynthesizing organisms and triggered

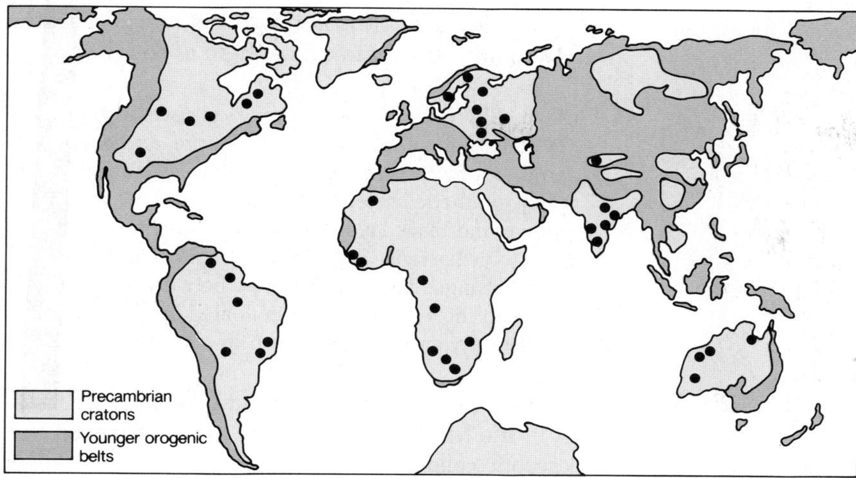


**Figure 9–24** Distribution of Precambrian banded iron formations. Note that all occur in Precambrian shield areas.

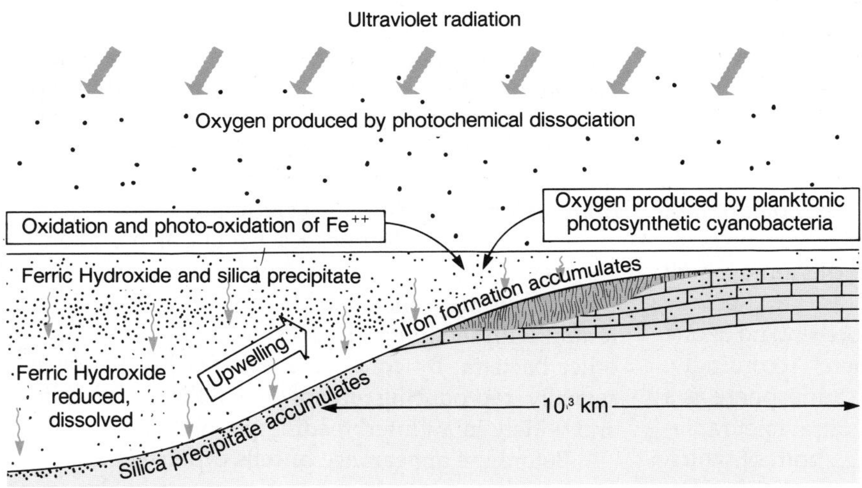


**Figure 9–25** Depositional model for the origin of Superior-type banded iron formations.

the precipitation of massive quantities of BIF between 2.5 and 2.0 billion years ago.

## Red Beds

Continental **red beds** first appeared about 1.8 billion years ago, following the deposition of the Proterozoic banded iron formations (Table 9–1). The color of red beds is caused by the presence of ferric oxide, usually as the mineral hematite ($Fe_2O_3$) that forms under oxidizing conditions. These deposits become increasingly abundant through the Proterozoic and are quite common in the Phanerozoic.

Figure 9–26 shows red beds from the Waterberg Group of South Africa, considered to be the oldest deposit of this type. A detailed study of these rocks shows that the red color is restricted to those deposits that accumulated in continental environments and particularly in

**Figure 9–26** Outcrop of cross-bedded, fluvial sandstone of the Waterberg Group, South Africa. These are the oldest known continental red beds. (Photo used with permission of Kenneth A. Eriksson, Virginia Polytechnic Institute and State University.)

stream systems, while those facies attributed to nearshore marine sedimentation are unpigmented. The iron responsible for the red color was derived from weathering of minerals within the deposits.

The appearance of continental red beds suggests an oxidizing atmosphere. But we stated earlier that the Proterozoic atmosphere contained little free oxygen, perhaps as little as 1 to 2 percent in the early part of that eon. Is this sufficient to account for oxidized iron in sediments? Probably not; other attributes of this atmosphere must be considered.

Before abundant free oxygen was present, no ozone ($O_3$) layer existed in the upper atmosphere. Accordingly, as free oxygen ($O_2$) was released to the atmosphere as a waste product during photosynthesis, ultraviolet radiation converted some of it into O and $O_3$, both of which oxidize surface materials more efficiently than $O_2$. Once an ozone layer became established in the upper atmosphere, most of the ultraviolet radiation failed to penetrate to the surface, and $O_2$ became the primary agent for oxidation of surface deposits.

## PROTEROZOIC LIFE

In the last chapter we noted that Archean fossils are not very common, and all are varieties of bacteria—unicellular prokaryotes. The Early Proterozoic is likewise characterized by these organisms. Apparently bacteria and cyanobacteria were the only life-forms present during the first 2 billion years of life history.

Lynn Margulis of Boston University has characterized the Early Proterozoic biosphere as follows:

> Two billion years ago, the land was covered not by forests, marshes, grasslands, tundra, and chaparral, but by the blue-green and purple scum of photosynthetic bacteria underlain by yellow, brown and black layers of nonphotosynthetic and anaerobic bacteria. Bacterial fruiting bodies and nets of fungus-like bacteria grew between soil particles. In the sea only bacteria floated and swam; in the air only spores were blown by the winds.[3]

Apparently little diversification had occurred up to this point in life history. In fact, cells more advanced than prokaryotic cells did not appear until about 1.4 billion years ago. Even the well-known, 1.8- to 2.1-billion-year-old Gunflint Formation of Ontario, Canada, with 12 species of microorganisms, contains only bacteria and cyanobacteria (Fig. 9–27).

Actually, the lack of organic diversity during the Archean and Early Proterozoic is not too surprising because prokaryotic cells reproduce asexually. Recall our discussion in Chapter 5 on the sources of variation in populations. Most variation results from sexual reproduction, during which the offspring receives half of its genetic makeup from each parent. Mutations may introduce new variations into any population, but the effects of mutations in prokaryotes are limited.

Consider this. Suppose a beneficial mutation occurred in a bacterium. How could this mutation spread throughout the population? It could not, since the bacterium with the mutation does not share its genes with other bacteria. By contrast, a beneficial mutation in a sexually reproducing organism could spread quickly and widely in an interbreeding population.

Before the appearance of cells capable of sexual reproduction, evolution was a comparatively slow process, accounting for the low organic diversity during the Archean and Early Proterozoic. This situation did not persist, however. In the Middle Proterozoic, cells appeared that reproduced sexually, and the tempo of evolution picked up markedly.

### A New Type of Cell Appears

The appearance of **eukaryotic cells** marks one of the most important events in the history of life. The degree of organizational complexity of eukaryotic cells is considerably greater than in prokaryotic cells, and eukary-

[3]L. Margulis. 1982. *Early Life,* New York: Van Nostrand Reinhold Co., p. 75.

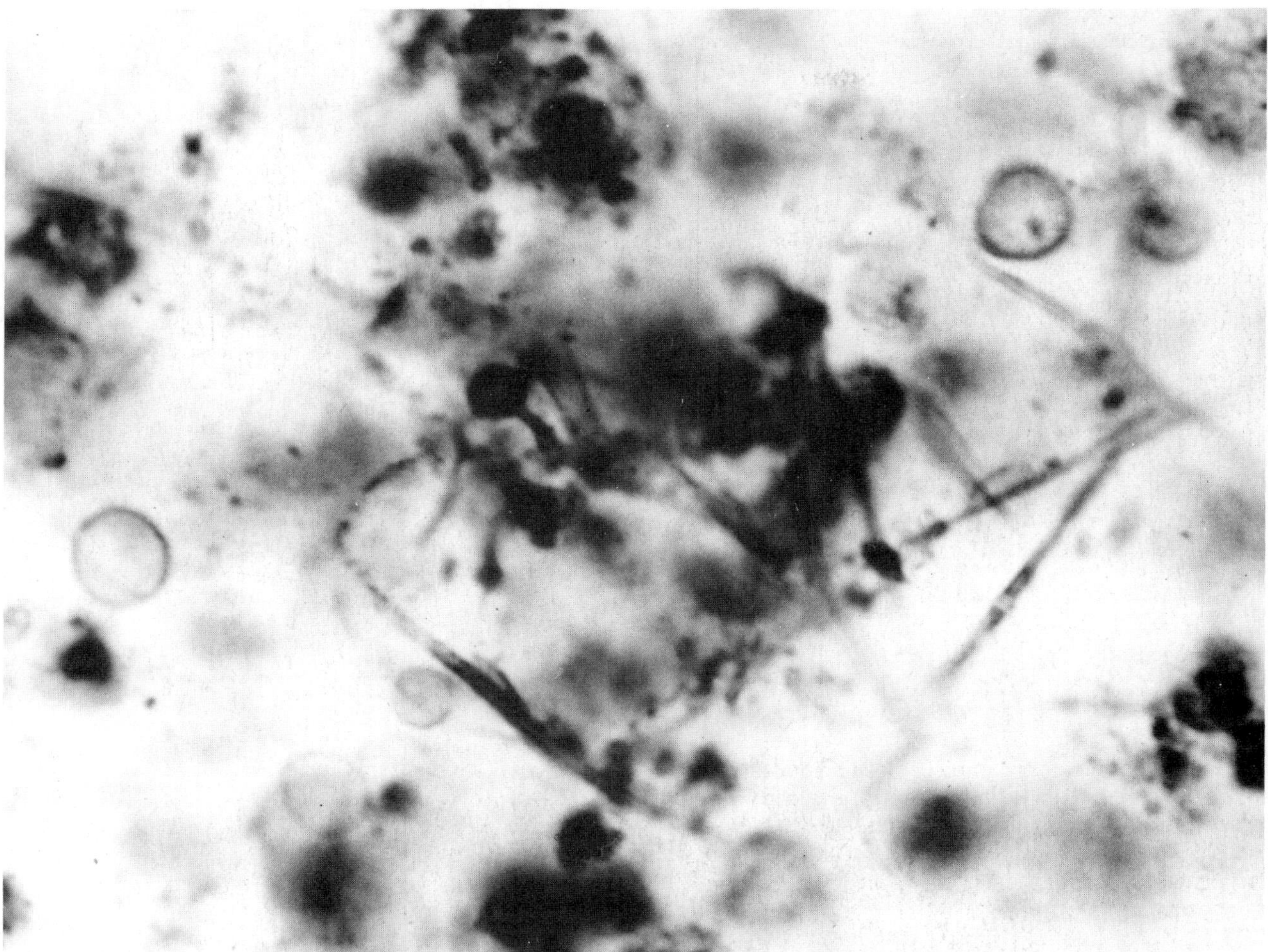

**Figure 9–27** Photomicrographs of spheroidal and filamentous microfossils from stromatolitic chert of the Gunflint Iron Formation, Ontario.

otic cells have a membrane-bounded cell nucleus that contains the genetic material. Differences between these two types of cells are summarized in Figure 9–28. Most eukaryotes are multicellular, although those of the kingdom Protoctista are unicellular (Appendix B). In marked contrast to prokaryotes most eukaryotes reproduce sexually, and most are aerobic, thus eukaryotes could not have appeared until some free oxygen was present in the atmosphere.

A number of Proterozoic fossil localities have yielded unicellular eukaryotes, but the oldest one appears to be in the 1.2- to 1.4-billion-year-old Beck Springs Dolomite of southeastern California (Fig. 9–29, Table 9–3). Fossils that appear to be unicellular algae from the 1.0-billion-year-old Bitter Springs Formation of Australia show evidence of both mitosis and meiosis, processes used only by eukaryotic cells.

Although the fossils of the Beck Springs Dolomite are commonly cited as the first known eukaryotes, this is a bit uncertain, because it is difficult to be sure that microscopic fossils are eukaryotes rather than prokaryotes.

The fossil evidence for the appearance of eukaryotic cells comes from size and relative complexity. Figure 9–28 shows that eukaryotic cells are larger, commonly much larger, than prokaryotic cells. Cells larger than 60 microns appear in abundance about 1.4 billion years ago. As for relative complexity, prokaryotic cells are

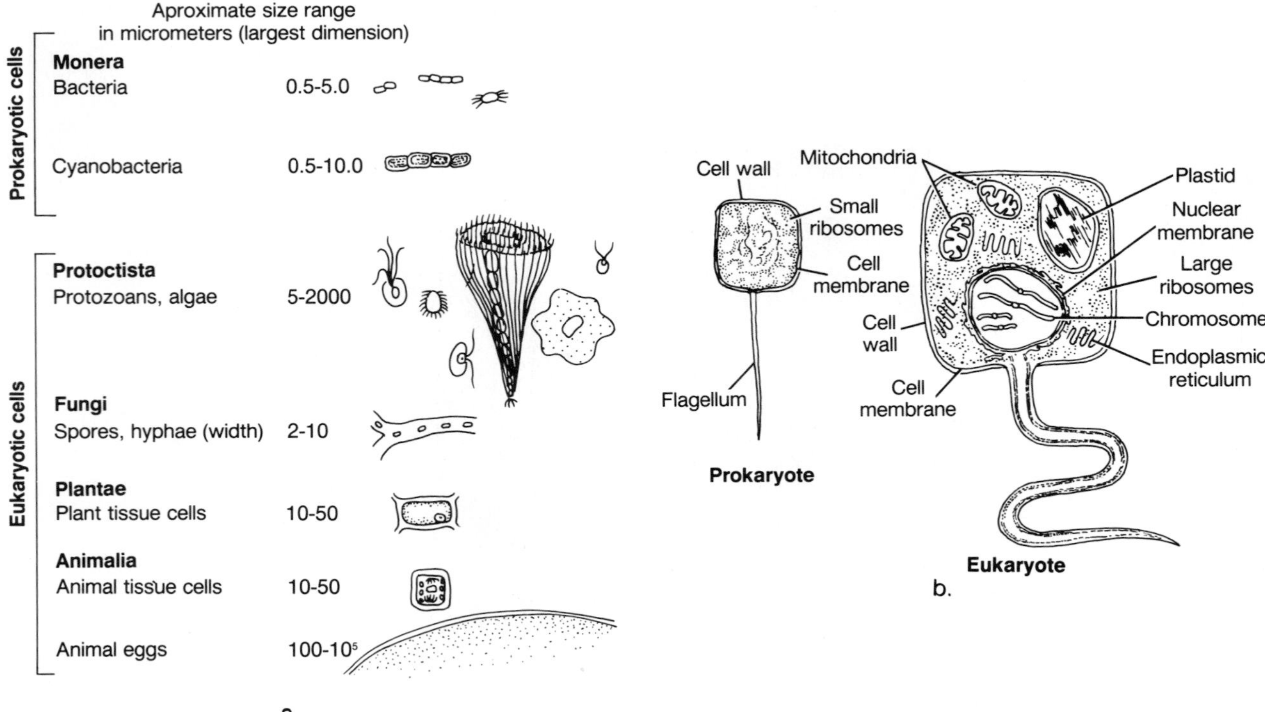


**Figure 9–28** Comparisons of prokaryotic and eukaryotic cells. a. Sizes of various prokaryotic and eukaryotic cells. b. Cell structure. Note that eukaryotes have a cell nucleus containing the genetic material and several organelles such as mitochondria and plastids.

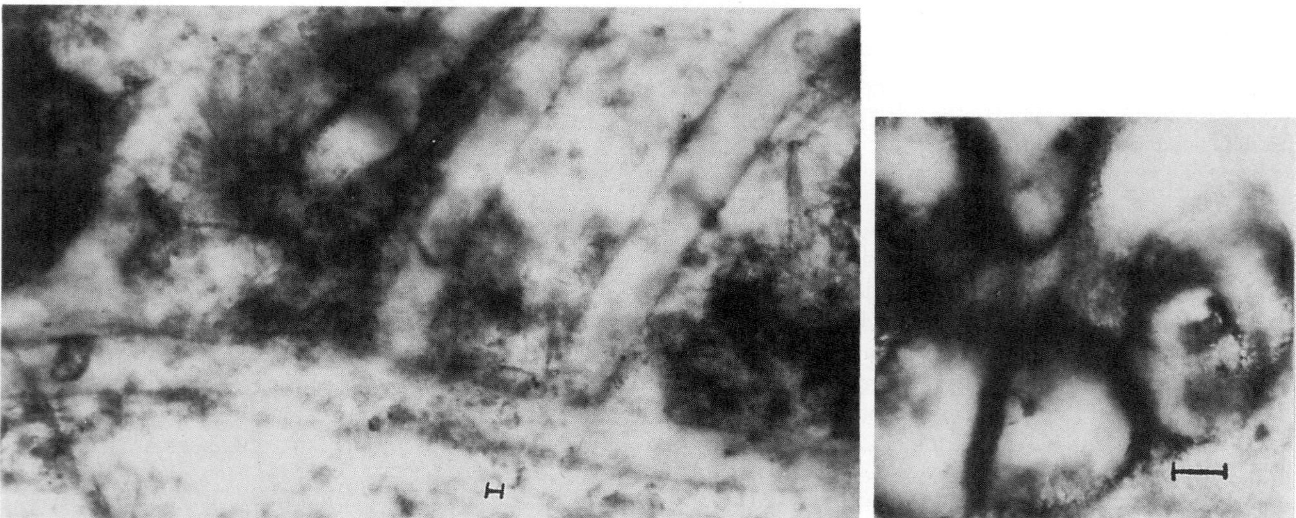

**Figure 9–29** Microfossils from the Beck Springs Dolomite, California. These fossils may be the oldest known eukaryotic cells. (Photos courtesy of Preston Cloud, University of California, Santa Barbara.)

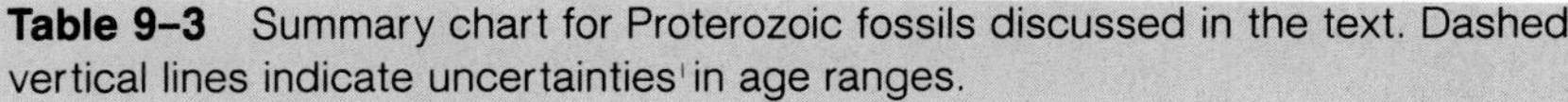

**Table 9–3** Summary chart for Proterozoic fossils discussed in the text. Dashed vertical lines indicate uncertainties in age ranges.

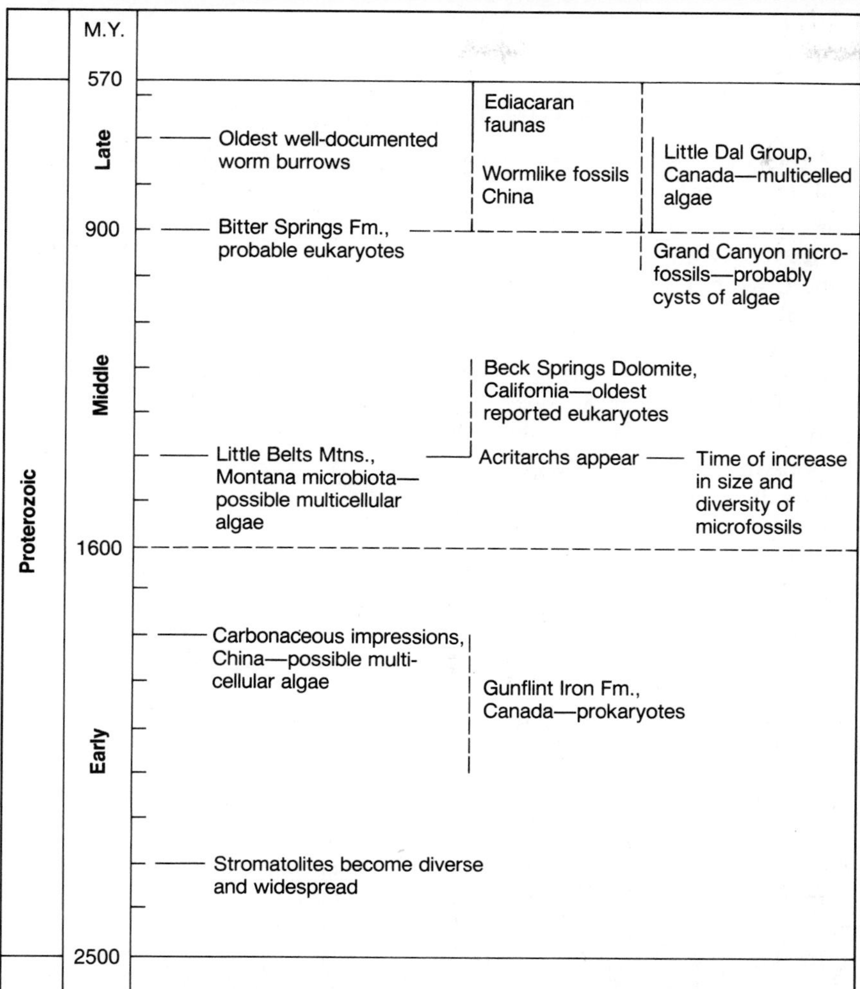


typically spherical or plate-like. Proterozoic fossils of branched filaments, flask-shaped organisms, and some containing what appear to be internal, membrane-bounded structures all suggest a eukaryotic level of organization.

Additional evidence for eukaryotes in the Proterozoic comes from a fossil group called **acritarchs**. These hollow fossils are probably the cysts of planktonic algae (Fig. 9–30), which became quite abundant during the Late Proterozoic but first appeared about 1.4 billion years ago. Numerous microfossils with vase-shaped skeletons have been recovered from Late Proterozoic rocks of the Grand Canyon (Fig. 9–30). These have been tentatively identified as cysts of some kind of algae.

The precise time of appearance of eukaryotes is debatable, but undoubtedly they were present by the Late Proterozoic. Perhaps they appeared as early as 1.4 billion years ago, the maximum age of the Beck Springs Dolomite. Most authorities agree that eukaryotes had appeared by at least 1.0 billion years ago.

## Multicellular Organisms

**Multicellular organisms** are composed of many cells, often billions. More importantly, multicellular organisms have cells specialized to perform specific functions such as reproduction, respiration, and food gathering. We know from fossils that multicellular organisms ap-

## Perspective 9–2

# Symbiosis and the Origin of Eukaryotes

Stratigraphic position and radiometric ages tell us that eukaryotic cells were present by the Middle Proterozoic. However, the fossil record reveals nothing about how they evolved from prokaryotic ancestors. For answers to that question, we must turn to the study of modern microorganisms.

A currently popular theory among evolutionary biologists holds that "several prokaryotes make a eukaryote."[1] In essence, this means that eukaryotic cells formed from several prokaryotic cells that had formed a symbiotic relationship. *Symbiosis*, the living together of two or more dissimilar organisms, is quite common among modern organisms. It may take the form of *parasitism*, in which one organism lives at the expense of another, or the two may coexist with mutual benefit. For example, lichens, once considered to be plants, are actually symbiotic fungi and algae.

In a symbiotic relationship, each symbiont must be capable of metabolism and reproduction, but the degree of dependence in some symbiotic relationships is such that one symbiont cannot live independently. Many parasites, for example, cannot exist outside a host organism. And so it may have been with Proterozoic prokaryotes. That is, two or more prokaryotes may have entered into a symbiotic relationship (Fig. 1). According to the theory that eukaryotes arose from such an association, the

[1]L. Margulis. 1981. *Symbiosis in Cell Evolution*, New York: W.H. Freeman and Co., p. 2.

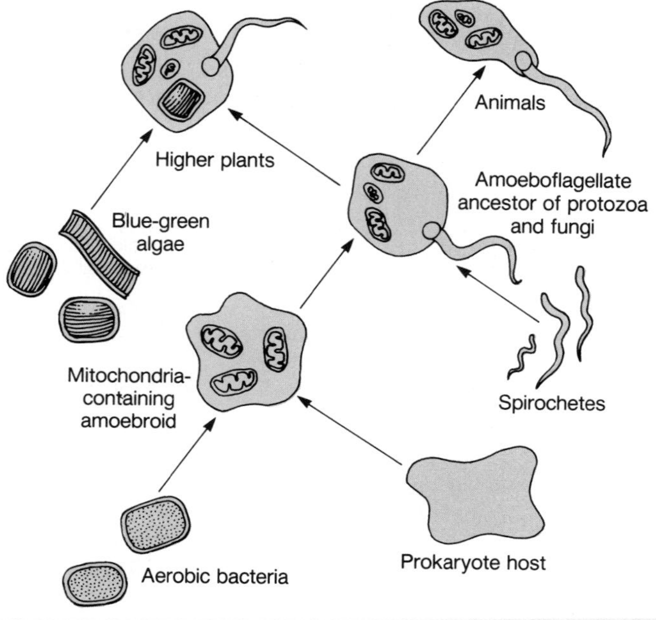


**Figure 1** Symbiosis theory of the origin of eukaryotic cells. An aerobic bacterium and a larger host of the kingdom Monera united to form a mitochondria-containing amoeboid. An amoeboflagellate was formed by a union of the amoeboid and a bacterium of the spirochete group; this amoeboflagellate was the direct ancestor of two kingdoms—Fungi and Animalia. Another kingdom, Plantae, was founded when this amoeboflagellate formed a union with blue-green algae (cyanobacteria) that became plastids.

peared in the Late Proterozoic, but we have no fossil evidence of the transition from their unicellular ancestors.

Once again, the study of modern organisms gives some clues as to how this transition may have occurred. Perhaps some unicellular organism divided and formed a group of cells that did not disperse but remained together as a colony. The cells in some colonies may have become somewhat specialized, similar to the situation in the living **colonial organisms** (Fig. 9–31). Further specialization might have led to simple multicellular organisms such as sponges that consist of cells specialized for reproduction, respiration, and food gathering. In other words, specialization of cells might have led to the development of organs with specialized functions.

Is there any particular advantage in being multicellular? After all, until the Late Proterozoic, life seems to have thrived even though all known organisms were unicellular. In fact, unicellular organisms are quite successful at what they do, but what they do is limited. For

interdependence of symbionts grew until the unit could exist only as a whole.

Supporting evidence for the symbiosis theory outlined here comes from the study of living eukaryotic cells. For example, eukaryotic cells contain internal structures called *organelles* that have their own complements of genetic material. Although these organelles, such as plastids and mitochondria, cannot exist independently today, it seems that they were once capable of reproduction as free-living organisms.

Another way of evaluating the symbiosis theory is to look at protein synthesis. Prokaryotic cells synthesize proteins, but they can be thought of as a single system, while eukaryotes are a combination of protein-synthesizing systems. That is, within eukaryotes, some of the organelles, such as mitochondria and plastids, are capable of protein synthesis. These organelles, then, with their own genetic material and protein-synthesizing capabilities, are thought to have been free-living bacteria that entered into a symbiotic relationship. With time, the interdependence of the various units grew until life was possible only as an integrated whole (Fig. 1).

Although the fossil record does not record the acquisition of organelles or symbiosis, one living eukaryote can give some idea of what the first eukaryotes may have been like. The living giant amoeba *Pelomyxa*, which lives in the mud of ponds, lacks mitochondria (Fig. 2). Two types and hundreds of individual bacteria, however, have a symbiotic relationship with *Pelomyxa* and perform the same function as mitochondria. Perhaps a similar relationship among Proterozoic prokaryotes was the first stage in the origin of eukaryotic cells.

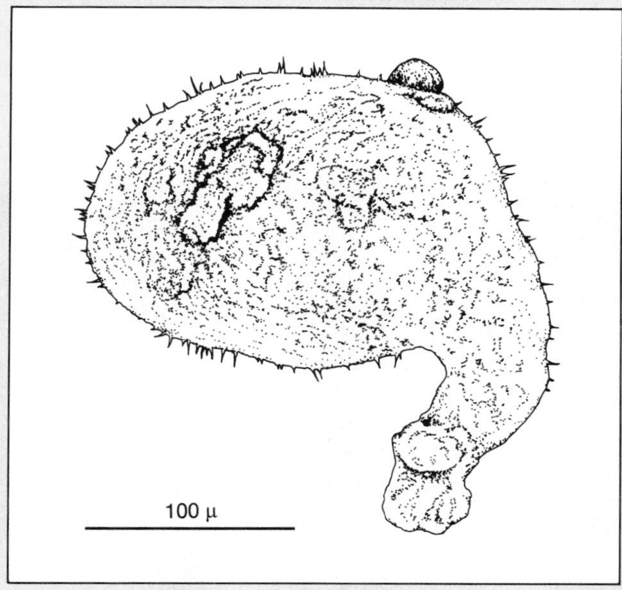


**Figure 2** The giant amoeba *Pelomyxa* lacks mitochondria but has a symbiotic relationship with bacteria that serve the same function as mitochondria.

example, they cannot become very large, because as size increases, proportionally less of the cell is exposed to the external environment in relation to its volume. In other words, as size increases, the amount of surface area compared to volume decreases; consequently, the process of transferring materials from the external environment into the cell becomes less efficient. Multicellular organisms live longer, since cells can be replaced; and more offspring can be produced, because some cells are specialized for reproduction. And finally, cells have increased functional efficiency as they become specialized into organs.

### Multicellular Algae

Carbonaceous impressions of probable multicellular algae are known from 1,000- to 700-million-year-old rocks in Spitzbergen, China, India, and the Little Dal Group of northwestern Canada (Fig. 9–32, Table 9–3). According to H.J. Hofmann, "The large size, the carbonaceous composition, the general shape, and the geologic

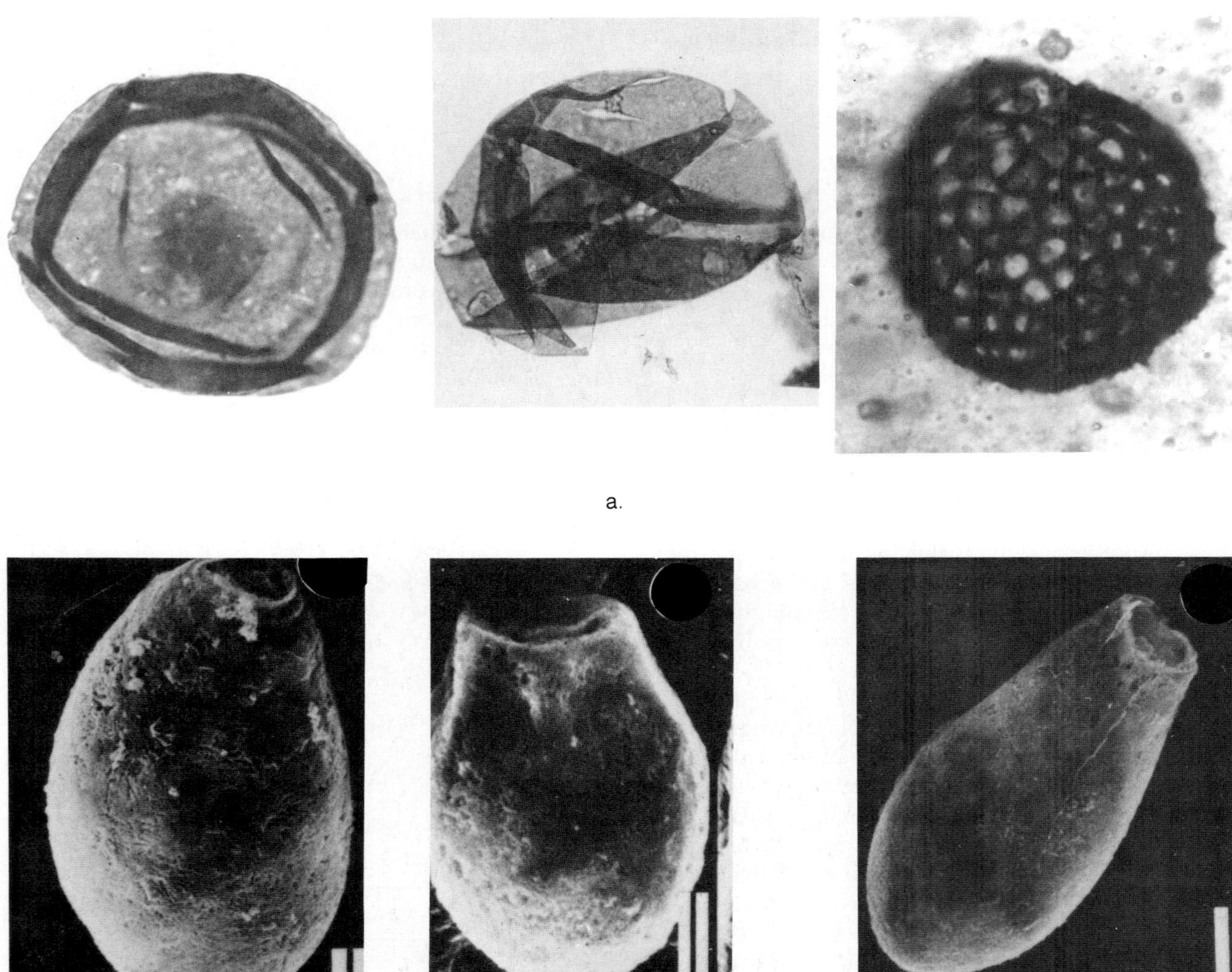

**Figure 9–30** Late Proterozoic microfossils. a. Acritarchs are probably the cysts of algae. b. Vase-shaped microfossils, probably cysts of some kind of algae. Both a and b are common in the Late Proterozoic and are thought to represent eukaryotic organisms. (Photo a courtesy of Andrew H. Knoll, Harvard University. Photo b courtesy of Bonnie Bloeser, University of Southern California.)

age suggest that these organisms were photosynthesizing eucaryotes, probably some kind of planktonic algae."[4] In fact, some organic sheets in the Little Dal Group closely resemble the modern green alga called *sea lettuce.*

Even older rocks contain carbonaceous impressions that may be multicellular algae, but this is uncertain. For example, the 1.4-billion-year-old microbiota from the Little Belt Mountains of Montana contains filaments and spherical forms of uncertain affinities (Fig. 9–32, Table 9–3). Carbonaceous macroscopic filaments from 1.8-billion-year-old rocks of China are also suggestive of multicellular algae.

[4]H. J. Hofmann. 1985. The Mid Proterozoic Little Dal Macrobiota, MacKenzie Mountains, North-West Canada. *Palaeontology,* 28, pt. 2: p. 351.

### The First Animals

In the "Prologue" we discussed Late Proterozoic multicellular animal fossils discovered in the Pound

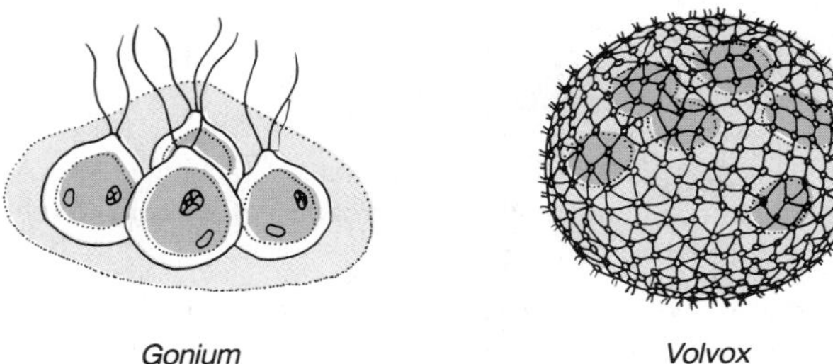


**Figure 9–31** Unicellular versus multicellular organisms. Although *Gonium* consists of as few as four cells, all cells are alike and can produce a new colony. *Volvox* has some cells specialized to perform different functions, so has crossed the threshold that separates unicellular from multicellular organisms.

Quartzite of the Ediacara Hills, South Australia (Fig. 9–1). Evidence for pre-Ediacaran animals is scarce, but there is some. For example, a jellyfish-like impression is known from rocks 2,000 m below the Pound Quartzite. And in many areas, burrows, presumably made by worms, occur in rocks at least 700 million years old. The presence of burrows implies that they were made by animals that, "had evolved hydrostatic skeletons, that is, fluid-filled body spaces that work against muscles, so that the animal could dig in the sea bed."[5]

Until very recently the Ediacaran faunas contained the oldest unquestioned animal fossils. However, wormlike fossils associated with fossil algae were recently discovered in 700- to 900-million-year-old strata in China

[5]J. W. Valentine. 1978. The Evolution of Multicellular Plants and Animals. *Scientific American,* 239, no. 3: 142.

a.

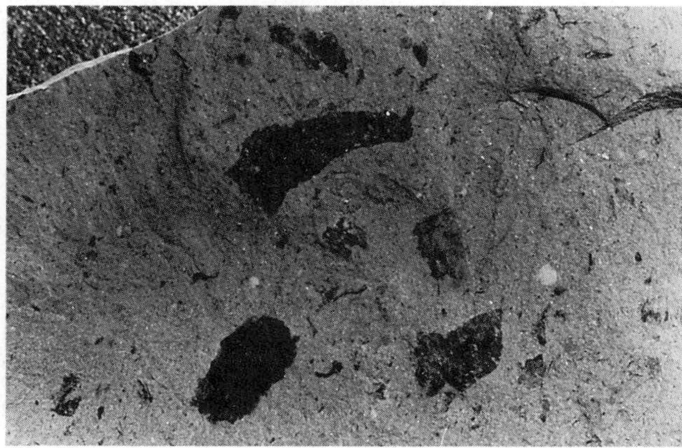
b.

c.

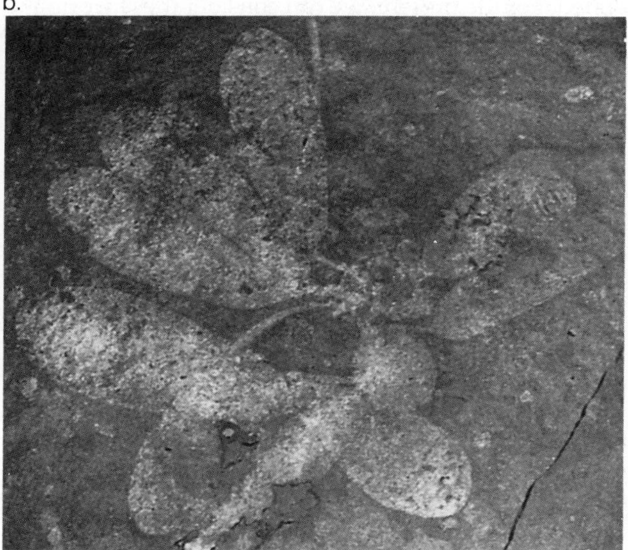
d.

**Figure 9–32** a and b. Carbonaceous impressions in Proterozoic rocks in the Little Belt Mountains, Montana. These may be impressions of multicellular algae, but this is uncertain. c and d. Carbonaceous impressions of multicellular algae from the Little Dal Group, Canada. (a and b courtesy of Robert Horodyski, Tulane University; c and d courtesy of H.J. Hofmann, University of Montreal.)

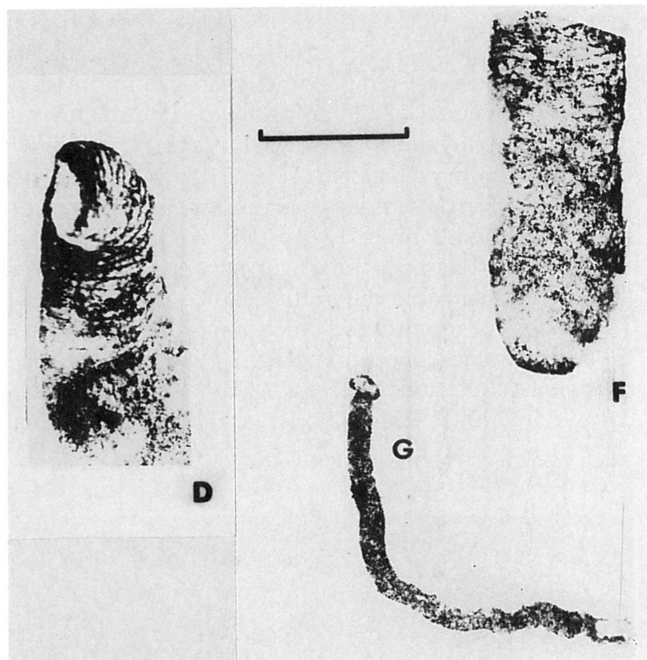


**Figure 9–33** Wormlike body fossils from the Late Proterozoic of China. (Photos courtesy of Sun Weigno, Nanjing Institute of Geology and Palaeontology, Academia Sinica, Nanjing, People's Republic of China.)

(Fig. 9–33, Table 9–3). Perhaps these are worms, but both their biological affinities and age have been questioned. For the present, we can consider these wormlike fossils as persuasive but not conclusive evidence for pre-Ediacaran animals.

All known Proterozoic animals were soft-bodied; that is, they lacked the durable exoskeletons that characterize many Phanerozoic invertebrates. There is some evidence, however, that the incipient stages of skeletonization occurred much earlier. For example, some Ediacaran animals may have had chitin, possibly a chitinous carapace, and others may have possessed some calcareous skeletal elements.

By the latest Proterozoic, several skeletonized animals probably existed. Evidence for this conclusion comes from minute scraps of shell-like material and denticles from larger animals, and spicules, presumably from sponges. Durable skeletons of chitin, silica, and calcium carbonate, however, appeared in abundance at the beginning of the Phanerozoic Eon 570 million years ago.

## CHAPTER SUMMARY

1. The crust-forming processes that characterized the Archean continued into the Proterozoic but at a considerably reduced rate.
2. Archean cratons served as nuclei about which Proterozoic crust accreted. One large landmass that formed by this process is called Laurentia. It consisted mostly of North America and Greenland.
3. The major events in the Proterozoic evolution of Laurentia were the Hudsonian orogeny, origin of the Colorado-Central Plains and Southwestern provinces, origin of granite-rhyolite provinces, the Grenville-Llano orogeny, and the Midcontinent rift.
4. By the Late Proterozoic there were probably three or more supercontinents. Laurasia consisted of Laurentia and what is now Europe and most of Asia. East and West Gondwana may have united in the latest Proterozoic to form Gondwana.
5. Greenstone belts formed during the Proterozoic but at a reduced rate, and they differ in detail from their Archean counterparts.
6. Ophiolite sequences, which mark convergent plate margins, are first well documented from the Early Proterozoic. A modern style of plate tectonics seems to have been established during the Early Proterozoic.
7. Quartzite-carbonate-shale assemblages are known from the Late Archean but become common in Proterozoic rocks. These rock assemblages were deposited on passive continental margins and in intracratonic basins.
8. Widespread glaciation occurred during the Early and Late Proterozoic.
9. The atmosphere became progressively richer in free oxygen through the Proterozoic. Photosynthesizing cyanobacteria were largely responsible for oxygenation of the atmosphere.
10. During the period from 2.5 to 2.0 billion years ago most of the world's iron ores were deposited as Superior-type banded iron formations.
11. The first continental red beds were deposited about 1.8 billion years ago. The widespread occurrence of oxidized iron in sedimentary rocks indicates an oxidizing atmosphere.

12. Unicellular prokaryotes are the only known life-forms from the Early Proterozoic. Eukaryotic cells first appeared during the Middle Proterozoic.
13. A symbiotic relationship among some prokaryotic cells may account for the origin of eukaryotes.
14. The oldest fossils of multicellular organisms are carbonaceous impressions, probably of algae, in rocks between 1 billion and 700 million years old.
15. The Late Proterozoic Ediacaran faunas include the oldest known animal fossils other than burrows. Animals were widespread at this time, but all were soft-bodied, so fossils are not common.

## IMPORTANT TERMS

acritarch
Algoma-type BIF
banded iron formation (BIF)
colonial organism
Colorado-Central Plains province
Ediacaran faunas
eukaryotic cell
granite-rhyolite province
Grenville province
Hudsonian orogeny
Laurasia
Laurentia
Llano province
Midcontinent rift
multicellular organism
ophiolite
quartzite-carbonate-shale assemblage
red beds
Southwestern province
Superior-type BIF
tillite

## REVIEW QUESTIONS

1. Summarize the major difference between the Archean and Proterozoic.
2. Discuss the events leading to the accretion of continental crust during the Hudsonian orogeny.
3. How did the granite-rhyolite provinces form, and why were they important in the evolution of Laurentia?
4. What is the Midcontinent rift, and what kinds of rocks does it contain?
5. Discuss the significance of ophiolites and quartzite-carbonate-shale assemblages in establishing that the Proterozoic was characterized by a modern style of plate tectonics.
6. What kind of evidence would confirm that a suspected Proterozoic tillite was actually a glacial deposit?
7. How are Superior-type BIFs thought to have been deposited?
8. The atmosphere seems to have been chemically oxidizing by 1.8 billion years ago. What evidence supports this statement?
9. What evidence indicates that eukaryotic cells appeared between 1.4 and 1.0 billion years ago?
10. Explain how a symbiotic relationship among Proterozoic prokaryotes may have given rise to eukaryotes.
11. Since the transition from unicellular to multicellular organisms is not recorded by fossils, how can we account for this event?
12. Briefly review the evidence for the presence of animals in the Late Proterozoic.

## ADDITIONAL READING

Condie, K. C. 1982. *Plate Tectonics and Crustal Evolution*. New York: Pergamon Press.

Glaessner, M. F. 1984. *The Dawn of Animal Life*. New York: Cambridge University Press.

Margulis, L. 1982. *Early Life*. New York: Van Nostrand Reinhold Co.

Mendaris, L. G., Jr., C. W. Byers, D. M. Mickelson, and W. C. Shanks. 1983. *Proterozoic Geology: Selected Papers from an International Symposium*. The Geological Society of America, Memoir 161.

Schopf, J. W. 1978. The Evolution of the Earliest Cells. *Scientific American,* 239 no. 3: 111–137.

Windley, B. F. 1984. *The Evolving Continents*. New York: John Wiley & Sons.

# 10

## CHAPTER OUTLINE



## PROLOGUE

*"The Grand Canyon is the one great sight which every American should see," declared President Theodore Roosevelt. "We must do nothing to mar its grandeur." And so, in 1908, he named the Grand Canyon a national monument to protect it from exploitation. In 1919 the Grand Canyon National Monument was upgraded to a National Park primarily because both the geology and scenery exposed in the canyon are unparalled.*

*When people think of the Grand Canyon, many think of the seemingly limitless time represented by the rocks exposed in the walls. For most people, staring down at the rocks in the 1.5 km deep canyon, is their only exposure to the concept of geologic time.*

*Major John Wesley Powell was the first geologist to explore the Grand Canyon region. Major Powell, a Civil War veteran who lost his right arm in the battle of Shiloh, led a group of hardy explorers down the unchartered Colorado River through the Grand Canyon in 1869. Without any maps or other information, Powell and his group ran the many rapids of the Colorado River in fragile wooden boats, hastily recording what they saw. Powell wrote in his diary that "all about me are interesting geologic records. The book is open and I read as I run."*

*From this initial reconnaissance, Powell led a second expedition down the Colorado River in 1871. This second trip included a photographer, a surveyor, and three topographers. This expedition made detailed topographic and geologic maps of the Grand Canyon area as well as the first photographic record of the region.*

*Probably no one has contributed as much to the understanding of the Grand Canyon as Major Powell. In recognition of his contributions, the Powell Memorial was erected on the South Rim of the Grand Canyon in 1969 to commemorate the hundredth anniversary of his first voyage.*

*When we stand on its rim and look down into the Grand Canyon, we are really looking far back in time, all the way back to the early history of our planet. More than one billion years of history are recorded in the rocks of the Grand Canyon, ranging*

# Geology of the Early Paleozoic Era

*from violent mountain building episodes to periods of incursions and retreats of shallow seas.*

*The oldest rocks exposed in the Grand Canyon record two major mountain-building episodes during the Proterozoic Eon. The first episode, represented by the Vishnu and Brahma schists, records a time of uplift, deformation, and metamorphism. This mountain range was eroded to a rather subdued landscape and was followed by deposition of approximately 4,000 m of sediments and lava flows of the Grand Canyon Supergroup. These rocks and the underlying Vishnu and Brahma schists were uplifted and formed a second Proterozoic mountain range. This mountain range was also eroded to a nearly flat surface by the end of the Proterozoic Eon.*

*The first sea of the Paleozoic Era transgressed over the region during the Cambrian Period, depositing sandstones, siltstones, and limestones. A major unconformity separates the Cambrian rocks from the Mississippian limestones exposed as the cliff forming Redwall Limestone. Another unconformity separates the Red Wall Limestone from the overlying Permian Kaibab Limestone, which forms the rim of the Grand Canyon.*

*The Grand Canyon in all its grandeur is a most appropriate place to start our discussion of the Paleozoic history of North America.*

Major John Wesley Powell who led the first geologic expedition down the Grand Canyon, Arizona.

## INTRODUCTION

Having reviewed the geologic history of the Archean and Proterozoic eons, we now turn our attention to the Phanerozoic Eon, comprising the remaining 12 percent of geologic time. First, however, let us briefly examine how the Archean and Proterozoic differ from the Phanerozoic.

Recall that during the Archean, crust-forming processes generated greenstone belts and granite-gneiss complexes, which were shaped into cratons. The most common Archean sedimentary rocks are graywackes and argillites, typical of tectonically active regions. While a modern style of plate tectonics did not begin until the Early Proterozoic, many geologists believe that the Archean style of plate tectonics involved numerous collisions of minicontinents, resulting in rapid crustal growth. The Proterozoic Eon can be characterized as a time of modernization in which the sedimentary-rock assemblages and plate tectonic style were like those of the Phanerozoic Eon. During the Proterozoic Eon, plate movements, extensive plutonism, and regional metamorphism thickened, stabilized, and increased the area of continental crust.

At the beginning of the Phanerozoic Eon, there were six major continental landmasses, four of which straddled the paleoequator. Plate movements during the Phanerozoic Eon created a changing panorama of continents and ocean basins whose positions affected atmospheric and oceanic circulation patterns and created new environments for habitation by the rapidly evolving biota.

The Paleozoic history of most continents involves major mountain-building activity along the continental borders and numerous shallow-water marine transgressions and regressions over their interiors. These transgressions and regressions were caused by global changes in sea level probably related to plate activity and glaciation.

The following chapters present the geologic history of North America in terms of those major transgressions and regressions rather than a period-by-period chronology. While we will focus on North American geologic history, we will endeavor to place those events in a global context.

## THE CONTINENTAL FRAMEWORK

During the Precambrian, continental accretion and orogenic activity occurred, forming sizable continents. One such continent was Laurentia which consisted mostly of North America and Greenland (Fig. 9–3). By the beginning of the Paleozoic Era, most of the continents consisted of two major components that differed in structure and history: the cratons and mobile belts.

### Cratons

**Cratons** are the relatively stable and immobile areas of continents. They are usually large and form the foundation on which Phanerozoic sediments were deposited. Cratons typically consist of two parts: a shield and a platform (Fig. 10–1).

The *shields* are composed of Precambrian rocks, predominantly igneous and metamorphic, and were extremely stable during the Phanerozoic Eon. Their stability during that time contrasts sharply with their Precambrian history of extensive orogenic activity. The platform portion of the craton is typically covered by relatively flat-lying Phanerozoic rocks. Like the shield, the platform was generally stable throughout most of the Phanerozoic.

Platform rocks include sandstones and carbonates that were deposited in shallow, transgressing seas. These widespread and shallow seas, called **epeiric seas**, were a common feature of most Paleozoic cratonic histories. They resulted from sea-level changes primarily caused by continental glaciation as well as plate tectonic activity.

While many of the Paleozoic platform rocks are still flat-lying, in some places they were gently warped into broad basins, domes, and arches which stood out as low islands during the Paleozoic Era and supplied sediments to the surrounding epeiric seas.

### Mobile Belts

Elongated areas that mark the sites of former mountain building are called **mobile belts**. Most are located along the margins of continents. Here sediments were deposited in the relatively shallow waters of the continental shelf and the deeper waters at the base of the continental slope. During plate convergence along these margins, the sediments were deformed and intruded by magma, creating mountain ranges.

Four mobile belts formed around the margin of the North American continent during the Paleozoic Era; these are known as the **Franklin, Cordilleran, Ouachita,** and **Appalachian mobile belts** (Fig. 10–1). Each was the site of mountain building in response to compressional forces at convergent plate boundaries that formed such mountain ranges as the Appalachians and Ouachitas.

## PALEOZOIC GLOBAL GEOGRAPHY

We are so familiar with the present-day configuration of the continents that the Earth's earlier geography seems totally foreign to us. One of the goals of historical geology is to provide paleogeographic reconstructions of the world for the geologic past. The paleogeographic history for the Mesozoic and Cenozoic eras is well-understood primarily because the pattern of magnetic-reversal stripes on the ocean floor allows geologists to reconstruct the positions of the continents with a high degree of certainty for this time interval.

The history of the Paleozoic Era, however, is not as precise, in part because geologists cannot use ocean-floor magnetic-reversal patterns because Paleozoic oceanic crust has been subducted and thus destroyed during the formation of a single continent at the end of the Paleozoic Era. Paleogeographic reconstructions for the Paleozoic Era are based on structural relationships and climate-sensitive sediments such as red beds, evaporites, and coals, as well as the distribution of plants and animals. Figure 10–2 shows the Paleozoic continents superimposed on a map of the modern world. Six major Paleozoic continents are recognized: **Gondwana** (South America, Florida, Africa, Antarctica, Australia, India, and parts of the Middle East and southern Europe), **Laurentia** (most of modern North America, Greenland, Scotland, and part of eastern Russia), **Baltica** (Russia west of the Ural Mountains, Scandinavia, Poland, and northern Germany), **Siberia** (Russia east of the Ural Mountains and Asia north of Kazakhstan and south of Mongolia), **Kazakhstania** (a triangular continent centered on Kazakhstan), and **China** (all of southeast Asia, including China, Indochina, part of Thailand, and the Malay Peninsula). The paleogeographic reconstructions that follow are based on the methods used to determine

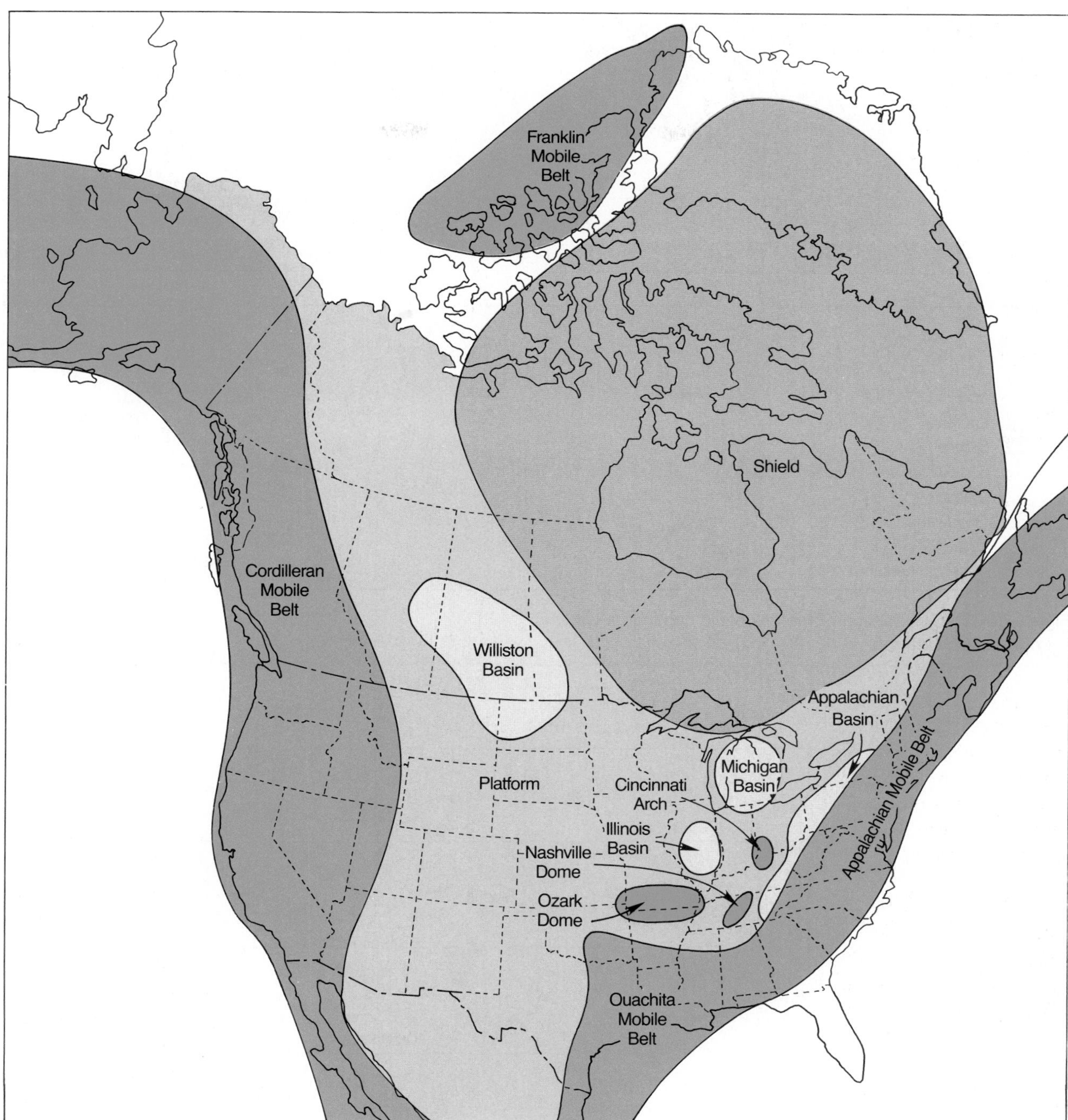


**Figure 10–1** Map of North America showing the mobile belts and major cratonic basins and domes that formed during the Paleozoic Era.

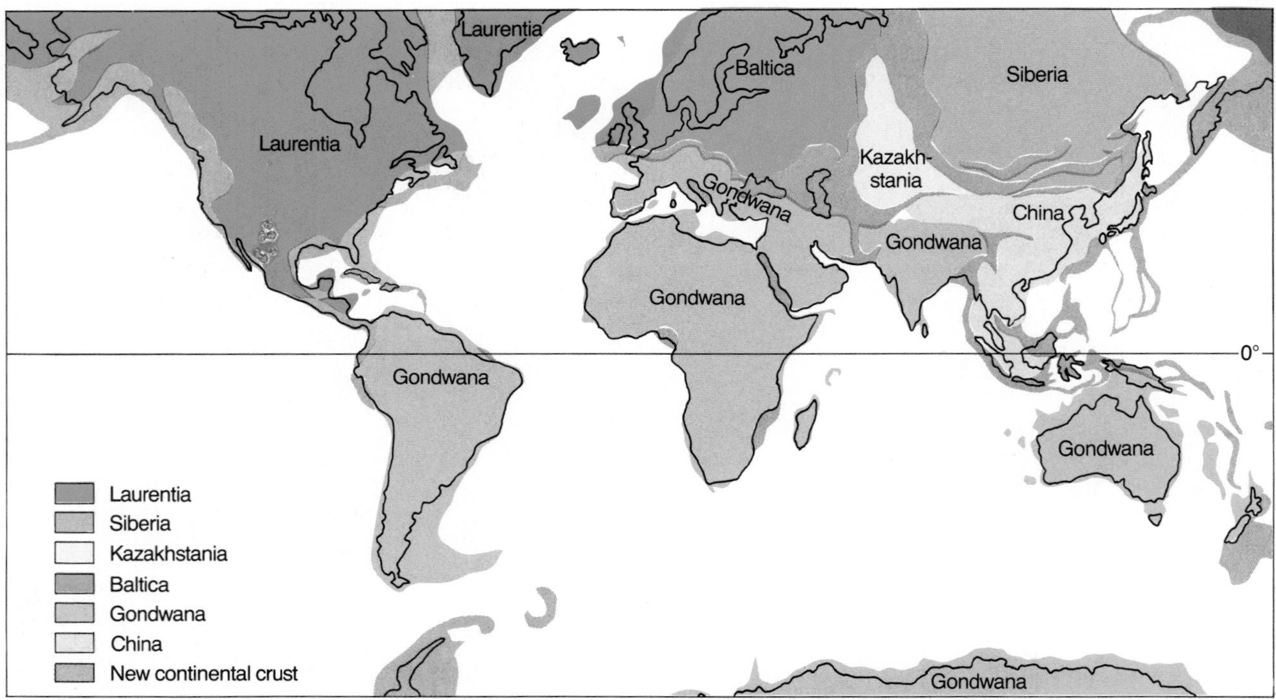


**Figure 10–2** Comparison of the Paleozoic continents with their present counterparts.

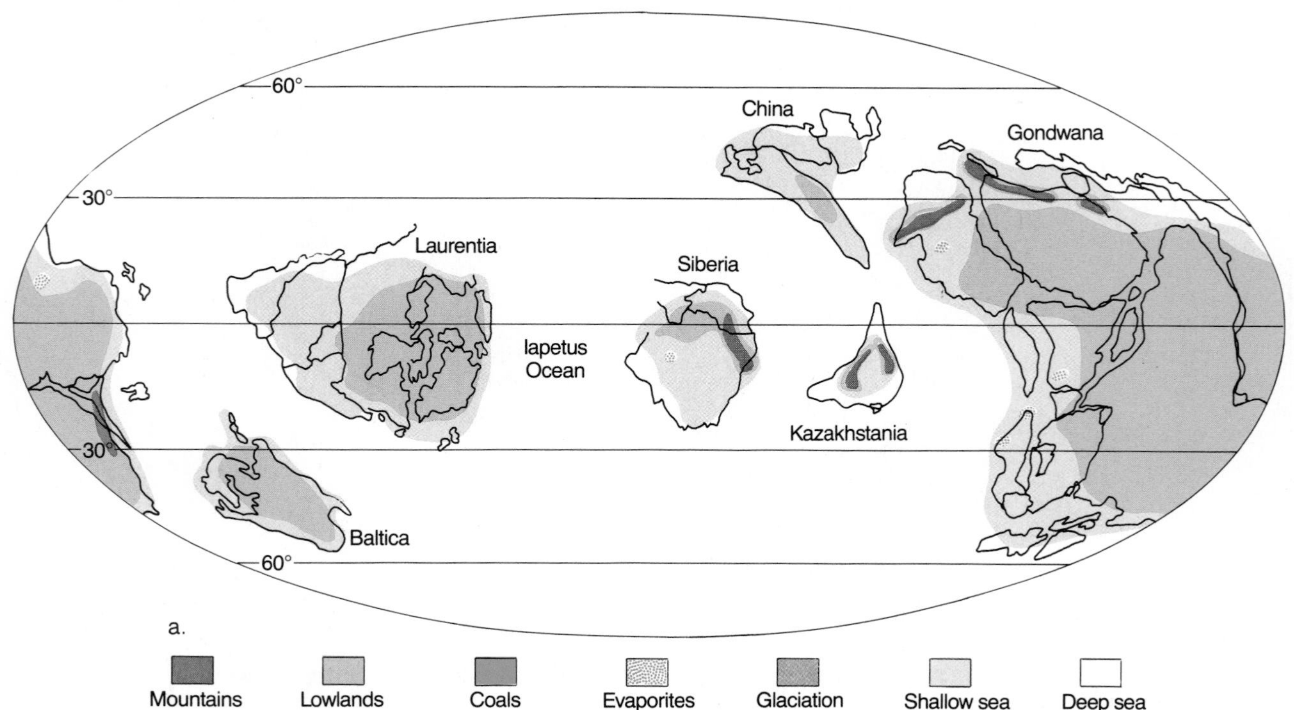


**Figure 10–3** Paleogeography of the world for the (a) Late Cambrian Period, (b) Middle Ordovician Period, and (c) Middle Silurian Period.

and interpret the location, geographic features, and environmental conditions on the paleocontinents (Chapter 4 and Perspective 10–1).

## Early Paleozoic Global History

In contrast to today's global geography (Fig. 10–2), the Cambrian world consisted of six isolated continents dispersed around the globe at low tropical latitudes (Fig. 10–3a). Water circulated freely among ocean basins, and the polar regions were apparently ice-free. By the Late Cambrian, epeiric seas had covered large areas of Laurentia, Baltica, Siberia, Kazakhstania, and China, while major highlands were present in northeastern Gondwana, eastern Siberia, and central Kazakhstania.

During the Ordovician and Silurian periods, plate movement played a major role in the changing global geography (Figs. 10–3b, 10–3c). Gondwana moved approximately 40 degrees south to a south polar location, as indicated by Late Ordovician tillites found today in

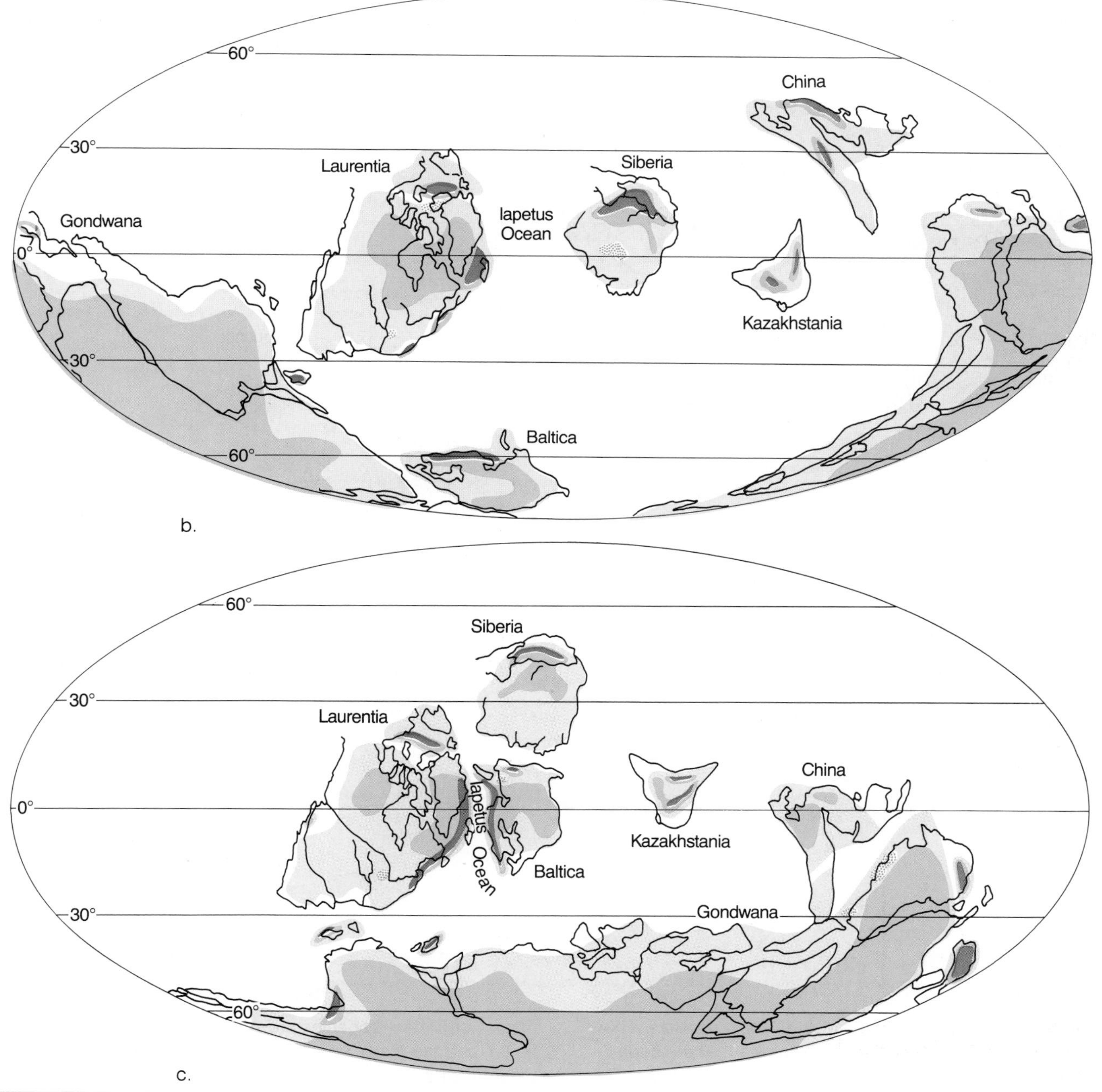

the Sahara Desert. Baltica, paralleling Gondwana's movement, initially moved southward, then separated from the Gondwana plate and moved northward. It eventually collided with eastward-moving Laurentia during the Late Silurian to form Laurasia (see Chapter 11). Siberia shifted from an equatorial position in the Cambrian Period to north temperate latitudes in the Silurian Period. These plate movements resulted in mountain building along the eastern margin of Laurentia (the Taconic orogeny) during the Middle and Late Ordovician Period, and along the western margin of Baltica during the Silurian Period (the Caledonian orogeny). With this plate tectonic overview in mind, we now focus our attention on North America and the role it played in the Early Paleozoic geologic history of the world.

## EARLY PALEOZOIC EVOLUTION OF NORTH AMERICA

The beginning of the Paleozoic Era was a relatively quiet time for North America in contrast to the Late Proterozoic tectonism that occurred during the Grenville-Llano orogeny (Chapter 9). While we know little about the Paleozoic ocean basins, because they have been destroyed by subduction, we do know there were major transgressions and regressions over the craton. Furthermore, as North America rotated counterclockwise and moved eastward in the lower latitudes, major mountain building took place along its eastern borders, while minor

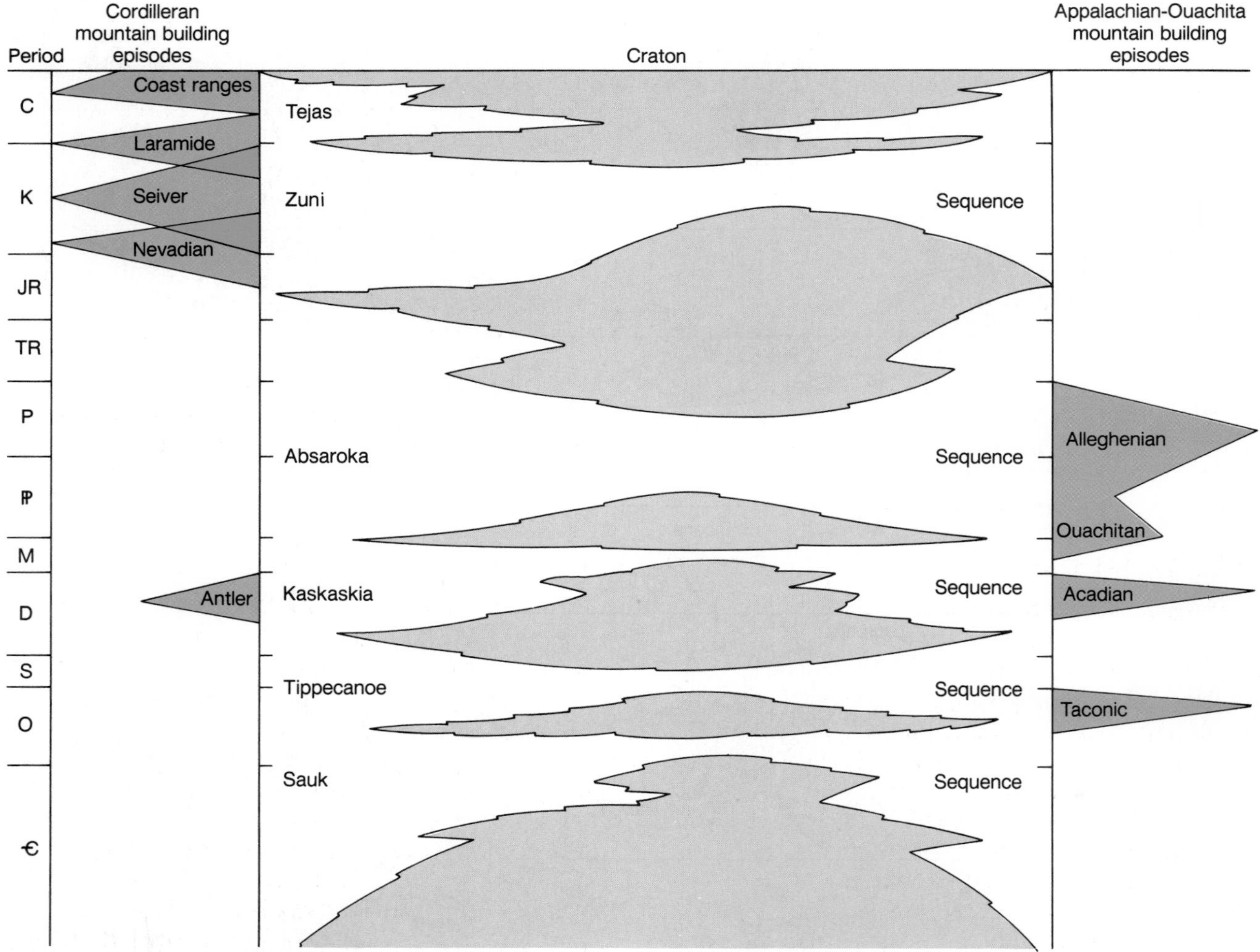


**Figure 10–4** Cratonic sequences of North America. The white areas represent sequences of strata that are separated by major unconformities shown as shaded areas. The major Cordilleran orogenies are shown on the left side of the figure, and the major Appalachian orogenies are shown on the right side.

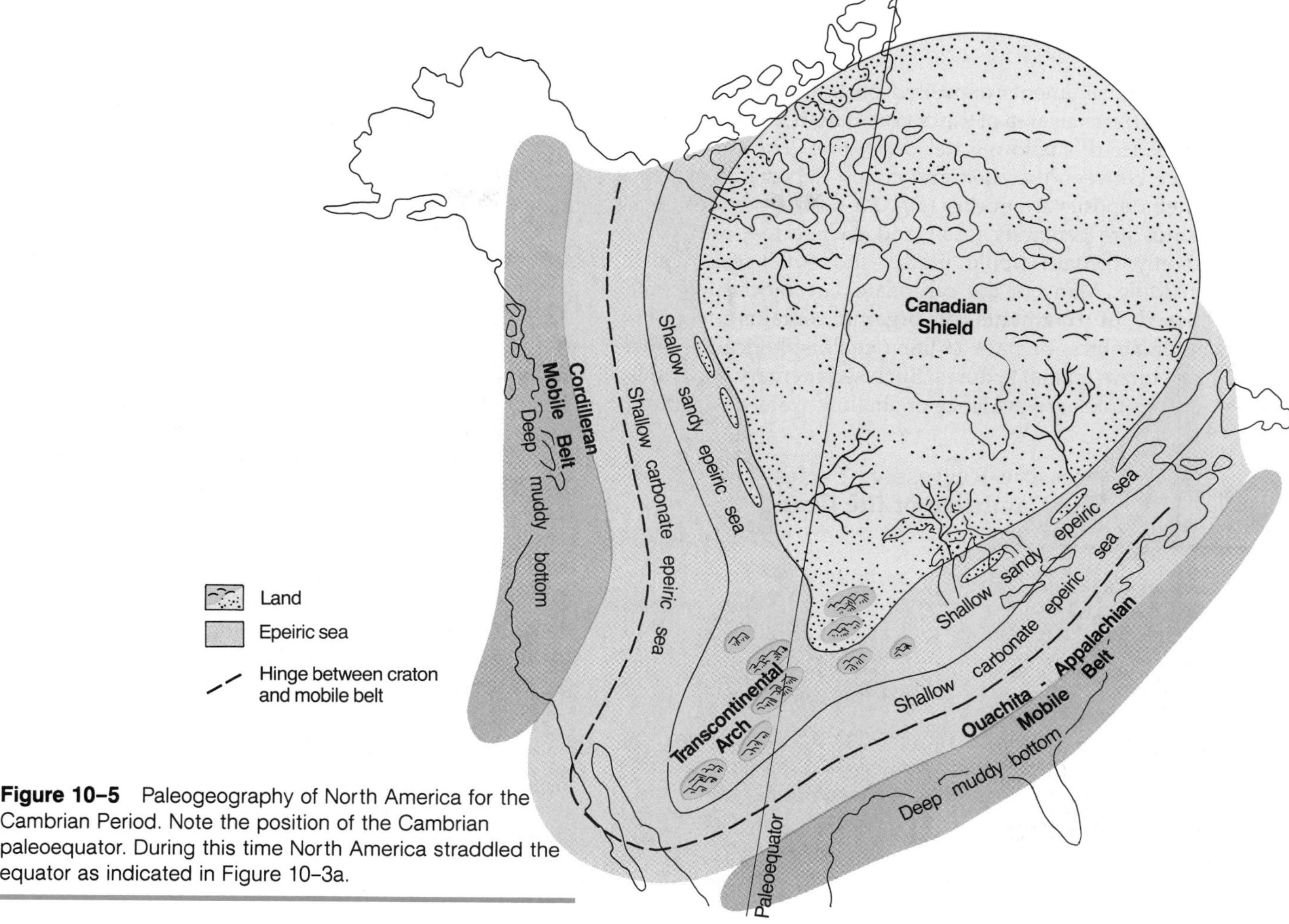


**Figure 10–5** Paleogeography of North America for the Cambrian Period. Note the position of the Cambrian paleoequator. During this time North America straddled the equator as indicated in Figure 10–3a.

orogenies occurred along its western margin. It is therefore convenient to discuss the history of the North American craton from a perspective of mobile belt formation and continental accretion.

## The North American Craton

The American geologist Laurence L. Sloss proposed in 1963 that the sedimentary-rock record of North America could be subdivided into six cratonic sequences. A **cratonic sequence** is a large-scale (greater than supergroup) lithostratigraphic unit representing a major transgressive-regressive cycle, and bounded by craton-wide unconformities (Fig. 10–4). The transgressive phase, which is usually covered by younger sediments, is commonly well preserved, while the regressive phase of each sequence is marked by an unconformity. Where rocks of the appropriate age are preserved, each of the six unconformities can be shown to extend across the various sedimentary basins of the North American craton and into the mobile belts along the cratonic margin.

Geologists have also recognized major unconformity bounded sequences in cratonic areas outside of North America. Such global transgressive and regressive cycles are caused by sea-level changes and are believed to result from major tectonic and glacial events.

# THE SAUK SEQUENCE

Rocks of the **Sauk sequence** record the first major transgression onto the North American craton (Fig. 10–4). Prior to this, most of the craton was above sea level. Over a period of at least 50 million years (m.y.), the Precambrian igneous and metamorphic rocks were deeply weathered and eroded as rivers swept across this barren, plantless landscape. As the transgressive phase of the Sauk began, the seas encroached onto the craton, producing an epeiric sea (see Perspective 10–2). By the Late Cambrian, the Sauk Sea covered most of North America, such that only a part of the Canadian Shield and a few large islands were above sea level (Fig. 10–5). These islands, collectively referred to as the **Transcontinental Arch**, extended from New Mexico to Minnesota and the Lake Superior region.

The sediments deposited on both the craton and along the shelf area of the craton margin show abundant evidence of shallow-water deposition. The only difference between the shelf and craton deposits is that the shelf deposits are thicker (Fig. 10–6). In both areas, the sands are generally clean and well-sorted, and commonly contain ripple marks and small-scale cross-bedding. Many of the carbonates are bioclastic (composed of fragments of organic remains), contain stromatolites, or have oolitic (small, spherical calcium carbonate grains) textures. Such sedimentary structures and textures are evidence of shallow-water deposition.

## A Transgressive-Regressive Facies Model: The Cambrian of the Grand Canyon Region

Recall from Chapter 3 that sediments get increasingly finer the farther away from land one goes. Therefore in a stable environment, where sea level remains the same, coarse clastic sediments are typically deposited in the nearshore environment, and finer-grained sediments are deposited in the offshore environment. Carbonates form farthest from land in the area beyond the reach of terrigenous sediments. During a transgression, these facies (sediments that represent a particular environment) migrate in a landward direction (Fig. 3–13).

The Cambrian rocks of the Grand Canyon region (see "Prologue") provide an excellent model for illustrating the sedimentation patterns of a transgressing sea. During the Late Proterozoic, the Sauk Sea was largely confined to the continental margins (continental shelves and slopes) so that most of the craton was above sea level. Beginning in the Early Cambrian, the Sauk Sea encroached onto the craton. In the Grand Canyon area, the Tapeats Sandstone represents the basal transgressive shoreline deposits that accumulated as marine waters slowly transgressed from the shelf onto the western margin of the craton during the Early Cambrian (Fig. 10–7). These sediments are clean, well-sorted sands, of the type one would find today on a beach. As the transgression continued into the Middle Cambrian, muds and silts of the Bright Angel Shale were deposited over the Tapeats Sandstone. By the Late Cambrian, the Sauk Sea

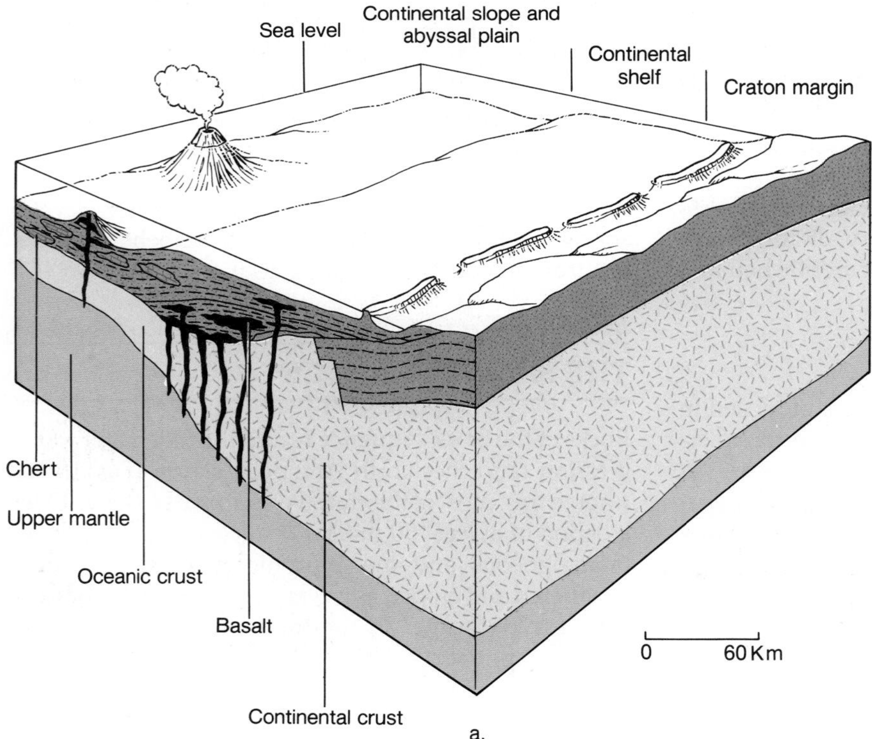


**Figure 10–6** Depositional setting for the Cordilleran margin for the Late Proterozoic and Early Cambrian in the northern Great Basin, Nevada and Utah. Wave-formed ripple marks, small scale cross-bedding, stromatolites, and oolitic and biocalstic limestones in these rocks indicate they were deposited in shallow water on the shelf and the craton.

## Perspective 10–1

# Paleogeographic Reconstructions and Maps

The key to any reconstruction of world paleogeography is the correct positioning of the continents in terms of latitude and longitude as well as orientation of the paleocontinent relative to the paleonorth pole. The main criteria used for paleogeographic reconstructions are paleomagnetic, biogeographic, tectonic, and climatic.

Paleomagnetic data are the only source of quantitative data on the orientations of the continents. For the Paleozoic Era, however, the paleomagnetic data are often inconsistent and contradictory. This is because secondary magnetizations may be acquired through the effects of metamorphism or weathering.

The distribution of faunas and floras provides a useful check on the latitudes determined by paleomagnetism, and can provide additional limits on longitudinal separation of continents. It is well-known that the distribution of plants and animals is controlled by both climatic and geographic barriers. Such information can be used to position continents and ocean basins in a way that accounts for the biogeographic patterns indicated by fossil evidence.

Tectonic activity is implied by deformed sediments associated with andesitic volcanic zones and ophiolite sequences. Such features allow geologists to recognize ancient mountain chains and zones of subduction. These mountain chains may have been separated by subsequent plate movement, and the recognition of large, continuous mountain chains provides important information about continental positions for times in the geologic past.

Climate-sensitive sedimentary rocks are used to interpret past climatic conditions. Some detrital sediments are indicative of humid environments. Desert dunes are typically well-sorted and cross-bedded on a large scale, and associated with other deposits indicating an arid environment. Coals form in freshwater swamps where temperature and rainfall are adequate for abundant plant growth. Evaporites result when evaporation exceedes precipitation, such as in desert regions or along hot, dry, shorelines. Tillites result from glacial activity and indicate cold, wet environments. Various animal and plant fossils can also be used to help determine ancient water and air temperatures. However, because paleoclimates involve the interaction between the atmosphere, hydrosphere, and lithosphere, it is the most difficult to accurately interpret.

Paleogeographic features can be determined by associations of sedimentary rocks and sedimentary structures. For example, large-scale cross-beds may indicate aeolian or windblown conditions such as in deserts. Delta complexes and deep-sea fans have characteristic internal features and three-dimensional forms that can be recognized in the rock record, just as coal and associated deposits usually follow a particular sequence. These features can be used to interpret such geographic features as lakes, streams, swamps, and shallow and deep marine areas.

Former mountain belts can be recognized by folded and faulted sedimentary rocks associated with metamorphic and igneous rocks. We have already mentioned the association of andesites and ophiolite sequences as evidence of former mountain building.

By combining all relevant geologic, paleontologic, and climatologic information, geologists can construct paleogeographic maps (see Figs. 4–41, and 10–3). Such maps are simply interpretations of the geography of an area for a particular time in the geologic past. The majority of paleogeographic maps show the distribution of land and sea, probable climatic regimes, and such geographic features as mountain ranges, swamps, or glaciers.

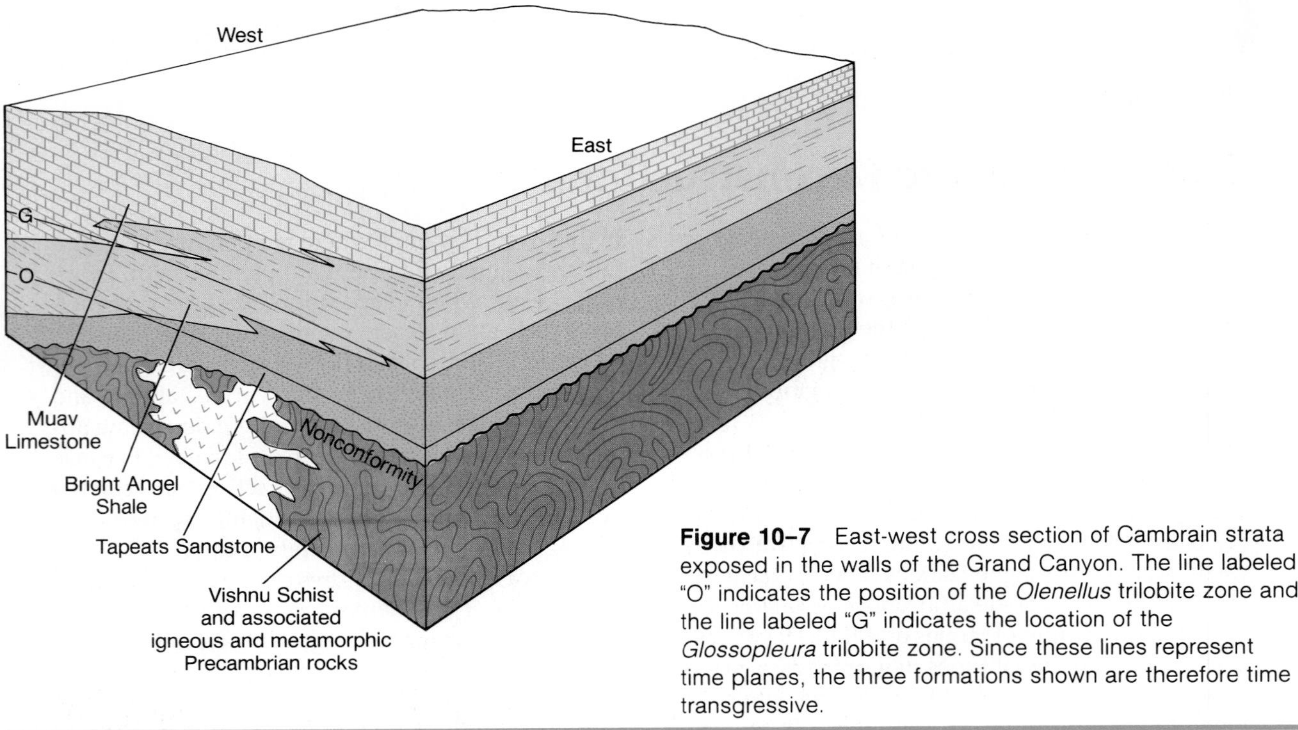


**Figure 10–7** East-west cross section of Cambrain strata exposed in the walls of the Grand Canyon. The line labeled "O" indicates the position of the *Olenellus* trilobite zone and the line labeled "G" indicates the location of the *Glossopleura* trilobite zone. Since these lines represent time planes, the three formations shown are therefore time transgressive.

has transgressed so far onto the craton that in the Grand Canyon area carbonates of the Muav Limestone were deposited over the Bright Angel Shale. Together, these three formations form a typical transgressive sequence that can be recognized by coarse deposits near the base of the section, followed by increasingly finer sediments near the top (Fig. 10–7). This vertical succession of sandstone (Tapeats), shale (Bright Angel), and limestone (Muav) records the craton-ward migration of progressively offshore facies through time.

Cambrian rocks of the Grand Canyon region also illustrate that formations are usually not the same age everywhere they occur; that is, they are time-transgressive. Detailed mapping of these three formations, including correlation of trilobite faunas, shows that deposition of the Muav Limestone had already started in the West before deposition of the Tapeats Sandstone had finished in the East. Correlations based on trilobites indicate the Bright Angel Shale is Early Cambrian in age in California, and Middle Cambrian in age in the Grand Canyon area. The age differences within the Bright Angel Shale illustrate the time-transgressive nature of formations and facies.

The same facies relationship just discussed also occurred elsewhere on the craton as the seas encroached from the Appalachian and Ouachita mobile belts onto the craton interior (Fig. 10–8). Carbonate deposition predominated as the Sauk transgression continued during the Early Ordovician, and the islands of the Transcontinental Arch were soon covered by the advancing Sauk Sea. By the end of Sauk time, the majority of the craton was submerged beneath a warm, equatorial epeiric sea (Fig. 10–3a).

## THE TIPPECANOE SEQUENCE

Near the end of the Early Ordovician, the Sauk Sea regressed from the craton, revealing a landscape of low relief. The rocks exposed were predominantly limestones, which experienced deep and extensive subaerial erosion, because North America was located in a tropical environment. That erosion produced a widespread unconformity that marks the boundary between the Sauk and Tippecanoe sequences.

Deposition of the **Tippecanoe sequence** began like the Sauk, with a major transgression onto the craton. This transgressing sea deposited clean quartz sands over a large area of the craton. The most famous of these Tip-

## Perspective 10-2

# Pictured Rocks National Lakeshore

Exposed along the south shore of Lake Superior between Au Sable Point and Munising in Michigan's Upper Peninsula is the beautiful and imposing wavecut sandstone called Pictured Rocks cliffs (Fig. 1). The rocks exposed in this area, part of which is designated a National Lakeshore, comprise the Upper Cambrian Munising Formation, which is divided into two members: The lower Chapel Rock Sandstone and the upper Miner's Castle Sandstone (Fig. 1). The Munising Formation unconformably overlies the Upper Proterozoic Jacobsville Sandstone and is unconformably overlain by the Middle Ordovician Au Train Formation. The reddish-brown, coarse-grained Jacobsville Sandstone was deposited in streams and lakes over an irregular erosion surface (Fig. 1). Following deposition, the Jacobsville was slightly uplifted and tilted.

As we have discussed previously, the Sauk Sea covered most of North America by the Late Cambrian Period, with only a part of the Canadian Shield and a few large islands still above sea level (Fig. 10–5). By the Late Cambrian, the transgressing Sauk Sea reached the Michigan area. During the first phase of this transgression, the Chapel Rock Sandstone was deposited. The principal source area for this unit was the Northern Michigan highlands, an area that corresponds to the present Upper Peninsula. Following deposition of the Chapel Rock Sandstone, the Sauk Sea retreated from the area.

The Miner's Castle Sandstone represents a second transgression of the Sauk Sea in the area. However, this second transgression covered most of the Upper Peninsula of Michigan and drowned the highlands that were the source for the Chapel Rock Sandstone.

The source area for the Miner's Castle Sandstone was the Precambrian Canadian shield area to the north and northeast. The Miner's Castle Sandstone contains rounder, better sorted, and more abundant quartz grains than the Chapel Rock Sandstone, indicating a different source area. A major unconformity separates the Miner's Castle Sandstone from the overlying Middle Ordovician Au Train Formation.

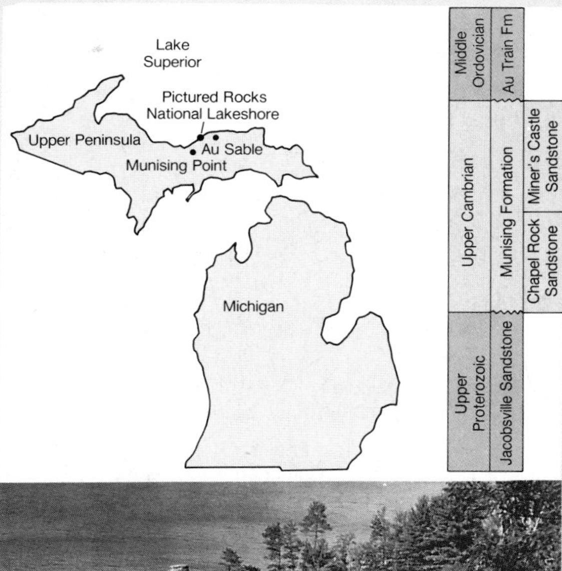

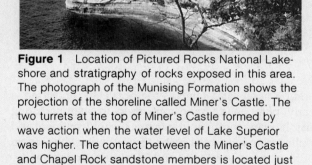

**Figure 1** Location of Pictured Rocks National Lakeshore and stratigraphy of rocks exposed in this area. The photograph of the Munising Formation shows the projection of the shoreline called Miner's Castle. The two turrets at the top of Miner's Castle formed by wave action when the water level of Lake Superior was higher. The contact between the Miner's Castle and Chapel Rock sandstone members is located just above lake level.

One of the most prominent features of Pictured Rocks is Miner's Castle (Fig. 1). Miner's Castle is a wavecut projection along the Pictured Rocks National Lakeshore. The lower sandstone unit at water level is the Chapel Rock Sandstone, while the rest of the feature is composed of the Miner's Castle Sandstone. The two turrets of the castle formed as sea stacks during a time following the Pleistocene when the water level of Lake Superior was much higher.

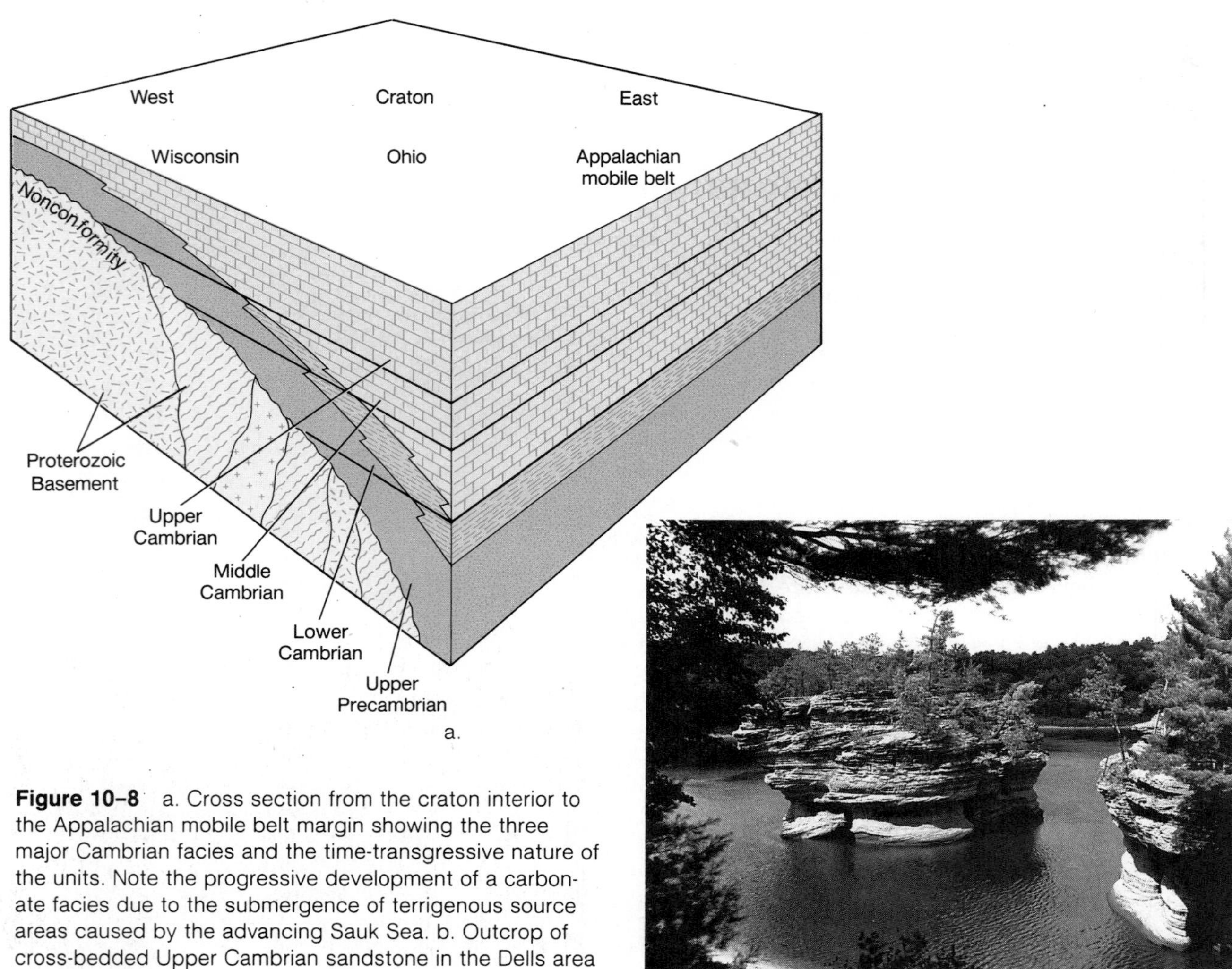


**Figure 10–8** a. Cross section from the craton interior to the Appalachian mobile belt margin showing the three major Cambrian facies and the time-transgressive nature of the units. Note the progressive development of a carbonate facies due to the submergence of terrigenous source areas caused by the advancing Sauk Sea. b. Outcrop of cross-bedded Upper Cambrian sandstone in the Dells area of Wisconsin.

pecanoe basal sandstones is the St. Peter Sandstone, a nearly pure quartz sandstone used in manufacturing glass. It occurs throughout much of the midcontinent and resulted from numerous cycles of weathering and erosion of Proterozoic sandstones, and Cambrian sandstones deposited during the Sauk transgression (Fig. 10–9).

The Tippecanoe basal sandstones were followed by the development of extensive limestones (Fig. 10–10). These limestones were predominantly the result of deposition by calcium carbonate–secreting organisms such as corals, brachiopods, stromatoporoids, and bryozoans. In addition to the limestones, there were also many dolomites. Most of the dolomites were formed as the result of substituting magnesium for some of the calcium within calcite, and in the process, converting the limestones into dolomites.

In the eastern portion of the craton, limestones and dolomites gradually merge into shales. The shales mark the farthest extent of clastic sediments derived from weathering and erosion of the highlands formed during the Taconic orogeny, a tectonic event we will discuss later.

## Tippecanoe Reefs and Evaporites

Widespread reefs and evaporite deposits are major features of the Tippecanoe sequence. Remember that during the Ordovician and Silurian periods, Laurentia was moving northward, yet still was located within tropical latitudes (Figs. 10–3b, 10–3c). Most of the North American craton during that time was covered by a warm epeiric sea. In the present-day Great Lakes region, reef-

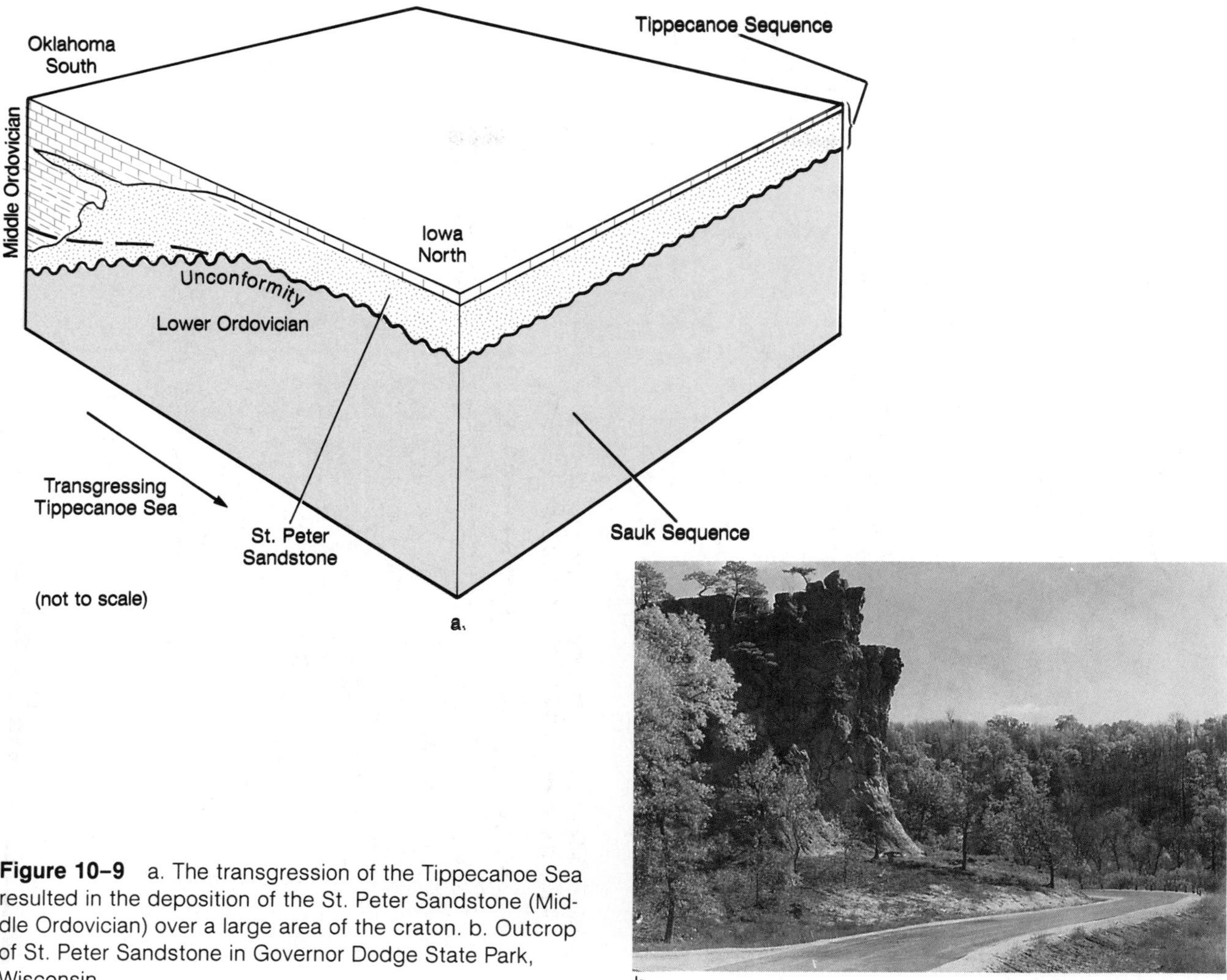


**Figure 10–9** a. The transgression of the Tippecanoe Sea resulted in the deposition of the St. Peter Sandstone (Middle Ordovician) over a large area of the craton. b. Outcrop of St. Peter Sandstone in Governor Dodge State Park, Wisconsin.

fringed basins developed. The evaporation of seawater in these basins resulted in the precipitation of thick evaporite mineral deposits such as gypsum and halite. Before discussing Silurian reefs and evaporites, we will briefly examine present-day reefs and their environments to better understand reef environments.

## General Characteristics of Modern Organic Reefs

**Organic reefs** are limestone structures constructed by living organisms, some of which contribute skeletal materials to the reef framework (Fig. 10–11). Today corals and calcareous algae are the most prominent reef builders, while in the geologic past, other organisms played a major role. Regardless of the organisms dominating reef communities, reefs appear to have occupied the same ecological niche in the past that they do today. Because of the ecological requirements of reef-building organisms, reefs today are confined to a narrow latitudinal belt between 30 degrees north and south of the equator. Corals, the major reef-building organisms today, require warm, clear, shallow water of normal salinity for optimal growth.

The size and shape of a reef is largely the result of the interaction between the reef-building organisms, the bottom topography, wind and wave action, and subsidence of the sea floor. Reefs also alter the area around them by forming barriers to water circulation or wave action.

Reefs are commonly long, linear masses that form a barrier between a shallow platform on one side and a

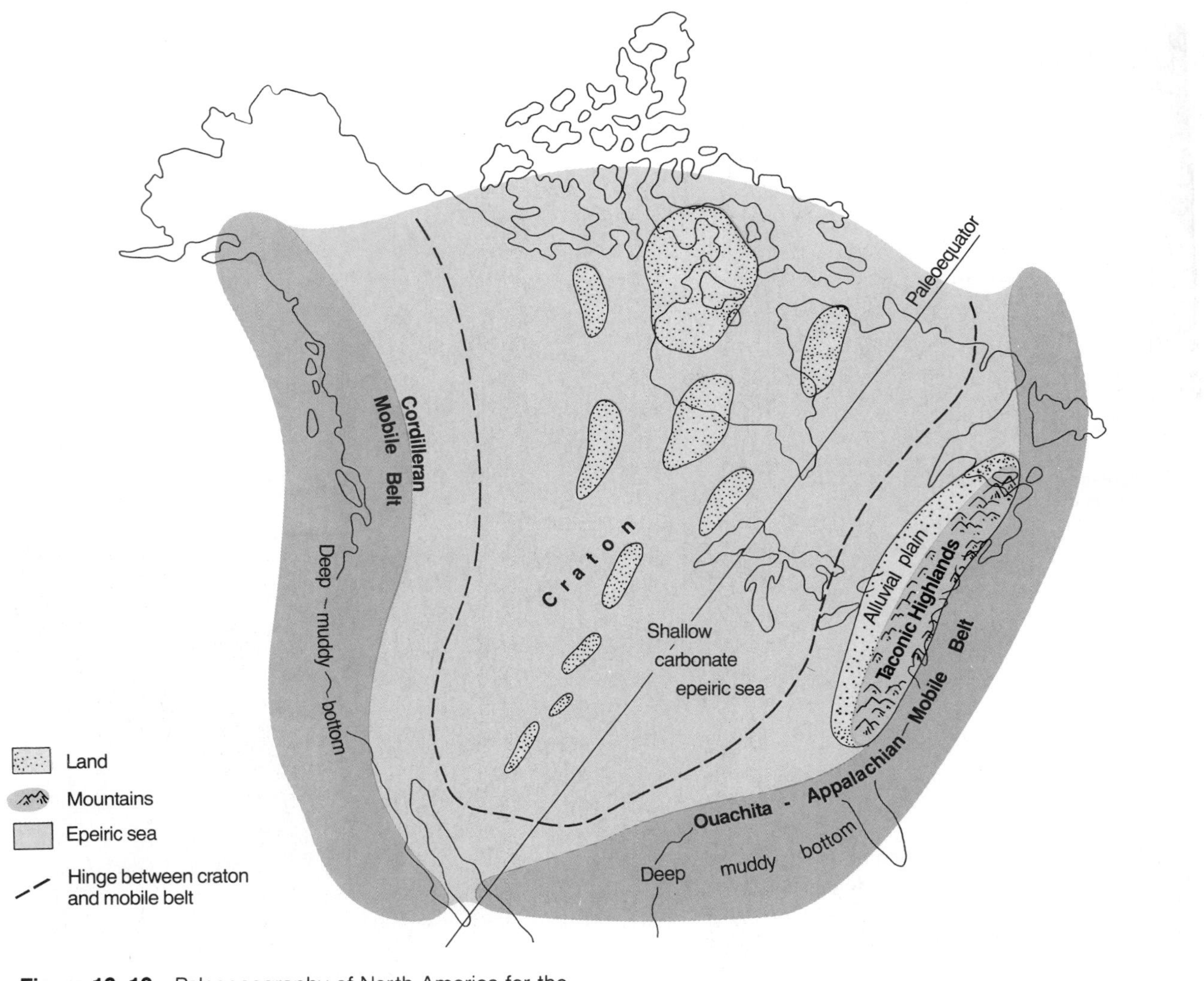


**Figure 10–10** Paleogeography of North America for the Ordovician Period. Note that the position of the equator has changed, indicating North America was rotating in a counterclockwise direction.

comparatively deep marine basin on the other side (Fig. 10–11). Reefs create and maintain a steep seaward front that absorbs incoming wave energy. As skeletal material breaks off from the reef front, it accumulates along a forereef slope. The reef barrier itself is porous and composed of reef-building organisms. The back-reef lagoon area is a low-energy, quiet-water zone where fragile and sediment-trapping organisms thrive. The back-reef lagoon area can also become the site of evaporitic deposits when circulation to the open sea is cut off.

Modern examples of barrier reefs are the Florida Keys, Bahama Islands, and Great Barrier Reef of Australia. Other types of reefs include circular fringing reefs that develop around islands, Pacific atolls that are built on submerged volcanic peaks, and small patch reefs, several meters in diameter, that form in a variety of settings.

Reefs have been common features in the geologic past and have been built by a variety of organisms. The first skeletal builders of reef-like structures were archaeocyathids. These conical-shaped organisms lived during the Cambrian Period and had double, perforated, calcareous shell walls. Archaeocyathids built small mounds which have been found on all continents except South

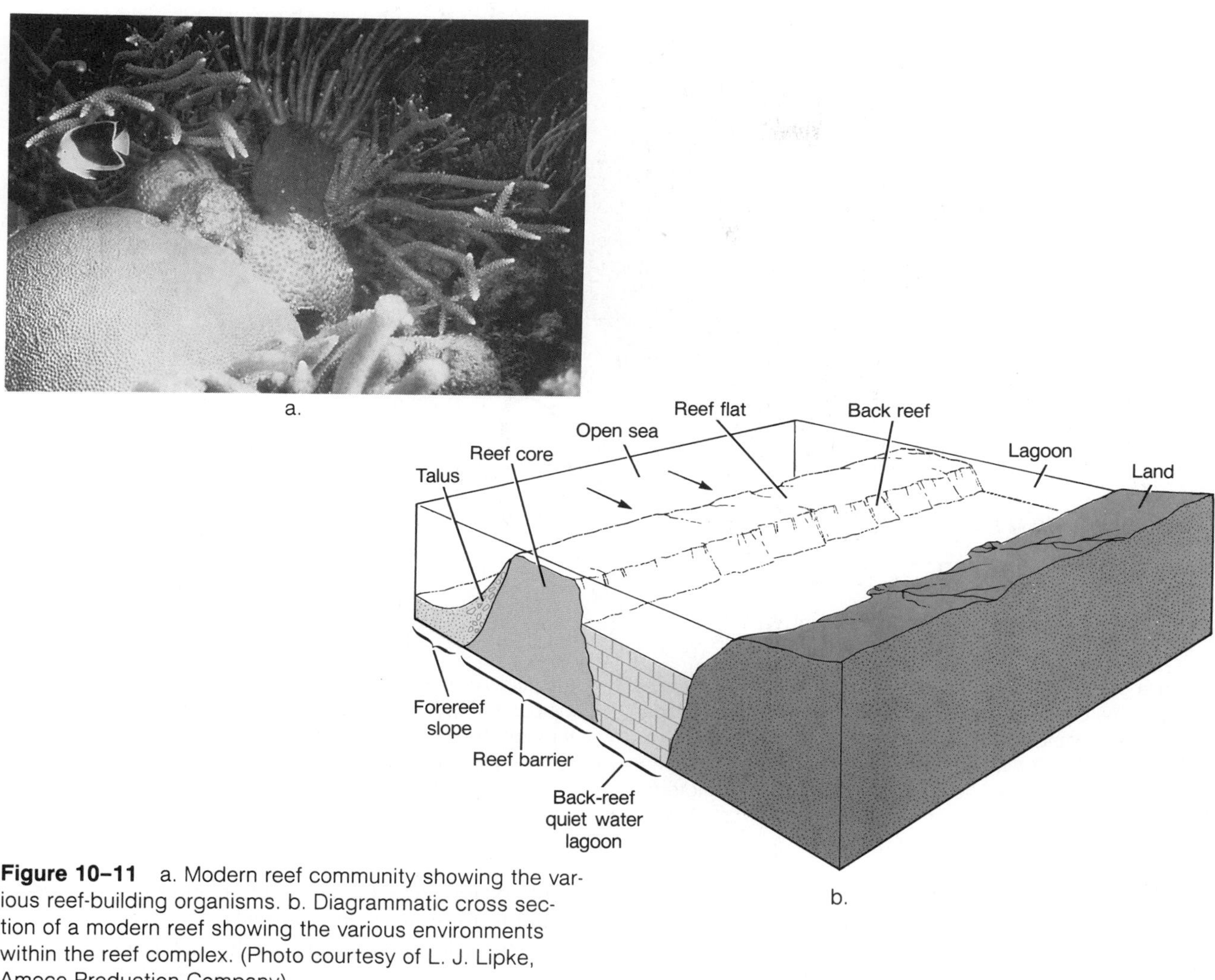


**Figure 10–11** a. Modern reef community showing the various reef-building organisms. b. Diagrammatic cross section of a modern reef showing the various environments within the reef complex. (Photo courtesy of L. J. Lipke, Amoco Production Company)

America (Fig. 12–9). Beginning in the Middle Ordovician, stromatoporoid-coral reefs became common in the low latitudes and remained so throughout the rest of the Phanerozoic Eon. The burst of reef building seen in the Late Ordovician through Devonian periods was probably in response to evolutionary changes triggered by the appearance of extensive carbonate sea floors and platforms beyond the influence of terrigenous sediments.

## Silurian Organic Reefs and Evaporite Facies

The Middle Silurian rocks (Tippecanoe sequence) of the present-day Great Lakes region are world-famous for their reef and evaporite deposits and have been extensively studied (Fig. 10–12). The most famous structure in the region is the Michigan Basin. It is a broad, circular basin surrounded by large barrier reefs. No doubt these reefs contributed to increasingly restricted circulation and the precipitation of Upper Silurian evaporites within the basin (Fig. 10–13).

Within the rapidly subsiding interior of the basin, another type of reef is found. *Pinnacle reefs* are tall, spindly structures up to 100 m high. They reflect the rapid upward growth needed to maintain themselves near sea level during subsidence of the basin (Fig. 10–13). In addition to the pinnacle reefs, bedded carbonates and thick sequences of salt and anhydrite are also found in the Michigan Basin.

As the Tippecanoe Sea gradually regressed from the craton during the Late Silurian, precipitation of evaporite minerals occurred in the Appalachian, Ohio, and

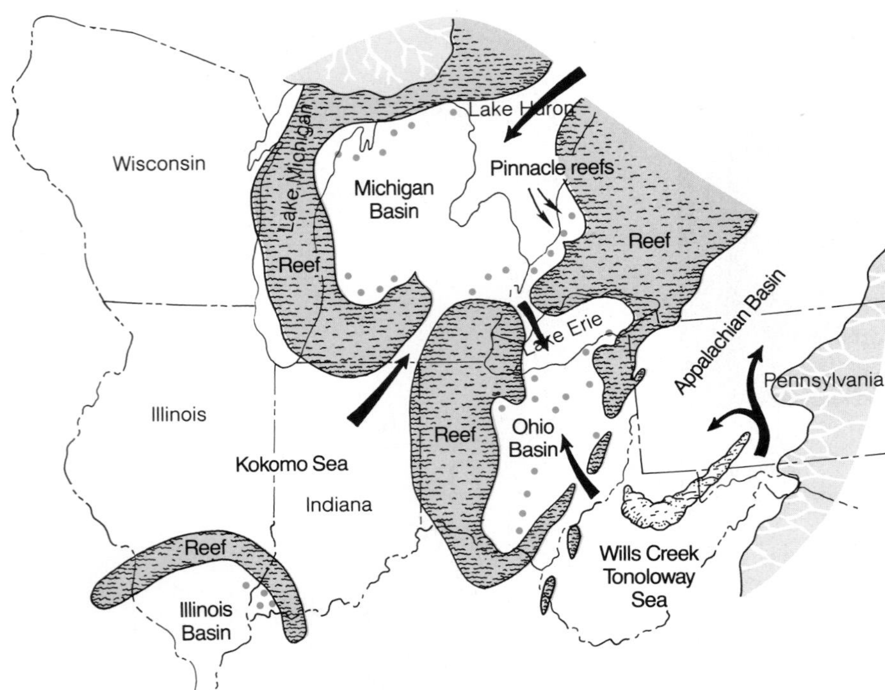


**Figure 10–12** Silurian paleogeography of the Great Lakes region. The basins present at this time were in the Michigan, Ohio, and Indiana-Illinois-Kentucky areas. Pinnacle reefs grew upward within these basins, while barrier reefs such as those surrounding the Michigan Basin probably restricted the inflow of marine water at times. During such times, evaporite deposits formed within the basin. Narrow inlets through the reefs allowed seawater to replenish the basin, resulting in carbonate deposition.

Michigan basins. In the Michigan Basin alone, approximately 1,500 m of sediments were deposited, nearly half of which are halite and anhydrite. How did such thick sequences of evaporites accumulate? One possibility would be a drop in sea level, bringing the top of the barrier reef up to or beyond sea level, and thus preventing the influx of new seawater into the basin. Evaporation of the basinal seawater would result in the precipitation of salts. A second possibility is that the reefs grew upward so close to sea level that they formed a sill or barrier that eliminated interior circulation (Fig. 10–14).

Since North America was still near the equator during the Silurian Period (Fig. 10–3c), temperatures were probably high. As circulation to the Michigan Basin was restricted, seawater within the basin evaporated, forming a brine. Because the brine was heavier, it concentrated near the bottom, and minerals were precipitated onto the basin floor. There was some replenishment of seawater over the sill and through channels cut in the barrier reefs, but this only added new seawater that later become concentrated as brine. In this way, the concentration of the brine in the basin increased until the salts could no longer be held in solution and therefore precipitated to form evaporite minerals.

The order and type of salts that precipitate from seawater depends on their solubility, original concentration of seawater, and local conditions of the basin. In general, different salts precipitate in order of least soluble to most soluble. Therefore, calcium carbonate usually precipitates out first, followed by gypsum*, and lastly halite. However, due to fluctuations in the amount of seawater entering the basin, as well as changing geologic conditions, there can be many lateral shifts and interfingering of the limestone, anhydrite, and halite facies (Fig. 10–15).

Periodic evaporation of seawater according to this model could account for the observed vertical and lateral distribution of evaporites in the Michigan Basin. However, associated with those evaporites are pinnacle reefs, and the organisms that constructed those reefs could not

*Recall from Chapter 4 that gypsum ($CaSO_4 \cdot 2H_2O$) is the common sulfate precipitated from seawater, but when deeply buried gypsum loses its water and is converted to anhydrite ($CaSO_4$).

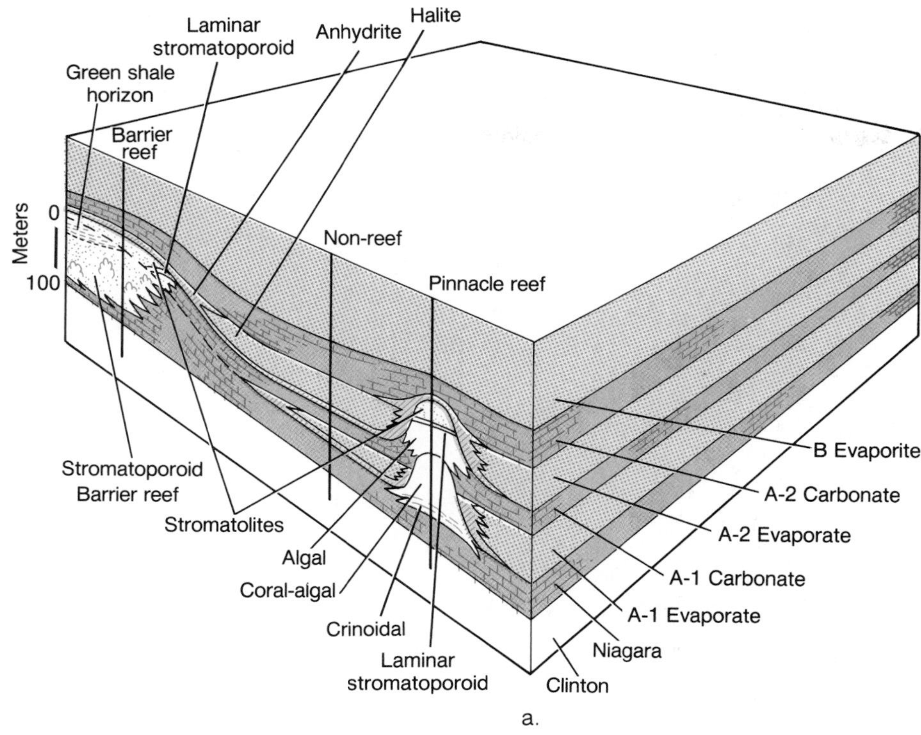

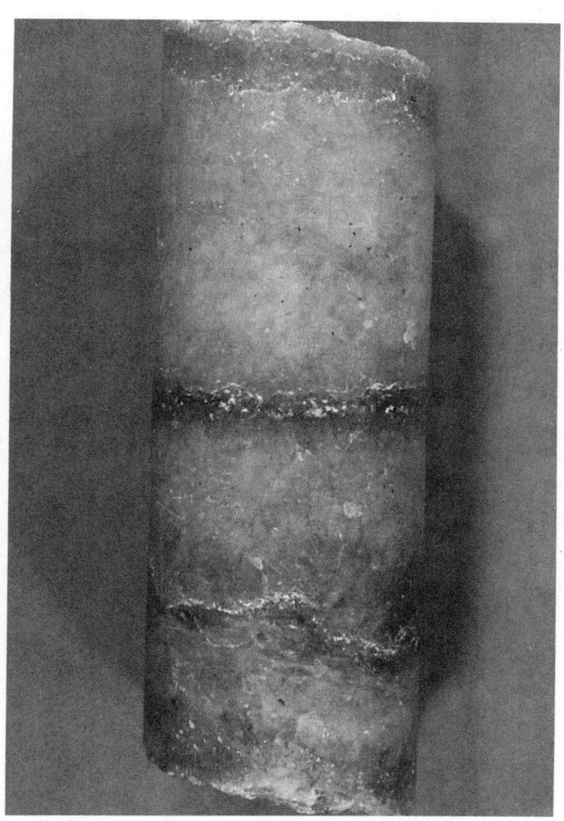


**Figure 10–13** a. Generalized cross section of the northern Michigan Basin during the Silurian Period. b. Stromatoporoid barrier-reef facies. c. Evaporate facies.

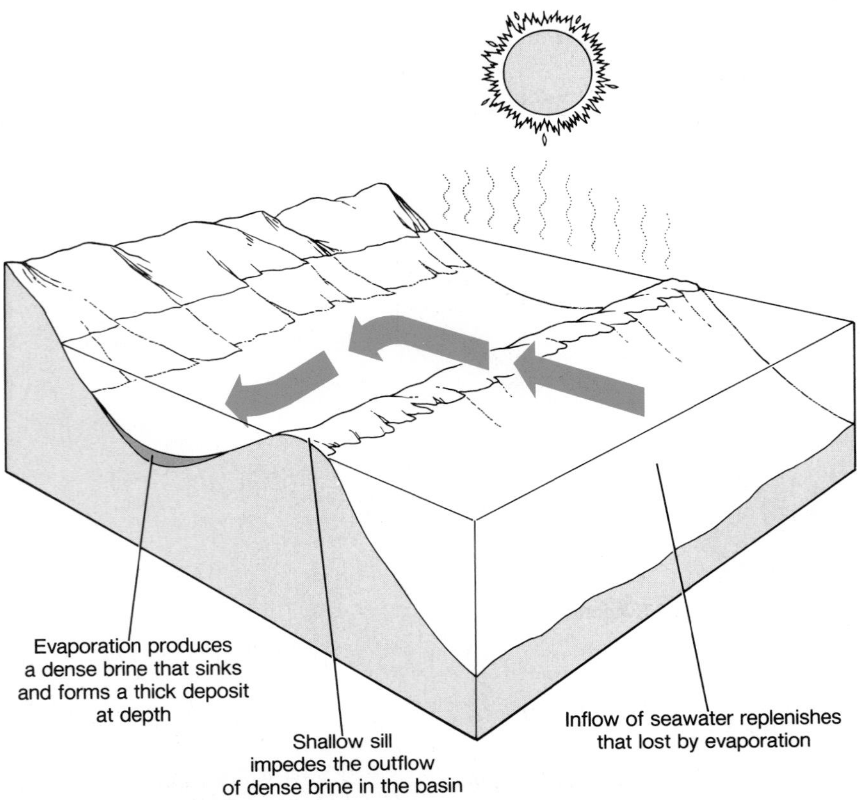
Evaporation produces
a dense brine that sinks
and forms a thick deposit
at depth
Shallow sill
impedes the outflow
of dense brine in the basin
Inflow of seawater replenishes
that lost by evaporation

**Figure 10–14** Silled basin model for evaporite sedimentation by direct precipitation from seawater. Vertical scale is greatly exaggerated.

have lived in such a highly saline environment (10–13). How, then, can such contradictory features be explained? Numerous models have been proposed, ranging from cessation of reef growth followed by evaporite deposition, to alternation of reef growth and evaporite deposition. Although the Michigan Basin has been studied extensively for years, no model yet proposed completely explains the genesis and relationship of its various reef, carbonate, and evaporite facies.

### The End of the Tippecanoe Sequence

By the Early Devonian, the regressing Tippecanoe Sea had retreated to the craton margin. During this regression, marine deposition was initially restricted to a few cratonic basins and finally to the craton margin. By the end of Tippecanoe time, most of the craton was a low-relief landscape.

During this last phase of regression (Early Devonian), the craton underwent a period of mild deformation resulting in many domes, arches, and basins. Between the regression of the Tippecanoe Sea and the transgression of the Kaskaskia Sea, the rising arches and subsiding basins were eroded. Their truncated edges were submerged by the succeeding transgressing Kaskaskia Sea, producing a regional unconformity that forms the boundary between the two sequences.

## THE APPALACHIAN MOBILE BELT AND THE TACONIC OROGENY

Having examined the Sauk and Tippecanoe geologic history of the craton, we turn our attention to the Appalachian mobile belt, where the first orogeny began during the Middle Ordovician.

Throughout Sauk time, the Appalachian region was a broad, structurally passive, continental margin. Sedimentation was closely balanced by subsidence as thick,

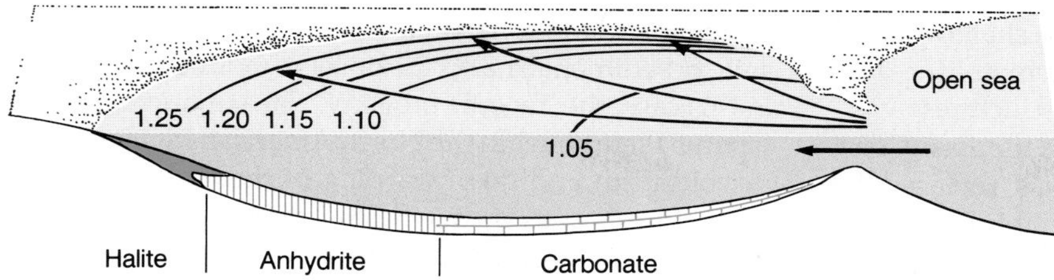


**Figure 10–15** A model for deposition of evaporites in a closed basin. Water removed from the basin by evaporation is replenished by seawater from the open sea through narrow inlets or over a sill. As the seawater in the basin becomes more concentrated by evaporation, the less soluble minerals precipitate, followed by the more soluble minerals. Thus, the evaporites form a succession of facies across the basin sea floor.

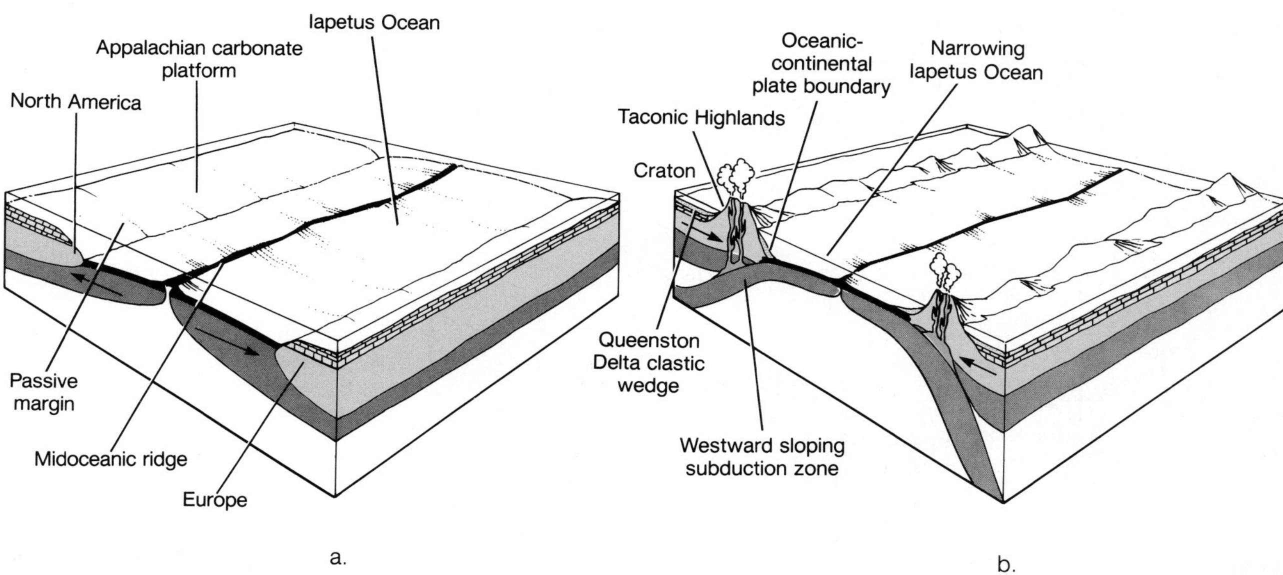


**Figure 10–16** Evolution of the Appalachian mobile belt from the Late Proterozoic to Late Ordovician. a. During the Late Proterozoic to Early Ordovician, the Iapetus Ocean was opening up along a divergent plate boundary. Both the east coast of North America and the west coast of Europe were passive plate margins where large carbonate platforms existed. b. Beginning in the Middle Ordovician, the North American and European passive margins became oceanic-continental plate boundaries resulting in orogenic activity.

shallow-marine sands were succeeded by extensive carbonate deposits. During this time, the **Iapetus Ocean** was widening as a result of movement along a divergent plate boundary (Fig. 10–16).

Beginning with the subduction of the Iapetus plate beneath North America (an oceanic-continental convergent plate boundary—see Chapter 6), the Appalachian mobile belt was born (Fig. 10–16). The resulting **Taconic orogeny,** named after the present-day Taconic Mountains of eastern New York, central Massachusetts, and Vermont, is the first of several orogenies to affect the Appalachian region.

Obviously, any record of mountain building is complex, and we cannot go into all the details or subdivisions in a book such as this. However, we can provide an overview of the event and explain its underlying causes in a plate tectonic framework as well as describe the evidence used by geologists to reconstruct such events.

The Appalachian mobile belt can be divided into two depositional environments. The first is the extensive, shallow-water carbonate platform that formed the eastern continental shelf and stretched from Newfoundland to Alabama. It represents the Sauk Sea transgression onto the craton, creating a large, shallow sea where carbonates accumulated (Fig. 10–17a). The shallow-water depth on the platform is indicated by stromatolites, desiccation cracks, and other sedimentary structures.

Carbonate deposition ceased along the East Coast during the Middle Ordovician and was replaced by deep-water deposits characterized by thinly bedded black shales, graded beds, coarse sandstones, graywackes, and associated volcanics (Fig. 10–17b). This suite of sediments marks the onset of mountain building, in this case, the Taconic orogeny. The subduction of the Iapetus Plate beneath the North American Plate resulted in volcanism and downwarping of the carbonate platform, forming an area where sediments accumulated (Fig. 10–17c). Throughout the Appalachian mobile belt,

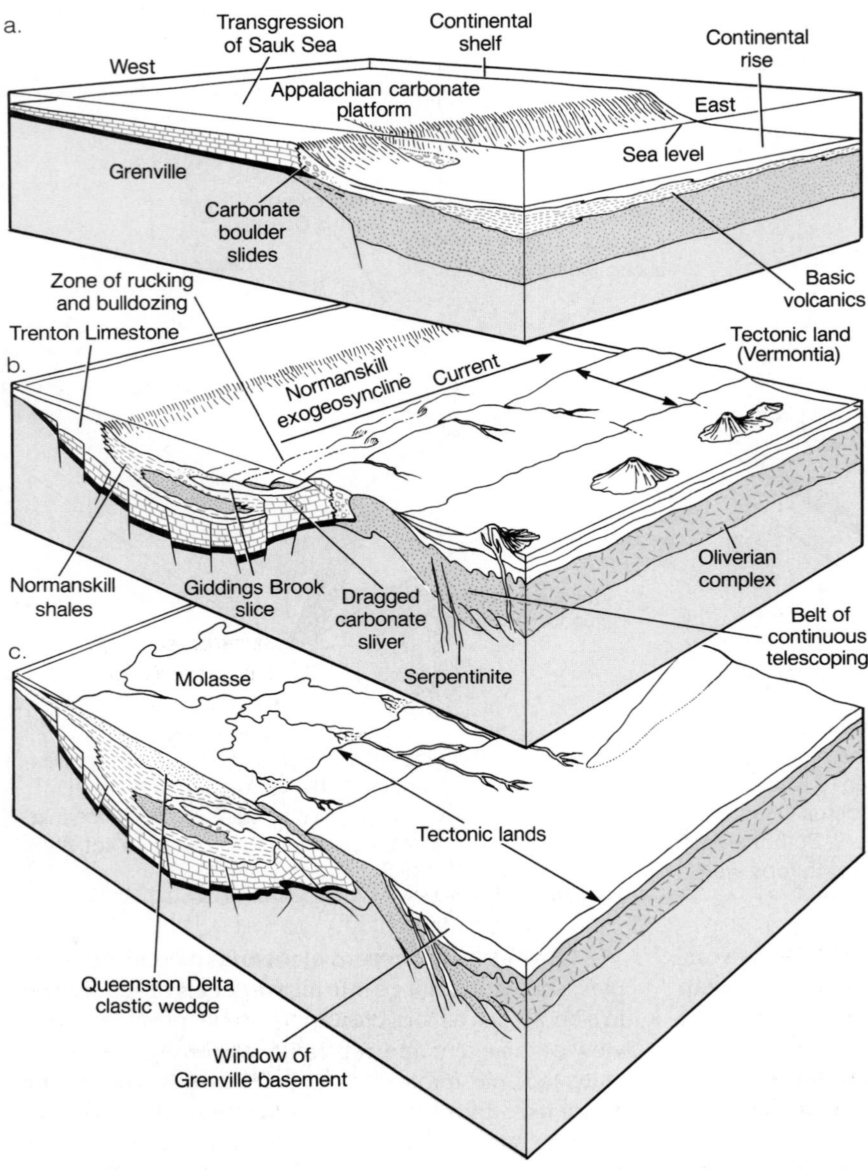


**Figure 10–17** Evolution of the Taconic orogeny along the continental margin of North America in Western New England. a. Pre-Taconic. b. Early phases of the Taconic orogeny. c. Late phases of the Taconic orogeny.

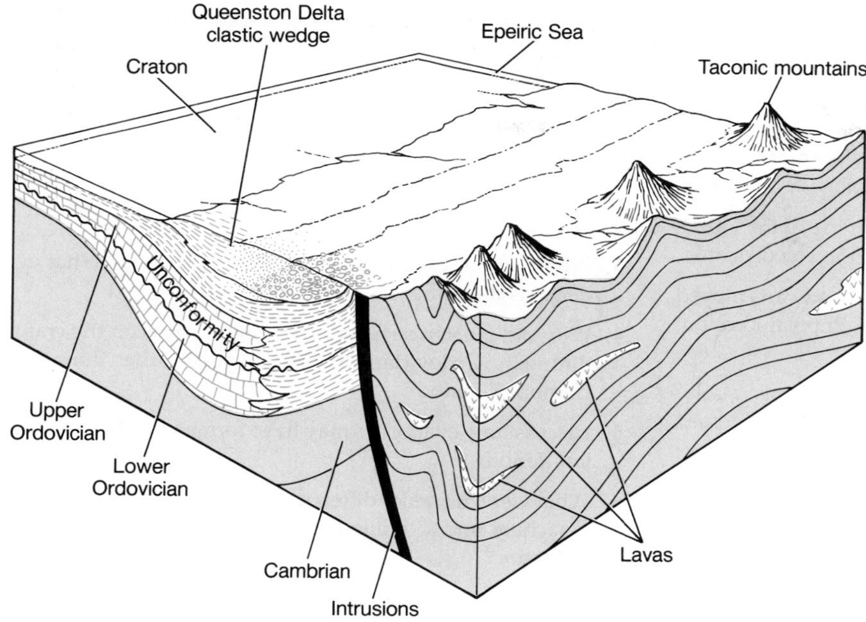


**Figure 10–18** Reconstruction of the Taconic Highlands and Queenston Delta clastic wedge. The clastic wedge consists of thick coarse-grained detrital sediments nearest the Highlands and thins laterally into finer-grained sediments on the craton.

facies patterns, paleocurrent, and sedimentary structures all indicate that these deposits were derived from the east, where the Taconic Highlands and associated volcanoes were rising (Fig. 10–17c).

Additional structural, stratigraphic, petrologic, and sedimentologic evidence has provided much information on the timing and origin of this orogeny. For example, at many locations within the Taconic belt, pronounced angular unconformities occur where steeply dipping Lower Ordovician rocks are overlain by gently dipping or horizontal Silurian and younger rocks.

Other evidence includes volcanic activity in the form of deep-sea lava flows, volcanic ash layers, and intrusive bodies in the area from present-day Georgia to Newfoundland. These igneous rocks show a clustering of radiometric ages between 480 to 440 m.y. ago. In addition, regional metamorphism coincides with the radiometric dates.

The final piece of evidence for the Taconic orogeny is the development of a large **clastic wedge**, an extensive accumulation of mostly clastic sediments that are deposited adjacent to an uplifted area. These deposits are thickest and coarsest adjacent to the highland area and become thinner and finer-grained away from the source area, eventually passing into the carbonate cratonic facies (Fig. 10–18). The clastic wedge resulting from the erosion of the Taconic Highlands is referred to as the **Queenston Delta**. Careful mapping and correlation of these clastic deposits indicates that more than 600,000 cubic km of rock were eroded from the Taconic Highlands. Based on this figure, geologists estimate the Taconic Highlands were at least 4,000 m high.

The Taconic orogeny marked the first pulse of mountain building in the Appalachian mobile belt and was a response to the subduction taking place beneath the east coast of North America. As the Iapetus Ocean narrowed and closed, another orogeny occurred in Europe during the Silurian. The Caledonian orogeny was essentially a mirror image of the Taconic orogeny and the Acadian orogeny (Chapter 11) and was part of the global mountain- building episode that occurred during the Paleozoic Era. Even though the Caledonian orogeny occurred during Tippecanoe time, we will discuss it in the next chapter because it was intimately related to the Acadian orogeny.

## CHAPTER SUMMARY

Table 10-1 provides a summary of the geologic history of the North American craton and mobile belt as well as global events and sea-level changes for the Early Paleozoic.

1. Six major continents existed at the beginning of the Paleozoic Era; four of them were located near the paleoequator.
2. While Laurentia was moving northward, Gondwana moved nearly halfway around the world to a south polar location, as indicated by tillite deposits.
3. Most continents consist of two major components: a relatively stable craton over which epeiric seas transgressed and regressed, surrounded by mobile belts in which mountain building took place.
4. The geologic history of North America can be divided into cratonic sequences that reflect craton-wide transgressions and regressions. The Sauk and Tippecanoe sequences were deposited during the latest Proterozoic to Early Devonian.
5. The first major transgression onto the craton was by the Sauk Sea. At its peak it covered the craton except for a series of large, northeast-southwest trending islands called the Transcontinental Arch.
6. The Tippecanoe sequence began with deposition of an extensive sandstone over the exposed and eroded Sauk landscape. During Tippecanoe time, extensive carbonate deposition occurred. In addition, large barrier reefs enclosed basins, resulting in evaporite deposition within these basins.
7. The eastern edge of North America was a stable carbonate platform during Sauk time. During Tippecanoe time an oceanic-continental convergent plate boundary formed, resulting in the Taconic orogeny, the first of several orogenies to affect the Appalachian mobile belt.
8. The newly formed Taconic Highlands shed sediments into the western epeiric sea, producing a clastic wedge called the Queenston Delta.

## IMPORTANT TERMS

| | |
|---|---|
| Appalachian mobile belt | Kazakhstania |
| Baltica | Laurentia |
| China | mobile belt |
| clastic wedge | organic reef |
| Cordilleran mobile belt | Ouachita mobile belt |
| craton | Queenston delta |
| cratonic sequence | Sauk sequence |
| epeiric sea | Siberia |
| Franklin mobile belt | Taconic orogeny |
| Gondwana | Tippecanoe sequence |
| Iapetus Ocean | Transcontinental Arch |

## REVIEW QUESTIONS

1. How can geologists determine the locations of continents for the Paleozoic Era?
2. Why are cratonic sequences a convenient way to study the geologic history of the Paleozoic Era?
3. Where was the Transcontinental Arch located? What evidence is there that it existed?
4. Why did greater carbonate deposition occur on the craton during the Ordovician and Silurian periods than during the Cambrian Period?
5. Discuss how evaporites may have formed during the Silurian Period?
6. What were the major differences between the Appalachian, Ouachita, and Cordilleran mobile belts during the Early Paleozoic?
7. What evidence is there in the rock record that the Taconic orogeny occurred?
8. What evidence indicates that the Iapetus Ocean began closing during the Middle Ordovician?

## ADDITIONAL READINGS

Bambach, R. K., C. R. Scotese, and A. M. Ziegler. 1980. Before Pangaea: The Geographies of the Paleozoic World. *American Scientist,* 68, no. 1: p. 26–38.

Burton, D. L., ed. 1971. *Cambrian of the New World.* London: Wiley-Interscience.

Stearn, C. W., R. L. Carroll, and T. H. Clark. 1979. *The Geological Evolution of North America, 3d ed.* New York: Ronald Press.

Stewart, J. H., C. H. Stevens, and A. E. Fritsche, eds. 1977. *Paleozoic Paleogeography of the Western United States.* Pacific Coast Paleogeography Symposium 1. Society of Economic Paleontologists and Mineralogists, Tulsa, Oklahoma.

**Table 10–1** Summary of Early Paleozoic events.

| Age (millions of years) | Geologic Period | Sequence | Changes of sea level in meters (Rising 300 – 0 – 300 Falling) | Cordilleran mobile belt | Craton | Ouachita mobile belt | Appalachian mobile belt | Major events outside North America |
|---|---|---|---|---|---|---|---|---|
| 408–438 | Silurian | Tippecanoe | | | Extensive barrier reefs and evaporites common. | | | Caledonian orogeny |
| 438–505 | Ordovician | Tippecanoe / Sauk | | | Queenston Delta clastic wedge.<br>Transgression of Tippecanoe Sea.<br>Regression exposing large areas to erosion. | | Taconic orogeny | Continental glaciation in Southern Hemisphere. |
| 505–570 | Cambrian | Sauk | Present Sea Level | | Canadian Shield and Transcontinental Arch only areas above sea level.<br>Transgression of Sauk Sea. | | | |

# 11

## CHAPTER OUTLINE



## PROLOGUE

*The unusual state of preservation of the Pennsylvanian-aged Mazon Creek biota of northeastern Illinois provides a significant insight regarding the soft-part anatomy of organisms rarely preserved in the fossil record. The biota is divided into marine and nonmarine components and contains the only known or oldest fossil representatives of several major animal groups.*

*The Mazon Creek fossils occur in spheroidal to elliptical shaped iron-carbonate concretions ranging from 1 to 30 cm long. Rapid burial and the formation of concretions around the organisms were primarily responsible for the excellent preservation of the organisms. Not only are the hard parts of organisms preserved, but also impressions and carbonaceous films of the soft-bodied animals, as well as plants. More than 320 species of animals and 350 species of plants have been described from this classic fossil assemblage.*

*The environment that these plants and animals lived in was a large delta in which sluggish southward-flowing rivers emptied into a subtropical epeiric sea that covered most of the present state of Illinois. Two major habitats are represented: a swampy forested lowland of the subaerial delta, and the shallow marine environment of the actively prograding delta.*

*A diverse marine fauna lived in the warm, shallow waters of the delta front and included coelenterates, mollusks, echinoderms, arthropods, worms and fish. From this fauna have come the only known fossils of several animal groups, including the lamprey, whose gills and liver can be discerned in the impressions.*

*More than 350 species of plants lived in the swampy lowlands surrounding the delta. Almost all of the plants were seedless vascular plants, typical of the kinds that comprised the Pennsylvanian coal-forming swamps of North America. Also found in the swampy lowlands were numerous insects, including millipedes and centipedes, as well as spiders and other animals such as scorpions and amphibians. In the ponds, lakes, and rivers were many fish, shrimp, horseshoe crabs, and ostracodes.*

# Geology of the Late Paleozoic Era

*Collecting of the Mazon Creek biota began in the mid-1800s by local farmers and townspeople. This was soon followed by scientific studies by such famous geologists and paleontologists as J. D. Dana, L. Lesquereux, and E. D. Cope. Study of the Mazon Creek fossils has been uneven through the years, with much of it concentrating on the descriptions of the many plant and animal species recovered. Today's research is still largely concerned with describing the complete assemblage, but also with determining the phylogenetic relationships of the organisms and the paleoecologic significance of the assemblage.*

a. b.
Seedless vascular plant fossils in iron-carbonate concretions from Pennsylvanian deposits, Mazon Creek locality, Will County, Illinois. a. *Neuropteris.* b. *Annularia.*

## INTRODUCTION

The Devonian, Carboniferous (Mississippian, Pennsylvanian), and Permian periods comprise the Late Paleozoic Era. During this time, the separate continents assembled to form the supercontinent **Pangaea**. This assemblage was the result of collisions along convergent plate boundaries. Coals, evaporites, and tillites record the varied climates that existed during this time period. Major glacial-interglacial intervals occurred over much of Gondwana during the Late Mississippian to Early Permian. The growth and retreat of these glaciers profoundly affected the world's biota as well as contributing to global sea-level changes. Additionally, mountain building resulting from continental collisions strongly influenced ocean and atmospheric circulation patterns.

## LATE PALEOZOIC GLOBAL GEOGRAPHY AND CLIMATE

The Late Paleozoic was a time of continental collisions, mountain building, fluctuating sea levels, and varied climates. These events resulted from plate interaction and major continental ice-sheet advances and retreats. We will initially examine the changing paleogeography and climates of the Late Paleozoic, and then place North American events into this global framework.

### The Devonian Period

By the Devonian Period, Baltica and Laurentia collided, forming the larger paleocontinent Laurasia (Fig. 11–1a). This collision initially caused the **Caledonian orogeny** in northwestern Baltica along a convergent plate boundary and continued with the **Acadian orogeny** in eastern Laurentia (northern Appalachian mobile belt). The resulting highlands were located in the equatorial zone, and vast amounts of reddish nonmarine fluvial sediments were eroded from them. These sediments covered large areas of northern Europe (Old Red Sandstone) and eastern North America (the Catskill Delta). Other Devonian tectonic events, probably related to the collision of Laurentia and Baltica, include the **Antler orogeny** in the Cordilleran mobile belt, the **Ellesmere orogeny** along the northern margin of Laurentia, and the change from a passive to an active compressive plate margin in the Uralian mobile belt of eastern Baltica. The

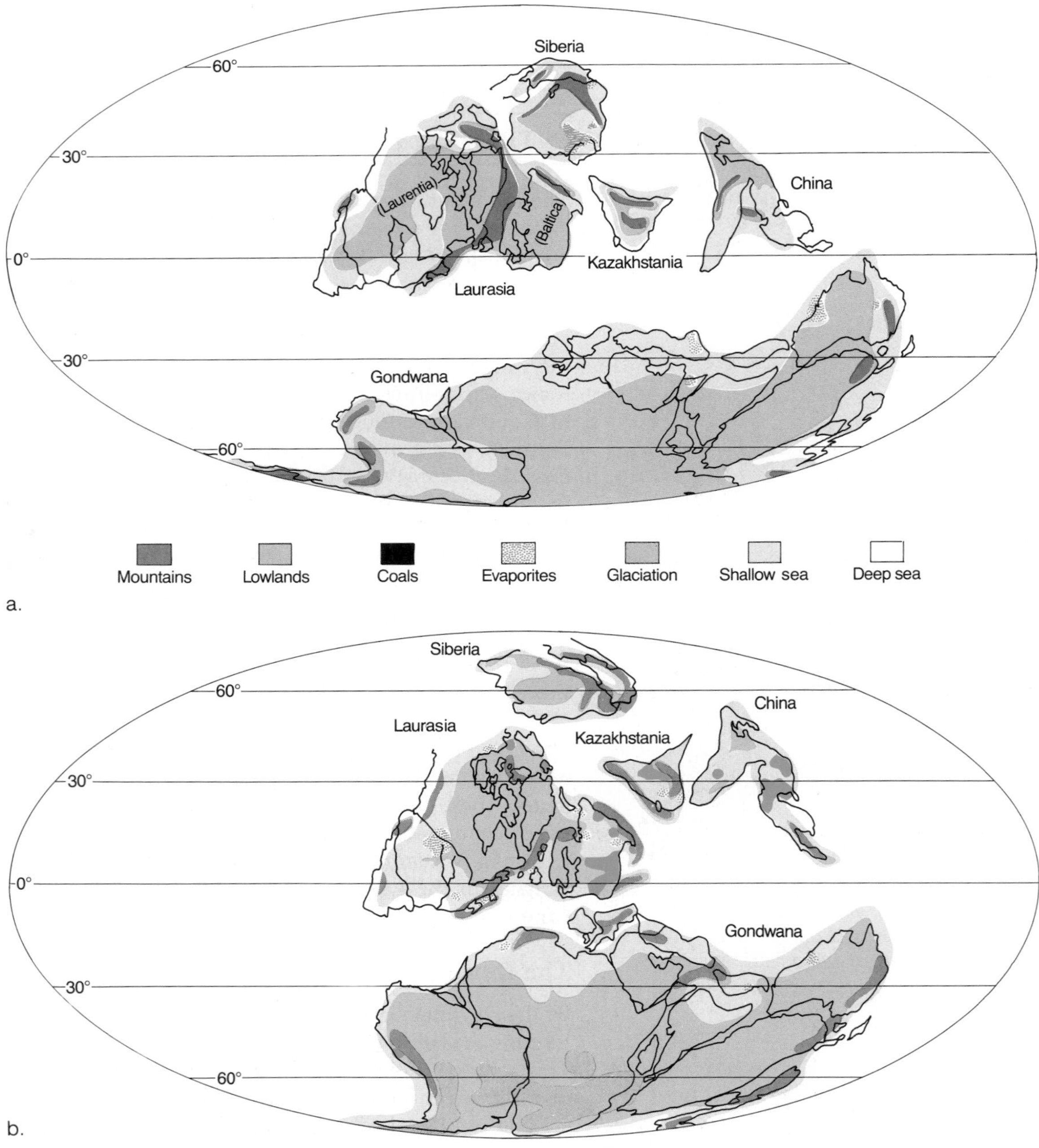


**Figure 11–1** Paleogeography of the world for the (a) late Early Devonian Period and (b) middle Early Carboniferous Period.

distribution of reefs, evaporites, and red beds, as well as similar land plants the world over, suggests the climates were rather uniform throughout the world during the Devonian Period.

## The Carboniferous Period

During the Carboniferous Period, Gondwana moved over the South Pole, resulting in extensive continental glaciation (Figs. 11–1b and 11–2a). The advance and re-

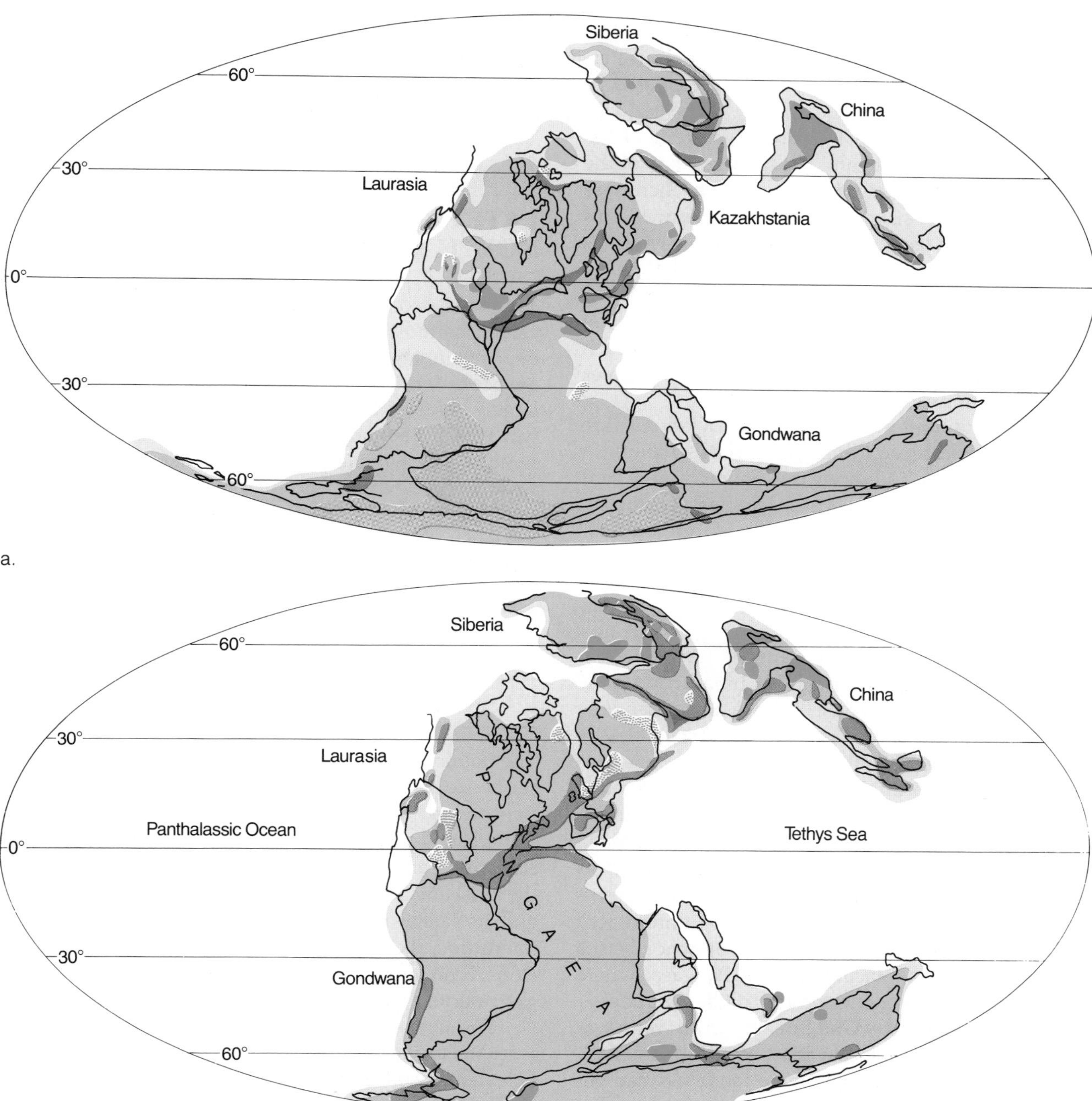


**Figure 11–2** Paleogeography of the world for the (a) middle Late Carboniferous Period and (b) early Late Permian Period.

treat of these glaciers produced global sea-level changes that affected sedimentation patterns on the cratons. As Gondwana continued to rotate in a clockwise fashion and move northward, it collided with southern Laurasia during the Late Carboniferous, as indicated by tectonism in the Ouachita, southern Appalachian, and Hercynian mobile belts. Elsewhere, China and Kazakhstania collided with Siberia. By the end of the Carboniferous Period, all the paleocontinents were tightly clustered as Pangaea began to take shape.

The Carboniferous coal basins of eastern North America, western Europe, and the Donetz Basin of the USSR all lay in the equatorial zone, where there was high rainfall and consistently warm temperatures. This is indicated by the lack of strong seasonal growth rings in fossil plants from these coal basins. The fossil plants found in the coals of Siberia and China, however, show well-developed growth rings, signifying seasonal growth with abundant rainfall and distinct seasons such as occur in the temperature zones (latitudes 40 degrees to 60 degrees north).

Glacial conditions in the high southern latitudes are indicated by widespread tillites and glacial striations in southern Gondwana, indicating movement of large continental ice sheets (Fig. 6–4). These ice sheets spread toward the equator and, at their maximum growth, extended well into the middle temperate latitudes.

### The Permian Period

With the collision between Laurasia and Siberia-Kazakhstania, and the movement of China farther to the northwest, the assemblage of Pangaea was essentially complete by the end of the Permian Period (Fig. 11–2b). Surrounding this supercontinent was an enormous, single ocean called **Panthalassa** that not only extended from pole to pole, but covered 300 degrees of longitude at the equator. Waters of this ocean must have circulated much more freely than today, resulting in more equitable water temperatures. For example, equatorial currents driven by trade winds would have flowed uninterrupted around five-sixths of the Earth's circumference before impinging on the east-facing coast of Pangaea, where they would have been deflected into higher latitudes—much as the modern- day Gulf Stream carries warm waters northward along the east coast of North America.

The formation of a single large landmass had climatic consequences for the terrestrial environment as well. Terrestrial Permian sediments indicate that arid and semi-arid conditions were widespread over Pangaea. The mountain ranges produced by the **Ouachita, Alleghenian, and Hercynian orogenies** were high enough to create a rain shadow that blocked the moist, subtropical, easterly winds—much as the southern Andes mountains do in western South America today. This produced very dry conditions in North America and Europe, as evident from the extensive Permian evaporites found in western North America, central Europe, and parts of Russia. Permian coals, indicative of abundant rainfall, were mostly limited to the northern temperate belts (latitude 40 degrees to 60 degrees north), while the last remnants of the Carboniferous ice sheets retreated to the mountainous regions of eastern Australia.

## LATE PALEOZOIC HISTORY OF NORTH AMERICA

The Late Paleozoic cratonic history of North America included periods of extensive shallow-marine carbonate deposition, large coal-forming swamps, as well as dry, evaporite-forming terrestrial conditions. Cratonic events largely resulted from sea-level changes due to Gondwanan glaciation and tectonic events related to the assemblage of Pangaea. Mountain building that began with the Ordovician Taconic orogeny continued with the Caledonian, Acadian, Ouachitan, and Alleghenian orogenies. These orogenies were part of the global tectonic process that resulted in the assemblage of Pangaea by the end of the Paleozoic Era.

## KASKASKIA SEQUENCE

The boundary between the Tippecanoe and **Kaskaskia sequence** is a major unconformity. As the Kaskaskia Sea transgressed over the craton during the late Early Devonian, the majority of the basal beds deposited consisted of clean, mature, quartz sandstones. Examples are the Oriskany Sandstone of New York and Pennsylvania and its lateral equivalents (Fig. 11–3). The Oriskany Sandstone, like the basal Tippecanoe St. Peter Sandstone, is an important glass sand as well as being a good gas-reservoir rock.

The source areas for the basal Kaskaskia sandstones were the eroding highlands of the Appalachian mobile belt area as well as the exhumed Cambrian and Ordovician sandstones along the margins of the Ozark Dome and of the Canadian Shield in Wisconsin. The complete absence of similar sands in the Silurian carbonate beds below the Tippecanoe-Kaskaskia unconformity, indicates these source areas were then submerged. Elsewhere on the craton, the basal units of the Kaskaskia

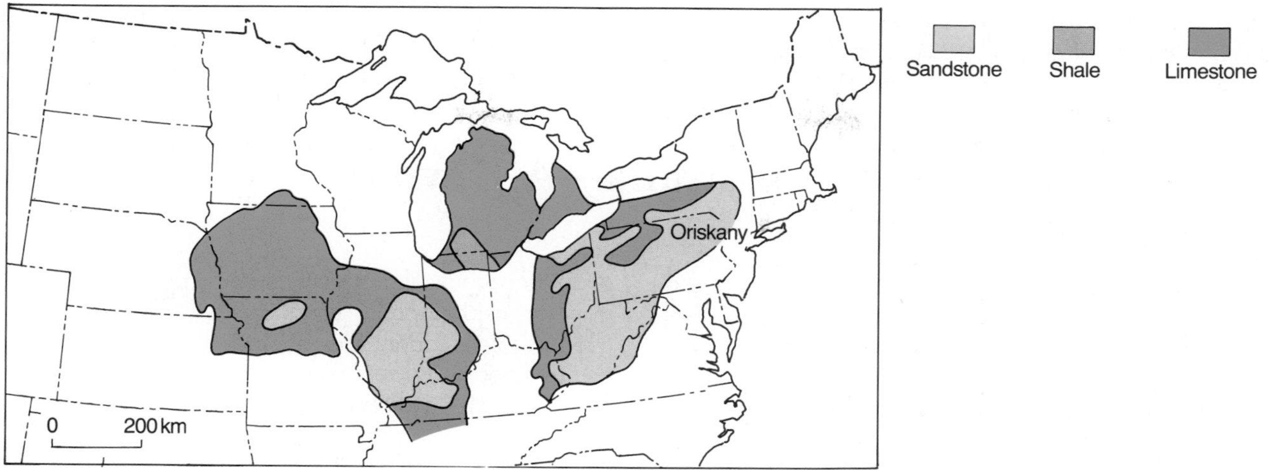


**Figure 11–3** Extent of the basal units of the Kaskaskia sequence in the eastern and north-central United States.

consist of carbonates that are frequently difficult to differentiate from the underlying Silurian carbonates unless they are fossiliferous.

Except for widespread Late Devonian and Early Mississippian black shales, the Kaskaskian rocks were mainly carbonates, reefs, and associated evaporite deposits. In many other parts of the world, such as southern England, Belgium, central Europe, Australia, and Russia, the Middle and early Late Devonian epochs were times of major reef building (see Perspective 11–1).

## Reef Development in Western Canada

The Middle and Late Devonian reefs of western Canada contain large reserves of petroleum and hence have been widely studied from outcrops and in the subsurface (Fig. 11–4). These reefs began forming as the Kaskaskia Sea transgressed southward in western Canada. By the end of the Middle Devonian, they had coalesced into a large barrier-reef system that restricted the flow of oceanic water into the back-reef platform, thus creating conditions for evaporite precipitation (Fig. 11–4). In the back reef, up to 300 m of evaporites were precipitated in much the same way as occurred in the Michigan Basin during the Silurian (Fig. 10–13). More than one-half of the world's potash, which is used in fertilizers, comes from these Devonian evaporites. By the middle of the Late Devonian, reef growth stopped in the western Canada region, although non-reef carbonate deposition still continued.

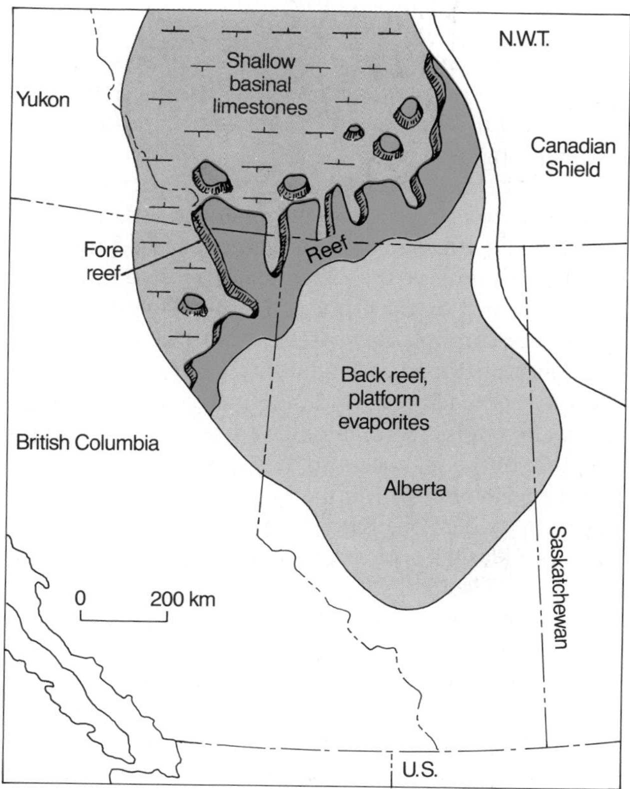


**Figure 11–4** Reconstruction of the extensive Devonian reef complex of western Canada. These extensive reefs controlled the regional facies of the Devonian epeiric seas.

## Perspective 11–1:

# The Canning Basin, Australia—A Devonian Great Barrier Reef

One of the largest and most spectacularly exposed fossil-reef complexes in the world is the Great Barrier Reef of the Canning Basin, Western Australia (Fig. 1). This barrier-reef complex developed during the Middle and Late Devonian Period when the Canning Basin was covered by a tropical epeiric sea (Fig. 11–1). The reefs are now exposed as limestone ridges that extend for some 350 km along the northern margin of the Canning Basin, but they probably continued around the present northern coastal region to join with similar reefs exposed in the Bonaparte Basin to the east.

The limestone reefs rise 50 to 100 m above the surrounding plains in much the same way they looked when the area was covered by the Devonian epeiric sea. The reason for this is that the shales and other soft sediments deposited in the open ocean in front of the reef complex (Fig. 10–11) have been eroded away, leaving the resistant limestone reefs standing as ridges.

The reefs themselves were constructed primarily by calcareous algae, stromatoporoids, and tabulate and rugose corals, which comprised the main components of the other major reef complexes in the world at that time. An interesting feature of these Canning Basin reefs is the contribution of column-shaped stromatolites which are found in the reef, back-reef, and marginal-slope areas of the reef complex. Stromatolites are an unusual component of the reef because they ceased to be an abundant element of marine faunas by the end of the Proterozoic Eon. Throughout the Phanerozoic Eon, stromatolites typically formed only in areas generally inhospitable to animals.

The outcrop along Windjana Gorge beautifully reveals the various features and facies of the Devonian Great Barrier Reef complex (Fig. 2). The reef core consists of unbedded limestones comprised predominantely of the aforementioned calcareous algae, stromatoporoids, and corals. The back-reef facies (Fig. 10–11) is bedded and makes up the major part of the total reef complex environment. A diverse and abundant fauna of calcaeous algae, stromatoporoids, various corals, some bivalves, gastropods, cephalopods, brachiopods, and crinoids lived in this lagoonal area behind the reef core.

In front of the reef core was the steep forereef slope (Fig. 10–11) that supported some organisms, including algae, sponges, and stromatoporoids. This facies contains considerable reef talus, an accumulation of debris eroded by waves from the reef front. The ocean-basin deposits contain the fossils of mainly nektonic and planktonic organisms such as fish, radiolarians, cephalopods, and conodonts.

Near the end of the Late Devonian, nearly all the reef-building organisms as well as much of the associated fauna of the Canning Basin Great Barrier Reef became extinct. As we will discuss in Chapter 12, few massive tabulate-rugose-stromatoporoid reefs are known from latest Devonian or younger rocks.

**Figure 1** Aerial view of Windjana Gorge showing the Devonian Great Barrier Reef exposed as a limestone ridge.

**Figure 2** Outcrop of Devonian Great Barrier Reef along Windjana Gorge. The talus of the forereef area can be seen on the left side of the picture sloping away from the reef core which is unbedded. To the right of the reef core is the back-reef facies which is horizontally bedded. (Photos courtesy of Geoffrey Playford, University of Queensland, St. Lucia, Australia)

a.

## Black Shales

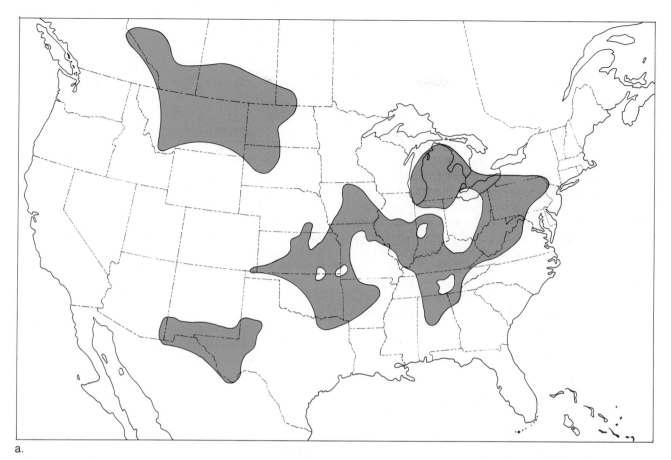

In North America, many areas of carbonate-evaporite deposition gave way to a greater proportion of shales and coarser clastics beginning in the Middle Devonian and continuing into the Late Devonian. This change to clastic deposition resulted from the formation of new source areas brought on by the mountain-building activity associated with the Acadian orogeny in North America.

As the Devonian Period ended, a marked change in sedimentation took place over the craton with the appearance of a very persistent black shale. This black shale is commonly called the Chattanooga Shale in the eastern United States but is known by a variety of local names elsewhere. While it is best developed from the Appalachian mobile belt to the Mississippi Valley, its correlatives can also be found in many western states and in Alberta, Canada (Fig. 11–5).

The Devonian and Mississippian epeiric black shales of North America are typically noncalcareous, thinly bedded, and usually less than 10 m thick (Fig. 11–5). If they are rich in land-plant remains, they are commonly

b.

**Figure 11–5** a. The extent of the Upper Devonian to Lower Mississippian Chattanooga Shale and its equivalent units in North America. b. Upper Devonian New Albany Formation, Button Mold Knob Quarry, Kentucky.

termed *carbonaceous*. If marine productivity was high, the shales are *bituminous* and contain mostly degraded algal material. Fossils are typically rare in the Devonian and Mississippian black shales, but some Devonian black shales do contain rich conodont faunas with large numbers of individuals. Because many black shales are relatively unfossiliferous, they are difficult to date and to correlate. In places where they can be dated, usually by conodonts (pelagic animals), acritarchs (pelagic algae), or plant spores, the lower beds are Late Devonian, and the upper beds are Early Mississippian in age.

While the origin of such extensive black shales is still being debated, the essential features required to produce them include undisturbed anaerobic bottom water, a reduced supply of clastic sediment, and high organic productivity in the overlying oxygenated waters. High productivity in the surface waters leads to a shower of organic material, which, decomposing on the undisturbed substrate, depletes the dissolved oxygen at the sediment-water interface. Several models have been proposed for the depositional environments of black shales, and all involve the creation of a quiet, anaerobic environment beneath an aerated one (Fig. 11–6).

The wide extent of such apparently shallow-water black shales in North America remains puzzling. Nonetheless, these shales are rich in uranium and are an important source rock of oil and gas in the Appalachian region.

### The Late Kaskaskia–A Return to Extensive Carbonate Deposition

Following deposition of these widespread black shales, carbonate sedimentation dominated the remainder of the Mississippian Period (Fig. 11–7). During this time, a variety of carbonate sediments were deposited in the epeiric sea, as indicated by the extensive deposits of crinoidal limestones (rich in crinoid fragments), oolitic limestones, and various other limestones (Fig. 11–8). These Mississippian limestones show sorting of the fossil fragments, cross-bedding, ripple marks, and scoured structures, all of which indicate shallow-water environments analogous to some of those observed on the present-day Bahama Banks. Finally, there were many small organic reefs growing throughout the craton that were all much smaller than the large barrier-reef complexes that dominated the earlier Paleozoic seas.

As the Kaskaskia Sea regressed from the craton during the Late Mississippian, probably in response to a growing Gondwanan ice sheet, large quantities of sand were deposited. The resulting sandstones, particularly in the Illinois Basin, have been extensively mapped and drilled, because they are excellent reservoirs for petroleum. With the retreat of the Kaskaskia Sea to the craton margin, the craton was once again exposed to erosion that resulted in one of the most widespread regional unconformities and karst topographies in the world.

## ABSAROKA SEQUENCE

The **Absaroka sequence** includes uppermost Mississippian through Lower Jurassic rocks. We will be concerned here with only the Paleozoic rocks of the Absaroka sequence. The extensive unconformity separating the Kaskaskia and Absaroka sequences serves to divide those strata equivalent to the European Carboniferous System into the North American Mississippian and Pennsylvanian systems. Not only are the rocks above and below the unconformity different, they were also products of quite different tectonic regimes resulting from plate collisions that sutured North America to western Europe and northwestern Africa.

The lowermost sediments of the Absaroka sequence are confined to the margins of the craton. In general, the deposits are thickest in the east and southeast, near the emerging highlands of the Appalachian and Ouachita mobile belts, and thin westward onto the craton. The lithologies also reveal lateral changes from nonmarine clastics and coals in the east, through transitional marine-nonmarine beds, to largely marine clastics and limestones farther west (Fig. 11–9).

### Cyclothems

One important aspect of these Pennsylvanian rocks is its repetitive nature, particularly the alternation of marine and nonmarine units, including coals. Such rhythmically, repetitive sedimentary sequences are called **cyclothems**. They result from numerous alternations of marine and nonmarine conditions, usually in areas of low relief. While seemingly simple, cyclothems are produced by a delicate interplay between nonmarine deltaic and shallow marine interdeltaic and shelf environments.

For purposes of illustration, we can look at a typical coal-bearing cyclothem from the Illinois Basin (Fig. 11–10. Such a cyclothem contains nonmarine units, capped by a coal, overlain by marine units. Figure 11–10 shows the environments of deposition that produce cyclothems. The initial units represent deltaic and fluvial deposits. Above them is an underclay that frequently contains root casts from the plants and trees that comprise

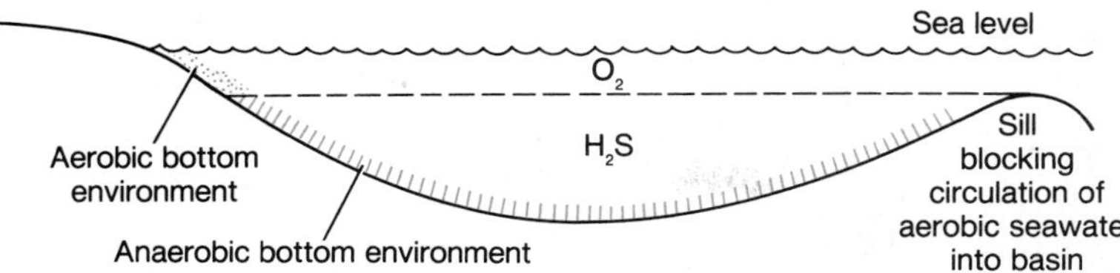


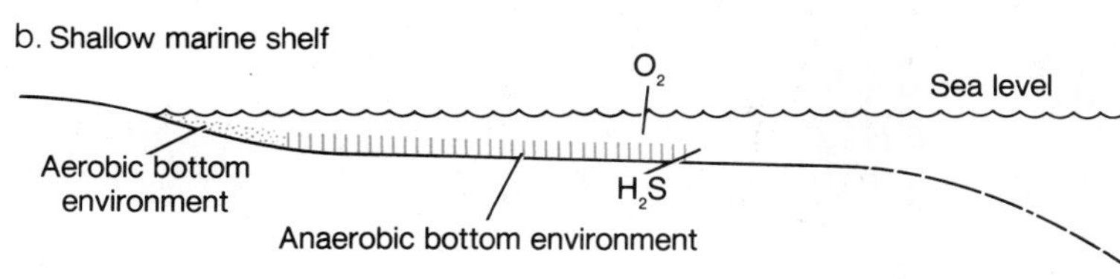


**Figure 11–6** Widespread deposition of black shales occurred during the Late Devonian to Early Mississippian and many models have been proposed for the depositional environments of these black shales. All of the models, however, involve a layer of quiet anaerobic water below an aerated layer. a. Silled basin environment. b. Shallow marine shelf environment.

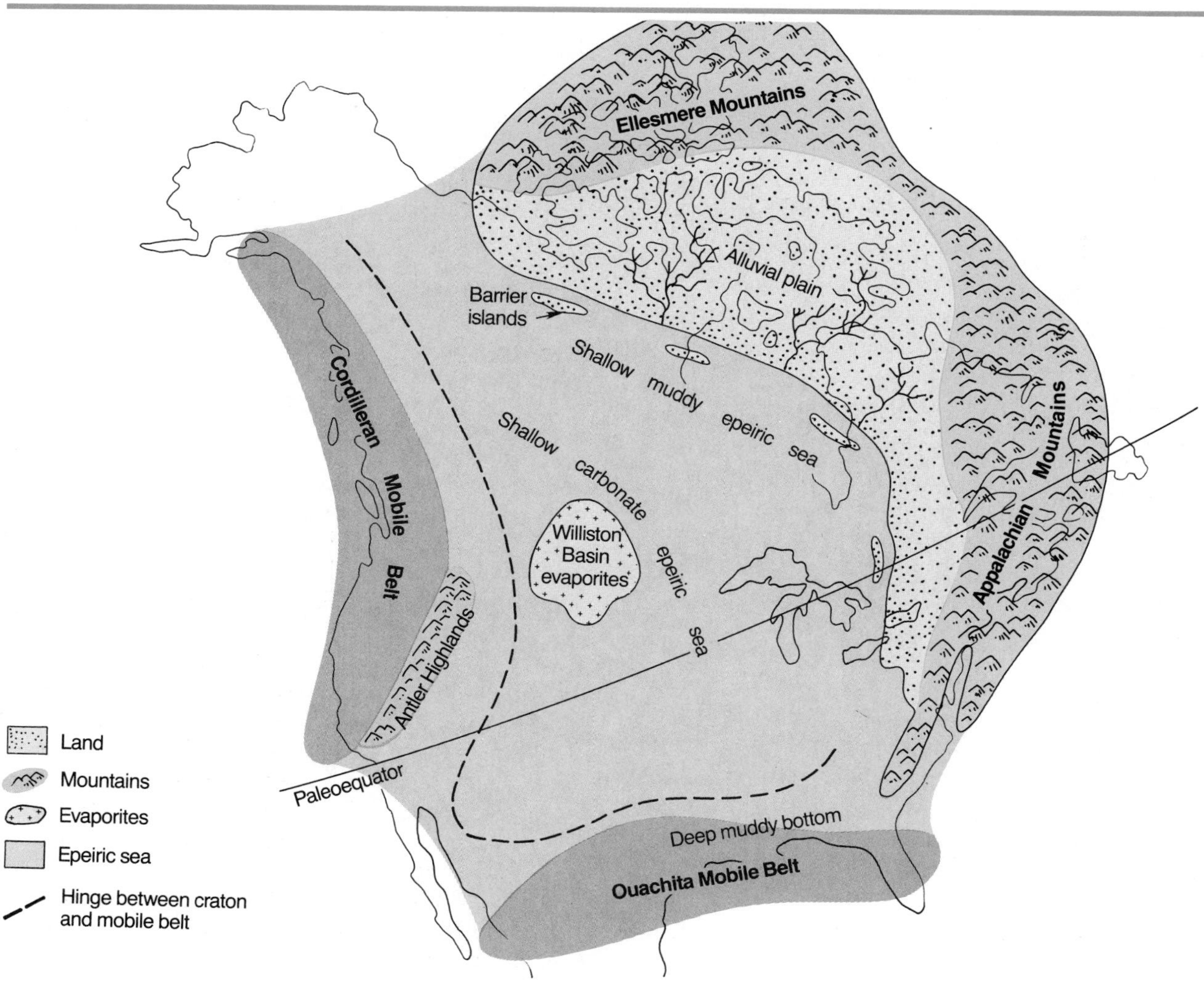

**Figure 11–7** Paleogeography of North America for the Mississippian Period.

a.

b.

**Figure 11–8** Mississippian limestones exposed near Bowling Green, Kentucky (a) and at Lewis and Clark Caverns, Montana (b). Extensive deposition of limestones occurrred in the Kaskaskia Sea. Many of these limestones contain sorted fossil fragments, cross-bedding, and ripple marks, all of which indicate deposition in shallow water.



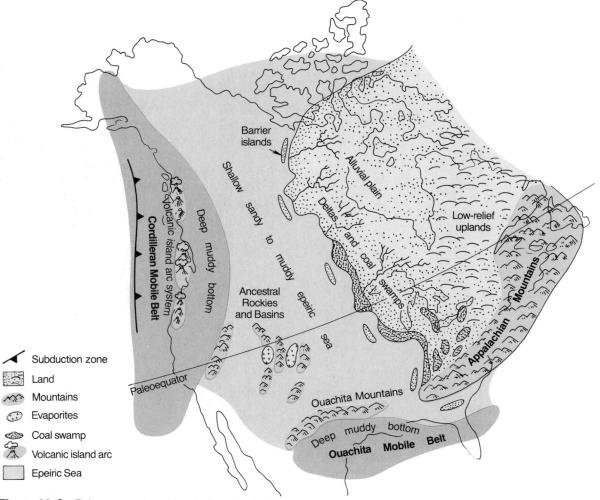

**Figure 11–9** Paleogeography of North America for the Pennsylvanian Period.

the overlying coal. The coal bed results from accumulations of plant material and is overlain by marine units of alternating limestones and shales, usually with an abundant marine invertebrate fauna. The marine cycle ends with an erosion surface. A new cyclothem begins with a nonmarine deltaic sandstone.

These cyclothems represent transgressive and regressive sequences with an erosional surface separating each cyclothem from the next. Thus, an idealized cyclothem passes upward from fluvial-deltaic deposits, through coals, to clastic shallow-water marine sediments, and finally to limestones, typical of an open marine environment, that represent the maximum transgression of the sea. It should be noted that Figure 11–10 represents ideal conditions. Frequently, all beds are not preserved because of abrupt changes from marine to

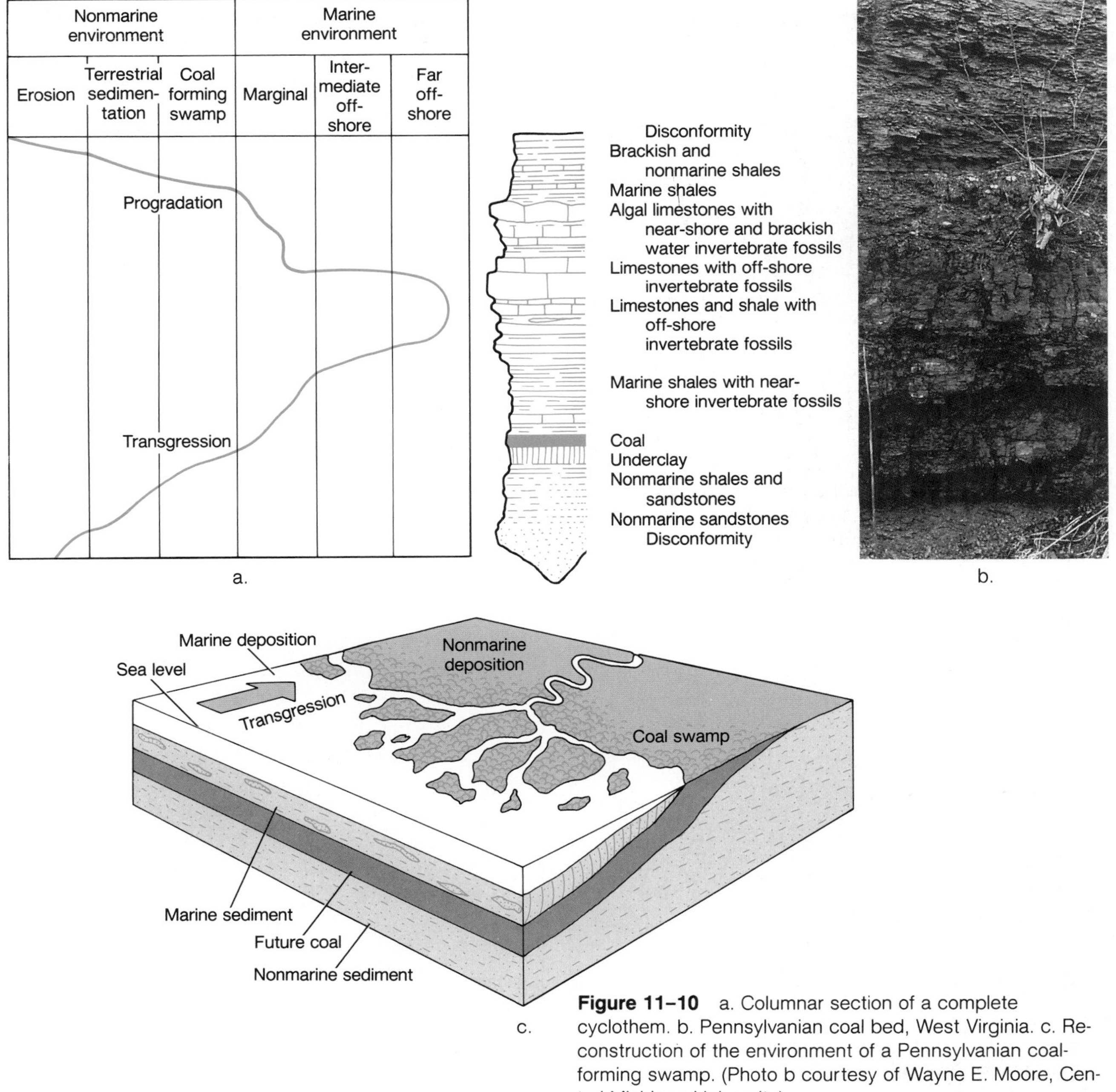


**Figure 11–10** a. Columnar section of a complete cyclothem. b. Pennsylvanian coal bed, West Virginia. c. Reconstruction of the environment of a Pennsylvanian coal-forming swamp. (Photo b courtesy of Wayne E. Moore, Central Michigan University)

nonmarine conditions or to removal of some units by erosion.

Such places as the Mississippi Delta and the Dutch lowlands represent modern coal-forming environments similar to those found in the Pennsylvanian System (Fig. 11–11). By studying these modern analogues, geologists can make reasonable deductions about conditions that existed in the geologic past.

The coal swamps that existed during the Pennsylvanian Period must have been large lowland areas neighboring the sea. In such cases, a very slight rise in sea level would have flooded large areas, while slight drops would have exposed large areas, resulting in alternating marine and nonmarine environments (Fig. 11–12). The same result could have been caused by rising sea level and progradation of a large delta, as occurs today in Louisiana (Fig. 11–11).

Such regularity and cyclicity in sedimentation over a large area requires an explanation, and the origin of the Pennsylvanian cyclothems has long been debated. In most cases, local cyclothems of limited extent can be explained by rapid but slight changes in sea level in a

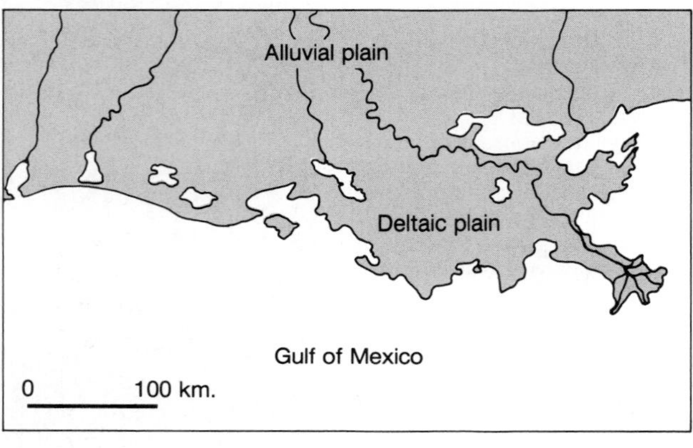


a.

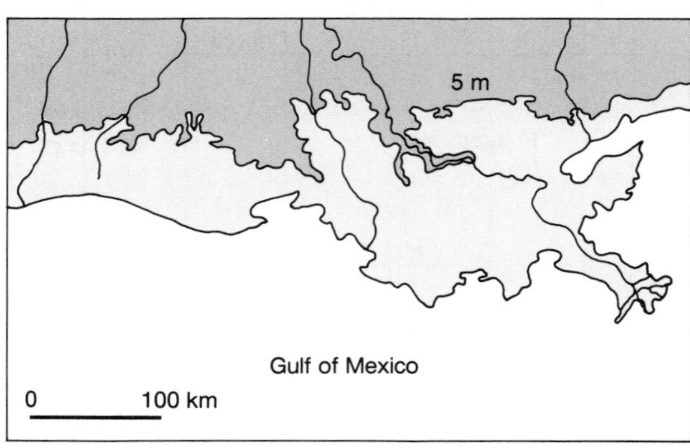


b.

**Figure 11–11** The Mississippi Delta and swampy alluvial plain is a modern example for certain Pennsylvanian deposits. a. The modern delta and surrounding swampy deltaic plain. b. The effect of a 5 m rise in sea level for comparison with the transgressive phase of a Pennsylvanian cyclothem.

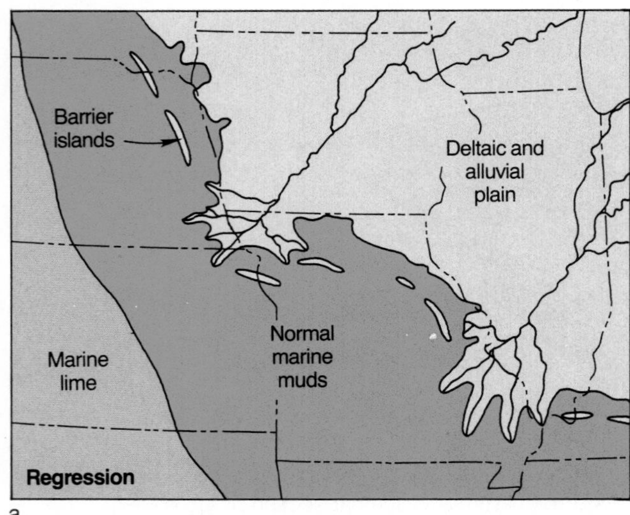


a.

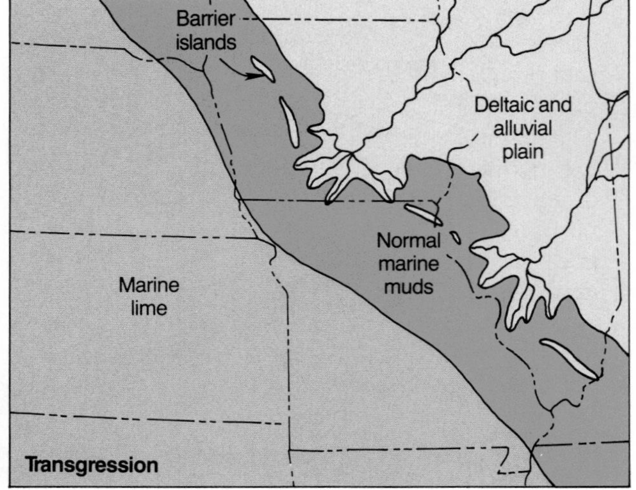


b.

**Figure 11–12** Fluctuations in environments during a regression (a) and transgression (b) for a typical Pennsylvanian cyclothem in the central United States. Compare this to the modern Mississippi Delta (Fig. 11–11).

## Perspective 11–2

# Brachiopod Migrations and Paleoclimatic Changes

An interesting study recently published by A. Raymond, P. H. Kelley, and C. B. Lutken,[1] relates articulate brachiopod migrations and changes in their latitudinal diversity gradient in the Northern Hemisphere to paleoclimatic changes brought about by the collision of Laurasia and Gondwana and the onset of continental glaciation in the Southern Hemisphere during the Carboniferous Period. This study traces the evolutionary history of articulate brachiopods, their migration patterns, and changes in the latitudinal diversity gradient to establish the timing and nature of paleoclimatic fluctuations during the Early and Middle Carboniferous.

It is well-known that today the diversity of organisms increases from the high latitudes to the equator. Such an increase results in a steep latitudinal diversity gradient. It is also well-established that when the climate was more equable from the pole to the equator in the past, the latitudinal diversity gradient was not as steep as it is today. Furthermore, it has been shown that when the environment changes, many organisms migrate to more suitable areas. This information was applied to an analysis of Carboniferous brachiopods to determine past climatic fluctuations.

The occurrences of 344 articulate brachiopod genera from nine stratigraphic intervals from the Early and Middle Carboniferous were tabulated. For purposes of determining migration patterns, the Northern Hemisphere was divided into four zones: high latitude (75 degrees to 50 degrees north latitude), middle latitude (35 degrees to 25 degrees north latitude), low latitude (25 degrees to 10 degrees north latitude), and equatorial (10 degrees to 0 degrees latitude).

Analysis of the data shows that during the late Early Carboniferous, 42 percent of all articulate brachiopod genera migrated north into a new zone. At the same time, the diversity of articulate brachiopods in the higher latitudes increased relative to the diversity of articulate brachiopods in the equatorial zone, producing a gentle latitudinal diversity gradient. The northern migration of articulate brachiopods and the gentle latitudinal diversity gradient indicate high-latitude warming. Early Carboniferous fossil land-plant distribution patterns corroborate the marine articulate brachiopod data, suggesting an Early Carboniferous terrestrial warming trend.

At the beginning of the Middle Carboniferous, 33 percent of the articulate brachiopod genera migrated into more southern zones, suggesting a high-latitude cooling trend. This southern migration continued during the Middle Carboniferous such that the latitudinal diversity gradient again steepened due to the loss of species in the high latitudes.

Having established that climatic changes occurred during the Early Carboniferous and Middle Carboniferous, let us consider what caused these changes. At the beginning of the Early Carboniferous, the east-west trending Tethys Sea separated Laurasia and Gondwana (Fig. 11–1b). A warm equatorial current encircled the globe and confined the warm waters to the equatorial region. By the end of the Early Carboniferous or during the Middle Carboniferous, Laurasia and Gondwana collided, forming Pangaea. At that time, the warm, equatorial current was deflected north and south along the eastern coast of Pangaea. Although the actual collision between Laurasia and Gondwana probably occurred during the Middle Carboniferous, significant amounts of warm water would have been deflected north and south as the Tethys Sea narrowed during the Early Carboniferous. Raymond, Kelley, and Lutken concluded that the deflection of warm waters north caused the high-latitude warming trend shown by the articulate brachiopod data.

The onset of continental glaciation in the Southern Hemisphere began in the Middle Carboniferous and was probably responsible for the high-latitude cooling indicated by the southward migration of articulate brachiopods and the Mid-Carboniferous marine mass extinction (see Chapter 12).

Analysis of the distribution of fossils such as articulate brachiopods provides geologists with an independent means of assessing paleoclimatic fluctuations and their possible causes, such as plate movements and continental glaciation.

[1]Raymond, A., P. H. Kelley, and C. B. Lutken. (In press). *Polar Glaciers and Life at the Equator: The History of Dinantian and Namaurian (Carboniferous) Climate.*

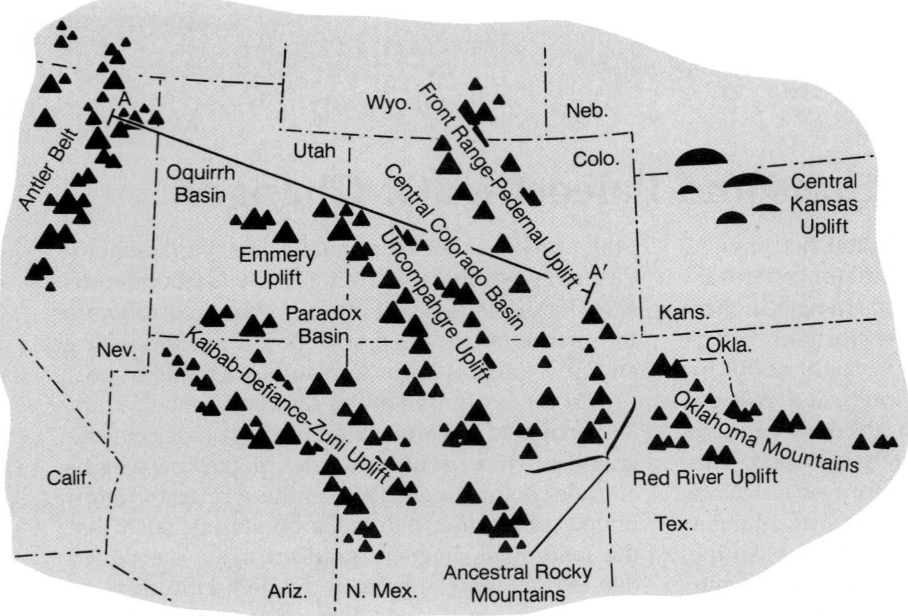


a.

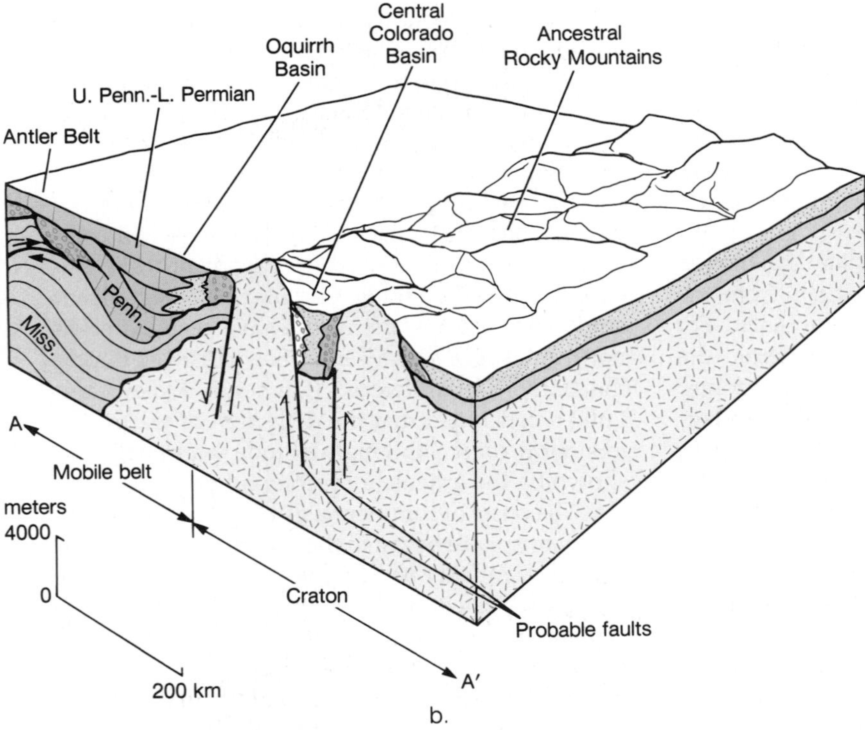


b.

**Figure 11–13** a. Location of the principal Pennsylvanian highland areas and basins of the southwestern part of the craton. b. Cross section of the Ancestral Rockies, which were elevated by faulting during the Pennsylvanian Period. Erosion of these mountains produced coarse, red-colored sediments that were deposited in the adjacent basins.

swamp-delta complex of low relief near the sea, or by localized crustal movement. Explaining widespread cyclothems is more difficult.

The hypothesis currently favored is a rise and fall of sea level related to advances and retreats of Gondwanan continental glaciers (see Perspective 11–2). When the Gondwanan ice sheets advanced, sea level dropped; and when they melted, sea level rose. Late Paleozoic cyclothem activity reasonably corresponds with Gondwanan glacial-interglacial cycles.

### The Late Absaroka—More Evaporite Deposits and Reefs

The Late Paleozoic was a time of mountain building in the Cordilleran, Ouachita, and Appalachian mobile belts. There was also uplift in the southwestern part of the craton, resulting in the **Ancestral Rockies** (Fig. 11–13). Ordinarily, cratonic areas are characterized by stability. However, the southwestern part of North America is an exception to this general rule. We will discuss the mobile belt activity in more detail later in this chapter; but for now, note that the uplift of those regions resulted in a regression of the Absaroka Sea, and by the Permian Period, the formerly extensive Paleozoic epeiric sea was restricted to parts of western North America.

The basins adjacent to the Ancestral Rockies contain both marine sediments and evaporites. The Paradox Basin, for example, is especially interesting because it provides a good example of the history of the southwestern part of North America (Figure 11–13).

This area was inundated during Early Pennsylvanian time by the Absaroka transgression. These marine deposits, predominantly black shales, were deposited over a karst topography developed on Upper Mississippian limestones. By the Middle Pennsylvanian, the sea had become restricted at its western access, resulting in cyclical deposits of salt, anhydrite, and gypsum. Fossiliferous and oolitic limestones were deposited around the periphery of the basin, while barrier and patch reefs grew abundantly along the margins. By the end of the Pennsylvanian Period, the Paradox Basin was filled mainly with arkosic sediments from the recently uplifted Uncompahgre Highlands, which formed part of the Ancestral Rockies (Fig. 11–13).

The Absaroka Sea occupied a narrow zone of the south central craton by the Early Permian (Fig. 11–14). Fossiliferous limestones, characteristic of Upper Pennsylvanian deposits, are overlain by Lower Permian unfossiliferous shales, red beds, and evaporites. The thick salt beds in Kansas and Oklahoma are evidence of the restricted nature of the sea and the evaporation that occurred when the seas retreated from the central craton.

The last deposits of the Paleozoic epeiric seas occurred in the western part of Texas and southeastern New Mexico where a remarkable sequence of interrelated lagoonal, reef, and open-shelf sediments formed (Fig. 11–15). In this region, three irregularly subsiding basins developed in and among shallowly submerged platforms. Limestones, shales, and sandstones were deposited in the basinal areas, while massive reefs developed around the basins' edges (Fig. 11–16). In the lagoonal areas behind the reefs, thin limestones, red beds, and evaporites formed. As the passageways between the basins became more restricted, evaporites gradually filled the individual basins with deposits up to 600 m thick.

Spectacular deposits that represent this history can be seen today in the Guadalupe Mountains of Texas and New Mexico where the Capitan Limestone forms the caprock of these mountains (Fig. 11–17). These reefs have been extensively studied because of the tremendous oil production that comes from this area.

By the end of the Permian Period, the Absaroka Sea had retreated from the craton, leaving continental red beds over much of the southwestern and eastern areas.

## HISTORY OF THE LATE PALEOZOIC MOBILE BELTS

Having examined the history of the craton for the Kaskaskia and Absaroka sequences, we now turn our attention to the orogenic activity in the mobile belts. The mountain building that occurred during this time had a profound influence on the climate and sedimentary history of the craton. In addition, it was part of the global tectonic regime that sutured the continents together, forming Pangaea by the end of the Paleozoic Era.

## CORDILLERAN MOBILE BELT

During the Late Proterozoic and Early Paleozoic, the Cordilleran area was a passive continental margin along which extensive continental shelf sequences developed. Thick marine sediments graded laterally into thin, cratonic units as the Sauk Sea transgressed onto the craton. Beginning in the Middle Paleozoic, an island arc (called the **Klamath Arc**) formed off the western margin

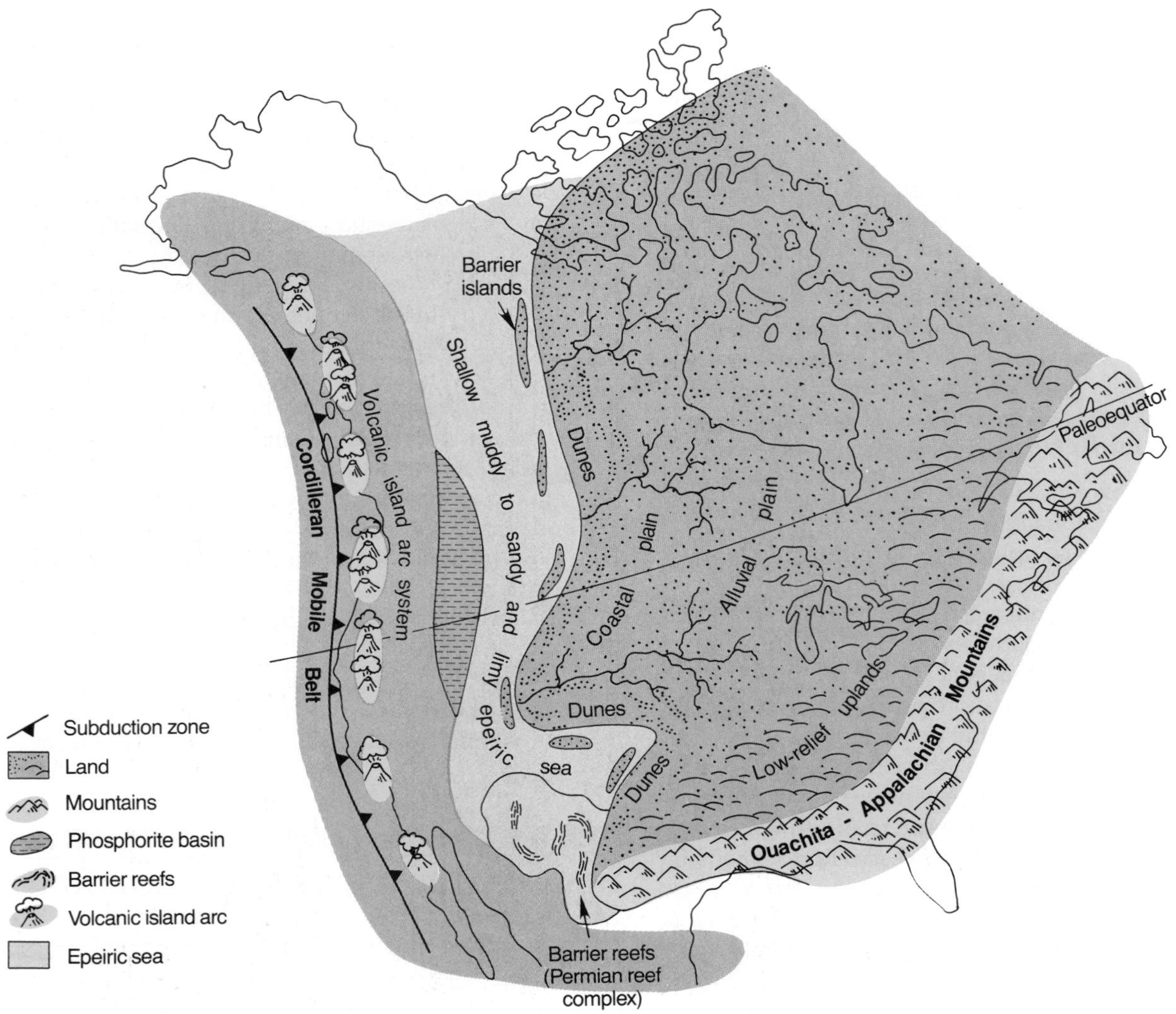


**Figure 11–14** Paleogeography of North America for the Permian Period.

of the craton (Fig. 11–18). During the Late Devonian and Early Mississippian, a collision between this eastward-moving island arc and the western border of the craton occurred (Fig. 11–19). In central Nevada, evidence of the collision can be seen along the Roberts Mountain thrust fault, where deep-water continental slope and rise deposits were thrust eastward as far as 160 km to come to rest overlying shallow-water continental shelf carbonates. Together, these deep-water and shallow-water rocks comprise a section more than 10 km thick (Fig.11–20).

As noted earlier, this orogenic event is called the **Antler orogeny**. It resulted from the closure of the basin between the Klamath Island Arc and the continent. The Antler orogeny was the first of a series of orogenic events to affect the Cordilleran mobile belt. During the Mesozoic and Cenozoic, this area was the site of major tectonic activity as a variety of terranes were accreted to the craton and a major mountain chain formed.

## CRATONIC UPLIFT—THE ANCESTRAL ROCKIES

During the Pennsylvanian Period, the uplift of the southwestern part of the North American craton resulted in several mountain ranges more than 1 km in elevation. These mountain ranges had diverse histories and did not all occur at the same time (Fig. 11–13). The group

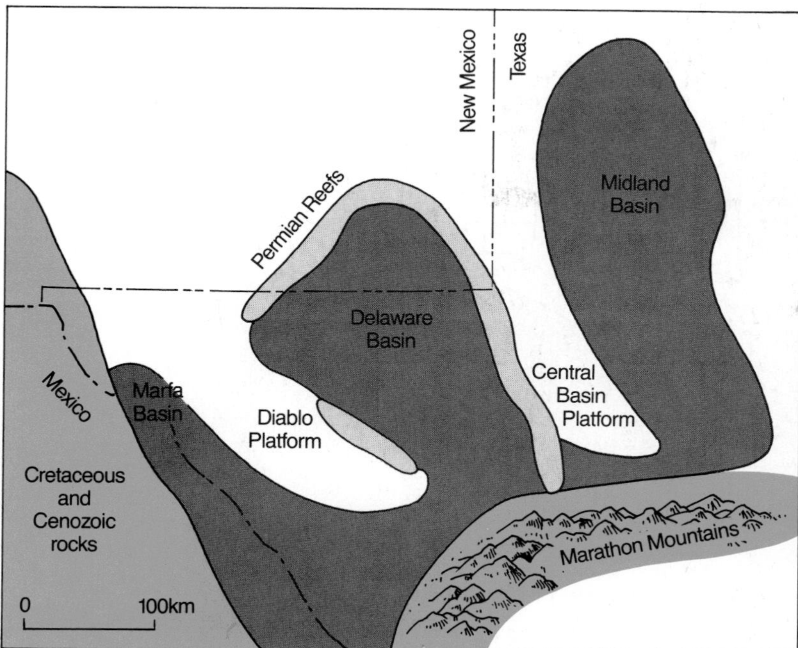


**Figure 11–15** Location of the West Texas Permian basins and surrounding reefs.

**Figure 11–16** Reconstruction of the Middle Permian Capitan Limestone reef environment. Shown are brachiopods, corals, bryozoans, and large glass sponges.

**Figure 11–17** The prominent light-colored Capitan Limestone forms the caprock of the Guadalupe Mountains. The Capitan Limestone is rich in fossil corals and associated reef organisms.

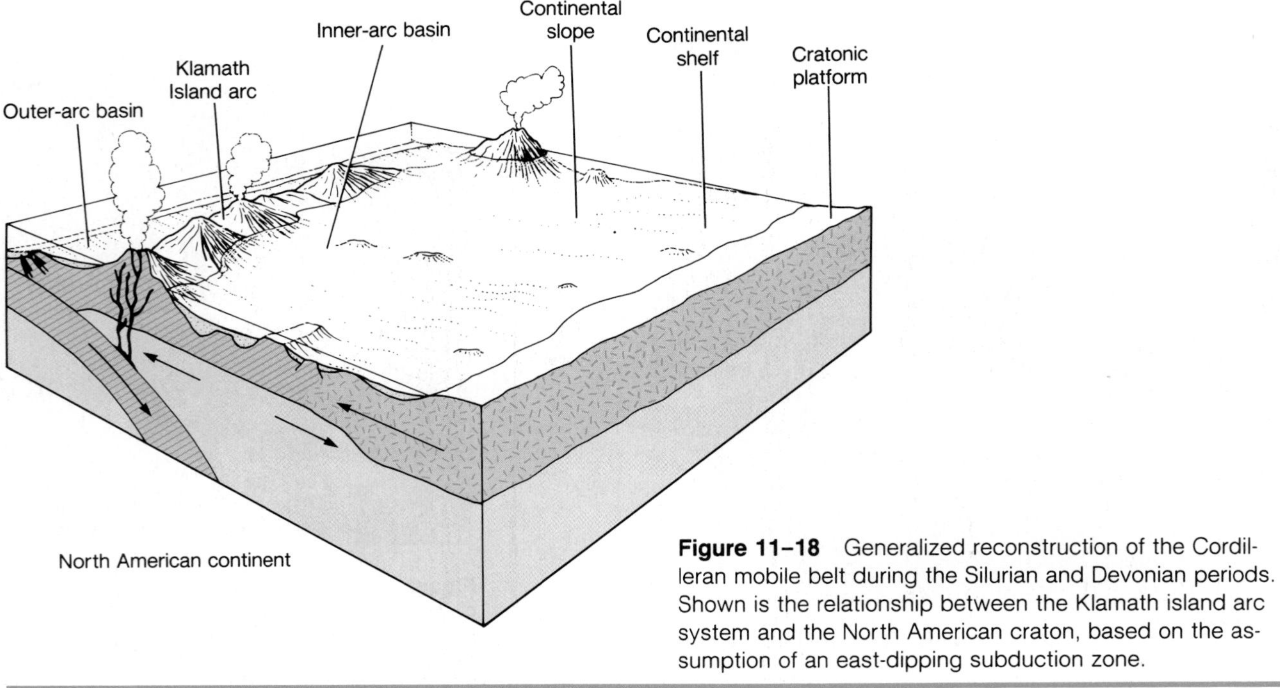


**Figure 11–18** Generalized reconstruction of the Cordilleran mobile belt during the Silurian and Devonian periods. Shown is the relationship between the Klamath island arc system and the North American craton, based on the assumption of an east-dipping subduction zone.

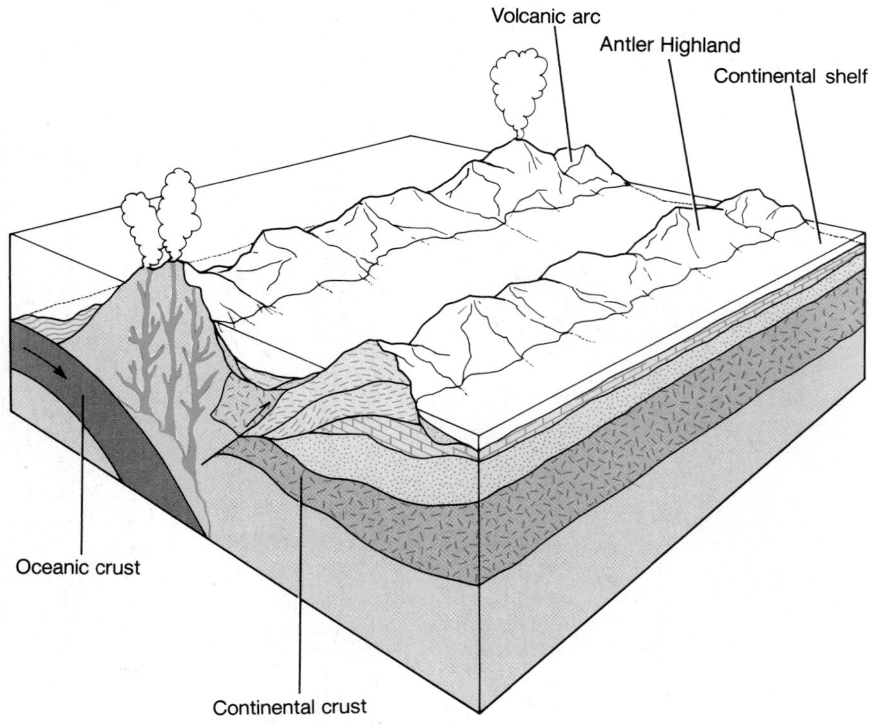


**Figure 11–19** Reconstruction of the Cordilleran mobile belt during the Early Mississippian, showing the effects of the Antler orogeny.

**Figure 11–20** Aerial photograph showing the Roberts Mountain thrust fault. View is south-southeast to the mouth of Cortez Canyon, from a point over Crescent Valley, Nevada.

**Figure 11–21** These spectacular exposures are the erosional remnants of Pennsylvanian arkoses that were later uplifted during the Laramide orogeny and are now exposed at Red Rock Canyon, Garden of the Gods, near Colorado Springs, Colorado.

of uplifts along the Texas-Oklahoma border and Texas Panhandle are called the Oklahoma Mountains; those across northeast Arizona, the Kaibab-Defiance-Zuni Uplift; while those in the Colorado, Utah, and New Mexico area, namely the Front Range-Pedernal and Uncompahgre uplifts, are referred to collectively as the *Ancestral Rocky Mountains* (mentioned earlier). Intracratonic mountain ranges are unusual, and these mountains apparently resulted from movement along near-vertical faults. Great stress was generated in this area as Gondwana collided with southern North America. Crustal adjustment in the form of faults resulted in uplift of cratonic blocks and downwarp of adjacent basins.

Growth of these mountains along faults elevated and exposed Precambrian basement rocks. As the mountains eroded, nearly 3 km of red arkosic sand and conglomerate were deposited in the surrounding basins (Fig. 11–13). Today they are preserved in the rocks of the Garden of the Gods near Colorado Springs (Fig. 11–21) and at the Red Rocks Amphitheatre near Morrison, Colorado.

## OUACHITA MOBILE BELT

The Ouachita mobile belt extends from Arkansas to north-central Mexico (Fig. 10–1). During the Late Proterozoic to Early Devonian, thousands of meters of sandstones, shales, and carbonates were deposited on the continental platform, while in the deeper-water portion of the mobile belt, bedded cherts and shales accumulated.

The Ouachita mobile belt changed from a passive plate margin to a convergent plate margin during the Mississippian Period (Fig. 11–22). It began receiving sediments from a volcanic source area to the south. Increasing sedimentation rates and volcanic tuffs interbedded with greywackes and black shale indicate development of a volcanic arc–subduction zone. The flood of clastics continued into the Pennsylvanian Period, forming a clastic wedge that thickens and becomes coarser to the south. The formation of the clastic wedge marked the beginning of uplift of the area and formation of a mountain range.

Thrusting and folding of sediments continued into the Late Pennsylvanian and Early Permian. These resulted from the compressive forces generated along the zone of subduction as Gondwana approached southern North America. Of the once lofty mountain range that formed during the Late Paleozoic, only the rejuvenated Ouachita Mountains of Arkansas and Oklahoma, and the Marathon Mountains of West Texas remain.

The Ouachita deformation was part of the general worldwide tectonic activity that united Gondwana with Laurasia. The Hercynian, Appalachian, and Ouachita mobile belts were continuous and marked the southern boundary of Laurasia (Fig 11–2). The tectonic activity that resulted in the uplift in the Ouachita mobile belt was very complex, involving South America and several microplates that eventually became part of Central America. The compressive forces impinging on the Ouachita mobile belt also affected the craton by causing epeirogenic uplift of the southwestern part of North America as previously discussed.

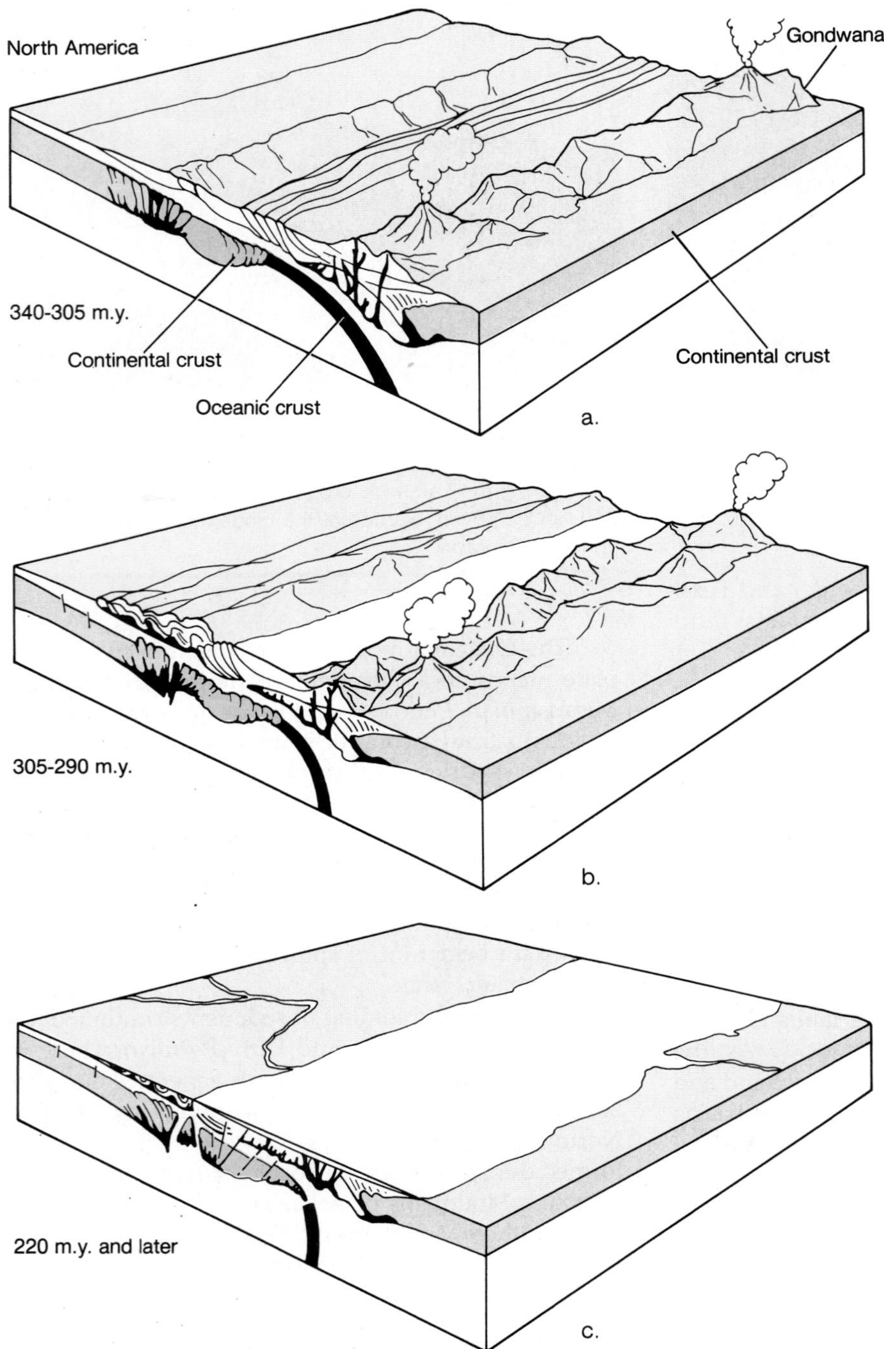


**Figure 11–22** Plate tectonic model for deformation of the Ouachita mobile belt. a. Incipient continental collision between North America and Gondwana began during the Mississippian to Pennsylvanian. b. Continental collision continued during the Pennsylvanian Period. c. Rifting, normal faulting, and fault-block tilting related to sea-floor spreading in the Gulf of Mexico Basin took place during the Mesozoic Era.

## APPALACHIAN MOBILE BELT

### Caledonian Orogeny

The Caledonian mobile belt extended along the northwest border of Europe and included Scotland, and Ireland, and Norway (Fig. 10–3). Beginning during the Middle Ordovician, subduction between the Iapetus Ocean and Europe began forming a mirror image of the convergent plate margin off the east coast of North America (Fig. 11–23).

The *Caledonian orogeny* (mentioned earlier) reached maximal uplift during the Late Silurian and Early Devonian, forming a highlands area from which clastics were shed. A large deltaic complex formed as thick red beds were deposited along the front of the Caledonian highlands. These deposits are known as the Old Red Sandstone.

### Acadian Orogeny

The third Paleozoic orogeny to affect North America and Europe began during the Late Silurian, culminating at the end of the Devonian Period. The *Acadian orogeny* (mentioned earlier) affected the Appalachian mobile belt from Newfoundland to Pennsylvania as sedimentary rocks were thrust northward and westward against the craton.

Like the Taconic and Caledonian orogenies, the Acadian orogeny occurred along a convergent plate boundary that started as an oceanic-continental boundary. It culminated in a continental-continental plate boundary as Europe and the northern part of North America collided (Fig. 11–24).

The Acadian orogeny was of greater magnitude than the Taconic orogeny, as indicated by more widespread regional metamorphism and granitic intrusions. Radiometric dates from those metamorphic and igneous rocks cluster between 350 to 400 million years ago. Furthermore, as with the Taconic orogeny, deep-water sediments were folded and thrust westward, producing profound angular unconformities that separate Upper Silurian from Upper Devonian rocks.

Weathering and erosion of the Acadian Highlands produced a thick clastic wedge called the **Catskill Delta**, named for the Catskill Mountains in upstate New York where it is well exposed. The Catskill Delta, composed of red, coarse conglomerates, sandstones, and shales contains nearly three times as much sediment as the Queenston Delta.

The Devonian rocks of New York are among the best studied on the continent. A cross section of the Devonian strata clearly reflects an eastern source (Acadian Highlands) for the Catskill facies (Fig. 11–24). These clastic rocks can be traced from eastern Pennsylvania, where the coarse clastics are approximately 3 km thick, to Ohio, where the deltaic facies are only

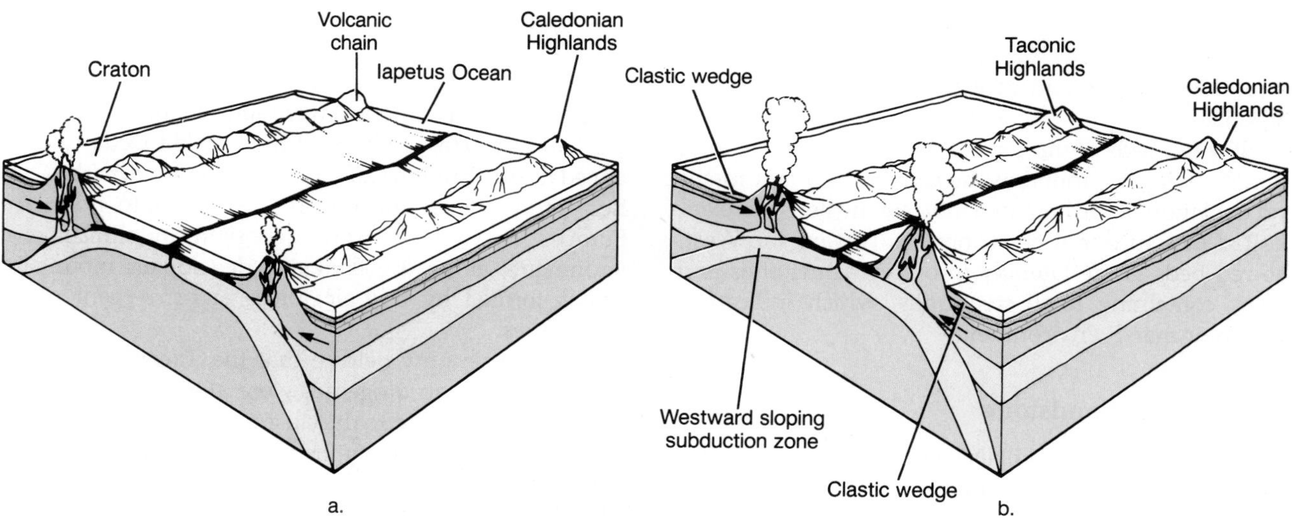


**Figure 11–23** The Caledonian orogeny began during the Middle Ordovician with subduction of the Iapetus Ocean plate beneath the European plate and reached maximum uplift during the Late Silurian to Early Devonian.

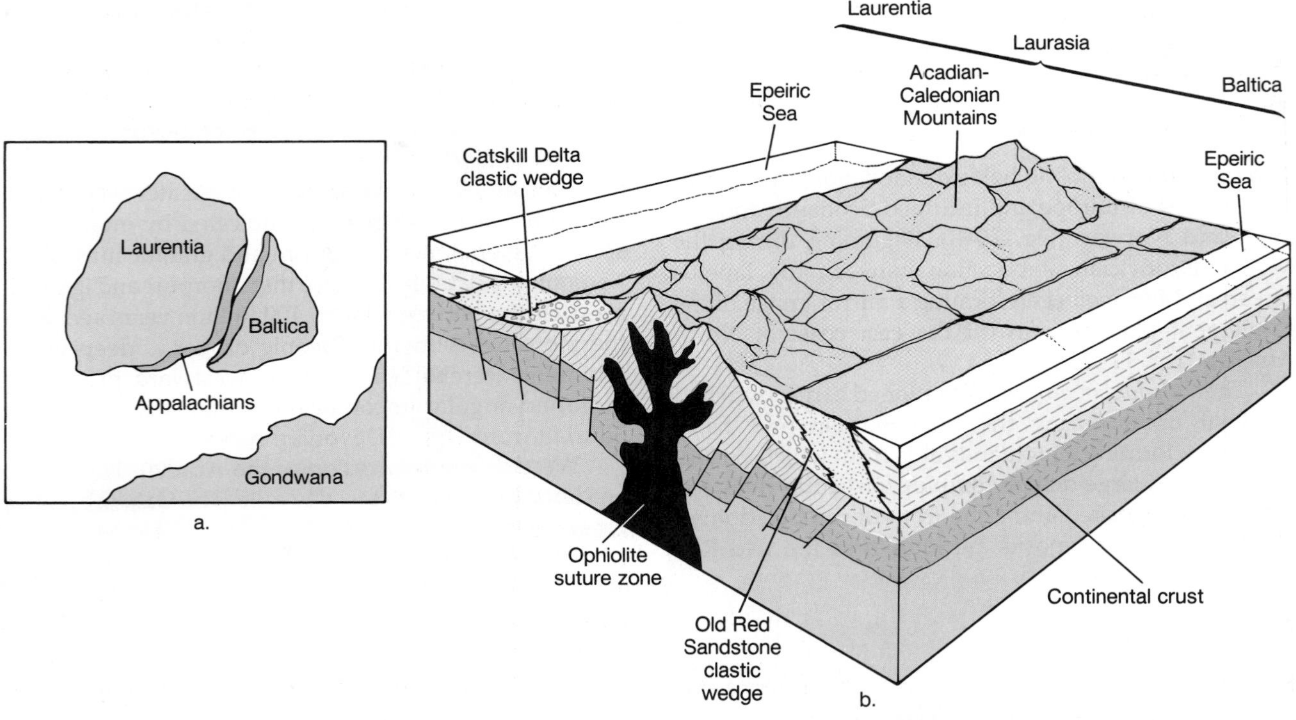


**Figure 11–24** a. Suturing of Laurentia to Baltica along a convergent plate boundary during the Middle to Late Devonian. b. Cross section of the collision between Laurentia and Baltica, showing the bilateral symmetry of the Catskill Delta clastic wedge and the Old Red Sandstone and their relationship to the Acadian and Caledonian Highlands.

about 100 m thick and consist of cratonic shales and carbonates.

The red beds of the Catskill Delta derive their color from the hematite found in the sediments. Plant fossils and oxidation of the hematite indicate the beds were deposited in a subaerial environment. Toward the West, the red beds grade laterally into grey sandstones and shales containing fossil treetrunks, which indicate a swamp or marsh environment.

## The Old Red Sandstone

The red beds of the Catskill Delta have a European counterpart in the Devonian Old Red Sandstone of the British Isles (Fig. 11–24). The Old Red Sandstone was a Devonian clastic wedge that grew eastward from the Caledonian Highlands. Like its North American Catskill counterpart, the Old Red Sandstone contains fresh-water fish, land plants, and early amphibian fossils. It also grades laterally into cratonic limestones.

By the end of the Devonian Period, Baltica and Laurentia (North America) were sutured together, forming Laurasia. The red beds of the Catskill Delta can be traced north, through Canada and Greenland, to the Old Red Sandstone of the British Isles and into northern Europe. These beds were deposited in similar environments along the flanks of developing mountain chains formed by tectonic forces at convergent plate boundaries.

Geologists now believe that the Taconic, Caledonian, and Acadian orogenies were all part of the same tectonic event related to the closing of the Iapetus Ocean (Fig. 11–25). This event began with paired oceanic-continental convergent plate boundaries during the Taconic and Caledonian orogenies, and culminated with a continental-continental convergent plate boundary during the Acadian orogeny as Laurentia and Baltica became sutured. Following this, and slightly further along the deformational belt, the Ouachita mobile belt became active, and it was in turn followed by the Hercynian-

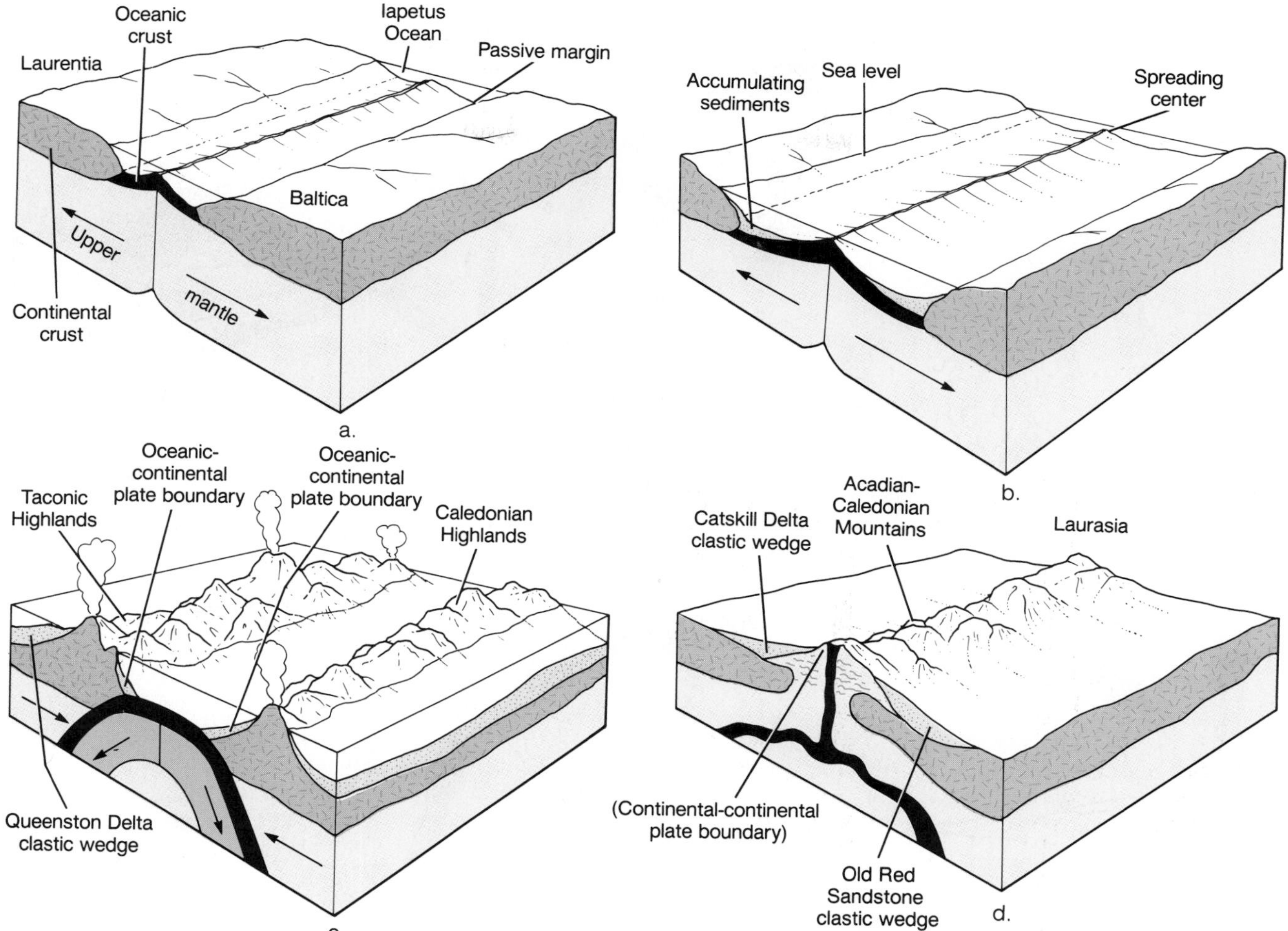


**Figure 11–25** Suggested evolution of the Taconic-Caledonian-Acadian mobile belt during the opening and closing of the Iapetus Ocean basin. a. Opening of the Iapetus Ocean basin during the Late Proterozoic. b. It continues to widen during the Cambrian Period. c. The Iapetus Ocean begins closing during the Middle Ordovician as subduction occurs on both sides. d. The Iapetus Ocean closes during the Devonian Period.

Alleghenian orogeny, thus completing the formation of Pangaea.

## Hercynian-Alleghenian Orogeny

The Hercynian mobile belt of southern Europe and the Appalachian and Ouachita mobile belts of North America mark the zone along which Europe (part of Laurasia) collided with Gondwana (Fig. 11–26). While South America and southern North America collided during the Pennsylvanian and Permian in the area of the Ouachita mobile belt, Europe and southeastern North America joined together with Africa as part of the *Hercynian-Alleghenian orogeny.*

Initial contact between southern Europe (Hercynian mobile belt) and Africa began during the Mississippian Period. The greatest deformation occurred during the Pennsylvanian and Permian periods. This event is referred to as the *Hercynian orogeny* (mentioned earlier). The central and southern parts of the Appalachian mobile belt (from New York to Alabama) were folded and thrust toward the craton as North America and Africa were sutured. This event in North America is referred to as the *Alleghenian orogeny* (also mentioned earlier).

**Figure 11–26** Pangaea, composed of its two major subcontinents, Laurasia and Gondwana, and the Paleozoic mobile belts where orogenic activity occurred.

These three Late Paleozoic orogenies (Ouachita, Hercynian, and Allegheny) represent the culmination of the joining together of Laurasia and Gondwana into a single massive continent called Pangaea.

## CHAPTER SUMMARY

Table 11–1 provides a summary of the geologic events occurring on the North American craton and mobile belts and shows how they relate to global events and sea-level changes for the Late Paleozoic.

1. During the Late Paleozoic, Baltica and Laurentia collided, forming Laurasia. Siberia and Kazakhstania collided and finally were sutured to Laurasia. Gondwana moved over the South Pole and underwent several glacial-interglacial periods, resulting in global sea-level changes and transgressions and regressions along the low-lying craton margins.
2. By the Late Permian, Laurasia and Gondwana collided, forming Pangaea, which was surrounded by the world ocean, Panthalassa.
3. The history of the North American craton can be deciphered from the rocks of the Kaskaskia and Absaroka sequences.
4. The basal beds deposited over the exposed Tippecanoe surface consisted of either mature sandstone, derived from the eroding Taconic highlands, or carbonate rocks.
5. Most of the Kaskaskia sequence was dominated by carbonates and associated evaporites. The Devonian Period was a time of major reef building in western Canada, southern England, Belgium, Australia, and Russia.
6. A persistent and widespread black shale, the Chattanooga Shale, was deposited over a large area of the craton during the Late Devonian and Early Mississippian.
7. The Mississippian Period was dominated by carbonate deposition.
8. Transgressions and regressions over the low-lying craton resulted in cyclothems and the formation of coals during the Pennsylvanian Period.
9. Cratonic mountain building occurred during the Pennsylvanian Period, and thick nonmarine clastics and evaporites were deposited in the intervening basins.
10. By the Early Permian, the Absaroka Sea occupied a narrow zone of the south central craton. Here, several large reefs and associated evaporites developed. By the end of the Permian Period, the Absaroka Sea had retreated from the craton.
11. The Cordilleran mobile belt was the site of a minor Devonian orogeny called the Antler orogeny during which deep-water sediments were thrust eastward over shallow-water sediments.
12. Mountain building occurred in the Ouachita mobile belt during the Pennsylvanian and Permian periods. This tectonic activity was partly responsible for the cratonic uplift that occurred in the southwest, producing the Ancestral Rockies.
13. The Caledonian, Acadian, Hercynian, and Alleghenian orogenies were all part of the global tectonic activity that produced Pangaea.

## IMPORTANT TERMS

Absaroka sequence
Acadian orogeny
Alleghenian orogeny
Ancestral Rockies
Antler orogeny
Caledonian orogeny
Catskill Delta
cyclothem
Ellesmere orogeny
Hercynian orogeny
Kaskaskia sequence
Klamath Arc
Ouachita orogeny
Pangaea
Panthalassa Ocean

## REVIEW QUESTIONS

1. What are cyclothems? How do they form? Why are they economically important?
2. What was the relationship between the Ouachita orogeny and the cratonic uplifts that occurred on the craton during the Pennsylvanian Period?
3. What is the evidence for glaciation on Gondwana during the Late Paleozoic?
4. What is the evidence for arid conditions in Laurentia during the Permian?
5. What do the various models proposed for the deposition of the Late Devonian to Early Mississippian black shales have in common? How do they differ? What is the evidence for a shallow-water environment of deposition?
6. How are the Caledonian, Acadian, Ouachita, Hercynian, and Alleghenian orogenies related to modern concepts of plate tectonics?
7. Compare and contrast the Taconic, Caledonian, and Acadian orogenies in terms of the tectonic forces that caused them and the sedimentary features that resulted.

**Table 11–1** Summary of Late Paleozoic events.

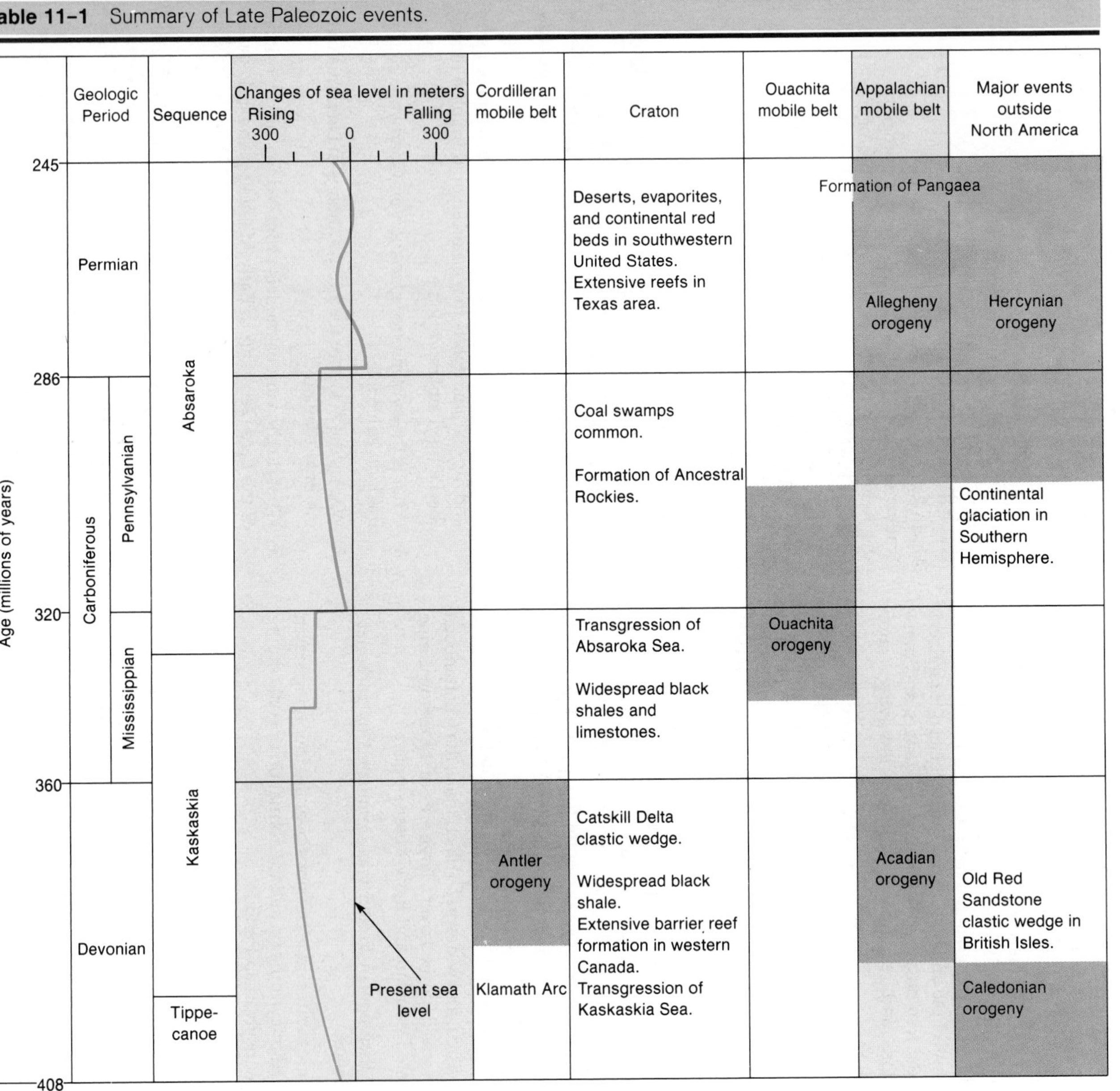


| Age (millions of years) | Geologic Period | Sequence | Cordilleran mobile belt | Craton | Ouachita mobile belt | Appalachian mobile belt | Major events outside North America |
|---|---|---|---|---|---|---|---|
| 245–286 | Permian | Absaroka | | Deserts, evaporites, and continental red beds in southwestern United States. Extensive reefs in Texas area. | | Formation of Pangaea; Allegheny orogeny | Formation of Pangaea; Hercynian orogeny |
| 286–320 | Carboniferous: Pennsylvanian | Absaroka | | Coal swamps common. Formation of Ancestral Rockies. | Ouachita orogeny | Allegheny orogeny | Hercynian orogeny; Continental glaciation in Southern Hemisphere. |
| 320–360 | Carboniferous: Mississippian | Absaroka; Kaskaskia | | Transgression of Absaroka Sea. Widespread black shales and limestones. | Ouachita orogeny | | |
| 360–408 | Devonian | Kaskaskia; Tippecanoe | Antler orogeny; Klamath Arc | Catskill Delta clastic wedge. Widespread black shale. Extensive barrier reef formation in western Canada. Transgression of Kaskaskia Sea. | | Acadian orogeny | Old Red Sandstone clastic wedge in British Isles. Caledonian orogeny |

8. How does the origin of evaporite deposits of the Kaskaskia sequence compare with the origin of evaporites of the Tippecanoe sequence?
9. How did the sedimentary environments of the Kaskaskia sequence differ from those of the Absaroka sequence in North America?
10. How did the formation of Pangaea and Panthalassa affect the world's climate at the end of the Paleozoic Era?

## ADDITIONAL READINGS

Bambach, R. K., C. R. Scotese, and A. M. Ziegler. 1980. Before Pangaea: The Geographies of the Paleozoic World. *American Scientist*, 68, no. 1: 26–38.

Dewey, J. F., and J. M. Bird. 1970. Mountain Belts and the New Global Tectonics. *Journal of Geophysical Research*, 75, no. 14: 2625–2647.

Fouch, T. D., and E. R. Magathan, eds. 1980. *Paleozoic Paleogeography of the West-Central United States*. Rocky Mountain Paleogeography Symposium 1. Rocky Mountain Section, Society of Economic Paleontologists and Mineralogists, Denver, Colorado.

Rodgers, J. 1970. *The Tectonics of the Appalachians*. New York: Wiley-Interscience.

Stearn, C. W., R. L. Carroll, and T. H. Clark. 1979. *The Geological Evolution of North America*, 3d ed. New York: John Wiley & Sons.

Stewart, J. H., C. H. Stevens, and A. E. Fritsche, eds. 1977. *Paleozoic Paleogeography of the Western United States*. Pacific Coast Paleogeography Symposium 1. Society of Economic Paleontologists and Mineralogists, Tulsa, Oklahoma.

Thomas, W. A. 1976. Evolution of Ouachita-Appalachian Continental Margin. *Journal of Geology*, 84, no. 3: 323–342.

Woodrow, D.L., and W. D. Sevan. 1985. The Catskill Delta. *Geological Society of America Special Paper* 201: 1–246.

# 12

## CHAPTER OUTLINE



## PROLOGUE

*In the fall of 1909, Charles D. Walcott was searching for fossils near Field, British Columbia, Canada, when one of the horses in his fossil-collecting party stumbled over a block of shale that had fallen on the trail. Upon splitting the shale, Walcott was amazed to find the impressions of a number of soft-bodied organisms preserved on the bedding surface. Walcott returned to the site the following summer and located the shale stratum that was the source of his fallen block of rock in the steep slope above the trail. He quarried the site and shipped back thousands of fossil specimens to the United States National Museum of Natural History, where he later studied and catalogued them.*

*The importance of Walcott's discovery is not that it was another collection of well-preserved Cambrian fossils, but rather that it allowed geologists a rare glimpse of a world previously almost unknown—that of the soft-bodied animals that lived some 530 million years ago. The beautifully preserved fossils from the Burgess Shale more nearly represent what a Middle Cambrian community was like than any deposit containing fossils that only had hard parts. Specifically, the Burgess Shale contains species of trilobites, sponges, brachiopods, mollusks, and echinoderms, all of which have hard parts and are characteristic of Cambrian faunas throughout the world. But in addition to the diverse skeletonized fauna, a large and varied fossil assemblage of soft-bodied animals is also present. In fact, 60 percent of the total fossil assemblage comprises soft-bodied animals, which usually are not preserved. In all, 107 genera of animals, 64 of which were soft-bodied and preserved as impressions, have been recovered from the Burgess Shale. This proportion of soft-bodied animals to those with hard parts is comparable to modern marine communities.*

*The Burgess Shale fauna reveals the evolutionary stage that marine life had reached by the Middle Cambrian: highly advanced organisms comprised complex communities as diverse in structure and adaptation as many modern marine communities. The Burgess Shale fauna makes speculation about the early evolution of metazoans difficult because it shows how little we know about early marine life. In fact, some of the organisms preserved may not represent any known phyla at all.*

# Life of the Paleozoic Era

*What conditions led to the remarkable preservation of the Burgess Shale fauna? When it was deposited, the Burgess Shale was located at the base of a steep submarine escarpment. The animals whose exquisitely preserved fossil remains are found in the Burgess Shale lived in and on mud banks that formed along the top of this escarpment. Periodically, this unstable area would slump and slide down the escarpment as a turbidity flow. At the base, the mud and animals carried with it were deposited in a deep-water anoxic environment devoid of life. In such an environment bacterial degradation could not destroy the buried animals, and they were compressed by the weight of the overlying sediments, eventually resulting in their preservation as carbonaceous impressions.*

Diorama of the environment and biota of the Phyllopod bed of the Burgess Shale. In the background is the vertical wall of the submarine escarpment with algae growing on it. The large cylindrical ribbed organisms on the muddy bottom in the foreground are sponges.

## INTRODUCTION

In this chapter we examine the history of Paleozoic life as a series of interconnected biologic and geologic events in which the underlying principles and processes of evolution and plate tectonics played a major role. The opening and closing of ocean basins, transgressions and regressions of epeiric seas, and the changing positions of the continents had a profound effect on the evolution of the marine and terrestrial communities.

A time of tremendous biologic change began with the appearance of skeletonized animals near the Precambrian-Cambrian boundary. Following this event, marine invertebrates began a period of adaptive radiation and evolution during which the Paleozoic marine invertebrate community greatly diversified. Indeed, the history of the Paleozoic marine invertebrate community was one of diversifications and extinctions. Vertebrates also evolved during the Paleozoic, first in the seas as fish, one group of which was ancestral to the first land animals, the amphibians. Before the end of the Paleozoic, reptiles had evolved from amphibians.

Plants preceded animals onto the land. Both plants and animals had to solve the same basic problems in making the transition from water to land. The method of reproduction proved to be the major barrier to expansion into environments for both groups. With the evolution of the seed in plants and the amniote egg in animals, this limitation was removed, and both groups were able to move into all the terrestrial environments.

The end of the Paleozoic Era was a time of major extinctions. The marine invertebrate community was greatly decimated, and many amphibians and reptiles on land became extinct.

## THE FIRST SHELLED FOSSILS

Several important questions that arise regarding the early history of life relate to the acquisition of a mineralized exoskeleton and its adaptive significance, the composition of the exoskeleton, and the rapid radiation of skeletonized animals beginning near the base of the Cambrian Period. Early geologists observed that the remains of skeletonized animals appeared rather abruptly in the fossil record. Charles Darwin addressed this problem in *On the Origin of Species* and observed that without a convincing explanation, such an event was difficult to reconcile with his newly expounded evolutionary theory.

Scientists generally recognize that because Cambrian life was so varied and complex, multicelled organisms must have had a long Precambrian history during which they lacked hard parts and thus did not leave a fossil record. As we discussed in Chapter 9, the oldest apparently multicelled organisms, a possible algae, are found in the 1.4-billion-year-old Belt Supergroup of Montana (Fig. 9–32), and recently wormlike fossils have been reported from rocks 700 to 900 million years old in China (Fig. 9–33).

Impressions of the first unequivocally multicelled animals belong to the widely distributed Ediacaran fauna, and are found in rocks between 570 to 670 million years old (Fig. 9–1). Associated with Ediacaran faunas in Namibia and southern China are the first shelled fossils (Fig. 12–1). These are small calcium carbonate tubes, presumably housing wormlike suspension feeding organisms. In addition, small organic tube-shaped fossils, also presumably housing wormlike suspension feeding animals occur with the calcareous tubes. By the latest part of the Proterozoic, several skeletonized animals had made their appearance, yet durable skeletons of chitin, silica, and calcium carbonate did not begin to appear in abundance until the beginning of the Phanerozoic Eon 570 million years ago.

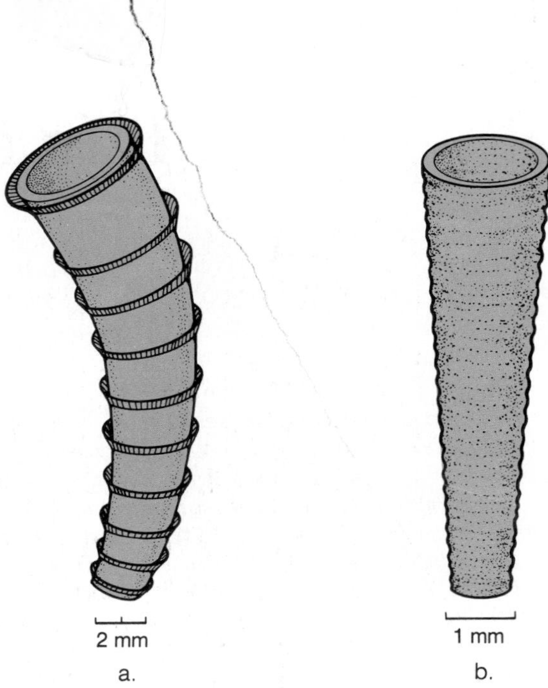


**Figure 12–1** The first shelled fossils. These small calcium-carbonate tubes presumably housed wormlike suspension-feeding ogranisms. They are found associated with Ediacaran faunas from Namibia and China. a. *Cloudina*. b. *Sinotubulites*.

## THE CAMBRIAN PERIOD AND THE EMERGENCE OF A SHELLY FAUNA

The Early Cambrian was characterized by a low-diversity shelly fauna consisting of animals that used both calcium carbonate and calcium phosphate to construct their skeletons. They included small worm tubes, mollusks and echinoderms, archaeocyathids, and brachiopods (Fig. 12–2). It is likely that this fauna was yet another "experiment," like the Ediacaran fauna of the Proterozoic Eon, but this experiment was very successful. By the Middle Cambrian, a large number of the major groups of invertebrate animals had appeared (Fig. 12–3). Many, such as the brachiopods, are still around today, while others, including the archaeocyathids and trilobites, are extinct. The Cambrian Period was also a time during which new body plans evolved (Fig. 12–4), and animals moved into new niches. As might be expected, the Cambrian witnessed a higher percentage of such experiments than any other period of geologic history.

The emergence of so many different organisms at this juncture in the Earth's history resulted in the development of new relationships and community structures as organisms underwent tremendous evolutionary change and filled previously unoccupied niches.

## THE ACQUISITION AND SIGNIFICANCE OF HARD PARTS

A striking aspect of the Early Cambrian fauna is that many animals already had fully developed features, that

**Figure 12–2** Three small Lower Cambrian shelly fossils. a. A conical sclerite (a piece of the armor covering) of *Lapworthella* from Australia. b. *Archaeooides*, an enigmatic spherical fossil from the Mackenzie Mountains, Northwest Territories, Canada. c. The tube of an anabaritid from the Mackenzie Mountains, Northwest Territories, Canada. (Photos courtesy of Simon Conway Morris and Stefan Bengtson, University of Cambridge, England)

is, their anatomies indicate an extended period of evolution before the evolution of hard parts. Furthermore, the major skeletonized animal groups did not all evolve at once, but rather evolved throughout the Cambrian and Ordovician Periods (Fig. 12–3). A preskeletonized period of evolution occurred during which members of a phylum evolved for millions of years as soft-bodied organisms.

The question remains as to why the invertebrate phyla initially acquired skeletons and why such acquisitions occurred over an extended period of time? The view long held by geologists was that the appearance of hard parts in the fossil record was rather sudden. However, recent evidence shows that the acquisition of hard parts by the major invertebrate phyla occurred over a period of perhaps 100 million years and came about from the need to eliminate mineral matter from their metabolic systems.

The formation of an exoskeleton confers several advantages. (1) It provides protection against ultraviolet radiation, allowing animals easier movement into shallower waters. (2) It helps prevent drying out in an intertidal environment. (3) It provides protection against predators. Previous assumptions that predators in Cambrian communities were not important have proven to be wrong. New evidence including actual fossils of predators and specimens of damaged prey, as well as antipredatory adaptations in some animals, indicates

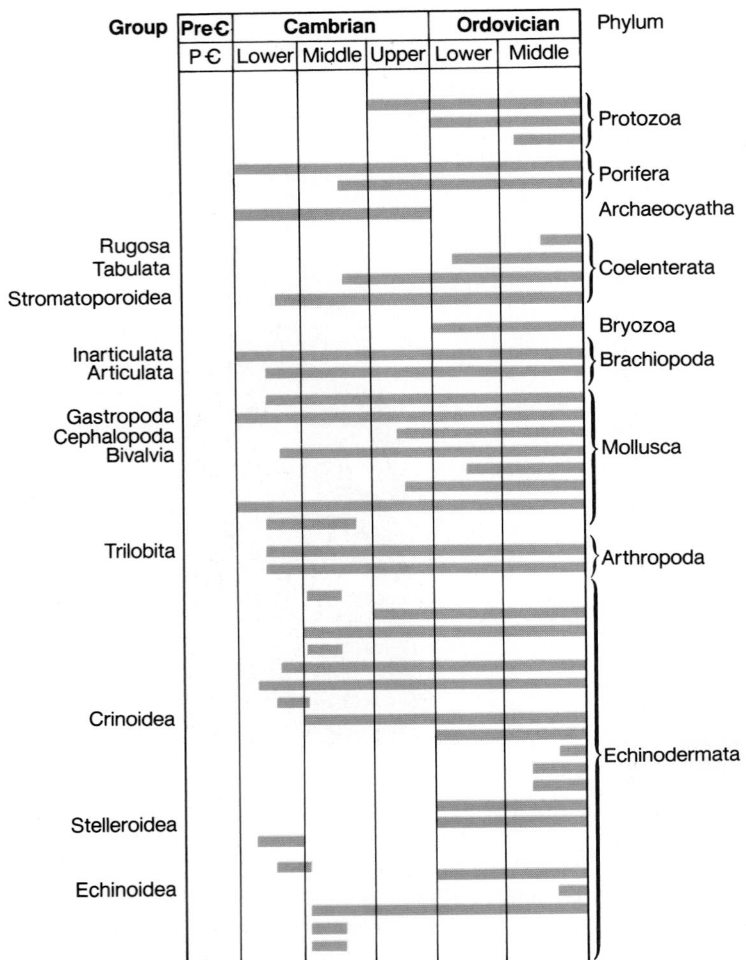


**Figure 12–3** The ranges of the earliest invertebrate phyla, showing the gradual evolution of the invertebrate phyla over a 100-million-year interval. Each line represents a major group in the phylum.

that the impact of predation was great (Fig. 12–5). With predators playing an important role in the Cambrian marine ecosystem, any mechanism or feature that protected an animal would certainly be advantageous and retained in the gene pool. (4) A supporting skeleton, whether an exo- or endoskeleton, allows animals to increase their size. (5) It also provides attachment sites for development of strong muscles, thus increasing locomotor efficiency in mobile animals.

Hard parts probably originated due to several factors rather than a single one. Whatever the reason, the acquisition of a mineralized skeleton was a major evolutionary innovation that allowed the invertebrates to successfully occupy many otherwise unavailable habitats of the marine environment.

## PALEOZOIC INVERTEBRATE MARINE LIFE

Having considered the origin, differentiation, and evolution of the Precambrian-Cambrian marine biota, we now examine the changes that occurred in the marine

**Figure 12–4** *Helicoplacus*, a primitive echinoderm that became extinct 20 million years after its first appearance about 510 million years ago. Such an organism (a representative of one of several short-lived echinoderm classes) illustrates the "experimental" nature of the Cambrian invertebrate fauna. (Photo by Porter M. Kier, courtesy of J. Wyatt Durham, University of California, Berkeley.)

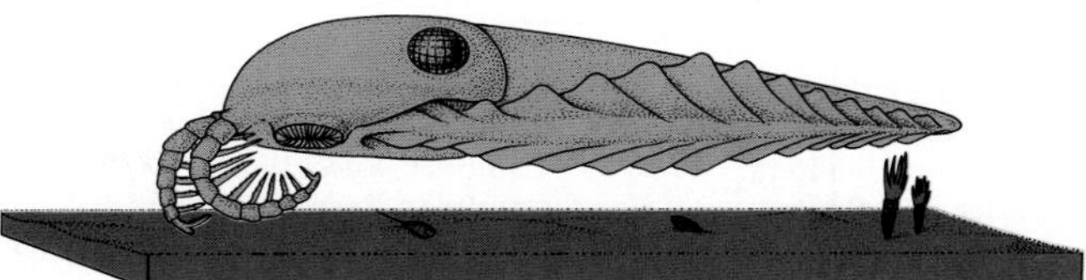

a.

b.

**Figure 12–5** a. *Anomalocaris*, a predator from the Early and Middle Cambrian Period was about 45 cm long and may have fed on trilobites. Its gripping appendages presumably carried food to its mouth. b. Wounds to the body of the trilobite *Olenellus robsonensis* (arrow). The wounds have healed, demonstrating that they occurred when the animal was alive and were not inflicted on an empty shell.

invertebrate community during the Paleozoic Era. Rather than focusing on the history of each invertebrate phylum (Table 12–1), we will survey the evolution of the marine invertebrate communities through time, concentrating on the major features and the changes that took place. To do that, we need to briefly examine the nature and structure of living marine communities so that we can make a reasonable interpretation of the fossil record.

## The Present Marine Ecosystem

In analyzing the present-day marine ecosystem, we must look at where organisms live and how they get around, as well as how they feed (Fig. 12–6). Organisms that live in the water column above the sea floor are called *pelagic*. They can be divided into two main groups: the floaters, or **plankton**, and the swimmers, or **nekton**. Plankton are mostly passive and go where the current carries them. Plant plankton such as diatoms, dinoflagellates, and various algae, are called *phytoplankton* and are mostly microscopic. Animal plankton are called *zooplankton* and are also mostly microscopic. Examples of zooplankton include foraminifera, radiolarians, and jellyfish. The nekton are swimmers and are mainly vertebrates such as fish; the invertebrate nekton include cephalopods.

Those organisms that live on or in the sea floor make up the **benthos**. They can be divided into those living on the sea floor, which are the *epifauna* (animals) or *epiflora* (plants), and those animals living in and moving through the sediments, which are the *infauna*. The benthos can be further divided into those organisms that stay in one place, called *sessile*, and those that move around on or in the sea floor, called *mobile*.

**Table 12–1** The Major Invertebrate Groups and Their Stratigraphic Ranges.

| | |
|---|---|
| **Phylum Protozoa** | Cambrian-Recent |
| Class Sarcodina | Cambrian-Recent |
| Order Foraminifera | Cambrian-Recent |
| Order Radiolaria | Cambrian-Recent |
| **Phylum Porifera** | Cambrian-Recent |
| **Phylum Archaeocyatha** | Cambrian |
| **Phylum Coelenterata** | Cambrian-Recent |
| Class Anthozoa | Ordovician-Recent |
| Order Tabulata | Ordovician-Permian |
| Order Rugosa | Ordovician-Permian |
| Order Scleractinia | Triassic-Recent |
| Class Hydrozoa | Cambrian-Recent |
| Order Stromatoporoida | Cambrian-Cretaceous |
| **Phylum Bryozoa** | Ordovician-Recent |
| **Phylum Brachiopoda** | Cambrian-Recent |
| Class Inarticulata | Cambrian-Recent |
| Class Articulata | Cambrian-Recent |
| **Phylum Mollusca** | Cambrian-Recent |
| Class Monoplacophora | Cambrian-Recent |
| Class Gastropoda | Cambrian-Recent |
| Class Bivalvia | Cambrian-Recent |
| Class Cephalopoda | Cambrian-Recent |
| **Phylum Annelida** | Precambrian-Recent |
| **Phylum Arthropoda** | Cambrian-Recent |
| Class Trilobita | Cambrian-Permian |
| Class Crustacea | Cambrian-Recent |
| Class Insecta | Silurian-Recent |
| **Phylum Echinodermata** | Cambrian-Recent |
| Class Blastoidea | Ordovician-Permian |
| Class Crinoidea | Cambrian-Recent |
| Class Echinoidea | Ordovician-Recent |
| Class Asteroidea | Ordovician-Recent |
| **Phylum Hemichordata** | Cambrian-Recent |
| Class Graptolithina | Cambrian-Mississippian |

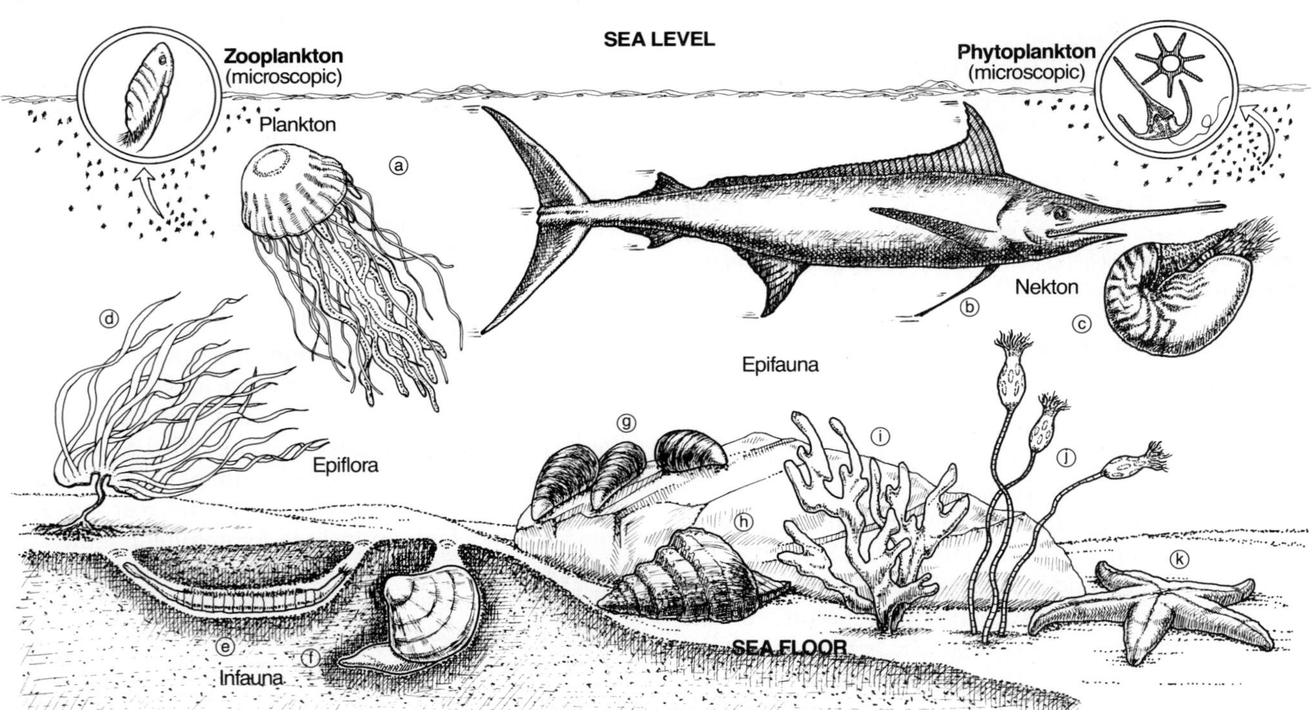


**Figure 12–6** Relationship of marine animals and plants to the surface and floor of the ocean. Plankton: jellyfish (a). Nekton: fish (b), and cephalopod (c). Benthos: (d through k). Sessile epiflora: seaweed (d). Sessile epifauna: bivalve (g), coral (i), crinoid (j). Mobile epifauna: starfish (k), gastropod (h). Infauna: worm (e), bivalve (f). Suspension feeders: bivalve (g), coral (i), crinoid (j). Herbivores: gastropos (h) Carnivores-scavengers: starfish (k). Sediment-deposit feeders: worm (i).

The feeding strategies of organisms are also important in terms of their relationships with other organisms in the marine ecosystem. There are basically four feeding groups; **Suspension-feeding** animals remove or consume microscopic plants and animals as well as dissolved nutrients from the water. **Herbivores** are plant eaters. **Carnivore-scavengers** are meat eaters, and **sediment-deposit feeders** ingest sediment and extract the nutrients from it.

We can define an organism's place in the marine ecosystem by where it lives and how it eats. For example, an articulate brachiopod is a benthonic, epifaunal suspension feeder, whereas a cephalopod is a nektonic carnivore.

Within an ecosystem there are several *trophic levels* of food production and consumption. The feeding hierarchy and hence energy flow in an ecosystem comprise a food web of complex interrelationships among the producers, consumers, and decomposers (Fig. 12–7). The **primary producers**, or *autotrophs*, are those organisms that manufacture their own food. Virtually all marine primary producers are phytoplankton. Feeding on the primary producers are the primary consumers, which are mostly suspension feeders. Secondary consumers feed on the primary consumers, and thus are predators, while tertiary consumers, which are also predators, feed on the secondary consumers. In addition to the producers and consumers, there are also transformers and decomposers. These are bacteria that break down the dead organisms that have not been consumed into organic compounds that are then recycled.

When we look at the marine realm today, we see a complex organization of organisms interrelated by trophic interactions and affected by changes in the physical environment. When one part of the system changes, the whole structure changes, sometimes almost insignificantly, other times catastrophically.

As we examine the evolution of the Paleozoic marine ecosystem, keep in mind how geologic and evolutionary changes can have a significant impact on the composition and structure of the ecosystem. For example, the major transgressions onto the craton opened up

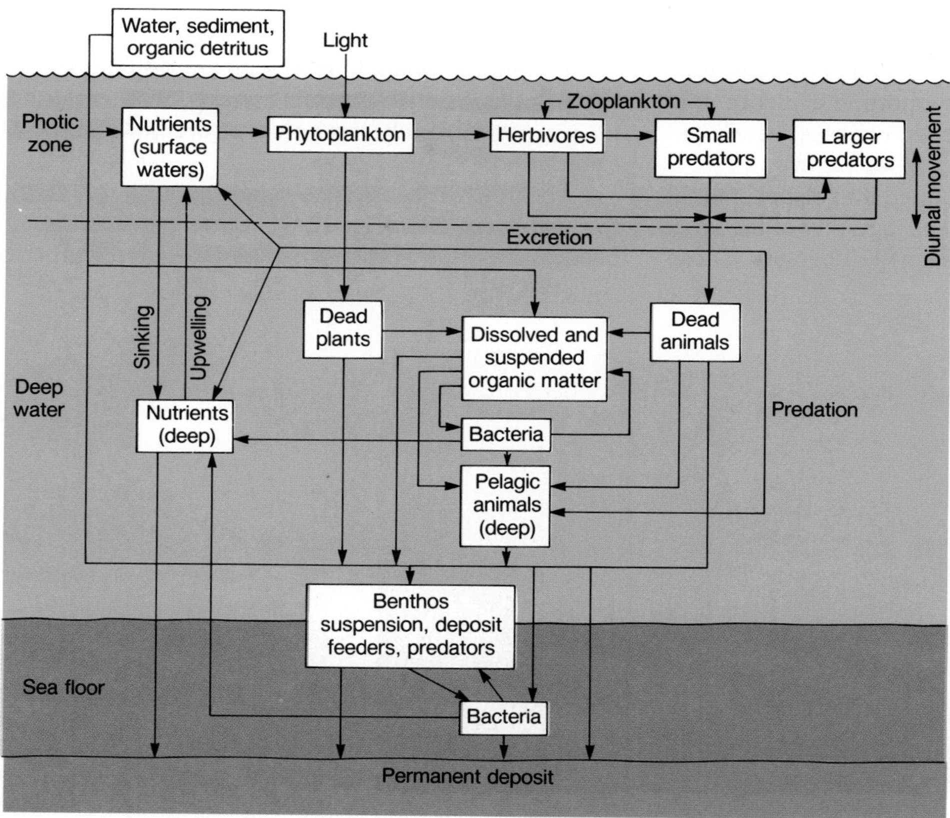


**Figure 12–7** Marine food web showing the relationships between the producers, consumers, and decomposers.

vast areas of shallow seas that could be inhabited. The movement of continents affected oceanic circulation patterns as well as causing environmental changes as the continents and their epeiric seas moved through different climatic zones.

## Cambrian Marine Community

Although almost all the major invertebrate phyla appeared in the fossil record during the Cambrian Period, many were represented by only a few species. While trace fossils were common, and echinoderms diverse, trilobites, brachiopods, and archaeocyathids comprised the majority of Cambrian skeletonized life (Fig. 12–8).

*Trilobites* were by far the most conspicuous element of the Cambrian marine invertebrate community and made up about half of the total fauna. Trilobites were benthonic, mobile sediment-deposit feeders that crawled or swam along the sea floor. They appeared in the Early Cambrian fully developed, reached their peak in the Late Cambrian, and then suffered mass extinctions near the end of the Cambrian Period from which they never fully recovered. While as yet no consensus exists on what caused the trilobite extinctions, it has been suggested that a sudden, temporary cooling of the seas may have played a major role. As supporting evidence, the trilobites living in the cool, deeper, offshore waters escaped the mass extinctions.

Trilobite faunas can be grouped into distinct faunal realms characteristic of the paleogeography of the time. Furthermore, trilobites are excellent guide fossils. Those of the Early Cambrian were cosmopolitan in distribution, while those of the Middle and Late Cambrian were more provincial.

**Figure 12–9** Reconstruction of a Cambrian reef-like structure built by archaeocyathids.

Cambrian *brachiopods* were mostly primitive types called *inarticulates*. They secreted a chitinophosphate shell, that is, one composed of the organic compound chitin combined with calcium phosphate. Inarticulate brachiopods also lacked a tooth-and-socket-arrangement along the hinge line. The *articulate* brachiopods, which have a tooth-and-socket arrangement, were also present but did not become abundant until the Ordovician Period.

The third major group of Cambrian organisms were the *archaeocyathids* (Fig. 12–9). These organisms were benthonic sessile suspension feeders that constructed

**Figure 12–8** Reconstruction of a Cambrian marine community showing floating jellyfish, swimming arthropods, benthonic sponges, and scavenging trilobites.

reef-like structures. The rest of the Cambrian fauna consisted of representatives of the other major phyla, including many organisms that were short-lived evolutionary experiments.

### The Burgess Shale Biota

No discussion of Cambrian life would be complete without mentioning one of the best examples in the world of a preserved soft-bodied biota, the Burgess Shale biota. As the Sauk Sea transgressed from the Cordilleran shelf onto the western edge of the craton, Early Cambrian sands were covered by a Middle Cambrian black, anoxic mud that allowed preservation of a diverse soft-bodied benthonic community. As we discussed in the "Prologue," these fossils were discovered in 1909 by C. D. Walcott near Field, British Columbia. They represent one of the most significant fossil finds of the century because they consist of impressions of soft-bodied animals and plants (Fig. 12–10), preservation of which is extremely rare in the fossil record. This discovery, therefore, allows us a valuable glimpse of organisms rarely preserved, as well as the soft-part anatomy of many extinct groups.

a.

b.

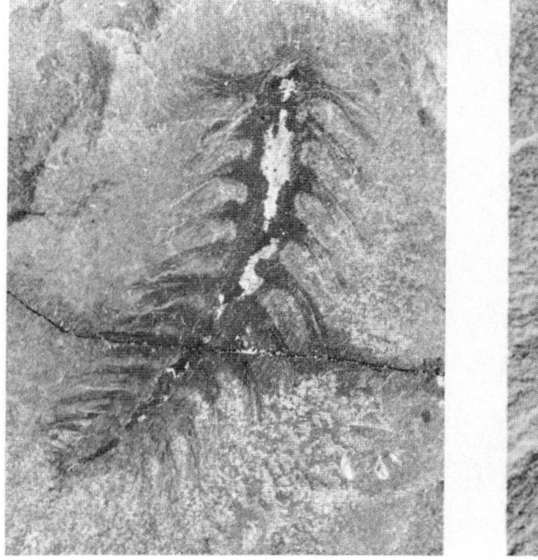

c.

d.

**Figure 12–10** Some of the fossil animals preserved in the Burgess Shale. a. *Burgessia bella*, an arthropod. b. *Waptia fieldensis*, another arthropod. c. *Burgessochaeta setigera*, an annelid worm. d. *Marrella splendens*, an arthropod that is the most abundant fossil animal in the Phyllopod bed of the Burgess Shale.

## Ordovician Marine Community

A major transgression began during the Middle Ordovician and near the end of the Ordovician culminated in the most widespread inundation of the craton ever known. Furthermore, warm, uniform climates reigned over the widespread epeiric seas during this time.

Not only did sedimentation patterns change dramatically from the Cambrian Period to the Ordovician Period, but changes in the composition of the fauna were equally striking. Whereas the Cambrian invertebrate community was dominated by three groups— trilobites, brachiopods, and archaeocyathids—the Ordovician Period was characterized by the adaptive radiation of many other animal phyla, (such as bryozoans and corals), with a consequent dramatic increase in the diversity of the shelly fauna. The Ordovician invertebrate community was dominated by epifaunal benthonic sessile suspension feeders (Fig. 12–11). The Ordovician Period was also a time of increased diversity and abundance of the acritarchs (organic-walled phytoplankton of unknown affinity), which were the major phytoplankton group of the Paleozoic Era and the primary food source of the suspension feeders (Fig. 12–12).

Whereas during the Cambrian Period, archaeocyathids were the main builders of reef-like structures, bryozoans, stromatoporoids, and tabulate and rugose corals assumed that role beginning in the Middle Ordovician. Many of these reefs were small patch reefs similar in size to those of the Cambrian but of a different composition, whereas others were quite large. As with modern reefs, Ordovician reefs exhibited a high diversity of organisms and were dominated by suspension feeders (Fig. 12–13).

Finally, three Ordovician fossil groups prove to be particularly useful for biostratigraphic correlation—the articulate brachiopods, graptolites, and conodonts (Fig. 12–14). The articulate brachiopods, present since the Cambrian, began a period of major diversification in the shallow-water marine environment during the Ordovician Period. They became a conspicuous element of the invertebrate fauna during the Ordovician and in succeeding Paleozoic periods.

Most graptolites were planktonic animals carried about by ocean currents. Combined with the fact that most individual species existed for less than a million years, fossil graptolites are excellent guide fossils and were especially abundant during the Ordovician and Silurian periods. Due to the fragile nature of their organic skeleton, graptolites are most commonly found in black shales.

**Figure 12–11** Reconstruction of a Middle Ordovician sea floor fauna. Cephalopods, crinoids, colonial corals, graptolites, trilobites, and brachiopods are shown.

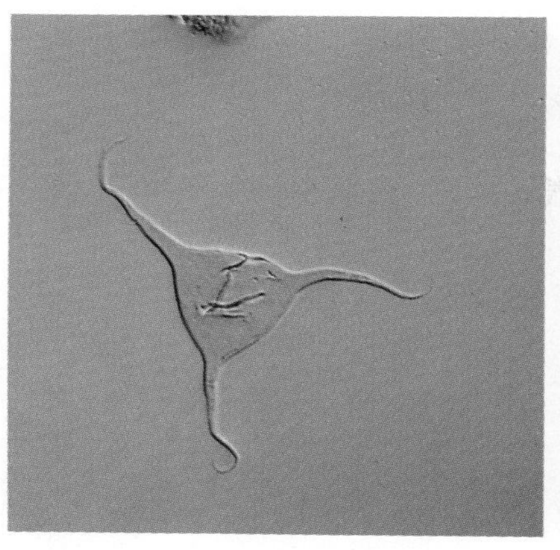

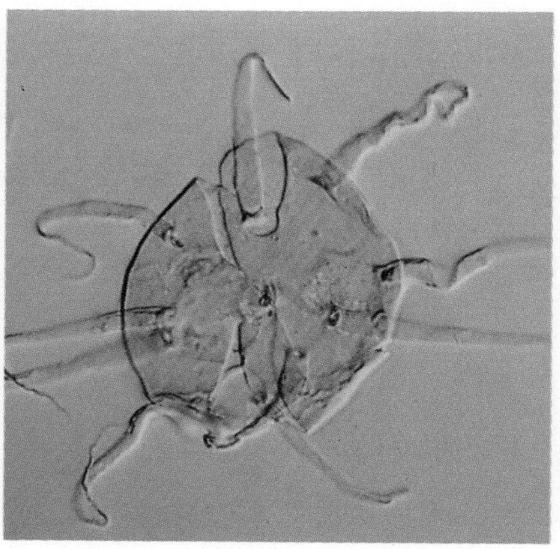

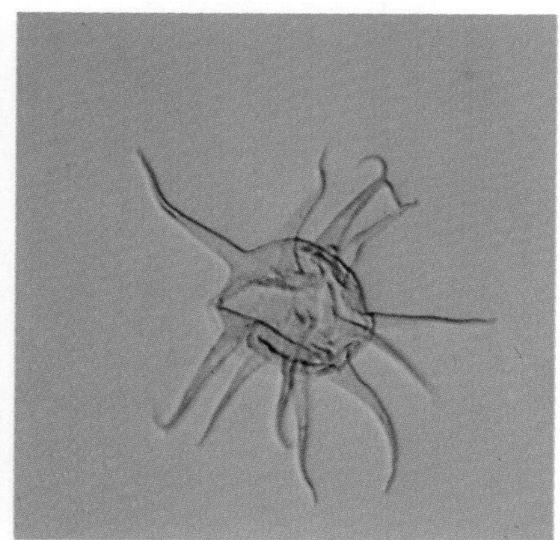

**Figure 12–12** Acritarchs from the Upper Ordovician Sylvan Formation, Oklahoma. Acritarchs are organic-walled phytoplankton and were the primary producers of the Paleozoic Era.

Carnivores
Cephalopods
Herbivores
Gastropods
Scavengers
Trilobites
Ostracods
Suspension feeders
Sponges
Stromatop-oroids
Bivalves
Brachiopods
Bryozoans
Corals
Echinoderms
Decomposers and transformers (bacteria)
Benthonic algae
Phytoplankton
Zooplankton
Organic matter
Nutrients

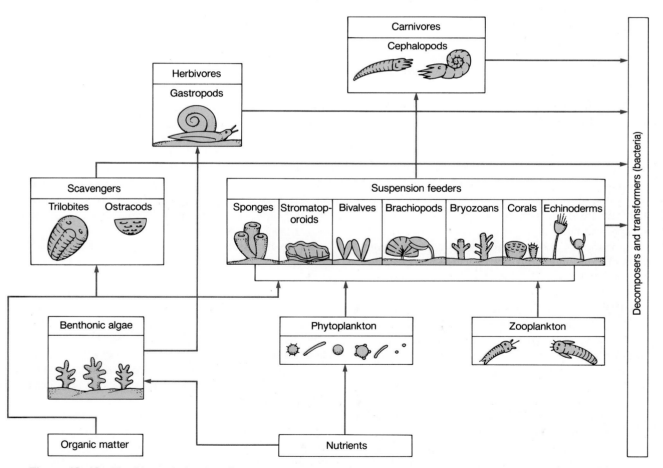

**Figure 12–13** Trophic analysis of an Ordovician reef community showing the relationship between the various organisms of the community. The phytoplankton and benthonic algae are primary producers, occupying the lowest trophic level. The primary consumers are the suspension feeders, making up the majority of the community. The highest trophic level is occupied by the carnivores, which in this community are the cephalopods.

**Figure 12–14** Representative brachiopods, graptolites, and conodonts. a. Brachiopods. b. Graptolites. *Phyllograptus angustifolius,* Norway. c. Conodonts *Cahabagnathus sweeti,* Copenhagen Formation (Middle Ordovician), Monitor Range, Nevada (upper left); *Phragmodus flexuosus,* Lenoir Limestone (Middle Ordovician), Friendsville, Tennessee (lower right); *Scolopodus* sp., Shingle Limestone, Single Pass, Nevada (right). (Photos (c) courtesy of Stig M. Bergstrom, Ohio State University.)

Conodonts are a group of well-known toothlike fossils whose wide distribution suggests that they represent elements of an organism that swam or floated. A recent discovery of conodonts within the carbonized impression of the conodont animal reveals the presence of fins, supporting the probability of a swimming mode of life for the conodont animal (Fig. 12–15). The wide distribution and short stratigraphic range of individual conodont species makes them excellent fossils for biostratigraphic zonation. Conodonts are easily recovered from limestones by dissolving the rock in acid—a process that does not attack the calcium phosphate material of which they are composed.

The end of the Ordovician Period was a time of mass extinctions in the marine realm. More than 100 families of marine invertebrates did not survive into the Silurian Period, and in North America alone, approximately one-half of the brachiopods and bryozoans died out. What caused such an event? Many geologists believe these extinctions were the result of the extensive glaciation that occurred in Gondwana at the end of the Ordovician Period (see Chapter 10).

Mass extinctions, those geologically rapid events in which an unusually high percentage of the fauna and/or flora becomes extinct, have occurred throughout geologic time (during the Ordovician, Devonian, Permian, and Cretaceous periods) and currently are the focus of much research and debate. (See Perspective 12–1.)

## Silurian and Devonian Marine Communities

The mass extinction at the end of the Ordovician Period was followed by rediversification and recovery of many of the decimated taxa. Brachiopods, bryozoans, gastropods, bivalves, corals, crinoids, and graptolites were just some of the groups that radiated again beginning during the Silurian Period.

As we discussed in Chapters 10 and 11, the Silurian and Devonian periods were times of major reef building. While most of the Silurian radiations of invertebrates represented repopulating of niches, organic reef builders diversified in new ways, building massive reefs larger than any produced during the Cambrian or Ordovician periods. This repopulation was probably due in part to renewed transgressions over the craton, and although there was a major drop in sea level at the end of the Silurian Period, the Middle Paleozoic sea level was generally high (Table 10–1).

The Silurian and Devonian reefs were dominated by tabulate and colonial rugose corals and stromatoporoids (Fig. 12–16). The growth of these reefs generally followed the characteristic ecologic succession in which small, stick-like tabulate and rugose corals initially colonized the subtidal sea floor. Following this, an intermediate stage began in which broad, mound-like tabulate and rugose corals dominated, forming low mounds. The final mature stage was reached when algae and stroma-

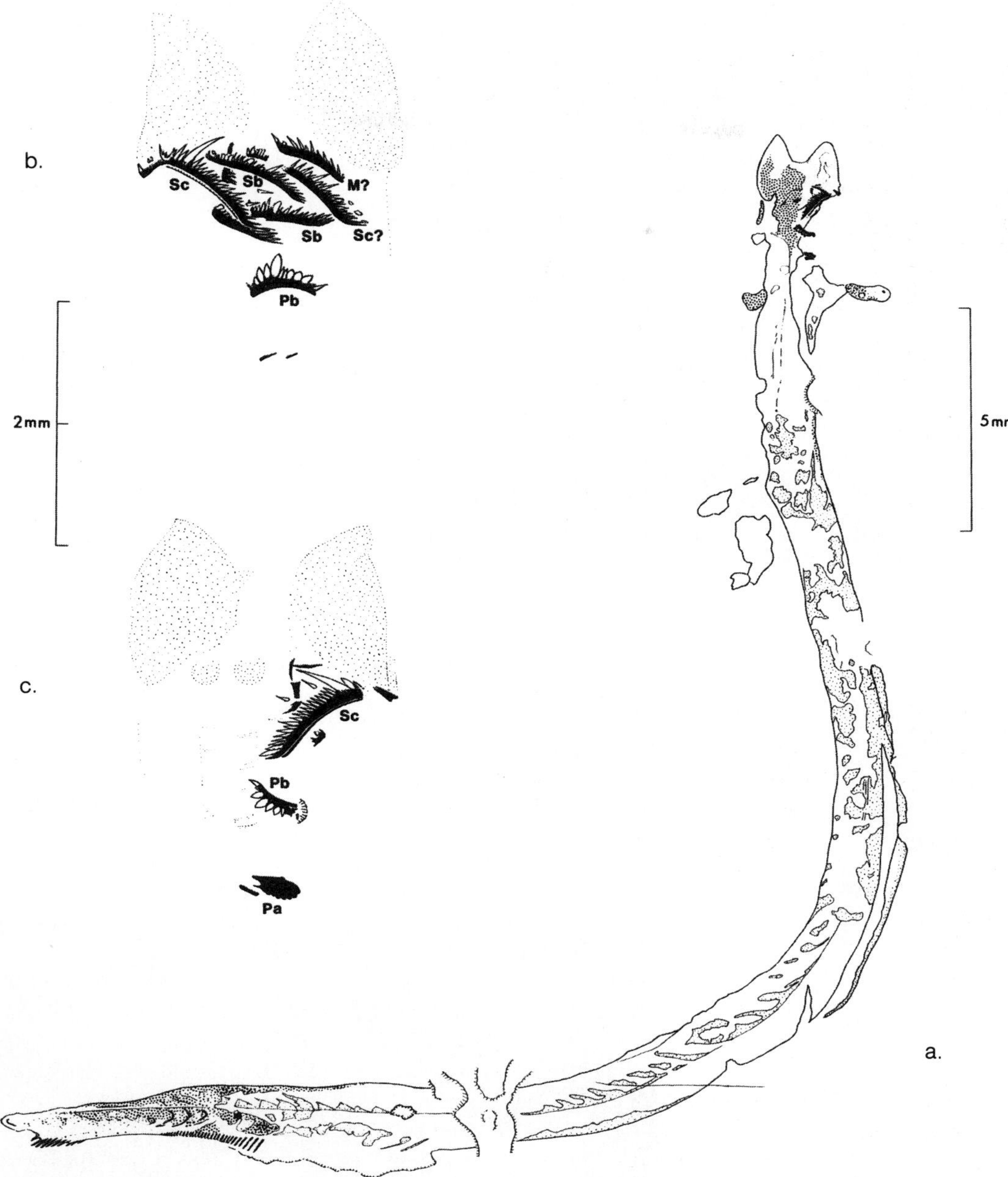


**Figure 12–15** a. Line drawing of *Clydagnathus*? cf. *cavusformis*, the conodont animal. b and c. The conodont elements found in this animal.

toporoids formed a ridge on the seaward side such that a quiet water zone and lagoon formed behind it. In the lagoon area, a wide variety of invertebrates flourished, including brachiopods, crinoids, bryozoans, and mollusks, among others. While the fauna of these Silurian and Devonian reefs was somewhat different from that of earlier reefs and reef-like structures, the general composition and structure are the same as in modern reefs.

The Silurian and Devonian periods were also the time when *eurypterids* (arthropods with scorpion-like bodies and impressive pincers) were abundant, especially in brackish and freshwater habitats (Fig. 12–17). *Ammonoids*, a subclass of the cephalopods, evolved from nautiloids during the Early Devonian and rapidly diversified. With their distinctive suture patterns, short stratigraphic ranges, and widespread distribution, ammonoids are excellent guide fossils for the Devonian through Cretaceous periods (Fig. 12–18).

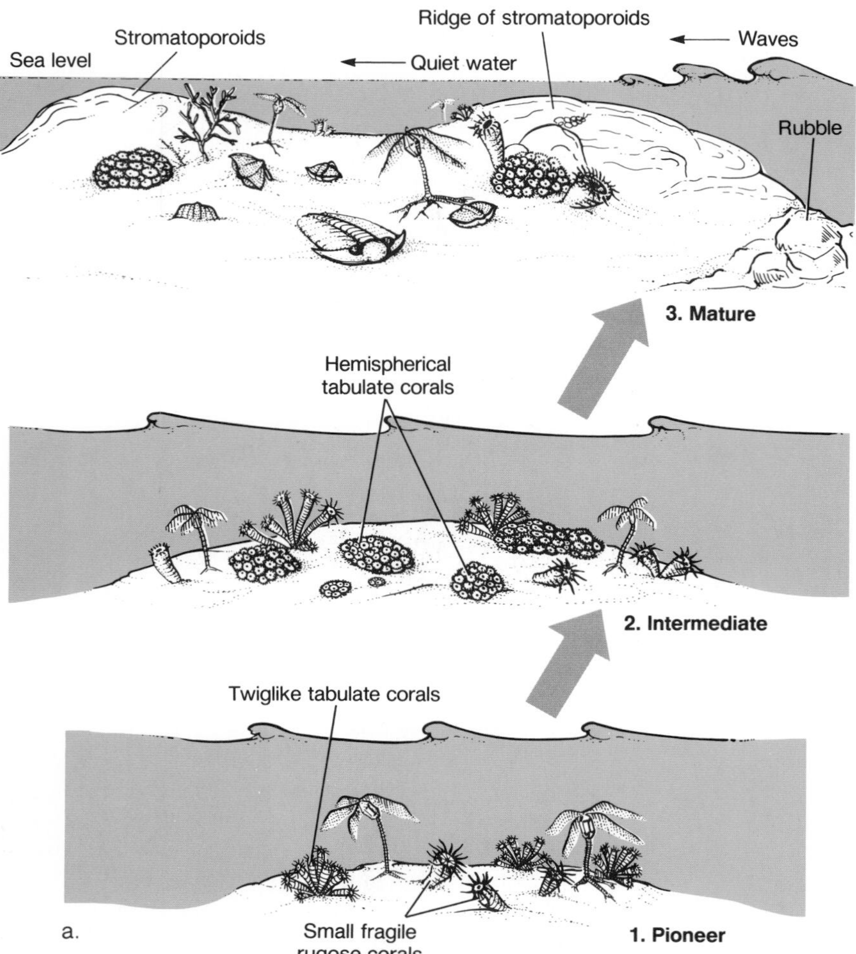

**Figure 12–16** a. Ecological succession of a typical Devonian reef. (1) Fragile, twiglike tabulate and rugose corals comprised the pioneer community. (2) Tabulate corals dominated the intermediate stage of development. (3) The reef grew close to sea level in the mature stage. A ridge of stromatoporoids absorbed the energy from breaking waves; behind them was a diverse community of species adapted to quieter water. b. Reconstruction of a Middle Devonian Reef from the Great Lakes area. It represents the area behind the stromatorporoid ridge, and shows corals, ammonoids, trilobites, and brachiopods.

**Figure 12–17** Reconstruction of a Silurian brackish-marine bottom scene near Buffalo, New York. Shown are algae, eurypterids, worms, and shrimp.

Another mass extinction occurred near the end of the Devonian Period. Geologists divide the Late Devonian Epoch into two ages—Frasnian and Famennian. This second great mass extinction occurred in Late Frasnian to Early Famennian time and resulted in a world-wide near total collapse of the large massive reef communities. On land, however, the seedless vascular plants were seemingly unaffected, although, the diversity of freshwater fish was greatly reduced.

Thus, in the marine realm, and particularly the reef and pelagic communities, extinctions were most extensive. Although the massive tabulate-rugose-stromatoporoid reefs declined during the Frasnian, they became virtually extinct by the end of the Famennian Age. Approximately 80 percent of the Frasnian brachiopod genera did not survive into the Famennian Age. Ammonoids experienced a similar decline, and many gastropods and bryozoans also became extinct during this time.

The acritarchs, the main food source of the suspension feeders, underwent a concurrent dramatic decline in abundance and diversity, perhaps contributing in part to the extinctions of many suspension-feeding invertebrates at this time.

The demise of the Middle Paleozoic reef community serves to highlight the geographic aspects of the Late Devonian mass extinction event. The tropical taxa

**Figure 12–18** *Imitoceras rotatorium* (DeKoninck), a goniatitic ammonoid cephalopod from the Lower Mississippian Rockford Limestone, near Rockford, Indiana. The distinctive suture pattern, short stratigraphic range, and wide distribution make ammonoids excellent guide fossils.

## Perspective 12–1

# Extinctions: Cyclical or Random?

Throughout geologic history, different plant and animal species have become extinct. In fact, extinction is a common feature of the fossil record, and the rate of extinction through time has fluctuated only slightly. Just as new species evolve, others become extinct. There have been, however, brief intervals in the past during which mass extinctions have eliminated large numbers of species (Fig. 1). Extinctions of this magnitude could only occur due to radical changes in the environment on a regional or global scale. The cause of these mass extinctions has been the subject of study and debate by geologists for many years.

When we look at the different mass extinction events that have occurred in the geologic past, there seem to be several common themes. The first is that mass extinctions have affected life both in the sea and on land. Second, there has been a preferential disappearance of tropical organisms during mass extinctions, especially in the marine realm. A third theme has been the tendency of certain animal groups to experience mass extinctions repeatedly. Such groups did not totally disappear during the first crises. The survivors diversified following the crisis, only to be depleted by another mass extinction. Three marine invertebrates display this characteristic: the trilobites, graptolites, and ammonoids. The fourth theme of mass extinctions is the alleged periodicity displayed during the Phanerozoic Eon. It has been proposed that mass extinctions have occurred every 26 million years.

When we look at the mass extinctions for the last 650 million years, we see that the first event involved only the acritarchs. Several extinction events occurred during the Cambrian Period, and these affected only marine invertebrates, mostly trilobites. Three other marine mass extinctions took place during the Paleozoic Era, one at the end of the Ordovician Period, involving many invertebrates; one near the end of the Devonian Period, affecting the major barrier reef–building organisms as well as the primitive armored fish, and the most severe one at the end of the Permian Period, when between 70 to 90 percent of the marine species became extinct. On land, a group of reptiles called pelycosuars also became extinct at the end of the Permian Period.

The Mesozoic Era experienced several mass extinctions, the most severe occurring at the end of the Cretaceous Period, when all large animals, including dinosaurs and sea-going animals such as plesiosaurs and ichthyosaurs, became extinct. The cause of the terminal Cretaceous extinction event is believed by many scientists to have resulted from a meteorite impact.

Several episodes of mass extinction occurred during the Cenozoic Era. The most severe was near the end of the Eocene Epoch.

While many scientists think of the marine mass extinctions as sudden events, they were rather gradual events, occurring over hundreds of thousands or even millions of years. Furthermore, instead of arguing for a catastrophic cause for the extinctions, many geologists believe that climatic changes are primarily responsible for the extinctions, particularly in the marine realm. Evidence of glacial episodes or other signs of climatic change have been correlated with the extinction events recorded in the fossil record.

---

were most severely affected; in contrast, the polar communities were seemingly little affected. Apparently, an episode of global cooling was largely responsible for the extinctions near the end of the Devonian Period. During such a cooling, the disappearance of tropical conditions would have had a severe effect on reef and other warm-water organisms. Cool-water species, on the other hand, could have simply migrated toward the equator. While cooling temperatures certainly played an important role in the Late Devonian extinctions, the closing of the Iapetus Ocean and the orogenic events of this time undoubtedly also played a role by reducing the area of shallow shelf environments where many marine invertebrates lived.

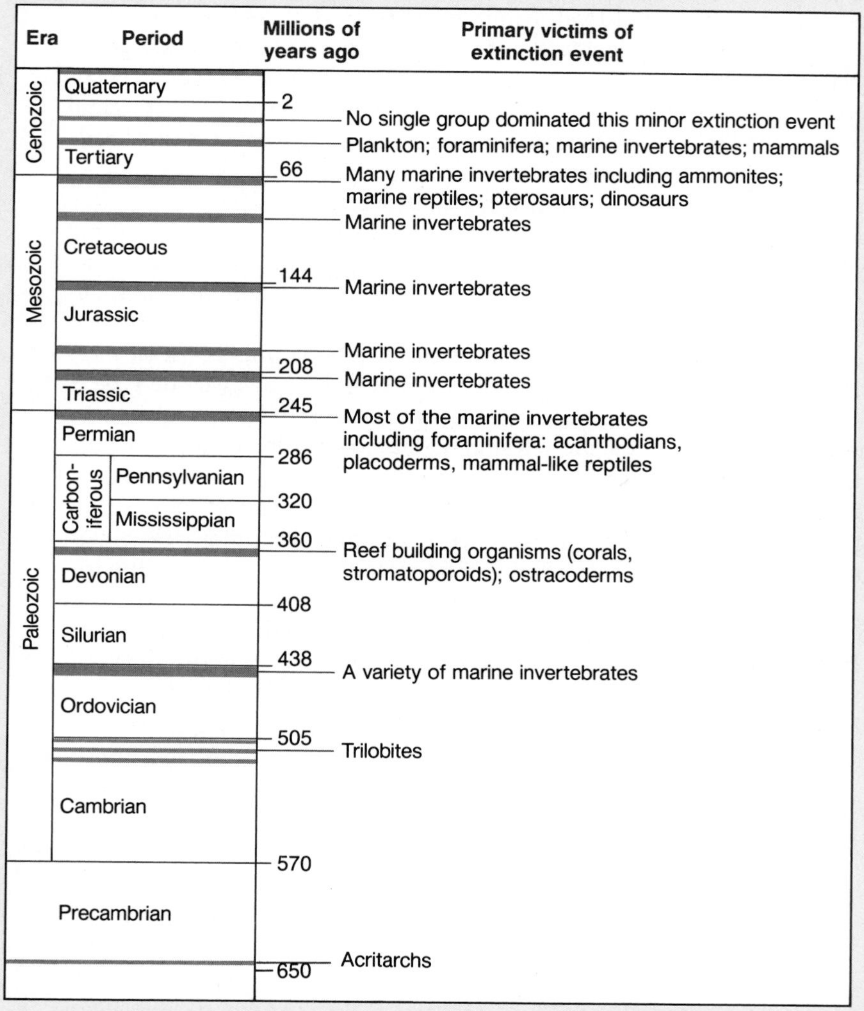

| Era | Period | Millions of years ago | Primary victims of extinction event |
|---|---|---|---|
| Cenozoic | Quaternary | 2 | |
| | | | No single group dominated this minor extinction event |
| | Tertiary | | Plankton; foraminifera; marine invertebrates; mammals |
| | | 66 | Many marine invertebrates including ammonites; marine reptiles; pterosaurs; dinosaurs |
| Mesozoic | | | Marine invertebrates |
| | Cretaceous | 144 | Marine invertebrates |
| | Jurassic | | Marine invertebrates |
| | | 208 | Marine invertebrates |
| | Triassic | 245 | Most of the marine invertebrates including foraminifera: acanthodians, placoderms, mammal-like reptiles |
| Paleozoic | Permian | 286 | |
| | Carbon-iferous: Pennsylvanian | 320 | |
| | Carbon-iferous: Mississippian | 360 | Reef building organisms (corals, stromatoporoids); ostracoderms |
| | Devonian | 408 | |
| | Silurian | 438 | A variety of marine invertebrates |
| | Ordovician | 505 | Trilobites |
| | Cambrian | 570 | |
| | Precambrian | 650 | Acritarchs |

**Figure 1** Times of mass extinctions and their primary victims.

## Carboniferous and Permian Marine Communities

The Carboniferous invertebrate marine community responded to the Late Devonian extinctions in much the same way the Silurian invertebrate marine community responded to the Late Ordovician extinctions, that is, by renewed adaptive radiation and rediversification. The brachiopods and ammonoids quickly recovered and again assumed important ecologic roles, while other groups, such as the lacy bryozoans and crinoids, reached their greatest diversity during the Carboniferous Period. With the decline of the stromatoporoids and the tabulate and rugose corals, large organic reefs like

## Perspective 12–1 (continued)

The fourth theme of mass extinctions is their apparent periodicity. There have been several proposals made that extinctions occur regularly through time. One of the most recent has been made by David Raup and John Sepkoski of the University of Chicago. They have suggested a periodicity of 26 million years since the end of the Paleozoic Era. According to them, the most recent mass extinction occurred about 11 million years ago. They originally graphed all family-level extinctions for marine invertebrates since the Middle Permian Period. They treated each family extinction as if it occurred at the end of a geologic age. They constructed their graph based on 39 ages since the Middle Permian. They found that their peaks representing times of increased extinctions approached a 26-million-year periodicity, and that some of the peaks, such as the one representing the end of the Cretaceous Period, coincided with recognized mass extinctions.

The idea of periodicity has remained controversial, in part because the dates used for the geologic ages are imperfect and vary among different workers. For instance, estimates for the end of the Oxfordian Epoch of the Jurassic Period range from 140 to 156 million years ago. Furthermore, some geologists question whether or not the plotted peaks represent anything more than what would occur randomly. In fact, three of the peaks in the original plot were only marginally above background levels of extinction.

What Raup and Sepkoski's data show is that the spacing of major extinction events does not occur precisely every 26 million years, but rather the occurrence is closer to 26-million-year intervals than would be expected for a random pattern. One possibility is that Mesozoic and Cenozoic mass extinctions may have been caused by some type of extraterrestrial event such as has been proposed for the terminal Cretaceous extinctions (see Chapter 14).

At this point, there does not seem to be strong enough evidence to attribute every major extinction event to an extraterrestrial cause. Based on the tremendous amount of knowledge that has been gained about extinctions during the past decade, we can only hope to get a clearer picture of the evolution and extinction of the Earth's biota during the next decade.

**Figure 12–19** Marine life of the Mississippian based on an Upper Mississippian fossil site at Crawfordville, Indiana. Invertebrate animals shown include crinoids, bastoids, lacy bryozoans, and small corals.

**Figure 12–20** Reconstruction of a Permian patch-reef community from the Glass Mountains of West Texas. Shown are algae, productid brachiopods, cephalopods, and corals.

those existing earlier in the Paleozoic virtually disappeared and were replaced by small patch reefs. These reefs were dominated by crinoids, blastoids, lacy bryozoans, brachiopods, and calcareous algae and flourished during the Late Paleozoic (Fig. 12–19). In addition, bryozoans, crinoids, and fusulinids (spindle-shaped foraminifera) contributed large amounts of skeletal debris to the formation of the vast bedded limestones that constitute the majority of Mississippian sedimentary rocks.

The Permian invertebrate marine faunas resembled those of the Carboniferous Period. However, because of the restricted size of the shallow seas on the craton and the reduced shelf space along the continental margins, they were more restricted in their distribution (Fig. 11–2b). The spiny and odd-shaped productids dominated the brachiopod assemblage and constituted an important part of the reef complexes that formed in the Texas region during the Permian Period (Fig. 12–20). The fusulinids (Fig. 12–21), which first appeared during the

a.

b.

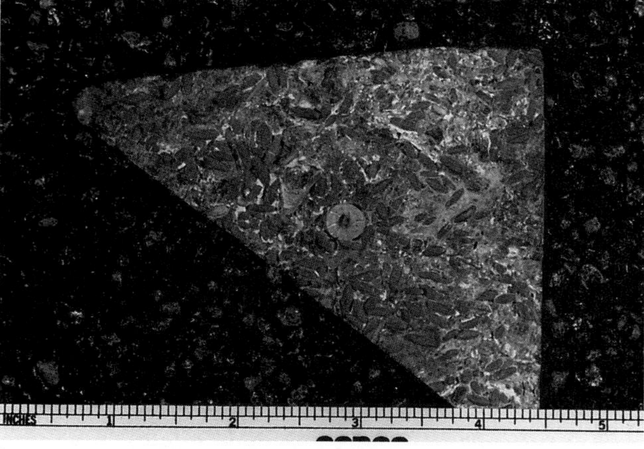


c.

**Figure 12–21** Fusulinids are a group of large benthonic foraminifera that are excellent guide fossils for the Pennsylvanian and Permian periods. a. *Leptotriticites tumidus*, Early Permian, Kansas. b. *Robustoschwagerian stanislavi*, Middle Permian, Texas. c. *Triticites*, Pennsylvanian, Kansas. (Photos courtesy of G. A. Sanderson, Amoco Production Co.)

Late Mississippian and greatly diversified during the Pennsylvanian Period, experienced a further diversification during the Permian; more than 5,000 species are known from the Permian Period alone. Because of their abundance, diversity, and worldwide occurrence, fusulinids are important guide fossils for Pennsylvanian and Permian strata. Bryozoans, sponges, and some types of calcareous algae also were common elements of the Permian invertebrate fauna.

## The Permian Marine Invertebrate Extinction Event

The greatest recorded mass extinction event to affect the marine invertebrate community occurred at the end of the Permian Period (Fig. 12–22). Before the Permian Period ended, roughly one-half of all marine invertebrate families and perhaps 90 percent of all marine invertebrate species became extinct. Fusulinids, rugose and tabulate corals, two bryozoan orders, and two brachiopod orders as well as trilobites and blastoids did not survive the end of the Permian Period. All of these groups had been very successful during the Paleozoic Era.

What caused such a crisis for the marine invertebrates? Many hypotheses have been proposed, but presently no completely satisfactory answer exists. Two currently discussed hypotheses are (1) a reduction of living area related to widespread regression of the seas and suturing of the continents and (2) decreased ocean salinity due to widespread arid climates. As we discussed earlier, there was a trend toward continental convergence during the Paleozoic, culminating in the formation of Pangaea by the end of the Permian Period. Such continental convergence resulted in regression of the epeiric seas from the cratons and loss of the shallow-water shelf area surrounding each continent. Decreased ocean salinity would affect those organisms with narrow salinity

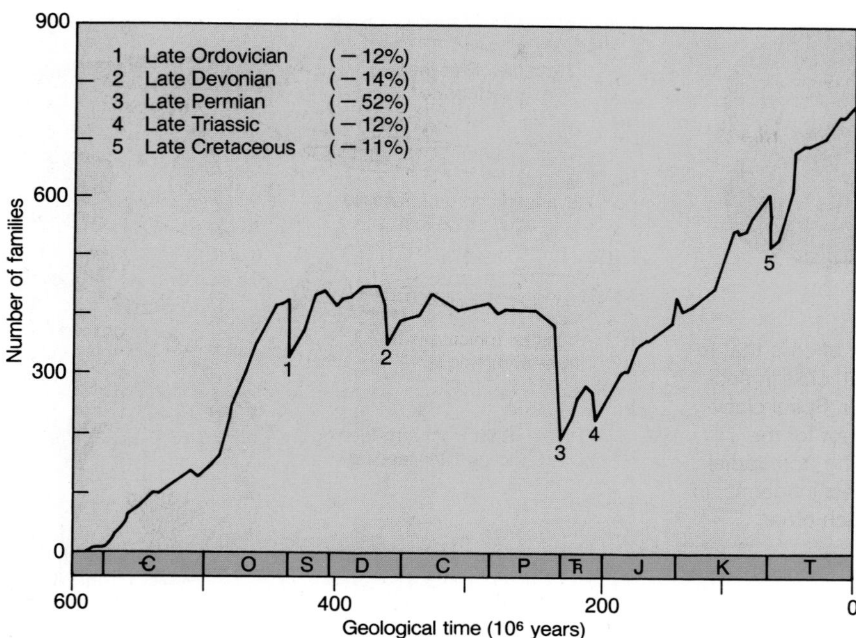


**Figure 12–22** Phanerozoic diversity for marine invertebrate and vertebrate families. Note the three episodes of Paleozoic mass extinctions, with the greatest occurring at the end of the Permian Period.

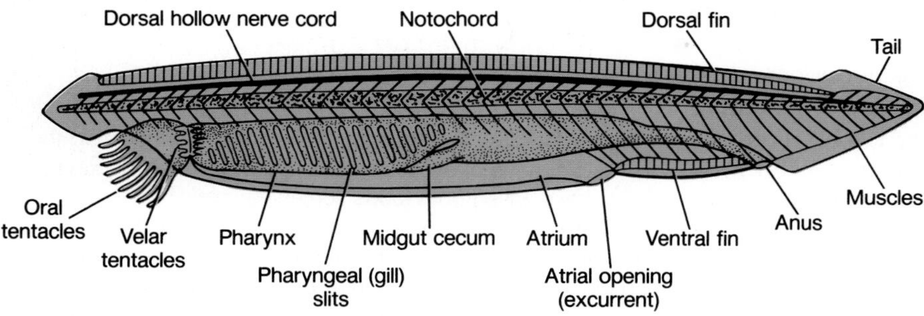


**Figure 12–23** Structure of the lancelet *Amphioxus* illustrating the three characteristics of a chordate: a notochord, dorsal hollow nerve cord, and gill slits.

tolerances, including most marine invertebrates. Extensive marginal marine evaporite deposits were formed during the Permian Period, and it is hypothesized that removal of the salts from the ocean to form these deposits lowered ocean salinity to lethal levels for many invertebrates. However, other calculations show that not enough salts could have been removed by this method to have significantly lowered the salinity level for the entire Panthalassic Ocean.

The Permian mass extinctions were probably caused by a combination of many interrelated geologic and biologic factors. In any case, the surviving marine invertebrate faunas of the Early Triassic were of very low diversity and were widely distributed around the world. This wide distribution indicates that either the taxa were tolerant of a wide temperature range, or the latitudinal temperature gradient was very gentle, enabling species of normal temperature tolerance to spread far and wide.

## VERTEBRATE EVOLUTION

A **chordate** is an animal that has, at least during part of its life cycle, a notochord, a dorsal hollow nerve cord, and gill slits (Fig. 12–23). **Vertebrates**, which are animals with backbones, are simply a subphylum of chordates.

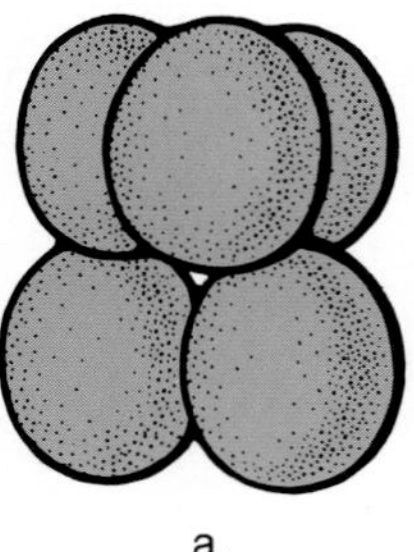


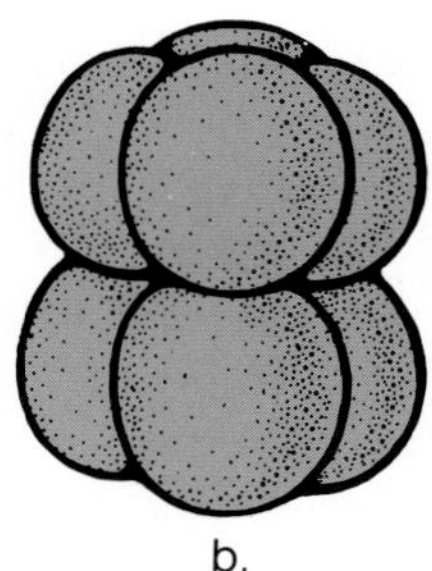


**Figure 12–24** a. Geometrical arrangement of cells that result from spiral cleavage. In this arrangement, cells in successive rows are nested between each other. Spiral cleavage is characteristic of all invertebrates except for the echinoderms. b. Arrangement of cells resulting from radial cleavage is characteristic of chordates and echinoderms. In this arrangement, cells are directly above each other.

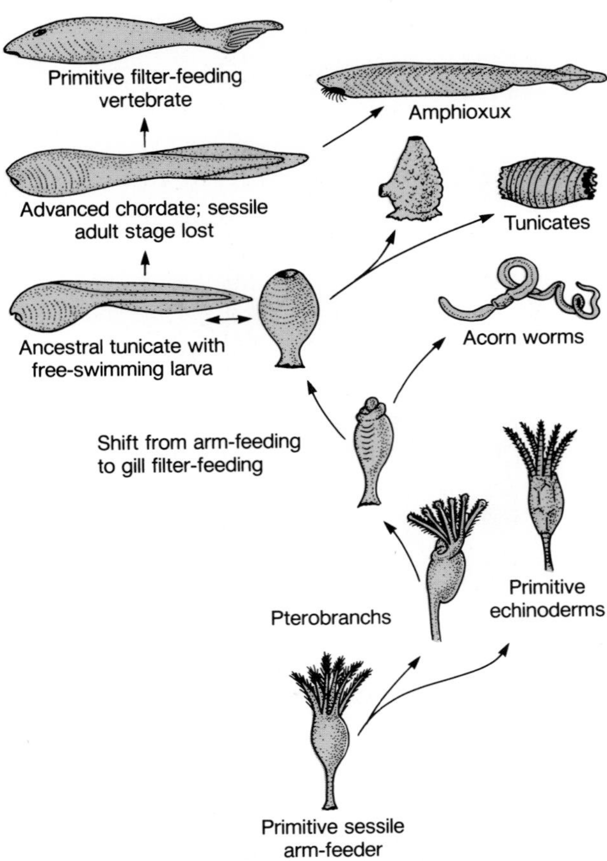


**Figure 12–25** A diagrammatic family tree suggesting the possible mode of evolution of vertebrates. The echinoderms may have arisen from forms somewhat similar to small pterobranchs; the acorn worm may have evolved from pterobranch descendants that had themselves evolved a gill-feeding system but were somewhat more advanced in other regards. Tunicates represent a stage in which, in the adult, the gill apparatus has become highly evolved. The important point though, is the development in some tunicates of a free-swimming larva with such advanced features as a notochord, nerve cord, and free swimming habit. In further progress to the lancelet *Amphioxus* and the vertebrates, the old sessile adult stage has been abandoned, and it is the larval type that has initiated the advance.

The ancestors and early members of the phylum Chordata were soft-bodied organisms that left few fossils. Consequently, we know very little about the early evolutionary history of the chordates or vertebrates. Surprisingly, a very close relationship exists between echinoderms and chordates. They may even have shared a common ancestor, since the development of the embryo is the same in both groups and differs completely from other invertebrates (Fig. 12–24). Furthermore, the biochemistry of muscle activity, blood proteins, and the larval stages are very similar in both echinoderms and chordates.

The evolutionary pathway to vertebrates may have begun with a sessile suspension-feeding animal with exposed cilia on its arms, somewhat like the one shown in Figure 12–25. Subsequently, these organisms evolved into nektonic gilled animals, perhaps looking somewhat like *Amphioxus* (Fig. 12–23). With modification of the notochord to vertebrae, the first true vertebrates evolved. (See Perspective 12–2.)

## FISH

The most primitive vertebrates are the fishes, and the oldest fish remains are found in the Upper Cambrian Deadwood Formation in northeastern Wyoming (Fig. 12–26). Here phosphatic scales and plates of *Anatolepis*, a primitive ostracoderm (armored jawless fish) have been recovered from marine sediments. All known Cambrian and Ordovician fish remains have been found in shallow, nearshore marine deposits. While this does not prove fish originated in the oceans, it does lend strong support to the idea. The earliest nonmarine fish remains are found in Silurian strata.

Figure 12–27 shows the geologic occurrence of the various major fish groups. The oldest and most primitive are the **ostracoderms**, whose name means "bony

## Perspective 12–2

# The Discovery of the Oldest Complete Fossil Fish

The oldest known vertebrate fossils are phosphatic plate and scale fragments of the jawless armored fish (ostracoderm) *Anatolepis*. These fragments are Late Cambrian (510 million years) and Early Ordovician (470 million years) in age and come from widely scattered areas in North America and Spitzbergen (Fig. 1). Far more complete remains of a similar fish (*Arandaspis*) are known from Early Ordovician aged rocks in Australia. In these specimens, the bone itself is not preserved but is represented by well-preserved molds in the rock showing detailed impressions of both the internal and external surface (Fig. 2).

Recently, two collections comprising 30 remarkably well-preserved ostracoderms were recovered from Lower Ordovician rocks (470 million years) by an international group of paleontologists during a collecting expedition in southern Bolivia. At least 10 of the specimens were virtually complete (Fig. 3).

David K. Elliott, vertebrate paleontologist at Northern Arizona University, Flagstaff, Arizona, said upon viewing the specimens, "This is one of the most exciting and important discoveries in lower-vertebrate studies in the last 50 years."

The fossils are up to 45 cm long and 15 cm wide. Bony plates cover the head region, while thin scales cover the body and tail. These fish lived in shallow marine waters and were probably poor swimmers. They are different enough from the Late Cambrian-Early Ordovician North American, Spitzbergen, and Australian fish to be placed in a new genus *Sacabambaspis*, named after a village near the fossil discovery.

The discovery of these fossils is significant because they are the oldest complete fossil fish specimens known. Furthermore, since fish are now known to have been widespread and diverse around 470 million years ago, they must have a longer history than paleontologists had previously assumed.

**Figure 1** Bony fragment of the ostracoderm fish *Anatolepis* from the Upper Cambrian Deadwood Formation, Wyoming. (Photo courtesy of John E. Repetski, U.S.G.S.)

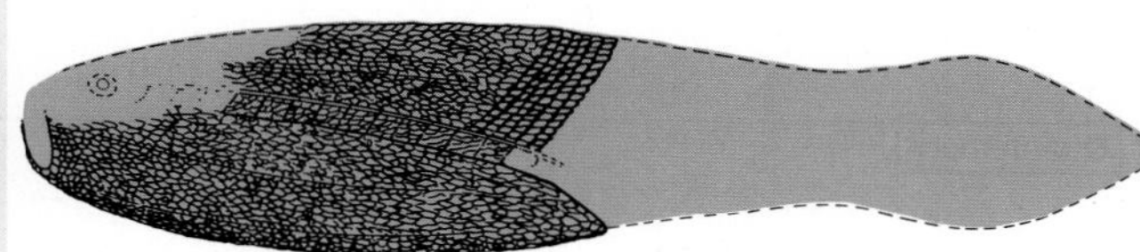

**Figure 2** Reconstruction of the Early Ordovician Australian fish *Arandaspis* based on well-preserved molds. The original length of this fish is estimated to be 12 to 14 cm.

**Figure 3** Fossil specimen of the newly discovered Bolivian ostracoderm fish *Sacabambaspis*.

**Figure 12–26** A fragment of a plate from *Anatolepis* cf. *A. heintzi,* the oldest known fish from the Upper Cambrian Deadwood Formation of Wyoming. (Photo courtesy of John E. Repetski, U.S.G.S.)

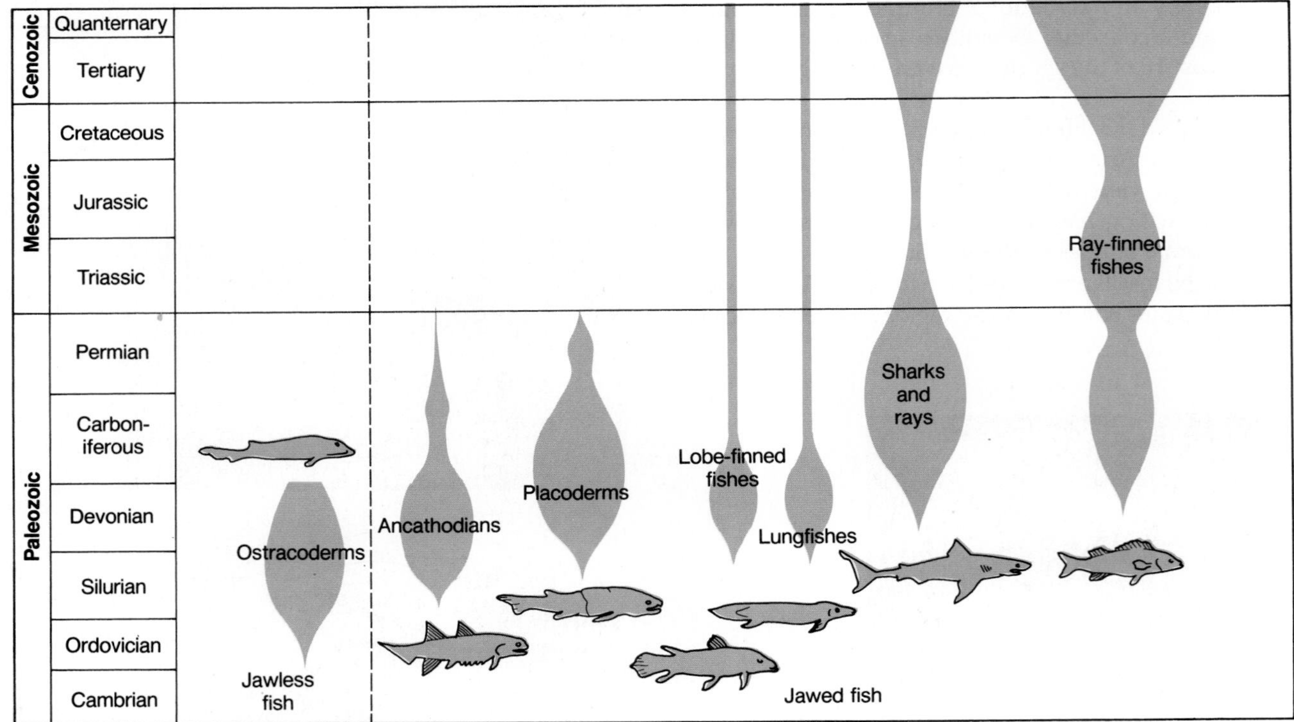


**Figure 12–27** Geologic occurrence of the major fish groups. All of the major groups of fishes were present during the Devonian Period. However, few placoderms and no ostracoderms survived beyond the Devonian Period.

skin." These are armored jawless fish that first appeared during the Late Cambrian, reached their zenith during the Silurian and Devonian periods, and then became extinct.

The majority of ostracoderms lived on the sea bottom. *Hemicyclaspis* is a good example of a bottom-dwelling ostracoderm (Fig. 12–28). It had a flattened underside, vertical scales, eyes on the top, and a mouth on the bottom of its head. The vertical scales allowed the fish to wiggle sideways, propelling itself along the sea floor, while the eyes on the top of its head allowed it to see such predators as cephalopods and jawed fish approaching from above. While moving along the sea bottom, it probably sucked up small bits of food and sediments through the jawless mouth on the bottom of its head. Another type of ostracoderm, represented by *Pteraspis* was more elongated and probably an active swimmer, although it also seemingly fed on small pieces of food it could suck up (Fig. 12–28).

One of the most important advances among primitive vertebrates was the evolution of jaws, which opened up new ecologic possibilities; fishes with jaws became herbivores and carnivores while their jawless ancestors could only feed on detritus. The evolution of the vertebrate jaw is an excellent example of evolutionary opportunism. The first three gill arches, which performed one function, were modified into a completely different structure (a simple jaw), which performed a different function (Fig. 12–29).

The remains of the first jawed fish, called **acanthodians** (Fig. 12–28), are found in Lower Silurian nonmarine rocks. Acanthodians are an enigmatic group of fish characterized by large spines, scales covering much of the body and reduced bony armor, jaws, and teeth. Their relationship to other fish has not been established. None of the known fossil ostracoderms are likely ancestors, and differences in the tail, spines, and teeth separate them from members of the bony fish. The acanthodians were most abundant during the Devonian, declined in importance through the Carboniferous, and became extinct during the Permian Period.

While we do not know how the acanthodians were related to other, more complex fish, we do know that during the Devonian Period a major adaptive radiation of jawed fish occurred. **Placoderms**, whose name means "plate-skinned," arose in the Late Silurian, and like acanthodians, reached their peak of abundance and

**Figure 12–28** Reconstruction of a Devonian sea floor fauna showing an ostracoderm (*Hemicyclaspis*) (a), placoderm (*Bothriolepis*) (b), acanthodian (*Parexus*) (c), and ray-finned fish (*Cheirolepis*) (d).

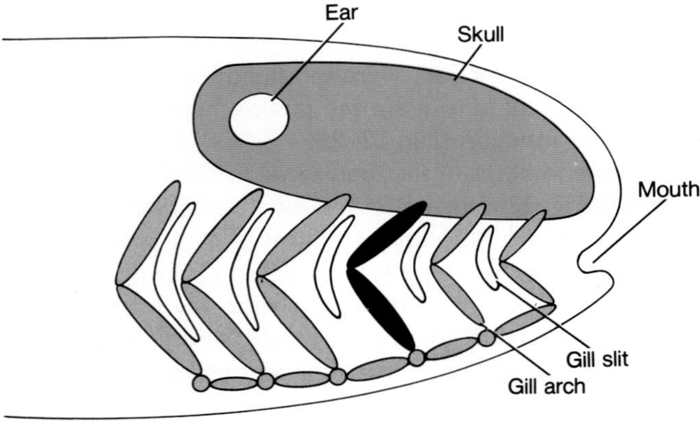


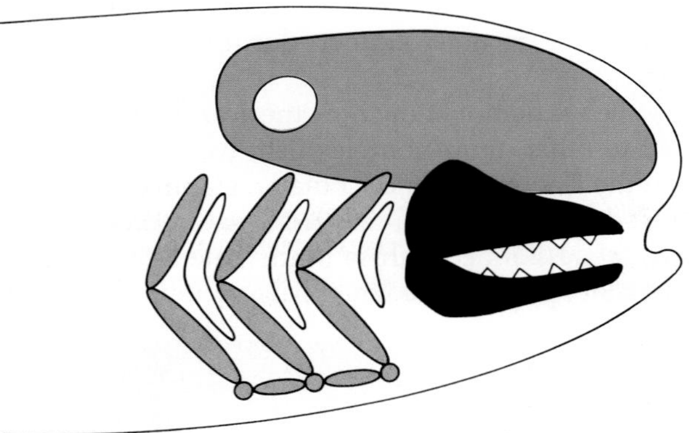


**Figure 12–29** Origin of the vertebrate jaw. Jaws (b) evolved in fishes by modification of the (a) anterior gill arch supports.

diversity during the Devonian Period. Placoderms were heavily armored jawed fish that lived in both fresh water and the ocean. Considerable variety existed among the placoderms, including small bottom-dwellers called *antiarchs* (Fig. 12–30), as well as *arthrodires,* which were the major predators of the Devonian seas. Arthrodires are best represented by *Dunkleosteus,* a Late Devonian fish that lived in the mid-continental North American epeiric seas (Fig. 12–30). It was by far the largest fish of the time, attaining a length of 10 m. It had a heavily armored head and shoulder region, a huge jaw lined with razor-sharp bony teeth, and a flexible tail, all features consistent with its status as a ferocious predator.

In addition to the abundant acanthodians, placoderms, and ostracoderms, other fish groups, such as the cartilaginous and bony fish, also made their debut during the Devonian Period. It is small wonder, then, that the Devonian is informally called the "Age of Fish," since all major fish groups were present during this period.

The **cartilaginous fish**, represented today by sharks, rays, and skates, first appeared during the Middle Devonian, and, by the Late Devonian, primitive marine sharks such as *Cladoselache* were quite abundant (Fig. 12–30). Cartilaginous fishes have never been as numerous nor as diverse as their cousins, the bony fishes, but they were, and still are, important members of the marine vertebrate fauna.

Along with the cartilaginous fish, the **bony fish** also first evolved during the Devonian Period. Because bony fish are the most varied and numerous of all the fishes, and because the amphibians evolved from them, their evolutionary history is of particular importance. The bony fish are divided into two groups, the familiar **ray-finned fish** and the **lobe-finned fish.**

The ray-finned fish first appeared in Devonian freshwater lakes and streams and from there rapidly expanded into the marine environment (Fig. 12–30). The term *ray-finned* refers to the way the fins are supported by thin bones that spread away from the body (Fig. 12–31). From a modest beginning during the Devonian, the ray-finned fish, which include most of the familiar marine and freshwater fishes such as trout, bass, perch, salmon, and tuna, rapidly diversified to dominate the Mesozoic and Cenozoic seas.

Present-day lobe-finned fish possess sturdy muscular fins and lungs. The fins do not have radiating bones, but rather articulating bones with the fin attached to the body by a fleshy shaft (Fig. 12–31). There are two major groups of lobe-finned fish, the *lung fish* and *crossopterygians.* The lung fish were fairly abundant during the Devonian Period, but today only three freshwater genera exist, one each in South America, Africa, and Australia. Their present-day distribution presumably reflects the Mesozoic breakup of Gondwana. Studies of modern lung fish indicate that lungs evolved from saclike bodies on the ventral side of the esophagus. These saclike bodies enlarged and improved their capacity for oxygen extraction, eventually evolving into lungs. When the lakes or streams in which lung fish live become stagnant and dry up, they breathe at the surface or burrow into the sediment to prevent dehydration. However, when the water is well oxygenated, lung fish rely upon gill respiration.

The **crossopterygians** are a second group of lobe-finned fish and a most important group because it was from them that the amphibians evolved. During the Devonian Period, two separate branches of crossopterygians evolved. One led to the amphibians, which we

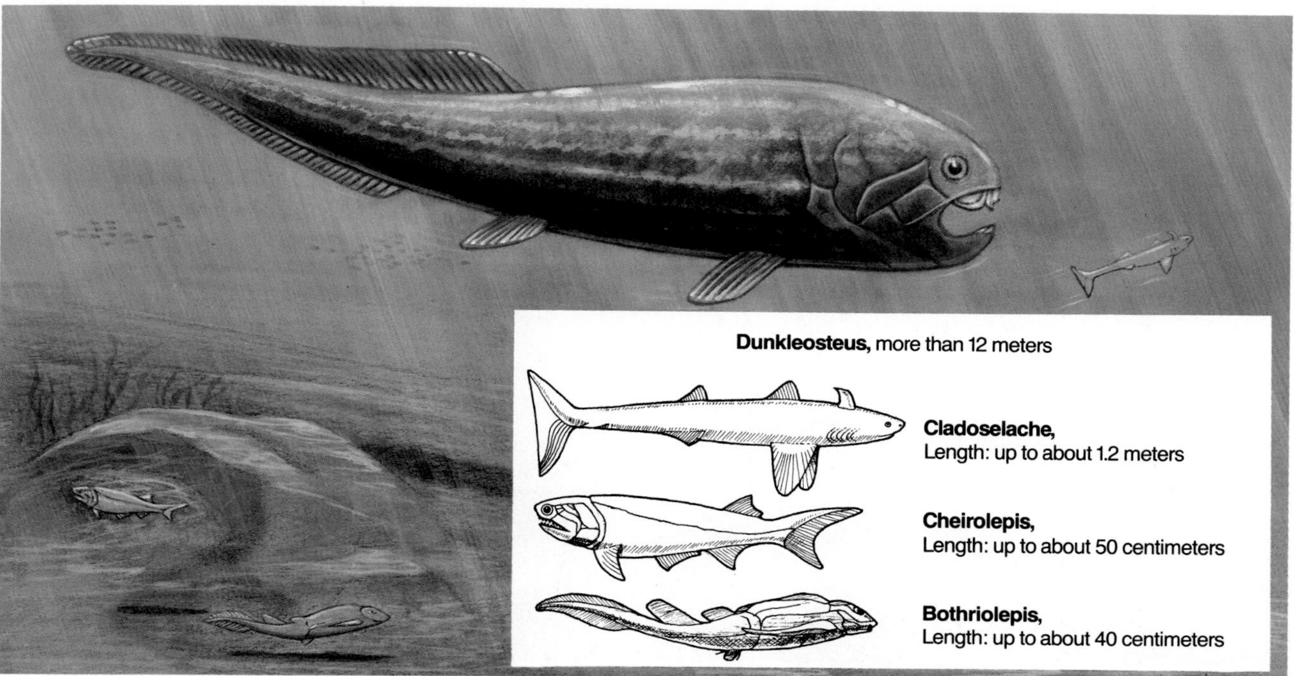


**Figure 12–30** Recreation of a Late Devonian marine scene. Shown is the giant placoderm *Dunkleosteus* in pursuit of the shark *Cladoselache*. Also shown is the bottom-dwelling placoderm *Bothriolepis* and the swimming ray-finned fish *Cheirolepis*.

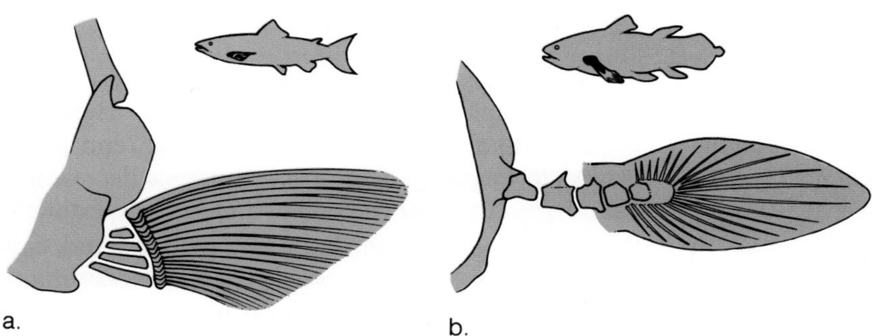


**Figure 12–31** Arrangement of fin bones for a typical ray-finned fish (a) and a lobe-finned fish (b). The muscles extend into the fin of the lobe-finned fish, allowing greater flexibility of movement than for the ray-finned fish.

will discuss shortly, while the other invaded the sea. This latter group, the *coelacanths*, were thought to have become extinct at the end of the Cretaceous. However, in 1938, fishermen caught a coelacanth in the deep waters off Madagascar (Fig. 5–27), and since then several more have been caught.

The group of crossopterygians that is ancestral to amphibians are called *rhipidistians*. These fish, attaining lengths of over 2 m, were the dominant freshwater predators of the Late Paleozoic. *Eusthenopteron*, a good example of a rhipidistian crossopterygian, had an elongate body that enabled it to move swiftly in the water, as

**Figure 12–32** *Eusthenopteron*, a member of the rhipidistian crossopterygians. The crossopterygians are the group from which the amphibians are thought to have evolved. *Eusthenopteron* had an elongate body and paired fins that could be used for moving about on land. It was about 28 cm long.

well as paired muscular fins that could be used for locomotion on land (Fig. 12–32). The structural similarity between crossopterygian fish and the earliest amphibians is striking and one of the better documented transitions from one major group to another (Fig. 12–33).

Before discussing this transition and the evolution of amphibians, it would be useful to place the evolutionary history of Paleozoic fish in the larger context of Paleozoic evolutionary events. Certainly the evolution and diversification of jawed fish as well as eurypterids and ammonoids had a profound effect on the marine ecosystem. Previously defenseless organisms either evolved defensive mechanisms or suffered great losses, possibly even extinction. As you will recall, trilobites suffered major extinctions at the end of the Cambrian Period, recovered slightly during the Ordovician, then declined greatly from the Silurian to their ultimate demise at the end of the Permian Period. Perhaps their lightly calcified external covering made them easy prey for the rapidly evolving jawed fish and cephalopods. Ostracoderms, although armored, would also have been easy prey for the swifter jawed fishes. Ostracoderms became extinct by the end of the Devonian Period, a time that coincides with the rapid evolution of jawed fish. The jawed placoderms also became extinct by the end of the Devonian Period while the acanthodians decreased in abundance after the Devonian, and became extinct by the end of the Paleozoic Era. On the other hand, cartilaginous and ray-finned bony fish expanded during the Late Paleozoic, as did the ammonoid cephalopods, the other major predator of the Late Paleozoic seas.

## AMPHIBIANS—THE VERTEBRATES INVADE THE LAND

Although amphibians were the first vertebrates to live on land, they were not the first land-living organisms. Land plants, which probably evolved from green algae, first evolved during the Silurian Period. Furthermore, animals, including early insects, millipedes, spiders, and even snails invaded the land before amphibians (see Perspective 12-3).

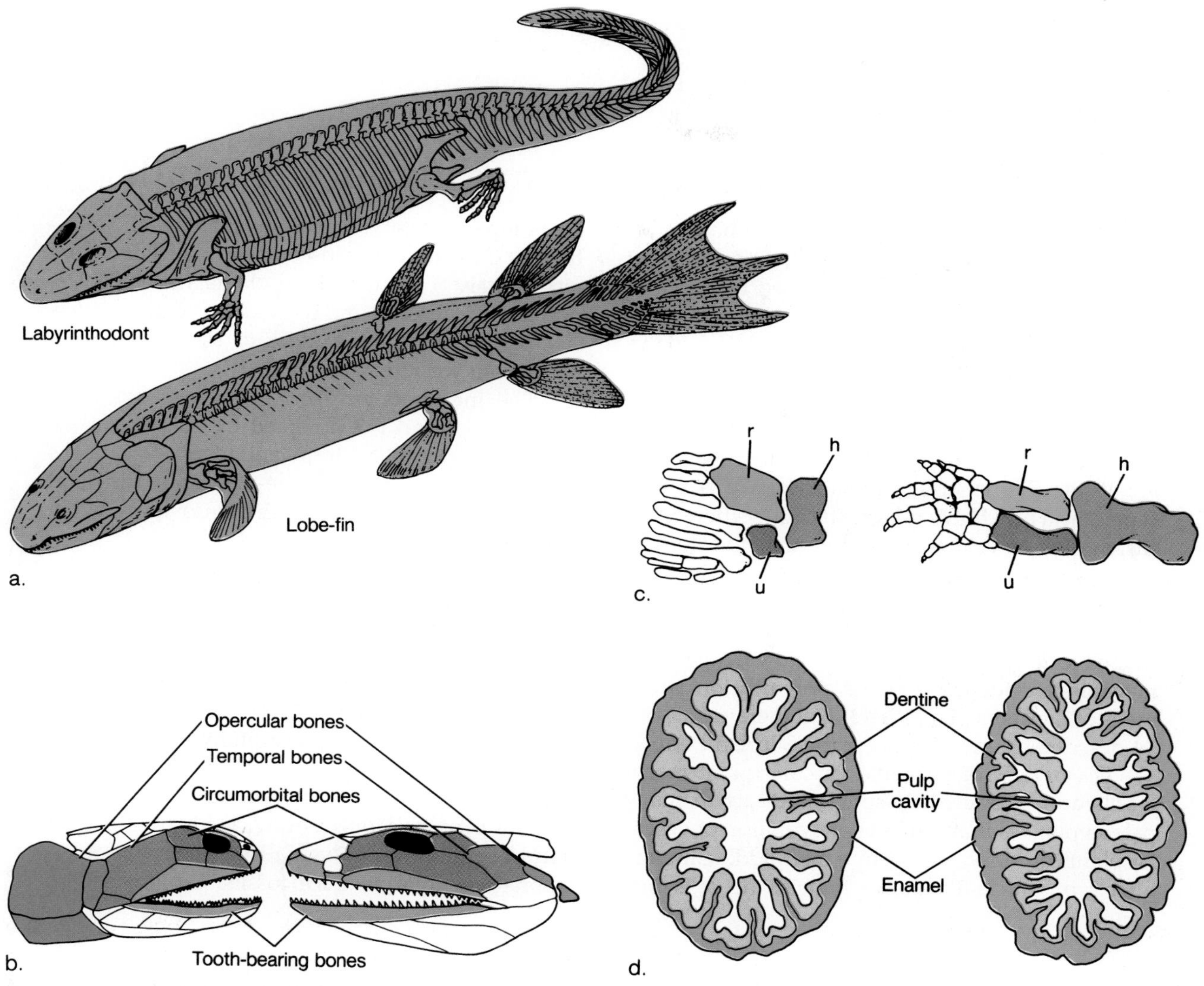


**Figure 12–33** Similarity between the crossopterygians and the earliest amphibians. a. Similarity of skeleton. b. Comparison of skulls and lower jaws of a crossopterygian (left) and an ichthyostegid amphibian (right). Shading identifies the bones the two skulls have in common. c. Comparison of the limb bones of a crossopterygian (left) and amphibian (right); shading identifies the bones (u = ulna, r = radius, h = humerus) that the two groups have in common. d. Comparison of cross sections of teeth showing the complex structure found in both the crossopterygians (left) and amphibians (right).

The transition from water to land required that several barriers be surmounted. The most critical for animals were desiccation, reproduction, the effects of gravity and extracting oxygen from the atmosphere by lungs rather than from water by gills. These problems were partly solved by the crossopterygians; they already had a backbone and limbs that could be used for walking, as well as lungs for extracting oxygen (Fig. 12-33).

The earliest amphibian fossils are found in the Upper Devonian Old Red Sandstone of eastern Greenland (see Perspective 12–4). These amphibians had streamlined bodies, long tails, and fins. In addition, they had four legs, a strong backbone, a rib cage, and pelvic and pectoral girdles, all of which were structural adaptations for walking on land (Fig. 12-34). The earliest amphibians appear to have many characteristics that were inherited from crossopterygians with little modification (Fig. 12-33).

Amphibians did not play a significant role in the terrestrial ecosystem of the Devonian Period. However, as

## Perspective 12–3

# Was There A Late Ordovician Invasion of the Land?

Recently there has been much discussion as to when the land was first colonized. There is excellent evidence that the land was colonized by vascular plants during the Silurian Period. Fragments of plant remains have been found in Middle Silurian rocks indicating primitive plants had invaded the land by that time. These early plants, like their Devonian descendents, had to live near bodies of water due to their reproductive requirements.

While Silurian plant fossil evidence is represented by a handful of geographically scattered fossils, the spore evidence for a Silurian land colonization is much more abundant and diverse. This may be due in part to the greater chance for preservation of spores, which have a resistant organic covering. In the North Atlantic region, Silurian vascular plant megafossil evidence is based primarily on specimens of *Cooksonia* (Fig. 12–42). Yet there are at least 14 described spore genera for the Silurian Period.

Discoveries of probable vascular plant megafossils and characteristic spores indicate to many paleontologists that the time of evolution of vascular plants occurred well before the Middle Silurian. Sheets of cuticle-like cells—that is, the cells that cover the surface of modern land plants—and tetrahedral clusters that closely resemble the spore tetrahedrals of primitive land plants, have been reported by Jane Grey and co-workers of the University of Oregon from Middle to Late Ordovician rocks from western Libya (Fig. 1).

The interpretation of these spores has been controversial and not accepted by all paleontologists. If they are in fact from land plants, this means that land plants had a long pre-Silurian record. And if land plants existed during the Ordovician, when did animals invade the land?

The first vertebrate animals made the transition from water to land during the Late Devonian. However, arthropods, including scorpions and flightless insects, evolved at least by the Early Devonian, based on their fossil representatives found in the Lower Devonian Rhynie Chert of Scotland. This unit is also famous for its beautifully preserved fossil plant flora.

A recent discovery of fossil burrows within a buried soil of the Upper Ordovician Juanita Formation in central Pennsylvania represents the oldest reported nonmarine trace fossils. While no hard

with other groups that moved into new and previously unoccupied niches, amphibians underwent rapid adaptive radiation and became abundant during the Carboniferous and Early Permian. The Late Paleozoic amphibians did not at all resemble the familiar frogs, toads, newts, and salamanders that make up the modern amphibian fauna. Rather they displayed a broad spectrum of sizes, shapes, and modes of life (Fig. 12–35). One such group of amphibians were the **labyrinthodonts**, so named for the labyrinthine wrinkling and folding of their teeth. Most labyrinthodonts were large animals, as much as two meters long. These typically sluggish creatures wallowed around in swamps and streams, eating fish, vegetation, insects, and other small amphibians (Fig. 12–35). The labyrinthodonts declined in abundance during the Permian Period, perhaps in response to changing climatic conditions and only a few survived into the Triassic Period.

## EVOLUTION OF THE REPTILES—THE LAND IS CONQUERED

Amphibians were limited in colonizing the land because they had to return to water to lay their gelatinous eggs. The evolution of the **amniote egg** (Fig. 12–36) freed reptiles from this constraint. In such an egg, the developing embryo is surrounded by a liquid-filled sac called the

parts or traces of organisms are associated with the burrows, their size, shape, and arrangement are consistent with a bilaterally symmetrical burrower that was resistant to desiccation. It is hypothesized that millipedes may have been the organism making these burrowers. Millipedes first appear in marine rocks of Early Silurian age, and some modern millipedes are active burrowers.

The existence of sizable burrowing organisms on dry land during the Late Ordovician implies there was some type of terrestrial vegetation for them to feed on. Burrowing animals could conceivably have fed on soil algae, now believed to have been in existence since the Late Precambrian. However, possible spores of primitive land plants have been reported from Middle to Upper Ordovician rocks. Such plants could have supported large populations of burrowing herbaceous arthropods as well as litter organisms. The discovery of Late Ordovician trace fossils in paleosoils is additional evidence for Late Ordovician terrestrial ecosystems, thus pushing back the colonization of the land further than had been believed.

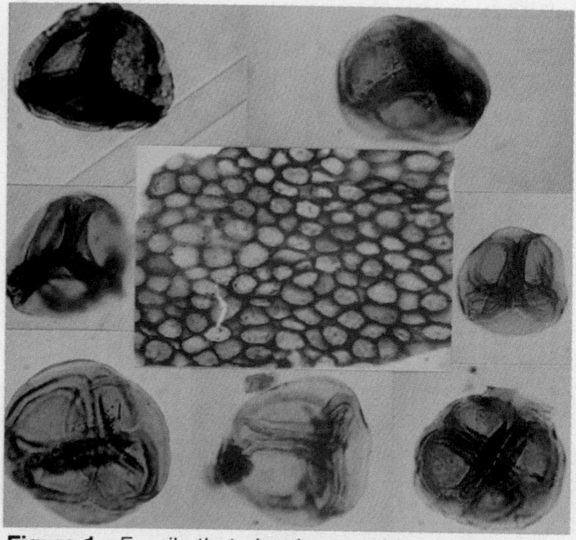

**Figure 1** Fossils that closely resemble the spore tetrahedrals of primitive land plants. The sheet of cuticlelike cells (center) is from the Upper Ordivician Melez Chograne Formation of Libya. The others are from the Upper Ordovician Djeffara Formation of Libya. (Photos courtesy of Jane Grey, Oregon State University.)

*amnion* and provided with both a yolk, or food sac, and an allantois, or waste sac. In this way the emerging reptile is, in essence, a miniature adult, bypassing the need for a larval stage in the water. The evolution of the amniote egg allowed vertebrates to completely colonize the land since they no longer had to return to the water as part of their reproductive cycle. The oldest known structure thought to be a reptilian egg comes from Lower Permian rocks of Texas, although some paleontologists question its authenticity. It is, however, generally acknowledged that the amniote egg evolved at the latest by the Early Pennsylvanian when the earliest reptile fossils are found.

Many of the differences between amphibians and reptiles are physiological and not preserved in the fossil record. However, there are enough differences found in the skull structure, limb construction, jaw and dental patterns, and vertebrae construction to suggest that reptiles evolved from labyrinthodont ancestors at least by the Early Pennsylvanian.

The oldest known reptiles are from the Lower Pennsylvanian Joggins Formation on the beach of the Bay of Fundy in Nova Scotia, Canada. Here, remains of *Hylonomus* are found in the sediments filling in rotted tree-trunks (Fig. 12–37). These earliest reptiles were small and agile and fed largely on grubs and insects. They belonged to the group of reptiles known as **captorhinomorphs**, the group from which all other reptiles evolved (Fig. 12-38). During the Permian Period, reptiles diversified and began to displace many amphib-

**Figure 12–34** Reconstruction of a Late Devonian landscape in the eastern part of Greenland. Shown is *Ichthyostega*, an amphibian that grew to a length of about 1 m. The flora of the time was diverse, consisting of a variety of small and large seedless vascular plants such as *Cyclostigma* (the large tree), and *Archaeopteris* (fern in front).

ians. The reptiles' success as land-dwelling animals can be accounted for by their advanced method of reproduction, their more advanced jaws and teeth, as well as their ability to move rapidly on land.

The **pelycosaurs**, or finback reptiles, evolved from the captorhinomorphs during the Pennsylvanian Period and were the dominant reptile group by the Permian Period. They evolved into a diverse assemblage of herbivores exemplified by *Edaphosaurus*, and carnivores such as *Dimetrodon* (Fig. 12–39). An interesting feature of the pelycosaurs is their sail. It was formed by vertebral spines that in life, were covered with skin. There have been numerous explanations for the function of the sail—a type of sexual display, protection, a display to look more ferocious—but current consensus seems to be that the sail served as some type of thermoregulatory device, raising the reptile's temperature by catching the sun's rays, or cooling it down by facing the wind. Because pelycosaurs are considered to be the group that the mammal-like reptiles evolved from, it is interesting that they may have had some sort of body-temperature control. Furthermore, the teeth of *Dimetrodon* were slightly differentiated, a condition characteristic of mammal-like reptiles and mammals.

The pelycosaurs became extinct during the Permian Period and were succeeded by the **therapsids**, mammal-like reptiles that evolved from the carnivorous pelycosaur lineage and rapidly diversified into herbivorous and carnivorous lines. Therapsids were small to medium-sized animals displaying the beginnings of many mammalian features: fewer bones in the skull due to fusion of many of the small skull bones, enlargement of the lower jawbone, differentiation of the teeth for various functions such as nipping, tearing, and chewing food, and a more vertical position of the legs for greater flexibility, as opposed to the way the legs sprawled out to the side in primitive reptiles (Fig. 12–39).

Furthermore, many paleontologists believe therapsids were endothermic, or warm-blooded, enabling them to maintain a constant internal body temperature. Such a condition allows exploitation of a variety of habitats, as indicated by their wide latitudinal distribution in Permian rocks (Fig. 12–40).

As the Paleozoic Era came to an end, the therapsids constituted about 90 percent of the known reptile genera and occupied a diverse range of ecological niches. The mass extinctions at the close of the Paleozoic that so greatly decimated the marine fauna had an even greater effect on the terrestrial population. By the end of the Permian Period, about 50 percent of the invertebrate marine families became extinct, compared with 75 percent of the amphibians and 80 percent of the reptile families. Plants, on the other hand, apparently did not experience as great a turnover as the animals.

**Figure 12–35** Reconstruction of a Carboniferous coal swamp showing the varied amphibian fauna of the time, including the large labyrinthodont amphibian *Eryops* (right), *Branchiosaurus* (left), and the serpent-like *Dolichosoma* (background).

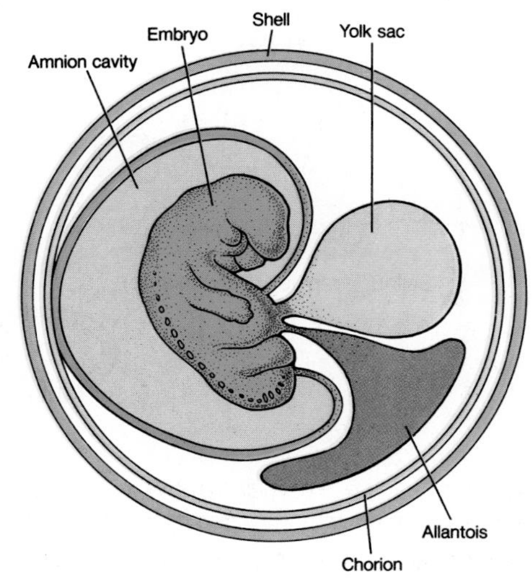


**Figure 12–36** The amniote egg. In such an egg, the embryo is surrounded by a liquid sac (amnion cavity) and provided with a food source (yolk sac), and waste sac (allantois). The evolution of the amniote egg freed reptiles from having to return to the water for reproduction and allowed them to completely conquer the land.

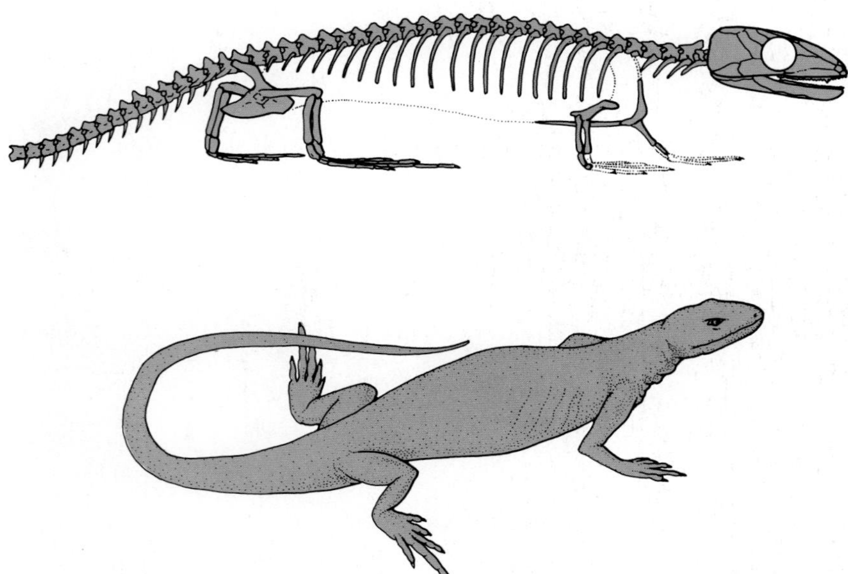

**Figure 12–37** Reconstruction and skeleton of the oldest known reptile, *Hylonomus lyelli* from the Pennsyvanian Period. Fossils of this animal have been collected from sediments that filled rotted tree stumps. *Hylonomus lyelli* was about 30 cm long.

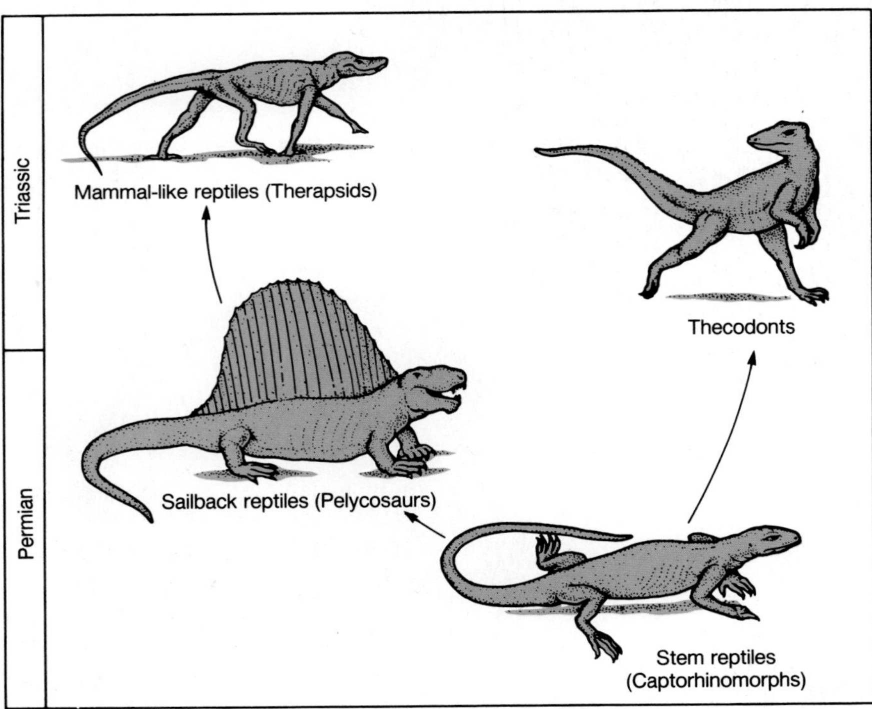


**Figure 12–38** Evolutionary relationships among the earliest reptiles.

**Figure 12–39** The pelycosaurs, or finback reptiles, are characterized by a sail on their back. One hypothesis explains it as a type of thermoregulatory device. Other hypotheses are that it was a type of sexual display or a device to make it look more intimidating. Shown here are the carnivore *Dimetrodon* (a) and the herbivore *Edaphosaurus* (b).

## PLANT EVOLUTION

When plants made the transition from water to land, they had to solve the same problems that animals did: desiccation, support, and the effects of gravity. Plants did so by the evolution of a variety of structural adaptations that were fundamental to the subsequent radiations and diversification that occurred during the Silurian, Devonian, and later periods (Table 12–2). Most experts agree that the ancestors of land plants first evolved in a marine environment, then moved into a freshwater environment and finally onto land. In this way, the problem of differences in osmotic pressures between salt and freshwater were overcome while the plant was still in the water.

Most land plants are *vascular*, which means they have a system of cells specialized for the movement of water and nutrients, as opposed to the *nonvascular* plants such as bryophytes and fungi, which do not have these specialized cells. The ancestor of terrestrial vascular plants was probably a green alga. While no fossil record exists of the transition from green algae to terrestrial vascular plants, comparison of their physiology reveals a strong link. Primitive *seedless vascular plants* (discussed further later on) such as the ferns resemble green algae in their pigmentation, important metabolic enzymes, and their reproductive cycle (Fig. 12-41). In addition, the green algae are one of the few plant groups to have made the transition from marine to fresh water.

The emergence of terrestrial vascular plants from an aquatic algal ancestry sometime during the Middle Silurian was accompanied by certain modifications that allowed them to occupy the new environment. (See Perspective 12-3 for a discussion of whether terrestrial vascular plants originated during the Ordovician Period.) The development of vascular tissue was an important step, as it allowed for the transport of food and water through the plant and provided some support for the plant body. Additional strength comes from the organic compounds *lignin* and *cellulose*, which are found throughout a plant's walls.

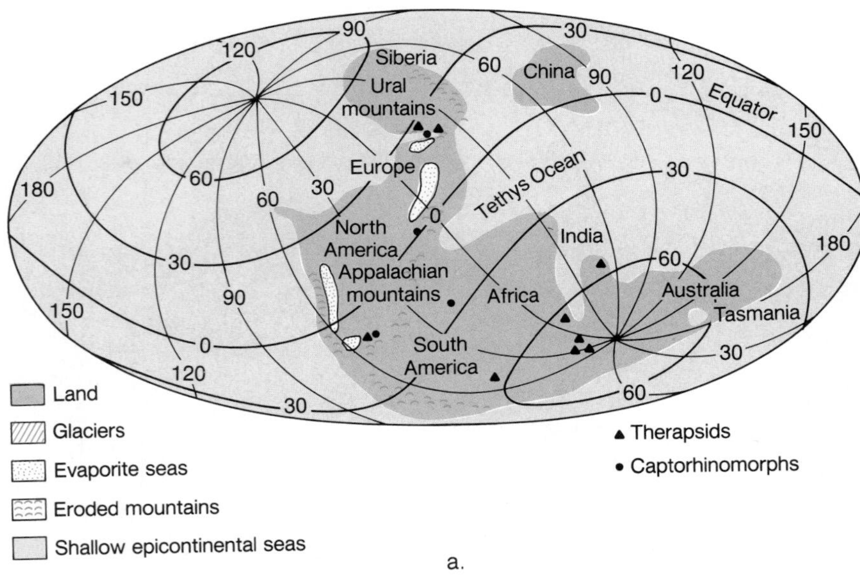

**Figure 12–40** a. Distribution of therapsids during the Permian. Notice that many of the localities are in the high southern latitudes of Gondwana where animals must have acquired the ability for high heat production and developed some insulation to have survived. b. Reconstruction of a Permian scene in Gondwana. Shown is the carnivorous therapsid *Cynognathus*. Many paleontologists believe therapsids were endothermic and may have had a covering of fur as shown here.

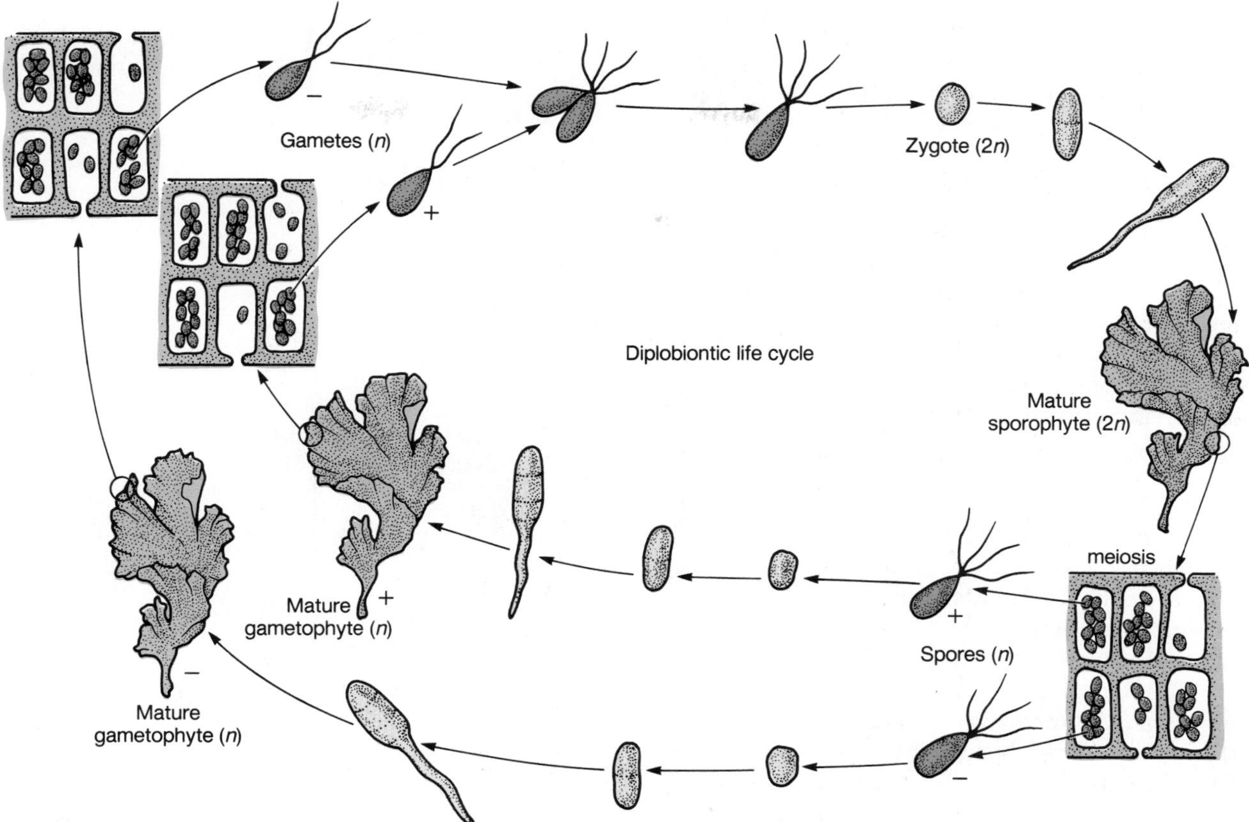


**Figure 12–41** The diplobiontic life cycle of the green algae *Ulva*. Here the zygote grows to form a multicellular sporophyte plant that produces spores by meiosis. These develop into a mature gametophyte plant that produces gametes that fuse to produce a zygote, which in turn grows into the sporophyte plant. This alternation of generations is the same method of reproduction found in vascular plants and is evidence that the vascular plants evolved from a green algae ancestor.

The problem of desiccation was circumvented by the evolution of *cutin*, an organic compound found in the outer-wall layers of plants. Cutin also provides additional resistance to oxidation, the effects of ultraviolet light, and the entry of parasites.

*Roots* evolved in response to the need to collect water and nutrients from the soil and to help anchor the plant in the ground. The evolution of *leaves* from tiny outgrowths on the stem or from branch systems provided plants with an efficient light-gathering system for photosynthesis.

## Silurian and Devonian Floras

Prior to the invasion of the land by vascular plants during the Silurian Period, the landscape was undoubtedly rather stark. There were certainly algae, and perhaps even land plants related to algae or bryophytes, but probably nothing else. Animals could not have survived until there was food for them. So, while the Ordovician and Silurian seas were teeming with life, the land was an uninviting and inhospitable place.

The earliest known vascular land plants are Y-shaped stems assigned to the genus *Cooksonia* from the Middle Silurian of Wales and Ireland. Together with Upper Silurian and Lower Devonian species from Scotland, New York state, Czechoslovakia, and the USSR, these earliest plants were small, simple, leafless stalks with a spore-producing structure at the tip (Fig. 12–42), and are known as **seedless vascular plants** because they did not produce a seed. They did not have a true root system, either, but anchorage and water uptake

## Perspective 12–4

# The Oldest Amphibians in North America

Geologists in Iowa reported in June 1988 the discovery of the oldest *tetrapod* (four-legged animal) fossil site in North America. The fossils were recovered from the Mississippian (335-million-year-old) St. Louis Formation, near Delta in southeast Iowa. Several hundred tetrapod fossils have been collected, and at least two amphibian and eight fish species have been discovered to date, with excavation of the site more than half completed.

Only 22 fossil sites of comparable age (Late Devonian and Mississippian) in the world are known, and most of those have yielded only fragmentary material. The Delta site represents the oldest amphibian fossils to be found in midcontinental North America.

The two amphibian species include the earliest temnospondyl (animals 1 to 2 m long, resembling salamanders) and a species informally called *protoanthracosaur* (Fig. 1). The temnospondyls had well-developed legs for walking on land, yet probably spent most of their time in the water. Protoanthracosaur combines features found in later amphibians with those found in the most primitive anthrocosaurs.

An important aspect of this find is that it will help paleontologists to better determine the relationship of the earliest amphibians to their fish ancestors. Furthermore, this fossil discovery will provide important information about the morphology and phylogenetic relationships among the earliest tetrapods.

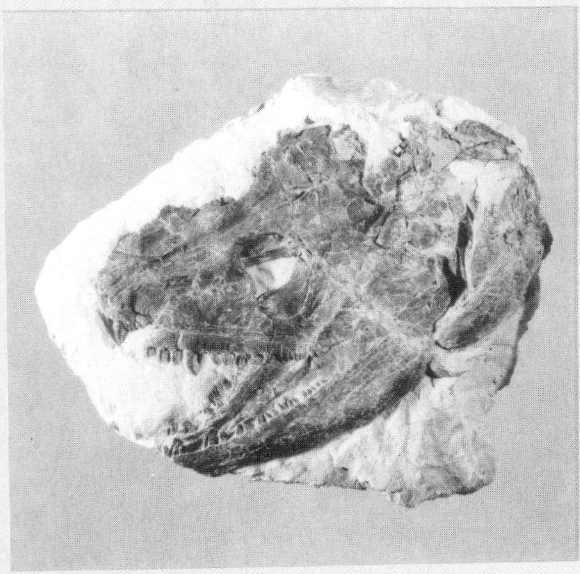

**Figure 1** Side view of fossil protoanthracosaur skull.

was accomplished by a *rhizome*, the underground part of the stem. The sedimentary rocks in which these plant fossils are found indicate that they lived in low, wet, marshy, freshwater environments.

An interesting parallel can be seen between seedless vascular plants and amphibians. Both were the first to make the transition from water to land and overcome the problems such a transition involved. Both groups, while very successful, were nonetheless still tied to a source of water required by their reproductive methods. In the case of amphibians, their gelatinous egg had to remain moist, while the seedless vascular plants required water for the sperm to travel through and reach the egg.

From this simple beginning, the seedless vascular plants evolved many of the major structural features characteristic of modern plants such as leaves, roots, and secondary growth. All of these features did not evolve simultaneously but rather at different times, a pattern known as *mosaic evolution*. This diversification and adaptive radiation took place during the Late Silurian and Early Devonian and resulted in a tremendous increase in diversity (Fig. 12–43). From the end of the Early Devonian to the end of the Devonian Period, the number of plant genera remained about the same, yet the composition of the flora changed. While the Early Devonian landscape was dominated by relatively small, low-

**Table 12–2** Major events in the evolution of land plants. The Devonian Period was a time of rapid evolution for the land plants. Major events were the appearance of leaves, heterospory, secondary growth, and the emergence of seeds.

| Period | Epoch | MYA | Events |
|---|---|---|---|
| Quarternary | | 2 | Flowering plants |
| Tertiary | | 60 | |
| Cretaceous | | 140 | |
| Jurassic | | 213 | Cycads |
| Triassic | | 248 | Conifer-type seed plants |
| Permian | | 286 | ferns |
| Carboniferous | | 360 | Seed ferns; Lycophytes |
| Devonian | Upper | 374 | Arborescence; Seeds; Megaphyllous leaves; Progymnosperms |
| | Middle | 387 | Secondary growth; Zosterophyllophytes |
| | Lower | 408 | Major diversification of vascular plants; Trimerophytes; Heterospory; Microphyllous leaves |
| Silurian | Pridoli | 414 | Rhyniophytes |
| | Ludlow | 421 | |
| | Wenlock | 428 | Tracheids; *Cooksonia* |
| | Llandovery | 438 | |
| Ordovician | Upper | 458 | First land plants |
| | Lower | 505 | |

lying, bog-dwelling types of plants, the Late Devonian witnessed forests of large tree-sized plants up to 10 m tall (Fig. 12–34).

In addition to the diverse seedless vascular plant flora of the Late Devonian, another significant floral event took place. The evolution of the seed at this time liberated land plants from their dependence on moist conditions and allowed them to completely spread over the land.

**Figure 12–42** The earliest known fertile land plant was *Cooksonia*, seen in this fossil from the Upper Silurian of South Wales. *Cooksonia* consisted of upright, branched stems terminating in sporangia (spore producing structures). It also had a resistant cuticle and produced spores typical of a vascular plant. These plants probably lived in moist environments such as mud flats. This specimen is 1.49 cm long. (Photo courtesy of Dianne Edwards, University College, England)

**Figure 12–43** Reconstruction of an Early Devonian landscape showing some of the earliest land plants. a. *Dawsonites.* b. *Protolepidodendron.* c. *Bucheria.*

Seedless vascular plants require moisture for successful fertilization because the sperm must travel to the egg on the surface of the gamete-bearing plant (gametophyte) to produce a successful spore-generating plant (sporophyte). Without moisture, the sperm would dry out before reaching the egg (Fig. 12–44). However, in the seed method of reproduction, the spores are not released to the environment as they are in the seedless vascular plants, but are retained on the spore-bearing plant, where they grow into the male and female forms of the gamete-bearing generation. In the case of the **gymnosperms**, or flowerless seed plants, these would be the male and female cones (Fig. 12–44). The male cone produces pollen, which contains the sperm and has a waxy coating to prevent desiccation, while the egg, or embyronic seed, is contained in the female cone. After fertilization, the seed then develops into a mature, cone-bearing adult plant. In this way the requirement of a moist environment for the gametophyte generation is solved. The significance of this development is that seed plants, like reptiles, were no longer restricted to wet areas, but were free to migrate into previously unoccupied dry environments. As will be discussed in Chapter 14, the evolution of the flowering plants (angiosperms) did not occur until the Cretaceous Period.

Before the seed plant could evolve, an intermediate evolutionary step was necessary. This was the development of *heterospory*, whereby a species produces two types of spores; a large one (megaspore) which gives rise to the female gamete-bearing plant, and a small one (microspore) that produces the male gamete-bearing plant. Prior to heterospory, plants were *homosporous;* that is, they produced spores all the same size that, upon germinating, grew into the gametophyte plant that produced both male and female gametes.

The earliest evidence of heterospory is found in the Early Devonian plant *Chaleuria cirrosa,* which produced spores of two distinct sizes (Fig. 12–45). The appearance of heterospory was followed several million years later by the emergence of progymnosperms—Middle and Late Devonian plants with fernlike reproductive habits and a gymnosperm anatomy—which gave rise in the Late Devonian to such other gymnosperm groups as the seed ferns and conifer-type seed plants. While the seedless vascular plants dominated the flora of the Carboniferous coal-forming swamps, the gymnosperms made

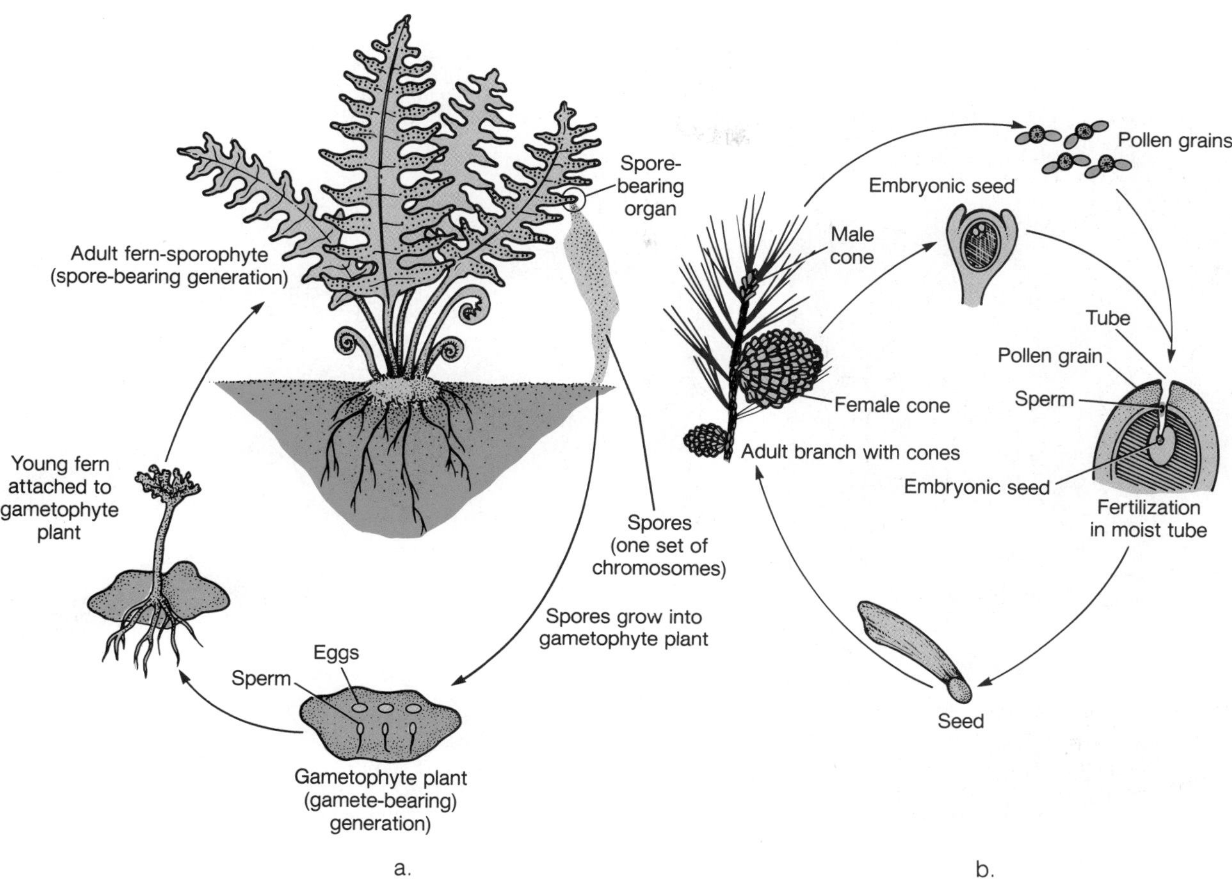


**Figure 12–44** a. Generalized life history of a seedless vascular plant. The sporophyte adult plant produces spores, which upon germination grow into small gametophyte plants that produce sperm and eggs. The fertilized eggs grow into the spore producing adult plant and the sporophyte-gametophyte life cycle begins again. b. Generalized life history of a gymnosperm plant. The adult plant bears both male cones that produce sperm-bearing pollen grains and female cones that contain embryonic seeds. Pollen grains are transported to the female cones by the wind. Fertilization occurs when the sperm, moving through a moist tube that grows from the pollen grain unites with the embryonic seed which then grows into a cone-bearing adult plant.

up an important element of the Late Paleozoic flora, particularly in the non-swampy areas.

## Late Carboniferous and Permian Floras

As discussed earlier, the rocks of the Pennsylvanian Period (Late Carboniferous) are the major source of the world's coal. Coal results from the alteration of plants living in low, swampy areas. The geologic and geographic conditions of the Pennsylvanian Period were ideal for the growth of seedless vascular plants, and consequently these coal swamps had a very diverse flora (Fig. 12–46).

It is evident from the fossil record that while the Early Carboniferous flora was similar to its Late Devonian counterpart, a great deal of evolutionary experimentation was occurring that would lead to the highly successful Late Paleozoic flora of the coal swamps and adjacent habitats. Among the seedless vascular plants, the lycopsids and sphenopsids were the most important coal-forming groups of the Pennsylvanian Period.

The lycopsids were present during the Devonian Period, chiefly as small plants, but by the Pennsylvanian Period they were the dominant element of the coal swamps, achieving heights up to 30m in such genera as

a.

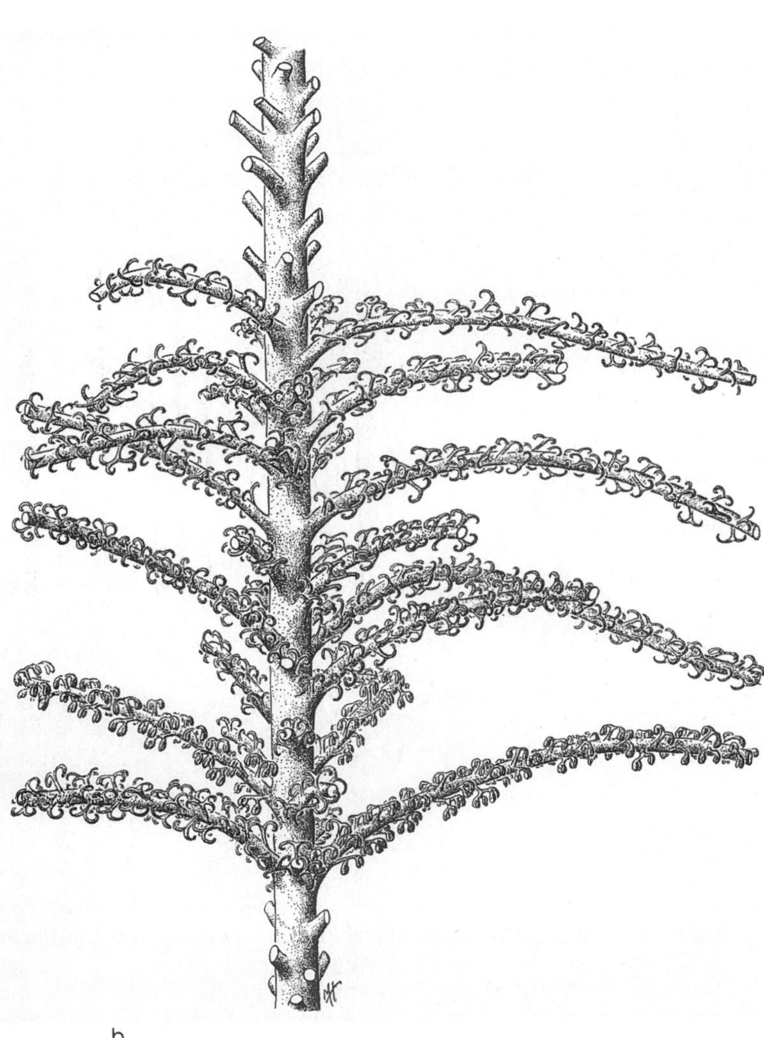

b.

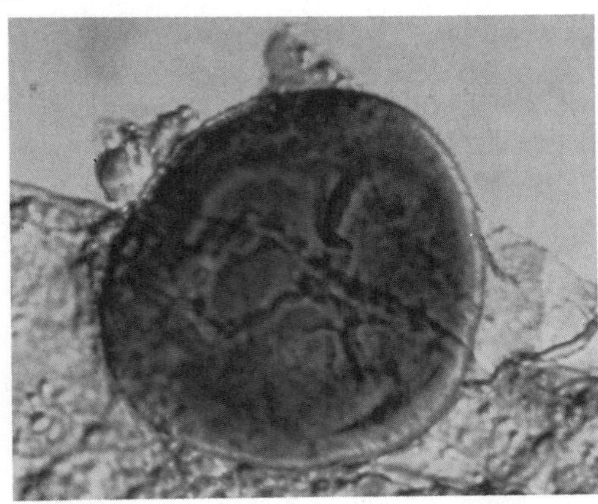

c.

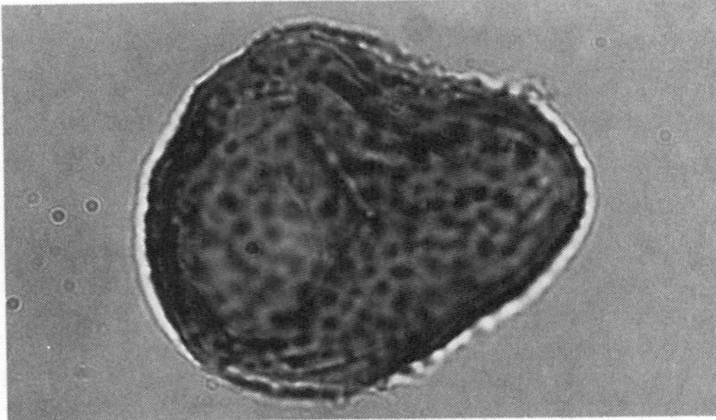

d.

**Figure 12–45** Specimen (a) and reconstruction (b) of the Early Devonian plant *Chaleuria cirrosa* from New Brunswick, Canada. This plant produced spores of two sizes; production of two sizes of spores by a plant is called heterospory. c. The larger spores are thought to have produced female

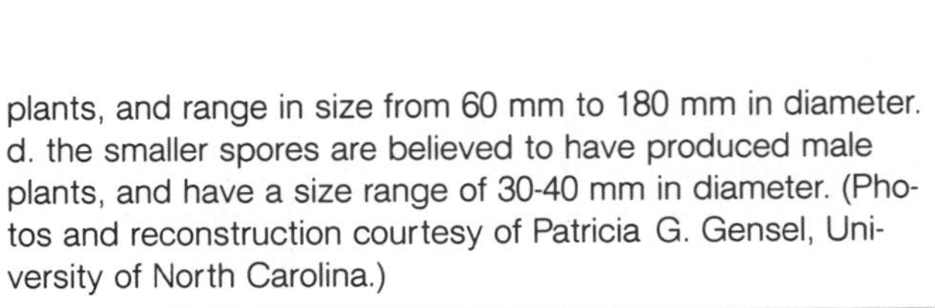

plants, and range in size from 60 mm to 180 mm in diameter. d. the smaller spores are believed to have produced male plants, and have a size range of 30-40 mm in diameter. (Photos and reconstruction courtesy of Patricia G. Gensel, University of North Carolina.)

**Figure 12–46** Reconstruction of a Pennsylvanian coal forest. Shown are the two common trees of this period *Lepidodendron* (large) and *Sigillaria* (small).

*Lepidodendron* and *Sigillaria* (Fig. 12–46). The Pennsylvanian Period lycopsid trees are interesting because they lacked branches except at their top. The leaves were elongate and similar to the individual palm leaf of today. As the trees grew, the leaves were replaced from the top, leaving prominent and characteristic rows or spirals of scars on the trunk. Today, the lycopsids are represented by small temperate-forest ground pines such as *Lycopodium*.

The sphenopsids were the other important coal-forming plant group and are characterized by being jointed and having horizontal underground stem-bearing roots. Many of these plants, such as *Calamites*, averaged 5 m to 6 m tall. Living sphenopsids include the horsetail (*Equisetum*), and scouring rushes (Fig. 12–47). Small seedless vascular plants and seed ferns formed a thick undergrowth or ground cover beneath these tree-like plants.

**Figure 12–47** Living sphenopsids include the horsetail *Equisetum*.

Not all plants were restricted to the coal-forming swamps, however. Among those plants occupying higher and drier ground were some of the *cordaites*, a group of tall gymnosperm trees that grew up to 50 m and probably formed large forests (Fig. 12–48). Another important non-swamp dweller was *Glossopteris*, the famous plant so abundant in Gondwana (Fig. 6–1), whose distribution is cited as critical evidence that the continents have moved through time.

The floras that flourished during the Pennsylvanian Period persisted into the Permian Period, but due to climatic and geologic changes resulting from tectonic events (see Chapter 11), they declined in abundance and importance. By the end of the Permian Period, the cordaites became extinct, while the lycopsids and sphenopsids were reduced to mostly small, creeping forms. The gymnosperms, whose life-style was more suited to the warmer and drier Permian climates, diversified and came to dominate the Permian, Triassic, and Jurassic landscapes.

**Figure 12–48** Reconstruction of a cordiate forest from the Late Carboniferous. Cordaites were a group of tall gymnosperm trees that grew up to 50 m tall.

## CHAPTER SUMMARY

Table 12–3 provides an overall summary of the major evolutionary events and their relationships to each other for the Paleozoic Era.

1. Soft-bodied multicelled organisms presumably had a long Precambrian history during which they lacked hard parts. The appearance of hard parts in different invertebrate groups occurred during the Cambrian and Ordovician periods and provided such advantages as protection against predators and support for muscles, enabling organisms to grow large and increase locomotor efficiency. Hard parts probably evolved as a result of several factors rather than just one.

2. The Cambrian invertebrate community was dominated by three major groups—the trilobites, brachiopods, and archaeocyathids. Little specialization existed among the invertebrates, and most phyla were represented by only a few species.

3. The Ordovician marine invertebrate community marked the beginning of the dominance by the shelly fauna and the start of large-scale reef building. The end of the Ordovician Period was a time of major extinctions for many of the invertebrate phyla.

4. The Silurian and Devonian periods were times of diverse faunas dominated by reefs, while the Caroniferous and Permian periods saw a great decline in invertebrate diversity. A major extinction occurred at the end of the Paleozoic Era, affecting the invertebrates as well as the vertebrates.

5. Chordates are characterized by a notochord, dorsal hollow nerve cord, and gill slits. The earliest chordates were soft-bodied organisms that were rarely fossilized. Vertebrates are a subphylum of the chordates.

6. The fish are the earliest known vertebrates. Their first fossil occurrence is in Upper Cambrian rocks. They have had a long and varied history, including jawless and jawed, armored forms (ostracoderms and placoderms), and cartilaginous and bony forms. The crossopterygians, a group of lobe-finned fish, gave rise to the amphibians.

7. The link between the crossopterygians and the earliest amphibians is very convincing and includes a close similarity of bone and tooth structures. During the Carboniferous Period, the labyrinthodont amphibians were the dominant terrestrial vertebrate animals.

8. The earliest record of reptiles is from the Late Carboniferous. The evolution of an amniote egg was the critical factor in the reptiles' ability to completely conquer the land.

9. The pelycosaurs were the dominant reptile group during the Early Permian, while the therapsids dominated the landscape for the rest of the Permian Period.

**Table 12–3** Major evolutionary events of the Paleozoic Era.

| Age (millions of years) | Geologic Period | | Invertebrates | Vertebrates | Plants | Extinction events |
|---|---|---|---|---|---|---|
| 245–286 | Permian | | Largest mass extinction event to affect the invertebrates. | Therapsids and pelycosaurs the most abundant reptiles. | Gymnosperms diverse and abundant. | Extinction of many groups of invertebrates. Acanthodians, placoderms, and pelycosaurs become extinct. |
| 286–320 | Carboniferous | Pennsyl-vanian | Fusulinids diversify. | Reptiles evolve. Amphibians abundant and diverse. | Coal swamps with flora of seedless vascular plants and gymnosperms. | |
| 320–360 | Carboniferous | Mississ-ippian | Crinoids, lacy bryozans, blastoids become abundant. Renewed adaptive radiation following extinctions of many reef builders. | | Gymnosperms appear (may have appeared during Late Devonian) | |
| 360–408 | Devonian | | Reef building continues. Eurypterids abundant. | Amphibians evolve. All major groups of fish present—Age of Fish. | First seed evolves. Seedless vascular plants diversify. | Extinctions of reef-building invertebrates. |
| 408–438 | Silurian | | Major reef building. Diversity of invertebrates remains high. | Ostracoderms common. Acanthodians, the first jawed fish, evolve. | Early land plants-seedless vascular plants. | |
| 438–505 | Ordovician | | Major adaptive radiation of all invertebrate groups. Suspension feeders dominant. | Ostracoderms begin to diversify. | Plants move to land? | Extinctions of a variety of marine invertebrates. |
| 505–570 | Cambrian | | Trilobites, brachiopods and archaeocyathids are most abundant. | Earliest vertebrates—jawless fish called ostracoderms. | | Mostly trilobites. |

10. Plants had to overcome the same basic problems as the animals, namely desiccation, reproduction, and gravity in making the transition from water to land.
11. The earliest seedless vascular plants were small, leafless stalks, with spore-producing structures on their tips. From this simple beginning, plants evolved many of the major structural features characteristic of modern plants.
12. By the end of the Devonian Period, forests with tree-sized plants up to 10 m had evolved. The Late Devonian also witnessed the evolution of the flowerless seed plants (gymnosperms) whose reproductive style freed them from having to stay near water.
13. The Carboniferous Period was a time of vast coal swamps, where conditions were ideal for the seedless vascular plants. With the onset of more arid conditions during the Permian Period, the gymnosperms became the dominant element of the world's flora.

## IMPORTANT TERMS

acanthodian
amniote egg
benthos
bony fish
captorhinomorphs
carnivore-scavenger
cartilaginous fish
chordate
crossopterygian
gymnosperm
herbivore
labyrinthodont
lobe-finned fish
nekton
ostracoderm
pelycosaur
placoderm
plankton
primary producer
ray-finned fish
sediment-deposit feeder
seedless vascular plant
suspension feeder
therapsid
vertebrate

## REVIEW QUESTIONS

1. Discuss the advantages hard parts, such as shells, confer on an organism.
2. Outline the major events in the evolution of the marine invertebrate community during the Paleozoic Era.
3. Why is the Burgess Shale fauna important, and what does it tell geologists about life during the Cambrian Period?
4. What were some of the probable causes for the various marine invertebrate extinctions during the Paleozoic Era?
5. Outline the evolutionary history of fish.
6. List and describe the problems that had to be overcome before organisms could inhabit the land?
7. Why were the reptiles so much more successful at living on land than the amphibians?
8. In what ways are therapsids more mammal-like than pelycosaurs?
9. What are the major differences between seedless vascular plants and gymnosperms? What is the significance of these differences in terms of occupying the terrestrial environment?

## ADDITIONAL READING

Bolt, J. R., R. M. McKay, B. J. Witzke, and M. P. McAdams. 1988. A New Lower Carboniferous Tetrapod Locality in Iowa. *Nature,* 333, no. 6175: 768–770.

Carroll, R. L. 1988. *Vertebrate Paleontology and Evolution.* New York: W. H. Freeman & Co.

Colbert, E. H. 1980. *Evolution of the Vertebrates.* New York: John Wiley & Sons.

Cowen, R. 1989. *History of Life.* Palo Alto, Calif.: Blackwell Scientific Publications.

Gensel, P. G., and H. N. Andrews, 1987. The Evolution of Early Land Plants. *American Scientist,* 75, no. 5: 478–489.

Lane, N. G. 1986, *Life of the Past,* 2d ed. Columbus, Ohio: Charles E. Merrell Publishing Co.

McMenamin, M. A. S. 1987. The Emergence of Animals. *Scientific American,* 256, no. 4: 94–102.

Thomas, B., and Spicer, R. 1987. *Evolution and Palaeobiology of Land Plants.* Portland, Oregon: Dioscorides Press.

Valentine, J. W., and E. M. Moores, 1974. Plate Tectonics and the History of Life in the Ocean. *Scientific American,* 230, no. 4: 80–89.

# 13

## CHAPTER OUTLINE



## PROLOGUE

*Approximately 150 to 210 million years after the emplacement of massive plutons created the Sierra Nevada Mountains (Nevadan orogeny), gold was discovered at Sutter's Mill (Fig. 13–1) on the South Fork of the American River, at Coloma, California. On January 24, 1848, James Marshall, a carpenter building a sawmill for John Sutter, found bits of the glittering metal in the mill's tailrace. Soon, settlements throughout the state were completely abandoned as word of the chance for instant riches spread throughout California. Within a year after the news of the gold discovery reached the East Coast, the Sutter's Mill area was swarming with more than 80,000 men, all seeking to make their fortune. In all, at least 250,000 men prospected the Sutter's Mill area, and while most were Americans, gold seekers came from almost every area of the world, including as far away as China. To most of them, it seemed that the gold must be simply waiting to be taken.*

*Of course, no one really thought of the consequences of so many people converging on the Sutter's Mill area, all intent on making easy money. In reality, only a small percentage of prospectors ever hit it big or were even moderately successful. The rest barely eked out a living until they eventually abandoned the dream and went home.*

*While it is true that some men dug $30,000 worth of gold dust a week out of a single claim and that gold was found practically on the surface of the ground, most of this easy gold was recovered very early during the gold rush. Most prospectors made only a living wage working their claims. Nevertheless, during the five years from 1848 to 1853 that constituted the gold rush proper, more than $200 million in gold was extracted.*

*Would-be prospectors could follow three basic routes to the gold fields. The most popular was the overland journey by wagon train from the East Coast to California—a route fraught with peril, including the threat of starvation, disease, and the crossing of the Sierra Nevada Mountains. Those that took this route and survived finally arrived at Sutter's Mill and dispersed from there to prospect. The Panama route to California was by ship to Chagres on the Isthmus of Panama. Here the potential prospector and his goods were transported to Panama City by boat and mules and then by ship to California. The men who came to California in this way were for the most part young and vigorous and without wives and children. They faced the usual*

# Geology of the Mesozoic Era

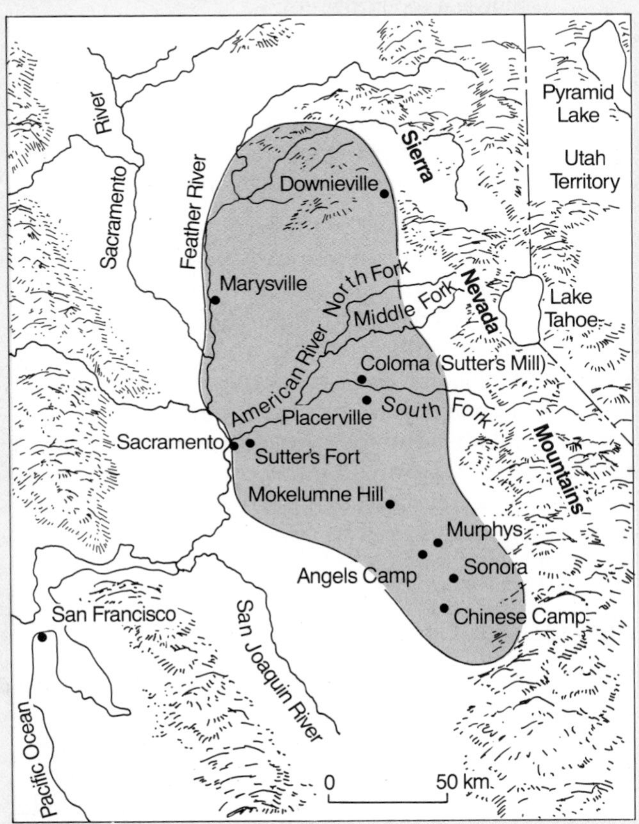


**Figure 13–1** California gold mining region. More than $200 million in gold was recovered from this region between 1848 to 1853.

DRY DIGGINGS

CALIFORNIA MINING ILLUSTRATED.

BARBER & BAKER, Publishers, cor. J & 3d sts, Sacramento.

WET DIGGINGS

QUARTZ MILL.

**Figure 13–2** "How the California Mines are Worked." From a letter-sheet of the early 1850s. Shown on the left is the way Dry or Hill Digging was done. The engraving on the right represents the manner of obtaining gold from the river beds. The Quartz Mill shown in the center was the means by which quartz rock containing gold was pulverized such that the minute particles of gold were gathered and amalgamated with quicksilver.

*hardships of steamy jungles, disease, and overcrowded ships, many of which were unseaworthy. The third route, by ship around the southern tip of South America, was the longest. However, it avoided the overcrowding encountered on the Panama route and the perils of the overland journey. Those that went by ship arrived at San Francisco, where many decided to stay, seeing that there were opportunities to make money in this new town.*

*The earliest prospectors came from the West Coast and were, for the most part, honest and hardworking. Those that arrived from elsewhere in 1849 and later were also mostly honest, hardworking men, but, unfortunately, some were thieves, outlaws, and murderers. Life in these mining camps was extremely hard and expensive. Frequently the shopowners and traders made more money than the prospectors did. Gambling halls and saloons sprang up all over since the men had little to do with their time except gamble and drink. Family life was virtually nonexistent as women comprised only 2 percent of the population.*

*The gold these prospectors sought was mostly in the form of placer deposits (Fig. 13–2). Gold is almost always associated with acidic igneous rocks, where it forms from mineral-rich solutions moving through cracks and fractures of the igneous body.*

*The gold- bearing igneous rocks, or mother lode, were never discovered, but weathering of the source rocks and mechanical separation of minerals by density during stream transport formed placer deposits. Many prospectors searched for the mother lode, but all of the gold recovered came from such placers.*

*One common means of mining these deposits was by panning, in which a prospector dipped a shallow pan into a stream bed, swirled the material around, and poured off the lighter material. Gold, being about six times heavier than most sand grains and rock chips, concentrates on the bottom of the pan and is picked out. An experienced panner could handle almost a cubic meter of material a day.*

## INTRODUCTION

The dawn of the Mesozoic ushered in a new era in Earth history. The major geologic event was the breakup of Pangaea, which affected oceanic and climatic circulation patterns as well as influencing the evolution of the terrestrial and marine biotas. While the Mesozoic Era is popularly known as the Age of Reptiles, it also was the time during which birds, mammals, and angiosperms (flowering plants) first evolved.

The Mesozoic Era (245 to 66 million years ago) is divided into three periods beginning with the Triassic (37 million years), followed by the Jurassic (64 million years), and finally the Cretaceous (78 million years). The stratotypes for the systems from which these periods derive their names are in the Hercynian Mountains of central Germany (Triassic), the Jura Mountains of northwest Switzerland (Jurassic), and the Paris Basin of northern France (Cretaceous).

In this chapter we emphasize the role of plate tectonics while examining Mesozoic geologic events. Since most of the Mesozoic geologic history of North America involves the continental margins, we focus our attention on the eastern, gulf, and western coastal regions. The transgressions and regressions of the final two epeiric seas and their depositional sequences (Absaroka and Zuni) will be incorporated into the geologic history of each of the aforementioned regions.

## BREAKUP OF PANGAEA

Any discussion of the history of the Mesozoic Era must include as its central theme the breakup of Pangaea. Recall that by the end of the Paleozoic Era all the land masses were sutured together into a single supercontinent called Pangaea (Fig. 11–26). Geologic, paleontologic, and paleomagnetic data indicate that the breakup of Pangaea occurred in four general stages:

1. Initial Triassic rifting between Laurasia and Gondwana;
2. Jurassic and Cretaceous separation of the various Gondwana continents;
3. Late Jurassic separation of South America from Africa; and
4. Cenozoic separation of Antarctica from Australia and rifting along the west side of Greenland, finally separating North America from Europe.

Figure 13–3 shows that by the beginning of the Mesozoic Era the continents were nearly equally divided between the northern and southern hemispheres. To achieve the present-day configuration (Fig. 13–3), in which about two-thirds of the world's landmass lies north of the equator, the Gondwana continents (except Antarctica) moved from the high southern latitudes toward the equator, and the Laurasian continents moved farther northward away from the equator. These movements profoundly affected the global climatic and oceanic regimes as well as the individual climates of continents.

### Evidence for Continental Breakup

The breakup of Pangaea can be documented by the extensional structures (fault-block basins or rift valleys) formed along the newly created continental margins. Modern-day analogs for the different stages of continental rifting can be found in various parts of the world today such as the East African Rift valleys (Fig. 6–15). As a continent begins to rift, fault-block basins are formed (Fig. 6–14). Associated with these tensional structures are basaltic dikes, sills, and lava flows, which can be radiometrically dated to determine the time of rifting. As the continents continue separating, and basaltic oceanic crust forms between them (Fig. 6–14), rates and direction of movement can be determined by radiometric and paleomagnetic dating of the oceanic crust (see Chapter 6 for a discussion of these techniques).

During the initial phase of plate fragmentation, structures called **triple junctions** may form where three plates join. While some are stable— such as the Galapagos triple junction at the boundary of the Pacific, Cocos, and Nazca plates—having maintained the same configuration for a long time (Fig. 1–19), most triple junctions

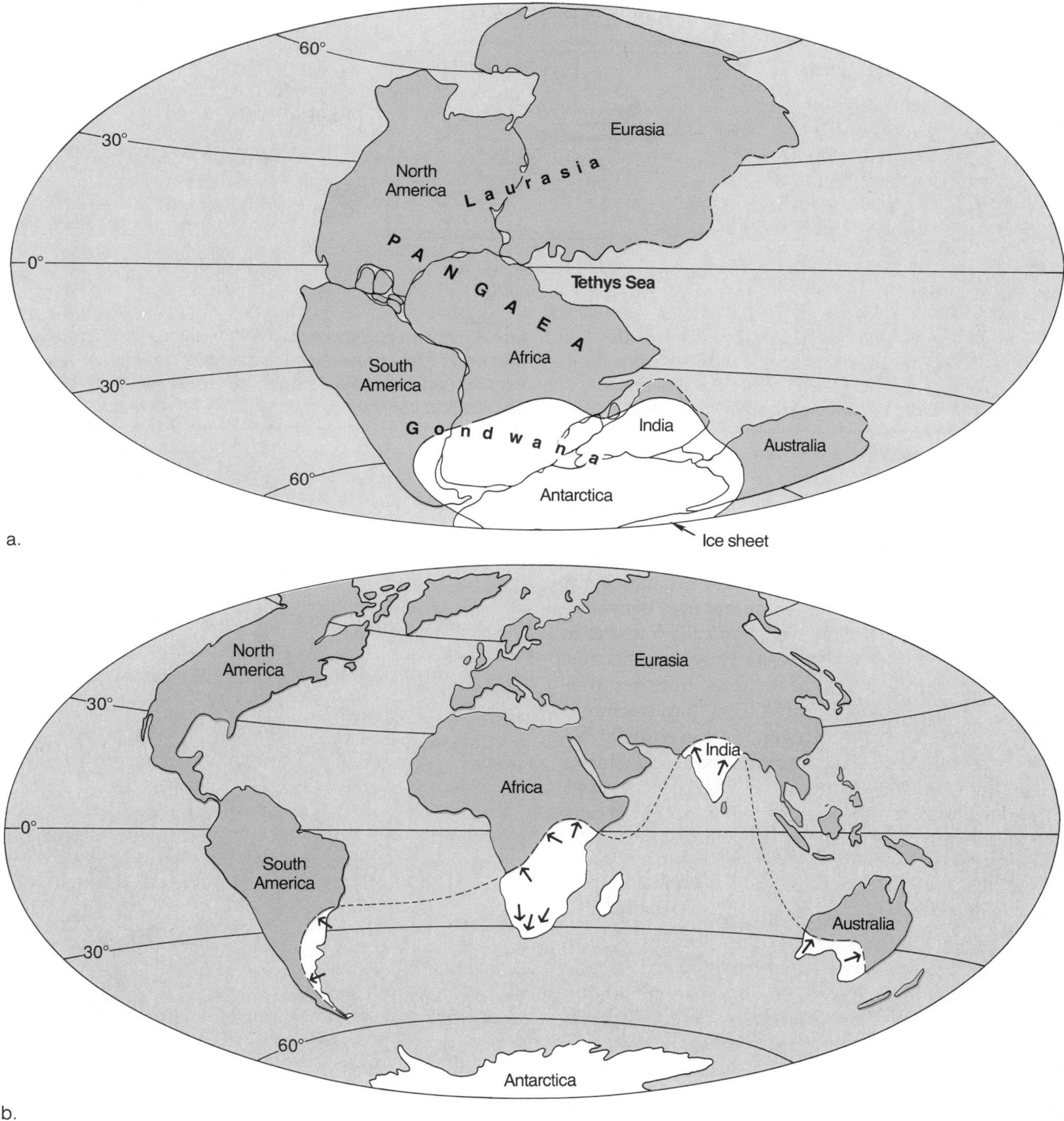


**Figure 13–3** a. Paleogeography of the world at the beginning of the Mesozoic Era. b. Present-day geography. White area is where the Gondwana ice sheet existed. Arrows indicate direction of Gondwana ice sheet movement.

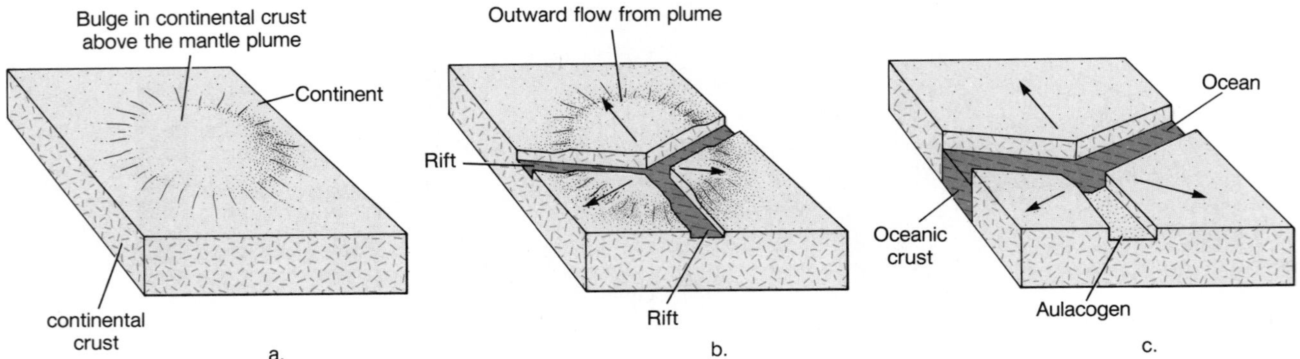


**Figure 13–4** Continental breakup and formation of an aulacogen caused by a mantle plume. a. A bulge forms over a rising mantle plume. b. Three radial rifts develop due to outward radial flow from mantle plume. c. The continent separates into two parts along two of the three rifts. New oceanic crust forms between the separating continents. The third rift becomes an aulacogen, or failed arm, and eventually fills with continental sediment.

are unstable and migrate laterally as plate boundaries change position.

The type of triple junction in which we are most interested involves three ridges, usually formed initially by a mantle plume. When a mantle plume rises beneath a continent, the crust is stretched and typically fractures in a three-pronged pattern (Fig. 13–4). However, the crust usually separates along two of the three fractures, thus leaving the third fracture inactive. This third fracture is called a failed arm, or **aulacogen**, and eventually becomes filled with sediment. Some of the world's major river systems, including the lower Niger, Mississippi, and Amazon, apparently occupy segments of failed rifts.

A modern example of a triple junction with an aulacogen is the East African Rift valleys–Red Sea–Gulf of Aden triple junction (Fig. 15–5). A Mesozoic example of the same type of triple junction involved the separation of South America from Africa during the Cretaceous Period (Fig. 13–5). Here, two arms separated to form the South Atlantic, while the third arm was abandoned and remains as the Benue Trough of Africa in which the Niger River flows.

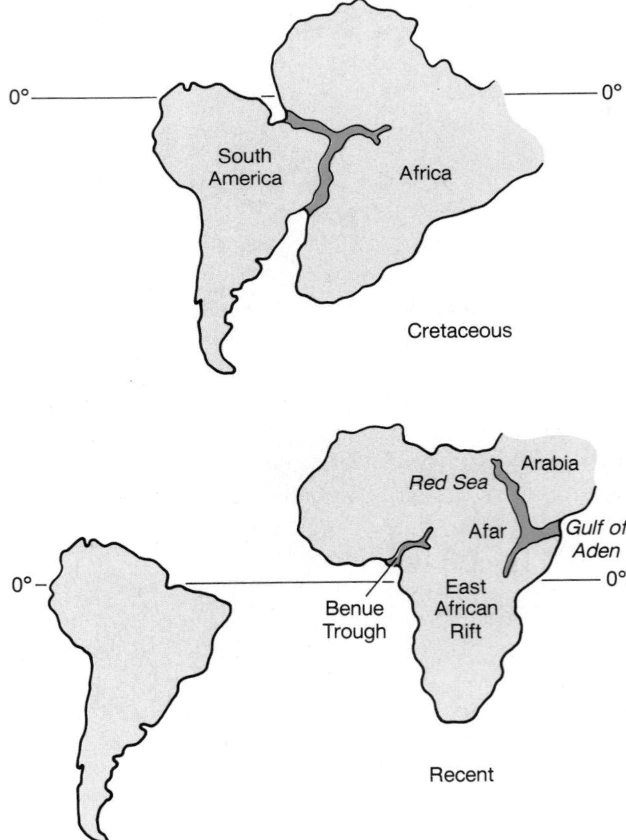


**Figure 13–5** The Benue Trough is an aulacogen that formed as a result of continental rifting between Africa and South America during the Cretaceous Period. The East African Rift valley may also be an aulacogen that formed when Africa and Arabia began separating 25 million years ago.

## The Breakup of Pangaea

The breakup of Pangaea began with rifting between Laurasia and Gondwana during the Late Triassic. By the end of the Triassic Period, North America had separated from Africa, and the Atlantic Ocean began expanding (Fig. 13–6). The initial rift between Laurasia and Gondwana was followed by separation of North America from South America sometime during the Late Triassic and Early Jurassic.

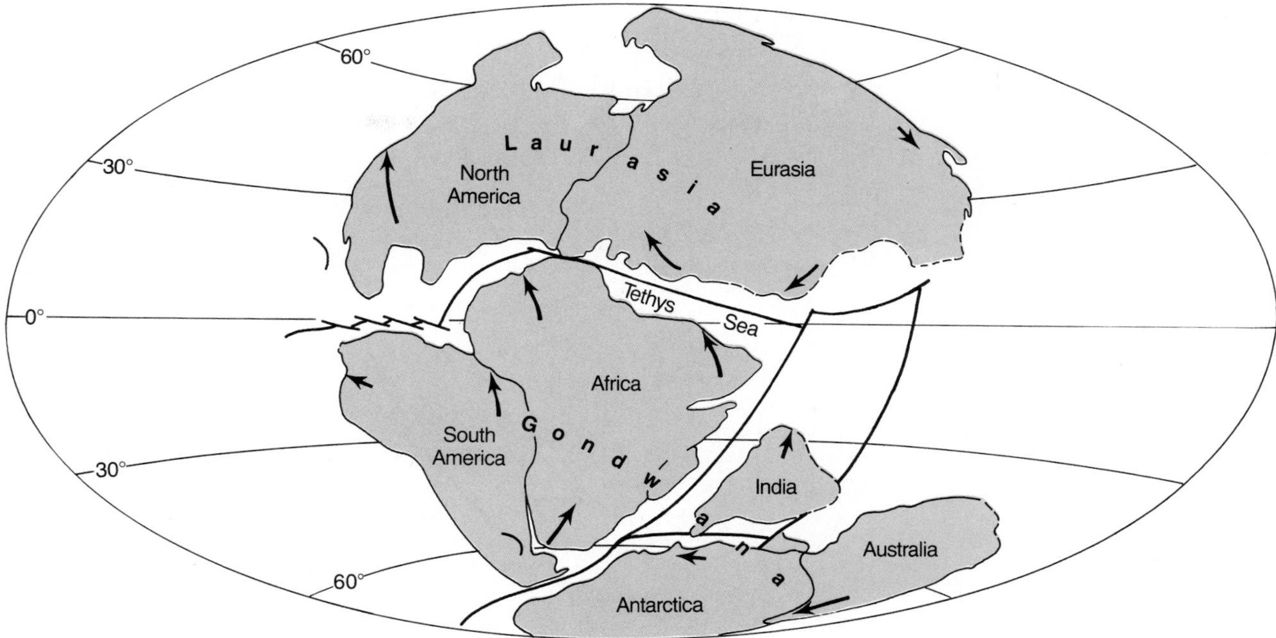


**Figure 13–6** Paleogeography of the world for the Late Triassic Period. Black arrows show the direction of movement for the continents.

This rifting and separation of the continents allowed water from the Tethys Sea to flow into the central Atlantic Ocean, and Pacific Ocean waters to flow into the newly created Gulf of Mexico, which was then little more than a restricted bay (Fig. 13–7). At that time, those areas were located in low tropical latitudes where high temperatures and high rates of evaporation were conducive to the formation of thick evaporite deposits.

The second stage in the breakup of Pangaea involved the separation of the various Gondawanan continents during the Jurassic and Cretaceous periods. A large rift separated Antarctica, Australia, and India from the southern ends of South America and Africa (Fig. 13–8). As rifting progressed, Antarctica and Australia, which remained sutured together, separated from South

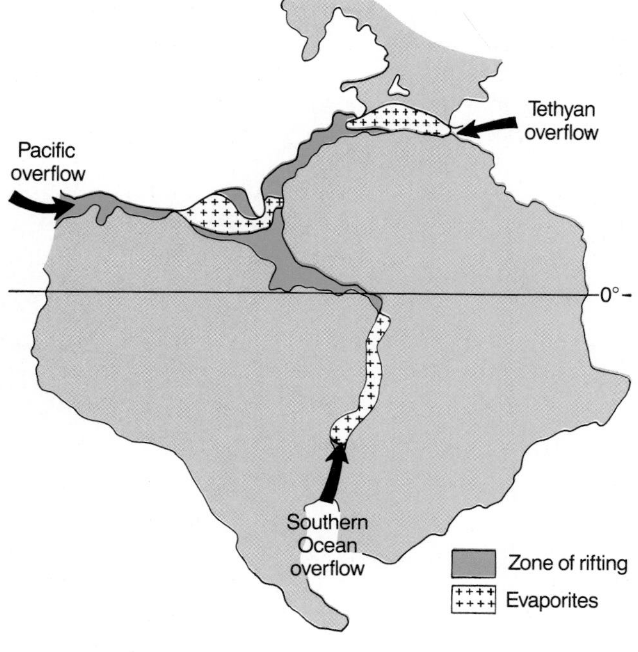


**Figure 13–7** Evaporites accumulated in shallow basins as Pangaea broke apart during the Early Mesozoic Era. Water from the Tethys Sea flowed into the central Atlantic Ocean, and water from the Pacific Ocean flowed into the newly formed Gulf of Mexico, and water from the south flowed into the southern Atlantic Ocean.

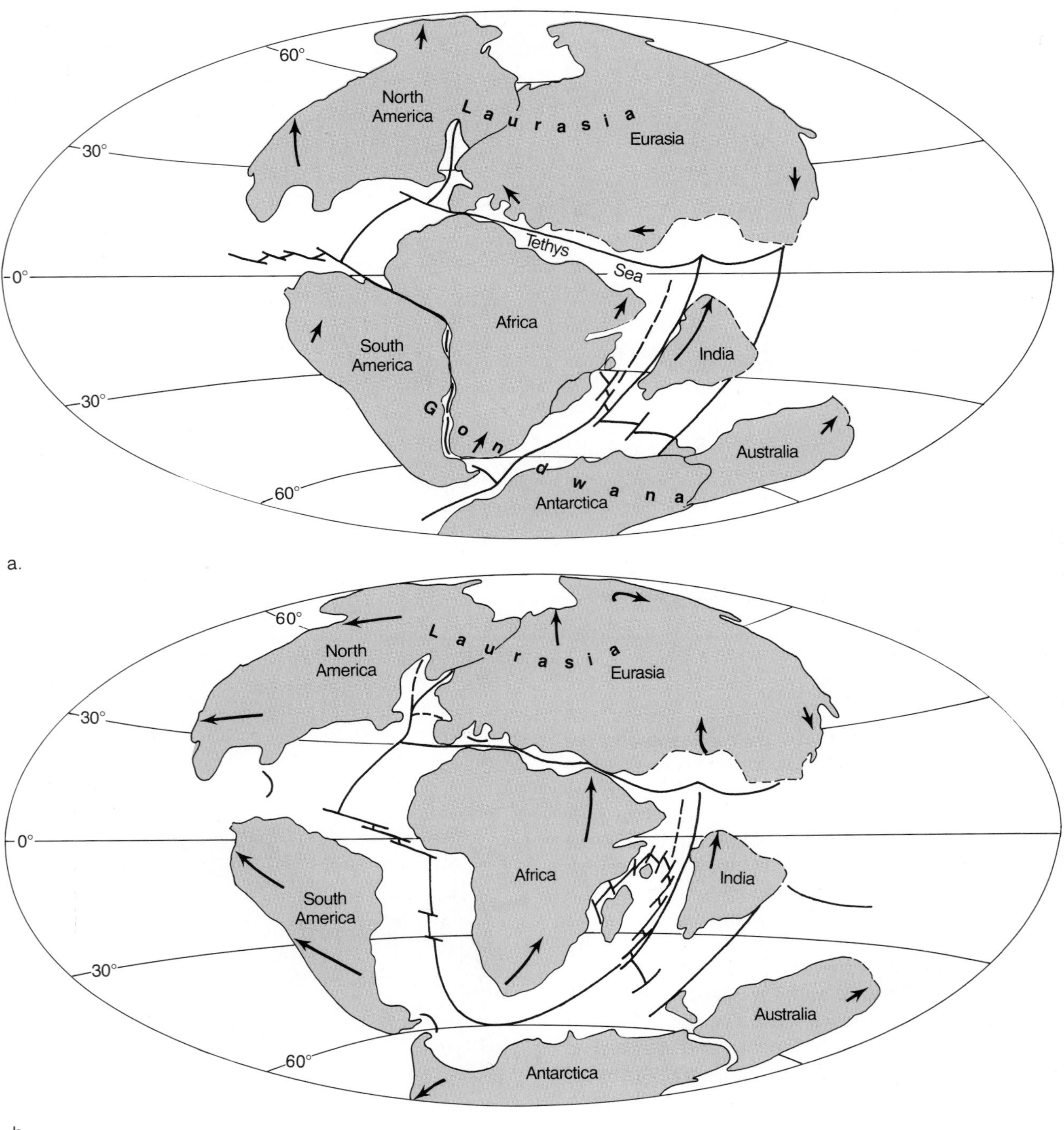


**Figure 13–8** Paleogeography of the world for the Jurassic Period (a) and the Cretaceous Period (b).

America and Africa, while India separated from all four continents and began moving northward to eventually collide with Asia during the Cenozoic, causing the formation of the Himalaya Mountains (see Chapter 15).

The third stage in the breakup of Pangaea began during the Late Jurassic, when South America separated from Africa (Fig. 13–8). The rifting and subsequent separation of these continents created another narrow evaporite basin, fed by southern ocean waters, where thick evaporite deposits accumulated (Fig. 13–7). During this time, the eastern end of the Tethys Sea began closing as a result of clockwise rotation of Laurasia and northward movement of Africa. This narrowed Jurassic and Cretaceous sea between Europe and Africa was the forerunner of the present Mediterranean Sea. Antarctica and Australia remained together at this time, while India continued its northward journey.

By the end of the Cretaceous Period, India had nearly reached the equator, South America and Africa were widely separated, and the eastern side of what is now Greenland began separating from Europe (Fig. 13–8). With the increased rifting of continents and growth of ocean ridges during the Cretaceous Period, a global rise in sea level resulted in a worldwide transgression. The transgression was caused by marine waters that were displaced onto the continents because of higher heat flow along the oceanic ridges and consequent expansion of the oceanic crust (Fig. 13–9). By the Middle Cretaceous, about one-third of the present land area was inundated by epeiric seas, and sea level probably stood as high as at any other time since the Ordovician.

The final stage in the breakup of Pangaea took place during the Cenozoic Era. It involved the separation of Antarctica from Australia and the rifting along the present west side of Greenland, such that North America was completely separated from Europe (Fig. 13–3).

### The Breakup of Pangaea and Its Effects on Global Climates and Ocean Circulation Patterns

By the end of the Permian Period the formation of a huge supercontinent was complete. Pangaea extended from pole to pole, covered about one-fourth of the Earth's surface, and was surrounded by Panthalassa, a global ocean that encompassed about 300 degrees of longitude (Fig. 11–2). Such a configuration exerted a tremendous influence on the world's climate and resulted in generally arid conditions over large parts of Pangaea.

The world's climates result from the complex interaction between wind and ocean currents and the location and topography of the continents. In general, dry climates occur on large land masses in areas remote from sources of moisture and where barriers to moist air form, such as mountain ranges. Wet climates occur near large bodies of water or where winds can carry moist air over land.

Past climatic states can be inferred from the distribution of climate-sensitive deposits such as evaporites, red beds, desert dunes, and coals. Evaporites require an environment where evaporation exceeds precipitation. Desert dunes and red beds are characteristic deposits of arid regions, and although both may form locally in humid regions, they are most widespread in dry areas. Coal forms in both warm and cool humid climates. Vegetation that is eventually converted into coal requires at least a good seasonal water supply; thus, coal deposits are indicative of humid conditions.

Widespread Triassic evaporites, red beds, and desert dunes in the low and middle latitudes of North and

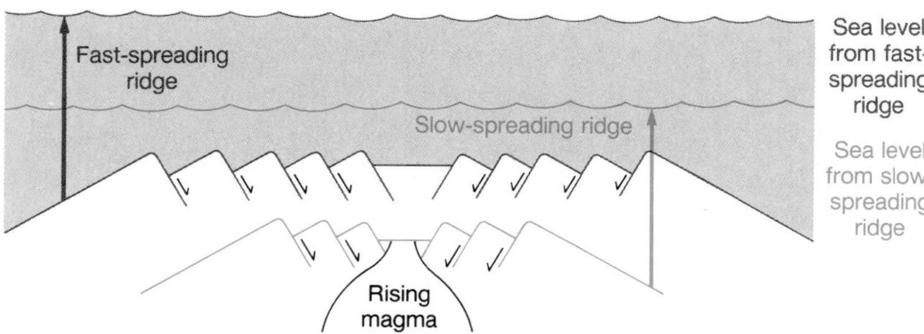


**Figure 13–9** The cross-sectional area of a mid-oceanic ridge depends on its spreading rate. The ridge forms due to expansion by heating. A fast-spreading ridge has a larger cross section and consequently displaces more seawater, resulting in a rise in sea level and transgressions onto the cratons. It is believed that faster spreading rates during the Cretaceous Period were responsible for the global rise in sea level during that time.

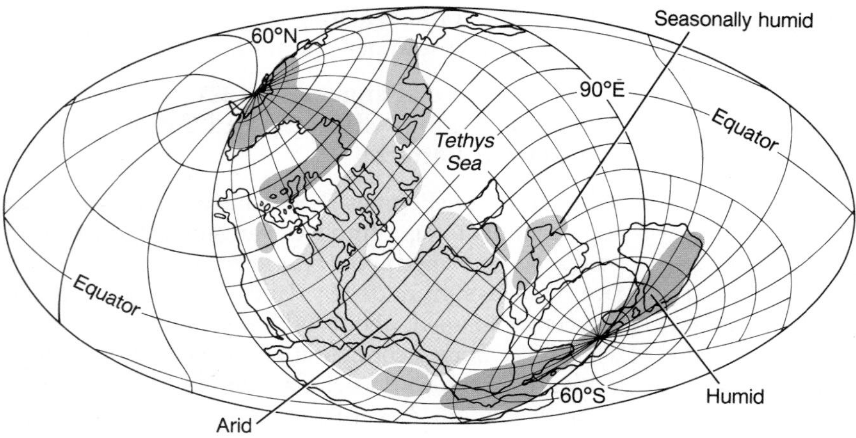


**Figure 13–10** Triassic climate zones inferred from the distribution of evaporites and coals.

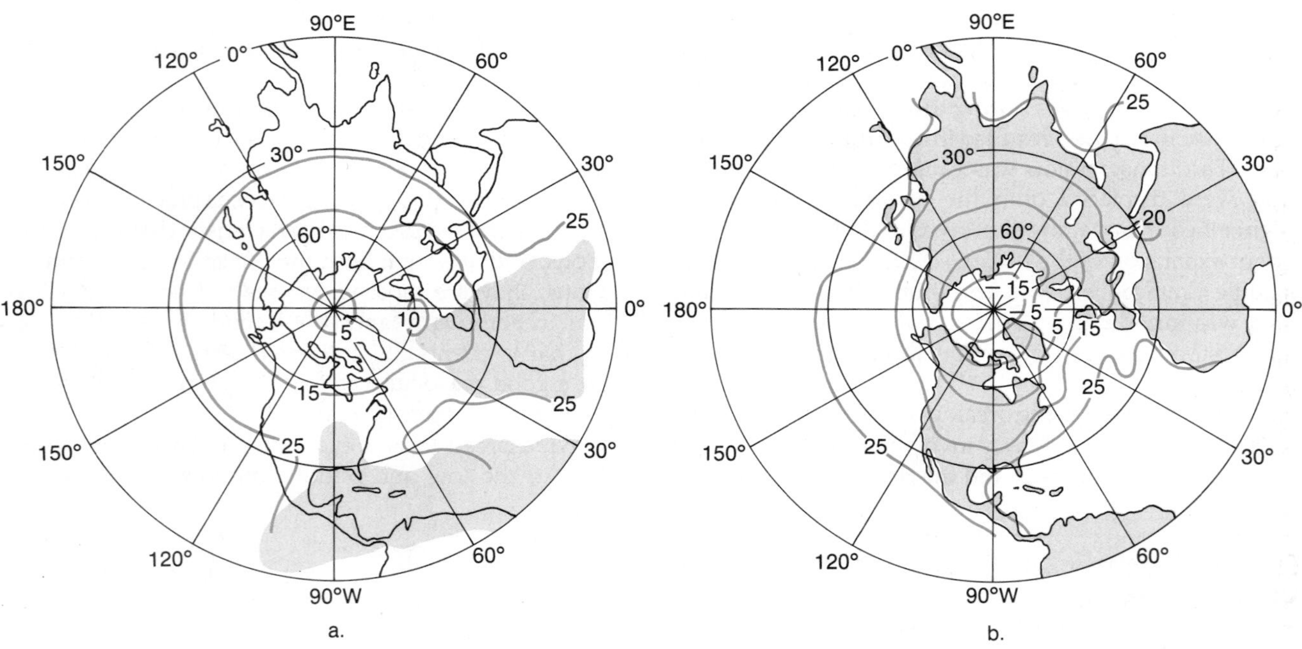


**Figure 13–11** North polar views of the Earth, showing computed temperature gradients for the Triassic Period (a) and the present (b). The positions of the Triassic continents (shaded areas) are superimposed on the present-day geography.

South America, Europe, and Africa indicate dry climates in those regions (Fig. 13–10). Coal deposits occurred mainly in the high latitudes, indicating humid conditions. These high-latitude coals are analogous to today's Scottish peat bogs or Canadian muskeg. The lands bordering the Tethys Sea were probably dominated by seasonal monsoon rains resulting from the warm moist winds and warm oceanic currents impinging against the east-facing coast of Pangaea (Fig. 13–10).

Another factor to consider in the reconstruction of global climates is the temperature gradient between the tropics and poles. The steeper the gradient, or the greater the temperature difference between the tropics and poles, the greater the circulation is of the oceans and atmosphere. Oceans absorb about 90 percent of the solar radiation they receive, while continents absorb only about 50 percent, and if they are snow-covered, they absorb only about 30 percent. The rest of the solar radiation is reflected back into space and lost.

Consequently, high latitudes dominated by seas are warmer than those dominated by land. Estimates of reflectivity can be made for various locations and times in the past if the geographic distribution of continents and oceans is known for those times. When such estimates are combined with heat-transfer rates between the ocean or land and the atmosphere, average annual temperatures can be calculated and *isotherms* (lines of equal temperature) plotted. Figure 13–11 shows that the northern hemisphere's temperature gradient for the Triassic Period had a calculated range of 20°C, while the present temperature gradient range is 41°C. This means the Triassic ocean and atmosphere circulation were more sluggish than they are today.

With the breakup of Pangaea during the Late Triassic, the global temperature gradient gradually increased as the northern-hemisphere continents moved further northward and displaced higher-latitude ocean waters. As temperature at the high latitudes decreased, circulation of the atmosphere and ocean was driven faster by the greater temperature differences between the tropics and the poles and the changing positions of the continents (Fig. 13–12). However, while the temperature gradient and seasonality on the land was increasing during the Jurassic and Cretaceous periods, the middle- and higher-latitude oceans were still quite warm, because warm waters from the Tethys Sea were circulating to the higher latitudes. This resulted in a relatively equable worldwide climate through the end of the Mesozoic Era.

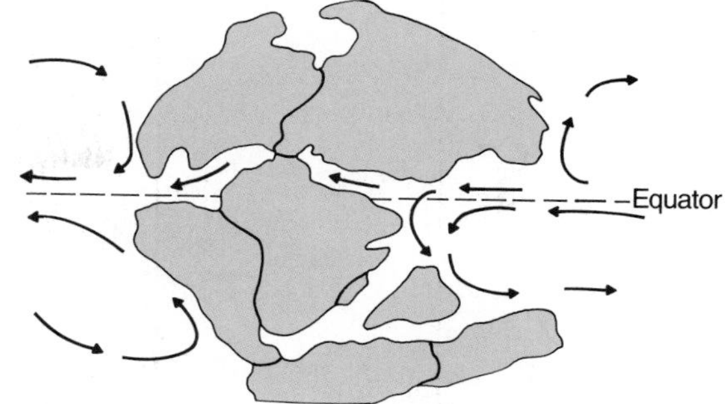


a. Jurassic Period (175 m.y.)

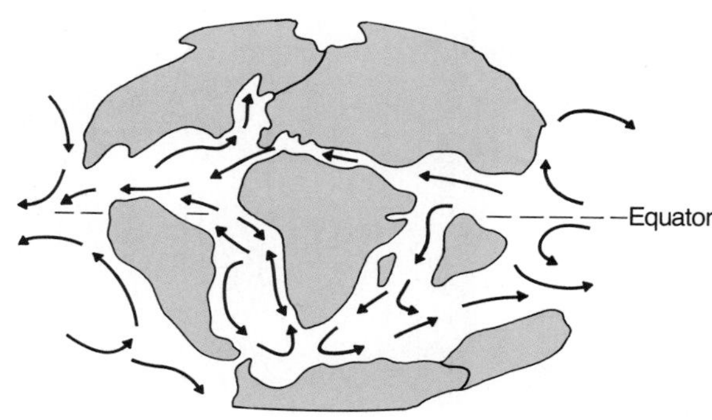


b. Late Cretaceous Period (70 m.y.)

**Figure 13–12** Oceanic circulation evolved from a simple pattern in a single ocean (Panthalassa) with a single continent (Pangaea) to a more complex pattern in the newly formed oceans of the Cretaceous Period. Ocean temperatures remained warm during the Cretaceous Period because warm waters from the Tethys Sea circulated to the higher latitudes.

## THE MESOZOIC HISTORY OF NORTH AMERICA

Having examined the global aspects of the breakup of Pangaea, we can now relate those events to the Mesozoic history of North America (Figs. 13–13, 13–14, 13–15).

In terms of tectonism and sedimentation, the beginning of the Mesozoic Era was essentially the same as the preceding Permian Period in North America (Fig. 11–14). Terrestrial red-bed sedimentation continued over much of the craton, while block faulting and basaltic igneous activity began in the Appalachian belt as North America and Africa began separating. The Gulf of Mexico was the site of extensive evaporitic deposition during the Late Triassic and Jurassic as North America separated from South America (Fig. 13–7). The Atlantic and Gulf coastal areas received thousands of meters of sediments during the Cretaceous Period as a result of the worldwide rise in sea level.

In western North America, marine deposition was continuous over much of the Cordillera but was interrupted by periods of mountain building. A volcanic island arc system that formed off the western edge of the craton during the Permian Period was sutured to North America sometime during Permian-Triassic time. This event, referred to as the *Sonoma orogeny*, (dicussed later in this chapter), was followed by relative tectonic quiescence. The entire Cordilleran area was then involved in a series of major mountain building episodes that resulted in the formation of the Rocky Mountains, the Sierra Nevada Mountains, and other lesser mountain ranges. While each orogenic episode has its own name,

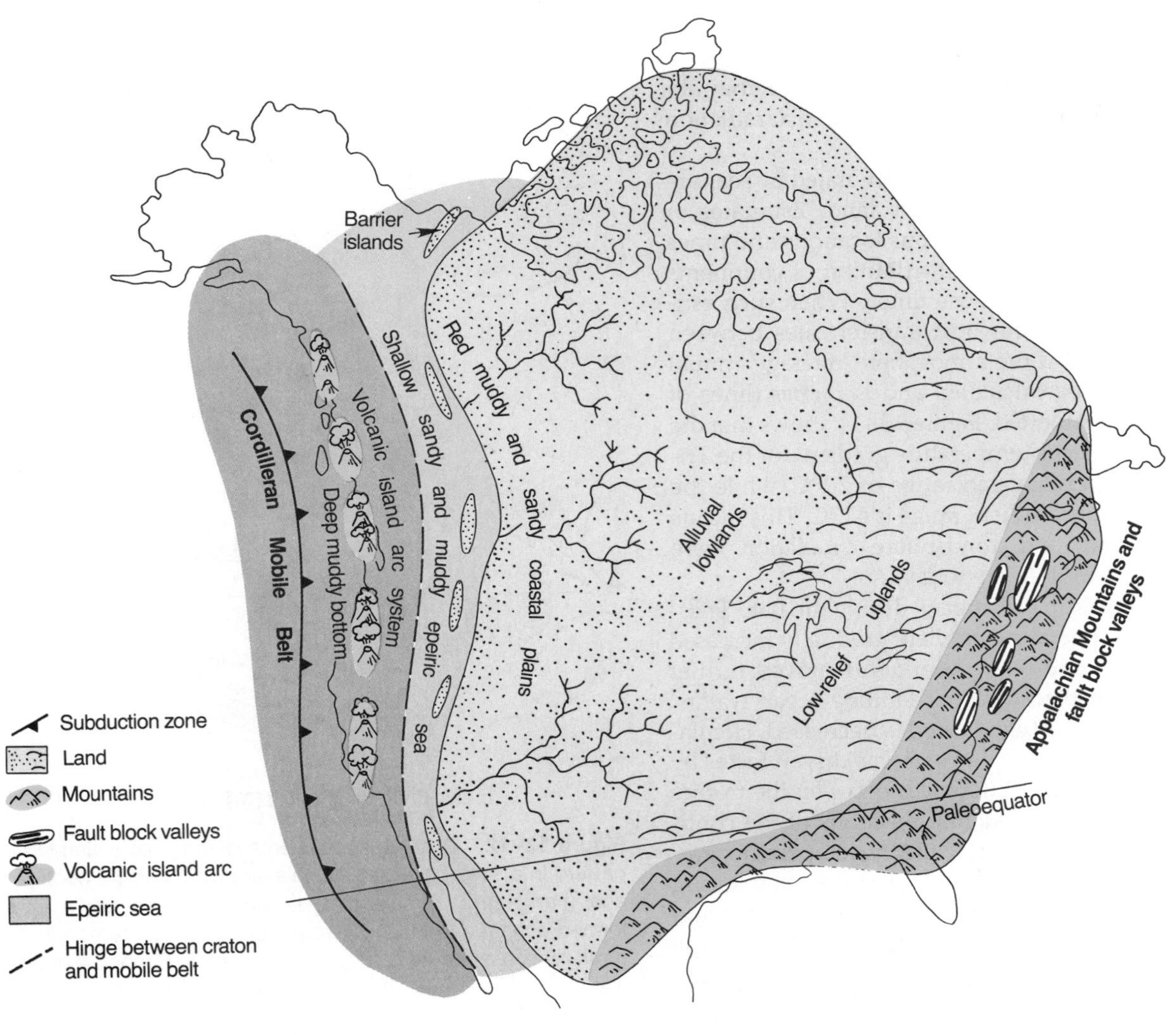


**Figure 13–13** Paleogeography of North America for the Triassic Period.

the entire mountain-building event is simply called the *Cordilleran orogeny* (also discussed later). With this simplified overview of the Mesozoic history of North America in mind, we will now examine the specific regions of the continent.

## CONTINENTAL INTERIOR

The history of the North American craton is recorded in unconformity-bound sequences that reflect advances and retreats of the epeiric seas over the craton (Fig. 10–4). While these transgressions and regressions played a major role in the Paleozoic geologic history of the continent, they were not as important during the Mesozoic Era. During the Mesozoic Era, most of the continental interior was above sea level and did not experience epeiric sea inundation. Consequently, the two Mesozoic cratonic sequences, the Absaroka (Late Mississippian to Early Jurassic) and **Zuni** (Early Jurassic to Early Paleocene) (Fig. 10–4), are incorporated as part of the history of the three continental margin regions of North America.

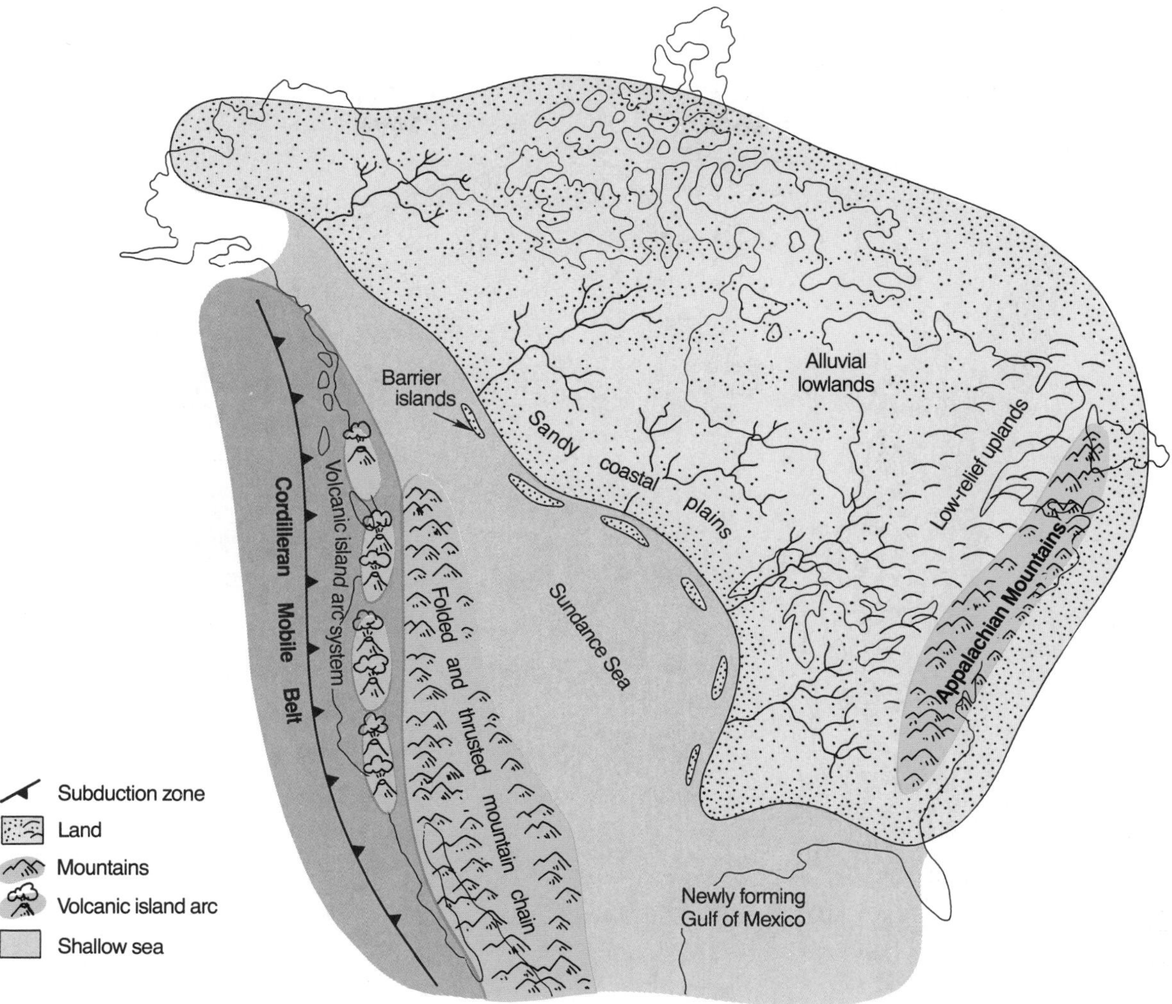


**Figure 13–14** Paleogeography of North America for the Jurassic Period.

## EASTERN COASTAL REGION

During the Early and Middle Triassic, coarse clastics derived from the erosion of the recently uplifted Appalachians (Allegheny orogeny) filled intermontane basins and other low areas. As erosion and basin filling continued, this once lofty mountain range was reduced to a low-lying plain. During the Late Triassic, the first stage in the breakup of Pangaea began, with North America separating from Africa. All along the eastern coast of North America from Nova Scotia to North Carolina, fault-block basins developed (Fig. 13–16), and these basins received great quantities (up to 6,000 m) of poorly sorted red-colored nonmarine clastic sediments known as the *Newark Group*. Lakes formed in the central parts of the basins; and along the margins of these lakes and adjacent riverbeds, reptiles roamed, leaving their footprints and trackways in the soft sediment (Fig 13–17). Curiously, although the Newark rocks contain numerous dinosaur footprints, they are almost completely devoid of dinosaur bones.

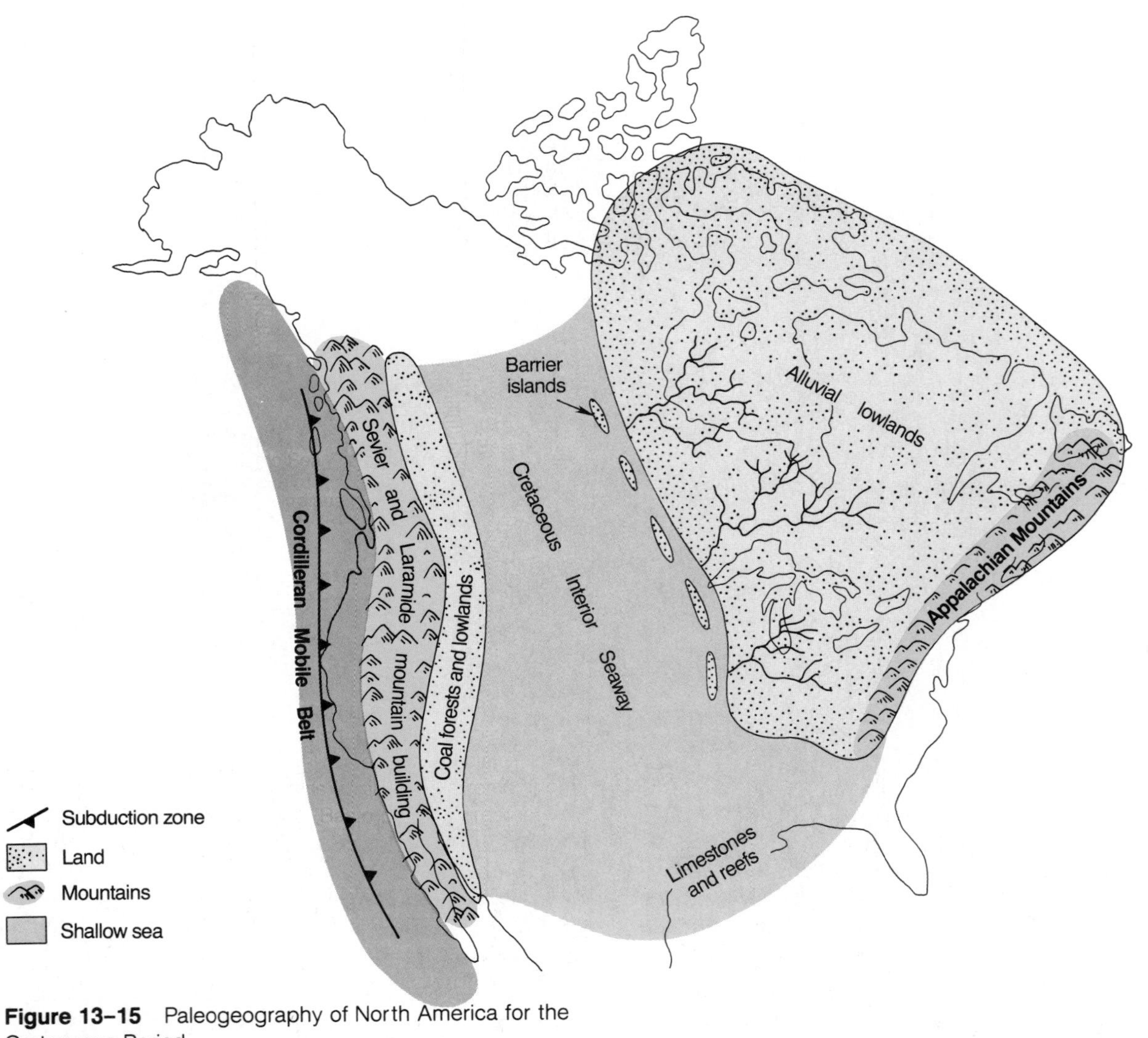


**Figure 13–15** Paleogeography of North America for the Cretaceous Period.

Concurrent with sedimentation in the fault-block basins were extensive lava flows that blanketed the basin floors as well as intrusions of numerous dikes and sills. The most famous intrusion is the prominent Palisades sill along the Hudson River in the New York–New Jersey area (Fig. 13–18). Radiometric dates obtained from the Palisades sill indicate it was intruded about 200 million years ago. While the Newark Group is considered Late Triassic in age, recent spore and pollen research indicates that deposition in some areas began during the Early Jurassic.

As the Atlantic Ocean grew, rifting ceased along the eastern margin of North America, and this once active plate margin became a passive, trailing continental margin. The fault–block mountains that were produced by this rifting eroded during the Jurassic and Early Cretaceous until only a broad, low-lying erosional surface remained. The sediments resulting from erosion contributed to the growing eastern continental shelf. During the Cretaceous Period, the Appalachian region was reelevated and once again shed sediments onto the continental shelf, forming a gently dipping, seaward-thickening wedge of rocks up to 3,000 m thick. These rocks are exposed in a belt extending from Long Island, New York to Georgia. In fact, the Atlantic coastline during the Cretaceous Period was probably similar to the present-day one and contained delta, beach, bay, lagoon, barrier-island, and marginal-shelf environments.

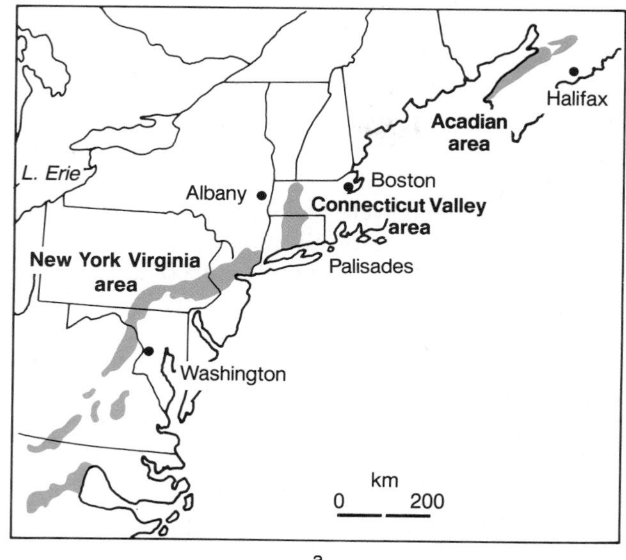


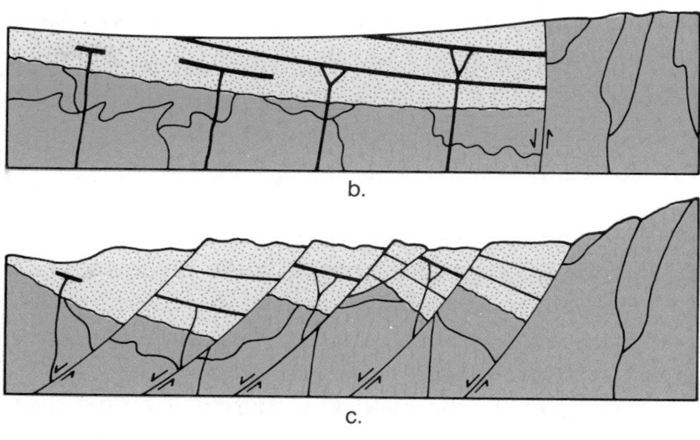


**Figure 13–16** a. Locations of Triassic fault-block basin outcrops in eastern North America. b. After the Appalachians were eroded to a low-lying plain by the Middle Triassic, fault-block basins formed as a result of Late Triassic rifting between North America and Africa. These valleys accumulated tremendous thicknesses of sediments and were themselves broken by a complex of normal faults during rifting (c).

**Figure 13–17** Reptile tracks in the Triassic Newark Group were uncovered during the excavation for a new state building in Hartford, Connecticut. Because the tracks were so spectacular, the building site was moved and the excavation was designated as a state park (Dinosaur State Park).

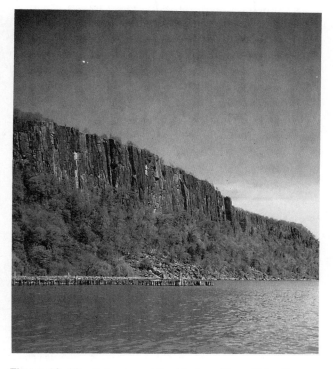

**Figure 13–18** Palisades of the Hudson River. This sill was one of many that were intruded into the Newark sediments during the Late Triassic rifting.

## GULF COASTAL REGION

Until the Late Triassic, the Gulf Coastal region was above sea level. However, as North America separated from South America during the Late Triassic, the Gulf of Mexico began forming (Fig. 13–7). With oceanic waters flowing into this newly formed, shallow, restricted basin, conditions were ideal for evaporite formation. More than 1,000 m of evaporites were precipitated at this time, and most geologists believe that these Jurassic evaporites are the source of the Tertiary salt domes found today in the Gulf of Mexico and southern Louisiana (Fig. 15–48). The history of these salt domes and their associated petroleum accumulations will be discussed in Chapter 15.

By the Late Jurassic, circulation in the Gulf of Mexico was less restricted, and evaporite depositional conditions had abated; normal marine limestones, shales, and sandstones accumulated in the alternately transgressing and regressing seas of the Gulf region (Fig. 13–19). These rocks and the underlying evaporites are not exposed because they are deeply buried beneath a thick cover of Cretaceous and Cenozoic sediments. Their thicknesses, lithologies, and facies relationships have been determined from geophysical evidence and from the vast amount of data accumulated during extensive drilling in this area by oil companies.

During the Cretaceous Period, the Gulf Coast region—like the eastern and western margins of the continent—was inundated by transgressing seas. The relatively low-lying Gulf coastal area was invaded by northward-moving marine waters during the Early Cretaceous. This can be seen in facies relationships in which nearshore sandstones are overlain by finer sediments characteristic of deeper waters. Following an extensive regression at the end of the Early Cretaceous, a major transgression began during which a wide seaway extended from the Arctic Ocean to the Gulf of Mexico (Fig. 13–20). Cretaceous sediments that were deposited in the Gulf Coast region formed a seaward-thickening wedge (Fig. 13–21).

Carbonate reefs were particularly extensive during the Cretaceous Period. Bivalves called *rudists* were the main constituent of many of these reefs (Fig. 14–4). Because of their high porosity and permeability, rudistoid reefs are excellent reservoir formations for oil. A good example of a Cretaceous reef complex is known from the Texas region of the Gulf Coast area (Fig. 13–22). Here the reef trend had a strong influence on the carbonate platform deposition of the region. The facies patterns

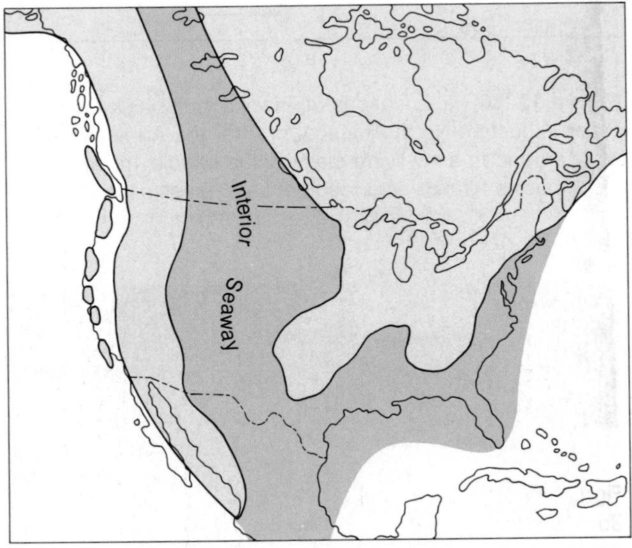


**Figure 13–20** Paleogeography of North America during the maximum Late Cretaceous transgression.

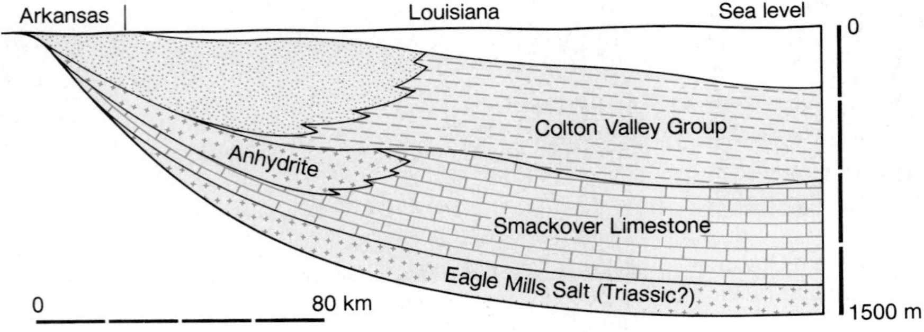


**Figure 13–19** Cross section of the Gulf Coast area for the Jurassic Period.

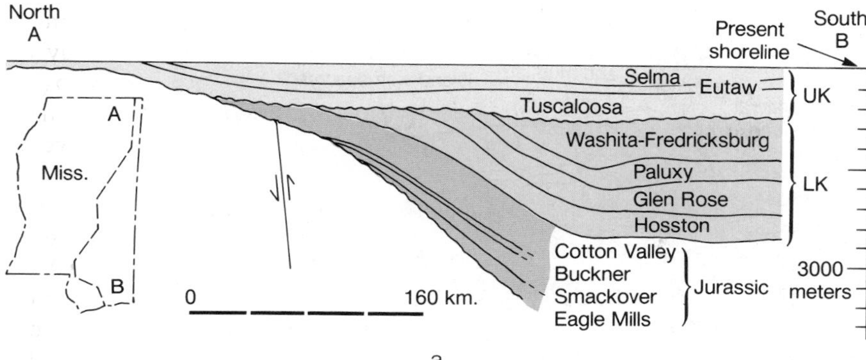


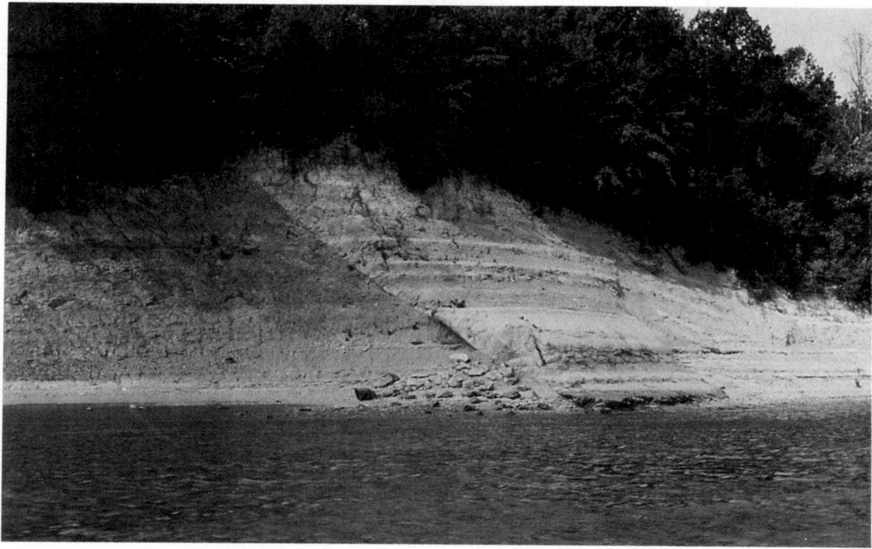


**Figure 13–21** a. Cross section of the Mississippi Gulf Coastal region showing how the Upper Cretaceous rocks form an angular unconformity with the underlying Jurassic and Lower Cretaceous rocks. b. Outcrop of the Demopolis Chalk (Selma Group) along the banks of the Tombigbee River near Demopolis, Alabama. (Photo courtesy of Mirza A. Beg, Geological Survey of Alabama)

of these carbonate rocks are equally as complex as are those found in the major barrier reef systems of the Paleozoic Era.

## WESTERN REGION

### Triassic Tectonics

The Cordilleran region had barely recovered from the Late Devonian and Early Mississippian suturing of the Antler island arc with the North American continent (Fig. 11–19) when it underwent yet another orogenic event, the **Sonoma orogeny**. While this orogeny is difficult to date precisely, most evidence places it at or near the Permian and Triassic boundary. Prior to this deformational event, an island arc and ocean basin analogous to the present-day Japanese Islands and Sea of Japan formed off the western North American craton and extended from Alaska to Nevada (Fig. 13–13). Like the preceding Antler orogeny, the Sonoma orogeny was caused by the collision of the island arc system with the southwestern margin of North America (Fig. 13–23).

This Permo-Triassic orogeny involved subduction of an oceanic plate under the island arc with subsequent thrusting of oceanic and island arc rocks eastward against the continental margin, creating the Sonoma Mountains (Fig. 13–23). The Triassic rocks of the far western Cordilleran include thick sections of volcanics

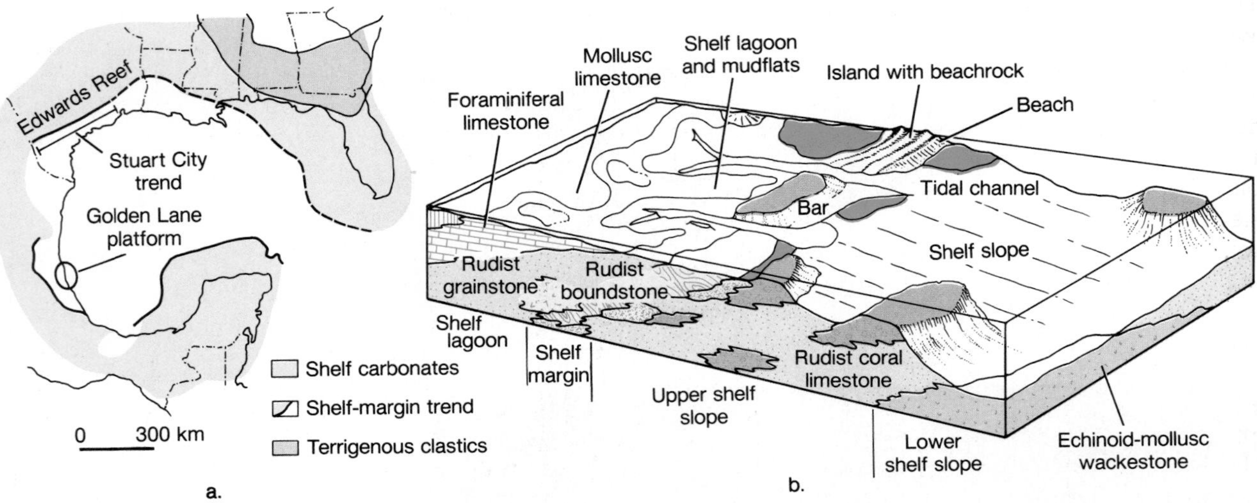


**Figure 13–22** a. Early Cretaceous shelf-margin facies around the Gulf of Mexico Basin. The reef trend is shown as a dark line. b. Reconstruction of the depositional environment and facies changes across the Stuart City reef trend, South Texas.

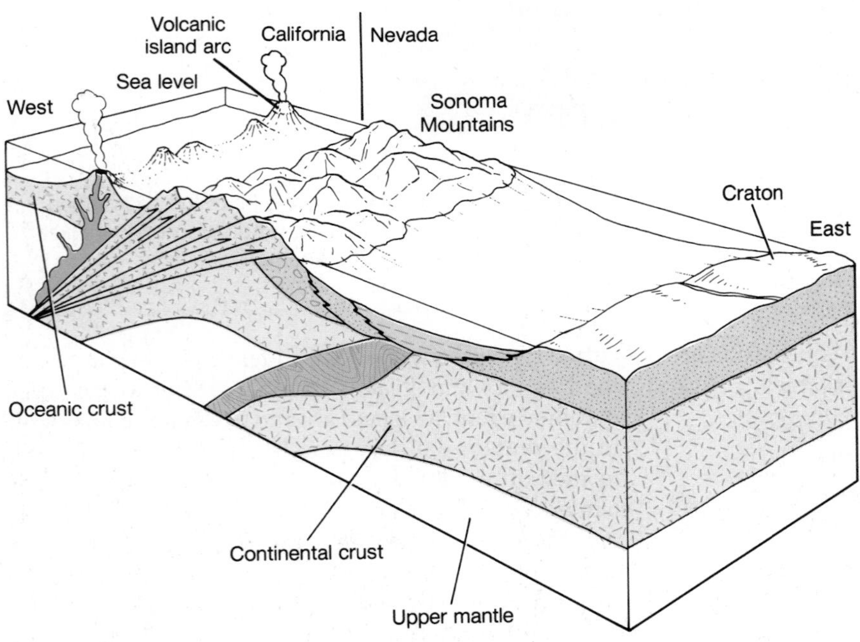


**Figure 13–23** Reconstruction of the tectonic activity that culminated in the Permian-Triassic Sonoma orogeny in western North America. The Sonoma orogeny was the result of a collision between the southwestern margin of North America and an island arc system.

and greywackes presumably derived from the island arc. However, some question exists as to whether these rocks were deposited where they are now found, or whether they represent parts of microplates that have been transported thousands of kilometers along with the Pacific plate and sutured onto the craton as part of a series of displaced terranes. Geologists now believe that the western part of the continent has grown as a result of accretion by many small microplates. The evidence strongly indicates that Alaska, British Columbia, portions of Washington, Oregon, and most of western California were added to North America by the collision of microplates and, in fact, represent displaced terranes. This topic will be covered in more detail at the end of the chapter.

Following the Early Mesozoic destruction of the volcanic island arc system during the Sonoma orogeny, the western margin of North America became part of an oceanic-continental convergent plate boundary similar to the one that now exists along the western margin of South America. This change came about because of the separation of North America from Africa during the Late Triassic. The intensity of orogenic events in the North American Cordillera is related to the sea-floor spreading rates in the Atlantic Ocean.

A steeply dipping subduction zone developed along the western margin of North America during the Late Triassic because of the westward movement of North America over the Pacific plate as the Atlantic Ocean basin opened. This newly created oceanic-continental plate boundary controlled Cordilleran tectonics for the rest of the Mesozoic Era and for most of the Cenozoic Era; this subduction zone marks the beginning of the modern circum-Pacific orogenic system.

## Triassic Sedimentation

While tectonism was occurring in the trench region of the Cordilleran mobile belt, Early Triassic sedimentation on the western continental shelf consisted of shallow-water marine sandstones and limestones. The Early Triassic climate of western North America remained essentially the same as it was during the Permian Period, namely arid to semi-arid, with periods of sufficient moisture to allow the growth of large trees. During the Middle and Late Triassic, the western shallow seas regressed further west, exposing large areas of former sea floor to erosion. Marginal marine and nonmarine Triassic rocks, particularly red beds, contributed to the spectacular and colorful scenery of such areas as the Painted Desert in western North America (see Perspective 13–1).

The Lower Triassic Moenkopi Formation of the southwestern United States is a succession of brick-red and chocolate-colored mudstones containing various sedimentary structures and fossils (Fig. 13–24). Such

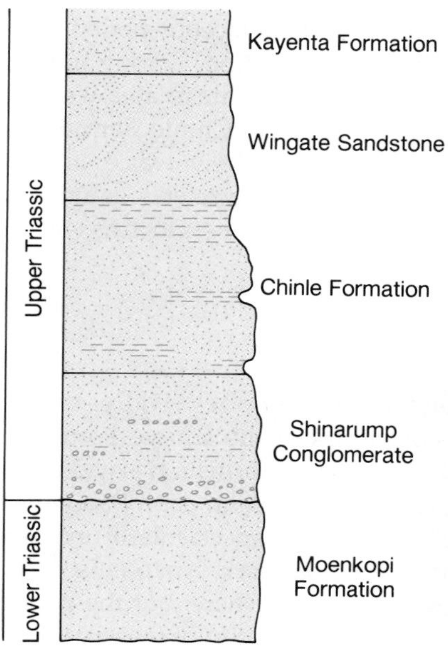

**Figure 13–24** a. Stratigraphic section of Triassic formations in the western United States. b. View of East Entrance of Zion Canyon, Zion National Park, Utah. The light colored massive rocks are the Jurassic Navajo Sandstone, while the slope forming rocks below the Navajo are the Upper Triassic Kayenta Formation.

## Perspective 13–1

# The Petrified Forest National Park

Petrified Forest National Park is located in eastern Arizona about 42 km east of Holbrook. The park consists of two sections: the Painted Desert, which is north of Interstate 40, and the Petrified Forest, which is south of the Interstate (Fig. 1).

The Painted Desert is a brilliantly colored landscape whose colors and hues constantly change throughout the day. The multicolored rocks of the Triassic Chinle Formation have been weathered and eroded to form a badland topography of numerous gullies, valleys, ridges, mounds, and mesas. The Chinle Formation is composed predominantly of different-colored shale beds. These shales and associated volcanic ash layers are easily weathered and eroded. Interbedded locally with the shales are lenses of conglomerates, sandstones, and limestones, which are more resistant to weathering and erosion than the shales and which form resistant ledges.

The Petrified Forest (Fig. 1) was originally set aside as a national monument to protect the large number of petrified logs that lay exposed in what is now the southern part of the Park. When the transcontinental railroad constructed a coaling and watering stop in Adamana, Arizona, passengers were encouraged to take excursions to "Chalcedony Park," as the area was then called, to see the petrified forests. In a short time, collectors and souvenir hunters hauled off tons of petrified wood, quartz crystals, and Indian relics. It was not until a huge rock crusher was built to crush the logs for the manufacture of abrasives that the area was declared a national monument and the petrified forests preserved and protected.

During the Triassic Period, the climate of the area was much wetter than today, with many rivers, streams, and lakes. About 40 different fossil plant species have been identified from the Chinle Formation. These include numerous seedless vascular plants such as rushes and ferns, as well as cycads. Such plants thrive in floodplains and marshes. Most of the logs are conifers and belong to the genus *Araucarioxylon*. Some of these trees grew more than 60 m tall and were up to 4 m in diameter. Apparently most of the conifers grew on higher ground or riverbanks. While many trees were buried in place, most appear to have been uprooted and transported by raging streams and rivers during times of flooding. Burial of the logs was rapid, and groundwater saturated with silica from the ash of nearby volcanic eruptions quickly permineralized the trees.

Deposition continued in the Colorado Plateau region during the Jurassic and Cretaceous periods, further burying the Chinle Formation. During the Laramide orogeny, the Colorado Plateau area was uplifted and eroded, exposing the Chinle Formation. Since the Chinle is mostly shales, it was easily eroded, leaving the more resistant petrified logs and log fragments exposed on the surface—much as we see them today.

sedimentary structures as desiccation cracks and ripple marks, as well as fossil amphibians and reptiles and their tracks, indicate deposition in a variety of continental environments, including stream channels, floodplains, and fresh and brackish water ponds. Thin tongues of marine limestones indicate three brief incursions of the sea. Local beds with gypsum and halite crystal casts attest to a rather arid climate.

Unconformably overlying the Moenkopi is the Upper Triassic Shinarump Conglomerate, a persistent unit generally less than 50 m thick that was derived from the newly uplifted areas in Arizona, western Colorado, and Idaho (Fig. 13–24).

Above the Shinarump are the vividly colored shales, silts, and sandstones of the Chinle Formation (Fig. 13–24). This Upper Triassic formation is widely exposed over the Colorado Plateau and is probably most famous for its petrified wood, spectacularly exposed in the Petrified Forest National Park, Arizona (see Perspective 13–1). Fossil ferns occur here, but the park is best known for its abundant, beautifully preserved logs of gymnosperms, especially conifers and plants called *cycads* (Fig. 14–11). Fossilization resulted from the silicification of the plant tissues. Weathering of volcanic ash beds interbedded with fluvial and deltaic Chinle sediments provided most of the silica for silicification. Some

**Figure 1** Petrified Forest National Park, Arizona. All of the logs here are *Araucarioxylon*, which is the most abundant tree in the park. The petrified logs have been weathered from the Chinle Formation and are mostly in the position they were when they were buried, some 200 million years ago.

trees were preserved in place, but most were transported during floods and deposited on sandbars and on floodplains, where fossilization occurred. After burial, silica-rich groundwater percolated through the sediments and silicified the wood.

While best known for its petrified wood, the Chinle Formation has also yielded fossils of labyrinthodont amphibians, phytosaurs, and small dinosaurs (see Chapter 14 for a discussion on the latter two animal groups).

The Wingate Sandstone, a desert dune deposit, and the Kayenta Formation, a stream and lake deposit, overlie the Chinle Formation. These two formations are well exposed in southwestern Utah and complete the Triassic stratigraphic succession in the Southwest (Fig. 13–24).

## Jurassic and Cretaceous Tectonics

The Mesozoic tectonic history of the North American Pacific coast area is very complex. It involved mountain-building activity resulting from Triassic oceanic-continental convergence, as well as accretion of microplates carried eastward by the subducting Pacific plate.

The fundamental cause of Mesozoic Cordilleran orogenic activity was the eastward subduction of the oceanic Pacific plate under the continental North American plate. This subduction varied in rate, inclination,

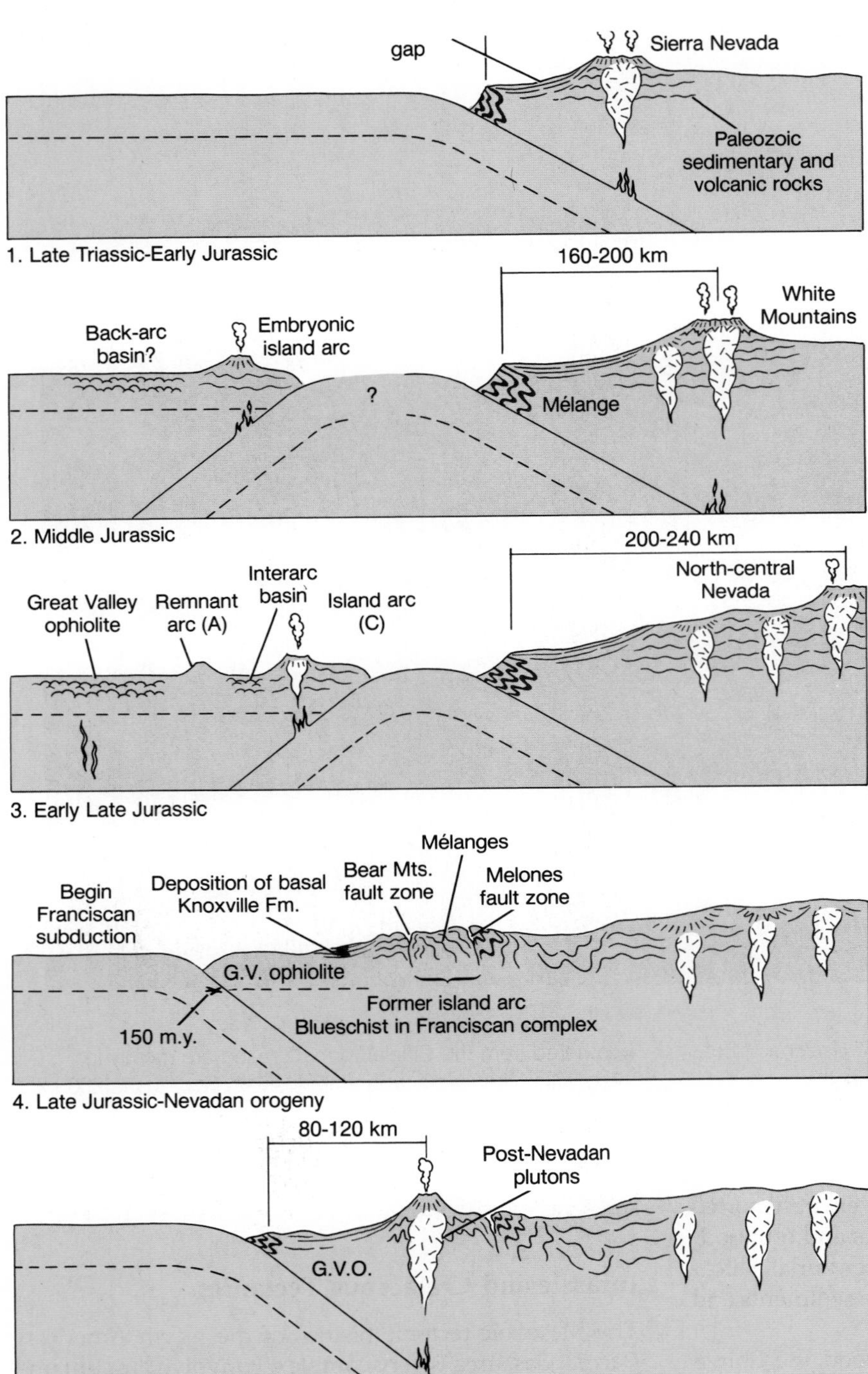


**Figure 13–25** Reconstruction of the tectonic evolution of the Sierra Nevada Mountains during the Mesozoic Era. Note the change from an oceanic-oceanic convergent plate margin to an oceanic-continental convergent plate margin with the beginning of the Nevadan orogeny during the Late Jurassic.

and, to a minor extent, direction. It resulted in an eastward shifting of deformation that progressively affected the trench, continental shelf, and finally the cratonic margin. Not only was the continental crust thickened during this time, but sedimentation patterns and regional geography were also affected (Figs. 13–14, 13–15).

Two subduction zones, each dipping in opposite directions, were apparently present off the west coast of North America during the Middle and early Late Jurassic (Fig. 13–25). As the westward-moving North American plate continued to override the oceanic Pacific plate, the more westerly subduction zone was eliminated, and an unusual rock assemblage, the Franciscan Formation, began accumulating in the remaining submarine trench (Fig. 13–26).

The Franciscan Formation is a thick (up to 7,000 m), chaotic mixture of rocks, predominantly greywackes. It also contains breccias, siltstones, black shales, cherts,

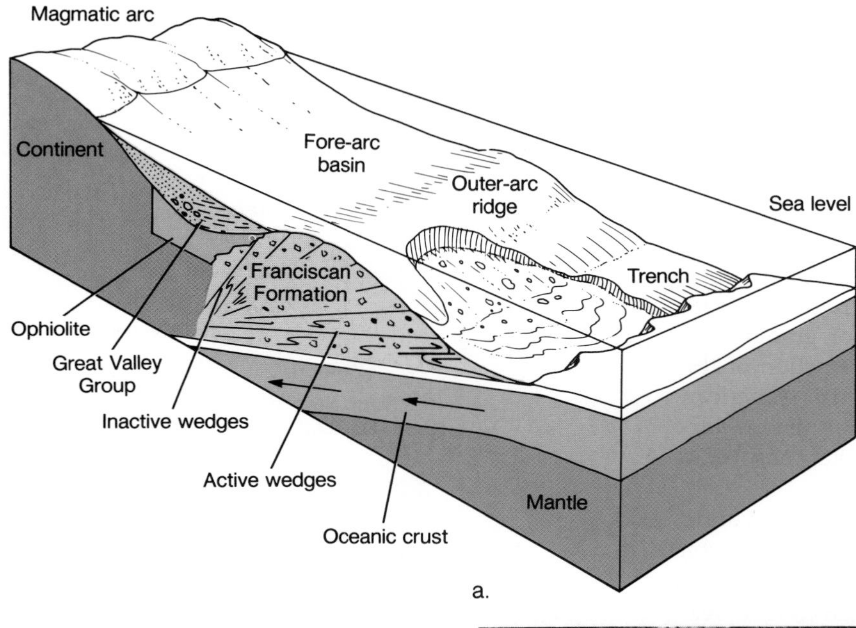

b.

**Figure 13–26** a. Reconstruction of the environment of deposition for the Franciscan Formation during the Late Jurassic and Cretaceous periods. b. Exposures of Franciscan pillow basalts. (Photo courtesy of J. William Guyton, California State University, Chico).

pillow basalts, volcanic breccias, and greenstones (Fig. 13–26). Fossils are extremely rare in the Franciscan, but radiolarians, clams, ammonites, and foraminifera indicate its age is Late Jurassic and Cretaceous. The various lithologies indicate that continental-shelf, slope, and deep-sea environments were brought together in a submarine trench when North America overrode the subducting Pacific oceanic crust. This subduction led in many instances to metamorphism of the various sediments, as attested to by the glauconite schists common in the Franciscan Formation.

East of the Franciscan Formation and presently separated from it by a major thrust fault is the Great Valley Group, consisting of more than 16,000 m of Cretaceous conglomerates, sandstones, siltstones, and shales. These sediments were deposited on the continental shelf and slope in a fore-arc basin setting at the same time the Franciscan deposits were accumulating in the submarine trench (Fig. 13–26).

In addition to the deformation of the rocks along the subduction zone, enormous volumes of granite and granodiorite were generated by the subduction of the Pacific plate under the continental North American plate. The result was repeated batholitic intrusions during the Jurassic and Middle and Late Cretaceous in the western Cordillera region (Fig. 13–27).

A major phase of deformation, which began during the Jurassic and continued into the Cenozoic, goes by the general term of **Cordilleran orogeny**. Just like the Paleozoic orogenic activity in the Appalachian mobile belt, the Cordilleran orogeny was a protracted event consisting of separate pulses manifesting themselves in different regions at different times. Much of this organic activity is related to the continued expansion of the Atlantic Ocean.

The initial phase of the Cordilleran orogeny, the **Nevadan orogeny** (Table 13–1) began during the Late Jurassic and continued into the Cretaceous as large volumes of granite-granodiorite were generated at depth beneath the western edge of North America. These granitic masses ascended as huge batholiths that are now recognized as the Sierra Nevada, Southern California, Idaho, and Coast Range batholiths (Fig. 13–27).

By the Late Cretaceous, most of the volcanic and plutonic activity had migrated eastward into Nevada and Idaho (Fig. 13–28). The probable cause for this migration was a change from high-angle to low-angle subduction, resulting in the subducting oceanic plate reaching its melting depth farther east (Fig. 13–29). Thrusting occurred progressively further east so that by Late Cretaceous time, it extended all the way to the Idaho-Washington border (Fig. 13–28).

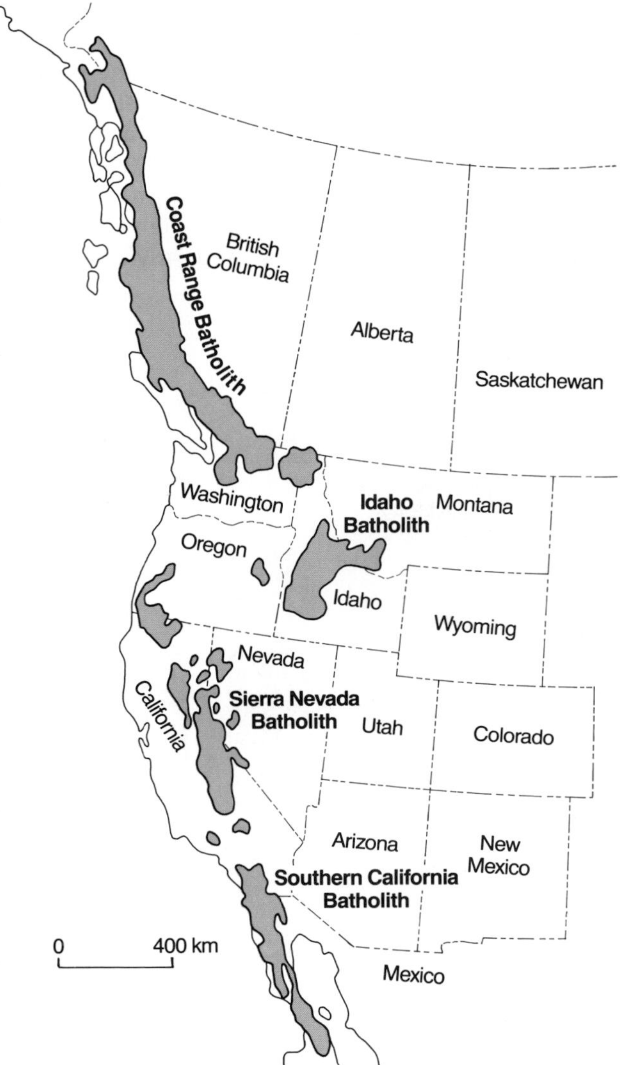


**Figure 13–27** Location of Jurassic-age and Cretaceous-age batholiths in western North America.

The second phase of the Cordilleran orogeny, the **Sevier orogeny** (Table 13–1), was mostly a Cretaceous event affecting the continental shelf and slope areas of the Cordilleran mobile belt. In response to the compressional forces generated in the subduction zone and transmitted eastward, shelf and slope strata were sheared from the underlying Precambrian rocks and broken along parallel planes of weakness to form multiple overlapping, low-angle thrust faults (Fig. 13–30). It is estimated that

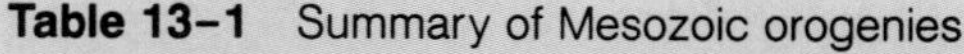

**Table 13–1** Summary of Mesozoic orogenies.

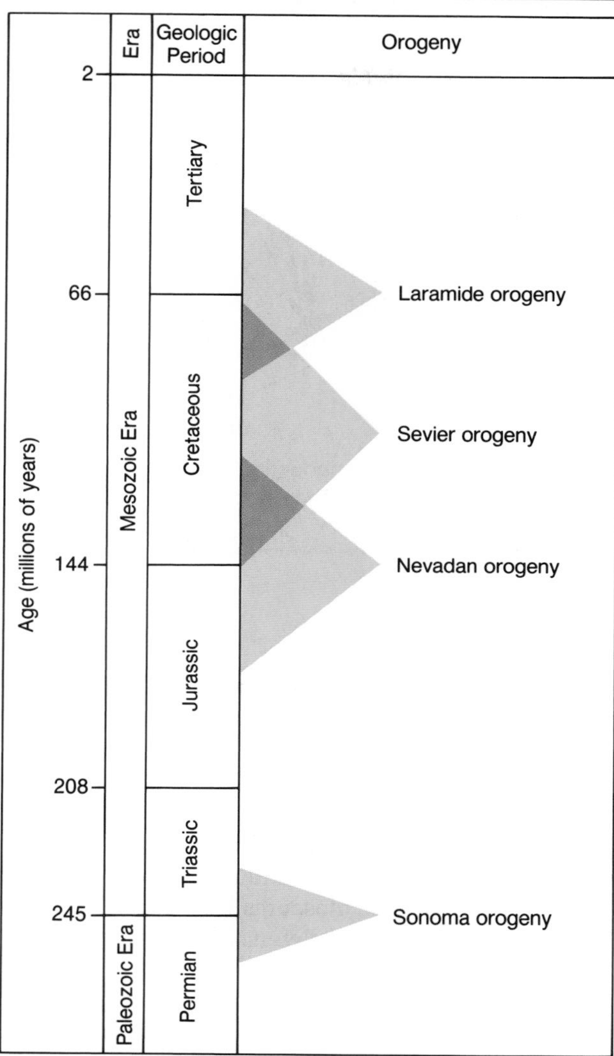


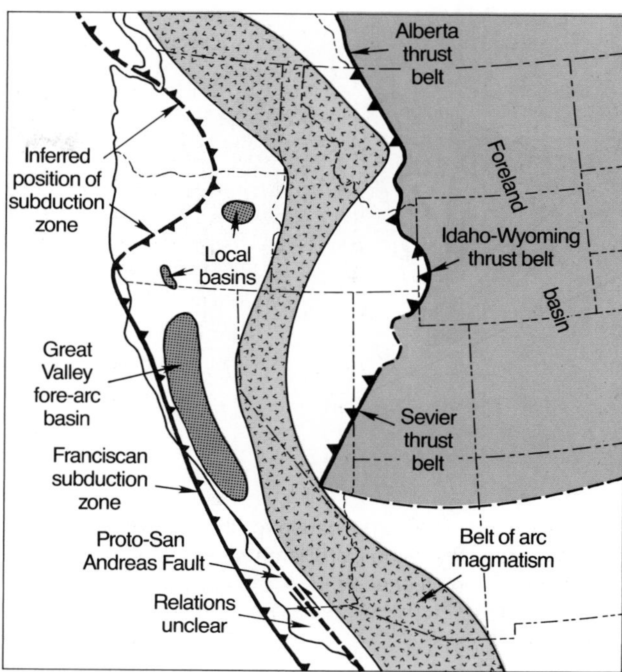


**Figure 13–28** Paleogeography of the Cordilleran region during the Late Cretaceous Period, 75 to 90 million years ago.

during the Sevier orogeny, these low-angle thrust faults resulted in more than 100 km of crustal shortening in the Nevada-Utah region.

Although the term *Sevier* is sometimes reserved for deformational events in southern California, Nevada, and Utah, similar deformation also occurred in Montana and western Canada. Here, each of the major mountain ranges consists of a block of Paleozoic shelf rock that has been thrust eastward along low-angle, westerly dipping fault surfaces (Fig. 13–30).

During the Late Cretaceous to Early Cenozoic, the final pulse of the Cordilleran orogeny occurred (Table 13–1). The **Laramide orogeny** developed east of the Sevier orogenic belt in the present-day Rocky Mountain areas of New Mexico, Colorado, and Wyoming. The Laramide orogeny differed considerably in tectonic style from the Nevadan and Sevier orogenies. While thrust faulting still occurred in the northern Rocky Mountain region, the more characteristic style of deformation in the middle and southern Rockies was arched domes, basins, and anticlines. Most of the structures of the present-day Rocky Mountains resulted from the Cenozoic phase of the Laramide orogeny, and for that reason, it will be discussed in Chapter 15.

## Jurassic and Cretaceous Sedimentation

Having just examined the Jurassic and Cretaceous orogenic events of western North America, we turn our attention to how the orogenic activity affected sedimentation patterns. The Early Jurassic deposits in a large part of the western region consist mostly of clean, cross-bedded sandstones indicative of windblown deposits. The thickest and most prominent of these is the Navajo Sandstone, a widespread cross-bedded sandstone that accumulated along the southwestern margin of the craton. The most distinguishing feature is its large-scale

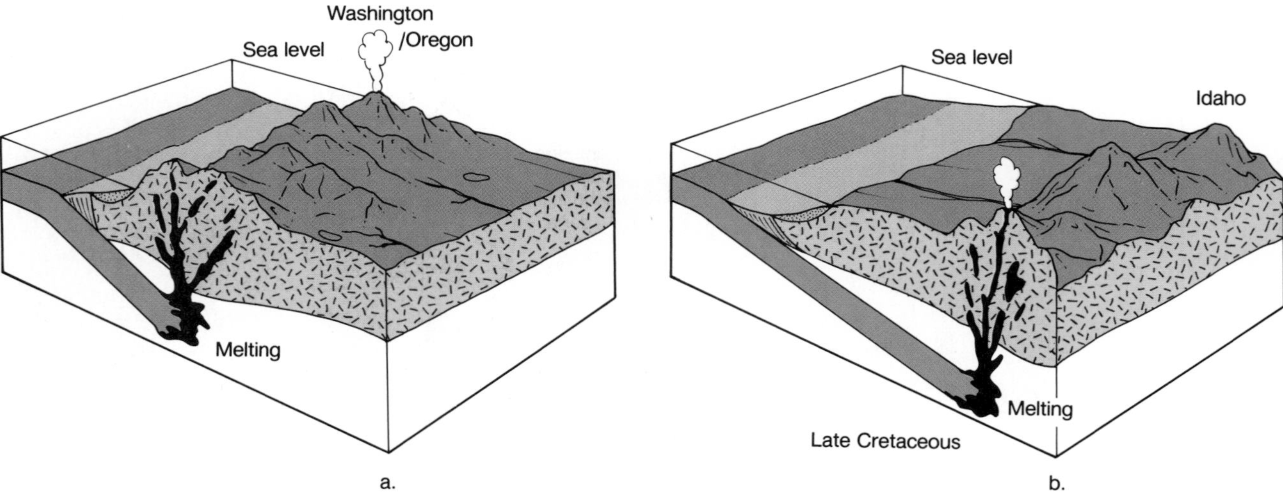


**Figure 13–29** A possible cause for the eastward migration of igneous activity in the Cordilleran region during the Cretaceous Period was a change from (a) high-angle to (b) low-angle subduction. As the subducting plate began to move downward at a lower angle, the depth of melting would be farther to the east.

cross-beds, some of which are more than 25 m high (Fig. 13–31). The upper part of the Navajo contains smaller cross-beds as well as dinosaur and crocodilian fossils. Based on its sedimentary structures, facies relationships, and fossil remains, the Navajo Sandstone probably represents a coastal dune environment.

Marine conditions became more widespread during the Middle Jurassic when the western interior of North America was flooded twice by a wide seaway called the **Sundance Sea** (Fig. 13–32). The resulting deposits, called the Sundance Formation, were largely derived from tectonic highlands to the west that paralleled the shoreline and were the result of intrusive igneous activity and associated volcanism that began during the Triassic.

During latest Jurassic time, the folding and thrust faulting that began as part of the Nevadan orogeny in Nevada, Utah, and Idaho formed a large mountain chain (Fig. 13–33). As the mountain chain grew and shed sediments eastward to form a clastic wedge, the Sundance Sea retreated northeastward (Fig. 13–33). A large part of the area formerly occupied by the Sundance Sea is now covered by the multicolored sandstones, mudstones, shales, and occasional lenses of conglomerates that comprise the world-famous Morrison Formation (Fig. 13–34).

The Morrison Formation contains the world's richest assemblage of dinosaur remains. Although most of the dinosaur skeletons are broken up, as many as 50 individuals have been found together in a small area. Such a concentration indicates they were brought together during times of flooding and deposited on stream-channel sandbars. Soils in the Morrison indicate that the climate was seasonably dry. The presence of turtle, crocodile, and fish fossils indicates that the lakes at the time were probably very shallow and saline.

While most major museums in the world have either complete dinosaur skeletons or at least bones from the Morrison Formation, the best place to see the bones still embedded in the rocks is at the visitor's center of Dinosaur National Monument near Vernal, Utah (Perspective 13–2).

Shortly before the end of the Early Cretaceous, Arctic waters spread southward over the craton, forming a large inland sea in the Cordilleran fore-basin area. By the beginning of the Late Cretaceous, this incursion joined the northward-transgressing waters from the Gulf area to create an enormous inland sea called the **Cretaceous Interior Seaway** that occupied the area east of the Sevier orogenic belt. Extending from the Gulf of Mexico to the Arctic Ocean, this seaway effectively divided North America into two large landmasses until just before the end of the Late Cretaceous (Fig. 13–35). Mid-Cretaceous transgressions also occurred on other continents and all were part of the global mid-Cretaceous rise in sea level that resulted from accelerated sea-floor spreading as Pangaea continued to fragment.

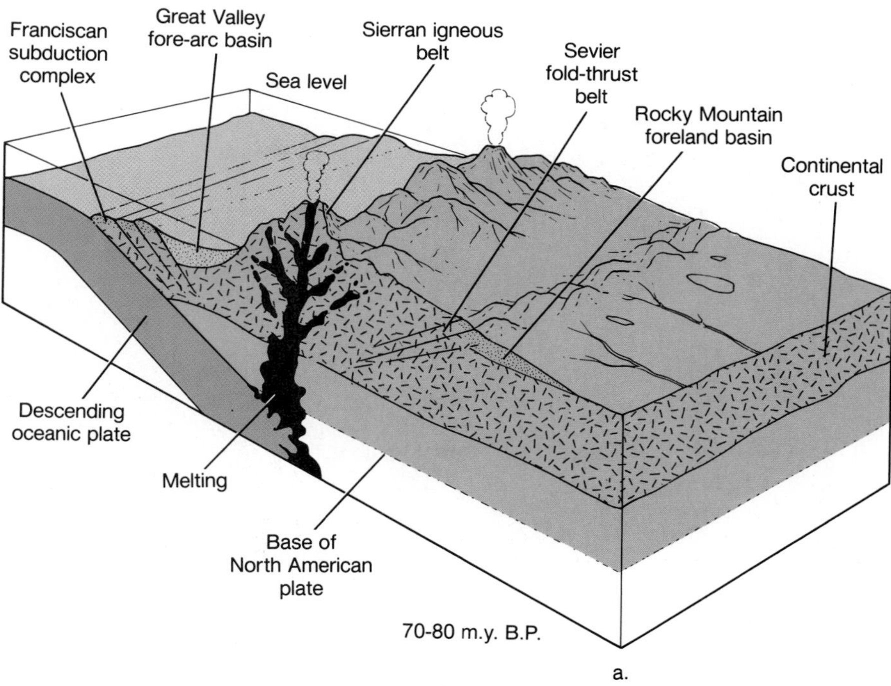

b.

**Figure 13–30** a. Reconstruction showing the associated tectonic features of the Late Cretaceous Sevier orogeny due to subduction of the Pacific plate under the North American plate. b. East face of Spring Mountains (looking northwest), west of Las Vegas, Nevada, showing the Keystone thrust fault, one of the major faults in the Sevier overthrust belt. The sharp boundary between the light colored Mesozoic rocks and the overlying dark colored Paleozoic rocks marks the trace of the Keystone thrust fault.

**Figure 13–31** Large cross-beds of the Jurassic Navajo Sandstone in Zion National Park, Utah.

The geologic history of the seaway is well documented from excellent stratigraphic correlations based on abundant fossil ammonoids as well as numerous ash beds, many of which can be radiometrically dated. These ash beds attest to continued volcanism in the western and central Cordillera. At its maximum extent, the seaway was more than 1,500 km wide and stretched from Utah to Iowa. Cretaceous deposits less than 100 m thick indicate the eastern margin of the seaway subsided slowly and received little sediment from the emergent, low-relief craton to the east. The western shoreline, however, shifted back and forth, primarily in response to fluctuations in the rate of sediment supply from the Cordilleran Sevier orogeny to the West. The facies relationships show lateral changes from conglomerate and coarse sandstone adjacent to the mountain belt through finer sandstones, siltstones, shales, and even limestones and chalks in the east (Fig. 13–35). During times of particularly active thrusting and uplift, these coarse clastic wedges of gravel and sand prograded even further east.

As the Cretaceous came to a close, renewed tectonic activity (Laramide orogeny), an influx of massive terrigenous clastics from the West, and a reduced rate of seafloor spreading, forced the withdrawal of this last great interior sea from the continent. As marine waters re-

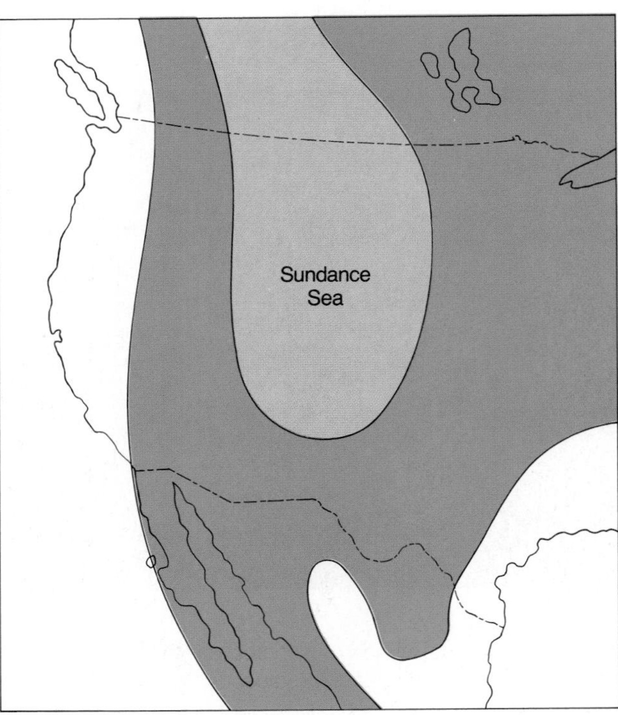


**Figure 13–32** Paleogeography of western North America showing the area covered by the Middle Jurassic Sundance Sea.

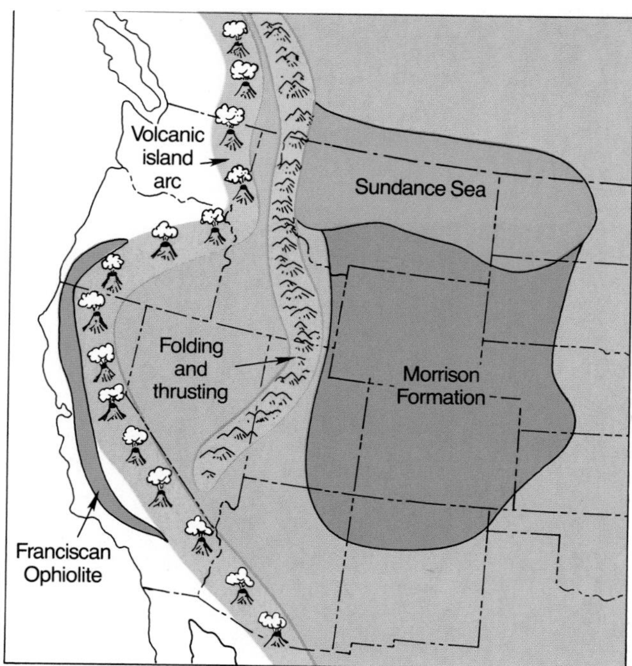


**Figure 13–33** Paleogeography of western North America for the Late Jurassic. A large inland mountain chain, approximately parallel to the coastline, extended for hundreds of kilometers. The Sundance Sea withdrew from the western interior and the nonmarine Morrison Formation accumulated in part of the area formerly occupied by the Sundance Sea.

treated to the North and South, marginal marine and continental deposition formed widespread coal-bearing deposits on the coastal plain.

## MICROPLATES AND THE GROWTH OF WESTERN NORTH AMERICA

In our discussion of the Cordilleran orogeny, we related the major regional events to subduction of the oceanic Pacific plate under the continental North American plate. Although subduction was the major controlling influence on the tectonic history and style of the Mesozoic Cordilleran mobile belt, it does not represent the complete story of mountain building and accretion in the area.

Several lines of evidence, introduced or clarified in the late 1970s and early 1980s, indicate that perhaps more than 25 percent of the entire Pacific Coast from Alaska to Baja California, and particularly Alaska and the Canadian Cordillera, consists of accreted lithospheric blocks, called **microplates** (Fig. 13–36). Some of these microplates are clearly of foreign origin, in that their fossils, stratigraphy, structural trends, and paleomagnetic properties differ completely from the rocks of the surrounding Cordilleran mobile belt and adjacent craton. Such microplates are referred to as **exotic terranes**. Some microplates of probable foreign origin, but

**Figure 13–34** Panoramic view of the Jurassic age Morrison Formation as seen from the Visitors Center, Dinosaur National Monument, Utah.

## Perspective 13–2

# Dinosaur National Monument

In 1909, Earl Douglass of the Carnegie Museum discovered "Dinosaur Ledge," a sandstone unit in the Upper Jurassic Morrison Formation that contained numerous dinosaur bones. After 13 years of excavating the layer, the Carnegie Museum had removed parts of 300 dinosaur specimens, two dozen of which were sufficiently complete to be reassembled (Fig. 1). Ten different dinosaur species were represented, as well as many other reptiles. It was by far the finest collection of dinosaur remains in the world. In the years that followed, the Smithsonian Institution and University of Utah also worked this rich quarry and recovered even more fossil specimens. Even more bones remained buried in an untouched part of the dipping sandstone ledge, and all that was needed to reveal them was to remove the overlying layers of shale and siltstone.

**Figure 1** *Camarasaurus lentus*, one of the best preserved and complete skeletons recovered from the Morrison Formation by the Carnegie Museum at Dinosaur National Monument.

Recognizing the scientific importance of this unit, the dinosaur quarry and 80 acres surrounding it were designated a national monument by President Wilson on October 4, 1915. Less than a year later it was included in the newly created National Park System. In 1938 Dinosaur National Monument was further expanded to 200,000 acres.

As far back as 1915, Earl Douglass envisioned making an exhibit in which the dinosaur bones would be exposed in relief in the tilted sandstone bed exactly where they came to rest 140 million years ago. Douglass stated in a letter to Dr. Charles Walcott, secretary of the Smithsonian Institution, "I hope that the Government, for the benefit of science and the people, will uncover a large area, leave the bones and skeletons in relief, and house them in. It would make one of the most astounding and instructive sights imaginable." However, it was not until 1953 that work began on constructing a truly unique museum around the still-buried sandstone ledge. The sediment overlying the remaining tilted sandstone ledge was removed, and the dinosaur bones were carefully exposed in bas relief. This quarry wall now forms the north wall of the visitors' center at Dinosaur National Monument (Fig. 2). The structure was completed and opened to the public in 1958.

not conclusively demonstrated to be exotic, are called **suspect terranes** to emphasize uncertainties about their origin. The discovery that mobile belts may also contain exotic and suspect terranes has resulted in a new field of research called *microplate tectonics.*

Many geologists believe these accreted lithospheric blocks, or terranes, formed elsewhere and were carried great distances on oceanic plates until they collided with other microplates or continental masses. They cite as evidence the many blocks of oceanic origin consisting of oceanic crust, islands, seamount chains, ridges, or island arcs. A few blocks clearly represent fragments of other continents.

The recognition of exotic terranes resulted from observations of abrupt changes in the paleontology, stratigraphy, paleomagnetic patterns, and structure of the

What was the landscape like during the Jurassic when dinosaurs roamed the area that is now Dinosaur National Monument? The land for miles around the present-day quarry was a low-lying desert during the Early Jurassic. This is indicated by the large cross-bedded dune sandstones of the Navajo Sandstone. Following deposition of the Navajo, a shallow sea transgressed from the west, depositing sandstone, siltstone, and limestones. During the Late Jurassic, the area from Mexico to Canada and from central Utah to the Mississippi River was above sea level. To the west was a mountain chain. Streams flowing from these mountains deposited sand and silt in the adjacent plains. Small lakes were numerous as well as some swamps. These stream, lake, and swamp deposits of the Jurassic coastal plain comprise the Morrison Formation.

Semitropical conditions prevailed in the area and forests of ginkos, cycads, and tree-ferns covered the land. In this setting pterosaurs glided through the air while dinosaurs roamed the landscape below. Crocodiles, turtles, and small mammals were also present. It was in the stream beds that most of the bones of the dinosaurs and other animals were deposited. During Cretaceous through Eocene time, the area was uplifted as part of the Laramide orogeny, and later the Green and Yampa rivers cut deep canyons in the area, exposing the rocks that make up Dinosaur National Monument.

a.

b.

**Figure 2** a. Visitors' center, Dinosaur National Monument. b. North wall of visitors' center showing dinosaur bones in bas relief, just as they were deposited 140 million years ago.

units in relation to the surrounding rocks. Presently, many geologists believe that microplate accretions to North America coincide and are associated with the deformation produced during the Sonoma, Nevadan, and Sevier orogenies. The accreted microplates are viewed as having been rafted along on oceanic plates. As the oceanic plates were subducted under the continent, island arcs, ridges, or microcontinents resisted subduction and were scraped off and accreted to the continent. It is estimated that more than 100 different-sized terranes have been added to the western margin of North America during the last 200 million years (Fig. 13–36).

The basic plate tectonic reconstruction for western North America remains unchanged. However, the details of such reconstructions in view of microplate tectonics are radically different. While the geologic history

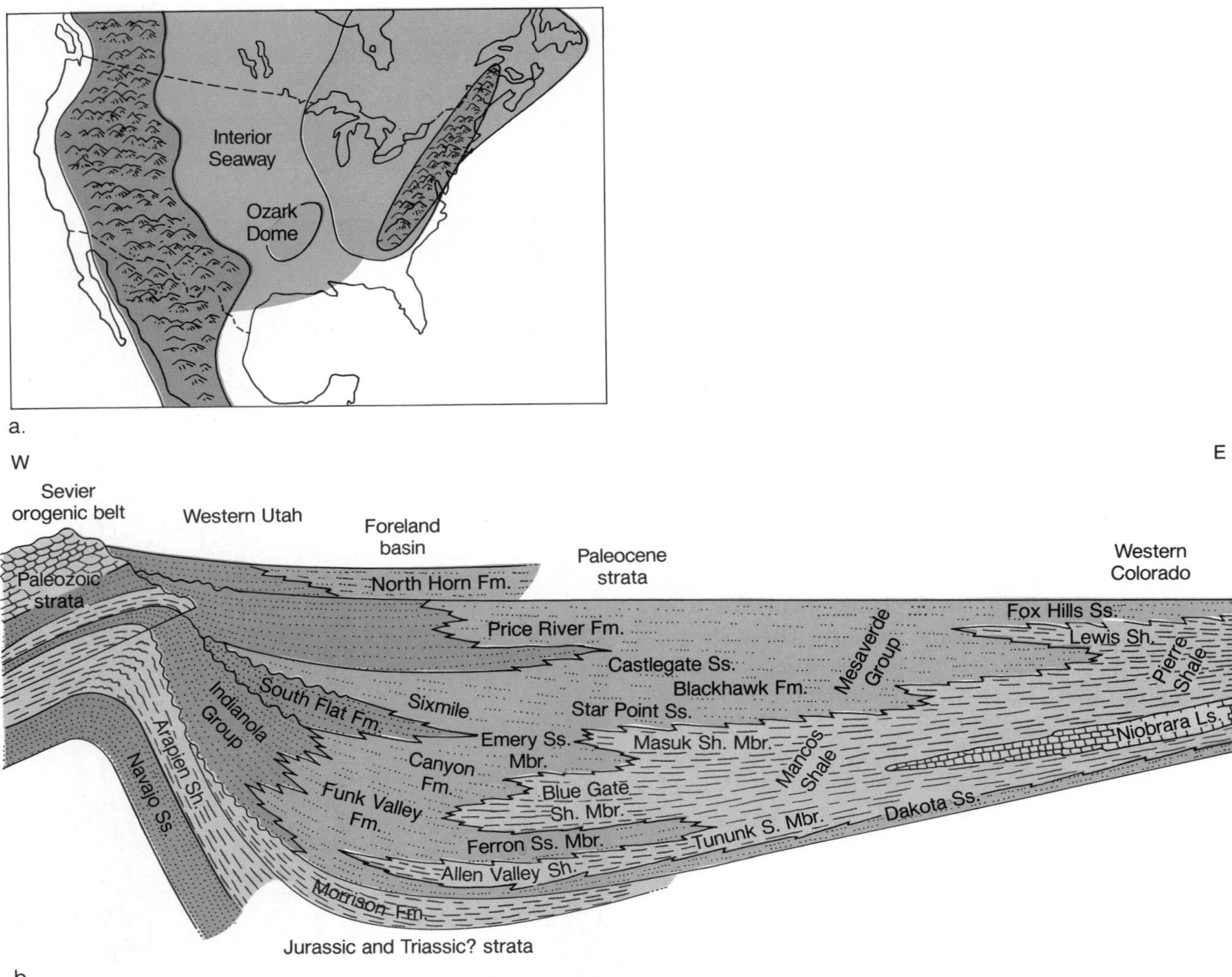


**Figure 13–35** a. Paleogeography of North America during the Late Cretaceous. The Cretaceous Interior Seaway stretched from the Arctic Ocean to the Gulf of Mexico and effectively divided North America into two large landmasses. b. Restored west-east cross section of Cretaceous facies of the western Cretaceous Interior Seaway showing their relationship to the Sevier orogenic belt.

of the Cordilleran mobile belt now appears more complex, microplate tectonics explains why active continental margins grow faster than passive trailing margins. Accreted terranes are indeed new additions to the continent rather than recycled older continental material. It has been estimated that western North America grew by more than 25 percent by microplate accretion during the Mesozoic Era.

We did not discuss microplates in relationship to the Paleozoic history of the Appalachian mobile belt, but a number of such plates are suspected to be present. However, many of them are difficult to define and not as easily recognized as those of the Mesozoic Cordilleran mobile belt because of greater deformation. Nonetheless, microplate tectonics provides a new way of viewing the Earth and of gaining a better understanding of the geologic history of the various continents.

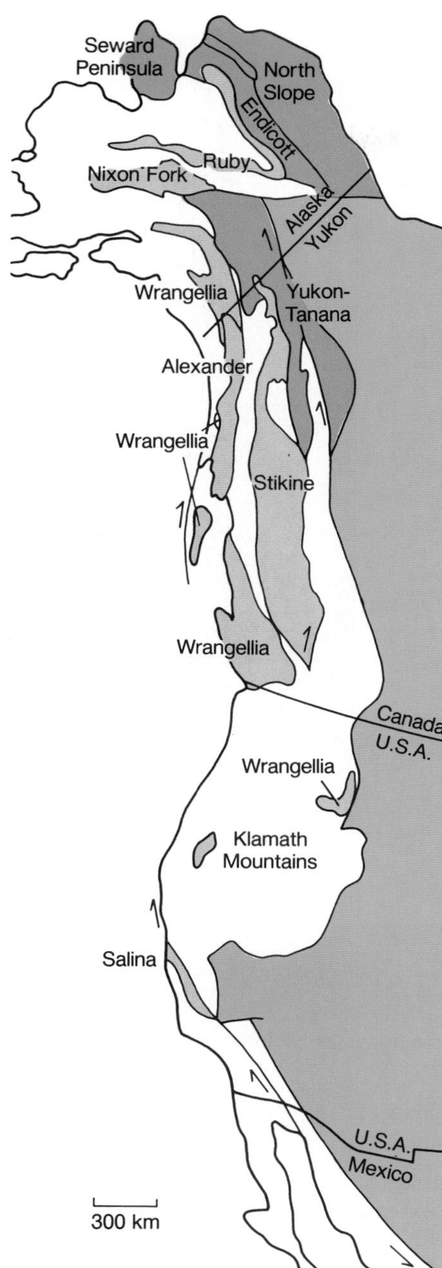


**Figure 13–36** Some of the accreted lithospheric blocks called microplates that form the western margin of the North American craton. The light brown blocks probably originated as parts of continents other than North America. The dark brown blocks are possibly displaced parts of North America. The North American craton is shown in green.

## CHAPTER SUMMARY

Table 13–2 provides an overall summary of the Mesozoic tectonic and sedimentation history of North America.

1. The breakup of Pangaea can be divided into four stages. The first stage involved the separation of North America from Africa during the Late Triassic, followed by the separation of North America from South America. The second stage involved the separation of Antarctica, India, and Australia from South America during the Jurassic and Cretaceous periods. During this stage India broke away from the still-united Antarctica and Australia landmass. During the third stage, South America separated from Africa, while Europe and Africa began to converge. In the last stage, Antarctica separated from Australia, and Greenland from North America.
2. The breakup of Pangaea influenced global climatic and atmospheric circulation patterns. While the temperature gradient from the tropics to the poles gradually increased during the Mesozoic, overall global temperatures remained equitable.
3. An increased rate of sea-floor spreading during the Cretaceous Period caused sea level to rise and transgressions to occur.
4. Except for incursions along the continental margin and two major transgressions (the Sundance Sea and the Cretaceous Interior Seaway), the North American craton was above sea level during the Mesozoic Era.
5. The Eastern Coastal Plain was the initial site of the separation of North America from Africa that began during the Late Triassic. During the Cretaceous Period, it was inundated by marine transgressions.
6. The Gulf Coast region was the site of major evaporite accumulation during the Jurassic Period as North America rifted from South America. During the Cretaceous Period, the Gulf Coast region was inundated by a transgressing sea, which, at its maximum, connected with a sea transgressing from the North to create the Cretaceous Interior Seaway.
7. Mesozoic rocks of the western region of North America were deposited in a variety of continental and marine environments. One of the major controls of sediment distribution patterns was tectonism.
8. Western North America was affected by four interrelated orogenies, the Sonoma, Nevadan, Sevier, and Laramide. Each involved batholithic intrusions as well as eastward thrust faulting and folding.
9. The cause of the Sonoma, Nevadan, Sevier, and Laramide orogenies was the changing angle of subduction of the

**Table 13–2** Summary of Mesozoic events.

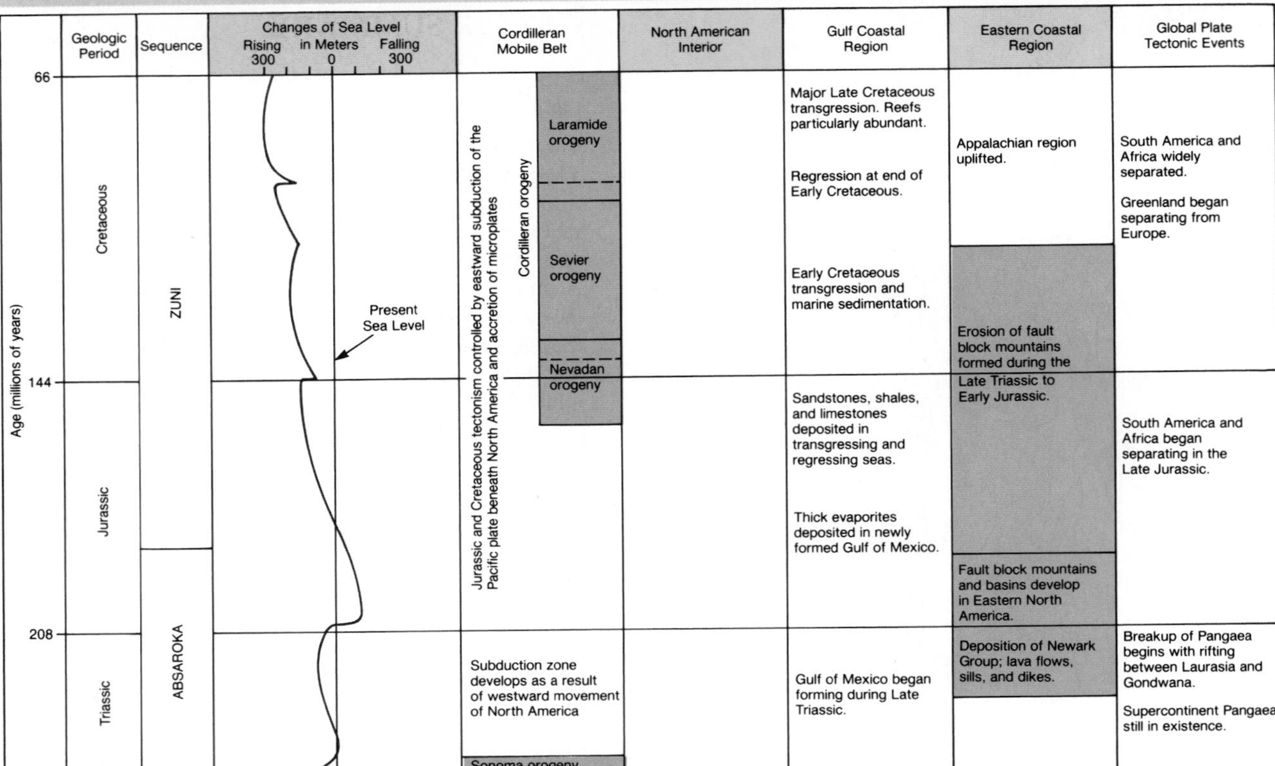


oceanic Pacific plate under the continental North American plate. The timing, rate, and to some degree the direction of plate movement was related to sea-floor spreading and the opening of the Atlantic Ocean.

10. Orogenic activity associated with the oceanic-continental convergent plate boundary in the Cordilleran mobile belt explains the structural features of the western margin of North America. However, it is believed that more than 25 percent of the North American western margin originated from the accretion of exotic and suspect terranes.

## IMPORTANT TERMS

aulocogen
Cordilleran orogeny
Cretaceous Interior Seaway
exotic terranes
Laramide orogeny
microplate
Nevadan orogeny
Sevier orogeny
Sonoma orogeny
Sundance Sea
suspect terrane
triple junction
Zuni sequence

## REVIEW QUESTIONS

1. Compare the tectonics of the Sonoma orogeny to the Antler orogeny.
2. How did the breakup of Pangaea affect oceanic and climatic circulation patterns?
3. How did the Mesozoic rifting that took place on the east coast of North America affect the tectonics in the Cordilleran mobile belt?
4. Discuss the tectonics of the Cordilleran mobile belt for the Mesozoic Era.
5. Using a diagram, explain how increased sea-floor spreading can cause a rise in sea level along the continental margins?
6. The American West is blessed with spectacular scenery. What geologic conditions led to such beauty?
7. Compare the depositional environments, tectonic setting, and depositional processes of the Triassic Newark Group in the East, to the Triassic Moenkopi, Shinarump, and Chinle Formations in the West.

8. Compare the tectonic setting and depositional environments of the Gulf of Mexico evaporites to the evaporite sequences of the Paleozoic Era.

9. How does microplate tectonics change our interpretations of the geologic history of the western margin of North America, and how does it relate to the Mesozoic orogenies that took place in that area?

## ADDITIONAL READINGS

Ben-Avraham, Z. 1981. The Movement of Continents. *American Scientist*, 69, no. 3: 291–299.

Bonatti, E. 1987. The Rifting of Continents. *Scientific American*, 256, no. 3: 96–103.

Coney, P. J., D. L. Jones, and J. W. H. Monger. 1980. Cordilleran Suspect Terranes. *Nature* 288: 329–333.

Dietz, R. S. and J. C. Holden. 1971. The Breakup of Pangea. In *Continents Adrift, Readings from Scientific American.* San Francisco: W. H. Freeman Co.

Jones, D. L., A. Cox, P. Coney, and M. Beck. 1982. The Growth of Western North America. *Scientific American*, 247, no. 5: 70–128.

Murray, G. E. 1961. *Geology of the Atlantic and Gulf Coastal Province of North America.* New York: Harper and Row.

Stearn, C. W., R. L. Carroll, and T. H. Clark. 1979. *The Geological Evolution of North America*, 3d ed. New York: John Wiley & Sons.

Wilhelm, O. and M. Ewing. 1972. Geology and History of the Gulf of Mexico. *Geological Society of America Bulletin*, 83, no. 3: 575–600.

# 14

## CHAPTER OUTLINE



## PROLOGUE

*About 80 million years ago, in what is now northern Montana, a wide coastal plain with rivers, swamps, and marshes sloped gently eastward to the sea. Unlike the semi-arid climate of today, this area was subtropical, probably much as southern Louisiana is now. Parts of the region were covered by dense vegetation including bald cypress, redwoods, and broad-leaf trees. Like the flora, the fauna was also varied. Fish, amphibians, turtles, and crocodiles lived in the streams and swamps, flying reptiles soared in the skies, and small primitive mammals scurried about.*

*Several types of dinosaurs lived in this area as well. Duck-billed dinosaurs were particularly abundant, but they shared their habitat with armored dinosaurs and herds of large, single-horned dinosaurs called* Monoclonius. *The large predator* Albertosaurus, *which looked much like* Tyrannosaurus, *and the small carnivore* Troödon *preyed upon the old, the weak, and unprotected juveniles.*

*Studies conducted by Princeton University and Montana State University under the direction of paleontologist John R. Horner have demonstrated that at least three dinosaur species used this area as a nesting ground. Numerous eggs, some unhatched, and a number of nests have been recovered. The eggs are of three types, the largest of which were laid by a duck-billed dinosaur, the genus* Maiasaura *(meaning "good mother lizard") (Fig. 14-1). These eggs are about 20 cm long, and the newborn measured 30 to 35 cm.*

*Eight maiasaur nests have been found at one site, each nest measuring 2 m in diameter and spaced about 7 m apart, a distance that equals the average length of an adult maiasaur. Twenty to 25 eggs were laid in a circular pattern in these bowl-shaped nests. It seems that these dinosaurs nested in colonies much as some modern birds and crocodiles do. Some nests contain the remains of juveniles up to 1 m long, which is considerably greater than their length at the time of hatching. This evidence implies that the young remained in the nest area for some time after hatching, and it further implies parental care. It seems likely that adults protected and fed the young.*

# Life of the Mesozoic Era

*Colonial nesting behavior seems to be well-documented by these discoveries. Additionally, smaller dinosaurs called* hypsilophodonts, *which laid intermediate-sized eggs, also in circular clutches, used the same nesting area repeatedly. Evidence for this conclusion is the fact that their nests occur in three superposed rock layers in a vertical sequence of strata 3 m thick. Hypsilophodont young did not remain in the nest as the maiasaur young did, but they did stay in the immediate nest area. Both hypsilophodont and maiasaur young stayed in the company of adults long after hatching. The smallest eggs found in this area were laid in two parallel rows, but the identity of the egg-layer is unknown.*

## INTRODUCTION

The Mesozoic Era is commonly referred to as the "Age of Reptiles," a phrase alluding to the fact that reptiles were the most diverse and abundant land-dwelling vertebrate animals. This is perhaps the most interesting chapter in the history of life because among the reptiles were the dinosaurs and their relatives, such as flying reptiles and marine reptiles. The Mesozoic diversification of reptiles

**Figure 14–1** Scene from the Late Cretaceous of northern Montana. A female *Maiasaura* guards her nest from the small, carnivorous dinosaur *Troödon. Albertosaurus*, a large predator, and the horned dinosaur *Monoclonius* are shown in the background.

was an important evolutionary event, but other equally important events also occurred. For example, mammals evolved from the mammal-like reptiles during the Triassic, while birds evolved from reptiles, probably small carnivorous dinosaurs, during the Jurassic.

Vast changes occurred in land-plant communities too. When the Mesozoic Era began, the dominant land plants were holdovers from the Paleozoic, but during the Cretaceous, the first flowering plants appeared. These plants diversified rapidly and soon became the most diverse and abundant land plants.

The Mesozoic was also a time of resurgence of marine invertebrates. Diversity among marine invertebrates had been drastically reduced as a result of the Permian extinction event, but survivors rapidly diversified, giving rise to increasingly complex marine invertebrate communities. Ammonites and planktonic microorganisms called *foraminifera* were particularly abundant and diverse.

The breakup of Pangaea that began during the Triassic continued throughout the Mesozoic. Nevertheless, the proximity of continents and mild Mesozoic climates allowed land animals and plants to occupy extensive geographic areas. As the fragmentation of Pangaea continued, however, some continents, Australia and South America especially, became isolated, and their faunas evolved independently.

Another mass extinction event occurred at the end of the Mesozoic. Once again organic diversity declined markedly as the dinosaurs and several other reptiles and some marine invertebrates died out. Since dinosaurs were victims of this extinction event, it has received more publicity than any other event although the Permian extinctions were of greater impact.

## MARINE INVERTEBRATES

Following the wave of extinctions that occurred at the end of the Paleozoic, the Mesozoic was a time of repopulation of the seas by the marine invertebrates. As the Atlantic and Indian oceans formed, and epeiric seas once again inundated the low-lying continental margins, invertebrate marine life returned with a flourish. Gone were many Paleozoic invertebrates such as the fusulinid foraminifera, blastoids, rugose corals, and trilobites.

The Early Triassic invertebrate marine fauna, while widely distributed, was not very diverse. By the Late Triassic, the seas were once again richly populated with invertebrates. The mollusks became increasingly diverse and abundant throughout the Mesozoic. The brachiopods, however, never completely recovered from their near extinction at the end of the Paleozoic, and while they diversified during the Triassic and Jurassic, they declined in numbers during the Cretaceous, remaining a minor invertebrate phylum during the Cenozoic. In areas of warm, relatively clear, shallow marine waters, corals again proliferated. These corals were of a new and familiar type, the *scleractinians* (Fig. 14–2). Echinoids, which were rare during the Paleozoic, greatly diversified during the Mesozoic (Fig. 14–3).

One of the major differences between Paleozoic and Mesozoic marine invertebrate communities was the increased abundance and diversity of burrowing organisms. Except for a few animals with hard parts, such as the inarticulate brachiopod *Lingula,* almost all Paleozoic burrowers were soft-bodied animals such as worms. The bivalves and echinoids, which were epifaunal elements during the Paleozoic, evolved various means of entering the infaunal habitats. This trend toward an infaunal existence may reflect an adaptive response to increasing predation from the rapidly evolving fish and cephalopods.

The end of the Mesozoic witnessed another major extinction event, this time involving mostly the large reptiles. The only extinctions of consequence in the marine invertebrate fauna were those affecting the ammonoid cephalopods, planktonic foraminifera, and a group of reef-dwelling *rudist* bivalves (Fig. 14–4).

We will now examine some of the major Mesozoic marine invertebrate and plant groups, beginning with the primary producers. The *coccolithophores* are an important group of living phytoplankton. Their remains, which are constructed of microscopic calcareous plates, collect in tremendous numbers on the ocean floor when the organisms die. Coccolithophore remains comprise many Cretaceous chalk beds such as the White Cliffs of Dover in England and the Demopolis Chalk of the southwestern United States (Fig. 14–5). Coccolithophores first appeared during the Jurassic and diversified tremendously during the Cretaceous. They continue to be abundant today and are one of the major primary producers supporting suspension feeders in today's oceans. *Diatoms* first appeared during the Cretaceous, but were more important as primary producers during the Cenozoic. Diatoms construct their shells out of silica. Today they are most abundant in cooler waters (Fig. 14–5). *Dinoflagellates* were common during the Mesozoic and today are the major primary producers in warm waters. Their remains are easily preserved as organic cysts (Fig. 14–5).

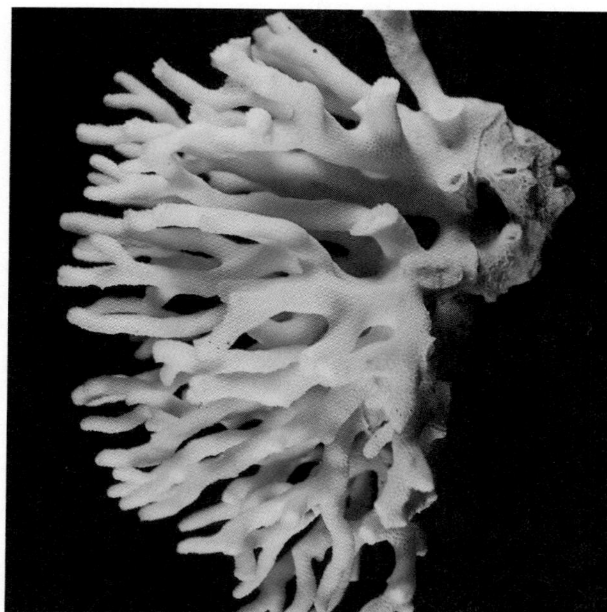

**Figure 14–2** Scleractinian corals appeared during the Triassic and proliferated in the warm, clear, shallow marine waters of the Mesozoic Era. Most living corals are scleractinians, represented here by the so-called "staghorn" coral.

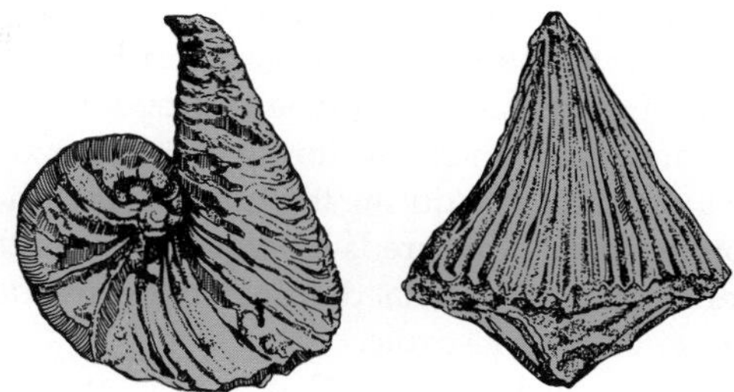

**Figure 14–4** Two different genera of Cretaceous reef-building rudistid bivalves.

**Figure 14–3** Echinoids, or sea urchins, were rare during the Paleozoic but became diverse and abundant during the Mesozoic.

**Figure 14–5** a. *Calcidiscus macintyrei,* a Miocene coccolith from the Gulf of Mexico (left); *Discoaster variabilis,* a Pliocene-Miocene coccolith from the Gulf of Mexico (right). b. *Actinoptychus senarius,* an Upper Miocene centric diatom from Java (left); *Cocconeis pellucida,* an Upper Miocene pinnate diatom from Java (right). c. *Deflandrea spinulosa,* an Eocene dinoflagellate from Alabama (left); *Spiniferites mirabilis,* a Neogene dinoflagellate from the Gulf of Mexico (right). (Scanning electron photomicrographs courtesy of (a) Merton E. Hill, Union Oil Company, (b) John Barron, United States Geological Survey, (c) John H. Wrenn, Amoco Production Company).

The mollusks, as previously noted, were the major invertebrate phylum of the Mesozoic. The mollusks include six classes, only three of which—the gastropods, bivalves, and cephalopods—are significant members of the marine invertebrate fauna. The gastropods increased in abundance and diversity during the Mesozoic, becoming most abundant during the Cretaceous, when the carnivorous forms appeared. However, it was the bivalves and cephalopods that dominated the invertebrate community of the Mesozoic.

Mesozoic bivalves diversified to inhabit many epifaunal and infaunal niches. Oysters and other clams became particularly diverse and abundant epifaunal suspension feeders and, in spite of a reduction at the end of the Cretaceous, continued to be important throughout the Cenozoic to the present (Fig. 14–6). The reef-forming rudists were a significant group of Mesozoic bivalves (Fig. 14-4). These epifaunal bivalves were geologically short-lived and so are excellent guide fossils for the Late Jurassic through the Cretaceous time interval. Rudists are also significant because they formed large tropical reefs, displacing corals as the main reef-builders. Bivalves also expanded into the infaunal niche during the Mesozoic. By burrowing into the sediment, they escaped predation from cephalopods and fish.

Cephalopods were one of the most important Mesozoic invertebrate groups. Their rapid evolution and nektonic life style make them excellent guide fossils (Fig. 14–7). Recall that two orders of cephalopods arose during the Paleozoic: the Nautiloidea, with simple sutures, and the Ammonoidea, with wrinkled sutures. The Ammonoidea are divided into three groups: the goniatites, ceratites, and ammonites. The ammonites, which are characterized by extremely complex suture patterns, were present during all three Mesozoic periods but were most prolific during the Jurassic and Cretaceous. Their tremendous success was a reflection of their ability to adapt to a wide range of marine environments. While most ammonites were coiled, some attaining diameters of 2m, others were uncoiled and led a near benthonic existence.

In spite of their successful adaptations, the ammonites became extinct at the end of the Cretaceous; explanations for their demise have ranged from competition with fish to extinction caused by a meteorite impact. Despite the fact the ammonites became extinct at or near the end of the Cretaceous, two other groups of cephalopods survived into the Cenozoic—the nautiloids and belemnoids, a group of squid-like cephalopods that was highly successful during the Jurassic and Cretaceous (Fig. 14–8).

Although the mollusks dominated the Mesozoic, other phyla were also important. The stalked echinoderms, which were abundant during the Paleozoic, became minor members of the Mesozoic marine invertebrate community. However, the echinoids, which were

**Figure 14–6** Bivalves, represented here by four Recent forms, were particularly diverse and abundant during the Mesozoic.

**Figure 14–7** Cephalopods were an important Mesozoic invertebrate group. The ammonites, which are characterized by extremely complex suture patterns, were particularly abundant and diverse during the Jurassic and Cretaceous and are excellent guide fossils for those periods. This specimen, *Scaphites preventricosus*, is from the Upper Cretaceous Colorado Formation, Toole County, Montana and shows the complex suture pattern characteristic of ammonites.

**Figure 14–8** Belemnoid cephalopods were abundant during the Jurassic and Cretaceous periods.

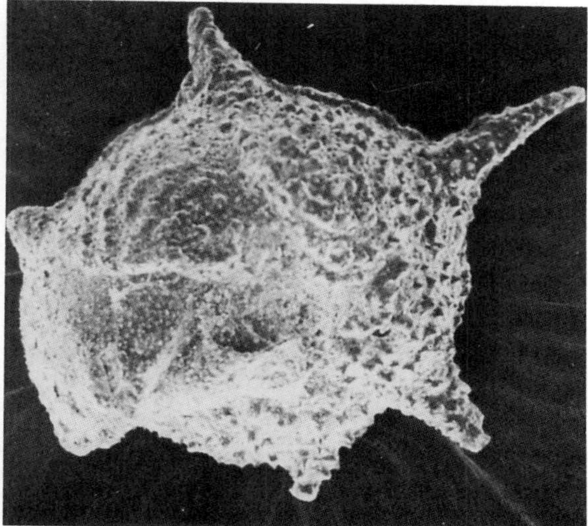

**Figure 14–9** Planktonic foraminifera, represented here by *Globotruncana calcarata* from the Cretaceous Pecan Gap Chalk of Texas, became diverse during the Jurassic and Cretaceous, but were affected by the extinction event at the end of the Cretaceous. (Photo courtesy of B.A. Masters, Amoco Production Co.)

exclusively epifaunal during the Paleozoic, branched out into the infaunal habitat and became very diverse and abundant (Fig. 14–3). Bryozoans, although rare in Triassic strata, diversified and expanded during the Jurassic and Cretaceous.

As is true today, where shallow marine waters were warm and clear, coral reefs proliferated. Mesozoic corals belong to the order Scleractinia (Fig. 14–2). Seleractinians use aragonite rather than calcite in constructing their skeletons, and most today have a symbiotic relationship with certain dinoflagellates. Because dinoflagellates require sufficient sunlight to function, they can live only in shallow waters. Whether scleractinian corals evolved from the rugose order or from an as yet unknown soft-bodied group that left no fossil record is still an unsettled question.

Lastly, the foraminifera underwent an explosive radiation during the Jurassic and Cretaceous that continued to the present. The planktonic forms (Fig. 14–9) in particular underwent rapid diversification, but they were also affected by the extinction event at the end of the Cretaceous. Most of the planktonic genera became extinct at the end of the Cretaceous, and only a few genera survived into the Cenozoic.

In general terms, we can think of the Mesozoic as a time of increasing complexity of the marine invertebrate community as it evolved from a simple one with low diversity and short food chains at the beginning of the Triassic to one that was highly complex with interrelated food chains near the end of the Cretaceous. This evolutionary history reflects the change in geologic conditions influenced by plate tectonic activity that we discussed in Chapter 13.

## FISHES AND AMPHIBIANS

Sharks and their relatives, the cartilaginous fishes, increased in abundance through the Mesozoic, but even so they never came close to matching the diversity of the bony fishes (Fig. 12–27). Nevertheless, sharks were, and still are, important elements of the marine fauna.

Among the bony fishes, the lung fishes and crossopterygians are represented by few Mesozoic genera and species. In fact, the crossopterygians declined and were almost extinct by the close of the era; only one living species is known (Fig. 5–27), and the group has no known Cenozoic fossil record.

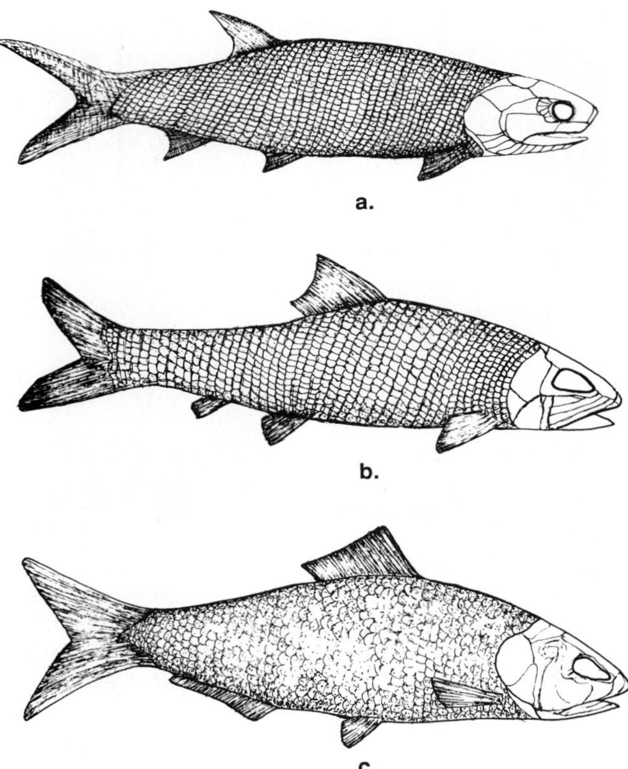


**Figure 14–10** Three stages in the evolution of bony fishes. The primitive chondrosteans (a) and the intermediate holosteans (b) are represented by few living species. In contrast, the advanced teleosts (c) include most species of living bony fishes.

The remaining bony fishes belong to three groups, which for convenience we can call primitive, intermediate, and advanced (Fig. 14–10). Superficially these fishes look much alike, but important changes occurred as one group replaced another (Table 14–1). By the end of the Cretaceous, the advanced group had become the dominant marine and freshwater fishes.

A few labyrinthodont amphibians persisted into the Mesozoic but died out by the end of the Triassic. Since the Pennsylvanian, the time of their greatest diversity, amphibians have made up only a small part of the total vertebrate fauna. Frogs and salamanders appeared during the Mesozoic, but their fossil records are poor.

## PRIMARY PRODUCERS ON LAND—PLANTS

As a prelude to our discussion of Mesozoic land animals, we will consider land-plant communities. After all, plants as photosynthesizers lie at the base of the food chain. In other words, they are the primary producers on land, while the animals, as consumers, are dependent upon plants.

Triassic and Jurassic land-plant communities, like those of the Late Paleozoic, were composed of seedless vascular plants such as ferns, horsetail rushes, and club mosses, and various gymnosperms (Table 14–2). Among the gymnosperms, however, the large seed ferns became extinct by the end of the Triassic. The *ginkgos* remained

a.

b.

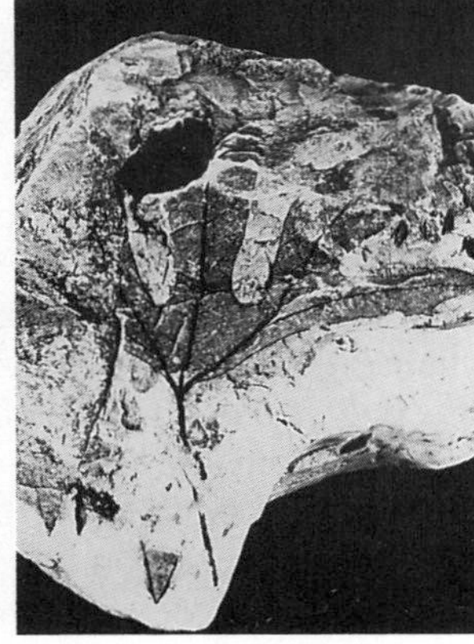

c.

**Figure 14–11** a. Living cycads, southern California. b and c. Fossil angiosperms from the Early Cretaceous Potomac Group of the eastern United States. b. *Sapindopsis*, Cecil County, Maryland. c. *Aralia* from New Jersey. (Photos b and c courtesy of Leo J. Hickey, Yale University).

**Table 14–1** Evolutionary trends in bony fishes.

| Primitive (Chondrostei) | Intermediate (Holostei) | Advanced (Teleostei) |
|---|---|---|
| Heavy, rhombic scales | Rhombic scales continued | Thin scales, of rounded shape |
| Internal skeleton, partly cartilaginous | Internal skeleton, partly cartilaginous | Internal skeleton, completely bony |
| Pelvic fins usually posterior | Pelvic fins usually posterior | Pelvic fins move forward in many forms |
| Lungs not transformed into swim bladder | Lungs transformed into swim bladder | Swim bladder completely hydrostatic |

**Table 14–2** Summary of biological and physical events for the Mesozoic.

| Geologic Period | Invertebrates | Vertebrates | Plants | Climate | Plate Tectonics |
|---|---|---|---|---|---|
| **Cretaceous** | Continued diversification of ammonites and belemnoids. Rudists become major reef-builders. Extinction of ammonites, rudists, and most planktonic foraminifera at end of Cretaceous. | Greatest diversity of dinosaurs<br>Extinctions of dinosaurs, flying reptiles and marine reptiles. Placental and marsupial mammals diverge. | Angiosperms appeared and diversified rapidly. Seedless plants and gymnosperms still common but less varied and abundant. | North-south zonation of climates more marked, but remained equable. Climate became more seasonal and cooler at end of Cretaceous. | Further fragmentation of Pangaea. South America and Africa had separated. Australia separated from South America but remained connected to Antarctica. North Atlantic continued to open. |
| **Jurassic** | Ammonite and belemnoid cephalopods increase in diversity. Scleractinian coral reefs common. Appearance of rudist bivalves. | First birds (may have appeared in Late Triassic). Time of giant sauropod dinosaurs. | Seedless vascular plants and gymnosperms only. | Much as Triassic. Ferns with living relatives restricted to tropics lived at high latitudes indicating mild climates. | Fragmentation of Pangaea continued, but close connections existed between all continents. |
| **Triassic** | The seas were repopulated by invertebrates that survived the Permian extinction event. Expansion of bivalves and echinoids into the infaunal niche. | Mammals evolved from cynodonts. Cynodonts became extinct. Thecodonts gave rise to dinosaurs. Flying reptiles and marine reptiles appeared. | Land flora of seedless vascular plants and gymnosperms as in Late Paleozoic. | Warm-temperate to tropical. Mild temperatures extended to high latitudes; polar regions may have been temperate. Local areas of aridity. | Fragmentation of Pangaea began in Late Triassic. |

abundant throughout the Mesozoic Era; ginkgos nearly disappeared during the Cenozoic and only one species survives today. *Conifers* also persisted from the Late Paleozoic, continued to diversify, and now are the most common land plants at high elevations and at the margins of the arctic. A new type of gymnosperm, the *cycads,* appeared during the Triassic. Cycads superficially resemble palm trees, and several varieties still exist in tropical and subtropical areas (Figs. 14–11, 14–12).

Marked changes in the composition of land-plant communities occurred during the Cretaceous. The long dominance of seedless plants and gymnosperms ended as many were replaced by **angiosperms,** or flowering plants (Figs. 14–11, 14–12 Table 14–2). The earliest angiosperms probably evolved from a specialized group of seed ferns. In any case, since the angiosperms appeared, they have adapted to nearly every terrestrial habitat from high mountains to low deserts. Some have even

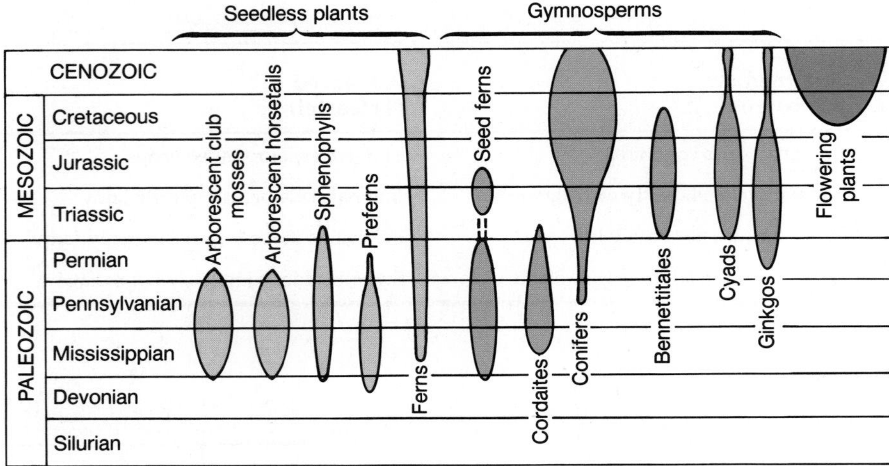


**Figure 14–12** Geologic ranges and relative abundance of land plants. Among the seedless vascular plants and gymnosperms, only ferns and conifers, respectively, are common elements in the modern flora. Flowering plants, the angiosperms, appeared during the Cretaceous and now account for more than 90 percent of all land-plant species.

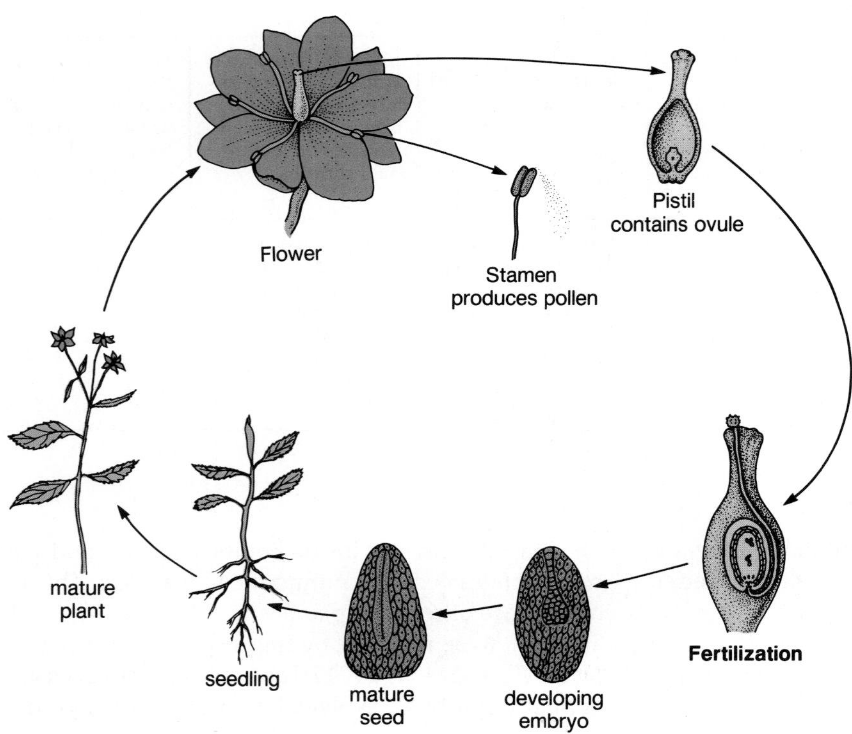


**Figure 14–13** The reproductive cycle in angiosperms.

adapted to shallow coastal waters, and a few varieties are carnivorous.

Several factors account for the phenomenal success of flowering plants, but chief among them is their method of reproduction (Fig. 14–13). There were two particularly important developments. One was the evolution of flowers, which attract animal pollinators, especially insects, and the other was the evolution of enclosed seeds. Wide dispersal of angiosperm seeds is accomplished in several ways—by the wind, carried in a fleshy fruit, or carried in burrs that become attached to animal fur.

Seedless plants and gymnosperms are important and still flourish in many environments; in fact, many botanists regard the ferns and conifers as emerging groups. Nevertheless, a measure of the angiosperms' success is that today they account for well over 90 percent of all land-plant species.

## REPTILES

The evolutionary relationships among the groups of fossil and living reptiles are summarized in Figure 14–14. Reptile diversification began during Pennsylvanian time with the appearance of the captorhinomorphs, apparently the first animals to lay amniote eggs (Chapter 12). From this basic stock of so-called *stem reptiles* all other reptiles arose. Birds and mammals, too, have their ancestors among the reptiles, so they are a part of this major evolutionary diversification.

Recall from Chapter 12 that the captorhinomorphs and pelycosaurs were the dominant land vertebrates of the Pennsylvanian and Permian periods. Here we continue our story of reptile diversification with a group called *thecodonts.*

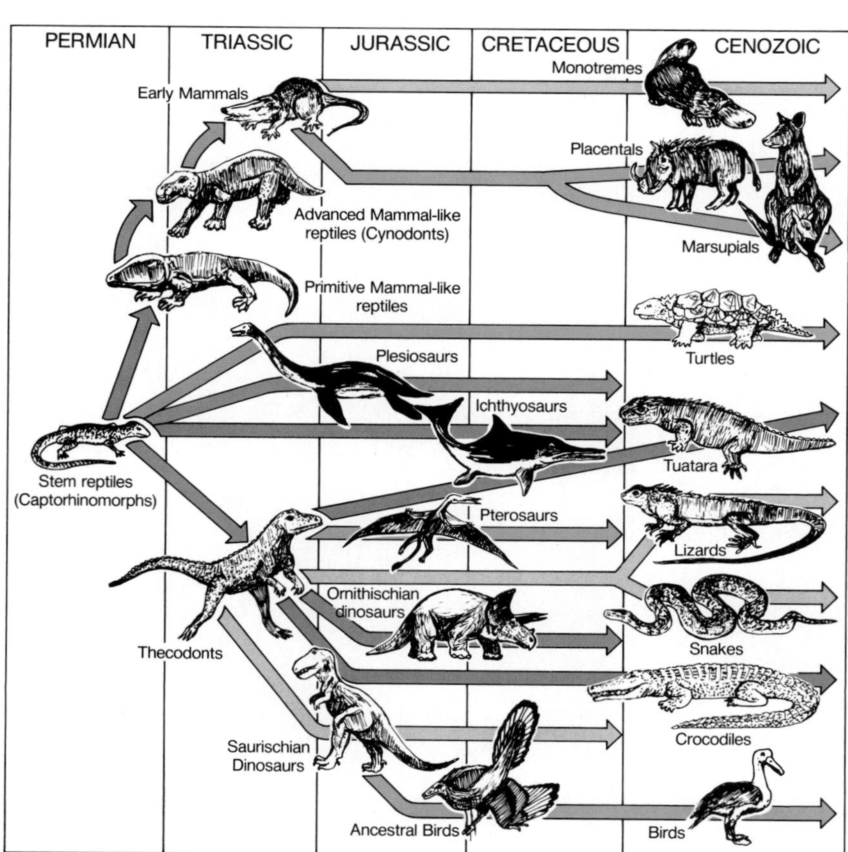


**Figure 14–14** Relationships among fossils and living reptiles, birds, and mammals.

## Thecodonts and the Ancestry of Dinosaurs

Compared with most other reptile orders, **thecodonts** had a comparatively brief history, existing for only 45 to 50 million years. Furthermore, their adaptive radiation was rather limited since it resulted in the origin of few genera and species. Thecodonts are nevertheless important because they were ancestors of the dinosaurs, flying reptiles, and three of the five orders of living reptiles (Fig. 14–14).

The thecodonts were Late Permian and Triassic reptiles characterized by teeth set into individual sockets, as are those of crocodiles, dinosaurs, mammal-like reptiles, and mammals. One group of thecodonts, the **pseudosuchians,** included lightly built carnivores 1 m or 2 m long (Fig. 14–15). These small predators had well-developed forelimbs and moved primarily on all four limbs; that is, they were **quadrupedal.** When running, however, they rose onto their hind limbs and moved in a **bipedal** fashion. The largest of the thecodonts were fully quadrupedal and covered with bony armor. Some of these were herbivores, while others—the phytosaurs, for example—were predators (Fig. 14–16). Phytosaurs were crocodile-like in appearance and probably lived much as modern crocodiles do.

Dinosaurs evolved from thecodonts during the Late Triassic, but their specific thecodont ancestor is not agreed upon. The traditional interpretation is that two distinct orders of dinosaurs were established when they first appeared, each of which may have had an independent origin from pseudosuchian thecodonts. Pelvic structure is the basis for recognizing the orders **Saurischia** and **Ornithischia** (Fig. 14–17, Table 14–3). Saurischian dinosaurs had a lizard-like pelvis, so are referred to as lizard-hipped dinosaurs. Ornithischians had a bird-like pelvis, hence are called bird-hipped dinosaurs. However, some investigators think that both orders of dinosaurs shared a common ancestor among the *lagosuchid* thecodonts, and that the subdivision of dinosaurs based on pelvic structure is not justified. In this book we follow the traditional approach and thus recognize the classification of dinosaurs shown in Figure 14–17 and Table 14–3.

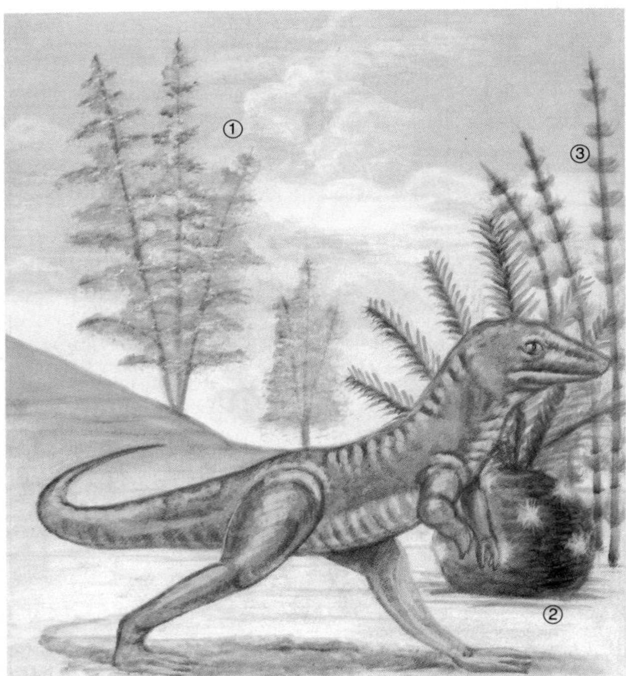

**Figure 14–15** Triassic bipedal thecodonts. a. Scene from the Early Triassic of South Africa showing the small thecodont *Euparkeria*. The vegetation consists of gymnosperms, including ginkgos (1) and cycads (2), and scouring rushes or horsetails (3), which are seedless vascular plants. b. Late Triassic scene from Europe. The classification of *Ornithosuchus* is uncertain. Some authorities call it an advanced thecodont while others believe it is a primitive dinosaur. In this scene some ferns are shown in addition to horsetails and cycads.

**Figure 14–16** Armored, quadrupedal thecodonts of the Triassic. *Desmatosuchus*, on the right, had a formidable appearance, but was a herbivore possessing weak jaws and teeth. The phytosaur *Rutiodon*, in the background, was a large carnivore that lived much as present day crocodiles do. A small dinosaur, *Coelophysis*, is shown in the left foreground. The vegetation includes ferns, horsetails, and gymosperms as in Figure 14–15.

## Dinosaurs

More than any other type of animal, the dinosaurs have inspired awe and have thoroughly captured the public imagination. As often as not, unfortunately, their popularization in cartoons, movies and many books has led to misunderstandings of dinosaurs in general. It is true that many were large animals, the largest animals ever to live on land. But not all were large. In fact, dinosaurs varied from giants to individuals no larger than a chicken (Fig. 14–18, Table 14–3).

A common but erroneous perception of dinosaurs is that they were poorly adapted animals that had trouble surviving. True, they became extinct, but to consider this a failure is to ignore the fact that for more than 140 million years they were the dominant land vertebrates. During their existence they diversified into numerous types and adapted to a wide variety of environments. Eventually the dinosaurs did die out, an event that then enabled mammals to become the predominant land vertebrates.

Neither were dinosaurs the lethargic beasts we often see portrayed in various media. Recent evidence suggests that at least some dinosaurs may have been very active and possibly warm-blooded. It also appears that some species cared for their young long after hatching, a behavioral characteristic most often associated with birds and mammals (see Prologue).

### Saurischian Dinosaurs

Two groups of saurischians are recognized, theropods and sauropods (Fig. 14–17, Table 14–3). Theropods were carnivorous bipeds that ranged in size from tiny *Compsognathus (Fig. 14–18) to Tyrannosaurus* (Fig. 14–19), the largest terrestrial carnivore known. A particularly interesting theropod was *Deinonychus,* which means terrible claw. It was a predator with large, sickle-like claws on the hind feet that were probably used to kill its prey (Fig. 14–20).

Included among the sauropods were the giant, quadrupedal herbivores such as *Apatosaurus, Diplodicus,*

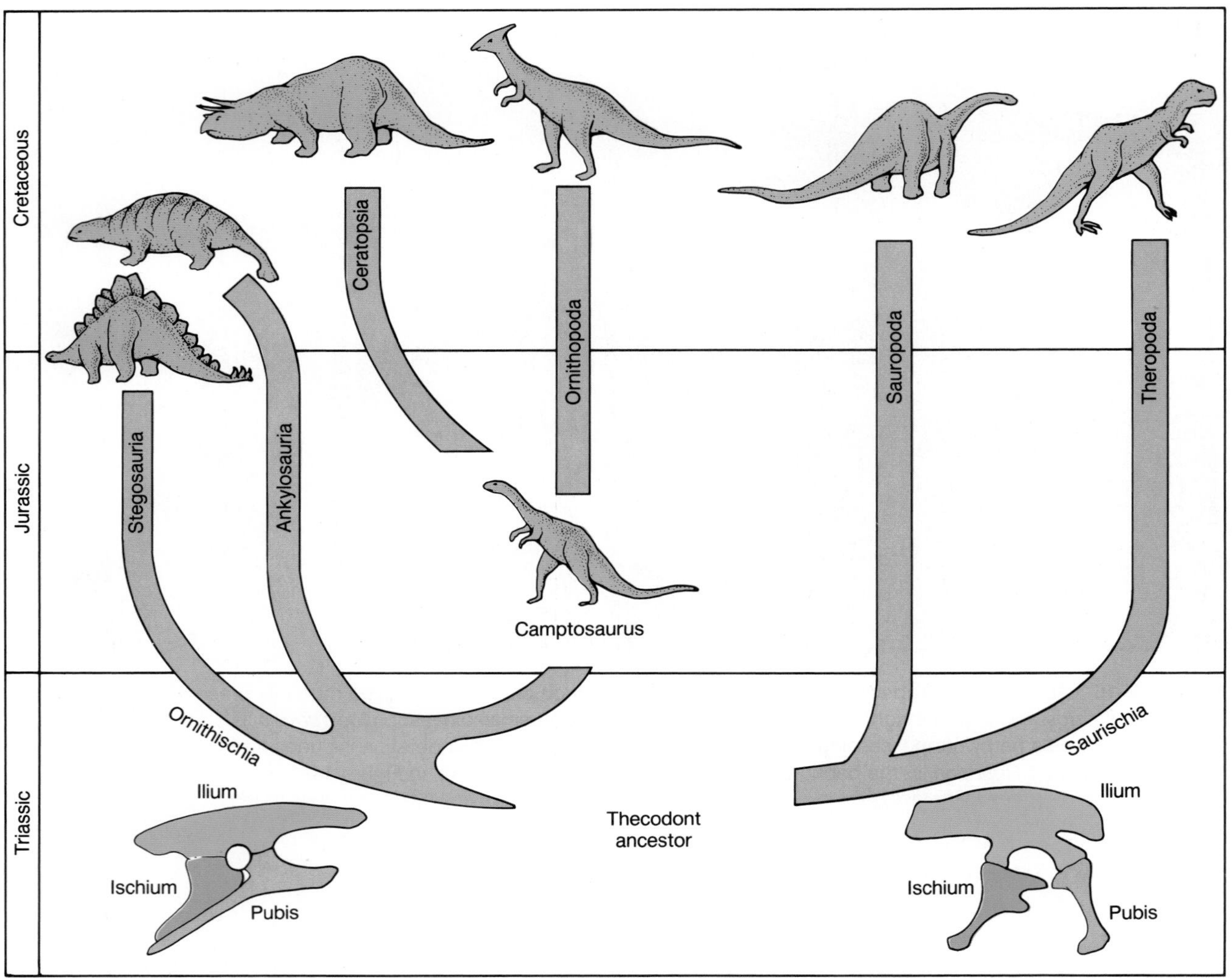


**Figure 14–17** Origin and relationships among dinosaurs. Both ornithischian and saurischian dinosaurs evolved from thecodonts, but each may have had an independent origin from that group. The pelvis of each order of dinosaurs is shown.

and *Brachiosaurus,* (Fig. 14–21), the largest known land animals of any kind. Recent discoveries of fossils in Colorado, New Mexico, and elsewhere suggest that even larger sauropods existed (Table 14–3) (Perspective 14–1). Evidence from fossil trackways indicates that sauropods moved in herds. Rather than speed, they depended on their size and herding behavior as their primary protection from predators.

### Ornithischian Dinosaurs

The great diversity of **ornithischians** is manifested by the fact that four distinct groups are recognized: ornithopods, ankylosaurs, stegosaurs, and ceratopsians (Fig. 14–17, Table 14–3). Ornithopods include the duck-billed dinosaurs, which had flattened, bill-like mouths (Fig. 14–22). These dinosaurs were particularly varied and abundant during the Cretaceous, and some species were characterized by head crests (Fig. 14–22). Such crests may have functioned as a means of species recognition, as display devices to attract mates, or as resonating chambers to amplify bellowing. Some duck-billed dinosaurs practiced colonial nesting and care of the young (see "Prologue"). All ornithopod genera except one were herbivores and primarily bipedal, but their

**Table 14–3** Summary chart of some of the dinosaurs mentioned in the text.

| Order | Suborder | Genus | Locomotion | Feeding | Length* | Weight* |
|---|---|---|---|---|---|---|
| Saurischia | Theropoda | *Compsognathus*<br>*Coelophysis*<br>*Deinonychus*<br>*Tyrannosaurus* | Bipedal | Carnivorous | 60 cm<br>3 m<br>2.4-4 m<br>14 m | 2-3 kg<br>29.5 kg<br>75 kg<br>7 tons |
| | Sauropoda | *Diplodocus*<br>*Apatosaurus*<br>*Brachiosaurus*# | Quadrupedal | Herbivorous | 27 m<br>21 m<br>23-27 m | 10.6 tons<br>30 tons<br>77 tons |
| Ornithischia | Ornithopoda | *Troödon*<br>Duck-bills† | Bipedal | Carnivorous<br>Herbivorous | 2.4 m<br>† | <br>† |
| | Ankylosauria | *Ankylosaurus* | Quadrupedal | Herbivorous | 6 m | 2-3 tons |
| | Stegosauria | *Stegosaurus* | Quadrupedal | Herbivorous | 9 m | 1.8 tons |
| | Ceratopsia | *Psittacosaurus*<br>*Triceratops* | Quadrupedal | Herbivorous | .8-1.5 m<br>9 m | Up to 23 kg<br>5.4 tons |

* Lengths and weights mostly from Lambert, D. 1983. A Field Guide to Dinosaurs, Avon Books, N.Y. Weights in metric tons (1 metric ton = 2204 lbs).

\# Partial remains of what appear to be even larger brachiosaurids have been discovered. For these, lengths greater than 30 m and weights up to 136 metric tons have been claimed.

† Duck-billed dinosaurs varied considerably in size: lengths ranged from 3.6 to 15m, and some larger species weighed more than 4 metric tons.

**Figure 14–18** *Compsognathus,* here compared with a cokeral, is one of the smallest known dinosaurs. Lizard bones have been found inside the rib cage of *Compsognathus* indicating that this small carnivore caught such speedy prey.

**Figure 14–19** Probably the best known theropod dinosaur is *Tyrannosaurus* shown here pursuing *Struthiomimus*, the "ostrich dinosaur." This scene from the latest Cretaceous of western North America also shows gymnosperms, a flowering plant on the left, and an early mammal called a multiturberculate in the right foreground.

well-developed forelimbs allowed them to walk in a quadrupedal fashion, too.

Ankylosaurs were heavily armored, quadrupedal herbivores, and some were quite large (Fig. 14–23, Table 14–3). Bony armor protected the back, flanks, and top of the head, and the tail ended in a large, bony club-like growth. No doubt a blow delivered by the powerful tail could seriously injure an attacking predator.

The stegosaurs, represented by the familiar genus *Stegosaurus*, (Fig. 14–23, Table 14–3), were quadrupedal herbivores with bony spikes on the tail that were undoubtedly used for defense, and bony plates on the back. The exact arrangement of these plates on the back is debatable, but many paleontologists believe they functioned as a device to absorb and dissipate heat.

The final group of ornithischian dinosaurs is the ceratopsian or horned dinosaurs. A rather good fossil record indicates that large, Late Cretaceous genera such as *Triceratops* evolved from small, Early Cretaceous ancestors (Fig. 14–24). The later ceratopsians were characterized by huge heads, a large bony frill over the top of the neck, and a large horn or horns on the skull. Fossil trackways indicate that these large, quadrupedal herbivores moved in herds.

We conclude our discussion of dinosaurs with a few general comments to put things into their proper perspective. Dinosaurs are fascinating animals, but they are not part of our experience, so we tend to overlook some other, equally fascinating aspects of the present-day animal world. For example, as large as some dinosaurs were, none are known to have been as large as living blue whales. One is awed by the massive armor of the ankylosaurs, but we may fail to realize that the most completely armored vertebrate animal to have ever lived is alive today—the turtle. The carnivorous dinosaurs are commonly portrayed as ferocious, aggressive beasts, and this is probably fairly accurate. Carnivores such as *Tyrannosaurus*, however, probably behaved much as large carnivores living today. They went for the easy kill, preying upon the old, the weak, and the young, or simply dined on carrion, if available.

### Warm-Blooded Dinosaurs?

All living reptiles are **ectotherms,** that is, cold-blooded animals whose body temperature varies in response to the outside temperature. **Endotherms,** warm-blooded animals, such as birds and mammals, are capable of maintaining a rather constant body temperature regardless of the outside temperature. Some investigators think that dinosaurs, or at least some dinosaurs, were endotherms too.

Proponents of dinosaur endothermy note that dinosaur bones are penetrated by numerous passageways which, when the animals were living, contained blood vessels. Typically, bones of endotherms have this structure, but considerably fewer of these passageways are found in bones of ectotherms. Modern crocodiles and turtles have this so-called endothermic bone structure, and in some small mammals the bone structure is more typical of ectotherms, yet we know that they are capable of maintaining a constant body temperature. It may be that bone structure is more related to body size and growth patterns than to endothermy.

Because endotherms have high metabolic rates, they must eat more than comparably sized ectotherms

**Figure 14–20** *Deinonychus* ("terrible claw") from the Early Cretaceous of Montana is shown here using its sickle-like claws to kill its prey. *Deinonychus* was a comparatively small theropod that measured 2.4 m to 4 m long.

**Figure 14–21** Scene from the Late Jurassic of western North America showing three giant, quadrupedal sauropods, *Diplodocus* (1), *Apatosaurus* (2), and *Brachiosaurus* (3). Although *Diplodocus* was the longest, the other two were much heavier. See Table 14–3 for sizes.

**Figure 14–22** Three duck-billed dinosaurs from the Late Cretaceous of western North America; *Anatosaurus* (1) and two of the several duck-billed dinosaurs with head crests, *Parasaurolophus* (2) and *Corythosaurus* (3). Notice that duck-billed dinosaurs had well-developed forelimbs and were capable of both bipedal and quadrupedal locomotion.

(Fig. 14–25). Consequently, endothermic predators require large prey populations. They would therefore constitute a much smaller proportion of the total animal population than their prey. In contrast, the proportion of ectothermic predators to their prey population is much greater. Where data seem to be sufficient, dinosaur predators appear to have made up 3 to 5 percent of the total population. These figures are comparable to modern mammalian populations. However, a number of uncertainties about the composition of fossil communities make this argument for endothermy unconvincing to many paleontologists.

Living endotherms have a large brain in relation to body size. A relatively large brain is not necessary for endothermy, but endothermy does seem to be a prerequisite for having a large brain, because a complex nervous system requires a rather constant body temperature. Some dinosaurs, particularly the small carnivores, did have a large brain in relation to their body, but many did not. The fact that the small carnivorous ones usually did seems to be a good argument for endothermy, but there is an even more compelling argument. The relationship of birds to small carnivores such as *Compsognathus* (Fig. 14–18) implies that these small dinosaurs were endothermic or at least trending in that direction. The large sauropods were probably not endothermic, but nevertheless may have been able to maintain their body temperatures within narrow limits as endotherms do. A large animal heats up and cools down slowly because it has a small surface area compared to volume. With proportionately less surface area to allow heat loss, sauropods probably retained body heat more efficiently than smaller dinosaurs.

One further point on endothermy in dinosaurs is that the flying reptiles, the *pterosaurs*, evolved from thecodonts as did the dinosaurs (Fig. 14–14). At least one species of pterosaur had hair or hair-like feathers. This is interesting because an insulating covering of hair or feathers is known only in modern endotherms. Furthermore, the physiology of active flight requires endothermy. Such evidence leads to the conclusion that perhaps both thecodonts and dinosaurs were endothermic.

a.

b.

**Figure 14–23** Armored and plated ornithischian dinosaurs. a. *Ankylosaurus* of the Late Cretaceous of western North America was the most completely armored dinosaur. b. *Stegosaurus* lived in western North America during the Late Jurassic. The large plates on the back may have functioned as a device to absorb and dissipate heat.

Obviously, considerable disagreement exists on dinosaur endothermy. In general, a fairly good case can be made for endothermic, small, carnivorous dinosaurs and pterosaurs, but for the others the question is still open.

## Flying Reptiles

The first vertebrate animals to fly are called **pterosaurs** (Fig. 14–26). Pterosaurs evolved from thecodonts during the Triassic and were abundant until their extinction at the end of the Mesozoic. Pterosaur flight adaptations include a wing membrane supported by an elongate fourth finger (Fig. 14–26), light hollow bones, and development of those parts of the brain associated with muscular coordination and sight. Size varied considerably. Some of the early species were sparrow- to robin-sized, while one Cretaceous pterosaur from Texas had a wingspan of at least 12 m. At least one pterosaur had a coat of hair or hair-like feathers (Perspective 14–2).

Were pterosaurs, particularly the larger ones, active, wing-flapping fliers or simply gliders? Experiments with scale models indicate that *Pteranodon* (Fig. 14–26) could actively fly, but once airborne probably took advantage of thermal updrafts to stay aloft and thereby ranged far out over the open sea, mostly by gliding. Perhaps much like the modern frigate bird, *Pteranodon* glided close to the surface and plucked fish from the water with its long beak. Most paleontologists agree that the small pterosaurs were wing-flapping fliers.

The function of the large crest on the head of *Pteranodon* is uncertain. It may have functioned aerodynamically in maneuvering or braking, or it may have served as a counterbalance for the long beak. But some specimens seem to lack this feature, indicating that it may have been a sexual characteristic of some sort. If so, whether males or females had crests is unknown.

a.

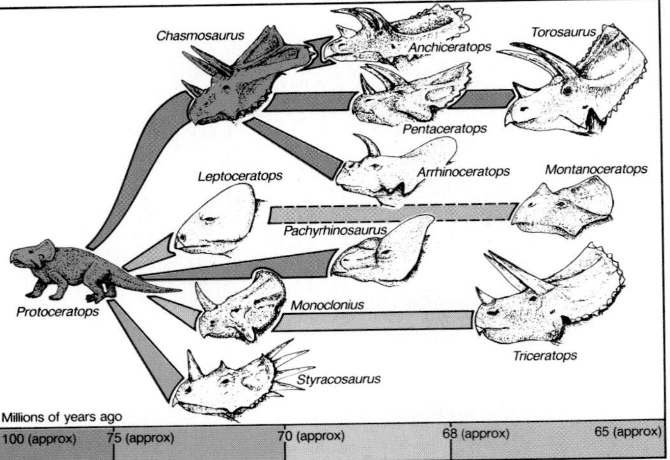


b.

c.

**Figure 14–24** Ceratopsian dinosaurs. a. *Psittacosaurus*, the parrot lizard from the Early Cretaceous of China and Mongolia, is the probable ancestor of all later ceratopsians. *Psittacosaurus* was semi-quadrupedal and measured .8 m to 1.5 m long. b. Evolutionary diversification of ceratopsians. An increase in size, change from semi-quadrupedal to quadrupedal stance, and development of horns and a large bony frill covering the top of the neck were the main evolutionary trends. The frill was probably for attachment of muscles that supported the large head. c. *Triceratops*, one of the best-known North American ceratopsians, was about 9 m long. The large brow horns were 1 m long.

Herbivorous

Carnivorous

**Figure 14–25** Evidence cited for endothermy in dinosaurs. If dinosaurs were endotherms, a large population of prey animals would have been necessary to support a low proportion of carnivores. This illustration shows the proportion of herbivores to carnivores for dinosaurs in Alberta, Canada. The proportions shown are consistent with populations of living endotherms, but some paleontologists question the accuracy of these data.

a.

b.

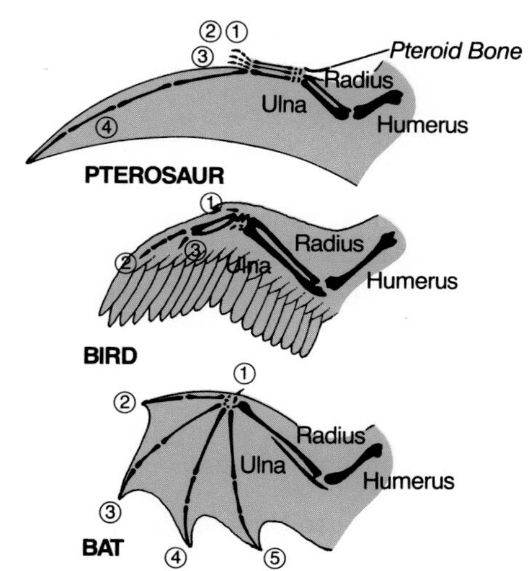


c.

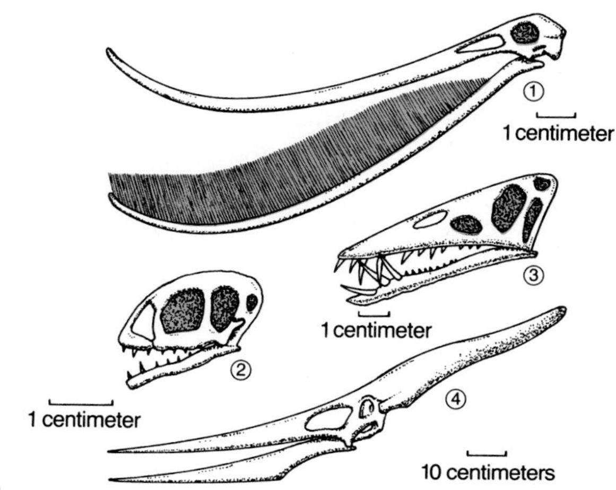


d.

**Figure 14–26** a. Long-tailed and short-tailed pterosaurs from the Jurassic of Europe. b. *Pteranodon,* a large, Cretaceous pterosaur. c. Comparison of the wings of a pterosaur (top), a bird, and a bat (bottom). In the pterosaur the fourth finger provides the support for the wing, while in the bird the second finger provides most of the support, and in the bat the second through fifth fingers support the wing. d. Skulls of pterosaurs: 1 had baleen-like teeth and may have strained plankton from seawater; 2 was a tiny pterosaur that may have eaten insects with its peg-like teeth, and 3 and 4 probably ate fish, since fish bones have been found within their rib cages.

## Perspective 14–1

# The Largest Dinosaurs

The largest living land animal, the African elephant, is about 4 m tall and weighs about 6 metric tons. In fact, living elephants weigh as much as, or more than, many dinosaurs. Nevertheless, it has been estimated that "more than half of the dinosaurs weighed more than 2 metric tons—a size attained by only about 2 percent of modern mammals.[1] Thus, many dinosaurs were very large animals, but just how large is "very large"? Certainly the giants among dinosaurs were the sauropods, especially those called *brachiosaurids* such as *Brachiosaurus* (Fig. 14–21); *Brachiosaurus* is estimated to have been 23 m to 27 m long, and may have weighed 77 metric tons!

It was once believed that the giant sauropods were too heavy to walk on land, hence they were commonly depicted partly or completely submerged in swamps and lakes (Fig. 1). However, recent studies indicate that the sauropods were quite capable of walking on dry land, although they no doubt visited swamps and lakes, where the chances of preservation of their bones was greater. Some paleontologists now think that many sauropods could even rear up on their hind legs in order to browse high up on trees (Fig. 2).

**Figure 1** It was once thought that the large dinosaurs, such as these sauropods, were too heavy to walk on dry land and must have been semiaquatic. Evidence from fossil footprints indicates that sauropods were quite capable of walking on land. Furthermore, if they had submerged themselves, as shown in this illustration, breathing would have been difficult because water pressure would prevent the chest from expanding.

Just how large the largest dinosaur was is unknown. However, partial remains from Colorado and New Mexico indicate that some were larger than *Brachiosaurus.* These giants among giants are known by such common names as supersaurus, ultrasaurus, and seismosaurus. Ultrasaurus may have been more than 30 m long and possibly weighed 136 metric tons (Fig. 3).

[1]Marx, J. L. 1978. Warm-Blooded Dinosaurs: Evidence Pro and Con. *Science,* 199: 1425.

## Marine Reptiles

To most people **ichthyosaurs** are probably the most familiar of the Mesozoic marine reptiles (Fig. 14–27). These animals were about 3 m long and were completely aquatic. Aquatic adaptations included a streamlined, somewhat fish-like body, a powerful tail for propulsion, and flipper-like forelimbs for maneuvering. The numerous sharp teeth indicate that ichthyosaurs were fish eaters.

Ichthyosaurs were so thoroughly aquatic that it is doubtful they could venture onto land at all. This poses a reproductive problem since reptile eggs will not survive if laid in water. Female ichthyosaurs probably retained the eggs in their bodies and gave birth to live young. Some fossils with young ichthyosaurs within the body cavity support this interpretation.

A second group of Mesozoic marine reptiles, the **plesiosaurs** occurred in two varieties, short-necked and

It may seem odd that ultrasaurus's estimated length is not much greater than that of *Brachiosaurus,* yet its estimated weight is more than 1.7 times as much. The reason is that the weight of an animal, or any object, increases by a factor of 8 if the linear dimensions are doubled. In other words, if the linear dimensions increase by a factor of 2, the weight increases eightfold. Thus, since ultrasaurus was perhaps 20 to 25 percent larger than *Brachiosaurus,* it probably weighed more than 1.7 times as much.

**Figure 2** Many paleontologists now think that sauropods could rear up and support their weight on their back legs and tail.

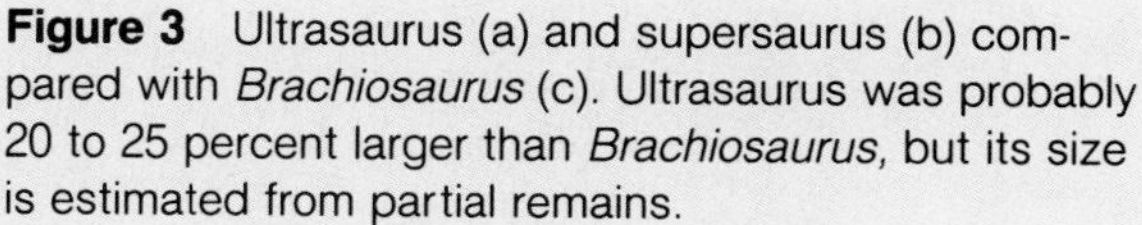

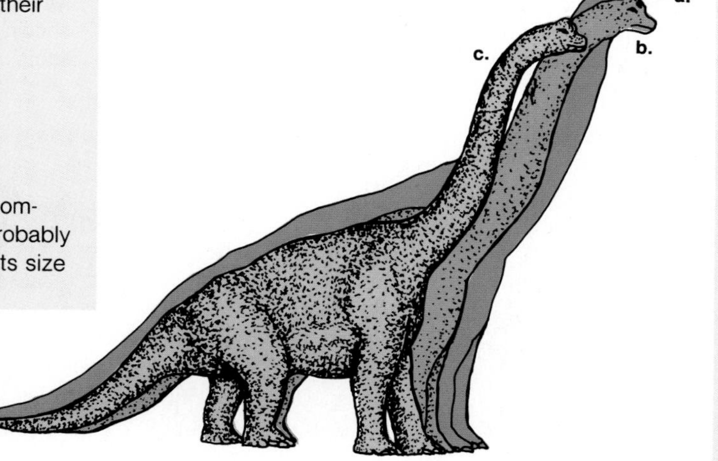


**Figure 3** Ultrasaurus (a) and supersaurus (b) compared with *Brachiosaurus* (c). Ultrasaurus was probably 20 to 25 percent larger than *Brachiosaurus*, but its size is estimated from partial remains.

long-necked (Fig. 14–27). Most plesiosaurs were between 3.6 m to 6 m long, but one species from Antarctica measures 15 m. Long-necked plesiosaurs were heavy-bodied animals with mouthfuls of sharp teeth and limbs specialized into oar-like paddles. They probably rowed themselves through the water and may have used their long necks in snake-like fashion to capture fish. Plesiosaurs probably came ashore to lay their eggs. No doubt they were clumsy on land but no more so than living walruses or seals.

## Crocodiles, Turtles, Lizards, and Snakes

By Jurassic time, crocodiles had replaced phytosaurs (Fig. 14–16) as the dominant freshwater predators. All crocodiles are amphibious, spending much of their time in water, but they are also well-equipped for walking on land. Overall, crocodile evolution has been conservative, involving changes mostly in size from a meter or so in Jurassic forms to 15 m in some Cretaceous species (Fig. 14–28).

**Figure 14–27** Mesozoic marine life. The short-necked and long-necked plesiosaurs *Peloneustes* (1) and *Muraenosaurus* (2) are shown along with two genera of ichthyosaurs, *Mixosaurus* (3) and *Eurhinosaurus* (4). Other marine reptiles shown are *Nothosaurus* (5), the probable ancestor of plesiosaurs, and *Placodus,* a genus of mollusk-eating placodont.

**Figure 14–28** A Late Cretaceous crocodile, 15 m long, *Phobosuchus,* is shown here attacking a small ornithopod.

Turtles, too, have been evolutionarily conservative since their appearance during the Triassic. The most remarkable feature of turtles is their heavy, bony armor; turtles are more thoroughly armored than any other vertebrate animal, living or fossil. Turtle ancestry is uncertain. One Permian animal had eight broadly expanded ribs, which may represent the first stages in the development of turtle armor.

Lizards and snakes are closely related to one another, and, in fact, lizards were ancestral to snakes. The limbless condition in snakes (some lizards are limbless, too) and skull modifications that allow snakes to open their mouths very widely are the main differences between these two groups. Lizards are known from Triassic strata, but they did not become abundant until the Late Cretaceous. Fossils of Late Cretaceous limbless reptiles may represent snakes, but they are perhaps limbless lizards. The first definite fossil snakes are from rocks of Early Cenozoic age.

## BIRDS

In Chapter 5 we discussed the origin of birds to illustrate how fossils provide evidence for evolution. We pointed out that *Archaeopteryx* (from Jurassic strata in Germany) has such avian features as a wishbone and feathers, but it otherwise more closely resembles the small carnivorous dinosaurs (Figure 5–38).

Until recently, *Archaeopteryx* was the only known pre-Cretaceous bird, but the discovery of fossils of two crow-sized individuals called *Protoavis* (Fig. 14–29) has perhaps changed that situation. These fossils are from Triassic rocks, so they predate *Archaeopteryx*, and some investigators think they were more bird-like than *Archaeopteryx*. *Protoavis* had hollow bones and the breastbone structure of birds, but since no impressions of feathers were found on these specimens, some investigators believe they may simply have been small carnivorous dinosaurs. If *Protoavis* proves to be a bird rather than a reptile, it would imply that birds evolved earlier than originally thought.

Other than *Archaeopteryx* and *Protoavis*, few other Mesozoic birds are known, but there are two that come from Late Cretaceous rocks in Kansas. One bird, *Hesperornis*, was a swimming bird with powerful kicking feet and vestigial wings (Fig. 14–30). *Ichthyornis* was an active flyer with well-developed wings, who seems to have lived much like modern terns. Neither of these two genera had the long-tail characteristic of older birds, but both seem to have had teeth. Teeth in birds were completely lost by the end of the Mesozoic, since no known Cenozoic bird had teeth.

**Figure 14–29** Comparison of *Archaeopteryx* (left) and *Protoavis*. Although some paleontologists have questioned the interpretation of *Protoavis* being a bird, others believe it was more bird-like than *Archaeopteryx*. Recovered bones of *Protoavis* are shown shaded.

## Perspective 14-2

# *Sordes Pilosus*—The Hairy Devil

Flying is a very strenuous activity and requires a high metabolic rate. Living vertebrate fliers, birds and bats, are endotherms, have efficient respiratory and circulatory systems to meet the high oxygen requirements of flight, and have an insulating coat to prevent loss of body heat. It seems reasonable to suspect that pterosaurs were similarly equipped.

In 1971 the Russian paleontologist A. G. Sharov discovered a pigeon-sized pterosaur of Jurassic age that showed clear evidence of a body covering of hair or hair-like feathers (Fig. 1). Sharov named this specimen *Sordes pilosus,* meaning "hairy devil," and he concluded that its entire body except for the tail was covered. The fact that *Sordes* had a hair- or feather-covered body and was a flier strongly suggests that this small pterosaur, and perhaps all pterosaurs, were endotherms.

One may wonder about the correct classification of *Sordes,* particularly if its body covering was actually feathers. Recall from Chapter 5 that *Archaeopteryx* (Fig. 5–38) had feathers and is classified as a bird, yet in most respects it more closely resembled small carnivorous dinosaurs. Similar reasoning would seem to dictate that *Sordes* also be called a bird, but it is classified as a reptile.

Choosing a single characteristic to define a class of organisms simply does not work very well. Furthermore, classification is more than naming organisms, it is intended to show evolutionary relationships (Table 5–1, Fig. 5–31). Perhaps *Sordes* was feathered, but it otherwise was quite unlike birds, and there is no evidence that *Sordes* or any other pterosaur was ancestral to birds.

*Archaeopteryx,* on the other hand, had a skeletal structure that indicates (1) a relationship to small carnivorous dinosaurs and (2) that with little modification it could have given rise to a more bird-like skeleton. The designation of *Archaeopteryx* as a bird is based mostly upon the presence of feathers. However, there is no reason why *Archaeopteryx* could not have been included among the reptiles.

It seems that *Sordes* and *Archaeopteryx* independently solved the same adaptive problem in very similar ways. The evolutionary trend in both was apparently toward endothermy, and both developed an insulating coat to preserve body heat.

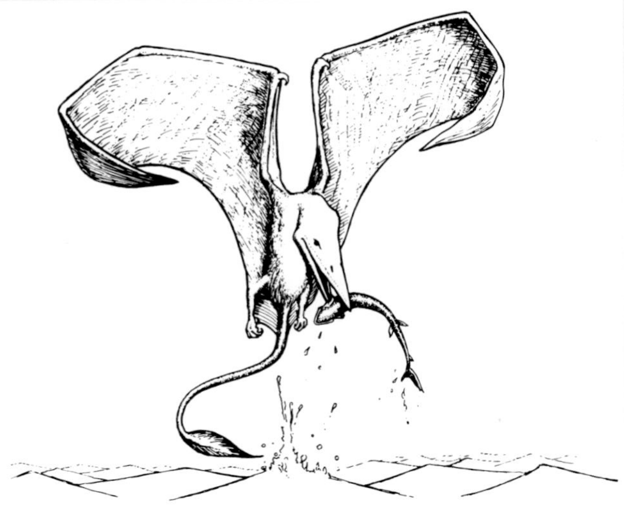

**Figure 1** *Sordes pilosus,* whose name means "hairy devil", was a small Jurassic pterosaur that had an insulating coat of hair or hair-like feathers.

## FROM REPTILE TO MAMMAL

**Therapsids,** or the advanced mammal-like reptiles, were briefly described in Chapter 12. These reptiles diversified into numerous species of herbivores, and carnivores and during the Permian they were the dominant terrestrial vertebrates. One particular group of carnivorous therapsids called **cynodonts** was the most mammal-like of all and by the Late Triassic gave rise to the class Mammalia (Fig. 14–14, Table 14–2).

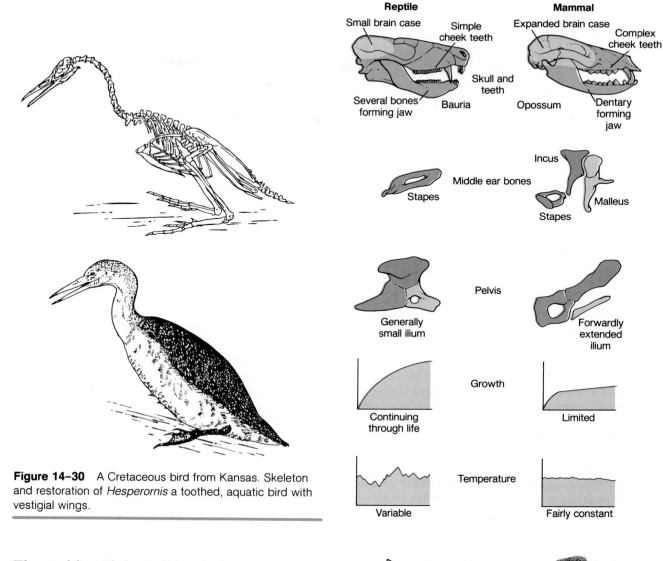

**Figure 14–30** A Cretaceous bird from Kansas. Skeleton and restoration of *Hesperornis* a toothed, aquatic bird with vestigial wings.

## Therapsids and the Origin of Mammals

The transition from cynodonts to mammals is well-documented by fossils and is so gradational that classification of some fossils as either reptile or mammal is difficult. We can easily recognize living mammals as those warm-blooded animals with hair or fur that have mammary glands and, except for the platypus and spiny anteater, give birth to live young (Fig. 14–31).

Obviously these criteria for recognizing living mammals are inadequate for classifying fossils. For them, we must use skeletal structure only. Several skeletal modifications characterize the transition from mammal-like reptiles to mammals, but distinctions between the two groups are based largely on details of the middle ear, the lower jaw, and the teeth (Table 14–4).

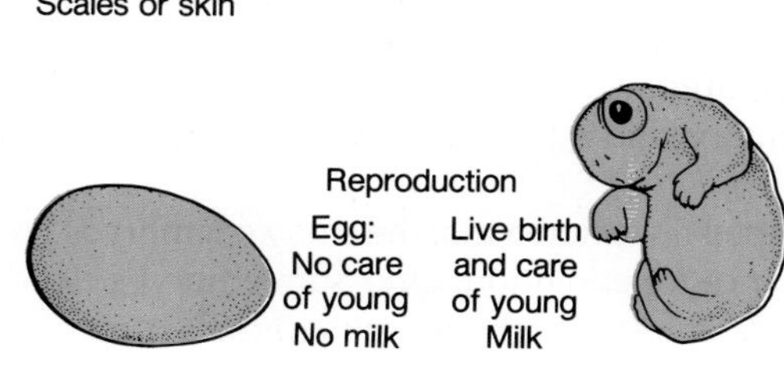


**Figure 14–31** Some contrasting characteristics of reptiles and mammals. Most living mammals can be easily identified by these characteristics. The platypus and spiny anteater, however, both lay eggs, but in most other respects are mammals.

**Table 14–4** Summary chart showing some of the characteristics and how they changed during the transition from reptiles to mammals.

| | Typical Reptile | Cynodont | Mammal |
|---|---|---|---|
| **Lower jaw** | Dentary and several other bones. | Dentary enlarged, other bones reduced. | Dentary bone only, except in earliest mammals. |
| **Jaw-skull joint** | Articular-quadrate | Articular-quadrate; some advanced cynodonts had both the reptile jaw-skull joint and the mammalian jaw-skull joint. | Dentary-squamosal |
| **Middle-ear bones** | Stapes | Stapes | Stapes, incus, malleus |
| **Secondary palate** | Absent | Partially developed | Well-developed |
| **Teeth** | No differentiation | Some differentiation | Fully differentiated |
| **Cold-vs. warm-blooded** | Cold-blooded | Probably warm-blooded | Warm-blooded |

Reptiles have only one small bone in the middle ear—the stapes—while mammals have three—the incus, the malleus, and the stapes. Also, the lower jaw of a mammal is composed of a single bone called the *dentary*, but a reptile's jaw is composed of several bones (Fig. 14–32). In addition, a reptile's jaw is hinged to the skull at a contact between the articular and quadrate bones, while in mammals the dentary contacts the squamosal bone of the skull (Fig. 14–32).

During the transition from cynodonts to mammals, the quadrate and articular bones that had formed the joint between the jaw and skull in reptiles were modified into the incus and malleus of the mammalian middle ear (Fig. 14–32). Fossils clearly document the progressive enlargement of the dentary until it became the only element in the mammalian jaw (Fig. 14–33). Likewise, a progressive change from the reptile to mammal jaw-joint is documented by fossil evidence. In fact, some of the most advanced cynodonts were truly transitional because they had a compound jaw joint consisting of (1) the articular and quadrate bones typical of reptiles, and (2) the dentary and squamosal bones as in mammals (Table 14–4).

In Chapter 5 we noted that the study of embryos provides evidence for evolution. Opossum embryos clearly show that the middle-ear bones of mammals were originally part of the jaw. In fact, even when opossums are born, the middle-ear elements are still attached to the dentary (Fig. 14-32), but as they develop further, these elements migrate to the middle ear, and a typical mammal jaw joint develops.

Several other aspects of cynodonts also indicate they were ancestral to mammals. Their teeth were somewhat differentiated into distinct types in order to perform specific functions. In mammals the teeth are fully differentiated into incisors, canines, and chewing teeth, but typical reptiles do not have differentiated teeth. Another mammalian feature, the secondary palate, was partially developed in advanced cynodonts (Fig. 14–34). This secondary palate is a bony shelf above the mouth that separates the nasal passages from the mouth cavity. It is an adaptation for eating and breathing at the same time, a necessary requirement for endotherms with their high demands for oxygen.

## MESOZOIC MAMMALS

Even though mammals appeared during the Late Triassic, their diversity remained low during the rest of the Mesozoic (Fig. 14–35, Table 14–5). A few Mesozoic mammals may have been as large as raccoons, but most were quite small—about the size of shrews or rats. Their fossil record, which consists of isolated teeth, jaw fragments, and fragmentary bones, is not particularly good, although a few skulls and even some fairly complete skeletons are known. Nevertheless, these fossils are important because two-thirds of mammalian history is recorded by Mesozoic fossils.

The first mammals retained several reptilian characteristics, but had mammalian features as well. For example, the Triassic triconodonts (Fig. 14–36) had the fully differentiated teeth typical of mammals, but they had both the reptile and mammal types of jaw joints. The earliest symmetrodonts retained several reptilian bones

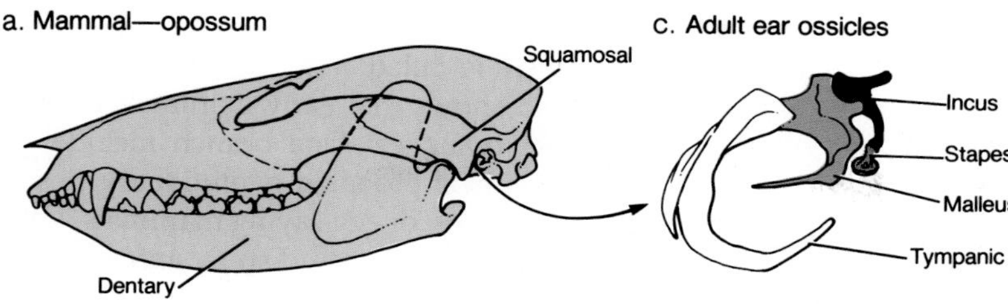


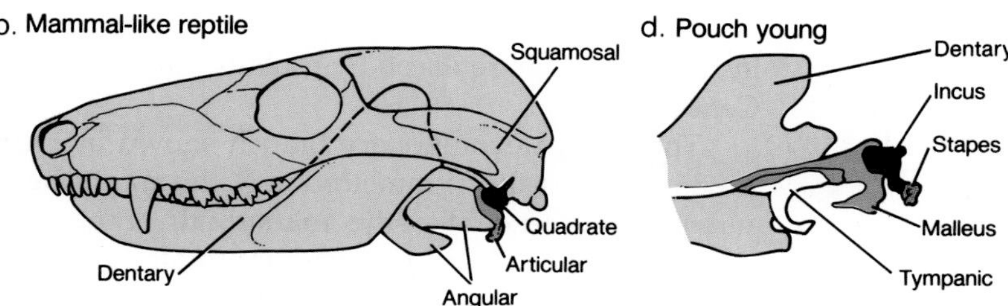


**Figure 14–32** a. The skull of an opossum showing the typical mammalian dentary-squamosal jaw joint. b. The skull of a cynodont shows the articular-quadrate jaw joint of reptiles. c. Enlarged view of an adult opossum's middle-ear bones. d. View of the inside of a young opossum's jaw showing that the elements of the middle ear are attached to the dentary during early development.

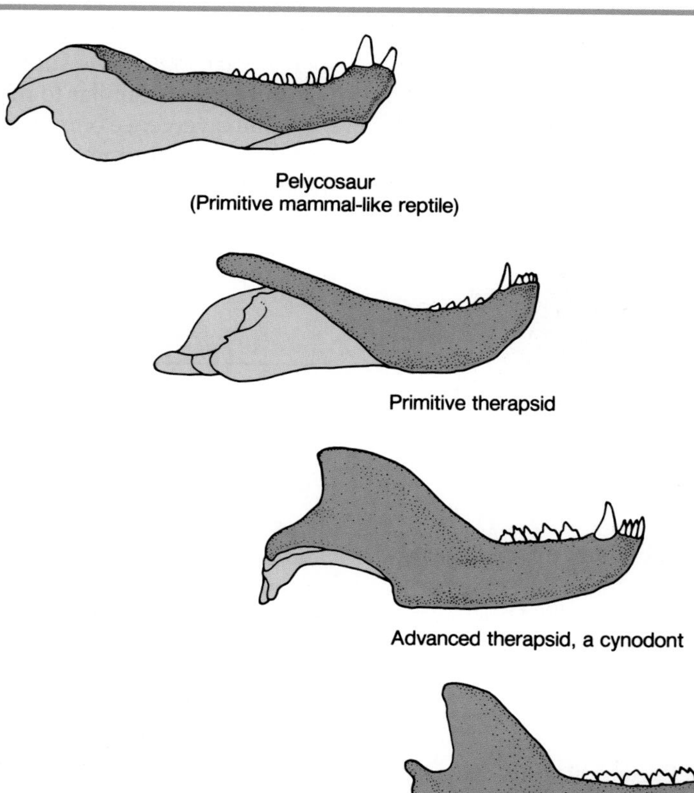


**Figure 14–33** During the transition from reptiles to mammals, the dentary bone (stippled) became progressively larger until it became the only bone in the lower jaw.

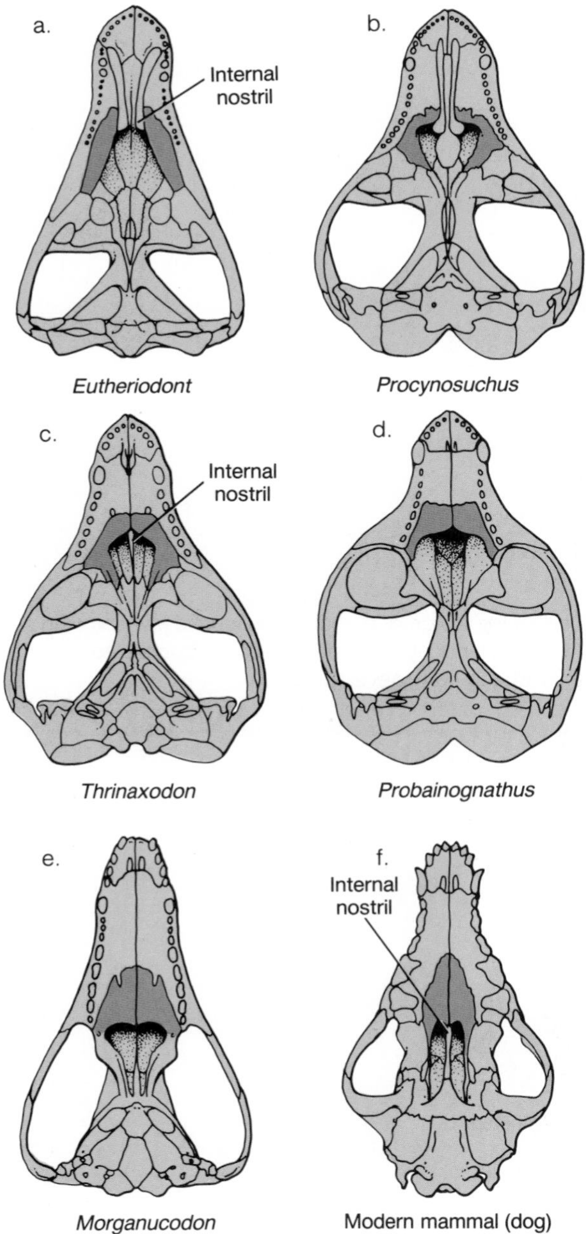


**Figure 14–34** Views of the bottoms of skulls of an early therapsid (a), three cynodonts (b through d), an early mammal (e), and a living mammal (f), showing the progressive development of the bony secondary palate (shaded).

in the lower jaw, but by the Cretaceous only the dentary was present (Fig. 14–37). In short, some mammalian features evolved more rapidly than others, thereby accounting for animals that possessed characteristics of both reptiles and mammals. Recall from Chapter 5 that such varying rates of evolution in different features of the same organism is called *mosaic evolution*.

Figure 14–35 shows that early mammals diverged into two distinct branches. One branch includes the triconodonts and their probable evolutionary descendants, the **monotremes,** or egg-laying mammals. Living monotremes are the platypus and spiny anteater of the Australian region. Also included in this branch are the *multituberculates,* the first mammalian herbivores (Figs. 14–19, 14–38); all other Mesozoic mammals probably preyed on insects, worms, and grubs. Multituberculates seem to have lived much like modern rodents, and, in fact, they were replaced by rodents during the Early Cenozoic.

The second evolutionary branch shown in Figure 14–35 includes the symmetrodonts, the **pantotheres,** and their descendants—the **marsupial,** or pouched, mammals, and the **placental mammals.** All living mammals except monotremes have ancestries that can be traced back through pantotheres and possibly to symmetrodonts.

Pantotheres were shrew-sized animals with poor fossil records, but details of their teeth indicate they were ancestral to both marsupial and placental mammals. Furthermore, pantotheres had bones projecting from the pelvis, that could have supported a pouch, so in this respect they were similar to modern marsupials (Fig. 14–39). The divergence of marsupials and placentals probably occurred during the Early Cretaceous, but undoubtedly both were present by the Late Cretaceous. The earliest placental mammals were members of the order *Insectivora* (Fig. 14–40), an order represented today by shrews, moles and hedgehogs.

## MESOZOIC CLIMATES AND PALEOGEOGRAPHY

The present continental positions and climatic patterns largely restrict the distribution of organisms. Land plants and animals have little opportunity to colonize distant areas because of physical barriers (especially the ocean basins) and climatic barriers. Mesozoic barriers to migration were apparently not as effective as they are now, because some Mesozoic organisms are known from areas that are now widely separated.

Fragmentation of the supercontinent Pangaea began by the Late Triassic and continues to the present, but during much of the Mesozoic, close connections existed between the various landmasses. The proximity of these landmasses, however, is not sufficient to explain Mesozoic biogeographic distributions, because climates are

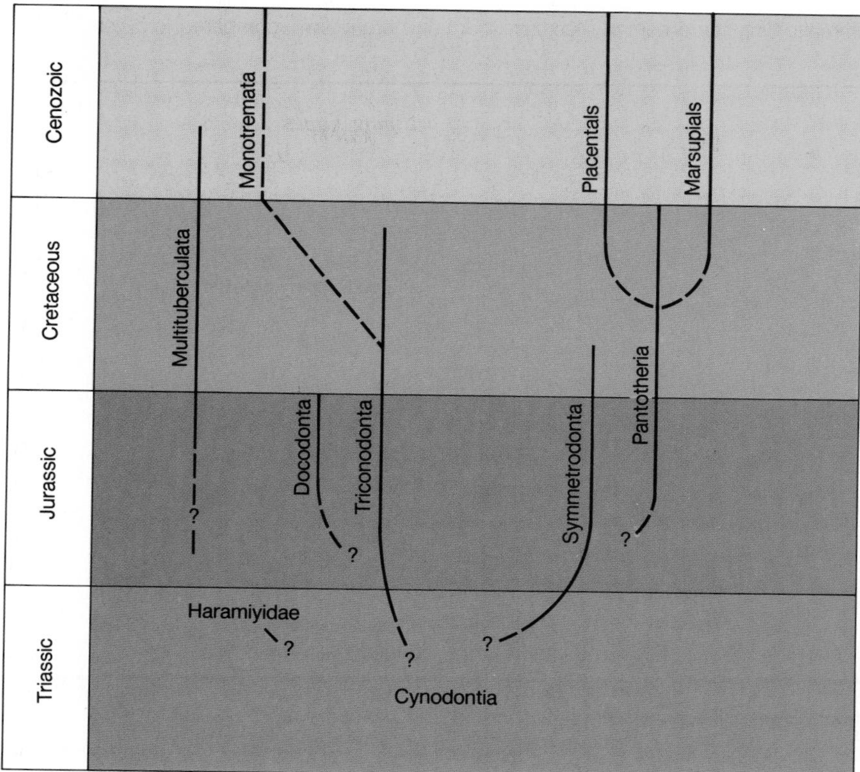


**Figure 14–35** Relationships among the early mammals. A number of relationships are uncertain, as shown by the question marks, but it seems that mammalian evolution proceeded along two branches. One branch led to the egg-laying mammals, the monotremes, which are represented by the living platypus and spiny anteater. The other branch led to marsupial and placental mammals.

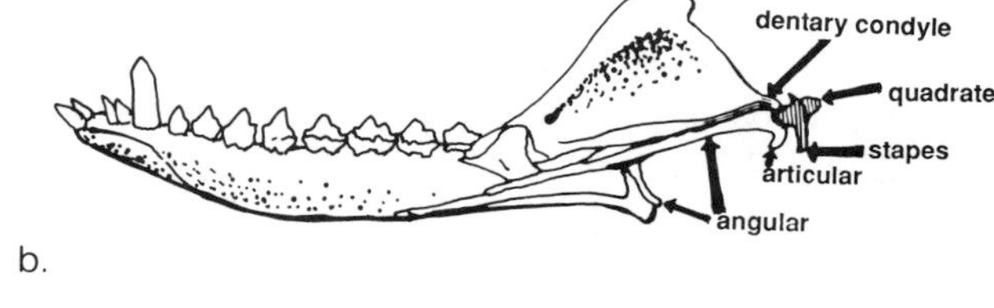


**Figure 14–36** a. One of the earliest mammals, the triconodont *Megazostrodon,* from the Early Jurassic of Africa, about 10 cm in length. b. Inside view of the jaw of the early mammal *Morganucodon.* The reptile jaw bones were still part of the lower jaw, but they were reduced in size.

**Table 14–5** Classification of Mesozoic mammals. Only ten orders of Mesozoic mammals are known, compared to 32 orders that existed during the Cenozoic.

| **Class Mammalia** | ***Living Types*** |
|---|---|
| Subclass Prototheria—Egg-laying mammals | |
| Order Docodonta | |
| Order Triconodonta | |
| Order Monotremata[a] | Platypus, spiny anteater |
| Subclass Allotheria | |
| Order Multituberculata | |
| Subclass Theria—Mammals that give birth to live young | |
| Order Symmetrodonta | |
| Order Pantotheria | |
| Order Marsupialia | Opossum, kangaroo, wombat |
| Order Creodonta | |
| Order Condylartha | |
| Order Insectivora | Shrew, mole, hedgehog |

[a] Until 1985 fossil monotremes were known only from Pleistocene deposits.

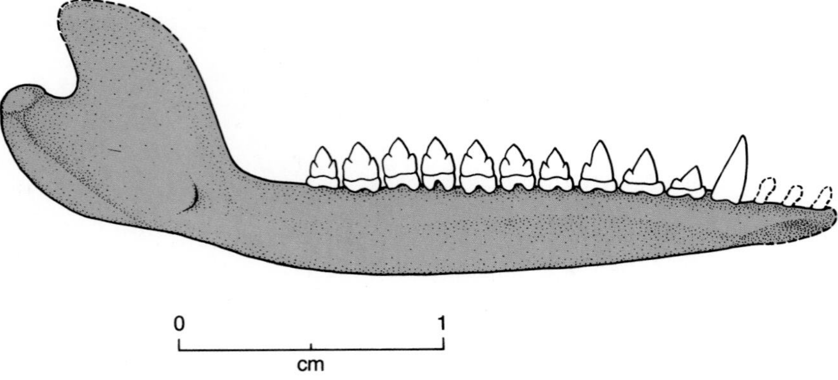


**Figure 14–37** Jaw of a symmetrodont mammal, showing the dentary as the only jaw bone. Note, too, that the teeth are differentiated into specific types.

also effective barriers to wide dispersal. During much of the Mesozoic, though, climates were more equable and lacked the strong north and south zonation characteristic of the modern world (Table 14–2). In short, Mesozoic plants and animals had greater opportunities to occupy much more extensive geographic ranges.

Pangaea persisted as a single unit through most of the Triassic. The Triassic climate was warm-temperate to tropical, although some areas, such as the present south-western United States, were arid. Mild temperatures extended 50 degrees north and south of the equator, and even the polar regions may have been temperate. The

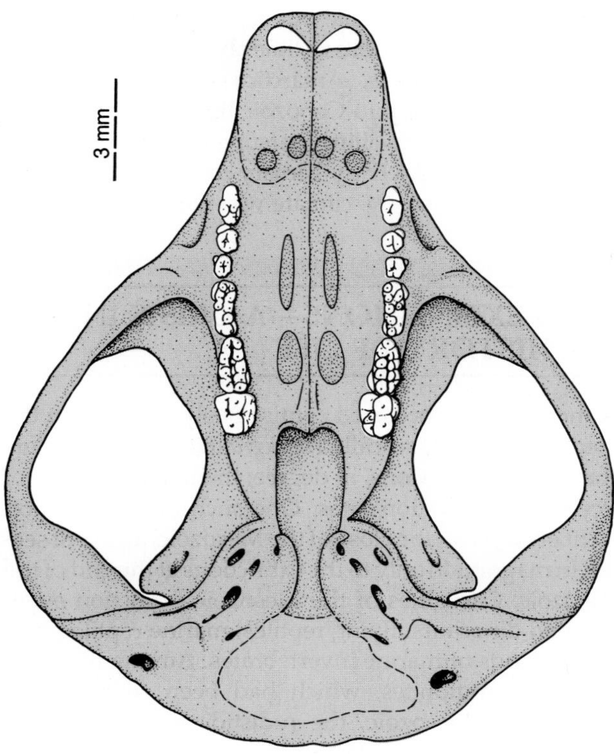


**Figure 14–38** Bottom view of the skull of a Cretaceous multituberculate from Mongolia. Multituberculates are so called because of the numerous rounded projections or tubercules on their molars.

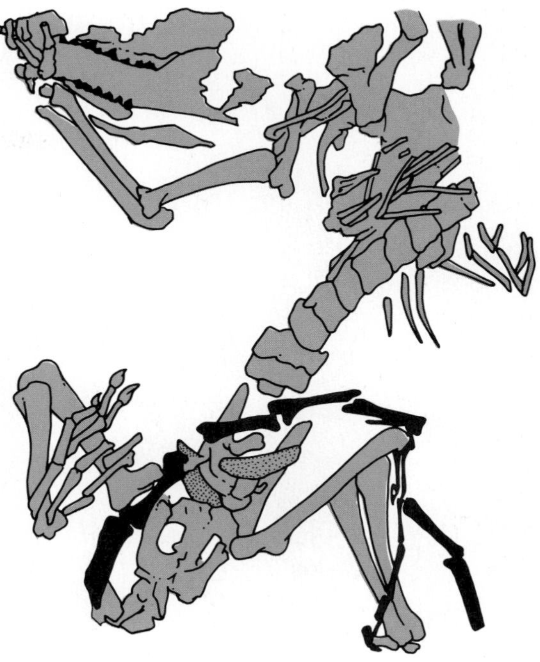

**Figure 14–39** Skeleton of a Jurassic pantothere from Portugal. The long tail (black) may have been used to grasp branches. The presence of bones (stippled) in the pelvic region may have supported a pouch as in living marsupials.

fauna was truly worldwide in its distribution. Phytosaurs (Fig. 14–16) were present in North America, Europe, and Madagascar. Some dinosaurs had continuous ranges across Laurasia and Gondwana. The peculiar gliding lizards (Fig. 14–41) occur in New Jersey and England.

By the Late Jurassic, Laurasia had become partly fragmented by the opening North Atlantic, but a connection still existed. The South Atlantic had begun to open so that a long, narrow sea separated the southern parts of Africa and South America. Otherwise the southern continents were still close together.

The mild Triassic climate persisted into the Jurassic. Ferns, whose living relatives are now restricted to the tropics of southeast Asia, are known from areas as far as 63 degrees south latitude and 75 degrees north latitude. Dinosaurs roamed widely across Laurasia and Gondwana. Specimens from the Morrison Formation in western North America and those in the Tendagura beds of eastern Africa are quite similar. For example, the giant dinosaur *Brachiosaurus* (Fig. 14–21) is known from both areas. Stegosaurs (Fig. 14–23) and some families of carnivorous dinosaurs lived throughout Laurasia and in Africa.

By the Late Cretaceous the North Atlantic had opened further, and Africa and South America were completely separated. South America remained an island continent until late in the Cenozoic. Its fauna, evolving in isolation, became increasingly different from faunas of the other continents. Marsupial mammals reached Australia from South America via Antarctica, but the South American connection was eventually severed. Placentals, other than bats and a few rodents, never reached Australia. This explains why the marsupials continue to dominate that continent's fauna even today.

Cretaceous climates were more strongly zoned by latitude, but they remained warm and equable until the close of that period. Climates then became more seasonal and cooler, a trend that persisted into the Cenozoic. Dinosaur and mammal fossils demonstrate that interchange was still possible, especially between the various components of Laurasia.

As one might expect, the paleontology of Antarctica is poorly known. Marine reptiles, including plesiosaurs 15 m long have been found on Seymour Island near the tip

**Figure 14–40** The oldest known placental mammals were members of the order Insectivora such as those in this scene from the Late Cretaceous. *Deltatheridium* (upper right) and *Zalambdalestes* probably fed on insects, worms and grubs.

of the Antarctic Peninsula, but since they were marine reptiles, their presence tells us little about continental connections. Nevertheless, Antarctica remained close to South America well into Cenozoic time, as indicated by the fact that a land-dwelling marsupial of Eocene age has been found there. An Early Tertiary coal bed on Seymour Island indicates that the climate was temperate.

## MASS EXTINCTIONS—A CRISIS IN THE HISTORY OF LIFE

Extinctions have occurred continuously throughout the history of life. This so-called *background extinction* differs from **mass extinctions,** in degree; the latter are times of accelerated extinction rates. One such mass extinction event occurred at the close of the Mesozoic, an event second in magnitude only to the extinctions at the end of the Paleozoic. Casualties of the Mesozoic extinction event included dinosaurs, flying reptiles, marine reptiles, and several kinds of marine invertebrates. Among the latter were the ammonites, which had been so abundant through the Mesozoic, the rudistid clams and some planktonic organisms (Fig. 5–28).

Numerous ideas have been proposed to explain Mesozoic extinctions, but most have been dismissed as improbable or untestable. A new proposal was made re-

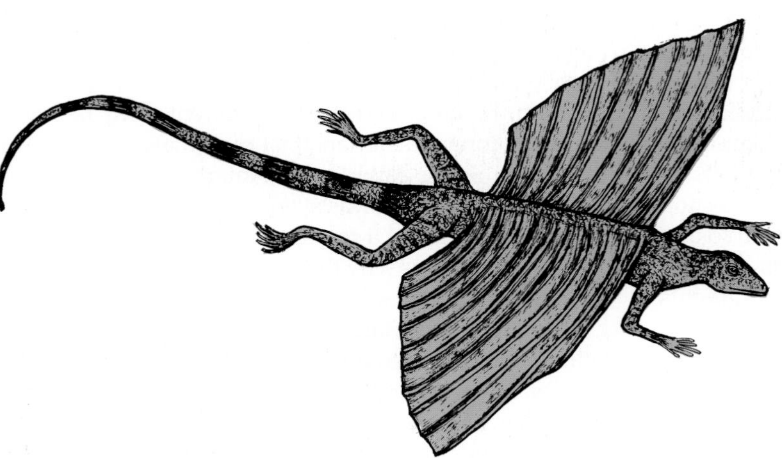

**Figure 14–41** The Late Triassic gliding lizard, *Kuchneosaurus*, is known from New Jersey and England. Its presence in these widely separated areas can be accounted for by the existence of Pangaea during the Triassic.

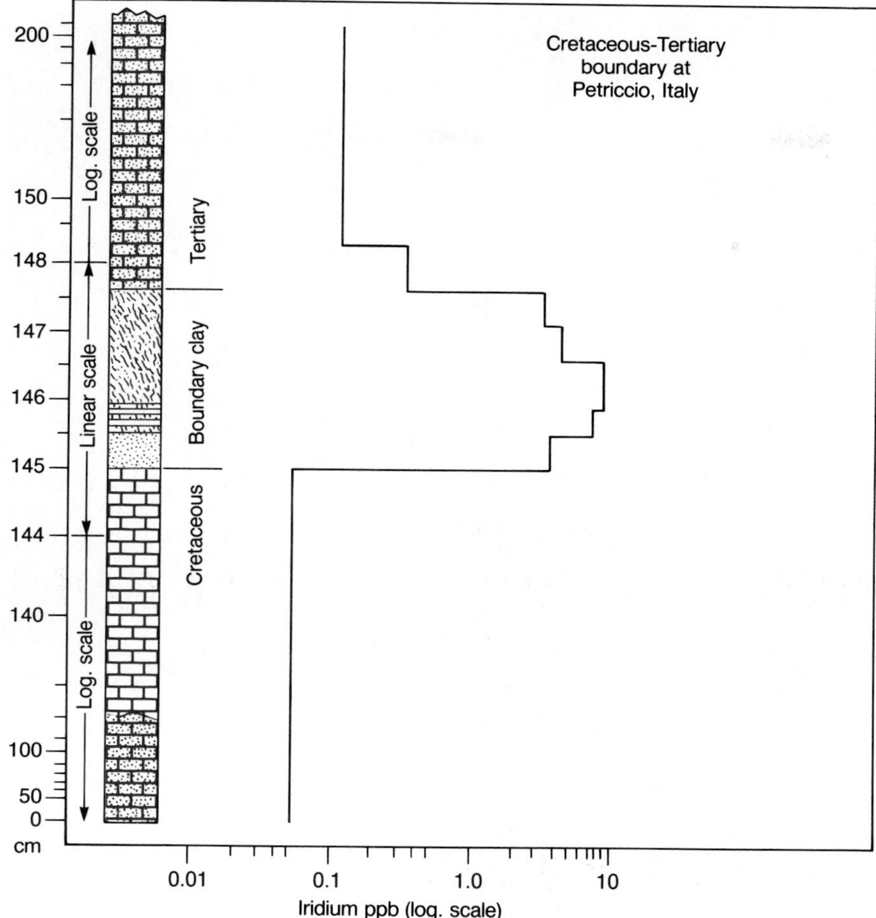


**Figure 14–42** Stratigraphy of the Cretaceous-Tertiary section in Italy showing the 2.5 cm thick boundary clay and the abundance of iridium. Even though iridium is present in crustal rocks in quantities measured in parts per billion (ppb), it shows a marked increase in abundance at the Cretaceous-Tertiary boundary.

cently based on a discovery at the Cretaceous-Tertiary boundary in Italy—a clay layer 2.5 cm thick, with an abnormally high concentration of the platinum group element iridium (Fig. 14–42). Since this discovery, high iridium concentrations have been identified at many other Cretaceous-Tertiary boundary sites. The significance of this discovery lies in the fact that iridium is rare in crustal rocks but occurs in much higher concentrations in some meteorites. Several investigators proposed a meteorite impact to explain this iridium anomaly, and they further postulated that the impact of a large meteorite, perhaps 10 km in diameter, set in motion a chain of events that led to extinctions (Fig. 14–43). Some Cretaceous-Tertiary boundary sites also contain soot and shock-metamorphosed quartz grains, both of which are cited as further evidence of an impact event. The meteorite-impact scenario goes something like this. Upon impact, about 60 times the mass of the meteorite was blasted from the Earth's crust high into the atmosphere, and the heat generated at impact started raging fires that added more particulate matter to the atmosphere. Sunlight was blocked for several months, causing a temporary cessation of photosynthesis; food chains collapsed, and extinctions followed. In addition, with sunlight greatly diminished, the Earth's surface temperatures were drastically reduced and could have added to the biologic stress.

The iridium anomaly is real, but its origin and significance are debatable. We know very little about the distribution of iridium in crustal rocks or how it may be

**Figure 14–43** Proposed meteorite impact at end of Cretaceous.

distributed and concentrated. Some geologists suggest that the iridium was derived from within the Earth by volcanism, but this is not conclusively supported by evidence.

Even if a meteorite did hit the Earth, did it lead to these extinctions? If so, both terrestrial and marine extinctions must have occurred at the same time. To date, strict time equivalence between terrestrial and marine extinctions has not been demonstrated. The selective nature of the extinctions is also a problem. In the terrestrial realm large animals were the most drastically affected, but not all dinosaurs were large, and crocodiles, close relatives of dinosaurs, were unaffected. Likewise, tropical plants seem to have suffered no ill effects from the supposed lower temperatures. This is puzzling because they have little resistance to cold.

Some paleontologists think that dinosaurs, some marine invertebrates, and many plants were already on the decline and headed for extinction before the end of the Cretaceous. A meteorite impact, if one actually occurred, may have simply hastened the process. There is even some evidence, although its significance is debatable, that indicates that dinosaurs survived into the Early Cenozoic, several tens of thousands of years after the proposed impact.

Investigators at the University of Chicago have proposed that the Mesozoic extinction event was only one of several to occur at 26-million-year intervals during the last 250 million years (Fig. 14–44). One possible cause of these cyclic extinctions is periodic meteorite showers. It has been suggested by some investigators that a companion star to the sun with a highly eccentric orbit could provide a mechanism for periodic meteor showers. In this scenario, when this star is close to the sun, it perturbs cometary orbits, thereby causing terrestrial impacts and mass extinctions.

If one mass extinction attributed to a meteorite impact has generated controversy, one can imagine just how controversial this new proposal is. The evidence for cyclic mass extinctions shown in Figure 14–44 seems compelling, but not all paleontologists are convinced. The critics of cyclic mass extinctions think that some minor extinction events have beeen overemphasized. Furthermore, they point out that if mass extinctions are caused by periodic meteorite impacts that iridium anomalies corresponding with extinction events should be present, but few such anomalies have been identified.

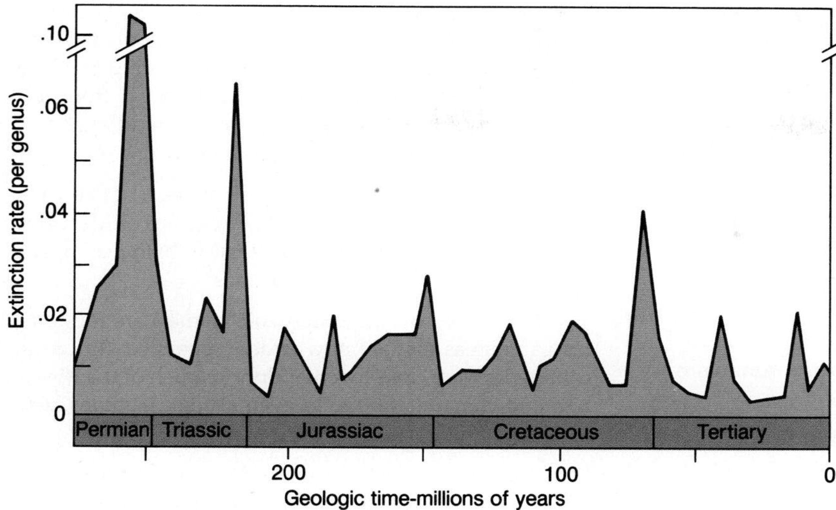


**Figure 14–44** Record of extinctions for 9773 genera of marine fossil animals determined by investigators at the University of Chicago. According to their calculations, mass extinctions have occurred at 26-million-year intervals shown by the peaks on the graph.

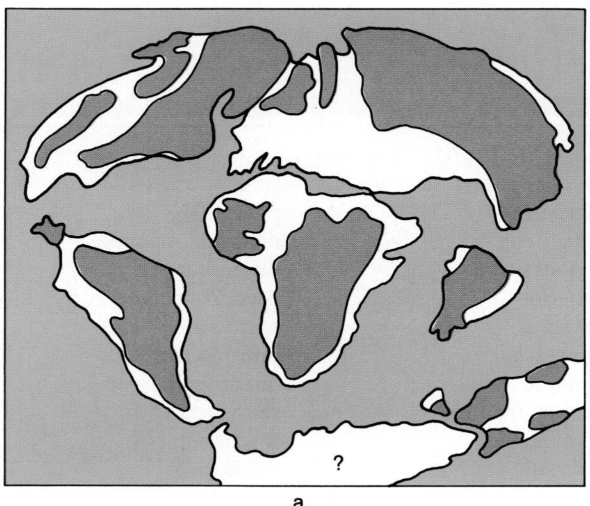


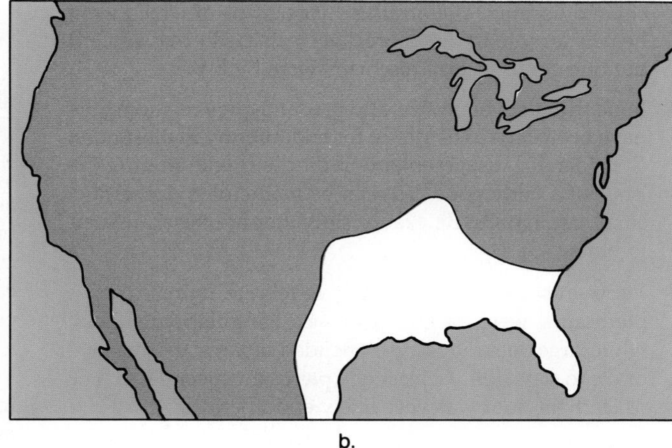


**Figure 14–45** Many paleontologists think that climatic changes and withdrawals of the epeiric seas at the end of the Cretaceous are sufficient to account for mass extinctions at that time. a. The areas of the continents occupied by epeiric seas during the Cretaceous. b. At the end of the Cretaceous the sea covered only a small part of North America. The epeiric seas had largely withdrawn from the other continents by this time as well.

In the final analysis we have no widely accepted explanation for Mesozoic extinctions. We do know that vast shallow seas occupied large parts of the continents during the Cretaceous, and that by the latest Cretaceous they had largely withdrawn (Fig. 14–45). We also know that the mild, equable climates of the Mesozoic became harsher and more seasonal by the end of the Mesozoic. Changes such as these seem adequate to many paleontologists to explain Mesozoic mass extinctions. But the fact remains that this extinction event was very selective, and no explanation accounts for all aspects of this crisis in the history of life.

## CHAPTER SUMMARY

1. Among the marine invertebrates, survivors of the Permian extinction event diversified and gave rise to increasingly complex Mesozoic marine invertebrate communities.
2. Some of the most abundant and diverse invertebrates were ammonites and foraminifera.
3. Triassic and Jurassic land-plant communities were composed of seedless plants and gymnosperms. Angiosperms, or flowering plants, appeared during the Early Cretaceous, diversified rapidly, and soon became the dominant land plants.
4. The Triassic thecodont reptiles included crocodile-like predators, armored herbivores, and small bipedal carnivores that were ancestral to dinosaurs.
5. Dinosaurs appeared during the Late Triassic, but were most abundant and diverse during the Jurassic and Cretaceous. Based on pelvic structure, two distinct orders of dinosaurs are recognized—Saurischia (lizard-hipped) and Ornithischia (bird-hipped).
6. Most carnivores and the giant quadrupeds were saurischian dinosaurs. The ornithischians, some of which were heavily armored, were more diverse than saurischians. All but one genus of ornithischians were herbivores.
7. Bone structure, brain size, and predator-prey relationships have been cited as evidence for endothermy in dinosaurs. Considerable disagreement on dinosaur endothermy exists, but a fairly good case can be made for endothermic, small, carnivorous dinosaurs, since they are closely related to birds.
8. Pterosaurs were the first flying vertebrate animals. Small pterosaurs were probably active, wing-flapping fliers, while large ones may have depended more on thermal updrafts to stay aloft. At least one pterosaur species had hair or feathers, so it was very likely endothermic.
9. The fish-eating, porpoise-like ichthyosaurs were thoroughly adapted to an aquatic life. Female ichthyosaurs probably retained eggs within their bodies and gave birth to live young. Plesiosaurs were heavy-bodied marine reptiles that probably came ashore to lay eggs.
10. During the Jurassic, crocodiles replaced phytosaurs as the dominant freshwater predators. Turtles and lizards were present through most of the Mesozoic. Snakes may have appeared during the Late Mesozoic, but this is not certain.
11. Birds almost certainly evolved from small carnivorous dinosaurs. The oldest known bird, *Archaeopteryx*, appeared during the Jurassic, but few other Mesozoic birds are known. Recent finds of *Protoavis* in Triassic rocks may represent a bird older than *Archaeopteryx*.
12. Among the mammal-like reptiles, the cynodonts show a transformation from the reptilian to the mammalian condition. The earliest mammals appeared during the Late Triassic, but they are difficult to distinguish from advanced cynodonts. Details of the teeth, the middle ear, and lower jaw are used to distinguish the two.
13. Several types of Mesozoic mammals existed, but all were small, and their diversity was low. A group of Mesozoic mammals called pantotheres gave rise to both marsupials and placentals during the Cretaceous.
14. The same types of some Mesozoic animals are found as fossils in areas that are now widely separated. Since the continents were close together during much of the Mesozoic and climates were mild even at high latitudes, animals and plants dispersed very widely.
15. Mesozoic mass extinctions account for the disappearance of dinosaurs, several other groups of reptiles, and a number of marine invertebrates. Paleontologists do not agree on the cause or causes for Mesozoic extinctions. One hypothesis holds that extinctions were caused by the impact of a large meteorite with the Earth. Many paleontologists reject the meteorite proposal and claim that withdrawal of epeiric seas and climatic changes can account for this extinction event.

## IMPORTANT TERMS

angiosperm
bipedal
cynodont
ectotherm
endotherm
ichthyosaur
insectivore
marsupial mammal
mass extinction
monotreme
pantothere
placental mammal
plesiosaur
pseudosuchian
pterosaur
ornithischian
quadrupedal
saurishcian
thecodont
therapsid

## REVIEW QUESTIONS

1. Briefly outline the major changes in the composition of Mesozoic marine invertebrate communities.
2. Discuss the changing aspects of land-plant communities during the Mesozoic.
3. What are the names of the two dinosaur orders, and how are they differentiated from one another?

4. Discuss several ways in which herbivorous dinosaurs protected themselves from predators.
5. What evidence indicates that some dinosaurs nested in colonies and used the same nesting sites repeatedly?
6. Briefly discuss the adaptations for flight seen in pterosaurs. Is there any evidence for endothermy in pterosaurs? If so, what?
7. In what ways was *Archaeopteryx* similar to and different from small carnivorous dinosaurs?
8. What skeletal features of cynodonts indicate that they were endotherms?
9. Explain what is meant when we say the earliest mammals show a mosaic evolution.
10. How does the breakup of Pangaea help us better understand the distribution of certain fossil plants and animals?
11. Briefly summarize the evidence for and against the proposal that a meteorite impact caused Mesozoic mass extinctions.

## ADDITIONAL READING

Bakker, R. T. 1986. *The Dinosaur Heresies.* New York: William Morrow and Company, Inc.

Colbert, E. H. 1983. *Dinosaurs: An Illustrated History.* Maplewood, New Jersey: Hammond Inc.

Horner, J. R. 1984. The Nesting Behavior of Dinosaurs. *Scientific American,* 250, no. 4: 130–137.

Hsü, H. J. 1986. *The Great Dying.* New York: Harcourt Brace Jovanovich, Publishers.

Lambert, D. 1983. *A Field Guide to Dinosaurs.* New York: Avon Books.

Langston, W. 1981. Pterosaurs. *Scientific American,* 244, no. 1: 122–136.

Lillegraven, J. A., Z. Kielan-Jaworowska, and W. A. Clemens, (eds.) 1979. *Mesozoic Mammals: The First Two-Thirds of Mammalian History.* Berkley, Calif.: University of California Press.

Savage, R. J. G. 1986. *Mammalian Evolution: An Illustrated Guide.* New York: Facts on File Publications.

Stewart, W. N. 1983. *Paleobotany and the Evolution of Plants.* New York: Cambridge University Press.

Wilford, J. N. 1985. *The Riddle of the Dinosaurs.*New York: Alfred A. Knopf, Inc.

# 15

## CHAPTER OUTLINE

PROLOGUE
INTRODUCTION
CENOZOIC PLATE TECTONICS—AN OVERVIEW
CENOZOIC OROGENIC BELTS
The Alpine-Himalayan Orogenic Belt
*The Alps*
*Roof of the World—The Himalayas*
The Circum-Pacific Orogenic Belt

PERSPECTIVE 15-1: THE PERILS OF LIVING NEAR A CONVERGENT PLATE MARGIN

THE NORTH AMERICAN CORDILLERA
Laramide Orogeny

PERSPECTIVE 15-2: INTERMONTANE BASIN DEPOSITION

Cordilleran Volcanism
Basin and Range Province
Colorado Plateau
The Pacific Coast
THE INTERIOR LOWLANDS
GULF COASTAL PLAIN
EASTERN NORTH AMERICA
Cenozoic Evolution of the Appalachians
The Atlantic Coastal Plain
CHAPTER SUMMARY

## PROLOGUE

*During the Eocene Epoch, several small plutons were emplaced in northeastern Wyoming. The best known of these, Devil's Tower, was established as our first National Monument by President Theodore Roosevelt in 1906. The tower rises nearly 260 m above its base and stands over 390 m above the floodplain of the nearby Belle Fourche River (Fig. 15–1). Visible from 48 km away, it served as a landmark for early travelers in the area.*

*The Cheyenne and Lakota Sioux Indians called the tower* Mateo Tepee, *which means "Grizzly Bear Lodge". It was also called "The Bad God's Tower," and, reportedly, "Devil's Tower"*

b.

**Figure 15–1** Devil's Tower, Wyoming. a. View to the North from several miles away. b. Close-up view. The rubble at the base of the tower is an accumulation of collapsed columns.

# Cenozoic History—Tertiary Period

*was a translation of this phrase. According to one Indian legend, the tower formed when the Great Spirit made it rise up from the ground. As the tower rose it carried several Indian children upward who were trying to escape from a gigantic grizzly bear. The bear, in its attempts to reach the children, left deep scratches in the tower's rocks (Fig. 15–2).*

*Geologists have a less picturesque explanation for the tower's origin. The near vertical striations (the bear's scratch marks) are simply the lines formed by the intersection of columnar joints. Columnar joints form in magma or lava as it cools and contracts, thus forming columns. Many of the columns are six-sided, but four-, five-, and seven-sided columns occur as well. The larger columns measure about 2.5 m across, and the pile of rubble at the tower's base is an accumulation of collapsed columns (Fig. 15–1).*

*Geologists agree that Devil's Tower was emplaced as a small pluton, and subsequent erosion exposed it in its present form. The type of pluton and the extent of its modification by erosion are debatable, however. Some geologists think Devil's Tower is the eroded remnant of a more extensive sill, or laccolith, while others think it is simply the magma that solidified in the neck of a volcano. Many similar, though less spectacular, small intrusions are found throughout the northern Black Hills of South Dakota. Bear Butte is of particular religious and historical significance to the Cheyenne and Lakota Sioux to this day.*

**Figure 15–2** Cheyenne legend account of the origin of Devil's Tower. (Photo of painting by Herbert Collins courtesy of Devil's Tower National Monument)

## INTRODUCTION

Traditionally, the Cenozoic has been divided into two periods, the **Tertiary** (66 to 2 million years ago) and the **Quaternary** (2 million years ago to the present). Both periods are futher divided into epochs (Fig. 15–3). The terms Tertiary and Quaternary are widely used among geologists, but a different scheme for designations of Cenozoic time is becoming increasingly popular. In this scheme, two periods are also recognized, the *Paleogene* and *Neogene*, but they do not correspond directly to the Tertiary and Quaternary periods (Fig. 15–3). In this book we will follow the traditional usage.

In Chapter 8 we noted that Precambrian time accounts for more than 87 percent of all geologic time. Thus, if all geologic time were represented by a 24- hour day, 21 hours would be the duration of the Precambrian. At 66 million years, the Cenozoic is brief, accounting for only about 20 minutes of our 24-hour day. Nevertheless, 66 million years is certainly long enough for significant evolution of the Earth and its biosphere to have taken place.

Many of the present features of the Earth have long histories, but the present distribution of land and sea and the present topographic expression of continents and their landforms are all the end products of Cenozoic processes. The Rocky Mountain region, for example, began its evolution during the Precambrian, but its present expression is largely the product of Cenozoic uplift and erosion. In short, the present distinctive aspect of the Earth developed very recently in the context of geologic time.

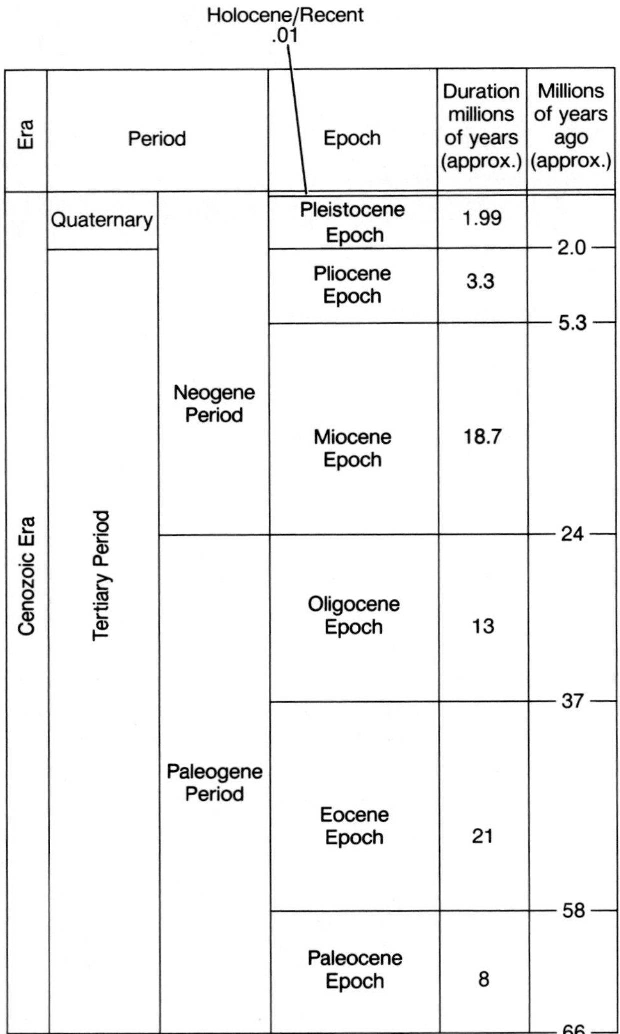

| Era | Period | | Epoch | Duration millions of years (approx.) | Millions of years ago (approx.) |
|---|---|---|---|---|---|
| Cenozoic Era | Quaternary | Neogene Period | Holocene/Recent | | .01 |
| | | | Pleistocene Epoch | 1.99 | 2.0 |
| | Tertiary Period | | Pliocene Epoch | 3.3 | 5.3 |
| | | | Miocene Epoch | 18.7 | 24 |
| | | Paleogene Period | Oligocene Epoch | 13 | 37 |
| | | | Eocene Epoch | 21 | 58 |
| | | | Paleocene Epoch | 8 | 66 |

**Figure 15-3** The geologic time scale for the Cenozoic Era.

## CENOZOIC PLATE TECTONICS—AN OVERVIEW

The Late Triassic fragmentation of the supercontinent Pangaea (Fig. 13–6) began an episode of plate motions that continues even now. As the Americas moved westward, the Atlantic Ocean basin opened until it attained its present dimensions, while the Pacific Ocean basin correspondingly decreased in size (Fig. 15–4).

Spreading centers such as the Mid-Atlantic Ridge and East Pacific Rise were established, and it is along these features that new oceanic crust is continuously generated. The age distribution of oceanic crust in the Pacific, however, is decidedly asymmetric because much of the crust in the eastern Pacific has been consumed at subduction zones along the western Americas (Fig. 6–12).

Another important event was the northward movement of the Indian plate and its collision with southern Asia (Fig. 15–4). India's long journey began during the Cretaceous when it separated from Gondwana and moved progressively north. In doing so, India, along with the northward-moving African plate, caused the closure of the Tethys Sea (Fig. 15–4). During the Early Tertiary, Australia separated from Antarctica and moved north to its present position.

The breakup of Pangaea and the drift of its various fragments account for the present geographic distribution of continents and oceans (Fig. 15–4). But continental rifting is not restricted to the Late Triassic. A triple junction (Chapter 6) is presently located at the junction of the East African Rift System, the rift in the Red Sea, and the rift in the Gulf of Aden (Fig. 15–5). Rifting in this region began during the Late Tertiary and continues at present.

Rifting in East Africa seems to be in its early stage, since the continental crust has not yet stretched and thinned enough for oceanic crust to form from below. In the Red Sea region, rifting was preceded by vast eruptions of basalt. In the succeeding rifting stage, a long, narrow sea formed, and by the Late Pliocene, oceanic crust began forming along the rift axis (Fig. 15-6). Rifting began even earlier in the Gulf of Aden than in the Red Sea. By the Late Miocene, the continental crust had stretched and thinned, and upwelling basaltic magma was sufficient to form oceanic crust.

Studies of this Late Cenozoic rifting event in Africa and adjacent regions are important for two reasons. First, they allow us to understand better the Late Triassic rifting of Pangaea and the origin of the Atlantic Ocean basin. Second, all three branches of these rifts remain active today, although rifting in East Africa is very slow. As a consequence of rifting in the Red Sea and Gulf of Aden, the Arabian plate has separated from Africa. Arabia's northward motion is responsible for a major north-south trending shear zone, the Dead Sea fault, and much of the present tectonic activity in the Middle East (Fig. 15–7).

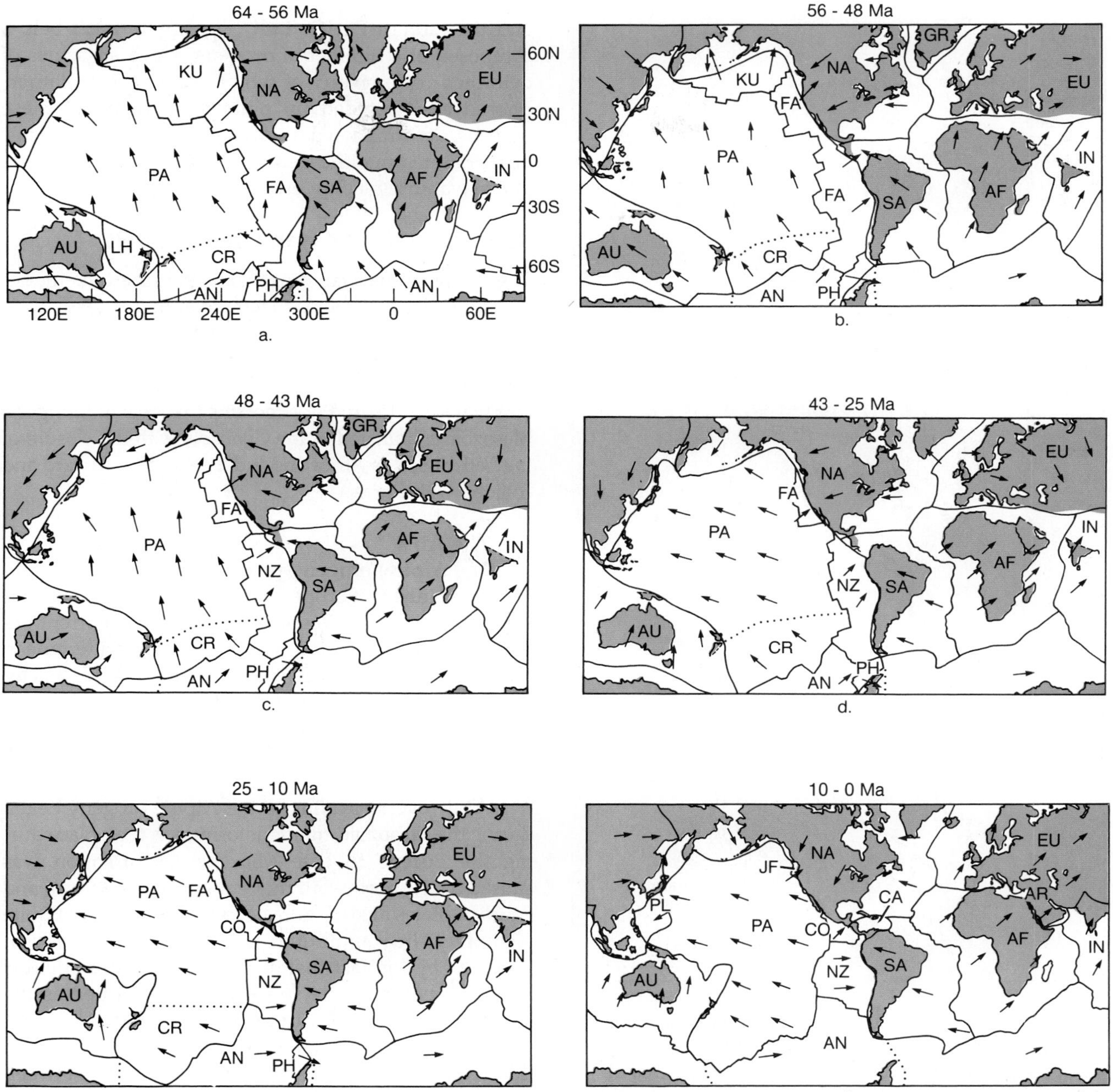


**Figure 15–4** Cenozoic plate movements. The Atlantic Ocean basin opened as the Americas moved westward from Europe and Africa, while the Pacific Ocean basin decreased in size. The Farallon plate was mostly consumed beneath the Americas; remnants of this plate are the Nazca, Cocos, and Juan de Fuca plates. Africa moved northward and partly closed the Tethys Sea. India also moved northward and collided with Asia. Australia moved northward to its present position. Abbreviations for plates: AF, African; AN, Antarctic; AR, Arabian; AU, Australian; CA, Caribbean; CO, Cocos; CR, Chatham Rise; EU, Eurasian; GR, Greenland; FA, Farallon; IN, Indian; JF, Juan de Fuca; KU, Kula; LH, Lord Howe; NA, North American; NZ, Nazca; PA, Pacific; PL, Philippine; PH, Phoenix; SA, South American.

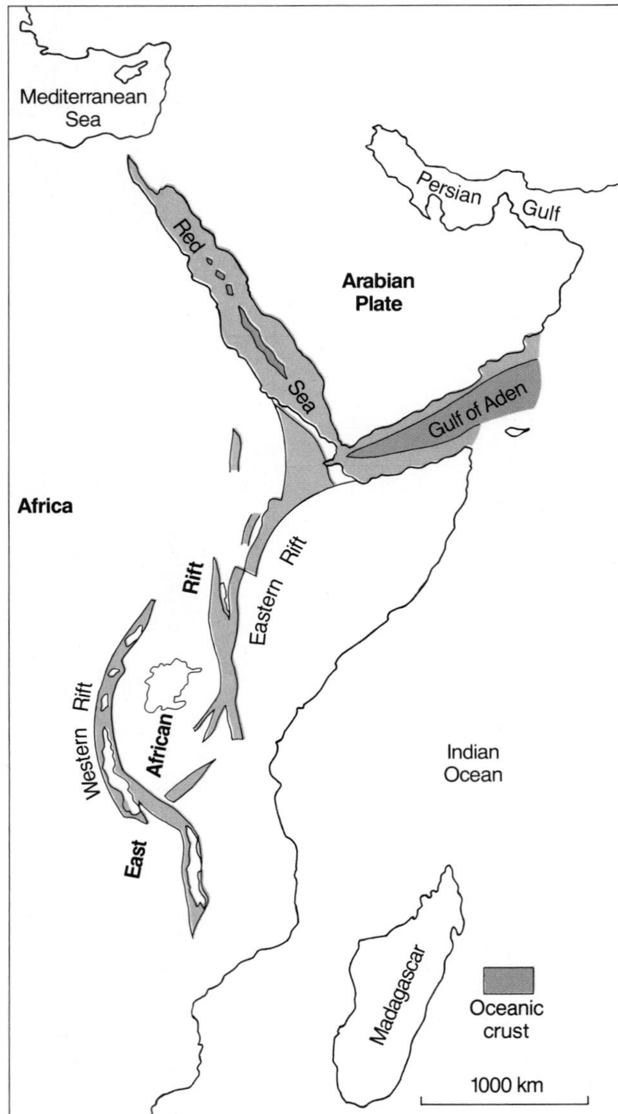


**Figure 15–5** A triple junction exists at the junction of the East African Rift System, and the rifts in the Red Sea and Gulf of Aden. Oceanic crust began forming in the Gulf of Aden about 10 million years ago. Rifting began later in the Red Sea, and oceanic crust is now forming. In East Africa, the continental crust has not yet stretched and thinned enough for oceanic crust to form from below.

## CENOZOIC OROGENIC BELTS

Cenozoic orogenic activity was largely concentrated in two major zones or belts, the **Alpine-Himalayan belt** and the **circum-Pacific belt** (Fig. 15–8). Each belt is composed of a number of **orogens** which are zones of deformed rocks, many of which have been metamorphosed and intruded by plutons. In many of these orogens, deformation dates well back into the Mesozoic but continued into the Cenozoic, and some, such as the Himalayan orogen, are orogenically active today.

### The Alpine-Himalayan Orogenic Belt

The Alpine-Himalayan orogenic belt includes the mountainous regions of the Mediterranean and extends eastward through the Middle East and India, and into southeast Asia (Fig. 15–8). Remember that the Tethys Sea separated much of Gondwana from Eurasia during Mesozoic time (Fig. 15–4). Plate motions that began during the Mesozoic culminated in the Cenozoic with the closure of the Tethys Sea as Africa and India moved northward and collided with Eurasia.

#### The Alps

The **Alpine orogeny** produced a zone of deformation in southern Europe extending from the Atlantic Ocean eastward to Greece and Turkey. Concurrent deformation also occurred south of the Mediterranean basin along the northwest coast of Africa (Fig. 15–8). Unraveling the complexities of this deformational event has been a long and arduous task, but the broad picture is now becoming clear, even though many details are still poorly understood.

While events leading to the Alpine orogeny began during the Mesozoic, major deformation took place during the Eocene to Late Miocene. Deformation was caused by the northward movements of the African and Arabian plates against Eurasia, but the story is not quite so simple. The complexities of Alpine geology are compounded by the fact that several small plates collided with Europe and were subsequently deformed as the larger Eurasian and African plates converged.

Deformation resulting from plate convergence in this region formed the Pyrenees Mountains between Spain and France, the Alps of mainland Europe, and the Apennines of Italy, and other mountain ranges (Fig. 15–9). The compressional forces generated by plate collisions produced complex thrust faults and large overturned folds called *nappes* (Fig. 15–10). Another result of plate convergence in this region was the formation of an isolated sea in the Mediterranean basin, which had formerly been a part of the Tethys Sea (Fig. 15–4) (See "Prologue" to Chapter 4).

The Atlas Mountains of northwestern Africa (Fig. 15–8) also formed as the African plate collided with Eurasia. Farther east in the Mediterranean basin, Africa is

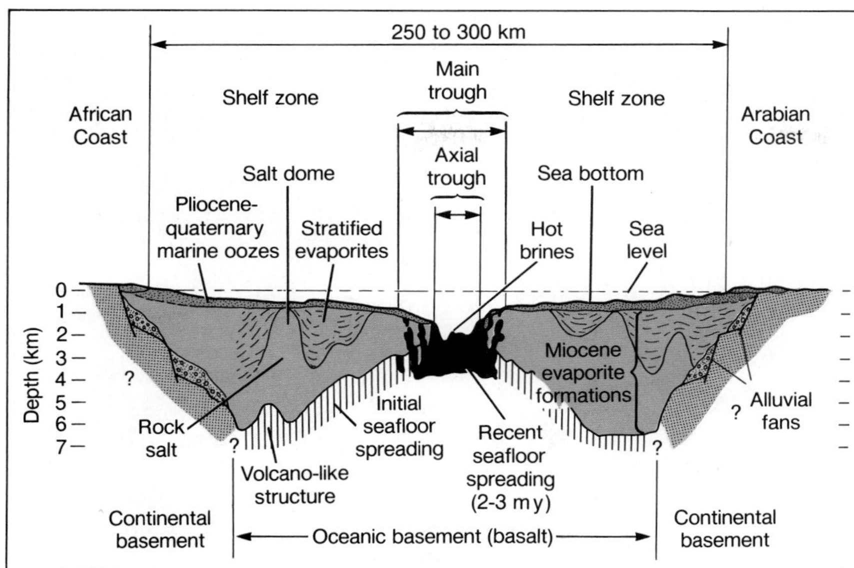


**Figure 15–6** Schematic north-south cross section of the Red Sea. Rifting that began during the Miocene Epoch has formed a long, narrow sea.

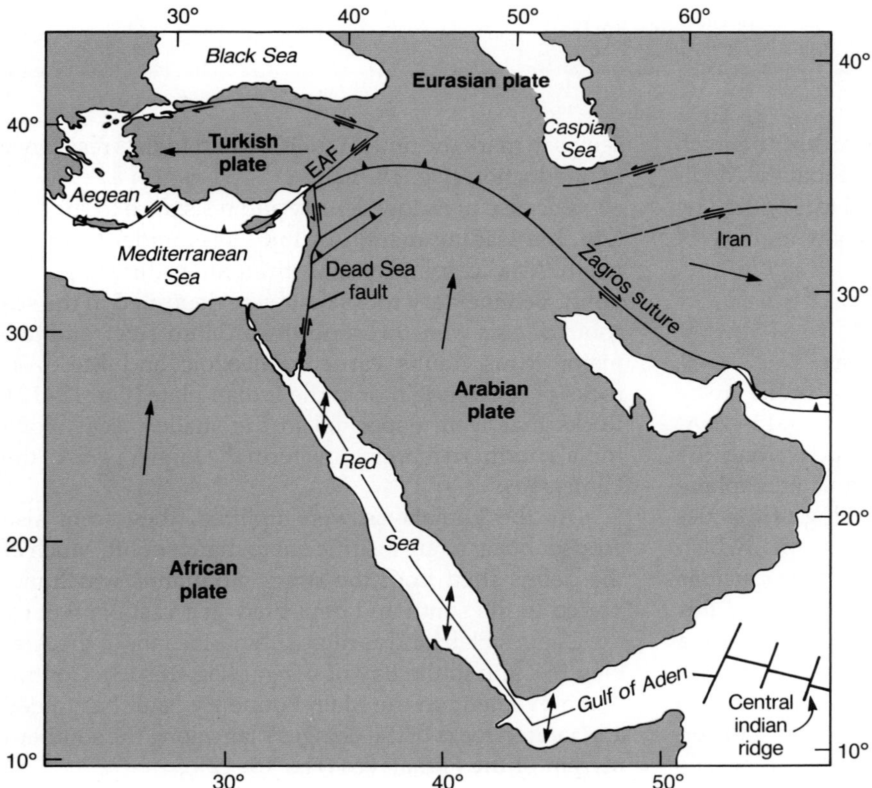


**Figure 15–7** The Arabian plate separated from the African plate as a consequence of rifting in the Red Sea and Gulf of Aden. Much of the tectonic activity in the Middle East is caused by the Arabian plate moving northward against Eurasia.

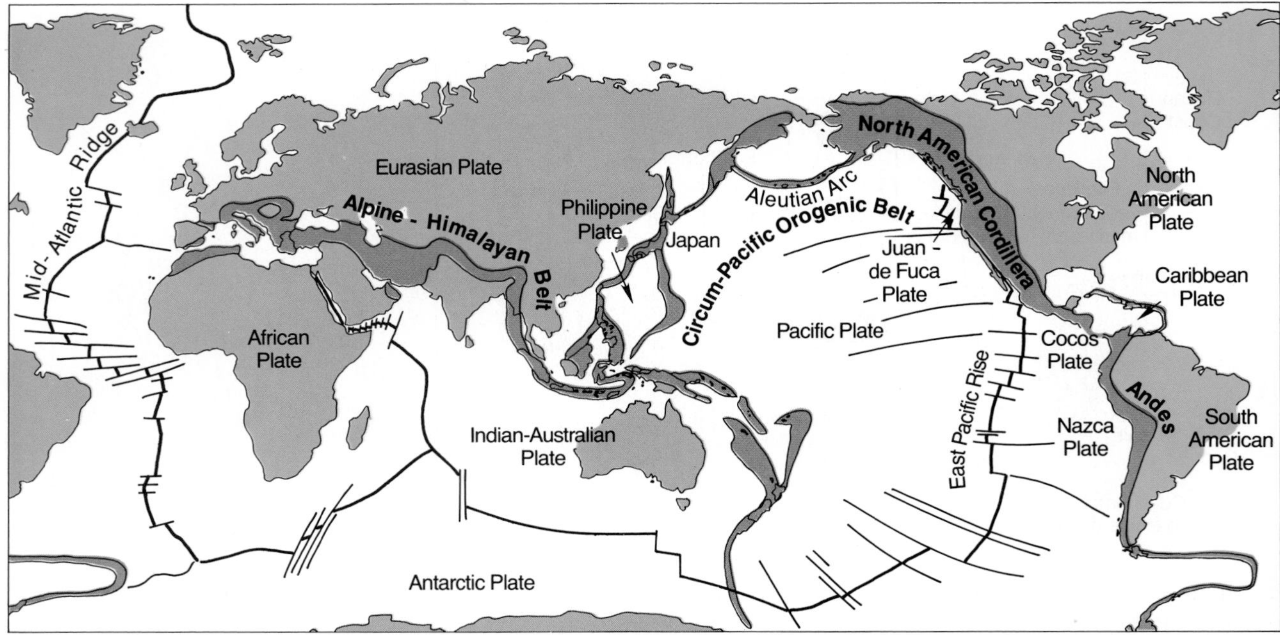


**Figure 15–8** Map showing the areas of Cenozoic orogenic activity. Orogenesis occurred mainly in two major belts—the Alpine-Himalayan belt and the circum-Pacific belt.

still forcing oceanic lithosphere northward beneath Greece and Turkey (Fig. 15–11). Evidence that the Mediterranean Basin remains geologically active is offered by active volcanoes in Italy and seismic activity in much of southern Europe.

### Roof of the World—The Himalayas

During Early Cretaceous time India broke away from Gondwana and began moving north (Fig. 15–4). At the same time, a subduction zone formed along the south-facing margin of Asia where oceanic lithosphere was consumed (Fig. 15-12). Partial melting of this descending oceanic lithosphere generated magma, which rose to form a volcanic chain and large granitic intrusions in what is now Tibet. India eventually approached this chain and destroyed it as it collided with Asia to form a **collision orogen**. As a result, two continental plates became sutured—India and Asia were now one.

The exact time of India's collision with Asia is uncertain, but sometime between 40 and 50 million years ago India's northward drift rate decreased abruptly, from between 15 and 20 centimeters per year to about five centimeters per year. Because continental lithosphere is not dense enough to be subducted, this decrease in rate seems to mark the time of collision and India's resistance to subduction (Fig. 15–12).

Because of its low density and resistance to subduction, the leading margin of India was underthrust beneath Asia, causing crustal thickening, thrusting, and uplift. Sedimentary rocks that were deposited in the sea south of Asia were thrust northward into Tibet, and two major thrust faults carried Paleozoic and Mesozoic rocks of Asian origin onto the Indian plate (Fig. 15–12). Rocks that were deposited in the shallow seas along India's northern margin now form the higher parts of the Himalayas.

As the Himalayas were uplifted, they were also eroded, but at a rate insufficient to match uplift. Much of the debris shed from the rising mountains was transported to the south and deposited as a vast blanket on the Ganges Plain and as huge submarine fans in the Arabian Sea and in the Bay of Bengal (Fig. 15–13). Continued movement on the Main Boundary Fault has folded and faulted rocks of the Ganges Plain along the southern margin of the Himalayas (Fig. 15–12).

Since its collision with Asia, India has been underthrust about 2,000 km beneath Asia! Presently, India continues moving north at about 5 cm per year. In other words, the Himalayas are still forming.

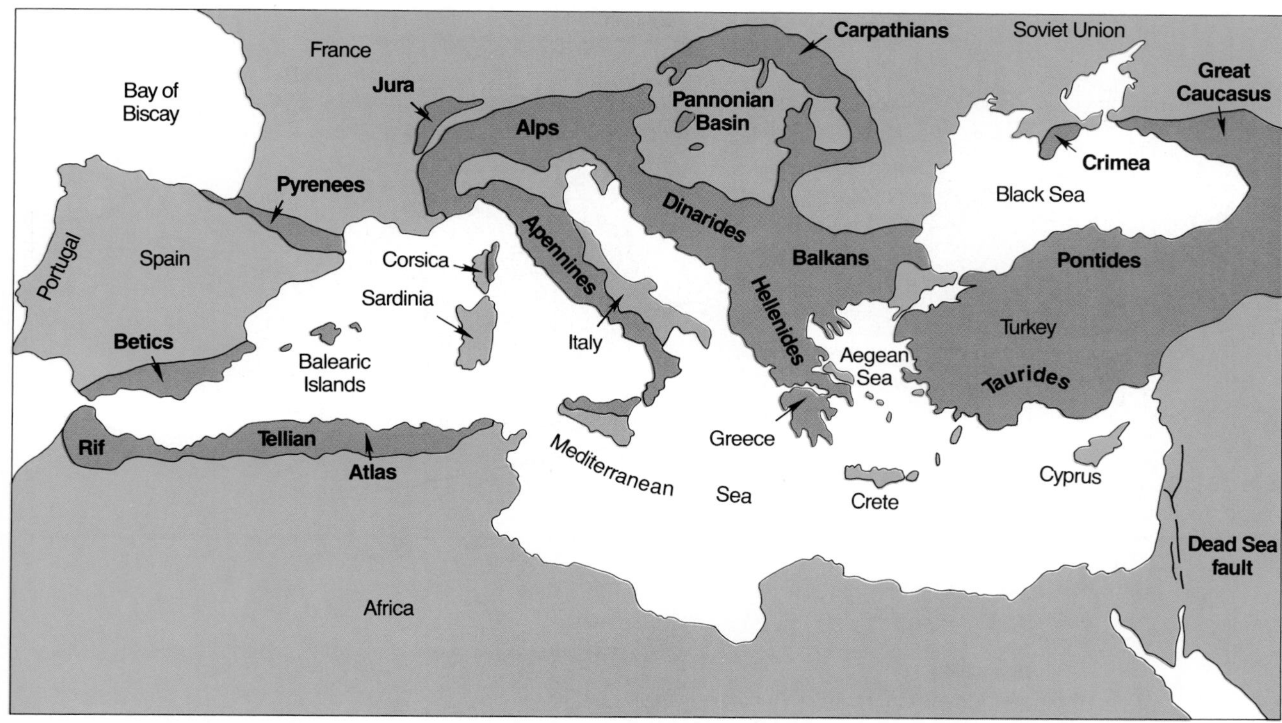


a.

b.

c.

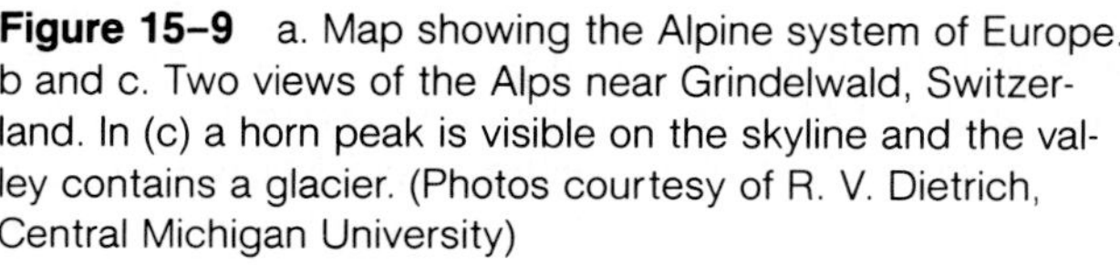

**Figure 15–9** a. Map showing the Alpine system of Europe. b and c. Two views of the Alps near Grindelwald, Switzerland. In (c) a horn peak is visible on the skyline and the valley contains a glacier. (Photos courtesy of R. V. Dietrich, Central Michigan University)

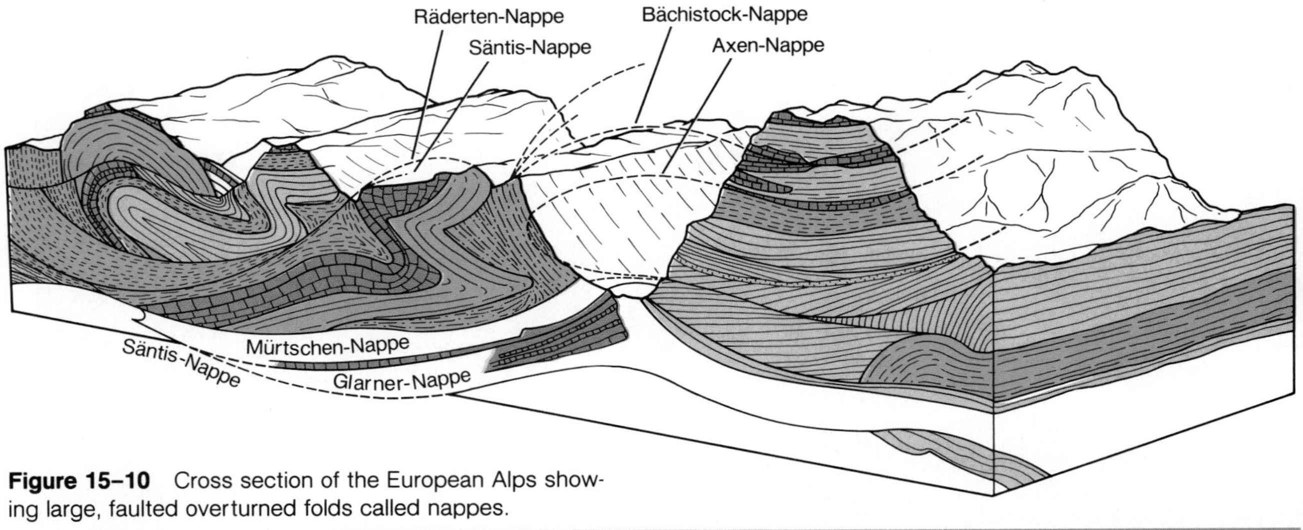


**Figure 15–10** Cross section of the European Alps showing large, faulted overturned folds called nappes.

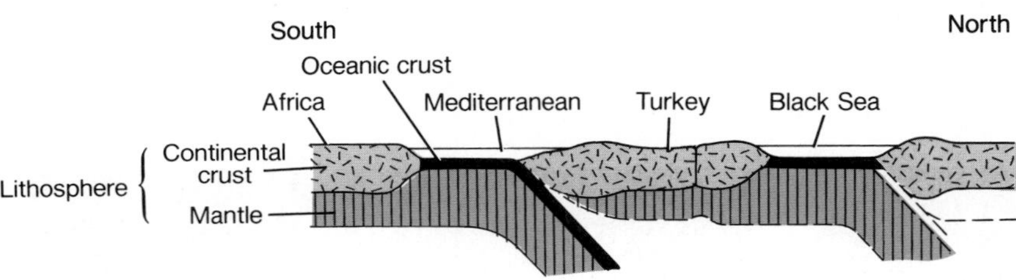


**Figure 15–11** In the eastern Mediterranean, Africa is approaching Eurasia, and oceanic lithosphere is being subducted beneath Turkey.

## The Circum-Pacific Orogenic Belt

The Pacific plate is being consumed at subduction zones along the western and northern margins of the Pacific Ocean basin (Fig. 15–14). This process has continued throughout the Cenozoic, giving rise to orogens in the Aleutians, the Philippines, Japan, and several other areas in the southwestern Pacific Ocean basin (Fig. 15–8).

Orogens of the western and northern Pacific are **arc orogens** characterized by subduction of oceanic lithosphere, deformation, and igneous activity. Japan, for example, is bounded on the east by the Japan Trench, where the Pacific plate is subducted. The Sea of Japan, a **backarc marginal basin** (Fig. 15–14), lies between Japan and the mainland of Asia. The origin of backarc marginal basins is not fully understood, but one theory holds that Japan was once part of mainland Asia and was separated as backarc spreading occurred (Fig. 15–15). Separation began sometime during the Cretaceous, Japan moved eastward over the Pacific plate, and oceanic crust formed in the Sea of Japan. Japan's geology is complex, and much of its deformation predates the Cenozoic. Nevertheless, considerable deformation, metamorphism, and volcanism occurred during the Cenozoic and, in fact, continues to the present.

Spreading at the East Pacific Rise is carrying the Cocos and Nazca plates eastward, where they are being subducted beneath Central and South America, respectively (Fig. 15–14). Volcanism and seismic activity indicate that the orogens in Central and South America remain active. We are reminded of this activity by events such as the 1985 earthquake that struck Mexico City and the tragic volcanic mudflows in Colombia in the same year (Perspective 15–1).

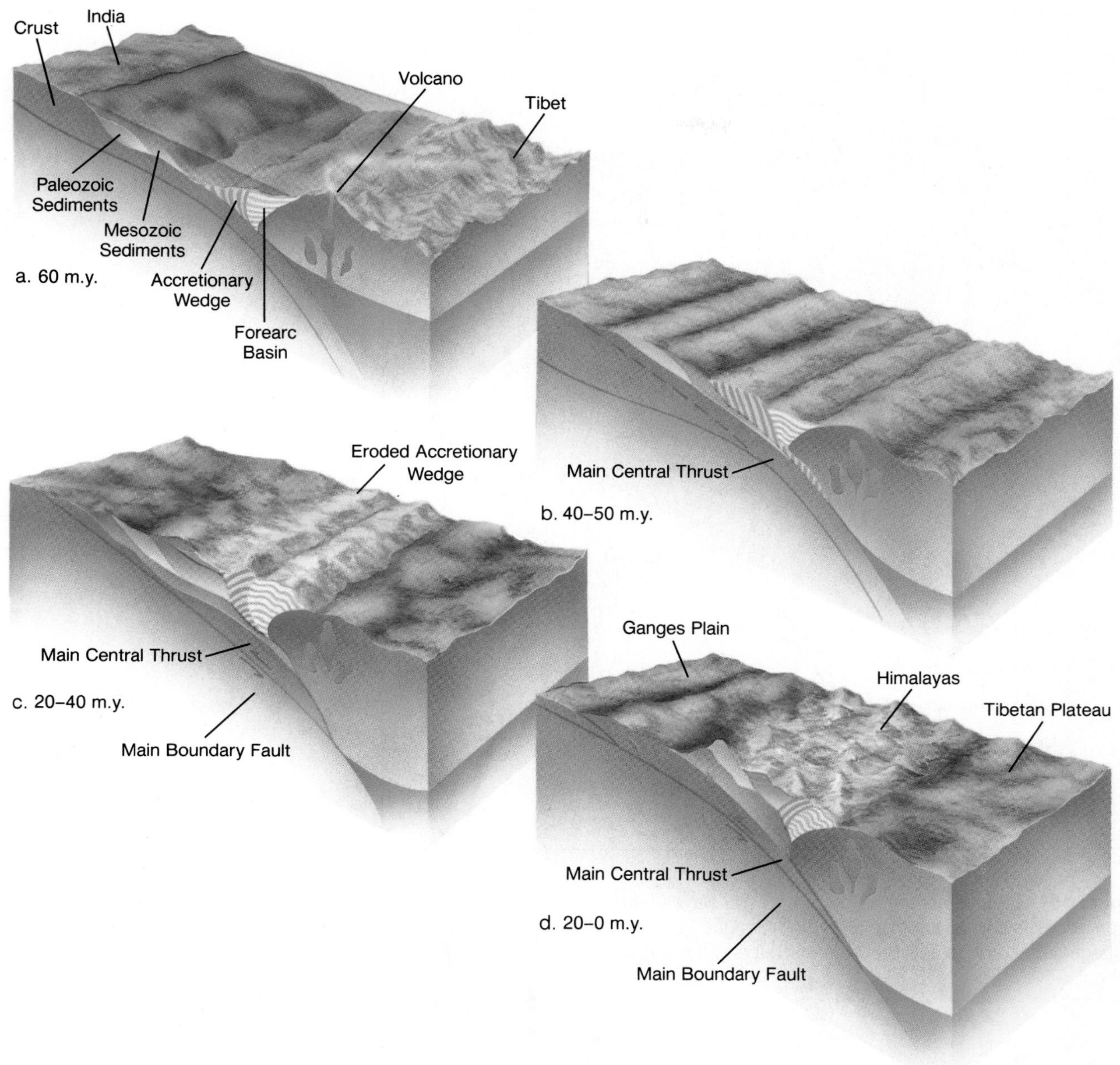


**Figure 15–12** Simplified cross sections showing the collision of India with Asia and the origin of the Himalayas. a. The northern margin of India before its collision with Asia. Subduction of oceanic lithosphere beneath southern Tibet as India approached Asia. b. About 40 to 50 million years ago India collided with Asia, but since India was too light to be subducted, it was underthrust beneath Asia. c. Continued convergence accompanied by thrusting of rocks of Asian origin onto the Indian subcontinent. d. Since about 10 million years ago India has moved beneath Asia along the Main Boundary Fault. Shallow marine sedimentary rocks that were deposited along India's northern margin now form the higher parts of the Himalayas. Sediment eroded from the Himalayas has been deposited on the Ganges Plain.

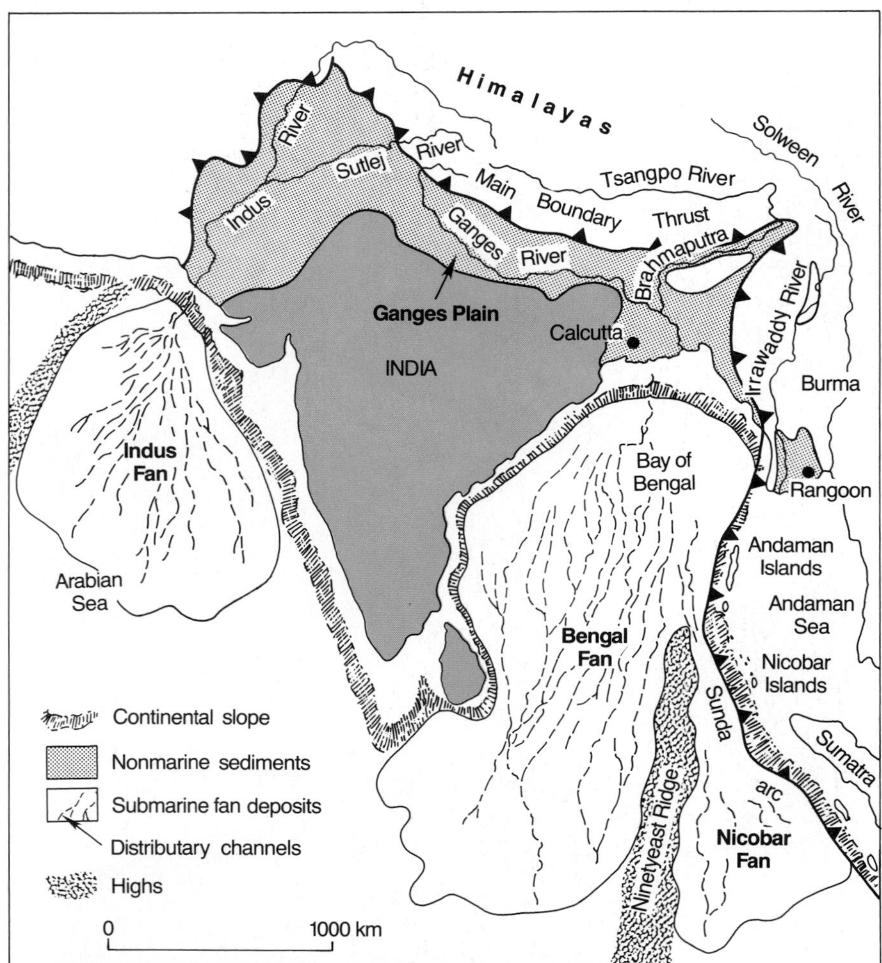


**Figure 15–13** Sediment eroded from the Himalayas has been deposited as a vast blanket on the Ganges Plain and as large submarine fans in the Arabian Sea and the Bay of Bengal.

In South America, the highest mountain range in the Western Hemisphere, the Andes (Fig. 15–8), resulted from Mesozoic-Cenozoic plate convergence. The formation of the modern Andes resulted from crustal thickening as sedimentary rocks were deformed, uplifted, and intruded by Mesozoic-Cenozoic granitic plutons. In addition, a major thrust fault carries the mountains over the Brazilian shield, further increasing the crustal thickness of South America (Fig. 15–16). Even though the Andes continue to be compressed at their margins, parts of the high Andes are subjected to tensional forces, resulting in downward movement of large blocks along normal faults.

Following its separation from Africa during the Mesozoic, South America was an island continent until the Late Tertiary. The land connection between North and South America was formed as a result of subduction at the Middle America Trench, together with arc magmatism (Fig. 15-17). We shall have more to say about this land connection between the Americas in Chapter 16, since the connection was an important one in the intercontinental migrations and extinctions of animals.

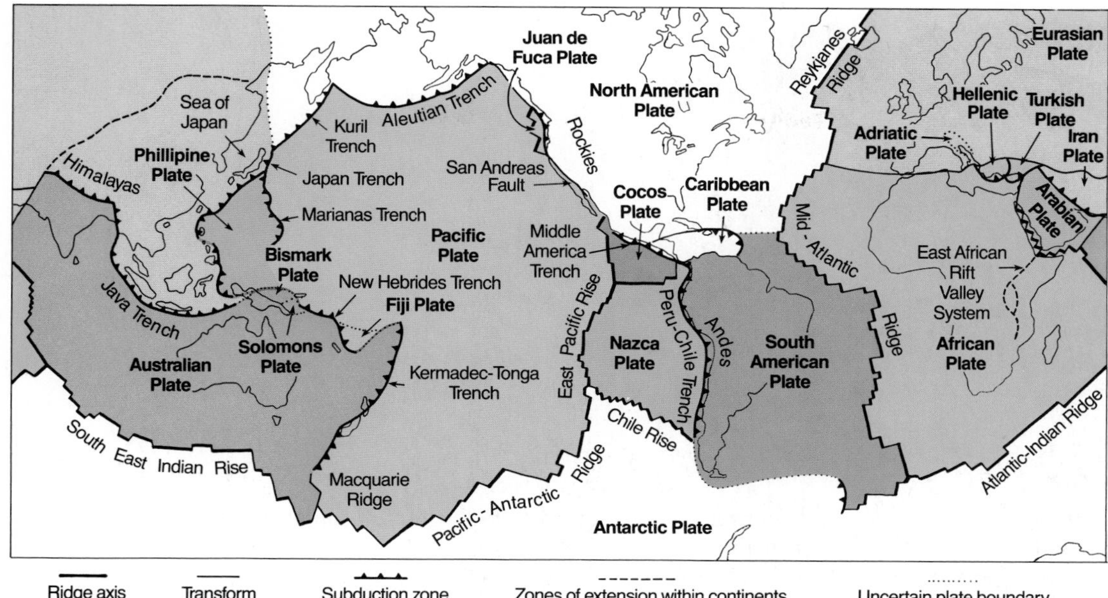


**Figure 15–14** The margins of the Pacific Ocean basin are bounded by subduction zones except along the west coast of North America. The Pacific plate is being consumed at subduction zones in the eastern and northern Pacific Ocean basin. The Nazca, Cocos, and Juan de Fuca plates, all remnants of the Farallon plate (see Fig. 15–4), are being consumed beneath the western Americas.

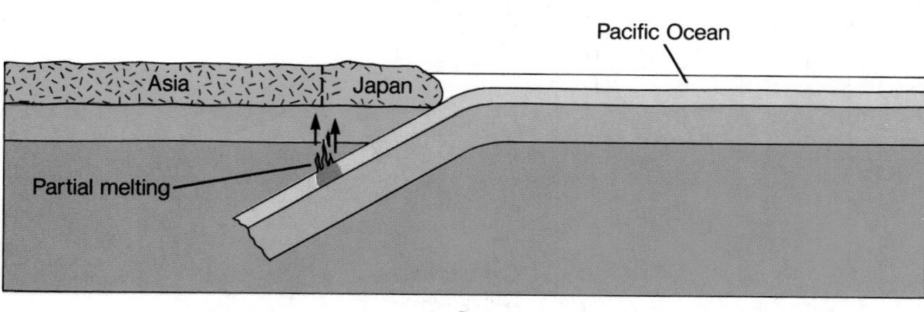


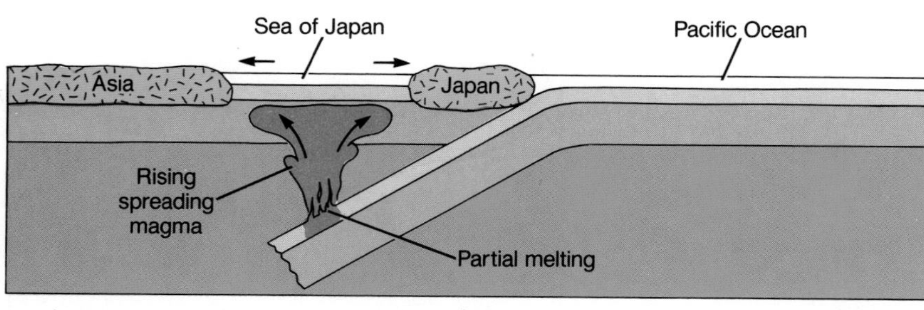


**Figure 15–15** The backarc marginal basin occupied by the Sea of Japan is thought to have formed by backarc spreading as shown in this model.

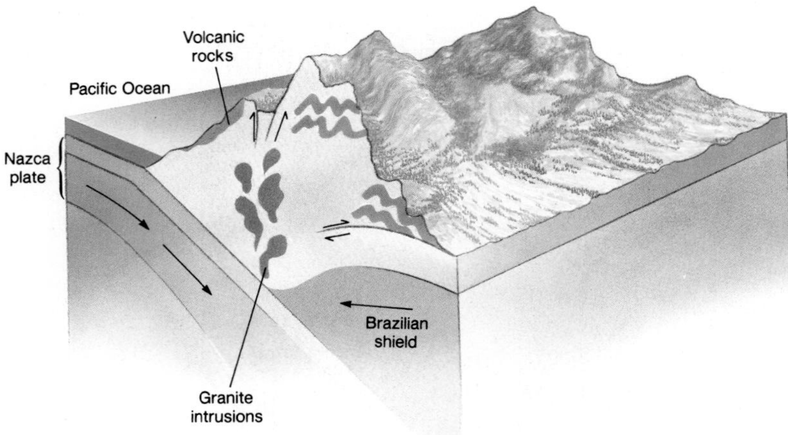


**Figure 15–16** The Andes of South America formed as a result of convergence of the Nazca and South American plates. A major thrust fault carries the mountains over the Brazilian shield.

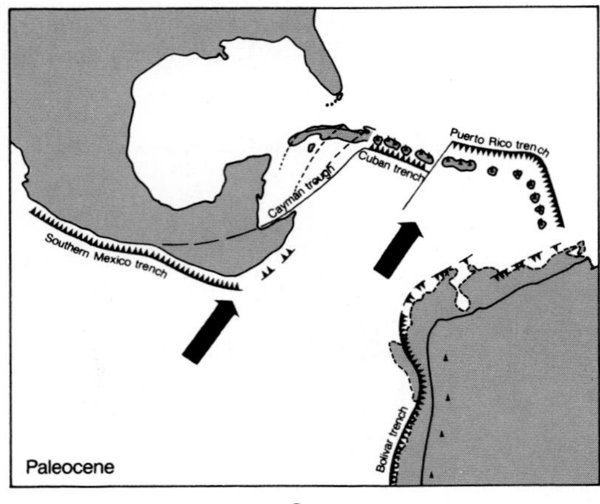


a.

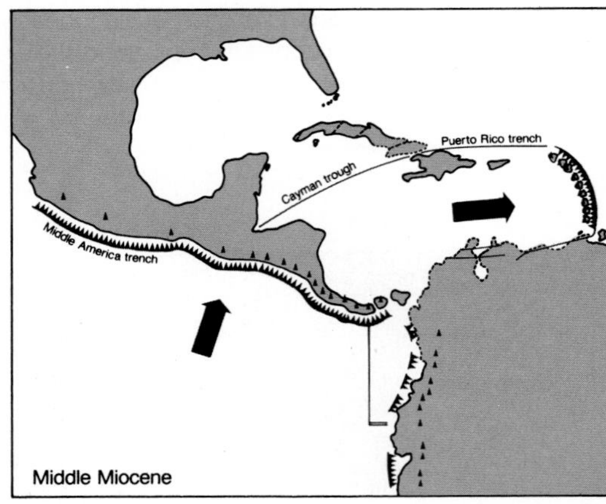


b.

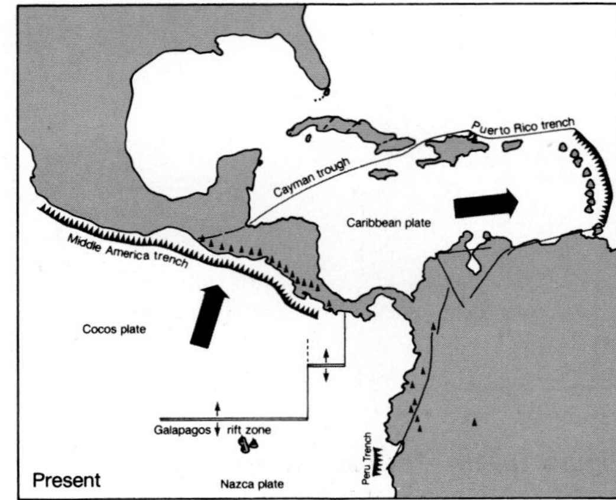


c.

**Figure 15–17** Origin of the land connection between North and South America. During the Paleocene, the Pacific plate was subducted at the Cuban and Puerto Rico trenches. From the Early Oligocene to present, however, subduction of the Pacific plate has occurred at the Midddle America Trench, resulting in arc magmatism and the origin of the land connection between North and South America.

## Perspective 15–1

# The Perils of Living Near A Convergent Plate Margin

The active state of the Central and South American orogens was demonstrated by two tragic events in 1985. One, an earthquake of 8.1 magnitude on the Richter Scale, claimed 9,000 lives in Mexico City. The other, a mudflow that engulfed Armero, Colombia and several villages, claimed 23,000 lives. In addition, several tens of thousands of people were injured or left homeless, and property damages of several billions of dollars resulted.

Even though earthquakes and mudflows are very different geologic phenomena, both of these events are related to convergent plate margin activities. The earthquake that struck Mexico City, the most populous urban area in the world, resulted from subduction of the Cocos plate at the Middle American Trench (Fig. 15–14). Sudden movement of the Cocos plate beneath Central America generated seismic waves that traveled outward in all directions. The violent shaking experienced in Mexico City and elsewhere, was caused by these seismic waves.

Much of the damage in Mexico City (Fig. 1) occurred because the city rests upon an unstable foundation. The underlying strata are mostly layers of sand, silt, and mud that accumulated in a large lake. In fact, Mexico City was built on the site once occupied by the Aztec capital city of Tenochtitlan, which was located on an island in this ancient lake. Unfortunately, unconsolidated sediments such as those beneath Mexico City amplify the shaking during earthquakes. Consequently, buildings on such unstable foundations are commonly more heavily damaged than those built on solid bedrock foundations.

Less than two months after the Mexico City earthquake, Colombia, South America, experienced its greatest recorded natural disaster. The Nazca plate is being subducted beneath South America (Fig. 15–14), and Nevado del Ruiz is one of several active volcanos where rising magma from the partially melted descending plate are erupted. A rather minor eruption of Nevado del Ruiz partially melted

a.

b.

**Figure 1** Earthquake damage in Mexico City. a. Collapsed building in Mexico City, with one floor piling upon another. b. Twisted railroad tracks near the epicenter of the earthquake. (Photos courtesy of C. Lomnitz, Universidad Nacional Autonoma de Mexico)

## Perspective 15-1 (continued)

the glacial ice on the mountain, and the meltwaters rushed down the valleys, picking up sediment as they went, and finally became viscous mudflows.

The city of Armero, Columbia, lies in the valley of the Lagunilla River, one of the river valleys inundated by mudflows. Of the 23,000 inhabitants of Armero, about 20,000 died, and most of the city was destroyed. Another 3,000 people were killed in other nearby valleys.

Natural disasters are not restricted to convergent plate margins—large earthquakes occasionally shake areas in continental interiors. Nevertheless, events such as these are common at convergent plate margins. Central and South America have experienced similar geologic upheavals in the past and can expect continuing activity of this type in the future.

## THE NORTH AMERICAN CORDILLERA

Bounded on the east by lowlands, and on the west by the Pacific Ocean, the **North American Cordillera** is a complex mountainous region extending from Alaska into Mexico (Fig. 15–8). In the western United States, the Cordillera widens to about 1,200 km (Fig. 15–18), and, according to William R. Dickinson of Stanford University, "Perhaps no segment of the circum-Pacific [orogenic belt] has undergone a more complex tectonic evolution during the Cenozoic."[1]

[1]1979, Cenozoic Paleogeography of the Western United States (p. 1), in *Cenozoic Paleogeography of the Western United States.* Pacific Section SEPM Paleogeography Symposium, v. 3.

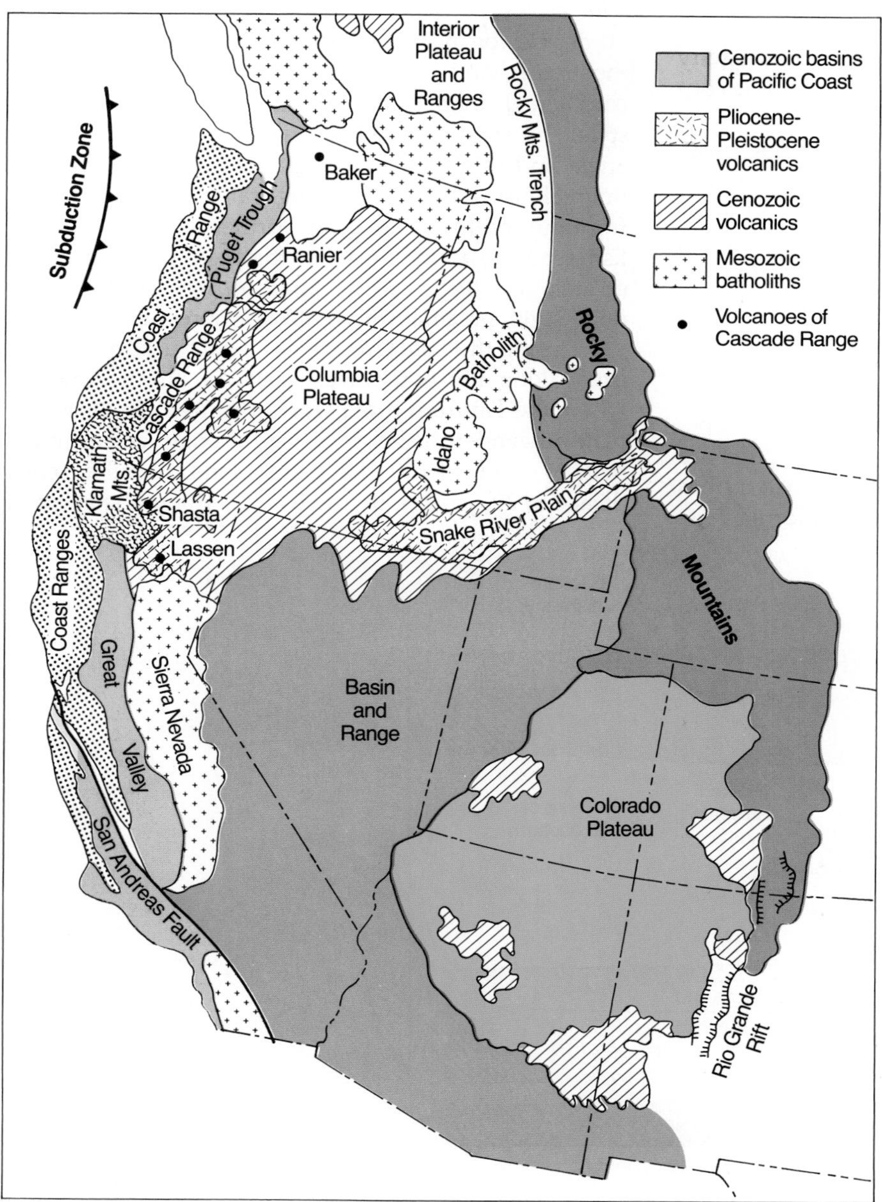


**Figure 15–18** Map of the Cordilleran region in the United States showing the main structural provinces.

The evolution of the North American Cordillera began during the Late Proterozoic when thick sequences of sediments were deposited along its west-facing continental margin (Fig. 9–16). Sediment deposition continued during the Paleozoic, and during the Devonian, part of the region was deformed during the Antler orogeny (Fig. 11–19). Beginning in the Late Jurassic and continuing into the Early Tertiary, deformation was more or less continuous as the Nevadan, Sevier, and Laramide orogenies progressively affected areas from west to east (Table 13–1). Both the Nevadan and Sevier orogenies were discussed in Chapter 13. The **Laramide orogeny** was a Late Cretaceous to Eocene event and will be considered in the following section.

Following the deformation of the Laramide orogeny, several other events occurred that contributed to the evolution of the Cordillera. For example, the structural style of the Basin and Range province (Fig. 15–18) developed as a result of extensional tectonics after the compressional deformation of the Laramide orogeny. Vast outpourings of rhyolitic and especially basaltic volcanics occurred in the Pacific Northwest, and volcanism continues in the Cascades of California, Oregon, and Washington. The Cordillera continues to evolve, especially along its western margin, where the Juan de Fuca plate is being subducted and where the **San Andreas transform fault** cuts through coastal California (Fig. 15–18).

## Laramide Orogeny

The Laramide orogen differs from other orogens in that deformation occurred much farther inland from the arc-trench system than is typical of these collisions. Also, deformation was not accompanied by significant batholithic intrusions. To account for these observations, geologists have modified the classic model for arc orogens. Figure 15–19 shows what probably happens when oceanic lithosphere is subducted beneath continental lithosphere; oceanic lithosphere descends at a steep angle, 30 degrees or more, arc magmatism occurs inland from the trench, and the thick sediments deposited on the continental margin are deformed. In the Laramide style, the subducted oceanic slab descends at a lower angle and moves nearly horizontally beneath the continental lithosphere, deforming cratonic crust far inland from the continental margin. Furthermore, arc magmatism seems to occur only when the descending plate penetrates as deep as the mantle, so in the Laramide style, magmatism would be supressed.

During the Late Cretaceous, a subduction zone existed along the entire west coast of North America, where the **Farallon plate** was consumed. The Farallon plate descended at about a 50 degree angle, and arc magmatism occurred 150 to 200 km inland from the trench (Figs. 15–20, 15–21). By the Early Tertiary, the angle of

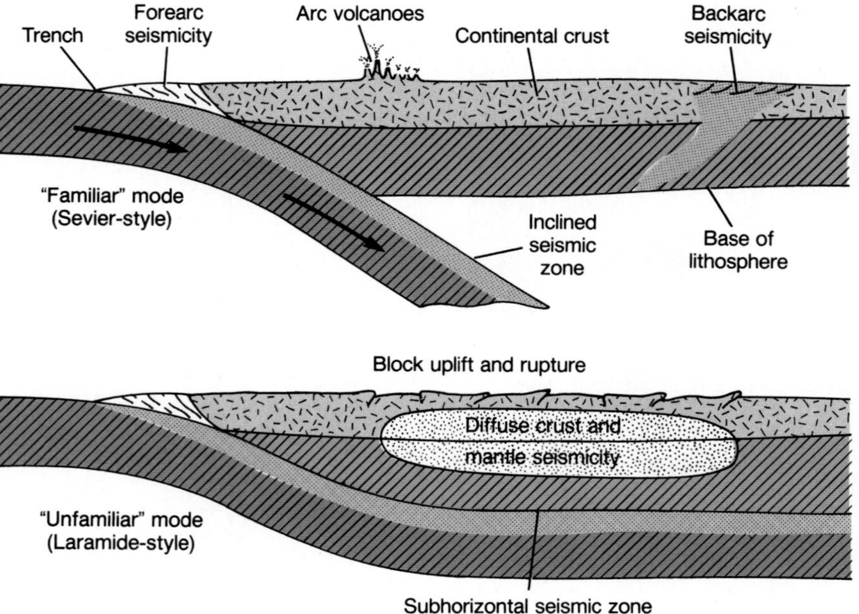


**Figure 15–19** Arc orogens resulting from steep and shallow subduction. In the shallow-subduction model, the subducted slab moves nearly horizontally beneath the continent, and arc volcanism ceases.

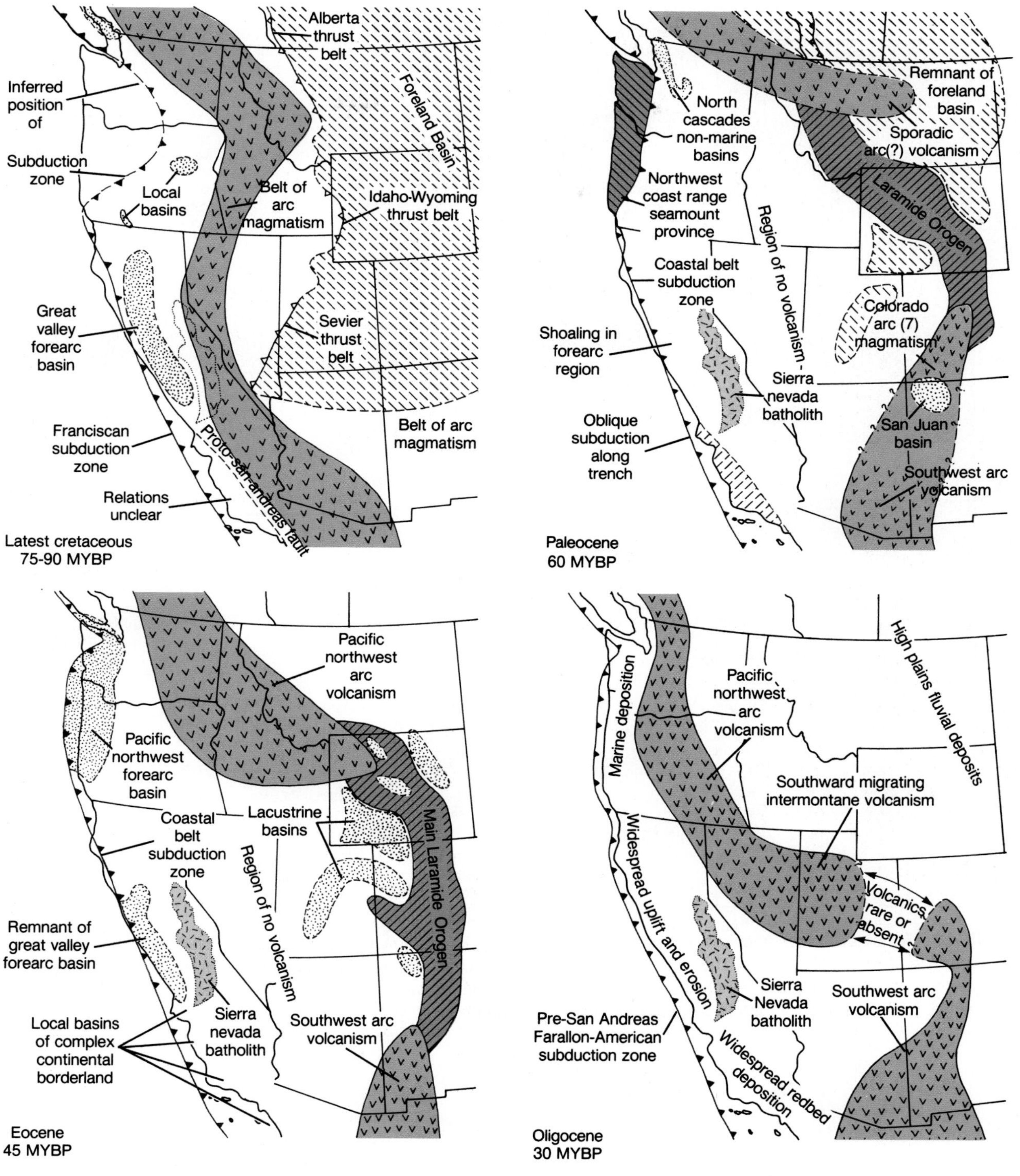


**Figure 15–20** Paleotectonic and paleogeographic maps for the Cordilleran of the United States from latest Cretaceous to Oligocene time. The Laramide orogen is located much farther inland than is typical of arc orogens. Also note that there was a region of little or no volcanism west of the Laramide orogen during the Paleocene and Eocene. Apparently, volcanism ceased in this area as the subducted Farallon plate moved nearly horizontally beneath the continent as shown in Figure 15–19.

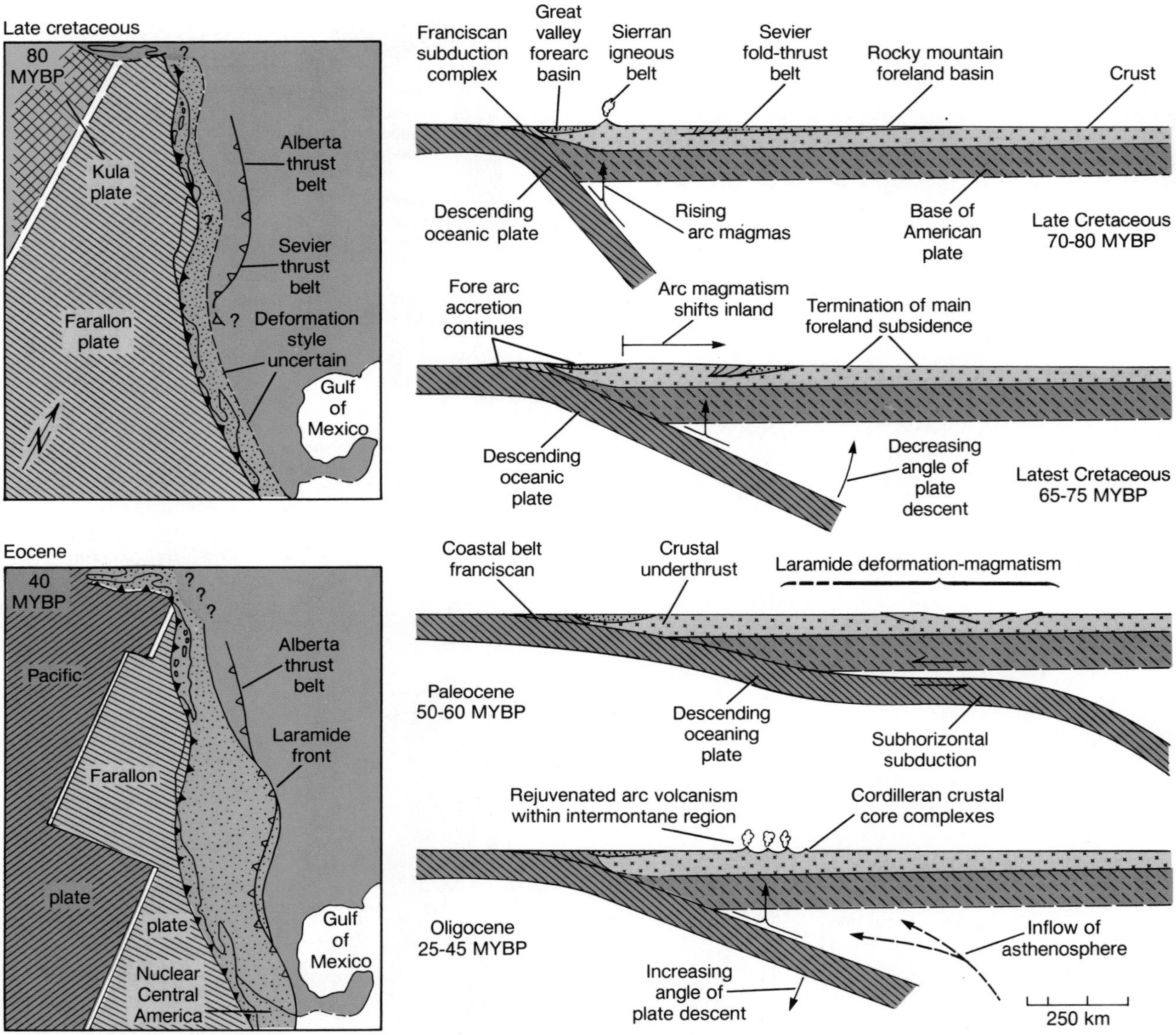

**Figure 15–21** Schematic cross sections and maps showing the Laramide style of deformation. The Laramide orogeny was caused by subduction of the Farallon plate beneath North America. Notice especially that as the angle of subduction decreases, arc magmatism shifts inland and eventually ceases. Also, deformation occurs much farther inland when the subducted plate moves nearly horizontally beneath the continent.

subduction apparently decreased, and the Farallon plate moved nearly horizontally beneath North America. As a result, arc magmatism shifted farther inland and eventually ceased, since the descending Farallon plate did not penetrate to the mantle. In fact, during Laramide deformation, there was a region of little or no magmatism west of the main Laramide orogen (Fig. 15–20).

Another consequence of the decreasing angle of subduction was a change in the tectonic style of deformation. The fold-thrust tectonic style of the Sevier orogenic belt gave way to large-scale buckling and fracturing, which produced fault-bounded uplifts (Fig. 15–21). Major mountain ranges with intervening intermontane basins formed along these faults (Fig. 15–22). These

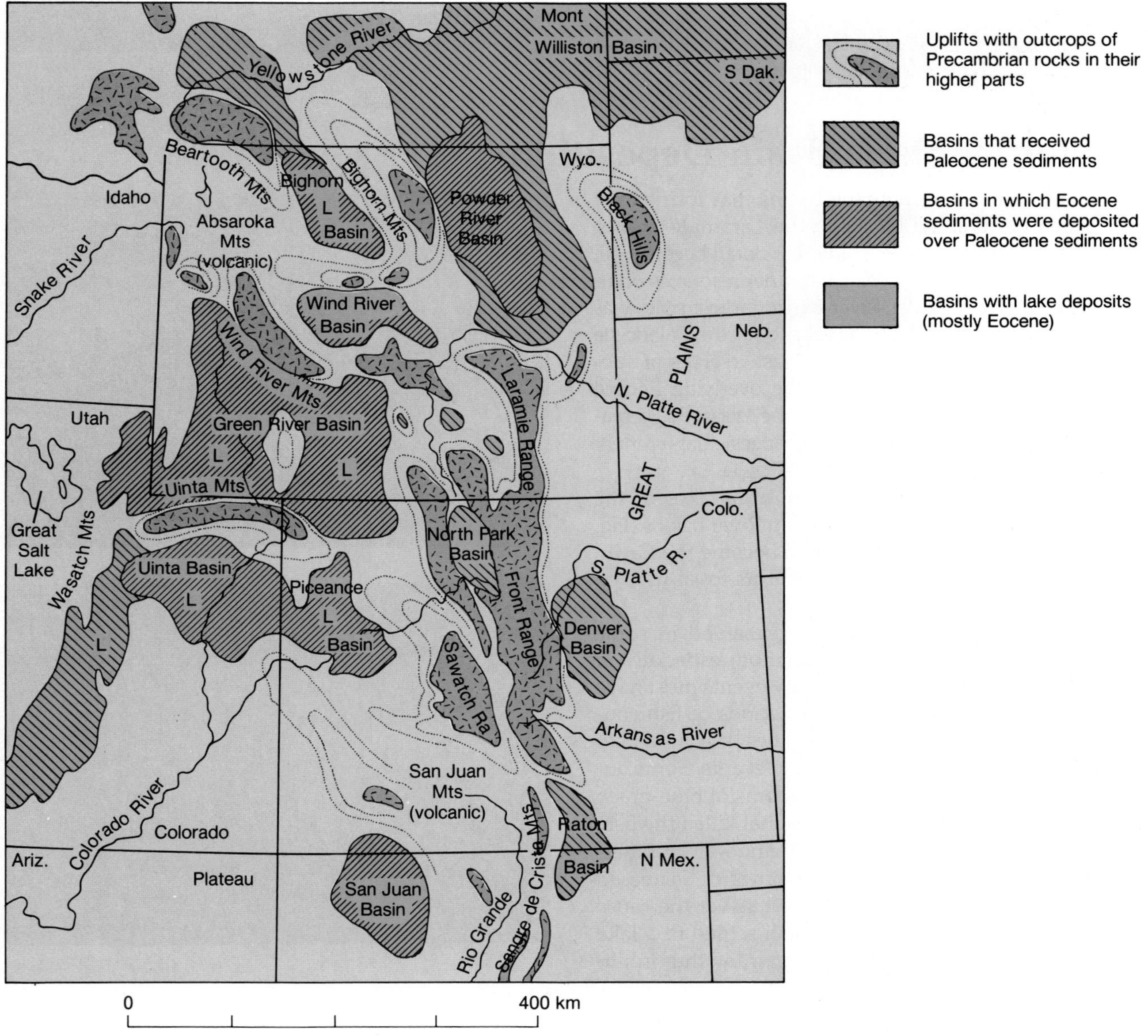


**Figure 15–22** Map showing Laramide uplifts and basins.

intermontane basins were the sites of Early Tertiary deposition of sediments that were eroded from the uplifted blocks (Perspective 15–2).

The Laramide orogen is centered mostly in the middle and southern Rockies in Wyoming and Colorado (Fig. 15–20). Nevertheless, Laramide compressional forces also deformed the Cordillera far to the north and south. In the northern Rockies of Montana and Alberta, Canada, large slabs of pre-Laramide strata were transported to the east along large overthrust faults. Along the Lewis overthrust of Montana, a large slab of Precambrian strata was moved at least 75 km eastward and now rests upon Cretaceous strata (Fig. 15–23). In the Canadian Rockies of Alberta, thrust sheets piled one upon another in shingle-like fashion, resulting in complex structural relationships.

Deformation also occurred south of the main Laramide orogen. In the Sierra Madre Oriental of east-central

## Perspective 15–2

# Intermontane Basin Deposition

Many of the intermontane basins that formed between the uplifted blocks of the Laramide orogen (Fig. 15–22) have been studied in detail because oil, natural gas, uranium, coal, and other resources occur in them. The Powder River Basin, for example, was the Early Tertiary depositional site for the Paleocene Fort Union Formation with its vast reserves of subbituminous coal (Fig. 1), and the overlying Eocene Wasatch Formation. Erosion of the Wasatch Formation in Utah accounts for the spectacular scenery in Bryce Canyon National Park (Fig. 1).

Sediments accumulated in large lakes in the Piceance Creek, Uinta, and Green River basins (Fig. 15–22). These sediments would become the Green River Formation, well-known for its fossil fish, oil shales, and evaporites.

Millions of fossil fish are preserved in some parts of the Green River Formation, especially in Wyoming (Fig. 2). Mass mortality events must have occurred to account for the thousands of fish concentrated on single bedding planes, but no agreement exists on what caused these events. Some authorities have suggested that blooms of blue-green algae produced toxic substances that killed the fish. Others propose that rapidly changing water temperatures or excessive salinity at times of increased evaporation were responsible. Whatever the cause of mass mortality events, the fish settled to a lake bottom that was deficient in oxygen and thus inhibited the process of decomposition.

The Green River Formation includes two types of deposits of particular interest—oil shale and

a.

b.

**Figure 1** a. Coal mine in the Fort Union Formation near Gillette, Wyoming. b. Exposures of the Wasatch Formation in Bryce Canyon National Park, Utah.

Mexico, Late Jurassic to Late Cretaceous shelf carbonates and clastic rocks are now part of a major fold-thrust belt (Fig. 15–24). Laramide structural trends can also be traced south into the area of Mexico City.

The Laramide style of deformation came to a stop during the Eocene. Apparently the cessation of Laramide deformation coincided with an increasing angle of descent of the Farallon plate (Fig. 15–21), and arc magmatism was reestablished as the descending slab penetrated the mantle. The uplifted, fault-bounded blocks of the Laramide orogen continued to be eroded, the debris filling the adjacent intermontane basins. By Late Tertiary time the rugged, eroded mountains had been nearly buried in their own erosional debris, forming a vast plain across which streams flowed (Fig. 15–25). During a renewed cycle of erosion these streams stripped away much of the younger basin-fill sediment and eroded downward, incising their valleys into the uplifted blocks. The deep canyons cutting through the present ranges are the products of this cycle of erosion

**Figure 2** Fossil fish from the Green River Formation. These fish average about 7.6 cm long.

evaporites. Oil shale is laminated, fine-grained sedimentary rock containing kerogen, an organic compound. Oil can be produced from oil shale by a process in which the rock is heated in the absence of oxygen, thus driving off the hydrocarbons and then recovering them by condensation. Currently this process is more expensive than recovering oil by conventional drilling and pumping, but the Green River oil shales constitute one of the largest untapped sources of oil in North America.

Figure 3 shows a depositional model for part of the Green River Formation. According to this model,

(Fig. 15–26). The present-day elevations of these ranges are the result of Late Tertiary uplift; in some areas uplift continues to the present.

## Cordilleran Volcanism

The vast batholiths of the Sierra Nevada Mountains in California, and in Idaho, and in the Coast Ranges of British Columbia were largely emplaced during the Mesozoic Era, but intrusive activity continued into the Tertiary. Numerous small plutons were emplaced, including copper- and molybdenum-bearing stocks in Utah, Nevada, Arizona, and New Mexico.

Tertiary extrusive volcanism occurred more or less continuously in the Cordillera, although it did vary in eruptive style and location (Fig. 15–27). Eocene lava flows and sedimentary rocks composed of volcanic rock particles accumulated in the Yellowstone National Park region of Wyoming. Further south in Colorado, lava flows, tuffs, and large calderas characterize the Oligo-

## Perspective 15–2 (continued)

coarse-grained alluvial fans formed at the basin margin and graded basinward into fine-grained mudflat and playa lake deposits. Oil shale probably formed from organic-rich ooze deposited in shallow, oxygen-deficient lake waters. Since little oxygen was present, the organic matter was not decomposed by oxidation.

Green River Formation evaporites are also interesting, and one, a sodium carbonate called *trona*, has been mined as a source of sodium compounds. The trona and associated evaporites such as halite are interpreted as deposits that accumulated in the central parts of lakes during times of intense evaporation (Fig. 3).

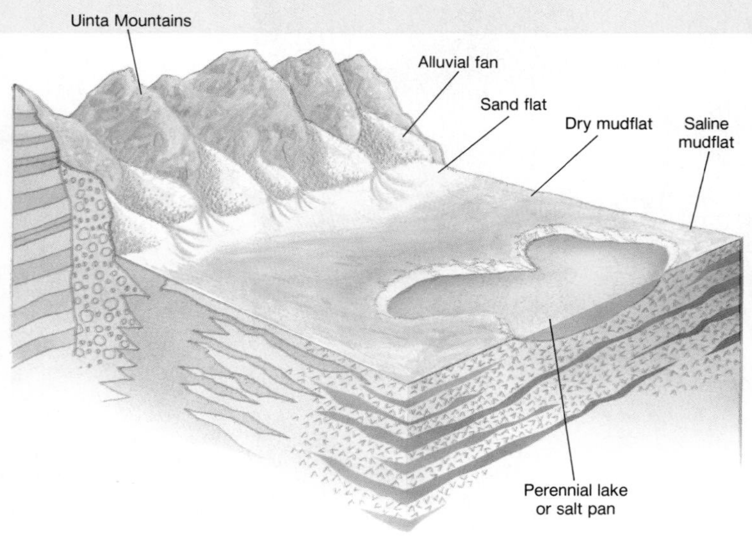


**Figure 3** Depositional model for part of the Green River Formation. Oil shales formed in shallow, oxygen-deficient lake waters, and evaporites were precipitated in the central part of the lake during times of intense evaporation.

cene San Juan volcanic field (Fig. 15–28). In Arizona, Pliocene to Quaternary volcanism built up the San Francisco Mountains. Here, volcanism may have ceased as recently as 1,200 years ago.

In the Pacific northwest a large area is underlain by about 200,000 km$^3$ of Miocene basalt flows, the Columbia River basalts (Fig. 15–29). These flows issued from long fissures and resulted in overlapping flows that are now well exposed in the walls of the deep gorges cut by the Snake and Columbia rivers (Fig. 15–30). The magnitude of some of these eruptive events is difficult to imagine. For example, the single Roza flow covers 40,000 sq

a.

b.

**Figure 15–23** a. The Lewis overthrust in Glacier National Park, Montana. Late Proterozoic strata of the Belt Supergroup rest upon Cretaceous strata. b. Chief Mountain, Montana, is an erosional remnant of the Lewis overthrust.

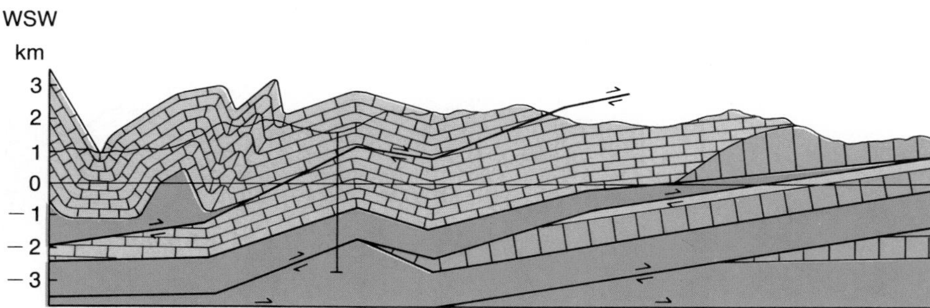


**Figure 15–24** Cross section showing part of the Laramide fold-thrust belt in the Sierra Madre Oriental Mountains of Mexico.

km and has been traced more than 300 km to the west from its source. Peter Hooper of Washington State University says:

> The enormity of this event is hard to visualize. A lava front about 30 m high, over 100 km wide, and at a temperature of 1100° C, advanced at an average rate of 5 km per hour.[2]

Most of the Columbia River basalts were erupted during a span of 3.5 million years and filled a basin with an aggregate thickness of 2,500 m at the basin center. However, despite all the detailed studies, the relationship of this phenomenal eruptive event to plate tectonics remains uncertain.

The Snake River Plain (Fig. 15–18) is actually a depression in the Earth's crust that has been filled mostly by Pliocene and younger basalts (Fig. 15–31). The fact that the volcanics are oldest in the southwestern part of the Snake River Plain and are progressively younger toward the northeast has led some geologists to propose that North America has migrated over a mantle plume. This plume, although its existence is debated, may now lie beneath Yellowstone National Park in Wyoming (Fig.

[2]1982. The Columbia River Basalts. *Science*, 215: 1466.

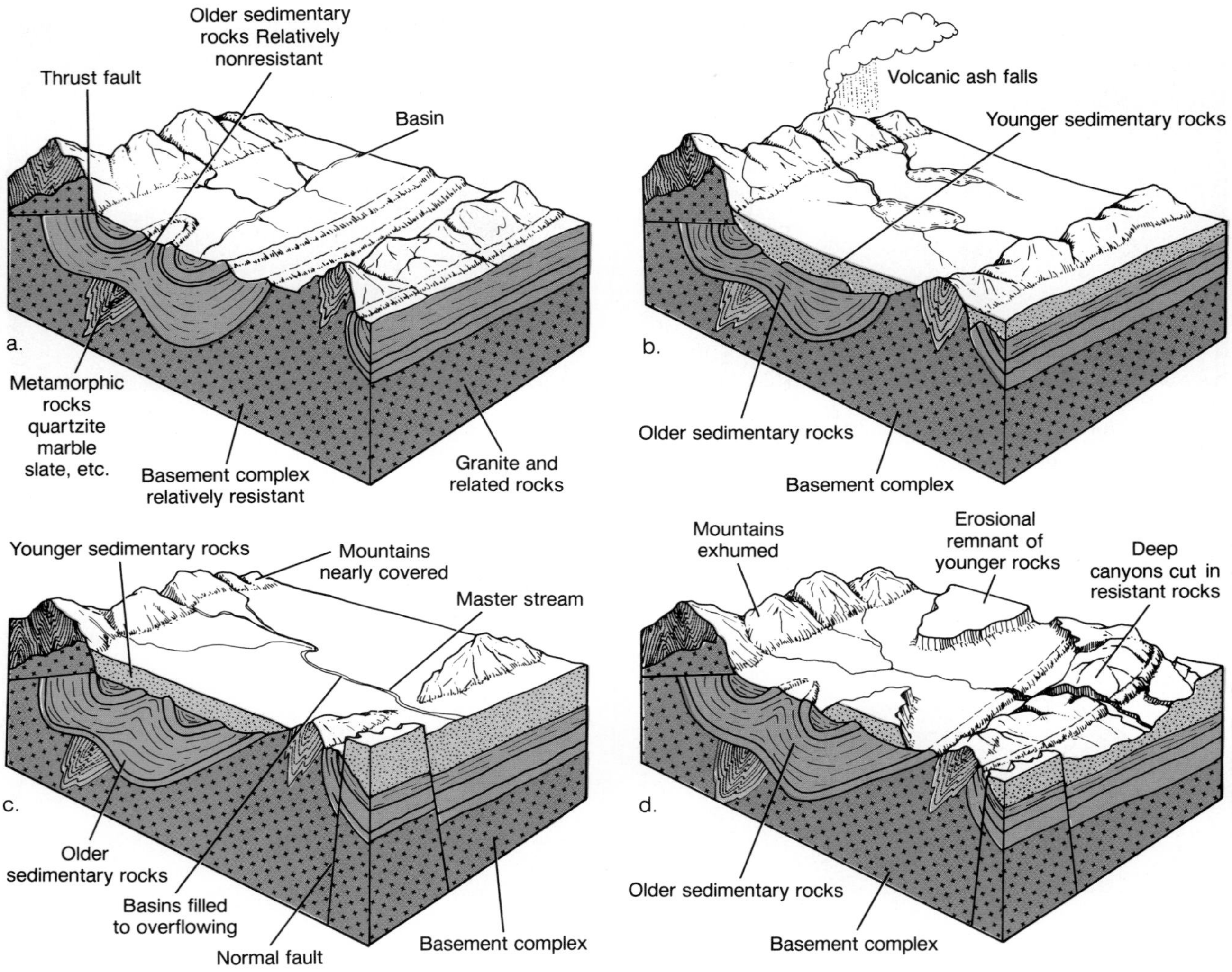

Thrust fault
Older sedimentary rocks Relatively nonresistant
Basin
a.
Metamorphic rocks quartzite marble slate, etc.
Basement complex relatively resistant
Granite and related rocks

Volcanic ash falls
Younger sedimentary rocks
b.
Older sedimentary rocks
Basement complex

Younger sedimentary rocks
Mountains nearly covered
Master stream
c.
Older sedimentary rocks
Basins filled to overflowing
Normal fault
Basement complex

Mountains exhumed
Erosional remnant of younger rocks
Deep canyons cut in resistant rocks
d.
Older sedimentary rocks
Basement complex

**Figure 15–25** a through c. Sediments eroded from the uplifted blocks of the Laramide orogen filled the intermontane basins. d. These basin sediments were partially excavated, and streams eroded deep canyons into the uplifted blocks.

**Figure 15–26** View of the Wind River Canyon, Wyoming. The canyon was eroded into the Owl Creek Mountains as shown in Figure 15-25d.

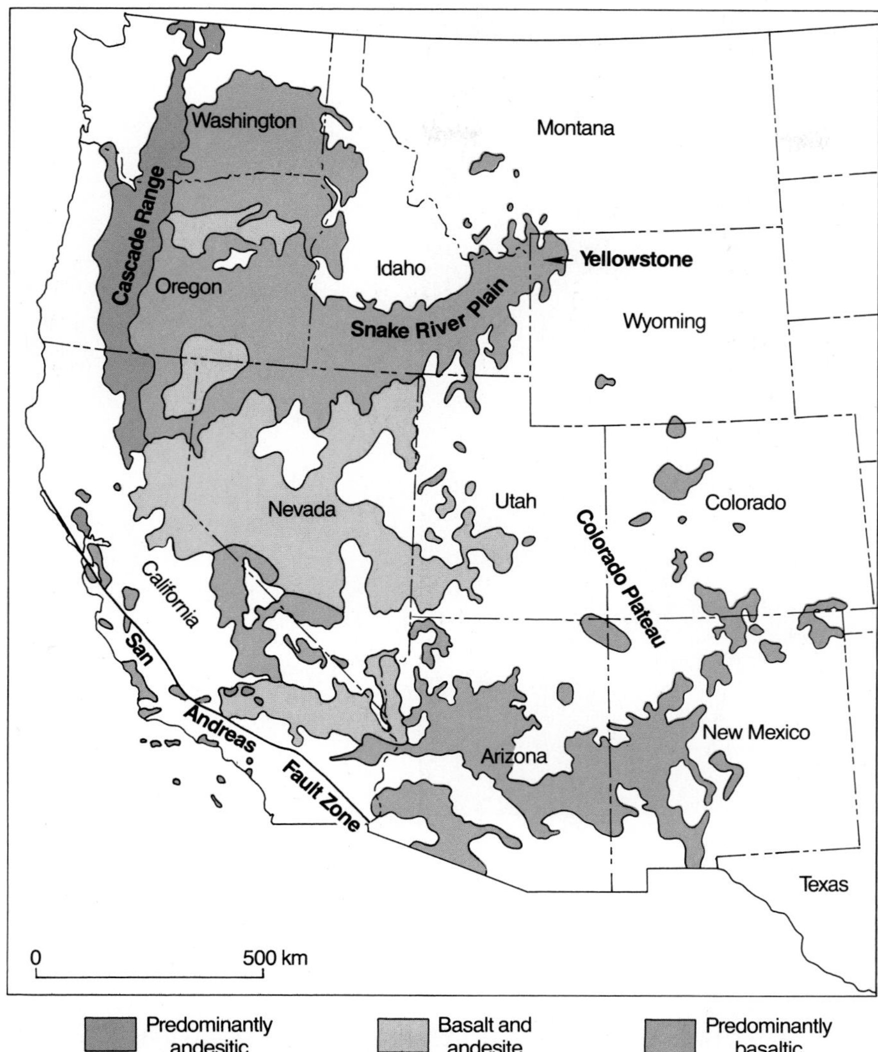


**Figure 15–27** Map showing the distribution of Cenozoic extrusive volcanics in the western United States.

15–18). Other geologists disagree with this plume hypothesis and propose that these volcanics were erupted along an intracontinental rift zone.

Bordering the Snake River Plain on the northeast is the Yellowstone Plateau (Fig. 15–18), an area of Late Pliocene and Quaternary rhyolitic and some basaltic volcanism (Fig. 15–32). As just noted, a mantle plume may be located beneath this area. Some source of heat at depth is indicated by the current hydrothermal activity, including one of the world's most famous geysers, Old Faithful (Fig. 15–33). Geophyscial evidence suggests, however, that the heat source may be only a body of intruded magma that has not yet completely cooled.

**Figure 15–28** Oligocene volcanic rocks of the San Juan Volcanic field in the San Juan Mountains, Colorado.

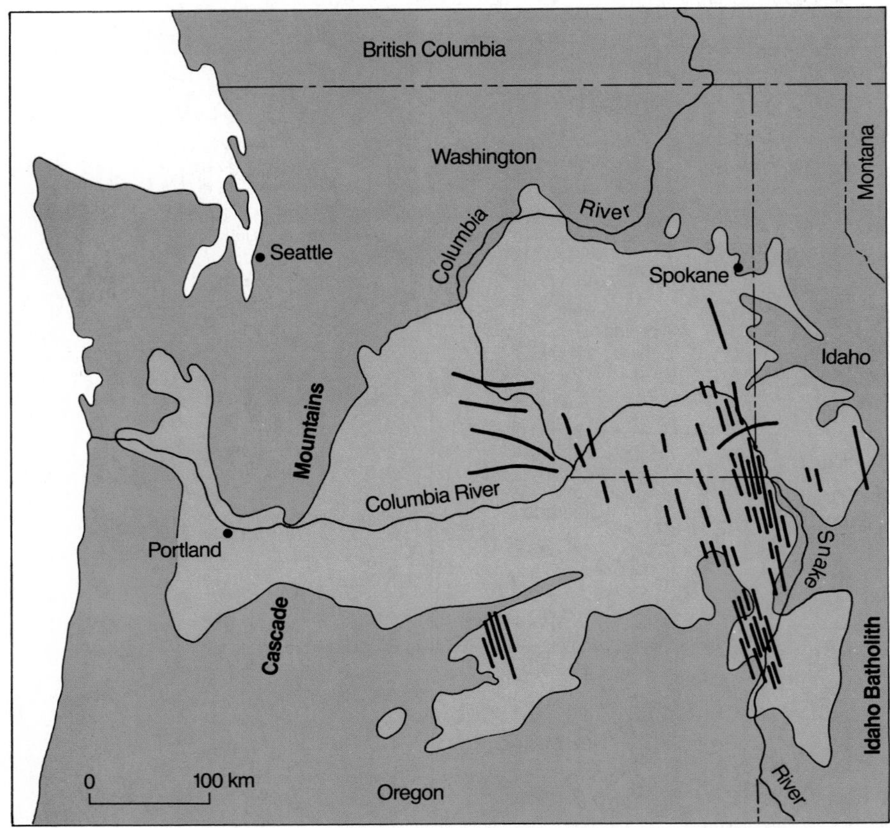


**Figure 15–29** Map showing the distribution of the Columbia River basalts. The fissures from which the basalts were erupted are shown by heavy lines.

**Figure 15–30** Columbia River basalts, Washington.

**Figure 15–31** Snake River Plain volcanic rocks. Basalt flow in Craters of the Moon National Monument, Idaho.

a.

b.

**Figure 15–32** Two basalt flows separated by the Tower Creek Conglomerate in Yellowstone National Park, Wyoming. b. The Grand Canyon of the Yellowstone River is eroded into the hydrothermally altered Yellowstone Tuff.

Some of the most majestic and highest mountains in the Cordillera are in the Cascade Range of northern California, Oregon, and Washington (Fig. 15–18, 15–34). Twice in this century Cascade volcanos have erupted. Mt. Lassen, in northern California, was active from 1914 to 1921, and more recently Mt. St. Helens, in Washington, devastated a large area when it erupted in May 1980. Cascade volcanism continues as a consequence of subduction of the Juan de Fuca plate (Fig. 15–18).

The modern volcanoes of the Cascade Range were built by andesitic volcanism during the Pliocene, Pleistocene, and Recent. The internal structure of one of these Cascade volcanos is exposed at Crater Lake, Oregon, where explosive eruptions about 6,600 years ago were followed by summit collapse and the formation of a large caldera (Fig. 15–34).

**Figure 15–33** Old Faithful geyser, Yellowstone National Park, Wyoming.

## Basin and Range Province

The continental crust of a large region centered in Nevada, but extending into adjacent states and northern Mexico, has been subjected to tensional forces and extended since the Late Miocene. Crustal extension in this region, the **Basin and Range province** (Fig. 15–18), has yielded north-south oriented, normal faults. Differential movements on these faults have produced a kind of topography that geologists have come to call *basin-and-range structure*, consisting of mountain ranges separated by broad valleys (Fig. 15–35). These mountain ranges consist of crustal blocks that moved relatively upward along parallel normal faults; fault-bounded crustal blocks that moved relatively downward formed the basins.

Before block-faulting began, the Basin and Range province was deformed in three successive orogenies: the Nevadan, the Sevier, and the Laramide. During the Early Tertiary, the entire area was an upland undergoing erosion; Early Miocene eruptions of rhyolitic lava flows and pyroclastics covered large parts of this region. Block-faulting and basaltic volcanism began during the Late Miocene. Large crustal blocks moved relatively upward and downward along normal faults, giving rise to the ranges and basins characteristic of this region. As the ranges were uplifted, they were eroded, and the erosional debris was transported into the adjacent basins

a.

b.

c.

**Figure 15–34** Vocanism in the Cascade Range of northern California, Oregon, and Washington. a. Mt. Lassen in northern California shown erupting in 1915 erupted from 1914 to 1921. b. Crater Lake, Oregon. Crater Lake is a caldera that formed when the summit of Mt. Mazama collapsed about 6600 years ago. The small cone within the caldera is a cinder cone called Wizard Island. c. May 18, 1980 eruption of Mt. St. Helens, Washington. (Photo courtesy of Lyn Topinka, U.S.G.S., David A. Johnston Cascades Volcano Observatory, Vancouver, Wash.)

where it was deposited as thick alluvial fans and playa-lake deposits.

At its western margin the Basin and Range province is bounded by a large escarpment that forms the east face of the Sierra Nevada Mountains (Fig. 15–36). The Sierra Nevada block was uplifted along normal faults during the Pliocene and Pleistocene and tilted to the west. Its crest now stands 3,000 m above the basins immediately to the east. Prior to uplift, the Basin and Range was in a subtropical climatic regime, but the rising Sierra Nevada Mountains created a rain shadow that caused the climate to become increasingly arid.

Several models have been proposed to account for basin-and-range structure. Among these are backarc spreading, spreading at the East Pacific Rise (which is thought to lie now beneath North America), a mantle plume beneath the area, and deformation related to the San Andreas transform fault (Fig. 15–37). There is little agreement on the mechanism, but we can be sure that whatever it is, it is still active.

## Colorado Plateau

The Colorado Plateau (Fig. 15–18) is a vast area of deep canyons, broad mesas, volcanic mountains, and brilliantly colored rocks (Fig. 15–38). Exposed strata range in age from Archean to Cenozoic. Remember from our discussions in Chapters 11 and 13 that the Colorado Plateau region was the site of deposition of extensive red beds during the Permian and Triassic. Many of these formations are now exposed in the uplifts and walls of canyons.

At the end of the Cretaceous, the Colorado Plateau was near sea level, as indicated by marine sedimentary

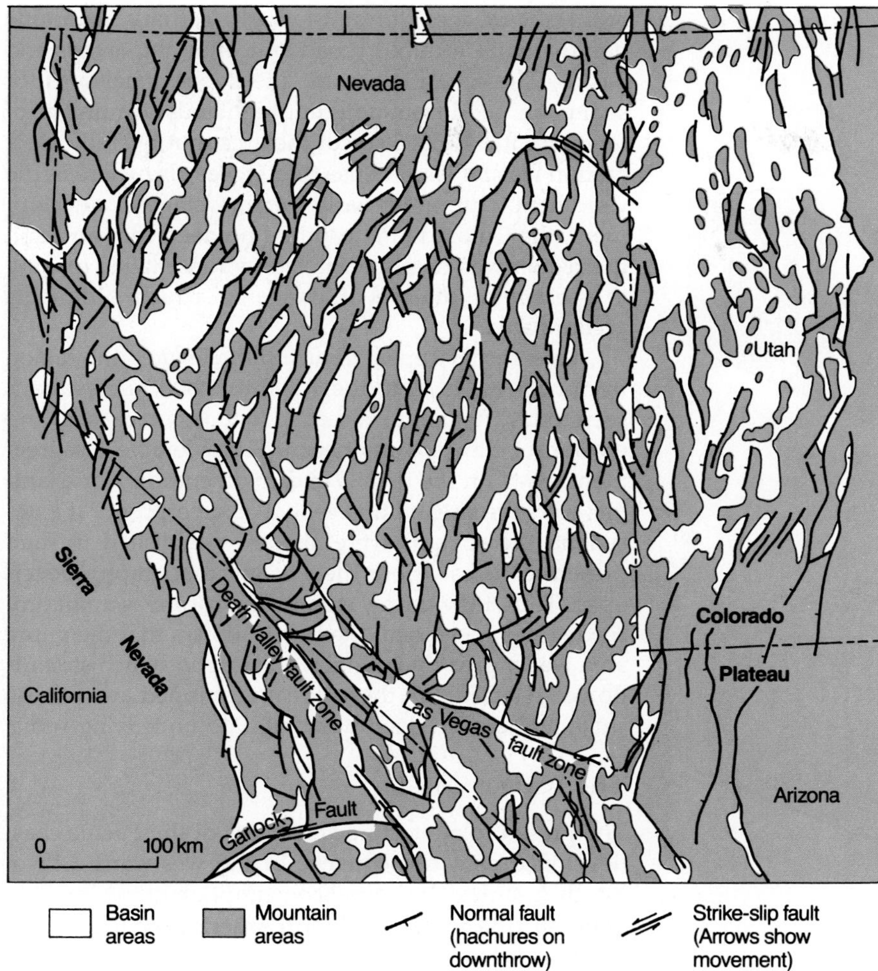


**Figure 15–35** Map of part of the Basin and Range province showing range-bounding faults and the trends of ranges and basins.

**Figure 15–36** View of the western margin of the Basin and Range province. The Sierra Nevada Mountains are bounded on the east by normal faults and rise 3,000 m above the basins to the east. (Photo courtesy of J. William Guyton, California State University, Chico)

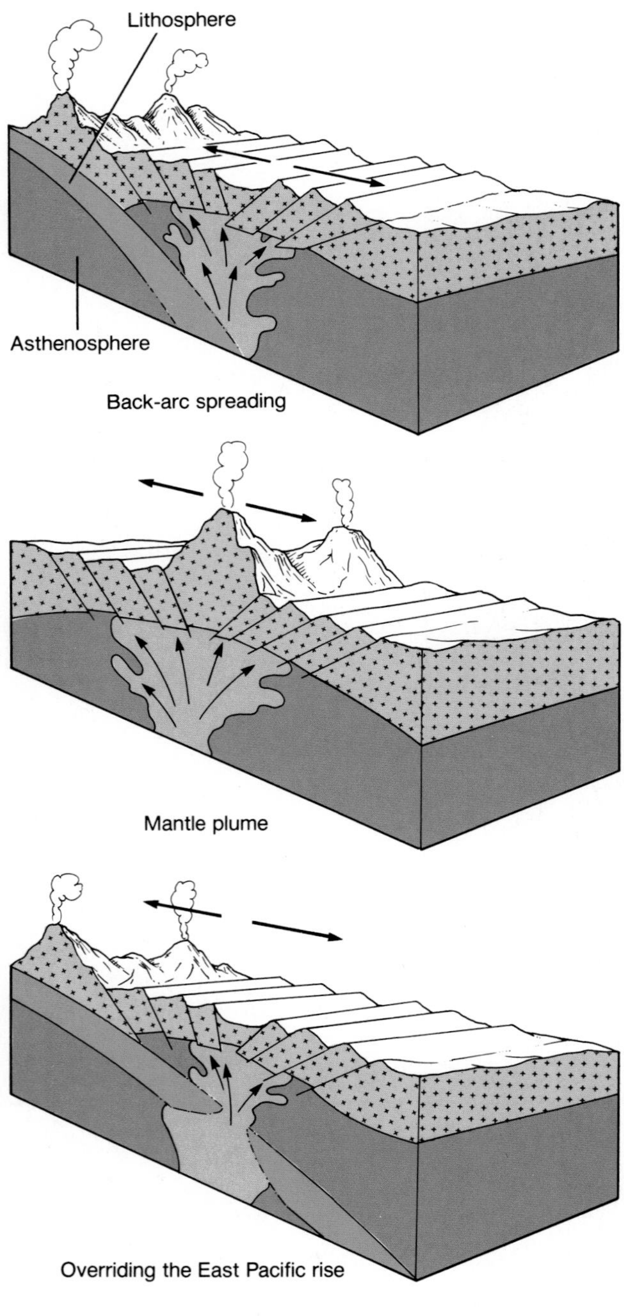


**Figure 15–37** Geologists agree that tensional forces caused by crustal extension are responsible for basin-and-range structure. There is no agreement, however, on what caused the crustal extension. These illustrations show three models that have been proposed to account for crustal extension and the resulting basin-and-range structure.

rocks of that age. During the Early Tertiary, Laramide deformation produced broad anticlines and arches and basins. These basins became the depositional sites for sediments shed from adjacent highlands. A number of large normal faults also cut the area, but overall deformation was far less intense than it was elsewhere in the Cordillera. During its Early Tertiary stage of development, the Colorado Plateau had no deep canyons, and the region was nearly encircled by mountains, making it a basin of internal drainage in which sediments accumulated.

Late Tertiary epeirogenic uplift elevated the region from near sea level to the 1,200 to 1,800 m elevations seen today. As uplift proceeded, deposition ceased, and erosion of the canyons began. Considerable disagreement exists on the details of canyon cutting. For example, were the streams antecedent or superposed? If antecedent, the streams were already established in their present courses before the existing topography developed. Thus, they simply eroded downward as uplift proceeded. If superposed, the implication is that the entire area was buried by younger strata upon which streams flowed. With uplift, these streams stripped away these strata and eroded downward into the underlying strata.

## The Pacific Coast

The present plate tectonic elements of the Pacific coast developed as a consequence of the westward drift of North America, the partial consumption of the Farallon plate, and the collision of North America and the Pacific-Farallon ridge. Prior to the Eocene, the entire Pacific coast was bounded by a subduction zone that stretched from Mexico to Alaska (Fig. 15–20). Most of the Farallon plate was consumed at this subduction zone, and now only two small remnants exist—the Juan de Fuca and Cocos plates. As discussed earlier, continuing subduction of these small plates accounts for seisimicity and volcanism in the Pacific northwest and Central America, respectively.

Westward drift of North America also resulted in its collision with the Pacific-Farallon ridge and the origin of the Queen Charlotte and San Andreas transform faults (Fig. 15–39). Since the Pacific-Farallon ridge was oriented at an angle to the margin of North America (Fig. 15–39), the continent-ridge collision occurred first in northern Canada during the Eocene and later during the Oligocene in southern California. In southern California, two triple junctions formed, one at the intersection of the North American, Juan de Fuca, and Pacific plates, and the other at the intersection of the North American, Cocos, and Pacific plates. Continued westward move-

a.

b.

c.

**Figure 15–38** Rocks of the Colorado Plateau. a. The Grand Canyon, Arizona. b. Arches National Park, Utah. Delicate Arch is shown in the distance. c. Shiprock is a volcanic neck 550 meters high in northeastern New Mexico.

ment of North America over the Pacific plate caused the triple junctions to migrate, one to the north and the other to the south, giving rise to the modern San Andreas transform fault (Fig. 15–39). A similar occurrence along the coast of Canada resulted in the origin of the Queen Charlotte transform fault (Fig. 15–39).

Where the San Andreas transform fault cuts through coastal California, it forms a complex zone of shearing. Additionally, faults such as this are seldom straight; they curve and branch into smaller faults (Fig. 15–40). Movements on complex fault systems of this type subject the crustal blocks within and adjacent to the shear zone to extensional and compressional stresses. Those areas subjected to extension form strike-slip basins, while the areas of compression are uplifted and supply sediments to the basins.

Cenozoic movements on the San Andreas transform fault account for the origin of numerous basins and

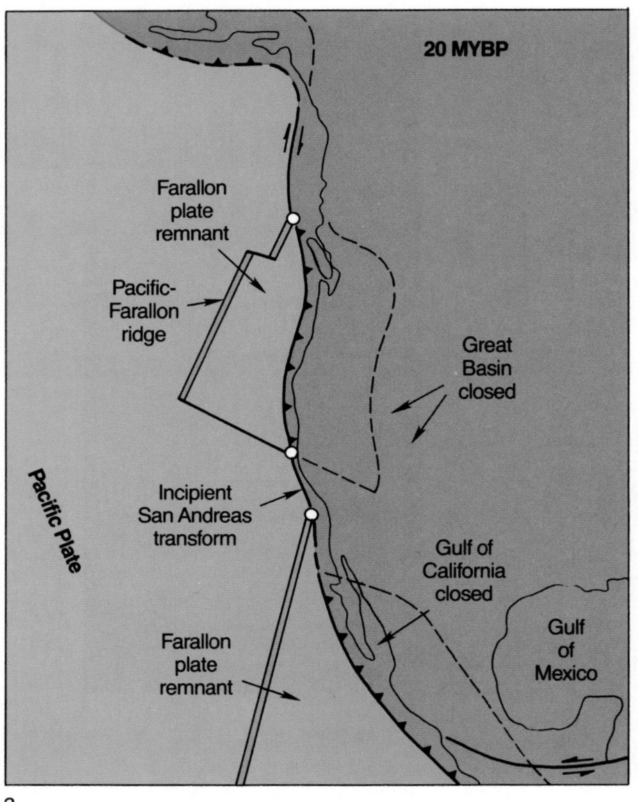


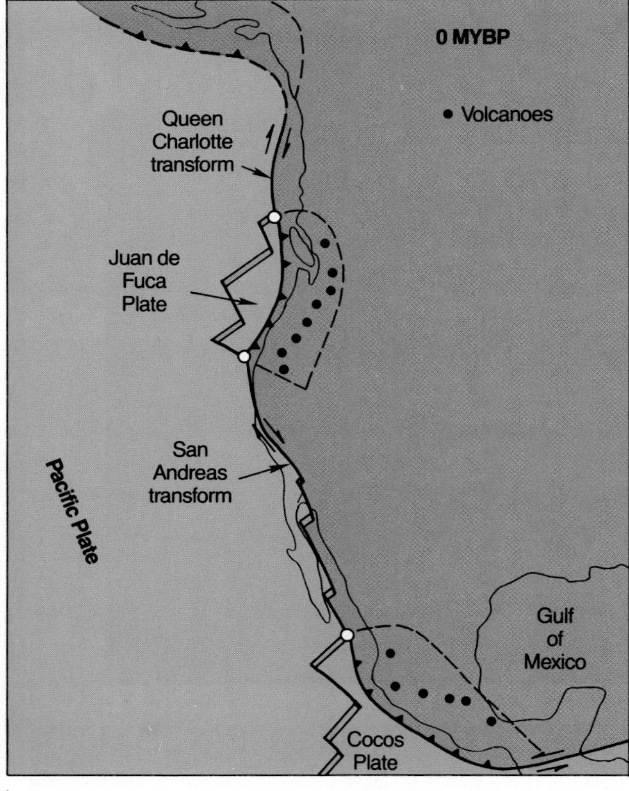


**Figure 15–39** a and b. Westward drift of North America caused it to collide with the Pacific-Farallon ridge. As North America overrode the ridge, the plate margin became bounded by transform faults rather than a subduction zone.

uplifts in California, and in adjacent areas off the southern California coast (Fig. 15–40). Many of these fault-bounded basins subsided below sea level and soon filled with turbidites and other deposits (Fig. 15–41). Many of these basins are areas of prolific oil and gas production.

## THE INTERIOR LOWLANDS

The Interior Lowlands is a vast region bounded by the Cordillera, the Canadian Shield, the Appalachians, the Ozark and Interior Low Plateaus, the Ouachita Mountains, and the Gulf Coastal Plain (Fig. 15–42). During the Cretaceous, most of the western part of this region, an area called the Great Plains, was covered by the Zuni epeiric sea. By Early Tertiary time, the Zuni Sea had withdrawn from most of North America, but a sizable remnant arm of the sea was still present in North Dakota during the Paleocene Epoch. Sediments derived from Laramide highlands located to the west and southwest were transported to this remnant sea where they were deposited in marginal marine and marine environments (Fig. 15–43).

Elsewhere within the western Interior Lowlands, sediment was also transported eastward from the Cordillera. These sediments were deposited in terrestrial environments, mostly in fluvial systems, and form large, eastward-thinning wedges that now underlie the Great Plains. The Black Hills, the easternmost uplift associated with Laramide deformation, were a sediment source for fluvial and lacustrine deposits, which were later intricately eroded into the White River Badlands of South Dakota (Fig. 15–44).

Igneous activity was not widespread in the western Interior Lowlands, but in some local areas it was signifi-

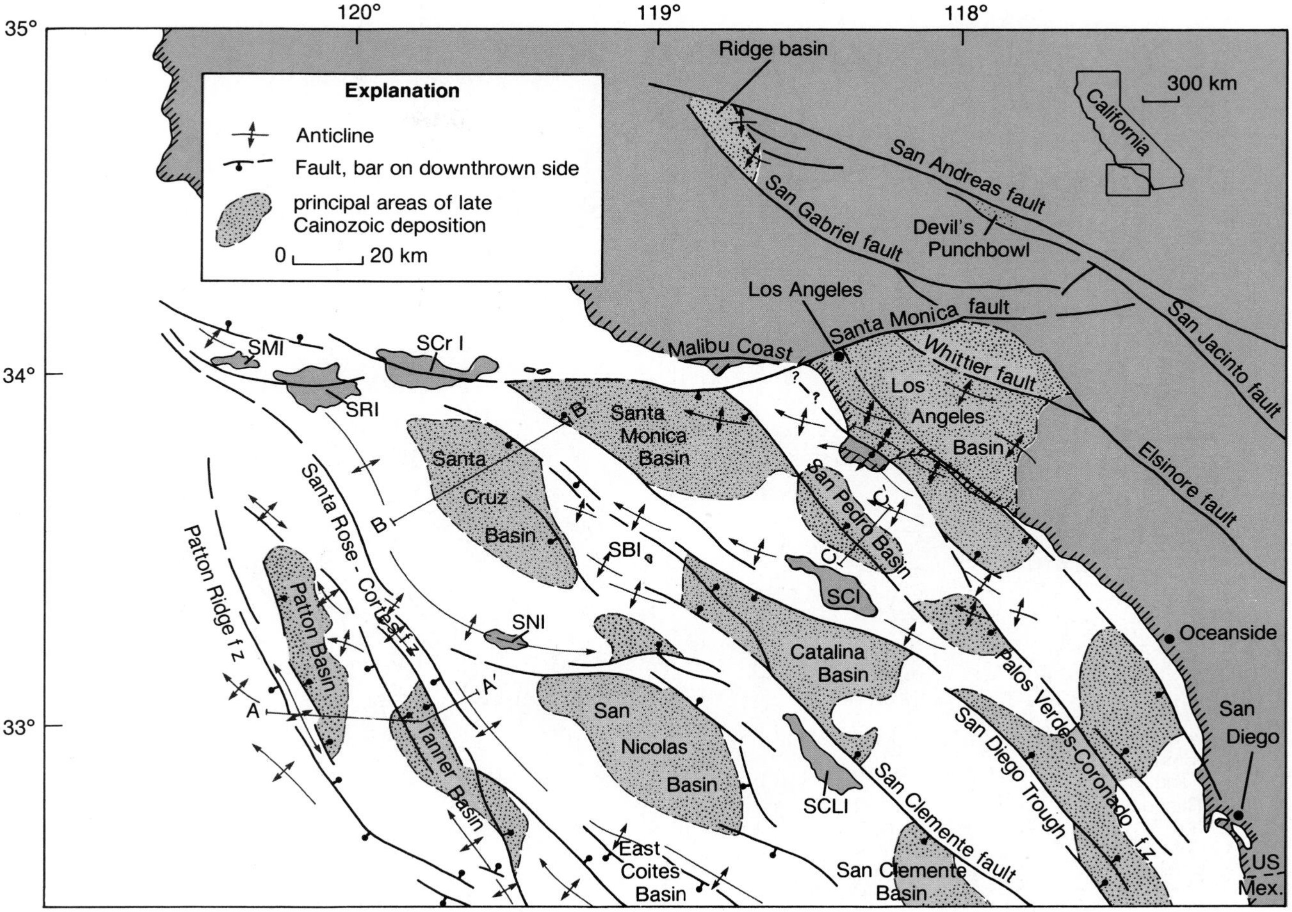

**Figure 15–40** Map showing major onshore faults in southern California and offshore Cenozoic basins. The principle areas of Late Cenozoic deposition are shown by the stipple pattern.

**Figure 15–41** Turbidite sandstone and siltstone beds of the Late Miocene Puente Formation of southern California.

cant. In northeastern New Mexico, for example, Late Tertiary extrusive volcanism produced volcanos and numerous lava flows. As a matter of fact, explosive volcanism in this area may have produced the volcanic ash that fell far to the northeast in Nebraska (see Perspective 3–1). A number of small intrusive bodies were emplaced in Colorado, Montana, South Dakota, and Wyoming. One of several in the Black Hills of South Dakota–Wyoming is called Devil's Tower (see "Prologue").

Eastward, beyond the Great Plains, Cenozoic deposits, other than those of Pleistocene glaciers (Chapter 17), are uncommon in the rest of the Interior Lowlands. Much of the Interior Lowlands was subjected to erosion during the Tertiary. Of course the eroded material had to be deposited somewhere, and that was on the Gulf Coastal Plain (Fig. 15–42).

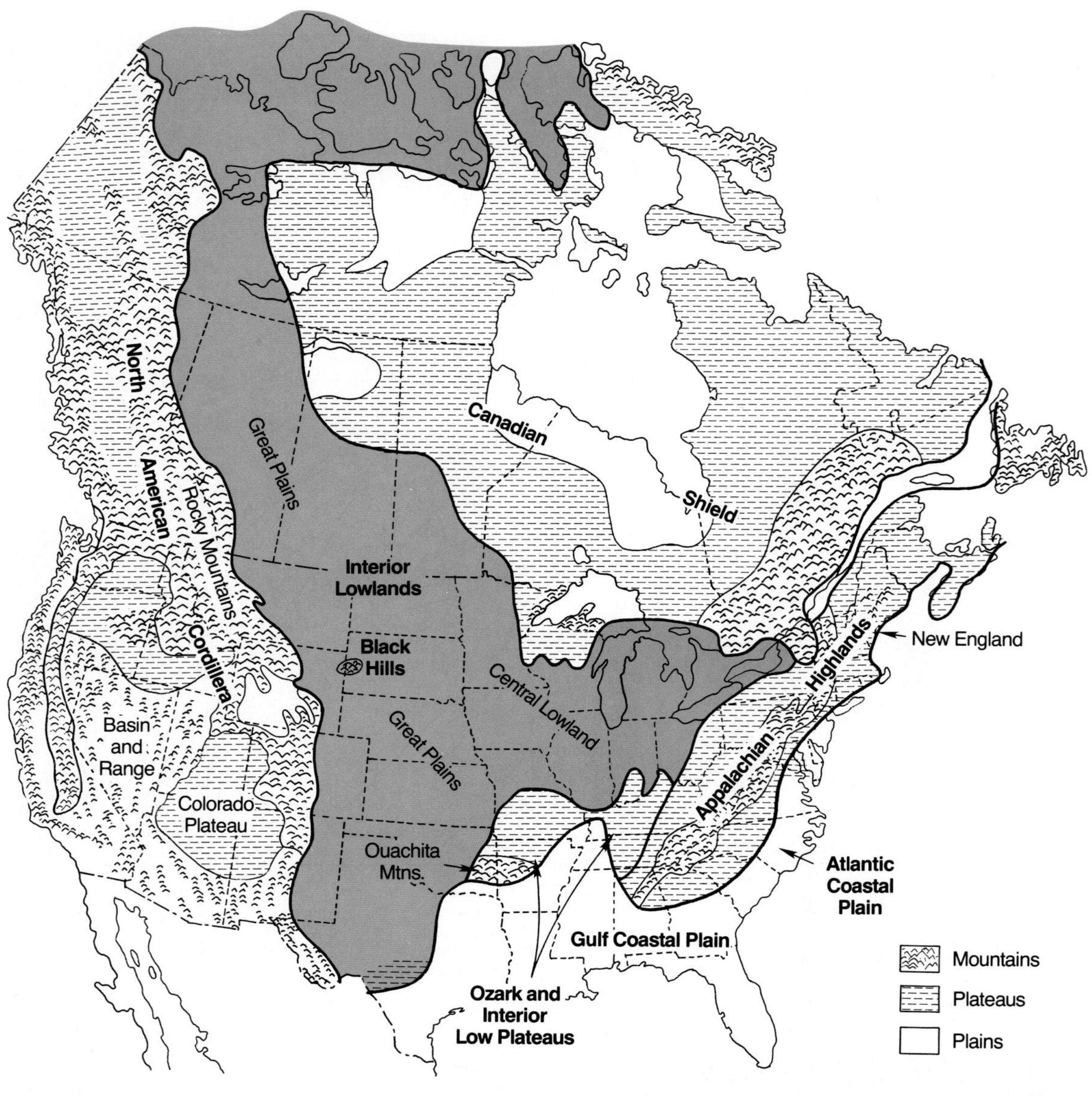


**Figure 15–42** Map showing the location of the Interior Lowlands of North America and other major physiographic provinces.

## GULF COASTAL PLAIN

Following the final withdrawal of the Cretaceous Zuni Sea, the Cenozoic **Tejas epeiric sea** (Fig.10–4) made a brief appearance on the continent. But even at its maximum extent, this sea was largely restricted to the Atlantic and Gulf Coastal plains and parts of coastal California (Fig. 15–45). In fact, its greatest incursion onto North America was in the area of the Mississippi Valley, where it extended as far north as southern Illinois. Epeiric seas were likewise restricted on the other continents except during the Early Tertiary of Europe.

Sedimentary facies development on the Gulf Coastal Plain was controlled largely by a regression of the Cenozoic Tejas epeiric sea (Fig. 10–4) . This sea extended far up the Mississippi River Valley during the

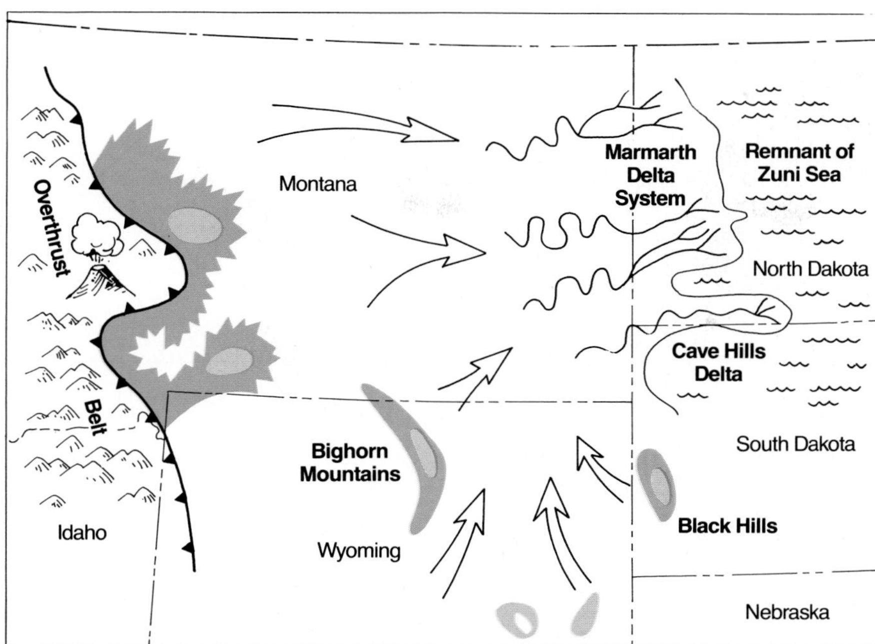


**Figure 15–43** Paleogeographic map showing some of the Laramide uplifts, and transport and deposition of sediment in the Paleocene remnant of the Cretaceous epeiric sea.

**Figure 15–44** The White River Badlands of South Dakota were formed by erosion of Cenozoic sedimentary rocks that were deposited east of the Black Hills.

Early Tertiary (Fig. 15–45), but then began its long withdrawal toward the Gulf of Mexico. Its regression, however, was periodically reversed by minor transgressions; eight transgressive-regressive episodes are recorded in Gulf Coastal Plain sedimentary rocks.

The general Gulf Coast sedimentation pattern was established during the Jurassic and persisted through the Cenozoic. Sediments were derived from the eastern Cordillera, western Appalachians, and Interior Lowlands and were transported toward the Gulf of Mexico. Deposition occurred in terrestrial, marginal marine, and marine environments. In general, the sediments form seaward-thickening wedges that grade from terrestrial facies in the north to progressively more offshore marine facies in the south (Fig. 15–46). They also clearly record the shoreward and offshore migration of the shoreline (Fig. 15–46).

Gulf Coastal Plain sediments—mostly sand, silt, shale, and lesser amounts of carbonate rocks—have been studied in considerable detail because many contain large quantities of oil and natural gas. And even though many of these rock units have no surface exposures, thousands of wells have been drilled through them, so their compositions, geometries, and stratigraphic relationships are well-known. Many of the hydrocarbon reservoirs are nearshore marine sands and deltaic sands that lie updip from the fine-grained source rocks that supply the hydrocarbons. Such oil and gas traps are called *stratigraphic traps* because they owe their existence to variations in the strata. The Wilcox Group of Eocene Age is an excellent example. Detailed studies of subsurface rocks and of Eocene-aged outcrops in Texas indicate deposition in fluvial, deltaic, barrier-island, and

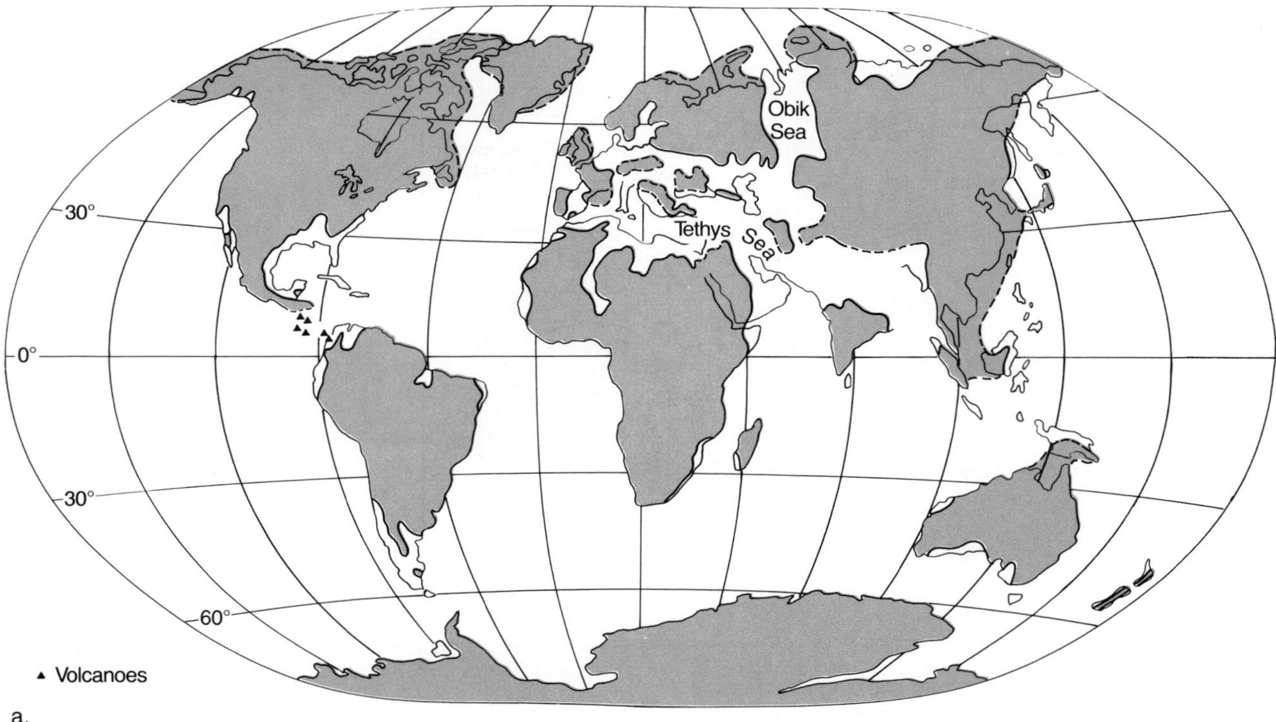


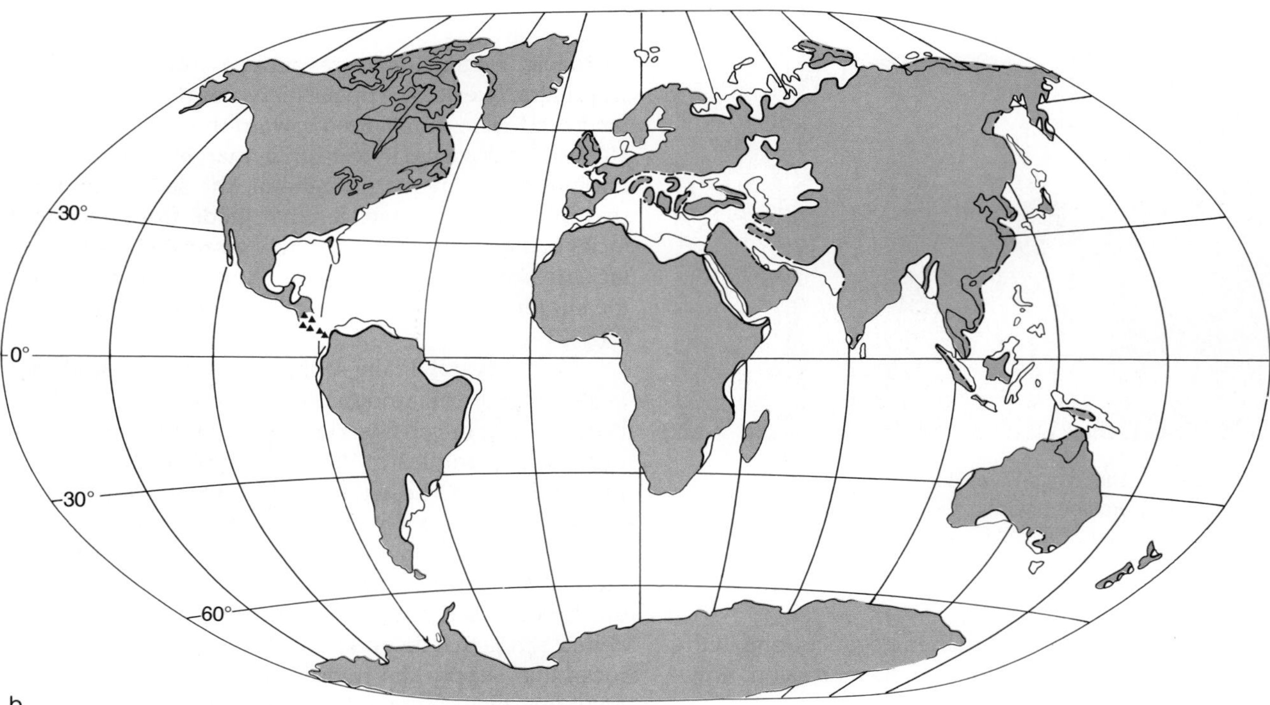


**Figure 15-45** Cenozoic epeiric seas. a. During the Early Cenozoic, the epeiric sea in North America covered the Atlantic and Gulf Coastal plains and extended far up the Mississippi River Valley. b. During Late Cenozoic time, the epeiric sea was restricted to the Atlantic and Gulf Coastal plains. By this time the epeiric seas had largely withdrawn from the other continents as well.

Quaternary
Upper Tertiary
Lower Tertiary
Cretaceous
Detrital shelf sediments
Carbonate shelf sediments
Deeper water sediments of continental slope and Sigsbee Deep

0 500km

Gulf of Mexico
Sigsbee Deep
Cuba
Yucatan

a.

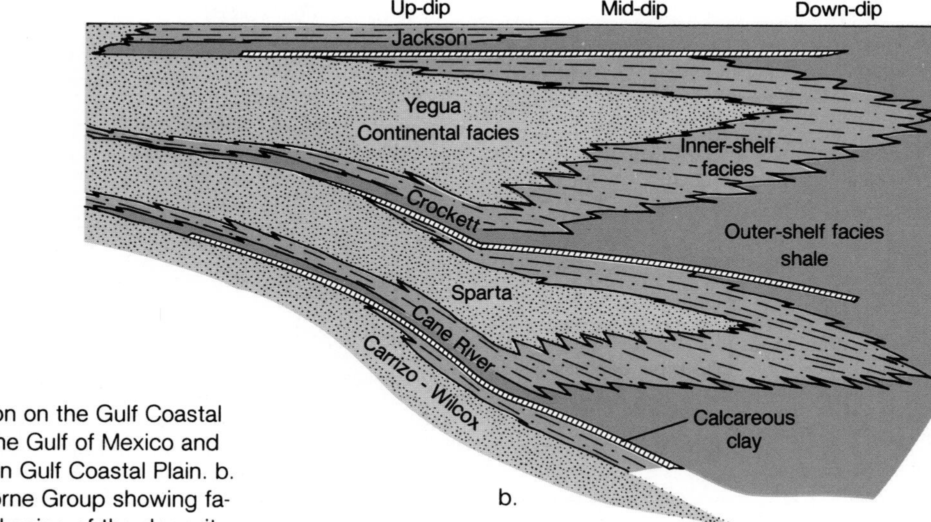

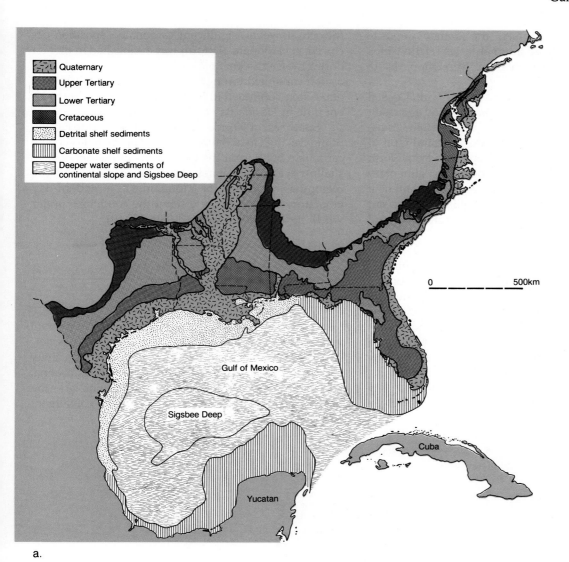


**Figure 15–46** Cenozoic deposition on the Gulf Coastal Plain. a. Depostional provinces in the Gulf of Mexico and surface geology of the northwestern Gulf Coastal Plain. b. Cross section of the Eocene Claiborne Group showing facies changes and the seaward thickening of the deposits.

beach environments. In fact, the deltaic deposits of the Wilcox Group have internal facies relationships much like those of the modern Mississippi Delta (Fig. 15–47).

*Structural traps* are common in the Gulf Coastal Plain, too, in which hydrocarbons are trapped as a result of folding, faulting, or both. In the Gulf Coast, the Jurassic Louann Salt was deposited in the ancestral Gulf of Mexico when North America first separated from North Africa (Fig. 13–7). Salt is a low-density sedimentary rock, and, when buried beneath more dense sediment, it tends to rise toward the surface in pillars called **salt domes** (Fig. 15–48). As the salt rises, it penetrates and deforms the overlying strata, creating structures that may trap oil or gas.

In addition to producing structures for the entrapment of oil and gas, the salt domes are natural resources in their own right. For example, the Avery Island salt dome in Louisiana has been mined for salt since the time of the Civil War. And a number of salt domes have a cap composed in part of sulfur, which is mined too. Salt domes have also been proposed sites for the storage of nuclear waste products.

Much of the Gulf Coastal Plain was dominated by detrital sediment deposition during the Cenozoic. In the Florida section of the coastal plain and the Gulf Coast of Mexico, however, significant carbonate deposition occurred. A carbonate platform was established in Florida during the Cretaceous, and shallow-water carbonate deposition continued through the Early Tertiary. Carbonate deposition continued in Florida, and occurs at the present time in Florida Bay and the Florida Keys.

Southeast of Florida, across the 80-km-wide Florida Strait, lies the Great Bahama Bank. This area has been a carbonate bank from the Cretaceous to present. Thick, shallow-water carbonates accumulated there, and the region has been used for a long time as a modern-day laboratory to help us understand the conditions under which limestone is deposited (Fig. 15–49).

## EASTERN NORTH AMERICA

The eastern seaboard has been a passive continental margin since Late Triassic rifting separated North America from North Africa and Europe. Some seismic activity still occurs there, (the 1886 Charleston, South Carolina, earthquake for example), but overall the region lacks the geologic activity characteristic of active, convergent margins.

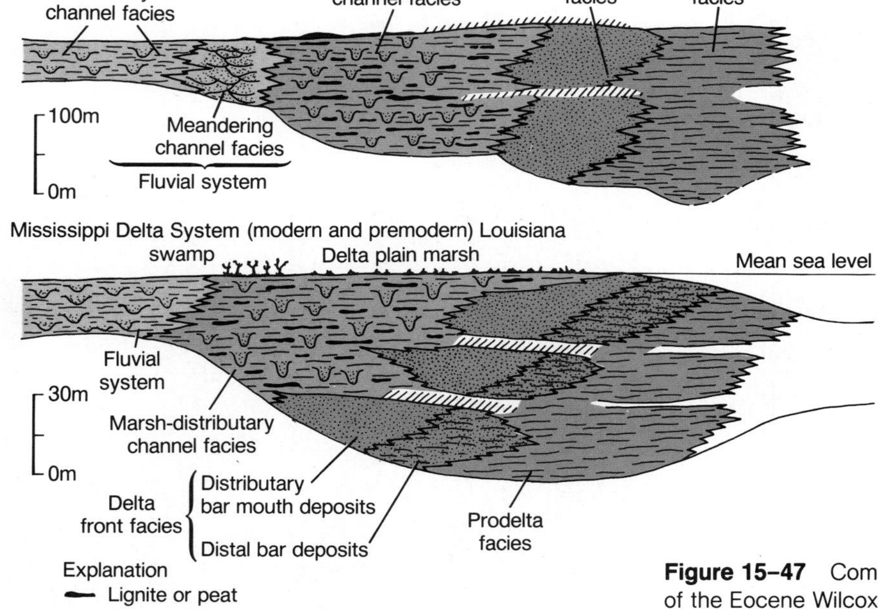


**Figure 15–47** Comparison of the Rockdale delta system of the Eocene Wilcox Group and the modern Mississippi Delta. Note that both depositional units contain similar facies.

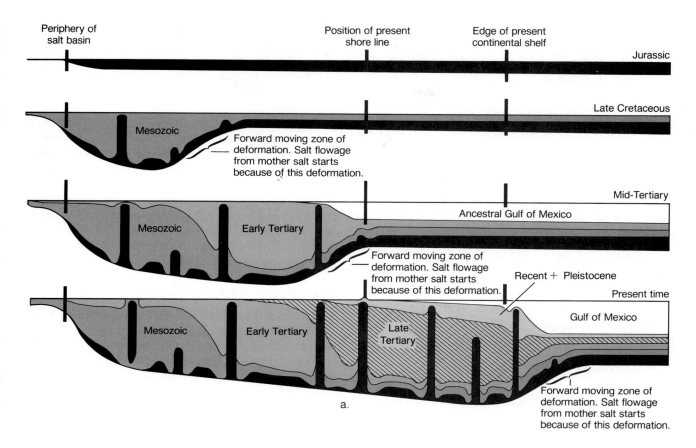

Periphery of salt basin

Position of present shore line

Edge of present continental shelf

Jurassic

Late Cretaceous

Mesozoic

Forward moving zone of deformation. Salt flowage from mother salt starts because of this deformation.

Mid-Tertiary

Ancestral Gulf of Mexico

Mesozoic

Early Tertiary

Forward moving zone of deformation. Salt flowage from mother salt starts because of this deformation.

Recent + Pleistocene

Present time

Gulf of Mexico

Mesozoic

Early Tertiary

Late Tertiary

Forward moving zone of deformation. Salt flowage from mother salt starts because of this deformation.

a.

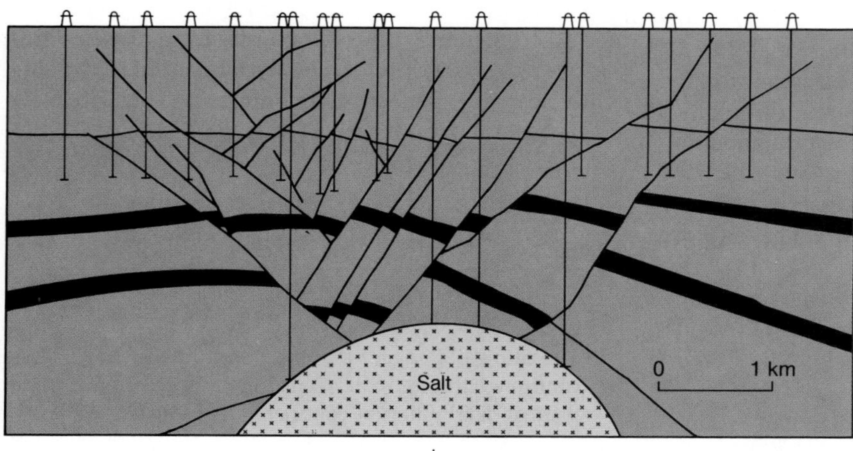


b.

**Figure 15–48** a. Illustrations showing four stages in the rise of Gulf Coast salt domes. b. Cross section of the top of the Heideberg salt dome, Mississippi, showing deformation of overlying strata.

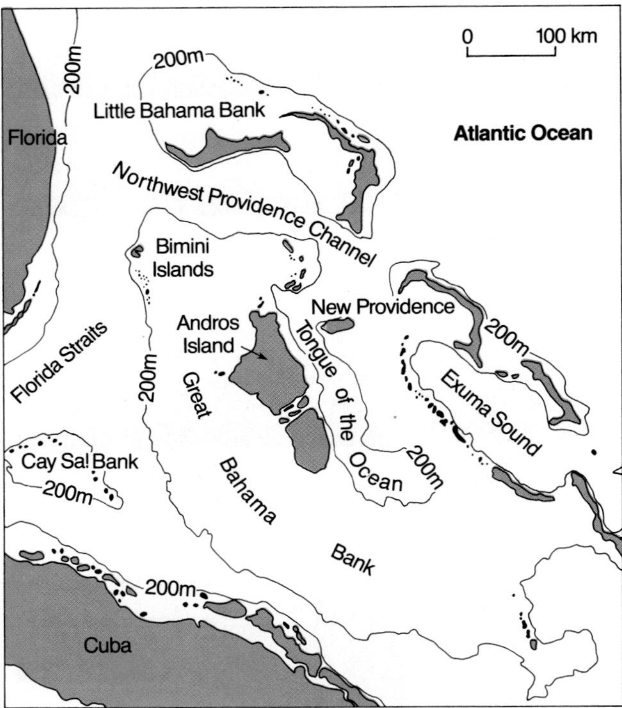


**Figure 15–49** Location of the Great Bahama Bank. This area is underlain by thick, shallow-water carbonate rocks, and carbonate depostion continues at present.

## Cenozoic Evolution of the Appalachians

The present distinctive topography of the Appalachian Mountains is the product of Cenozoic epeirogenic uplift and erosion. Recall that the Appalachians originally formed during the Late Paleozoic closure of the Iapetus Ocean, an event called the Allegheny orogeny. Block-faulting began during the Late Triassic in response to extensional forces related to the breakup of Pangaea (Fig. 13–6). In short, in what is now the eastern part of North America, an episode of rifting began that was probably very similar to the rifting that is now occurring in the Red Sea and Gulf of Aden (Fig. 15–5). Since that time, the area has been a passive continental margin.

By the end of the Mesozoic, the Appalachian Mountains had been eroded to a plain (Fig. 15–50). Cenozoic epeirogenic uplift rejuvenated the streams, which responded by renewed downcutting. As the streams cut, they were superposed on resistant strata and cut large canyons across these strata. For example, the distinctive topography of the Valley and Ridge province is the product of Cenozoic erosion and preexisting geologic structures. It consists of northeast-southwest trending ridges of resistant upturned strata and intervening valleys eroded into less resistant strata (Fig. 15–50). Large canyons were cut directly across the ridges, the result of superposition of streams.

The origin of erosion surfaces at different elevations in the Appalachians has been debated among geologists. Some geologists think these erosion surfaces are evidence of uplift followed by extensive erosion and renewed uplift. Others are convinced that they represent differential response to weathering and erosion. According to this view, a low-elevation erosion surface developed on softer strata and was eroded more or less uniformly. Higher surfaces represent the weathering and erosion response of more resistant strata.

## The Atlantic Coastal Plain

Sediments of the Atlantic Coastal Plain overlie rifted and thinned continental crust and oceanic crust (Fig. 15–51). Deposition began during the Jurassic, and even though sediments of this age are known only from a few deep wells, they are presumed to underlie the entire coastal plain. The distribution of Cretaceous and Cenozoic sediments is better known because both crop out in the coastal plain, and both have been penetrated by wells on the continental shelf.

Other than Cretaceous shelf-edge carbonates, most Atlantic coastal plain sediments are Cenozoic sandstones and shales. These sediments were derived from the Appalachian Mountains, transported east, and deposited in fluvial, nearshore marine and shelf environments. In general, the sediments form a seaward-thickening wedge, with sedimentary units dipping gently seaward. Beneath the continental shelf off New Jersey these sediments are as thick as 14 km.

## CHAPTER SUMMARY

1. The rifting of Pangaea that began during the Late Triassic continued through the Cenozoic and accounts for the present distribution of continents and oceans.

2. Cenozoic orogenic activity occurred mostly in two major belts—the Alpine-Himalayan orogenic belt and the circum-Pacific orogenic belt. Each belt is composed of smaller units called orogens.

3. The Alpine orogeny resulted from convergence of the African and Eurasian plates. Mountain building occurred

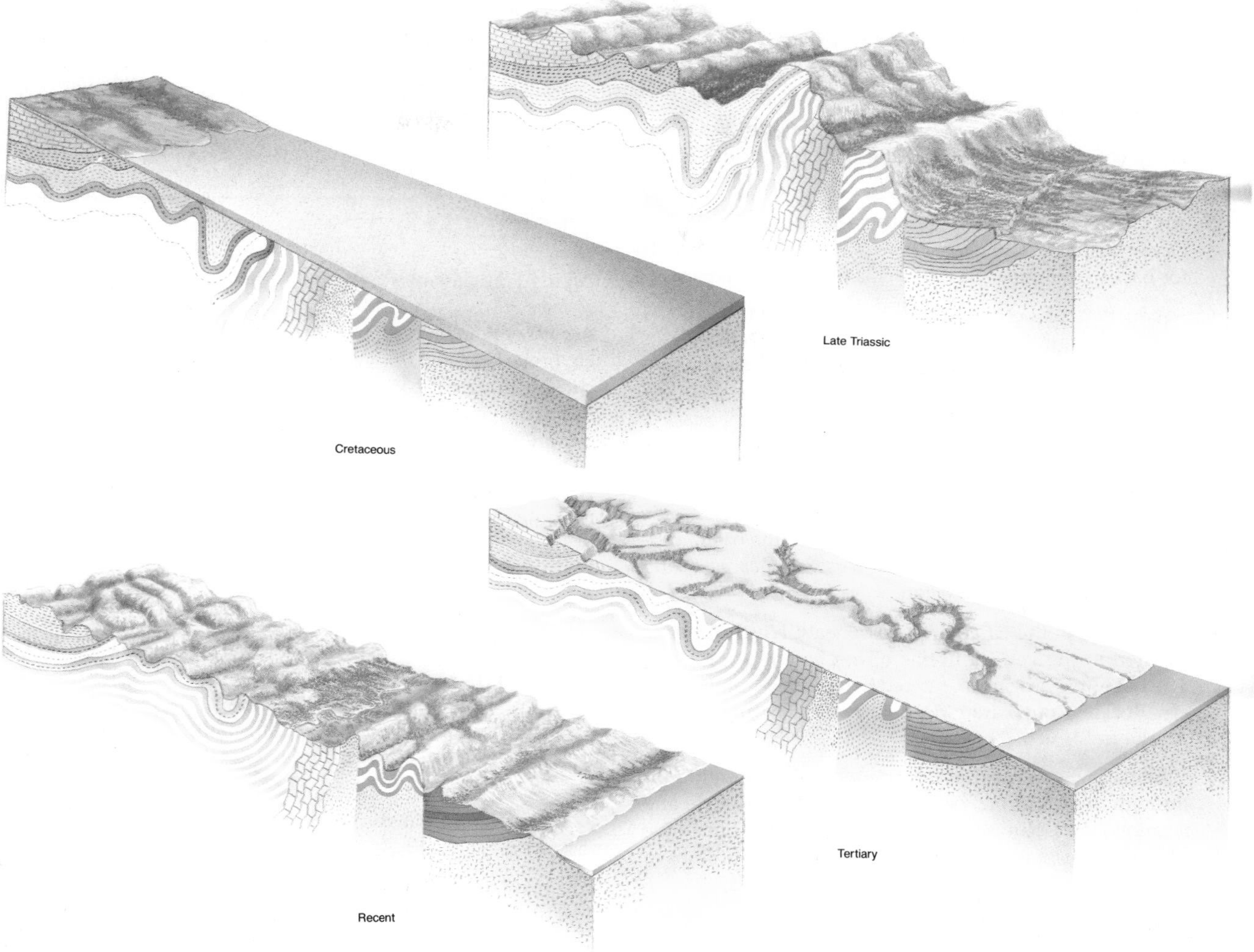

Late Triassic

Cretaceous

Tertiary

Recent

**Figure 15–50** Diagrams illustrating the origin of the present topography of the Appalachian Mountains. Erosion in response to Cenozoic uplift accounts for the present topography.

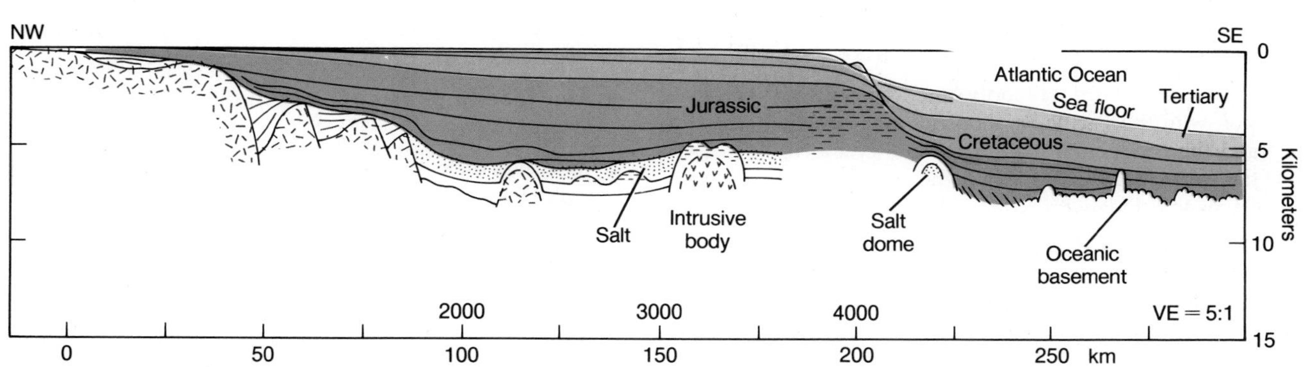


**Figure 15–51** The continental margin in eastern North America. The coastal plain and the continental shelf are covered mostly by Cenozoic sandstones and shales. Beneath these sediments are Cretaceous-aged and probably Jurassic-aged sedimentary rocks.

in southern Europe, the Middle East, and northern Africa. Plate motions also caused the closure of the Mediterranean Basin, which became a site of evaporite deposition.

4. India separated from Gondwana, moved north, and eventually collided with Asia, causing the uplift of the Himilayas.
5. Arc orogens characterize the western and northern Pacific Ocean basin. Backarc spreading appears to be responsible for backarc marginal basins such as the Sea of Japan.
6. Subduction of oceanic lithosphere occurred along the western margins of the Americas during much of the Cenozoic. This process continues beneath Central and South America, but the North American plate is now bounded mostly by transform faults.
7. The North American Cordillera is a complex mountainous region extending from Alaska into Mexico. Its Cenozoic evolution included deformation during the Laramide orogeny, extensional tectonics that formed the basin-and-range structures, intrusive and extrusive volcanism, and epeirogenic uplift and erosion.
8. Subduction of the Farallon plate beneath North America resulted in the vertical uplifts of the Laramide orogeny. The Laramide orogen is centered in the middle and southern Rockies, but Laramide deformation occurred from Alaska to Mexico.
9. Cordilleran volcanism was more or less continuous through the Cenozoic, but it varied in eruptive style and location. The Columbia River basalts represent one of the world's greatest eruptive events. Volcanism continues in the Cascades of the northwestern United States.
10. Tensional tectonics in the Basin and Range Province yielded north-south oriented, normal faults. Differential movement on these faults produced uplifted ranges separated by broad, sediment-filled basins.
11. The Colorado Plateau was deformed less than other areas in the Cordillera. Late Tertiary uplift and erosion were responsible for the present topography of the region.
12. The westward drift of North America resulted in its collision with the Pacific-Farallon ridge. Subduction ceased, and the continental margin became bounded by major transform faults, except where the Juan de Fuca plate continues to collide with North America.
13. Sediments eroded from Laramide uplifts were deposited in intermontane basins, on the Great Plains, and in a remnant of the Cretaceous epeiric sea in North Dakota.
14. Gulf Coastal Plain facies patterns were controlled by transgressions and regressions of the Cenozoic epeiric sea. A seaward-thickening wedge of sediments pierced by salt domes on the Gulf Coastal Plain contains large quantities of oil and natural gas.
15. Cenozoic epeirogenic uplift and erosion were responsible for the present topography of the Appalachian Mountains. Much of the sediment eroded from the Appalachians was deposited on the Atlantic Coastal Plain.

## IMPORTANT TERMS

Alpine-Himalayan orogenic belt
Alpine orogeny
arc orogen
backarc marginal basin
Basin and Range province
circum-Pacific orogenic belt
collision orogen
Farallon plate
Laramide orogeny
North American Cordillera
orogen
Quaternary
salt dome
San Andreas transform fault
Tejas epeiric sea
Tertiary

## REVIEW QUESTIONS

1. Explain how rifting in East Africa, the Red Sea, and Gulf of Aden helps us better understand the origin of the Atlantic Ocean basin.
2. Briefly summarize the events responsible for the Alpine orogeny.
3. How can you account for granitic intrusions in Tibet and the fact that the highest parts of the Himalayas are composed of shallow-marine sedimentary rocks?
4. What is an arc orogen? How do backarc marginal basins form? Give an example of a backarc marginal basin.
5. What accounts for the fact that the North American Cordillera in the United States is more complex than other segments of the circum-Pacific orogenic belt?
6. How does the Laramide orogen differ from typical arc orogens?
7. Why was there no intrusion of batholiths and no extrusive volcanism west of the Laramide orogen during the Early Tertiary?
8. How can the Basin and Range province be explained in the context of plate tectonics?
9. What sequence of events was responsible for the origin of the Queen Charlotte and San Andreas transform faults?
10. Laramide highlands shed large quantities of sediment. Where was this sediment deposited? Give specific examples of depositional sites.

11. Describe the sedimentary facies of the Gulf Coastal Plain. What event or process controlled facies patterns?
12. How do salt domes form, and why are oil and gas sometimes associated with salt domes?
13. Briefly outline the Cenozoic history of the Appalachian Mountains.

## ADDITIONAL READING

Armentrout, J. M., M. R. Cole, and H. Terbest, Jr. 1979. Cenozoic Paleogeography of the Western United States. *Pacific Coast Paleogeography Symposium 3*. Society of Economic Paleontologists and Mineralogists, Los Angeles, California.

Bonatti, E. 1987. The Rifting of Continents. *Scientific American*, 256, no. 3: 96–103.

Dickinson, W. R. 1978. Plate Tectonics of the Laramide Orogeny. In *Laramide Folding Associated with Basement Block Faulting in the Western United States*, ed. V. Matthew. *Geological Society of America Memoir* 151: 355–365.

Flores, R. M., and S. S. Kaplan, eds. 1985. Cenozoic Paleogeography of the West-Central United States. *Rocky Mountain Paleogeography Symposium* 3. Society of Economic Paleontologists and Mineralogists, Denver, Colorado.

Hooper, P. R. 1982. The Columbia River Basalts. *Science*, 215: 1463-1468.

Hsü, K. J. 1972. When the Mediterranean Dried Up. *Scientific American*, 227, no. 6: 27–46.

McBirney, A. R. 1978. Volcanic Evolution of the Cascade Range. *Annual Review of Earth and Planetary Sciences*, 6: 437–456.

Molnar, P. 1986. The Geologic History and Structure of the Himalaya. *American Scientist*, 74, no. 2: 144–154.

# 16

## CHAPTER OUTLINE



## PROLOGUE

*Streams flowing eastward from the Black Hills, South Dakota, and from mountain ranges in Wyoming deposited a vast blanket of Late Eocene to latest Oligocene sediments. Deposition occurred in meandering stream channels, on floodplains, and in small lakes. These deposits cover a large area but are best exposed in Badlands National Park, South Dakota (Fig. 16–1).*

*Badlands develop in dry areas where vegetation is sparse, and the rocks are nearly impermeable yet easily eroded. Rain falling on such unprotected rocks rapidly runs off and intricately dissects the surface by forming numerous small gullies, thus yielding the sharp angular slopes that characterize badlands. The Brule Formation (Fig. 16–1) is an example of such intricate dissection. In marked contrast, the underlying Chadron Formation erodes to form smooth, rounded surfaces (Fig. 16–1). Another striking feature of the rocks in the Badlands National Park is their color banding in shades of gray, red, and orange. This banding resulted from alternating episodes of deposition, soil formation, and ash-fall activity.*

*The rugged, scenic topography is reason enough to visit the park, but it has even more to offer; the rocks contain the most complete succession of mammal fossils known anywhere in the world. The Park Service has left a number of these fossils exposed but protected for viewing by park visitors. Among these mammals were various hoofed animals, including horses, camels, rhinoceroses, giant pig-like animals, and such extinct groups as oreodonts and titanotheres. Dog-like carnivores called creodonts and saber-toothed cats roamed the area, and small mammals such as rabbits, rodents, and insectivores scurried about (Fig. 16–2). In addition to mammals, fossil turtles, frogs, alligators, fish, and aquatic snails have been recovered as well.*

*Detailed studies of the fossils, ancient soil horizons* (paleosols)*, and depositional environments indicate that the climate of the area became progressively drier. During the Late Eocene, a humid forest habitat existed, but by latest Oligocene time, it was so dry that open grasslands had replaced the forests, and*

# Life of the Tertiary Period

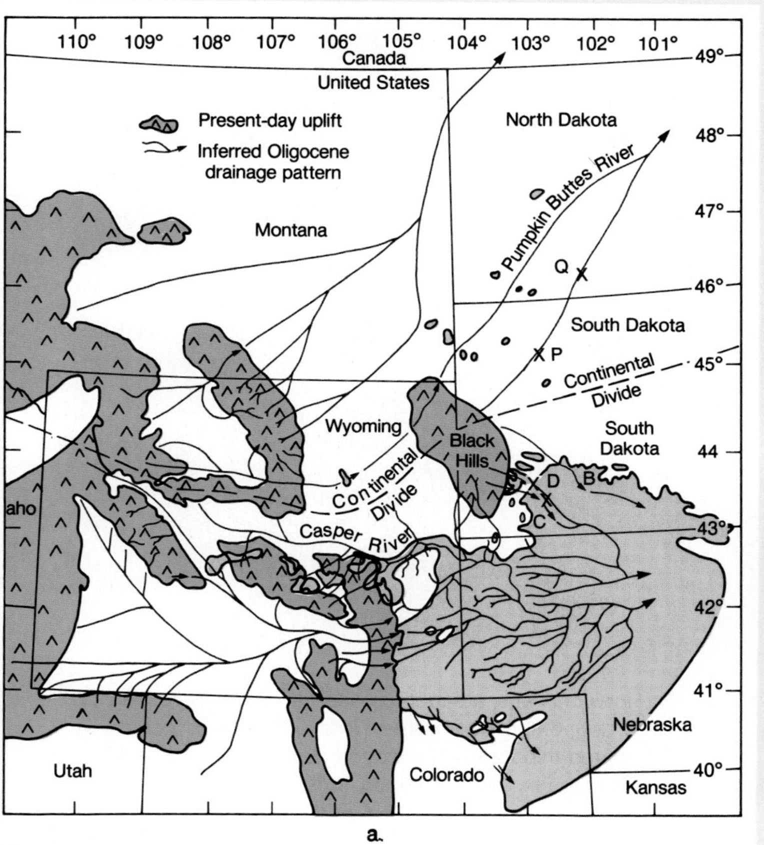

b.

c.

**Figure 16–1** a. Inferrred Oligocene drainage pattern and present surface and subsurface distribution of Oligocene rocks. b. Outcrops of the Oligocene Brule Formation (on skyline) and Chadron Formation (foreground) in Badlands National Park, South Dakota. Notice that erosion of the Brule Formation yields sharp, angular slopes while smooth, rounded slopes develop on the Chadron Formation. c. Color banding is a distinctive feature of the Oligocene rocks in Badlands National Park.

*only scattered groves of trees remained along stream margins. Concurrent changes occurred in the mammalian faunas as the Late Eocene forest-woodland fauna were replaced by mammals adapted to an open-grassland habitat. Once the mammals adapted to grasslands appeared, however, their evolutionary lineages persisted throughout the remaining time of deposition in this area.*

## INTRODUCTION

The Tertiary Period was a time during which the world flora and fauna continued to change as more familiar types of plants and animals appeared. In this chapter we

**Figure 16–2** Scene from the Early Oligocene of South Dakota showing some of the mammals that inhabited the region. Mammals shown include: 1. the rhinoceros-sized titanothere *Brontotherium*; 2. *Archaeotherium*, a giant pig-like mammal; 3. the hyena-like predator *Hyaenodon*; 4. *Hoplophoneus*, a saber-toothed cat; 5. *Hesperocyon*, an early dog; 6. the small camel *Poebrotherium*; 7. a small, flat-footed rhinoceros *Hyracodon*; 8. the pig-like oreodonts represented by *Merycoidodon*; 9. the squirrel-like rodent *Ischyromys*; and 10. the three-toed horse *Mesohippus*.

are concerned mainly with the adaptive radiation of mammals, and especially some of the more familiar types such as carnivores, elephants, and hoofed mammals. Recall from Chapter 14 that mammals evolved from cynodonts during the Late Triassic but were small and not very diverse through the rest of the Mesozoic. However, following the Mesozoic extinctions, mammals diversified and soon became the most abundant land-dwelling vertebrate animals.

Although we emphasize mammalian evolution in this chapter, one should be aware of other important events. The flowering plants continued to dominate land-plant communities, modern birds appeared early in the Tertiary, and some marine invertebrates continued to diversify. By the end of the Tertiary Period, many of the plant and animal genera living today were present.

## MARINE INVERTEBRATES AND PHYTOPLANKTON

The Tertiary marine ecosystem was populated mostly by those plants, animals, and single-celled organisms that survived the terminal Mesozoic extinction event. Gone were the ammonites, rudists, and most of the planktonic foraminifera. The major Cenozoic invertebrate groups that were especially prolific were the foraminifera, radiolarians, corals, bryozoans, mollusks, and echinoids. For the marine invertebrate community in general, there was increased provincialism during the Cenozoic because of changing ocean currents and latitudinal temperature gradients. In addition, the Cenozoic marine invertebrate faunas became more familiar in appearance.

Entire families of phytoplankton became extinct at the end of the Cretaceous. Only a few species in each major group survived into the Tertiary. These species diversified and expanded during the Cenozoic, perhaps because of decreased competitive pressures. The coccolithophores, diatoms, and dinoflagellates all recovered from their Late Cretaceous reduction in numbers to flourish during the Cenozoic. The diatoms were particularly abundant during the Miocene, probably because of the increased volcanism during this time. The volcanic ash provided increased dissolved silica in sea water, needed for diatoms to construct their skeletons. Massive Miocene diatomites have been found in California (Fig. 16–3).

**Figure 16–3** Outcrop of diatomite from the Miocene-aged Monterey Formation, Newport Lagoon, California.

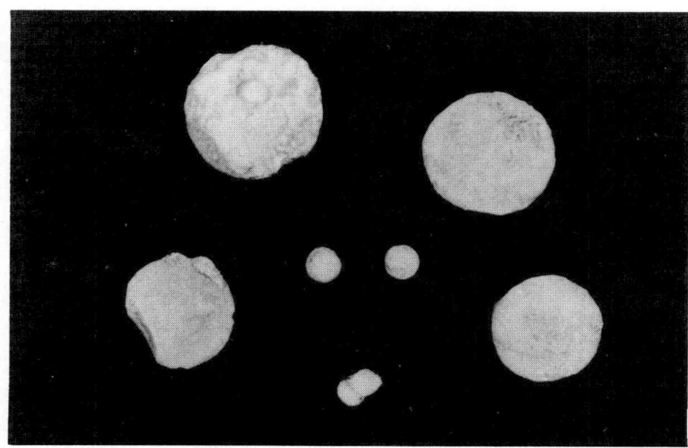

**Figure 16–5** The larger foraminifera such as *Nummulites* have shells several centimeters in diameter. Nummulitid forams were particularly abundant during the Early Tertiary and their shells make up thick beds of limestone in Europe, Asia, and North Africa. (Photo courtesy of Paul Brenckle, Amoco Research Co.)

The foraminifera comprised a major component of the Cenozoic marine invertebrate community. Though dominated by relatively small forms (Fig. 16–4), it included some exceptionally large forms that lived in the warm waters of the Cenozoic Tethys Sea. Shells of these larger forms accumulated to form thick limestones, some of which were used by the ancient Egyptians to construct the Sphinx and the Pyramids of Gizeh (Fig. 16–5).

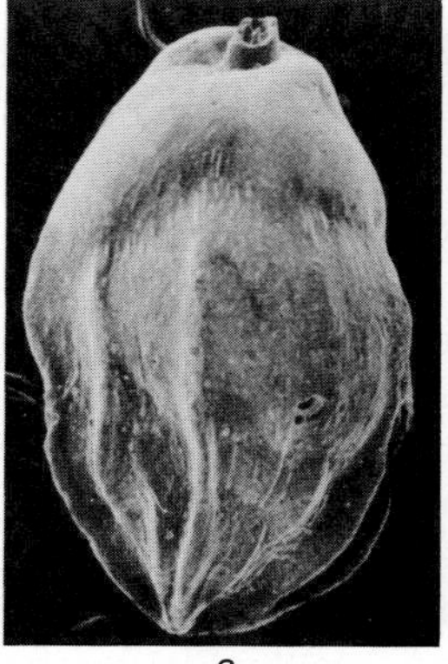

a.

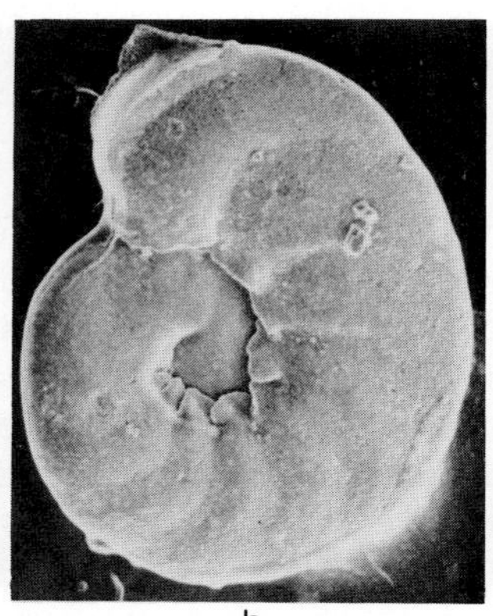

b.

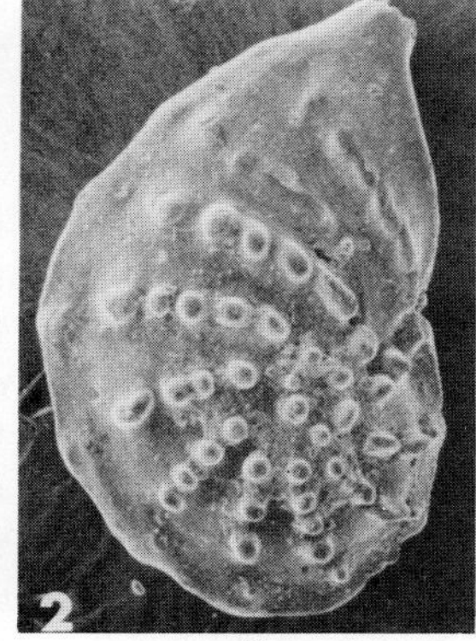



c.

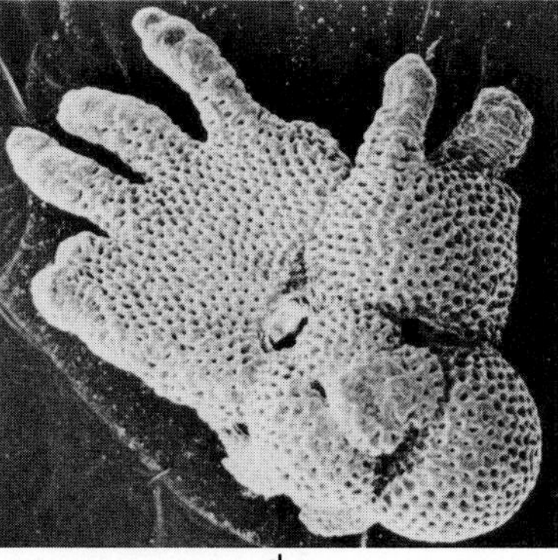

d.

**Figure 16–4** Foraminifera of the Cenozoic Era. a through c are benothonic forms. a. *Uvigerina cubana*, Late Miocene, California. b. *Cibicides americanus*, Early Miocene, California. c. *Lenticulina mexicana*, Eocene, Louisiana. d. A planktonic form, *Globigerinoides fistulosus*, Pleistocene, South Pacific Ocean. (Photos courtesy of B. A. Masters, Amoco Production Co.)

The corals were perhaps the main beneficiary of the terminal Cretaceous extinctions. Having relinquished their reef-building role to the rudists in the mid-Cretaceous, corals again became the dominant reef-builders during the Cenozoic. They formed extensive reefs in the warm waters of the Cenozoic oceans and were particularly prolific in the Caribbean and Indo-Pacific regions (Fig. 16–6).

Other suspension feeders such as the bryozoans and crinoids were also abundant and successful during the Tertiary as well as the Quaternary. The bryozoans, in particular, were very abundant (Fig. 16–7). Perhaps the least important of the Cenozoic marine invertebrates were the brachiopods, with fewer than 60 genera surviving today.

Just as during the Mesozoic, bivalves and gastropods were two of the major groups of marine invertebrates during the Tertiary, and they had a markedly modern appearance. Following the extinction of the ammonites and belemnites at the end of the Cretaceous, the Cenozoic cephalopod fauna consisted of nautiloids and shell-less cephalopods such as squids and octopuses.

The echinoids continued their expansion in the infaunal habitat and were particularly prolific during the Tertiary. New forms such as sand-dollars evolved during this time from biscuit-shaped ancestors (Fig.16–7).

a.

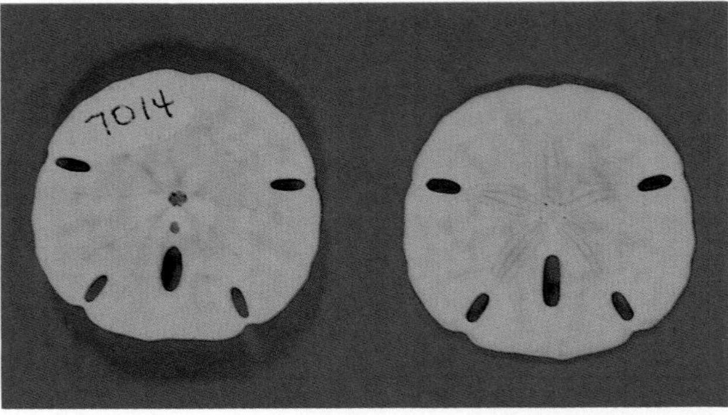


b.

**Figure 16–7** The suspension feeding bryozoans were a successful and abundant group during the Cenozoic. Echinoids were particularly abundant during the Tertiary and new infaunal forms such as sand-dollars evolved from their Mesozoic biscuit-shaped ancestors. a. Upper view of *Lunnulites*, a bryozoan from the Eocene of Alabama. b. Sand-dollars, a type of echinoid that evolved during the Tertiary. (Photo a courtesy of Alan H. Cheetham, Smithsonian Institution)

## TERTIARY BIRDS

Fully modern birds appeared during the Early Tertiary, including the first members of many living orders and families such as owls, hawks, ducks, penguins, and vultures. Today, birds vary considerably in size and adaptations, but the basic skeletal structure of birds has not changed significantly throughout the Tertiary. This uniformity is not surprising, since most birds are fliers, and adaptations for flying impose limitations on variations in structure.

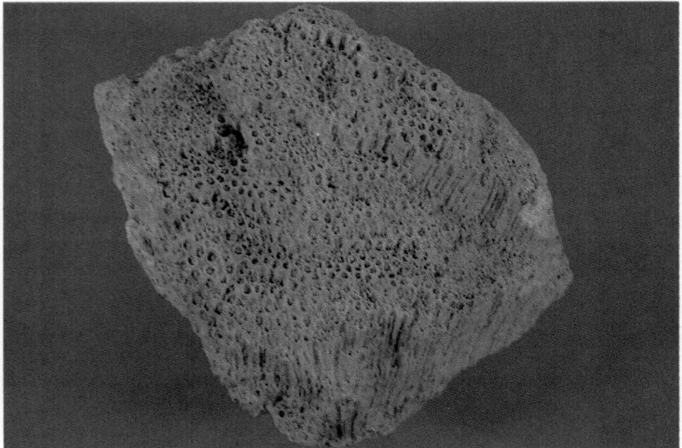

**Figure 16–6** Corals such as this colonial scleractinian were the dominant reef-builders during the Cenozoic Era.

Birds adapted to numerous habitats and increased in diversity through the Tertiary and into the Pleistocene. Since then, diversity of birds has decreased slightly. One of the more remarkable early adaptations was the development of large, flightless predatory birds. One of these flightless predators, *Diatryma*, was more than 2 m tall,

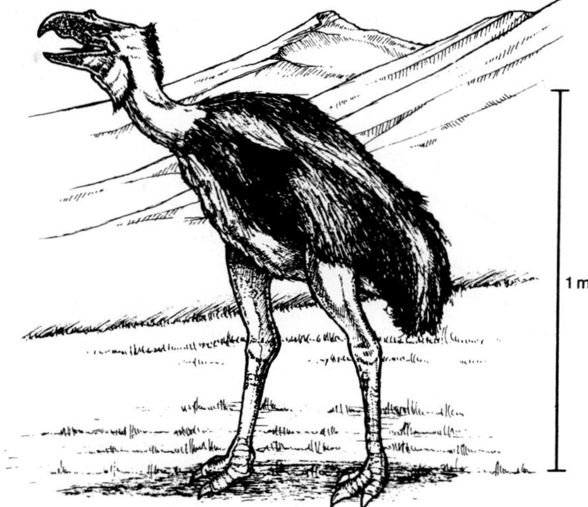


**Figure 16–8** Large, flightless predatory birds of the Early Tertiary. a. *Diatryma*, Eocene, North America. b. *Phororhacus*, Oligocene-Miocene, South America.

and had a huge head and beak, toes with large claws, and small vestigial wings (Fig. 16–8). Its legs were massive and short, indicating that *Diatryma* did not move very fast, but the early mammals upon which it preyed were slow-moving as well. *Diatryma* and related genera were widespread during the Early Tertiary of North America and of Europe, but eventually became extinct; apparently, they were replaced by the developing mammalian carnivores.

Large flightless birds still exist, but all are herbivores, and all are now restricted to the southern continents. These birds, ostriches, rheas, and others, probably evolved from a common ancestor sometime before Gondwana split into its individual parts, which would account for their seemingly anomalous present distribution (Fig. 16–9).

Flightless birds notwithstanding, the true success among birds belongs to fliers. They did not undergo much structural change during the Cenozoic, but a bewildering array of adaptive types arose. In fact, birds have been as successful as mammals; they have exploited the aerial habitat as fully as the mammals have adapted to terrestrial habitats.

## THE AGE OF MAMMALS BEGINS

For more than 100 million years mammals coexisted with dinosaurs; yet, their fossil record indicates that during this entire time they were neither diverse nor abundant (Table 14–5). Even during the Late Cretaceous, very near the end of the Age of Reptiles, only eight families of marsupial and placental mammals existed. This situation was soon to change. Mesozoic extinctions eliminated the dinosaurs and many of their relatives, thereby opening up numerous adaptive opportunities that were quickly exploited by the mammals. The Age of Mammals had begun.

The evolutionary history of the mammals is better known than the history of any of the other classes of vertebrates. Two factors account for this. First, Cenozoic terrestrial deposits are more common than Mesozoic and Paleozoic deposits, and many of these deposits are accessible at the surface or in the shallow subsurface (Fig. 16–1). Accordingly, mammals have a comparatively good fossil record.

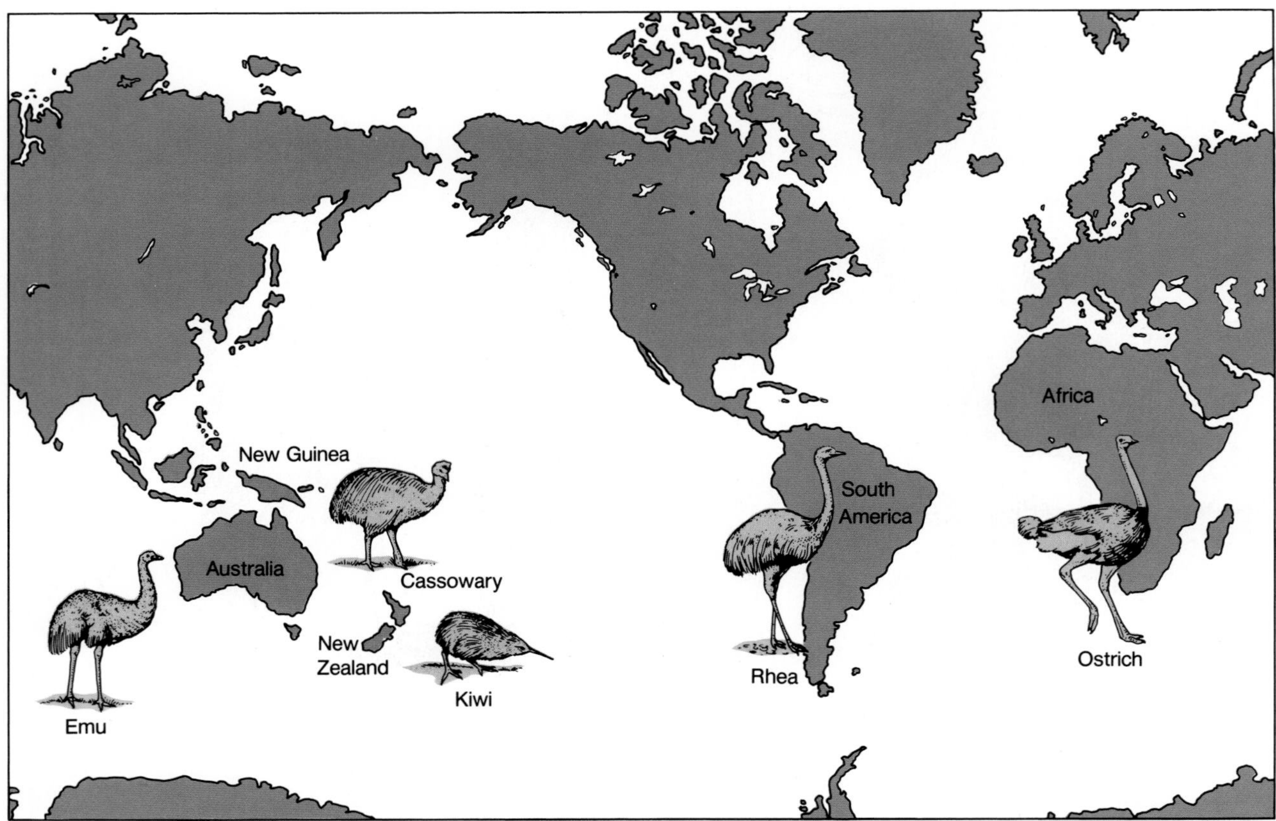


**Figure 16–9** Geographic distribution of flightless birds. All except the kiwi are large birds. The ancestors of these birds probably evolved when the Southern Hemisphere continents were assembled into a single landmass, Gondwana. When the continents separated, these ancestral birds were isolated from one another and each followed its own evolutionary course.

The second factor accounting for our greater understanding of mammalian evolution is that mammal fossils are easier to identify. In Chapter 14 we noted that differentiation of teeth was a trend established in the therapsids (Table 14–4). In true mammals the teeth are fully differentiated into distinctive types, and the chewing teeth, called **molars**, differ in each of the mammalian orders. In fact, a single mammal molar is commonly sufficient to identify the genus from which it came.

## PLACENTAL MAMMALS

Among living mammals only **monotremes** lay eggs, while marsupials and placentals give birth to live young. **Marsupials** are born in a very immature, almost embryonic condition, and then undergo further development in the mother's pouch. **Placentals**, on the other hand, have developed a different reproductive method. In these animals the amnion of the amniote egg (Fig. 12–36) has fused with the walls of the uterus, forming a *placenta.* Nutrients and oxygen are carried from mother to embryo through the placenta, permitting the young to develop much more fully before birth.

The phenomenal success of placental mammals is related in part to their reproductive method. A measure of this success is the fact that well over 90 percent of all mammals, fossil and living, are placentals. Recall from Chapter 14 that placental mammals of the order **Insectivora** were present during the Late Cretaceous. The great adaptive radiation of placental mammals began with shrew-like members of this order (Fig. 16–10).

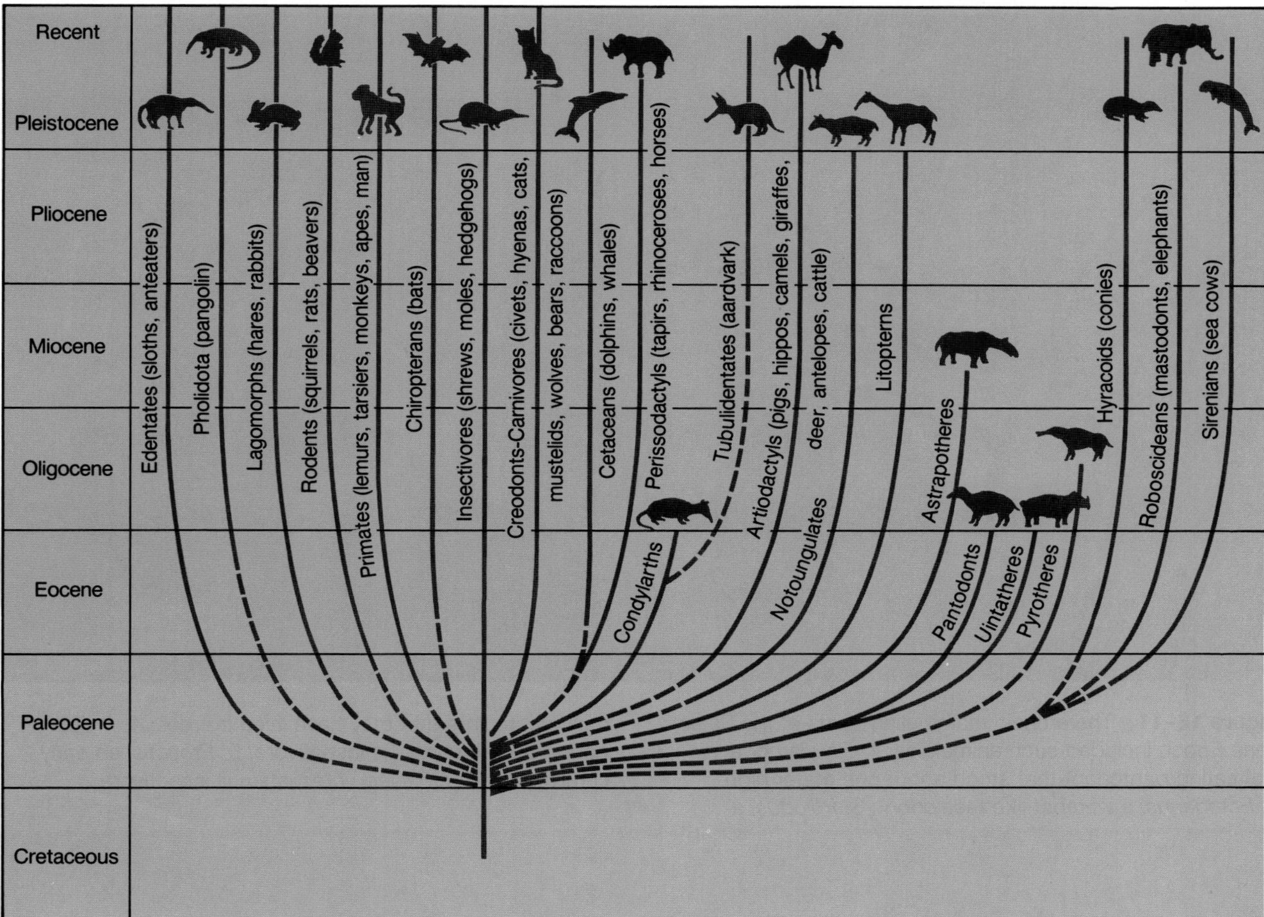


**Figure 16–10** The adaptive radiation of placental mammals began with shrew-like members of the order Insectivora. Diversification began during the Late Cretaceous and continued through the Cenozoic.

## DIVERSIFICATION OF PLACENTAL MAMMALS

A major adaptive radiation of mammals began during the Paleocene and continued through the Cenozoic Era. Paleocene mammalian faunas are considered archaic because they were composed of primitive mammals, some of which—including marsupials, insectivores, and the rodent-like multituberculates—were holdovers from the Mesozoic Era (Fig. 16–11). Another Mesozoic holdover, the order Condylartha, was particularly important because **condylarths** were ancestors of several other mammalian orders (Fig. 16–10, Perspective 16–1).

Thirteen new orders of mammals first appeared during the Paleocene (Fig. 16–12, Table 16–1). Among these were the first rodents, rabbits, primates, and carnivores, but many of the other new orders soon became extinct. Most of these Paleocene mammals, even those assigned to living orders, had not yet become clearly differentiated from their insectivore ancestors, and the differences between herbivores and carnivores were slight. Large mammals did not evolve until the Late Paleocene, and the first giant terrestrial mammals appeared during the Eocene (Fig. 16–13).

**Figure 16–11** The archaic mammalian fauna of the Paleocene Epoch included such animals as: 1. *Barylambda,* a leaf-eating pantodont that stood about one meter high; 2. *Ectoganus,* a wombat-like taeniodont; 3. *Ptilodus,* a multituberculate that may have lived in trees; 4. *Prodiacodon,* a hedgehog-like insectivore; 5. *Protictis,* an early carnivore; and 6. *Palaeoryctes,* a small insectivore.

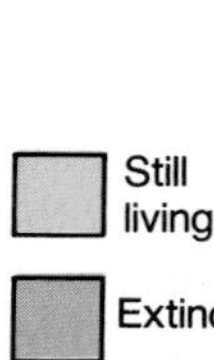


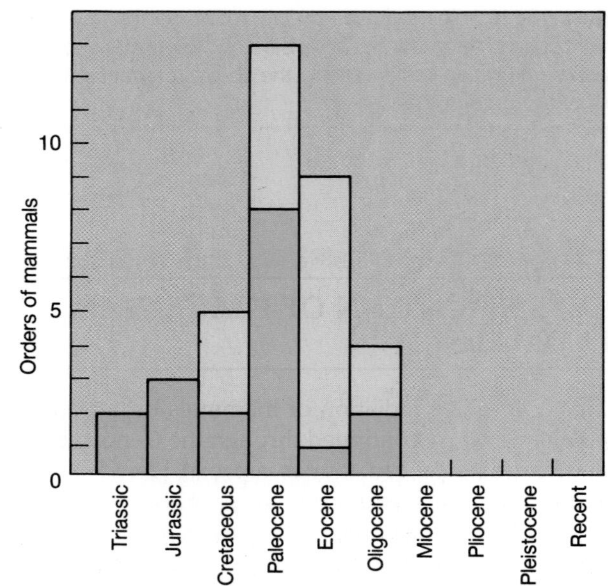


**Figure 16–12** Times during which the mammalian orders first appeared in the fossil record. Most of the living orders appeared during the Paleocene and Eocene, but many Paleocene orders became extinct. All living orders of mammals had evolved by Oligocene time.

**Figure 16–13** Middle Eocene mammals of western North America. The uintatheres (1) were rhinoceros-sized mammals with three pairs of bony protuberances on the skull and saber-like upper canine teeth. Some modern antlerless deer have enlarged canines used to strip edible bark from trees. Perhaps uintatheres used their canines for the same purpose. Other mammals shown include: 2. *Trogosus*, a bear-like mammal with clawed feet; 3. *Sinopa*, a fox-sized carnivore; 4. *Orohippus*, a small three-toed horse; 5. *Palaeosyops*, an early titanothere; and 6, *Patriofelis*, a bear-sized carnivore.

Diversification continued during the Eocene, when nine more orders evolved; all but one of these orders still exist (Fig. 16–12, Table 16–1). Most of the modern mammalian orders were present by Eocene time; yet if we could somehow go back and visit the Eocene, we would probably not recognize many of these animals. Some would be at least vaguely familiar to us, but the horses, camels, rhinoceroses, and elephants, for example, would bear little resemblance to their living descendants.

By Oligocene time, all of the living orders of mammals had evolved (Fig. 16–12, Table 16–1). Nevertheless, diversification continued, but it was within the existing orders as more familiar families and genera appeared (Fig. 16–14). Miocene and Pliocene mammals were mostly animals that we could easily identify. Perhaps a few would be rather odd-looking to us, but they would nevertheless be easily recognized as horses, deer, cats, elephants, and so on. A few unfamiliar types of mammals still existed during the Pliocene, but many of the genera then present were the direct ancestors of Pleistocene and Recent forms.

## Rodents, Rabbits and Bats

The adaptive success of rodents is manifested by the fact that the order Rodentia alone accounts for more than 40

**Table 16–1** The orders of fossil and living mammals and their geologic ranges.

| *Classification* | *Living species* | *Common names* | Triassic | Jurassic | Cretaceous | Paleocene | Eocene | Oligocene | Miocene | Pliocene | Pleistocene | Recent |
|---|---|---|---|---|---|---|---|---|---|---|---|---|
| Class Mammalia | | | | | | | | | | | | |
| Subclass Prototheria | | | | | | | | | | | | |
| Order Docodonta | | | x | x | | | | | | | | |
| Order Triconodonta | | | x | | | | | | | | | |
| Order Monotremata | 3 | Spiny anteater, platypus | | | x | x | x | x | x | x | x | x |
| Subclass Allotheria | | | | | | | | | | | | |
| Order Multituberculata | | | | x | x | x | x | | | | | |
| Subclass Theria | | | | | | | | | | | | |
| Order Symmetrodonta | | | | x | x | | | | | | | |
| Order Pantotheria | | | | x | x | | | | | | | |
| Order Credonta | | | | | x | x | | | | | | |
| Order Condylartha | | | | | x | x | x | x | | | | |
| Order Marsupialia | 242 | Opossum, kangaroo, koala | | | x | x | x | x | x | x | x | x |
| Order Insectivora | 406 | Shrew, mole, hedgehog | | | x | x | x | x | x | x | x | x |
| Order Xenungulata | | | | | | x | | | | | | |
| Order Taeniodonta | | | | | | x | x | | | | | |
| Order Tillodontia | | | | | | x | x | | | | | |
| Order Dinocerata | | | | | | x | x | | | | | |
| Order Pantodonta | | | | | | x | x | | | | | |
| Order Astraptheria | | | | | | x | x | x | x | | | |
| Order Notoungulata | | | | | | x | x | x | x | x | x | |
| Order Litopterna | | | | | | x | x | x | x | x | x | |
| Order Rodentia | 1687 | Beaver, squirrel, mouse, rat, porcupine, gopher | | | | x | x | x | x | x | x | x |

percent of all living mammal species. Rodents are extremely diverse and have adapted to a wide range of environments. Following their appearance during the Paleocene, they rapidly diversified and soon occupied many of the microhabitats unavailable to larger animals (Fig. 16–15).

Superficially, rabbits resemble rodents, but they differ in details of their anatomy and have an independent evolutionary history. Rabbits are known from the Paleocene but did not become abundant until the Oligocene. Long, powerful hind limbs for hopping and speed is the most obvious evolutionary trend in rabbits.

The oldest known fossil bat is a specimen from the Eocene Green River Formation of Wyoming (Fig. 16–16). Apart from forelimbs modified into wings, bats differ little from their insectivore ancestors. In fact, the bat's wing is simply the basic vertebrate forelimb modified to support a wing membrane. As opposed to pterosaurs and birds, bats evolved to use the entire hand for wing support (Fig. 16–16; compare with Fig. 14–26).

## Primates

The order **Primates** includes the tarsiers, lemurs, and lorises (the so-called lower primates) and living monkeys,

| *Classification* | *Living species* | *Common names* | Triassic | Jurassic | Cretaceous | Paleocene | Eocene | Oligocene | Miocene | Pliocene | Pleistocene | Recent |
|---|---|---|---|---|---|---|---|---|---|---|---|---|
| Subclass Theria | | | | | | | | | | | | |
| Order Lagomorpha | 63 | Pika, rabbit, hare | | | | x | x | x | x | x | x | x |
| Order Primates | 166 | Lemur, tarsier, loris, monkey, human | | | | x | x | x | x | x | x | x |
| Order Edentata | 31 | Anteater, sloth, armadillo | | | | x | x | x | x | x | x | x |
| Order Carnivora | 284 | Dog, cat, bear, skunk, seal, weasel, hyena, raccoon, panda | | | | x | x | x | x | x | x | x |
| Order Pyrotheria | | | | | | | x | x | | | | |
| Order Chiroptera | 853 | Bats | | | | | x | x | x | x | x | x |
| Order Dermoptera | 2 | Flying lemurs | | | | | x | x | x | x | x | x |
| Order Cetacea | 84 | Whale, dolphin, porpoise | | | | | x | x | x | x | x | x |
| Order Tubulidentata | 1 | Aardvark | | | | | x | x | x | x | x | x |
| Order Perissodactyla | 16 | Horse, rhinoceros, tapir | | | | | x | x | x | x | x | x |
| Order Artiodactyla | 171 | Pig, hippo, camel, deer, elk, bison, cattle, sheep, antelope | | | | | x | x | x | x | x | x |
| Order Proboscidea | 2 | Elephant | | | | | x | x | x | x | x | x |
| Order Sirenia | 5 | Dugong, sea cow, manatee | | | | | x | x | x | x | x | x |
| Order Embrithopoda | | | | | | | | x | | | | |
| Order Desmostyla | | | | | | | | x | x | | | |
| Order Hyracoidea | 11 | Hyrax | | | | | | x | x | x | x | x |
| Order Pholidota | 8 | Scaly anteater | | | | | | x | x | x | x | x |

apes, and humans (the higher primates) (Fig. 16–17). Much of the primate story is better told in Chapter 17 where we consider human evolution, so in this chapter we shall be brief. Primitive primates may have evolved by the Late Cretaceous, but they were undoubtedly present by the Early Paleocene. Most Paleocene primates were mouse-sized, long-snouted animals that looked much like the insectivores from which they evolved. By Eocene time, large primates had appeared (Fig. 16–18), and fairly modern-looking lemurs and tarsiers are known from Asia and North America. By Oligocene time, primitive New and Old World monkeys had appeared in South America and Africa, respectively. The hominoids, the group containing apes and humans, appeared during the Miocene (Chapter 17).

## Carnivores

Many land-living carnivorous mammals depend on speed, agility, and intelligence to catch their prey and have teeth modified for a diet of meat. The canine teeth are well-developed because they are used to make the kill, and carnivores have specialized shearing teeth called **carnassials** (Fig. 16–19). During the Paleocene, carnivorous mammals called **creodonts** and **miacids**

## Perspective 16-1

# Condylarths

A large part of the archaic Paleocene mammalian faunas consisted of members of the order Condylartha. Condylarths evolved from insectivores during the Late Cretaceous, diversified throughout the Paleocene, then dwindled to extinction during the Miocene. During their existence condylarths occupied all of the continents except Antarctica and Australia.

At the beginning of the Paleocene, most condylarths were small, but by the end of that epoch some had attained bear-size, and one Eocene genus was huge—its skull measured 1 m long. Both herbivorous and carnivorous condylarths are recognized (Fig. 1). None, however, showed the strong specializations that characterize true mammalian carnivores (order Carnivora), nor the specializations of the hoofed herbivorous mammals.

Our discussion of condylarths, a seemingly obscure, extinct group, is not without a purpose; condylarths are the probable ancestors of many other orders of mammals (Fig. 16–10). Whales, elephants, and sea cows appear to have evolved from condylarths, as well as several extinct orders of South American mammals. In addition, the small Eocene condylarth *Hyopsodus* (Fig. 2), if not directly ancestral to the even-toed hoofed mammals (order Artiodactyla), was very close to that ancestry. And, the sheep-sized, Paleocene-Eocene genus *Phenacodus* (Fig. 2) was the likely ancestor of the odd-toed hoofed mammals (order Perissodactyla).

Condylarths became extinct, but they were successful nevertheless. They existed for more than 50 million years, and were one of the few Early Cenozoic mammalian orders to become diverse, abundant, and widespread. If the number of descendant groups is a measure of evolutionary success, then condylarths were certainly successful.

**Figure 1** a. *Meniscotherium* from the Paleocene was a racoon-sized, herbivorous condylarth. b. The carnivorous condylarth *Mesonyx* was about the size of a large woverine and lived from Eocene to Oligocene time.

**Figure 2** Scene from the Early Eocene showing the condylarths *Hyopsodus* (bottom) and *Phenacodus* (top). *Hyopsodus* had clawed feet and may have been a tree climber. This animal or one very much like it was probably the ancestor of even-toed hoofed mammals (order Artiodactyla). *Phenacodus*, the likely ancestor of odd-toed hoofed mammals (order Perissodactyla), was sheep-sized and had hoofed, five-toed feet.

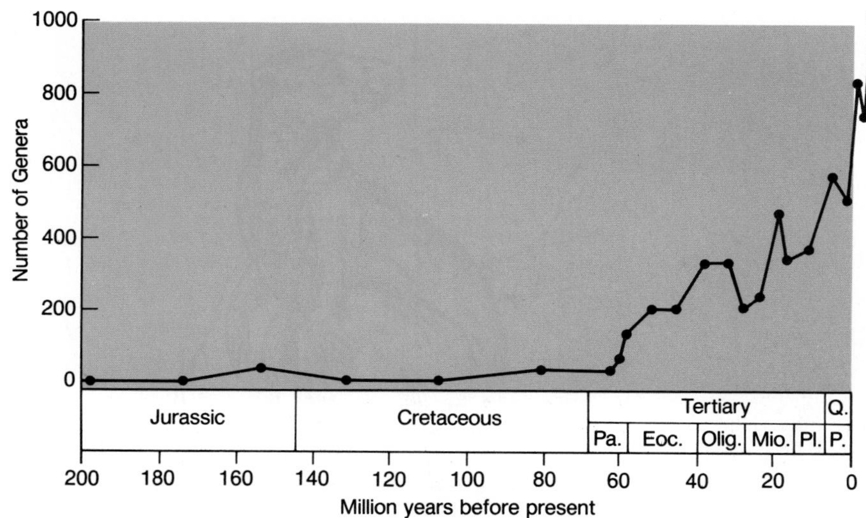


**Figure 16–14** Generic diversity of mammals. Diversity was low throughout the Mesozoic but increased markedly beginning in the Paleocene. Diversification continued since the Paleocene, although there were periodic reversals in this trend, particularly at the end of the Eocene when a number of archaic mammals became extinct.

**Figure 16–15** a. The Early Oligocene rodent *Paracricetodon* was the probable ancestor of modern rats and mice. b. One of the more bizarre-looking mammals was the Miocene horned rodent *Epigaulus*.

made their appearance. Most were rather small animals with short, heavy limbs (Fig. 16–20). They were not particularly fast runners, but, then, neither were their prey. Miacids were ancestral to all later members of the order Carnivora (Fig. 16–21). These weasel-like animals had well-developed carnassials and a brain larger than that of creodonts. But they retained such primitive features as short limbs. One of the groups of carnivores that arose

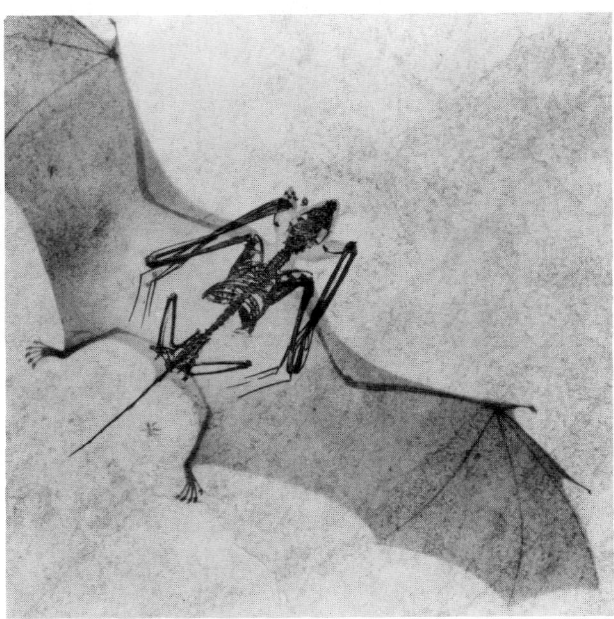

**Figure 16–16** The oldest known fossil bat, from the Eocene Green River Formation, Wyoming. The bat's wings are shown as they would have appeared when the bat was alive. Notice that the wing is supported by four elongated fingers.

**Figure 16–18** *Plesiadapis* was an Early Eocene squirrel-like primate that measured about 30 cm long.

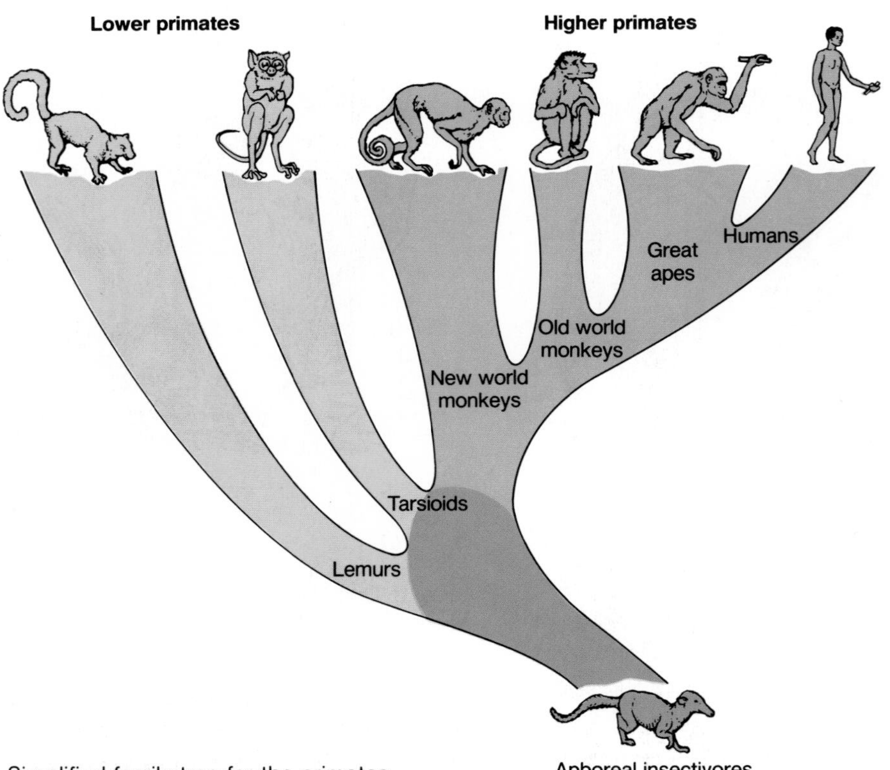


**Figure 16–17** Simplified family tree for the primates.

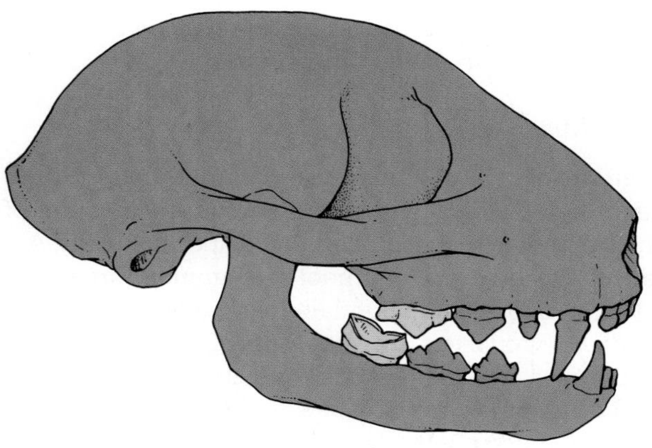

**Figure 16–19** Jaws and teeth of a cat. All members of the order Carnivora have large canine teeth and a specialized pair of shearing teeth called *carnassials* (shaded).

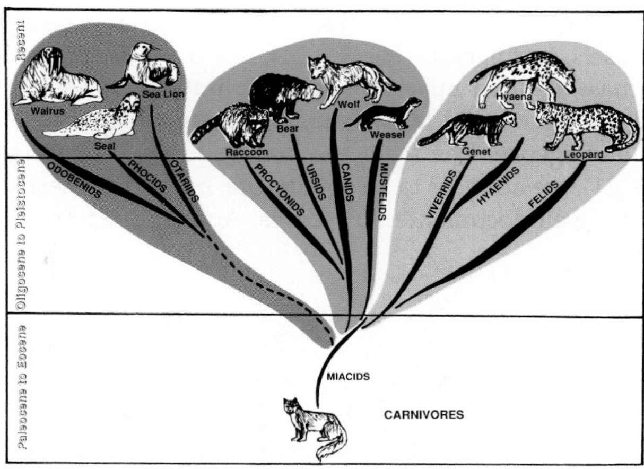


**Figure 16–21** Evolution of the canivorous mammals. Miacid ancestors gave rise to three major groups of carnivores, the aquatic carnivores, the racoon-bear-dog-weasel group, and the viverrid-hyena-cat group.

a.

b.

**Figure 16–20** a. Creodonts were Late Cretaceous to Paleocene carnivorous mammals. These restorations show the fox-sized creodont *Sinopa* (bottom), *Patriofelis* (top), which was bear-sized, and *Oxyaena*, which was about the size of a wolverine. b. The miacid carnivores, here represented by the genus *Miacis*, were the probable ancestors of all later members of the order Carnivora.

from miacids includes the cats, hyenas, and viverrids (mongooses, civet cats, and genet cats) (Fig. 16–21). Despite the obvious differences among these animals, studies of the fossil record and chromosomes of modern species clearly indicate their close relationships. Cats have not changed much since they first appeared, although there have been some remarkable developments in the canine teeth (Perspective 5–2).

The group of carnivores that includes dogs, bears, weasels, and pandas (Fig. 16–21), had better developed carnassials than miacids, and their limbs were longer. The braincase was larger, too, indicating a higher level of intelligence. The seals, sea lions, and walruses (Fig. 16–21), are adapted to an aquatic life, but their ancestry is not well documented by fossils. Aquatic adaptations include a streamlined body, a layer of blubber for insulation, and limbs modified into paddles.

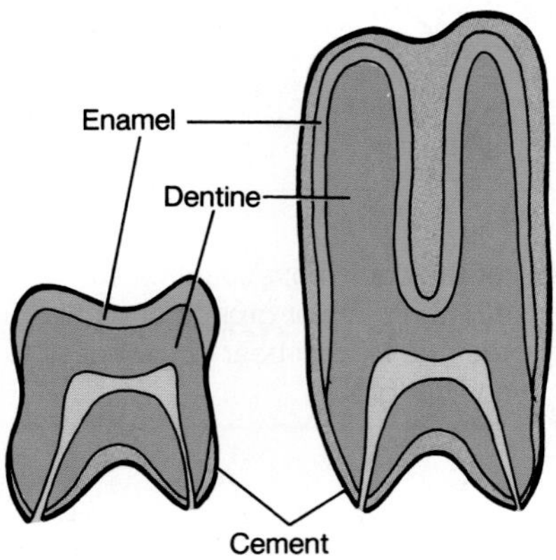


**Figure 16–22** Comparison of low-crowned and high-crowned teeth. Grazing ungulates developed high-crowned chewing teeth as an adaptation for eating abrasive grasses. The cusps of a high-crowned tooth are elevated into tall, slender pillars, and the entire tooth is covered by enamel and cement, both of which are hard substances.

## Ungulates

**Ungulates** are hoofed mammals, a group that includes the orders Artiodactyla and Perissodactyla, both of which evolved from condylarths (Perspective 16–1). **Artiodactyls**, the even-toed hoofed mammals, are by far the most diverse and abundant living ungulates; representative artiodactyls are cattle, sheep, goats, swine, antelope, deer and camels. **Perissodactyls**, the odd- toed hoofed mammals, consist of only 16 living species of horses, rhinoceroses, and tapirs (Table 16–1).

All ungulates are herbivores. The chewing teeth in mammals are the premolars and molars, but the ungulates show an evolutionary trend in the premolars called **molarization**. That is, the premolars have become enlarged to the size of molars; this gives the ungulates a continuous series of molar-like teeth for grinding vegetation. Some ungulates are **grazers** and have a grass diet. However, grasses contain tiny particles of silicon dioxide and are very abrasive, so they wear down teeth. Accordingly, the grazing ungulates have developed high-crowned teeth that are more resistant to abrasion (Fig. 16–22). In contrast, those ungulates characterized as *browsers*, which eat the tender shoots, twigs, and leaves of trees and shrubs, did not develop high-crowned teeth.

Many ungulates live in open-grassland habitats and depend on speed to escape predators. Adaptations for

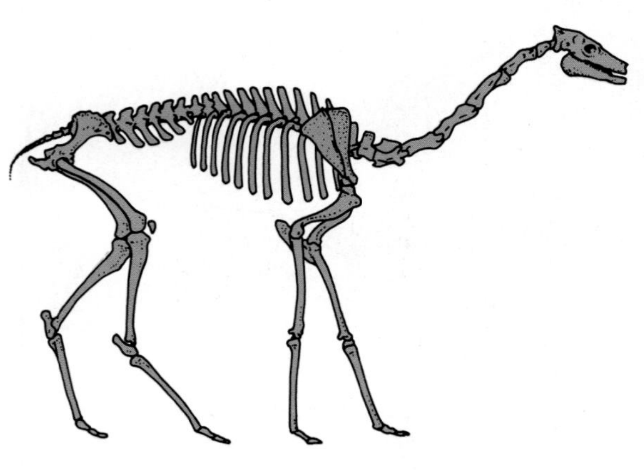

a.

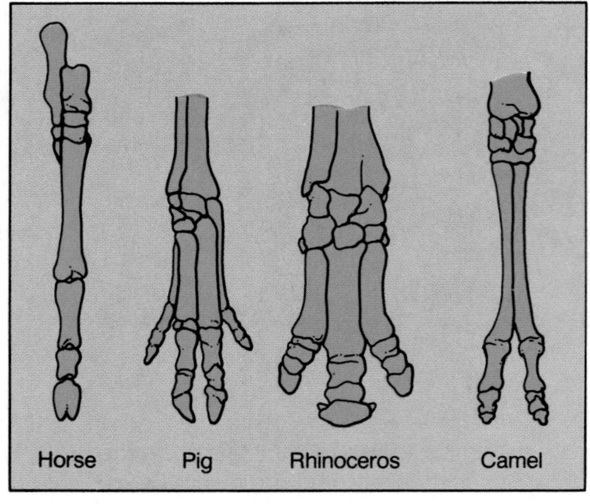


b.

**Figure 16–23** Modifications in the limbs of ungulates. a. *Oxydactylus*, a Miocene camel, shows the evolutionary trend of limb elongation. The bones located between the wrist and toes and between the ankle and toes became longer, thereby increasing the length of ungulate limbs. b. The ancestors of ungulates, the condylarths, had five-toed feet, but as shown here ungulates show a trend toward reduction in the number of toes; horses retain only one toe, pigs have four toes but with reduced side toes, rhinoceroses are three-toed, and camels retain two toes.

running include elongation of the bones of the palm and sole, which increases the length of the limbs (Fig. 16–23). Also associated with running is the trend to reduce the number of bony elements in the limbs, especially toes. The limbs of running ungulates are long and slender, ideally suited for speed. These animals walk and run on their toes, and each toe is covered by a hoof.

Not all ungulates are speedy runners. Some became very large, and their bulk alone was protection enough from predators. In this case massive limbs developed to support their great weight. Nevertheless, some large ungulates—rhinoceroses, for example—can run surprisingly fast, at least for short distances.

### Artiodactyls-Even-Toed Hoofed Mammals

Rabbit-sized ancestral artiodactyls appeared during the Early Eocene, but at this early stage of development they differed little from their ancestors, the condylarths (Perspective 16–1). From these tiny ancestors artiodactyls rapidly diversified into numerous families, many of which are now extinct (Fig. 16–24). Among these extinct families were the pig-like oreodonts (Fig. 16–2), which

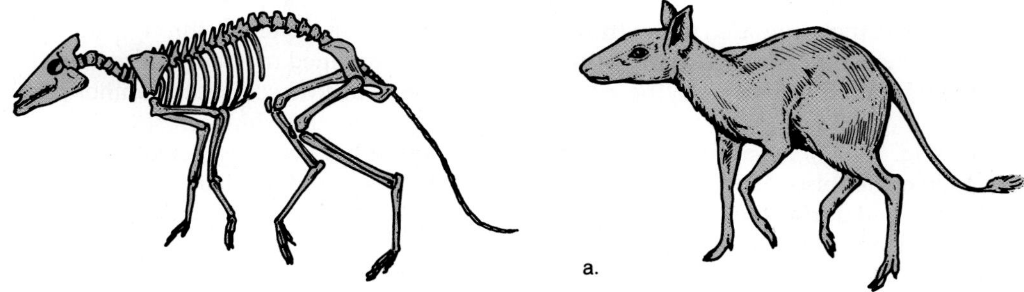

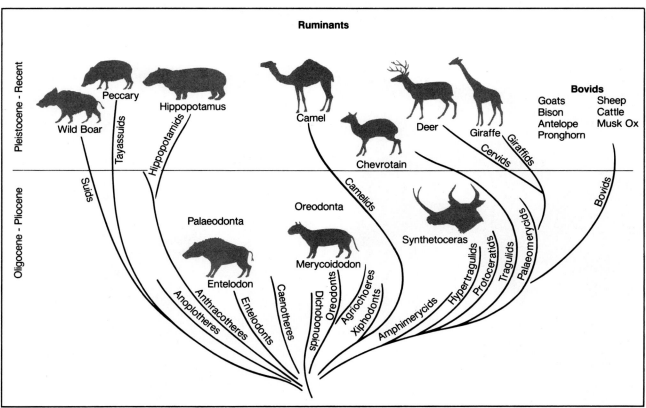


**Figure 16–24** a. Skeleton and restoration of the oldest known artiodactyl, *Diacodexis*. This animal was about the size of a rabbit. b. Simplified family tree of the artiodactyls. Early in their history artiodactyls split into two major groups; one group includes the pigs, hippopotamuses, and extinct giant hogs, and the other group is composed of ruminants or cud-chewing animals.

were common in North America until their extinction during the Pliocene, and a peculiar group of deer-like animals with horns on their noses (Fig. 16–25).

Small, four-toed, ancestral camels appeared early in the diversification of artiodactyls (Fig. 16–26). By the Oligocene all camels were two-toed, and during the Miocene they diversified into several distinctive types. Among these types were giraffe-like camels, gazelle-like camels, and giant camels standing 3.5 m at the shoulder (Fig. 16–26). Camels were abundant from Eocene to Pleistocene times in North America, and most of their evolution occurred on this continent. During the Pliocene Epoch camels migrated to South America, and Asia, where their descendants survive. In North America, camels became extinct near the end of the Pleistocene.

The *bovids* (Fig. 16–24), the most diverse living artiodactyls, include cattle, bison, sheep, goats, and antelopes. Bovids appeared during the Miocene, but most of their diversification took place in the Pliocene. The most common Tertiary bovid of North America was the pronghorn, which roamed the western interior in vast herds (Fig. 16–27). Bovids evolved on the northern continents but have since migrated to southern Asia and Africa, where they are most common today.

### Perissodactyls—Odd-Toed Hoofed Mammals

The living perissodactyls are the horses, rhinoceroses, and tapirs. These animals do not look much alike, but they and the extinct titanotheres and chalicotheres are united by several shared characteristics, and the fossil record indicates they all have a common ancestor among the condylarths (Fig. 16–28, Perspective 16–1).

Perissodactyls appeared during the Eocene and increased in diversity through the Oligocene, but they have declined markedly since then (Fig. 16–29). Today, perissodactyls constitute a minor part of the fauna, and it appears that rhinoceroses are on the verge of extinction. It has been suggested that competition with artiodactyls could explain the decline of perissodactyls. Most artiodactyls are called **ruminants**, meaning "cud-chewing." They are herbivores with complex three- or four-chambered stomachs. Ruminants use the same resources as perissodactyls, but they use them more efficiently.

a.

b.

**Figure 16–25** During the adaptive radiation of artiodactyls numerous families evolved, but many of these became extinct including the two shown here. a. The entelodonts, represented here by *Entelodon*, were two-meter-long herbivores that resembled wild boars. b. the peculiar protoceratids lived in North America until the Pliocene.

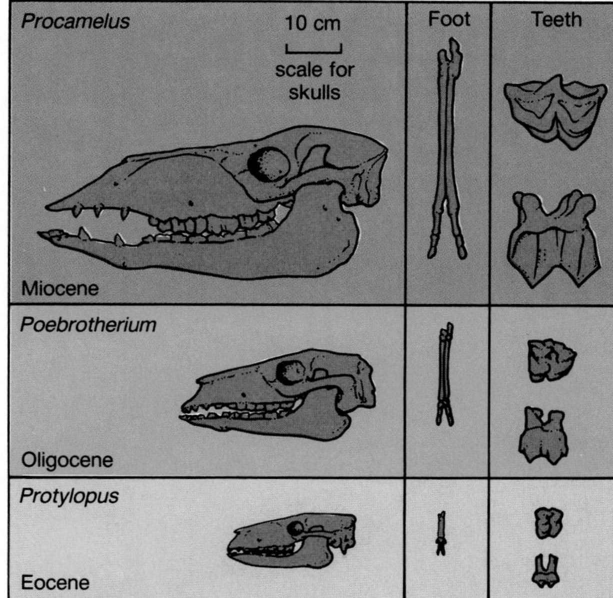


a.

b.

**Figure 16–26** a. These three genera of camels show the evolutionary trends of increase in size, elongation of the legs along with reduction of lateral toes, and development of high-crowned teeth. b. Two of the genera of North American Miocene camels. *Oxydactylus* (left) was a delicately built camel measuring about 1.4 m at the shoulder. *Alticamelus* (right) is called the giraffe camel because its legs and neck are so long.

**Evolution of Horses.** The earliest member of the horse family was ***Hyracotherium*** (Fig. 16–30) from the Early Eocene of Great Britain and western North America. About the size of a fox, it had four-toed front feet and three-toed hind feet. Each toe, however, was covered by a small hoof. *Hyracotherium* has few of the specializations we associate with horses (Table 16–2).

Since *Hyracotherium* was so unhorse-like, how can we be sure it belongs to the horse family at all? Fortunately, the fossil record of horses is exceptionally good, and this record clearly shows that *Hyracotherium* is linked to the modern horse, *Equus*, by a series of intermediates (Fig. 16–31). However, the various trends in horse evolution (Table 16–2) did not all occur at the same rate. For example, molarization of the premolars was complete by the Oligocene, but the reduction of toes to one was not complete until the Pliocene.

We can recognize two distinct branches in the adaptive radiation of horses (Fig. 16–31). Both have their ancestry in *Hyracotherium;* but one branch led to three-toed browsing horses, all of which are now extinct, and the other led first to three-toed grazing horses and finally to one-toed grazers. Horse evolution occurred almost totally in North America, but a few genera migrated to the Old World, and in the Pliocene one genus even reached South America.

The appearance of grazing horses during the Miocene coincides with the appearance of the open-prairie, grassland habitat. Like all habitats this one had its special problems. For one thing, as mentioned earlier, grass has silica and is very abrasive to the teeth, wearing them down rapidly. Horses and other ungulates adapted by developing high-crowned teeth that are more resistant to abrasion (Fig. 16–22). Also, speed was essential to escape predators in this habitat; thus the limbs became longer, and the number of toes was reduced (Fig. 16–31).

*Merychippus* is a good example of the early grazing horses (Fig. 16–31). It was about the size of a modern pony and had teeth well-suited for grazing. Its limbs were long, and most of the weight was borne on the third toe, although side toes were still present. By Pliocene time, one-toed grazers had evolved (Fig. 16–31). Compared with *Merychippus*, these horses were larger, had longer limbs, and higher-crowned teeth. Modern *Equus* evolved during the Late Pliocene in North America and migrated to the Old World, where it still lives in the wild. In North America, however, horses and a number of other large mammals became extinct at the end of the Pleistocene (Chapter 17). Modern horses were reintroduced into North America by Spanish explorers during the sixteenth century.

**Other Perissodactyls.** Other than horses, the only living perissodactyls are tapirs and rhinoceroses.

**Figure 16–27** The most common Tertiary bovid in North America was the pronghorn represented in this scene from the Early Pliocene by the extinct genus *Merycodus* (1). Other mammals shown include: 2. the horned rodent *Epigaulus*; 3. *Amebelodon*, a shovel-tusked mastodon; 4. the long-legged rhinoceros *Aphelops;* 5. the one-toed horse *Pliohippus;* and 6, *Megatylopus*, a giant camel.

Both had Eocene ancestors that looked much like *Hyracotherium* (Fig 16–32). In fact, all of the earliest perissodactyls, including the extinct titanotheres and chalicotheres, are so similar that it is difficult to differentiate among them.

Both tapirs and rhinoceroses increased in size during the Cenozoic, and both became more abundant, diverse, and widespread than they are now. An Oligocene-Miocene hornless rhinoceros of Asia, *Baluchitherium,* was the largest land mammal ever to exist (Fig. 16–33). Most rhinoceros evolution occurred in the Old World, but North American rhinoceroses were quite common until they became extinct during the Late Pleistocene (Fig. 16–33).

In addition to horses, rhinoceroses, and tapirs, small *Hyracotherium*-like ancestors also gave rise to titanotheres and chalicotheres (Fig. 16–28). Titanotheres existed only from the Early Eocene to Early Oligocene, but they evolved from small ancestors to giants about 2.5 m high at the shoulder (Figs. 16–1, 16–34, 5–25). The fossil evidence suggests that chalicotheres were never very abundant, although the group existed from the Early Eocene to the Pleistocene. Miocene and later chalicotheres were large animals that superficially resembled horses,

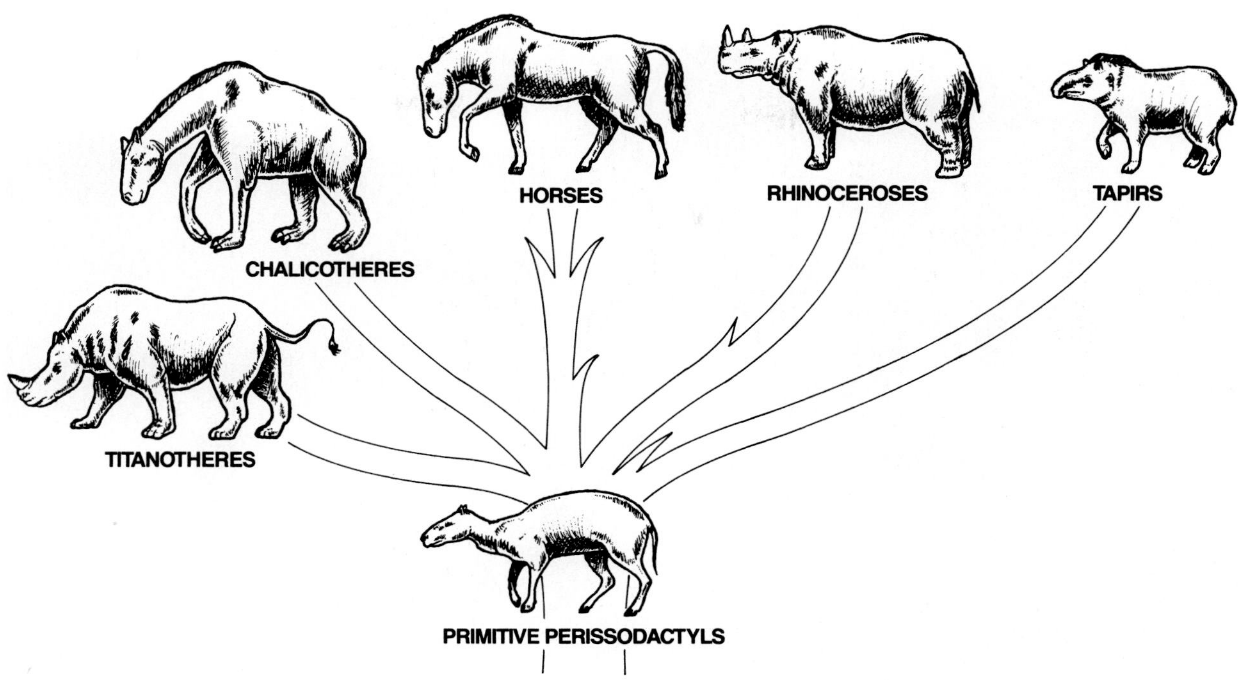


**Figure 16–28** Evolution of perissodactyls. Adaptive radiation and divergence from a common ancestor account for the fact that perissodactyls share several characteristics yet differ markedly in appearance.

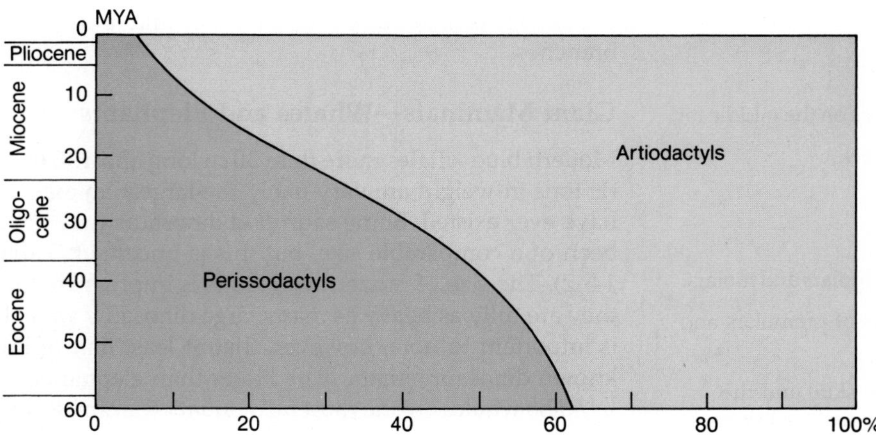


**Figure 16–29** Perissodactyls and artiodactyls were relatively abundant during the Cenozoic. At the beginning of the Eocene, perissodactyls were dominant, but they have steadily declined in abundance and now make up less than 10 percent of ungulate faunas.

**Figure 16–30** The oldest known horse, *Hyracotherium*, was a fox-sized animal with four-toed front feet and three-toed back feet. *Hyracotherium* probably browsed on leaves in the swampy forests where it lived.

**Table 16–2** Trends that characterize the evolution of horses during the Cenozoic Era.

1. Increase in size.
2. Lengthening of legs and feet.
3. Reduction of lateral toes, with emphasis on the middle toe.
4. Straightening and stiffening of the back.
5. Widening of the incisor teeth.
6. Molarization of the premolars
7. Increase in height of the crowns of premolars and molars.
8. Progressive complexity of the crowns of premolars and molars.
9. Deepening of the front portion of the skull and the lower jaws to accommodate the high-crowned premolars and molars.
10. Lengthening of the face in front of the eye, also to accommodate the high-crowned premolars and molars.
11. Increase in size and complexity of the brain.

but their feet had claws rather than hoofs (Fig. 16–35). The purpose of these claws is unknown. The prevailing opinion favors the interpretation that they were used to dig up roots and tubers or to hook and pull down branches.

## Giant Mammals—Whales and Elephants

Modern blue whales more than 30 m long and 130 metric tons in weight are very likely the largest animals to have ever existed. Some sauropod dinosaurs may have been of a comparable size, but this is uncertain (Table 14–2). The size of modern elephants is impressive, too; they are fully as heavy as many large dinosaurs were. It is important to note, however, that at least half of the known dinosaur genera were larger than elephants.

The whales are the most fully aquatic mammals. Evolutionary trends that occurred during the transition to the aquatic habitat included modification of the front limbs into flippers, loss of the hind limbs, migration of the nostrils to the top of the head to form the blowhole, and development of a large horizontal tail fluke used for propulsion.

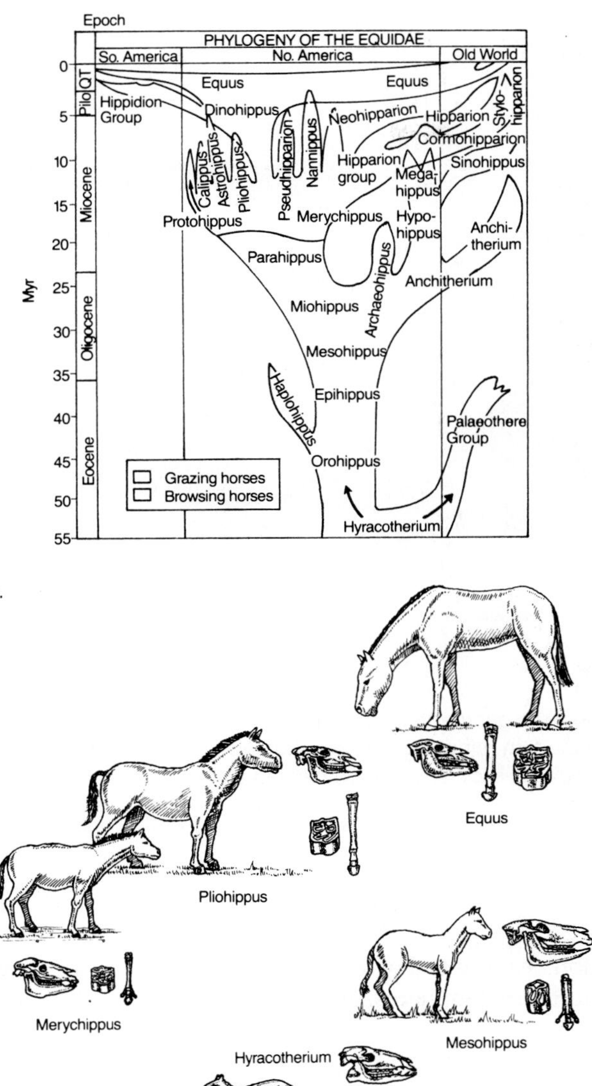


b.

**Figure 16–31** Evolution of horses. a. Summary chart showing the recognized genera of horses and their evolutionary relationships. Note that during the Oligocene, two separate lines emerged, one leading to three-toed browsing horses and the other to one-toed grazers. b. Simplified diagram showing some of the evolutionary trends from *Hyracotherium* to the modern horse, *Equus*. Important trends shown here include an increase in size, changes in skull proportions, loss of toes, and development of high-crowned teeth with complex chewing surfaces.

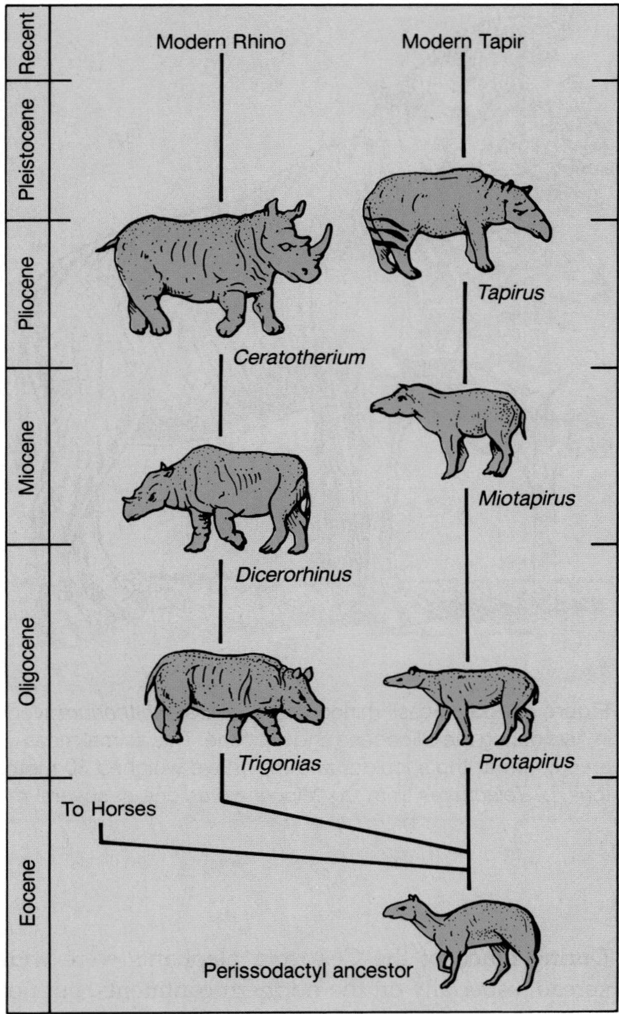


**Figure 16–32** Simplified chart showing the evolution of rhinoceroses and tapirs from a common ancestor. Both have increased in size.

Whales appeared during the Early Eocene and by the Late Eocene had become diverse and widespread, and some had achieved large size (Fig. 16–36). Eocene whales retained remnants of the hind limbs, had teeth resembling those of their condylarth ancestors, and were differently proportioned than living whales. By the Oligocene Epoch, both groups of living whales, the toothed whales and baleen whales, had evolved.

Living elephants are represented by only two genera and two species and are the dwindling remnants of the once diverse **proboscideans** (order Proboscidea).

a.

b.

**Figure 16–33** Fossil rhinoceroses. a. *Baluchitherium* lived in Asia during the Oligocene and Miocene. This animal measured 5.4 m at the shoulder and may have weighed 30 metric tons. b. *Teleoceras* from the Miocene was one of several rhinoceros genera to inhabit North America. About the size of a modern rhinoceros, it had short limbs and a hippopotamus-like body; it probably led a semi-aquatic life.

During much of the Cenozoic, elephants were widespread, especially on the northern continents, but now they are restricted to Africa and southeast Asia.

One of the earliest elephants, *Moeritherium* from the Eocene, had small tusks but most closely resembled a hippopotamus and probably lived a semiaquatic life (Fig. 16–37). By Oligocene time, elephants clearly showed the trend to large size and the development of a long proboscis and large tusks (Fig. 16–37). Mastodons with teeth adapted for browsing had evolved by Miocene time. They originated in Africa, but during the Miocene and Pliocene epochs they spread to the Northern Hemisphere continents. One genus reached South America during the Pleistocene. Tusk size and shape varied considerably among all mastodons (Fig. 16–38).

The last major evolutionary event to affect the proboscidians was the divergence of the modern elephant and mammoth lines during the Pliocene and Pleistocene (Fig. 16–37). Mammoths were no larger than modern elephants and, in fact, many were smaller, but they had the largest tusks of all the elephants. Until their extinction near the end of the Pleistocene, they lived on all the northern continents, as well as in Africa and India.

## TERTIARY VEGETATION AND CLIMATE

Angiosperms, or flowering plants, arose during the Cretaceous Period (Figs. 14–11, 14–12), and their diversification continued into the Tertiary. Some gymnosperms, especially conifers, remained abundant, and seedless vascular plants still occupied many habitats. Overall, the Cenozoic was a time when more and more familiar types of plants appeared.

Many of the Tertiary plants would be quite familiar to us, but their geographic distribution was markedly different from what it is now (Perspective 16–2). It has long been known that plant distribution is strongly controlled by climate, and for the Tertiary we see evidence of changing climatic patterns accompanied by shifting

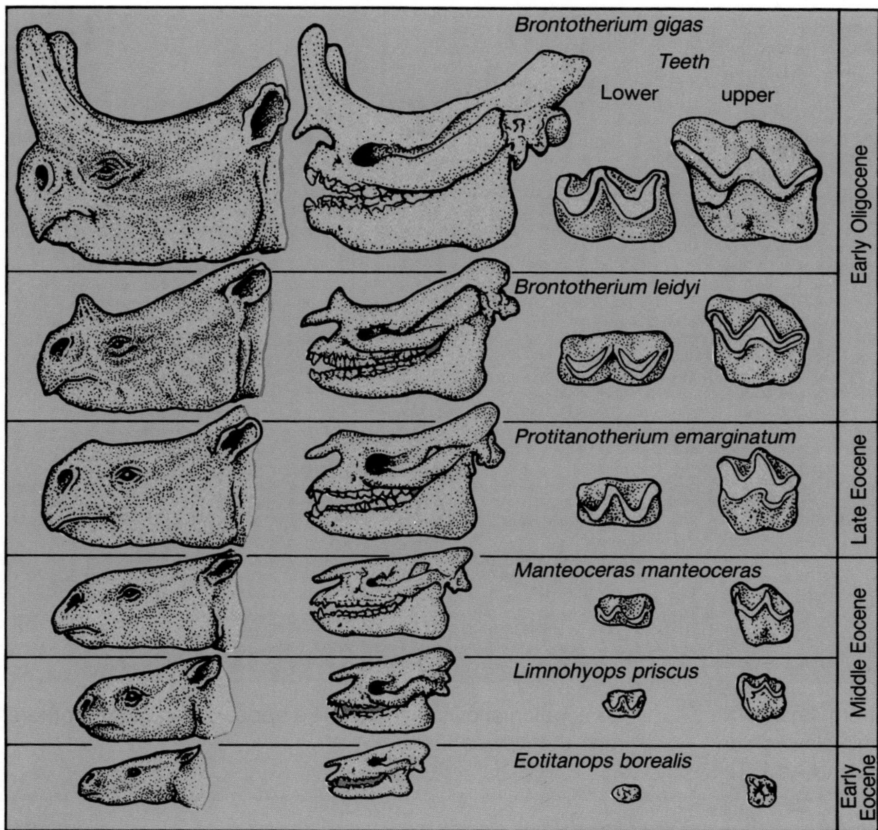


**Figure 16–34** Early Eocene to Early Oligocene skulls and molars of six species of North American titanotheres. Skull size increased and large nasal horns developed, but the teeth retained a constant pattern.

distributions of plants. Mean annual temperatures during the Paleocene and Eocene were high, but a marked temperature decrease occurred at the end of the Eocene (Fig. 16–39).

Leaf structure is a good climatic indicator and is the basis for the inferences shown in Figure 16–39. Leaves with entire or complete margins characterize areas of high annual precipitation and high mean annual temperatures, whereas leaves with incised margins typify cooler, less humid regions. If a fossil flora has a high percentage of entire-margined leaves, the climate was probably wet and warm.

The types of plants found in fossil floras are also good climatic indicators. For example, Paleocene strata in the western interior of North America contain fossil ferns and palms, which are common to warmer, more humid climates. In fact,

> No known Paleocene flora in North America indicates a climate cooler than warm-temperate, and subtropical climates existed at least as far north as latitude 62° in Alaska during part, if not all, of the Paleocene.[1]

Subtropical conditions continued into the Eocene of North America, probably the warmest of all the Tertiary

[1]J. A. Wolfe, and D. E. Hopkins. 1967. Climatic Changes Recorded by Tertiary Land Floras in Northwestern North America. In *Tertiary Correlations and Climatic Changes in the Pacific,* 11th Pacific Science Congress, p. 69.

**Figure 16–35** Chaliocotheres evolved from small, *Hyracotherium*-like ancestors into large animals such as *Moropus* of the Miocene of North America and Asia. This restoration shows the horse-like appearance of chalicotheres and the clawed feet.

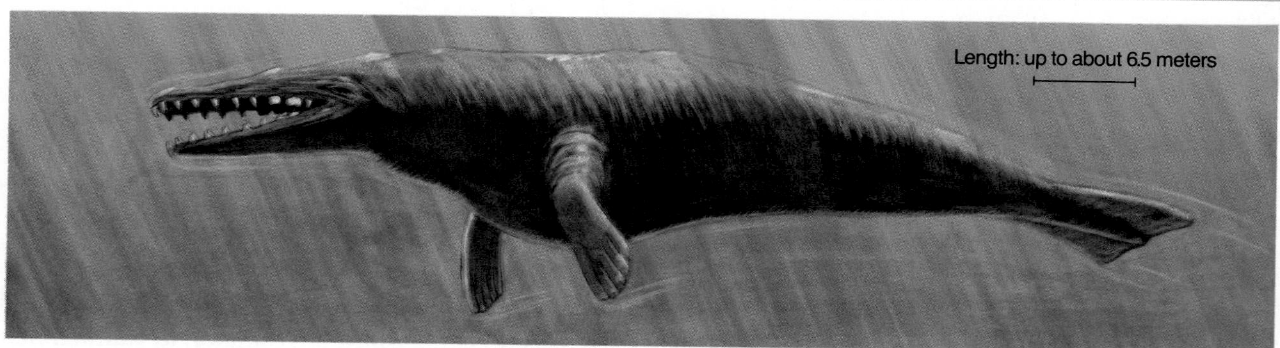


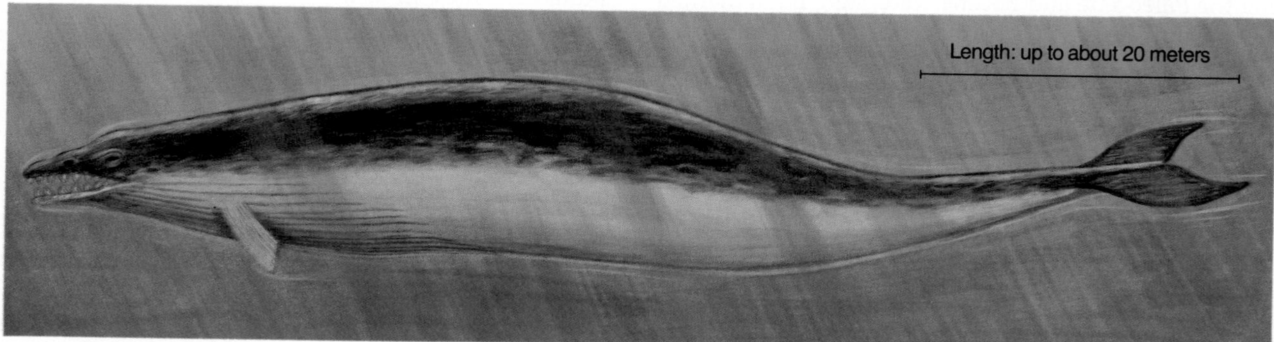


**Figure 16–36** Restoration of two Eocene whales. *Protocetus* was about 6.5 m long, and had teeth that closely resembled those of carnivorous condylarths. *Basilosaurus* was a giant whale, 20 m long, with a long, slender body and a very small head.

a.

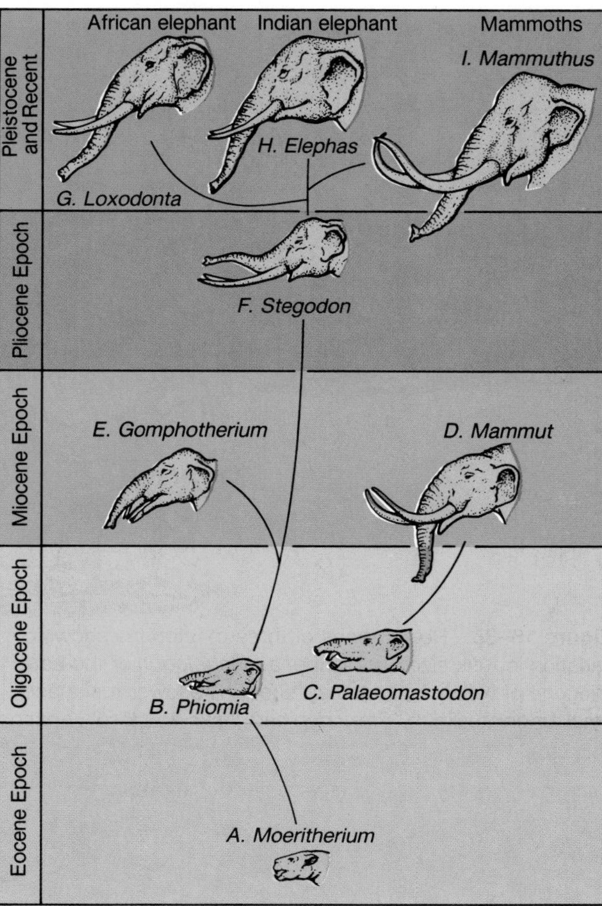


b.

**Figure 16–37** Evolution of elephants. a. *Moeritherium*, one of the oldest known ancestors of elephants, is shown here as it probably lived. This animal was the size of a large hog and had upper and lower incisors modified into small tusks. b. Simplified family tree of the elephants. Increase in size, and development of large tusks and a long proboscis were some of the evolutionary trends in elephants.

epochs. The fossil flora of the Eocene beds of the John Day Basin in eastern Oregon includes tropical ferns, figs, and laurels (Fig. 16–40), all of them plants found today only in the humid parts of Mexico and Central America.

A major climatic change occurred at the end of the Eocene Epoch (Fig. 16–39). Mean annual temperatures dropped as much as 7° C in about 3 million years. From the Early Oligocene on, mean annual temperatures have varied somewhat worldwide but overall have not changed much in the middle latitudes except during the Pleistocene.

A general decrease in precipitation over the last 25 million years took place throughout the Midcontinent of North America (Fig. 16–41). As the climate became drier, the vast forests of the Oligocene gave way first to *savannah* conditions (grasslands with scattered trees) and finally to *steppe* environments (short-grass prairie of the desert margin). Herbivorous mammals quickly adapted to the savannah habitat (Fig. 16–41) by developing high-crowned teeth suitable for a diet of grass. Among the horses, particularly, grazers became more common and browsers declined to extinction.

## INTERCONTINENTAL MIGRATRIONS

The mammalian faunas of North America, Europe, and northern Asia had many similarities throughout the Cenozoic. Even today Asia and North America are only narrowly separated at the Bering Strait, and several times during the Cenozoic the Bering Strait formed a land corridor across which mammals migrated (Fig. 16–42). During the Early Cenozoic, a land connection existed between Europe and North America (Fig. 16–42), freeing mammals to roam across all the northern continents, and many did; camels and horses are only two examples.

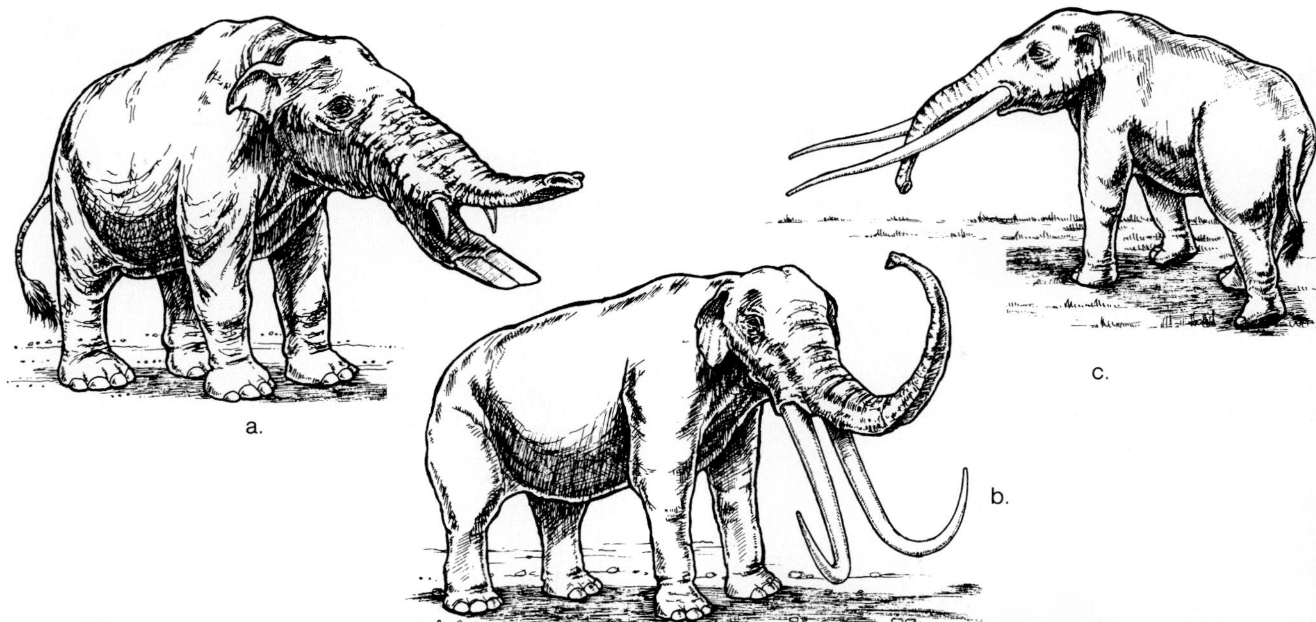


**Figure 16–38** Restorations of three mastodons showing variation in tusk size and shape. a. *Ambelodon* of the Late Miocene of North America had shovel-like lower tusks and small upper tusks. b. *Stegodon* had tusks so closely spaced that there was no space between them for the trunk. This mastodon lived during the Pliocene and Pleistocene. c. *Anancus* from France was a Pliocene-Pleistocene genus with straight tusks 3 m long.

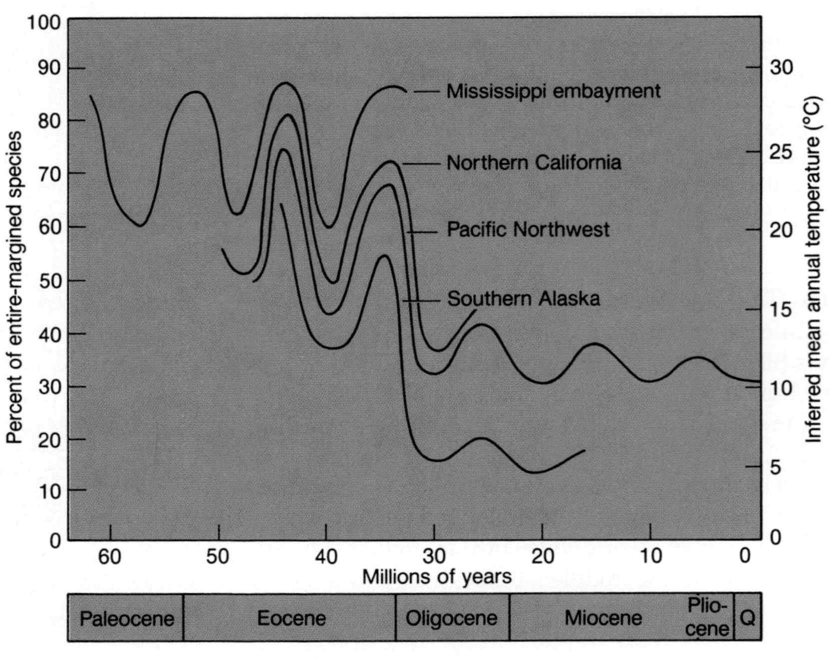


**Figure 16–39** Cenozoic climate. a. Adaptations to the environment are shown by leaf structure, so leaves are good indicators of climate. Plants adapted to cool climates typically have small leaves with incised margins (top), while plants of humid, warm habitats have larger, entire-margined leaves, and many have pointed drip-tips (bottom). b. Inferred climatic trends for four areas in North America based on percentages of plant species with entire-margined leaves. Notice the sharp drop in mean annual temperature at the end of the Eocene.

On the other hand, the southern continents were largely separate island continents during much of the Cenozoic. Africa remained fairly close to Eurasia, and, at times, faunal interchange between those two continents was possible. For example, elephants first evolved in Africa, but they migrated to all of the northern continents (Fig. 16–43).

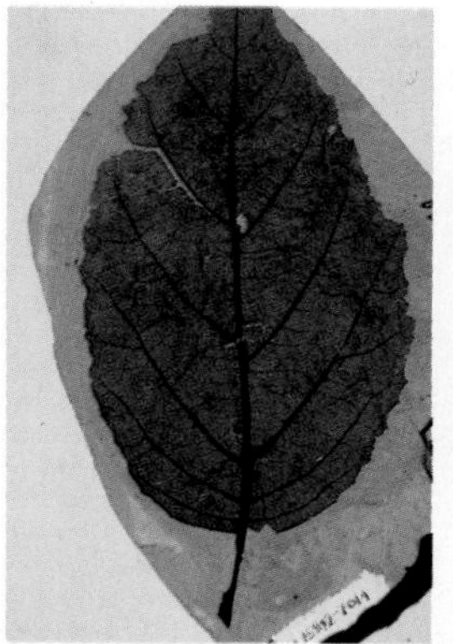

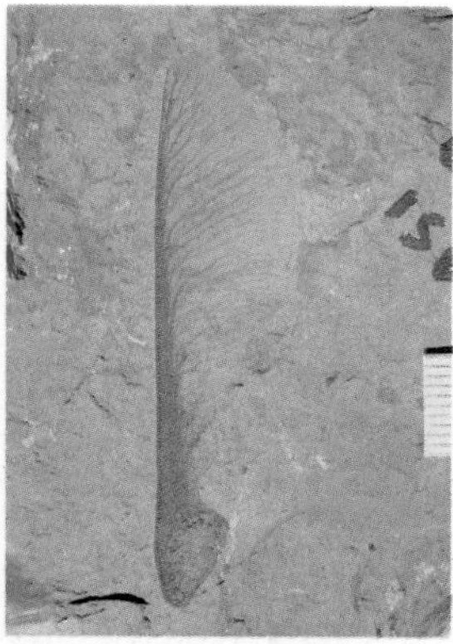

**Figure 16–40** Fossil alder leaf (left) and maple fruit from the Oligocene John Day beds, Oregon. (Photos courtesy of Steven R. Manchester, Indiana University, Bloomington)

South America was an island continent from the Late Cretaceous until a land connection with North America was established about 5 million years ago (Fig. 15–17). At that time, the South American mammalian fauna consisted of various marsupials, and several orders of placental mammals that lived nowhere else in the world (Fig. 16–44). These animals thrived in isolation and showed remarkable convergence with North American placentals (Fig. 5–22).

When the Isthmus of Panama formed, however, migrants from North America soon replaced most of the indigenous South American mammals. Among the marsupials only opossums survived, and most of the placentals died out too. Even though the land connection allowed migrations in both directions, very few South American mammals successfully migrated northward (Fig. 16–45).

Most of the living species of marsupials are restricted to the Australian region. Recall from Chapter 14 that marsupials occupied Australia before its complete separation from Gondwana, but apparently placentals, other than bats and a few rodents, never did.

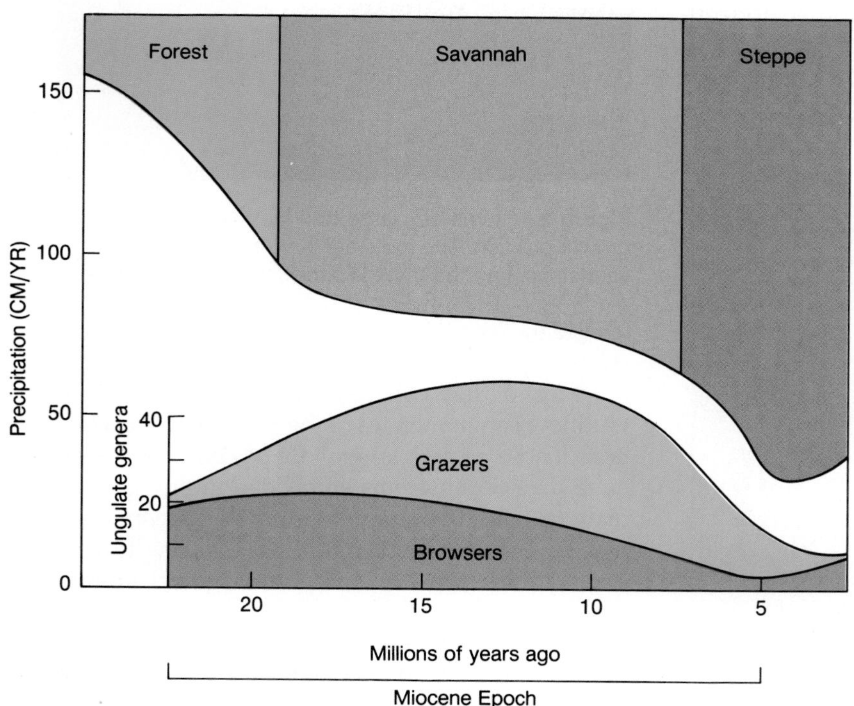


**Figure 16–41** Miocene climatic changes and changes in the ungulate fauna of midcontinental North America. The climate became drier and the vegetation changed from forest to grasslands with scattered trees (savannah) to short-grass prairie of the desert margin (steppe). Browsing ungulates were dominant during the Early Miocene, but grazers became more abundant when the climate became drier and the grasslands spread.

## Perspective 16-2

# Fossil Forests—Yellowstone National Park

Yellowstone National Park, Wyoming, does not now have a climate in which avocado, magnolia, laurel, and fig trees could grow. Yet, during the Eocene Epoch, these kinds of trees, along with spruce, fir, oak, redwoods, and pines, grew there. In fact, fossils of all of these tree types have been found in the same beds. Because magnolias and spruce, for example, do not now grow under the same climatic conditions, finding them in the same fossil beds led some investigators to conclude that Eocene plants had a wider range of environmental tolerances than they do today.

Fossil forests of similar mixtures occur at a number of sites in the northeastern part of Yellowstone Park. During the Eocene, these sites were in a valley located between two belts of volcanoes (Fig. 1). Recent studies document the following: many trees were buried in place and in their growth positions; some trees were transported and deposited horizontally as logs or as stumps that came to rest in a vertical position (these only *appear* to be in a growth position). As a consequence of transport, there was considerable mixing of ecologically different types of trees.

**Figure 2** Stump of a large tree that was buried in its growth position. The roots are in volcanic ash-rich sandstone, and the trunk is enclosed in conglomerate.

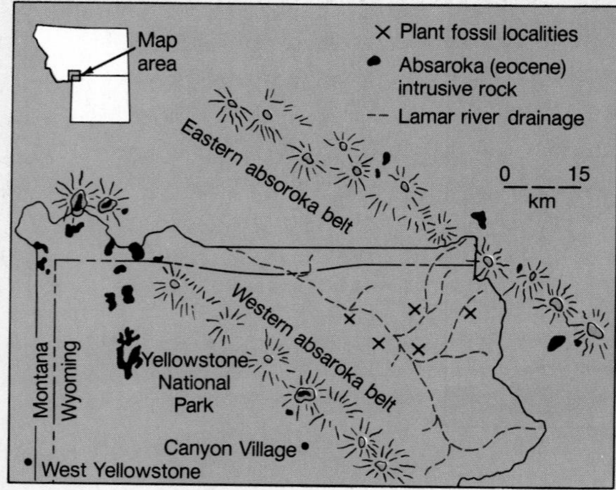


**Figure 1** Index map showing the locations of plant-fossil localities in Yellowstone National Park, Wyoming. During the Eocene, these localities were in a valley between two belts of volcanos.

Most of the fossil trees are buried in fluvial and mudflow conglomerates of the Lamar River Formation. Those buried in growth position, however, have their root systems embedded in fine-grained material and their upper parts in conglomerate (Fig.2). Streams and mudflows simply flowed around these trees and deposited gravel around their trunks.

Figure 3 is a depositional model for burial of trees. Mudflows and streams flowed down the flanks of the volcanoes, carrying logs and upright stumps down to lower elevations, where they were buried along with the trees still in growth position.

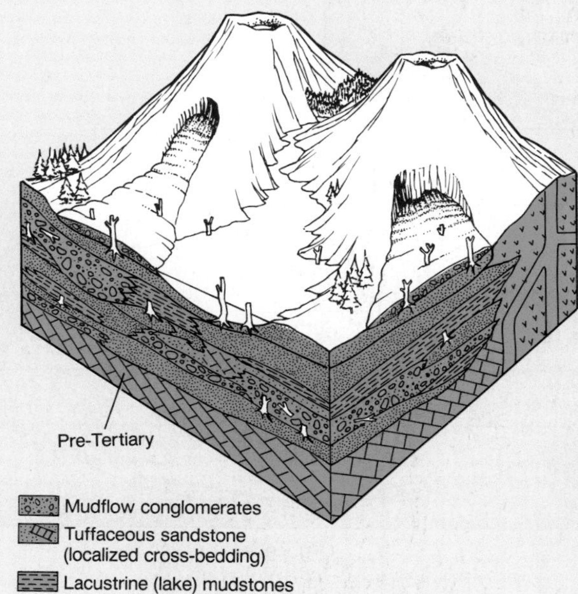


**Figure 3** Depositional model accounting for the burial of trees. Mudflows moved down the slopes of volcanoes, carrying trees from high-elevation to lower-elevation habitats. Some trees were buried in growth position. Some stumps floating in lakes came to rest in a vertical position and appear to be in growth position.

Transported stumps coming to rest in a vertical position are not unusual. Following the May 1980 eruption of Mt. St. Helens, Washington, stumps were observed floating vertically in Spirit Lake; some were grounded and came to rest in this position. This recent event also resulted in the burial of trees in growth position, as well as transport and deposition of logs. In fact, some layers of logs in Yellowstone Park look like log jams similar to the floating mass of logs observed in Spirit Lake (Fig. 4).

One recent investigator concluded that nine to twelve forest levels are represented in Yellowstone Park. Transport of stumps and logs from high-elevation, cool-temperature sites to tropical valleys accounts for the mixing of trees that were from different climatic zones. Nevertheless, the presence of fossil magnolias and avocados in Wyoming strata is strong evidence that the climate was considerably warmer than it is now.

**Figure 4** Following the 1980 eruption of Mt. St. Helens, Washington, Spirit Lake had a floating mat of logs and stumps.

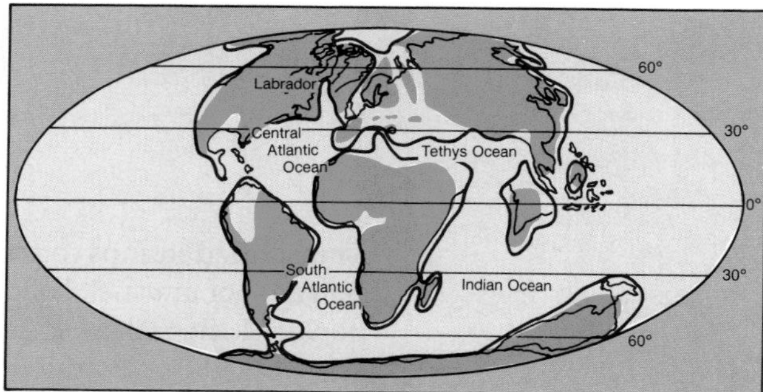

Paleocene

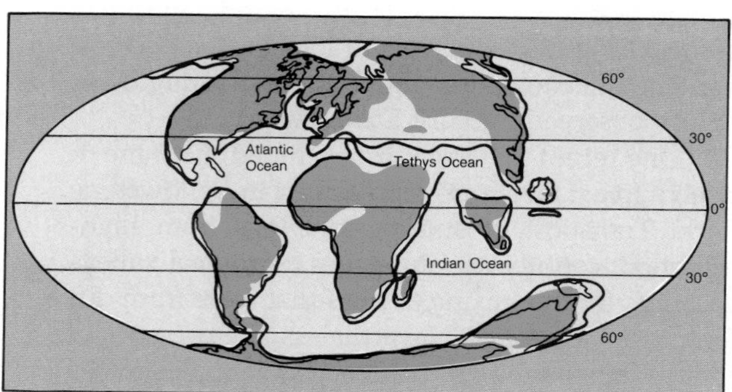

Eocene

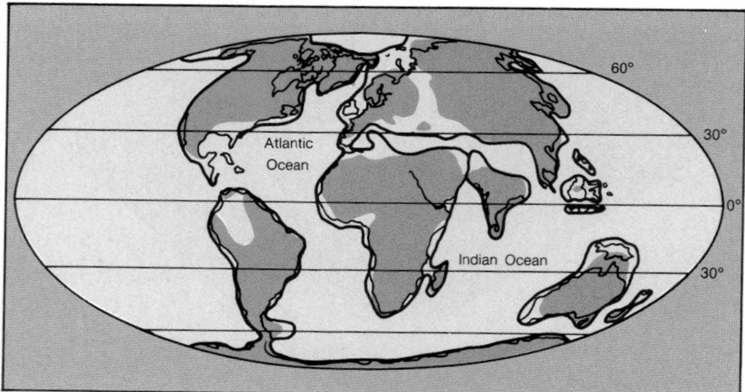

Oligocene

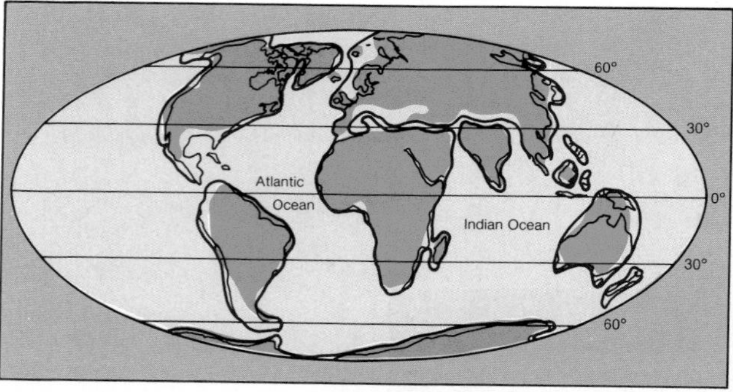

Miocene

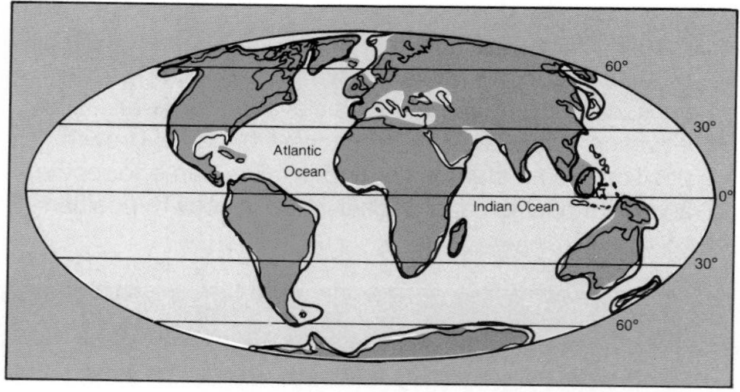

Pliocene

**Figure 16–42** Generalized paleogeographic maps showing positions of the continents and land connections between continents at various times during the Tertiary Period.

## CHAPTER SUMMARY

1. Marine invertebrate groups that survived the Mesozoic extinctions continued to expand and diversify during the Tertiary.
2. Birds belonging to living orders and families appeared during the Early Tertiary Period.
3. The Paleocene mammalian fauna was composed of Mesozoic holdovers and a number of new orders. This was a time of diversification among mammals, and several orders soon became extinct. Most living mammalian orders were present by the Eocene.
4. Mammals have a good fossil record, and because of their varied teeth mammal fossils are more easily identified than are fossils of other vertebrates, so their evolutionary history is known in considerable detail.

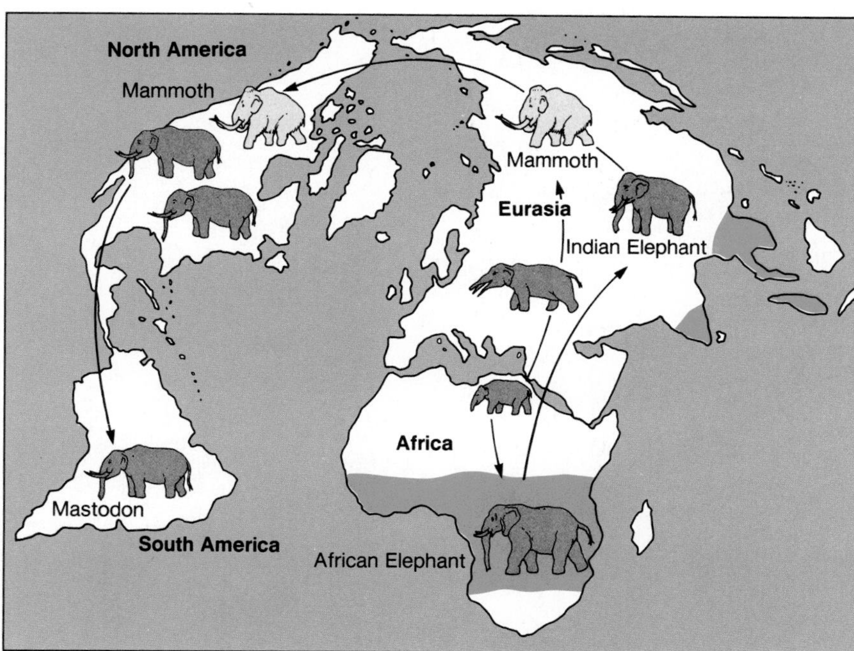


**Figure 16–43** Elephants evolved in Africa during the Eocene, but by Miocene time had spread to all the northern continents. Mastodons reached South America by the Late Pliocene. The shaded areas represent the present ranges of the African and Indian elephants.

**Figure 16–44** South America was an island continent during most of the Cenozoic Era, and its mammalian fauna consisted of numerous marsupials and several orders of placental mammals that lived only on that continent. In this scene from the Miocene of Argentina two carnivorous marsupials are shown; *Cladosictis* (1) measured about two meters long excluding the tail, and *Borhyhaena* (2) was about the size of a bear. Placental mammals shown are *Thoatherium* (3), an antelope-like member of the order Litopterna, and two members of the order Notoungulata, *Homalodotherium* (4) and *Protypotherium* (5). *Homalodotherium* was a two-meter-long herbivore, and *Protypotherium* was a rabbit-like herbivore.

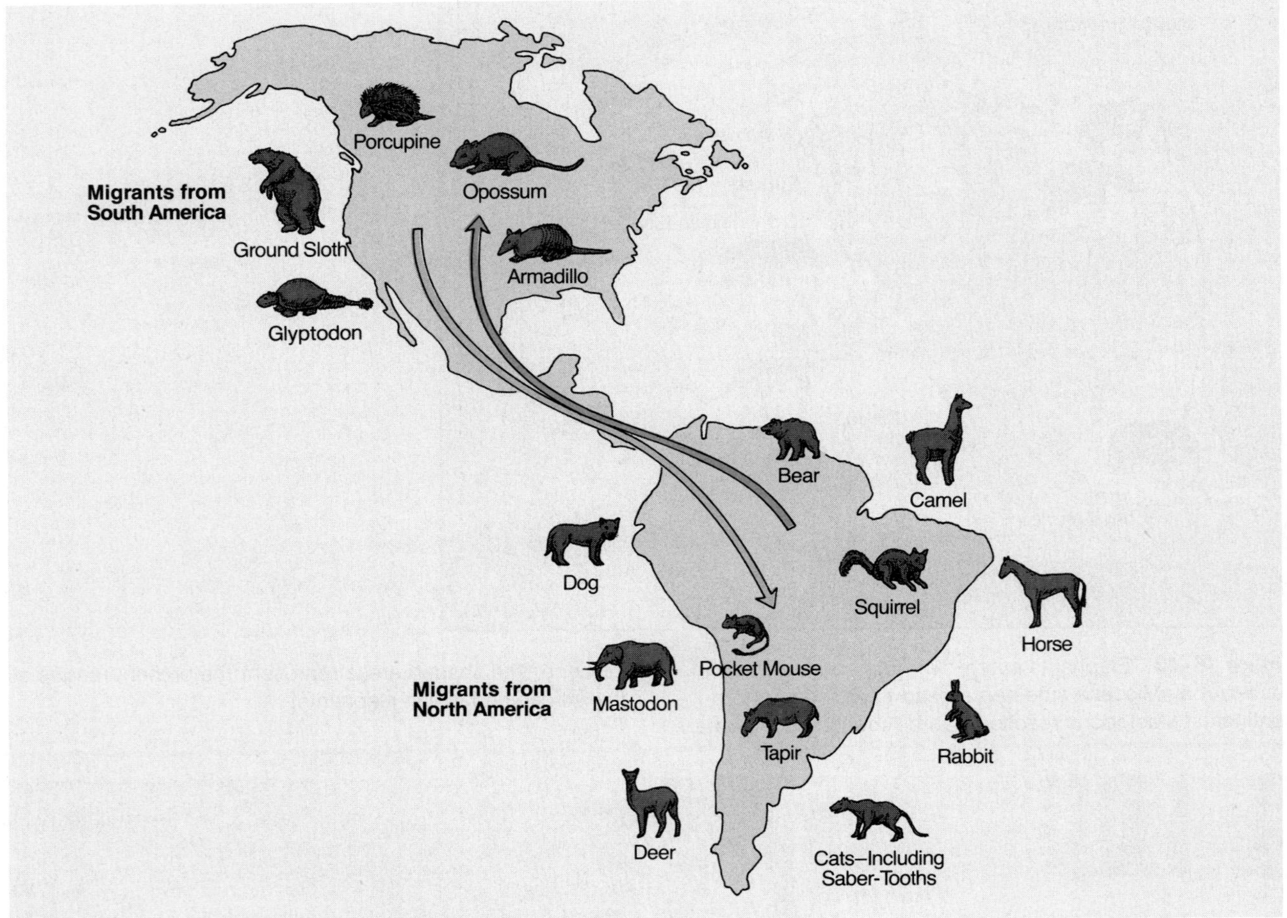


**Figure 16–45** South America existed as an island continent during most of the Cenozoic Era. Its fauna evolved in isolation and consisted of numerous marsupials and placental mammals that were restricted to that continent. When the Isthmus of Panama formed during the Late Pliocene (see Fig. 15–17), migrations between North and South America occurred. Many types of placental mammals migrated south and many South American mammals soon became extinct. Only a few mammals migrated north and successfully occupied North America.

5. Placental mammals owe much of their success to their method of reproduction. The placental order Insectivora appeared during the Cretaceous and was the ancestral stock for the placental adaptive radiation of the Cenozoic.

6. Small mammals, especially rodents, occupy microhabitats unavailable to larger animals. Bats, the only flying mammals, have forelimbs modified for wing support, but otherwise they differ little from their insectivore ancestors.

7. Paleocene carnivorous mammals called miacids were ancestral to all later placental carnivores. Adaptations in carnivores include minor modifications of the limbs, well-developed canine teeth, and specialized shearing teeth called carnassials.

8. The ungulates, perissodactyls and artiodactyls, evolved from condylarths during the Eocene. Ungulate adaptations include molarization of premolars for grinding up vegetation and limb modifications for speed.

9. Artiodactyls diversified throughout the Cenozoic and are now the most abundant ungulates. Perissodactyls were abundant and diverse during the earlier Tertiary but have declined to only 16 living species.

10. The evolutionary history of horses is particularly well documented by fossils. The earliest horse, *Hyracotherium*, and the modern horse, *Equus*, differ considerably, but a continuous series of intermediate fossils shows that they are related.

11. Horses, rhinoceroses, tapirs, titanotheres and chalicotheres diverged from a common ancestor during the Eocene.
12. Studies of fossils and living animals indicate that whales evolved from land-living mammals, probably condylarths. Elephants evolved from pig-sized ancestors, became diverse and abundant, especially on the northern continents, then dwindled to only two living species.
13. By the Pliocene Epoch, most mammals were of familiar types, and many of these were the ancestors of Pleistocene and Recent species.
14. Flowering plants continued their dominance in land-plant communities. Subtropical to tropical climates prevailed through the Paleocene and Eocene in North America. Temperatures declined at the end of the Eocene, and during the Oligocene-Miocene epochs the midcontinent region became increasingly arid.
15. Horses, camels, elephants and other mammals spread across the northern continents because land connections between these continents existed at various times during the Cenozoic. The southern continents were mostly isolated, and their faunas evolving in isolation were unique.

## IMPORTANT TERMS

artiodactyl
carnassials
condylarth
creodont
grazer
*Hyracotherium*
insectivore
marsupial
miacid
molar
molarization
monotreme
perissodactyl
placental
primate
proboscidian
ruminant
ungulate

## REVIEW QUESTIONS

1. Although numerous adaptive types of birds exist, there is little variation in basic structure. Why?
2. Give a brief summary of important Paleocene and Eocene evolutionary events among mammals.
3. Marsupials seem to have been successful only in Australia and South America. Explain. What happened to the South American marsupials near the end of the Cenozoic?
4. How do placental mammals differ from marsupials and monotremes?
5. Compare and contrast the wing structures of bats, birds, and pterosaurs.
6. How were the limbs and teeth of placental carnivores modified for a predatory life?
7. Why is the evolutionary history of mammals more fully known than the history of any other vertebrate class?
8. What are ungulates? How have their limbs been modified for speed?
9. What effect did the appearance of widespread grasslands during the Miocene have on the evolution of horses?
10. Explain how the fossil record demonstrates that animals as different as horses and rhinoceroses share a common ancestor.
11. Give a brief account of the evolutionary history of elephants.
12. How can leaf structure of fossil plants be used to make inferences about ancient climates.

## ADDITIONAL READING

Adams, D. B. 1981. The Nine Lives of the Sabercat. *Science 81*, 2, no. 1: 42–47.

Colbert, E. H. 1980. *Evolution of the Vertebrates.* New York: John Wiley & Sons.

Gingerich, D. D. et al. 1983. Origin of Whales in Epicontinental Remnant Seas: New Evidence from the Early Eocene of Pakistan. *Science*, 220: 403–406.

Marshall, L. G. 1988. Land Mammals and the Great American Interchange. *American Scientist,* 76: 380–388.

Savage, R. J. G. 1977. Evolution of the Carnivorous Mammals. *Paleontology*, 20: 237–271.

Savage, R. J. G. 1986. *Mammal Evolution: An Illustrated Guide.* New York: Facts on File Publications.

Simpson, G. G. 1951. *Horses.* Oxford: Oxford University Press.

Szalay, F. S., and E. Delson, 1979. *Evolutionary History of Primates.* New York: Academic Press.

Wolfe, J. A. 1978. A Paleobotanical Interpretation of Tertiary Climates in the Northern Hemisphere. *American Scientist,* 66: 694–702.

# 17

## CHAPTER OUTLINE



## PROLOGUE

*About 10,000 years ago almost all of the large terrestrial mammals of North America, South America, and Australia became extinct. Extinction events, both large and small, are part of the history of life and have been occurring since the Proterozoic Eon. This latest extinction event was modest by comparison to earlier ones. However, it was unusual in that it affected, with relatively few exceptions, only large terrestrial mammals weighing more than 40 kg.*

*During the Late Pleistocene, North America lost approximately 33 out of 45 genera (73 percent) of its large terrestrial mammals, South America 46 out of 58 genera (80 percent), Australia 15 out of 16 genera (94 percent), Europe 7 out of 23 genera (30 percent), and Africa, south of the Sahara, only 2 out of 44 genera (5 percent). Clearly, Europe and Africa, south of the Sahara, were less affected by the extinction event.*

*What caused this latest extinction, and why did it seemingly affect only large mammals? The debate over this particular extinction event rages between those who believe that the large mammals became extinct because they could not adapt to the rapid climatic changes resulting from the termination of the last glaciation, and those who believe the mammals were killed off by human hunters, a hypothesis known as* prehistoric overkill. *The question is, did climate or humans cause the extinctions of the large mammalian fauna at the end of the Great Ice Age?*

*Those researchers who favor a climatic cause for the extinctions point to the rapid changes in climate and vegetation that occurred over much of the Earth's surface during the Late Pleistocene. As the various glaciers began retreating, the North American and northern Eurasian open-steppe tundras were replaced by conifer and broadleaf forests as warmer and wetter conditions prevailed. The Arctic region flora changed from a productive herbaceous one that supported a variety of large mammals, including mammoths, to a relatively barren waterlogged tundra that supported a far sparser fauna. The southwestern United States region also changed from a moist area with numerous lakes, where sabertooths, giant ground sloths, and mammoths roamed, to a semi-arid environment unable to support a diverse large mammalian fauna.*

# Geology of the Quaternary Period and the Evolution of Humans

*While rapid changes in climate and their effect on vegetation can certainly result in changes in animal populations, there are several problems with the climatic hypothesis as an agent of extinction. First, why didn't the large mammals migrate to more suitable habitats as the climate and vegetation changed? After all, many other animal species did. For example, reindeer and the Arctic fox lived in southern France during the last glaciation and migrated to the Arctic when the climate became warmer. Why didn't the mammoths and other large cold climate-adapted animals simply migrate north?*

*The second argument against the climatic hypothesis is the apparent lack of correlation between extinctions and the earlier glacial advances and retreats throughout the Pleistocene Epoch. Previous changes in climate were not marked by episodes of mass extinctions.*

*Paul Martin of the University of Arizona at Tuscon, the leading proponent of the prehistoric overkill hypothesis, argues that the mass extinctions in North and South America and Australia coincided closely with the arrival of humans in each area. According to Martin, hunters had a tremendous impact on the faunas of North and South America about 11,000 years ago because the animals had no previous experience with humans. The same thing happened much earlier in Australia soon after people arrived about 40,000 years ago. The reason there were no large-scale extinctions in Africa and most of Europe is because animals in those regions had a long familiarity with humans.*

*There are also several arguments against the prehistoric-overkill hypothesis. One problem is that archeological evidence indicates that the early human inhabitants of North and South America, as well as Australia, probably lived in small, scattered communites, gathering food and hunting. It is hard to imagine how a few hunters could have decimated so many species of large mammals. On the other hand, humans have caused major extinctions on oceanic islands. For example, in a period of about 600 years after arriving in New Zealand, humans exterminated several species of the large, flightless birds called moas.*

*A second problem is that modern hunters concentrate on smaller, abundant, and less dangerous animals. The remains of horses, reindeer and other small animals are found in many prehistoric sites in Europe, while large mammoth and woolly rhinoceras remains are scarce.*

*Finally, there are few human artifacts found among the remains of extinct animals in North and South America, and there is usually little evidence that the animals were hunted. Countering this argument is the assertion that the impact on the previously unhunted fauna was so swift as to leave little evidence.*

*The reason for the extinctions of large mammals around 10,000 years ago is still unresolved and probably will be for some time. It may turn out that the extinction was a combination of many different circumstances. Populations that were already under stress from climatic changes were perhaps more vulnerable to hunting when humans occupied new areas.*

## INTRODUCTION

The Quaternary Period is divided into two epochs, the Pleistocene and Holocene (or Recent). Together they represent the past 2 million years in Earth history. From the perspective of 4.6 billion years, these two epochs are but a mere moment in geologic time. If we scale down the geologic history of the Earth to a 24-hour day, this 2 million-year episode would represent only the last 38 seconds of the day! However, from our perspective, these last 38 seconds (2 million years) are an extremely significant interval for two major reasons. First, our own species, *Homo sapiens*, evolved. Second, the Pleistocene Epoch was a time when large ice sheets advanced and retreated over the world's land surface, forming much of our present-day topography (Fig. 17–1). From evidence preserved in the rock record, times of glaciation in the geologic past appear to be rare events; thus, the study of the Pleistocene may provide clues about the causes of previous glaciations.

It seems hard to believe that so many competent naturalists of the last century were skeptical that continental glaciation occurred. Many naturalists invoked the biblical flood to explain the occurrence throughout Europe of large boulders far from their source. Others believed the boulders were rafted to their present location by icebergs floating in floodwaters. It was not until 1837 that the Swiss naturalist Louis Agassiz argued that the

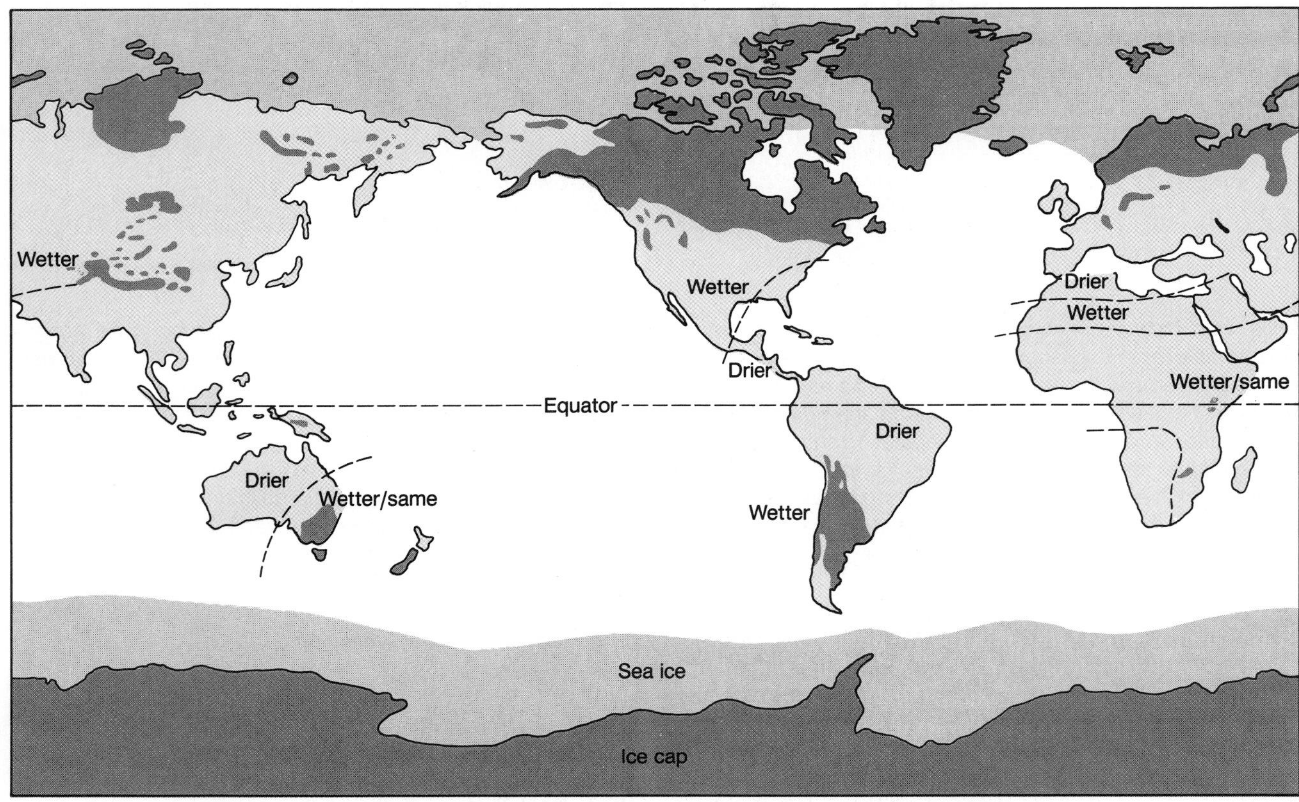


**Figure 17–1** Paleogeography of the world during the last glacial maximum, about 18,000 to 20,000 years ago. Large areas of the Norhtern Hemisphere were covered with ice caps, while many tropical and subtropical areas had mountain glaciers. In nonglaciated areas the pattern of rainfall was much different than it is today.

large displaced boulders (called *erratics*), coarse sedimentary debris, polished and striated bedrock, and U-shaped valleys found throughout parts of Europe were the result of huge ice masses moving over the land. Based on Agassiz's observations and arguments, geologists soon accepted that an Ice Age had indeed occurred during the recent geologic past.

We know today that the last Ice Age began about 2 million years ago and consisted of several intervals of glacial expansion separated by warmer interglacial periods. The present interglacial period began about 10,000 years ago, and geologists do not know whether we are still in an interglacial period or whether we are entering another colder glacial interval.

### The Onset of the Ice Age

The onset of glacial conditions really began around 40 million years ago when surface ocean waters in the high southern latitudes suddenly cooled, and the world's deep ocean basins quickly filled with water about 10° C colder than before. As the Tethys Sea gradually closed during the Oligocene Epoch, the supply of warm water to high latitudes diminished further. By the Middle Miocene a true Antarctic ice cap had formed, accelerating the production of very cold waters. Following a brief warming trend during the Pliocene Epoch, ice caps began forming in the Northern Hemisphere about 2 to 3 million years ago, and the Ice Age was underway (Table 17–1).

## PLEISTOCENE PALEOGEOGRAPHY AND CLIMATE

### Paleogeography

While the Pleistocene Epoch is well known for glaciation, it was also a time of tectonic unrest. Folding, fault-

**Table 17–1** Summary of the evolution of the Ice Age.

| Time (million years ago) | Events |
|---|---|
| 3 | Ice Age begins.<br>Icecaps form in Northern Hemisphere. |
| 5-3 | Gulf Stream intensified.<br>Isthmus of Panama closes. |
| 6 | Rapid cooling occurs. |
| 15 | Present abyssal oceanic circulation established.<br>Iceland-Faroe Ridge sinks; North Atlantic water wells up around Antarctica; more moisture and copious snowpack.<br>Antarctic icecap develops. |
| 30-25 | Antarctica much colder but still no icecap.<br>Circum-Antarctic current established. |
| 35-30 | Partial circulation around Antarctica.<br>Barriers in the western Pacific and closing of Tethys Sea in Middle East and Near East breaks up circum-equatorial circulation pattern. |
| 38 | First glaciers on Antarctica.<br>Deep-water circulation speeds up.<br>Rapid cooling of surface water in the south and of deep water everywhere. |
| 48-45 | Slight cooling takes place. |
| >50 | No glaciers or icecap on Antarctica.<br>Deep water much warmer than presently; circulates slowly.<br>Uniform climate and a warm ocean.<br>Ocean waters flow freely around the world at the equator. |

ing, and uplifts were common occurrences, resulting from Pacific and North American plate interaction along the San Andreas fault system. Many of California's coastal oil- and gas-producing fold structures formed during the Pleistocene Epoch. Elevated marine terraces covered with Pleistocene sediments testify to repeated uplifts during this time (Fig. 17–2).

Much of the Pleistocene volcanic activity in New Mexico, Arizona, Idaho, and Mexico resulted from continued subduction along the western continental margin of North America. Crater Lake, Oregon (Fig. 15–34), which occupies the collapsed caldera of Mt. Mazama; Yellowstone Lake; and Craters of the Moon National Monument, Idaho (Fig. 17–3), are three well-known examples of North American Pleistocene volcanism. Elsewhere, volcanic activity was occurring around the rim of the Pacific Ocean basin such as along South America's western margin and in the East Indies, as well as other locations such as Iceland, Spitzbergen, and the Himilayas.

**Figure 17–2** Wave-cut terraces on the west side of San Clemente Island, California. Each terrace represents a period when that area was at sea level. The highest terraces are now about 400 m above sea level.

**Figure 17–3** Volcanoes in Craters of the Moon National Monument, Idaho. (Photo courtesy of Wayne E. Moore, Central Michigan University).

Pleistocene crustal uplifts occurred in the Sierra Nevada Mountains, the Grand Teton Mountains, and portions of the central and northern Rocky Mountains. These uplifts, particularly the Sierra Nevada, influenced weather conditions by creating rain shadows that led to the present-day deserts located on the landward sides of these mountains. Pleistocene uplifts also occurred in the Alpine-Himalayan orogenic belt (see Chapter 15).

## Climate

The results of oxygen isotope research indicate that world climates grew progressively cooler from the beginning of the Eocene Epoch through the Pleistocene Epoch (Fig. 17–4). The culmination of this cooling trend was not a sudden plunge into frigidity, but rather oscillations between glacial and interglacial stages. Oxygen isotope data from deep-sea cores in the Atlantic, for instance, identify about 20 major warm-cold cycles for the last 2 million years. Variations in mean Earth temperatures during these cycles ranged from 6° C to 10° C. There were also numerous minor temperature fluctuations within each of these major cycles.

During times of glacier growth, those areas in the immediate vicinity of the glaciers experienced very cold conditions. Areas outside the glaciated region experienced varied climates. Because lower ocean temperatures reduced evaporation, most of the world was drier than it is today. However, some areas that are arid today were much wetter during times of glacial growth. For example, since the temperate, subtropical, and tropical zones were compressed toward the equator by the expanding cold belts, the rain that presently falls on the Mediterranean shifted so that it fell over the Sahara region. California and the arid Southwest were also much wetter during glacial times than today because a high-pressure zone over the northern ice cap deflected Pacific winter storms southward.

**Pollen analysis** is an especially useful tool in determining past climates (Fig. 17–5). Produced by the male reproductive bodies of seed plants, pollen possesses a hard waxy coating that resists destruction by chemical weathering. Most pollen is transported by wind and some settles in lakes, ponds, and bogs, where it becomes part of the bottom sediments. The pollen recovered from the sediments can commonly be identified to the family or type of plants or trees that produced it. By identifying, counting, and statistically analyzing the pollen present, one can infer the general floral composition of the area and make some general climatic interpretations.

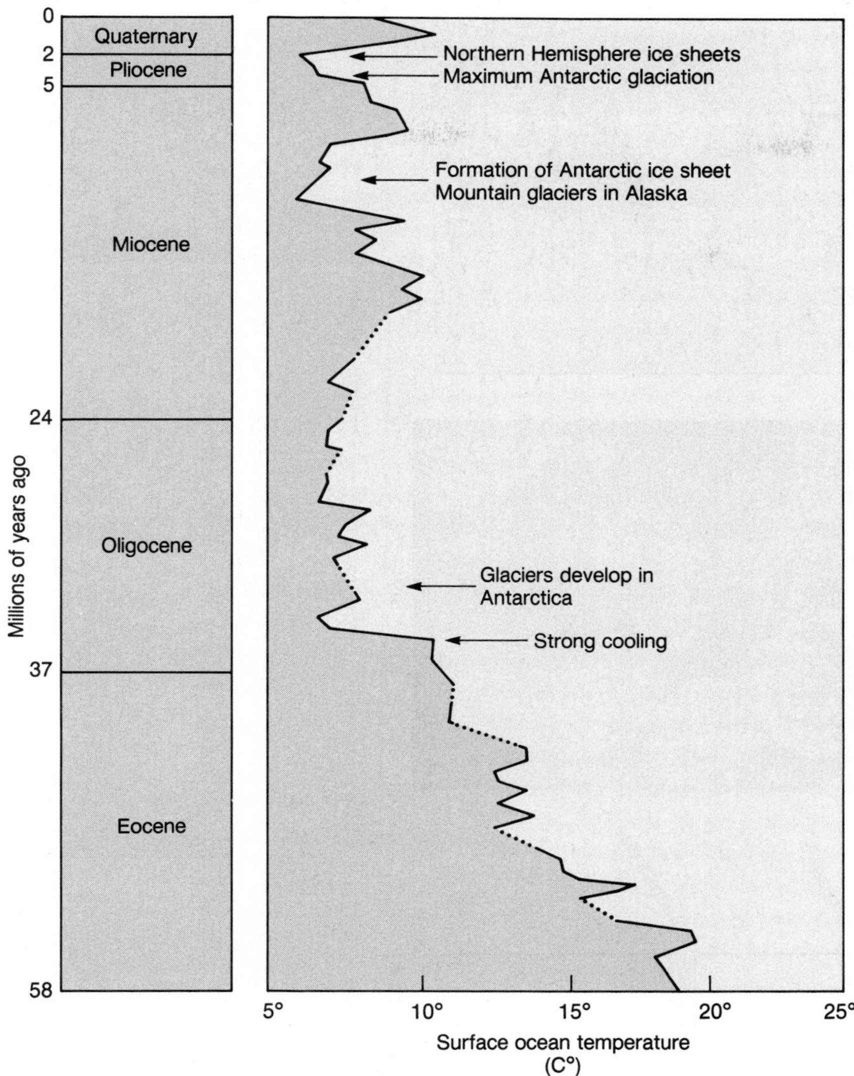


**Figure 17–4** Oxygen isotope curve from a sediment core in the western Pacific Ocean. Fluctuations in the recorded ratio of $0^{18}$ to $0^{16}$ reveal changes in surface ocean temperatures. A change from warm surface waters to colder conditions occurred around 35 million years ago.

Pollen diagrams (Fig. 17–5), tree-ring analysis (see Chapter 2), and measurements of fluctuations of mountain glaciers have produced a wealth of data for reconstructing the climatic history of the Northern Hemisphere since the last major ice sheets retreated 10,000 years ago (Fig. 17–6). Pollen data indicate a continuous trend toward warmer, more temperate climatic conditions until around 6,000 years ago. In many areas the interval between 8,000 and 6,000 years ago was a time of very warm temperatures. Following this warm period, conditions gradually became cooler. This change to a cooler, moister climate favored the growth of mountain glaciers in the Northern Hemisphere continents, known as the *Neoglaciation.* Careful studies of glacial deposits along the margins of present-day glaciers reveal that during the last 6,000 years several glacial expansions have occurred. The last expansion, between the middle thirteenth and middle nineteenth centuries, is marked

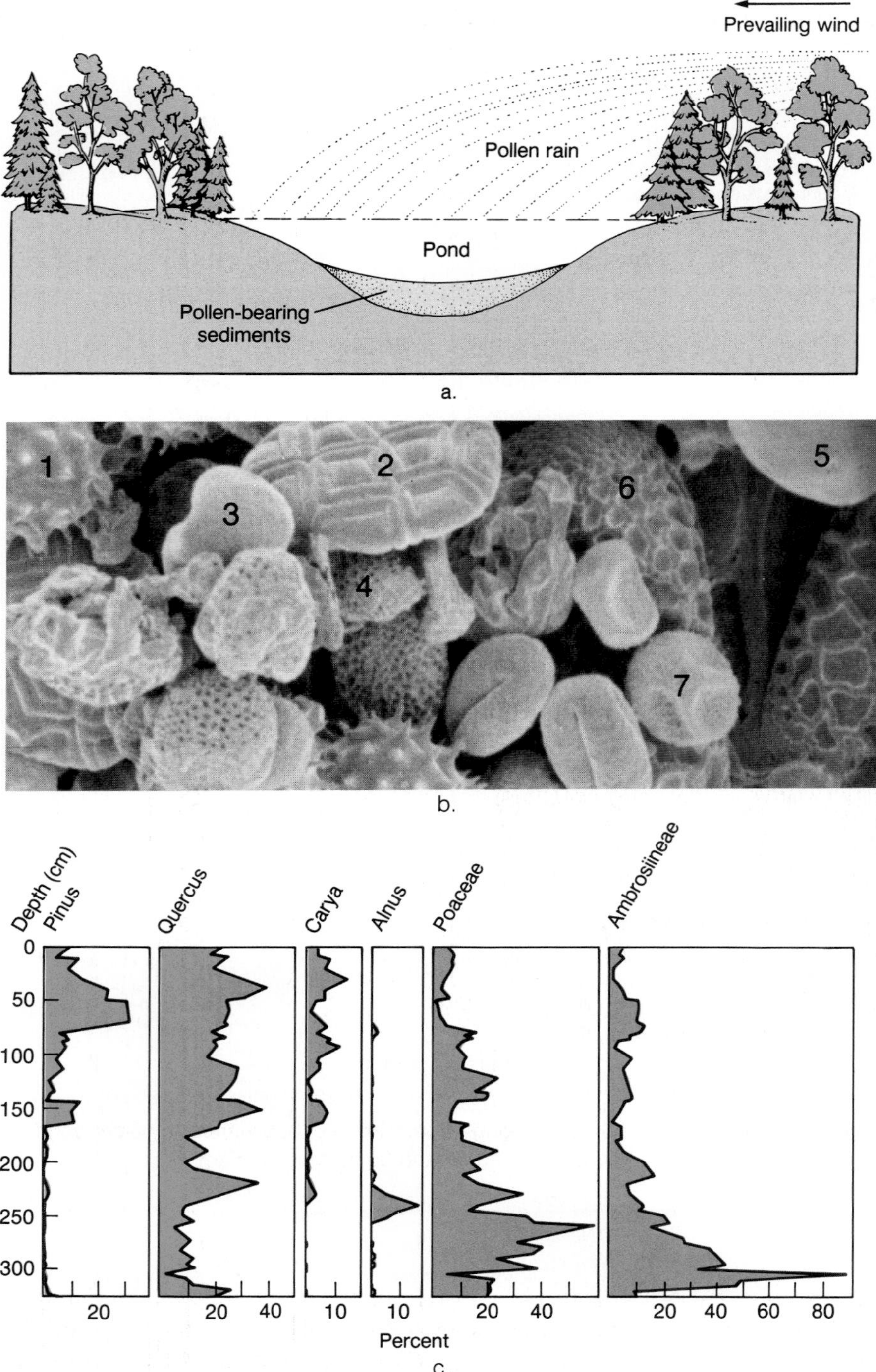


**Figure 17–5** Fossil pollen can be used to reconstruct past vegetation and climate. a. Windborne pollen grains from nearby trees fall into a pond where they are incorporated into the accumulating sediment. b. Scanning electron micrograph of modern pollen grains. Included are sunflower (1), guajillo (2), live oak (3), white mustard (4), little walnut (5), lechuguilla (6), and ash juniper (7). (Photo courtesy of Vaughn M. Bryant, Jr., Texas A and M University) c. Pollen diagram showing pollen abundance for six different trees. The pollen was recovered from samples taken in the Ferndale Bog, Atoka County, Oklahoma. Changes in the pollen abundance reflects changes in the climate during the past 12,000 years at this locality. (Diagram courtesy of Richard G. Holloway, Eastern New Mexico University)

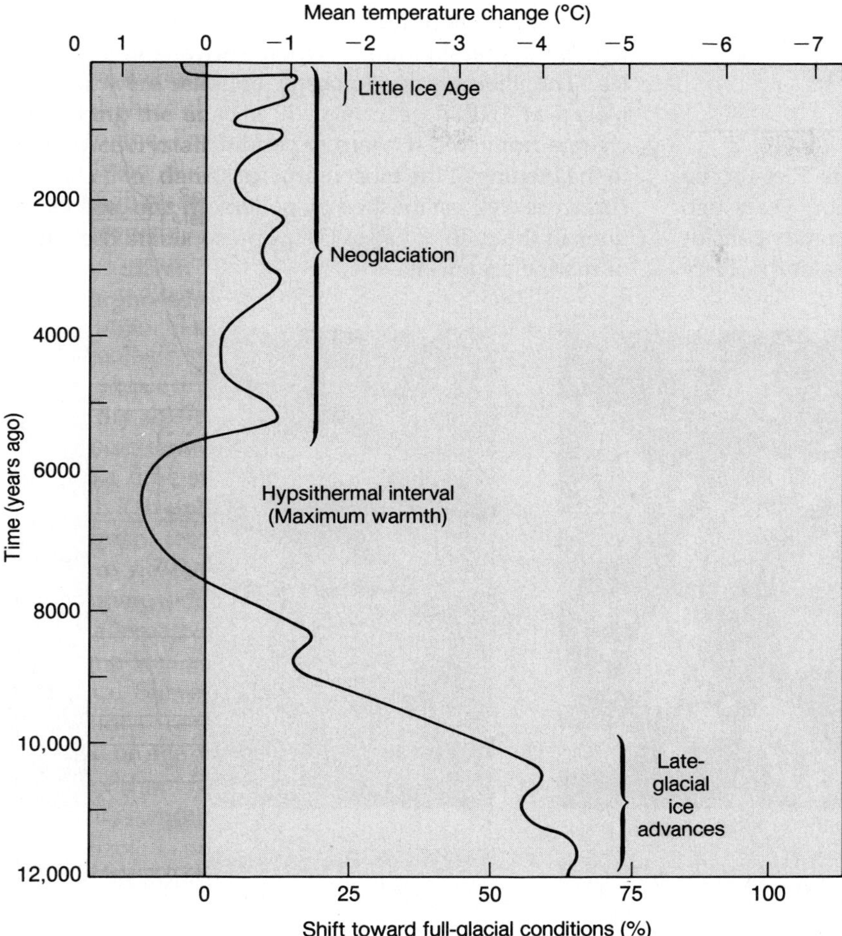


**Figure 17–6** Generalized curve of temperature changes in the Northern Hemisphere for the past 12,000 years. These temperature changes are based on snowline fluctuations and changes in plant-community composition in the mountains.

by such impressive glacial activity that it is called the *Little Ice Age*. During the Little Ice Age, there were frequent episodes of extremely snowy winters and cool, wet summers. These led to many crop failures throughout much of western Europe, resulting in large numbers of people dying of starvation and of diseases related to malnutrition.

Little Ice Age conditions persisted until the middle of the last century when a gradual warming trend caused mountain glaciers to retreat. Although minor temperature fluctuations have taken place since then, the overall warming trend has led to favorable conditions for crop production in the middle latitudes at a time when the human population was expanding rapidly and entering the Industrial Age.

This brief overview of climatic conditions reveals just how complex climatic patterns were for the last 2 million years. Large-scale warm-cold cycles had intermediate fluctuations superimposed on them, and those in turn were composed of even shorter variations, such as the Little Ice Age, which had profound effects on the social and economic fabric of human society.

## PLEISTOCENE AND HOLOCENE CHRONOLOGY

Estimated dates for the beginning of the Pleistocene Epoch have ranged from 0.4 to 4 million years ago (m.y.a.). A section in Calabria, Italy, is currently considered the most suitable for a Pleistocene stratotype. Here a boundary date of 2.0 ± 0.1 million years has been obtained by potassium-argon whole-rock analysis of pumice. The Pleistocene-Holocene boundary is currently placed at 10,000 years ago. It is based on a climatic change from cold to warmer conditions corresponding to the melting of the most recent ice sheets. This climatic change is well-established by pollen changes and variations in the ratio of $O^{18}$ to $O^{16}$ as preserved in the shells of marine organisms.

**Figure 17–7** Features characteristic of glaciated areas. a. Glacial till is the unsorted sediment deposited by a glacier. Till from the Mt. Cook area, South Island, New Zealand. b. Glacial striations are scratches in a rock caused by the abrasion of a moving glacier. Glacial striations and polish on the surface of a basalt flow, Devil's Postpile, California. c. Drumlins are long, streamlined hills composed of till and elongated parallel with the direction of glacial movement. Drumlins in Antrim County, Michigan. d. Erratics are ice-transported boulders that are carried by glaciers far from their original source. Precambrian erratic on top of Upper Cambrian sediments, South Hammond, New York. (a and d courtesy of R.V. Dietrich, c courtesy of B.M.C. Pape, Central Michigan University).

## Stratigraphy of Terrestrial Deposits

About 30 years after Louis Agassiz first proposed his theory for glaciation, research focused on outlining Ice Age history. This work involved recognizing and mapping terrestrial glacial features and placing them in a stratigraphic sequence. From glacial features such as the distribution of moraines, erratic boulders, and drumlins, (Fig. 17–7) geologists recognized that Pleistocene glaciers had covered about 27 million sq km, or about three times the area they now cover, and that the glacial advances and retreats were cyclical.

Mapping the distribution of terminal moraines and correlating the till deposits reveals at least four major episodes of Pleistocene glaciation for North America. Each of these advances was followed by retreating glaciers and warmer climates. The four glacial stages, the Wisconsin, Illinoian, Kansan, and Nebraskan, are named for the states representing the farthest advance where deposits are well-exposed; the interglacial stages, the Sangamon, Yarmouth, and Aftonian, are named for localities of well-exposed interglacial soil and other deposits (Fig. 17–8). However, detailed studies of glacial deposits have indicated that there were an as yet indeterminate number of pre-Illinoian glaciations, and that the history of glacial advances and retreats in North America is more complex than previously believed. While glacial stratigraphers have not yet determined just how many pre-Illinoian glacial advances and retreats occurred, it appears that the traditional four-part subdivision will have to be modified.

While at least four major Pleistocene episodes of glaciation are recognized in North America, six or seven major glacial advances and retreats are recognized in Europe, and at least 20 major warm-cold cycles can be detected in deep sea-cores. Why isn't there better correlation among the different areas if glaciation was a world-wide event? Part of the problem is that glacial deposits are typically chaotic mixtures of coarse materials, making correlation difficult. Furthermore, advances and retreats of the ice sheets usually destroy the sediment left by the previous advance, thus obscuring earlier chronological evidence. Even within a single major glacial advance, several minor advances and retreats may have occurred. For example, a detailed study of the glacial deposits of the Wisconsin Glaciation reveals that in Michigan and Illinois at least four distinct fluctuations of the ice margin occurred during the last 70,000 years (Fig. 17–9).

Geologists using sophisticated radiometric dating techniques and detailed pollen studies can now construct an accurate and detailed chronology for terrestrial Pleistocene glacial and interglacial episodes. However, because of the complexity of glacial deposits and the difficulty of separating similar-appearing terrestrial sediments, they cannot always recognize small-scale fluctuations in climate and correlate them with deposits in other areas.

| North America | Alpine region of Europe |
|---|---|
| Wisconsin | Würm |
| Sangamon | Riss-Würm |
| Illinoian | Riss |
| Yarmouth | Mindel-Riss |
| Kansan | Mindel |
| Aftonian | Gunz-Mindel |
| Nebraskan | Gunz |
| Pre-Nebraskan | Pre-Gunz |

a.

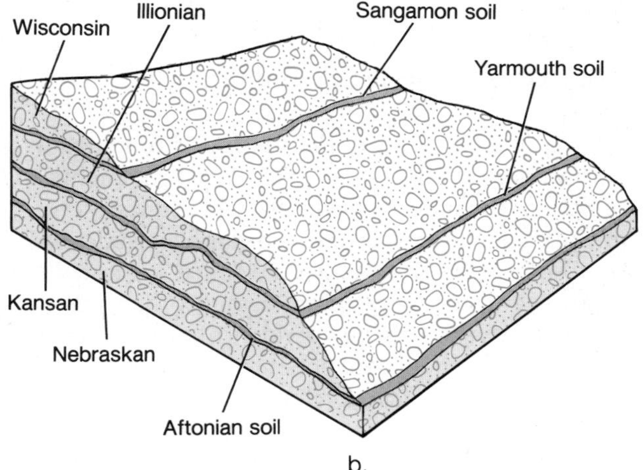


b.

**Figure 17–8** a. Standard Pleistocene nomenclature for glacial and interglacial stages. b. An idealized reconstruction showing the succession of deposits of the four glacial stages and the interglacially developed soils.

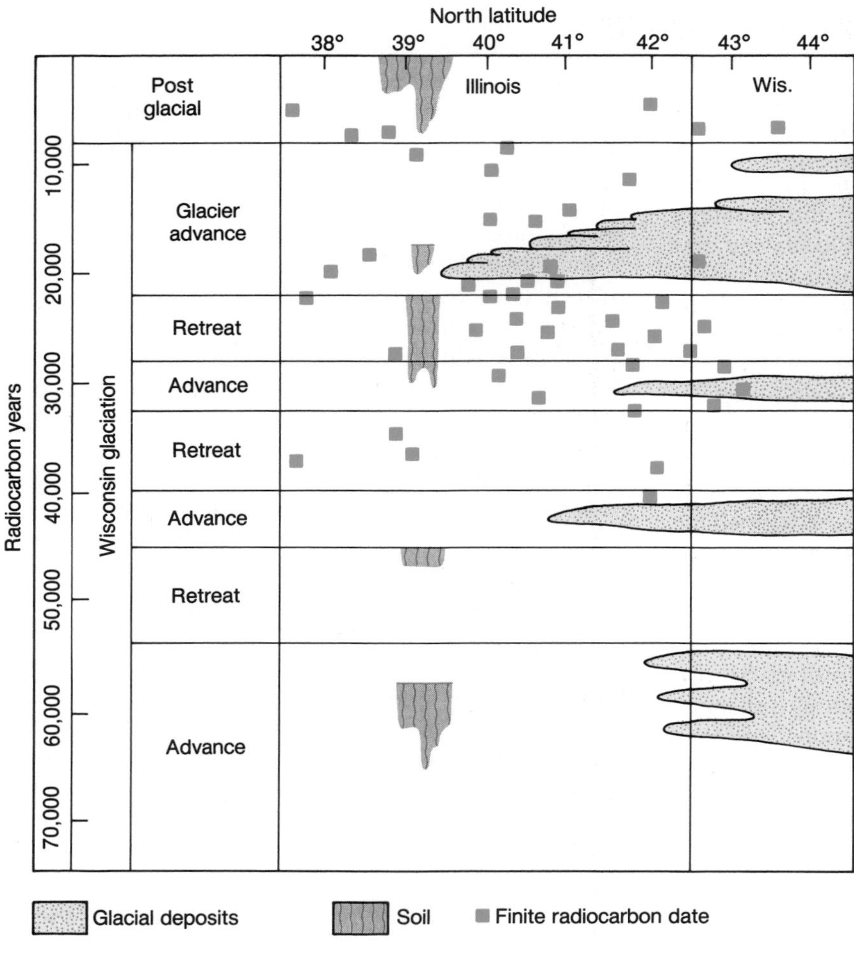


**Figure 17–9** The complex fluctuations of climate and glaciers in Illinois and Wisconsin during the last major period of glaciation are preserved in glacial deposits and intervening nonglacial sediments.

## Pleistocene Deep-Sea Sediments

The traditional view of Pleistocene chronology was based on terrestrial sequences of glacial sediments and seemed to indicate four major glacial advances in North America and perhaps six or seven glacial episodes in Europe. In the early 1960s, new evidence suggested that Pleistocene history is far more complex than was previously believed. This new evidence from ocean sediments reflected changes in ocean temperature and ocean chemistry that indicated climatic fluctuations. Such changes are recorded in the shells of planktonic foraminifera, which after dying and sinking to the sea floor, accumulate as sediment. These shells can be recovered for study by coring the ocean sediments. After the sediments are washed through screens, the foraminiferal shells are then concentrated and analyzed for coiling directions and for $O^{18}$ to $O^{16}$ ratios, as well as to determine the different species present.

One way to determine past changes in ocean surface water temperatures is simply to identify the planktonic foraminifera present to determine whether they were warm- or cold-water species. Many planktonic foraminifera are sensitive to changes in temperature and migrate to different latitudes when the surface water temperature changes. For example, the tropical species *Globorotalia menardii* is present or absent within Pleistocene cores depending on the surface water temperature. During times of cooler climate it is found only near the equator, while during warming times its range extends into the higher latitudes. Other planktonic foraminifera species change their coiling direction in response to temperature fluctuations. The Pleistocene species *Globorotalia truncatulinoides* predominantly coils to the right in water temperatures above 10° C, while in water below 8 to 10° C, it predominantly coils to the left (Fig. 17– 10). On the basis of changing coiling ratios, detailed climatic curves can be constructed (Fig. 17–10).

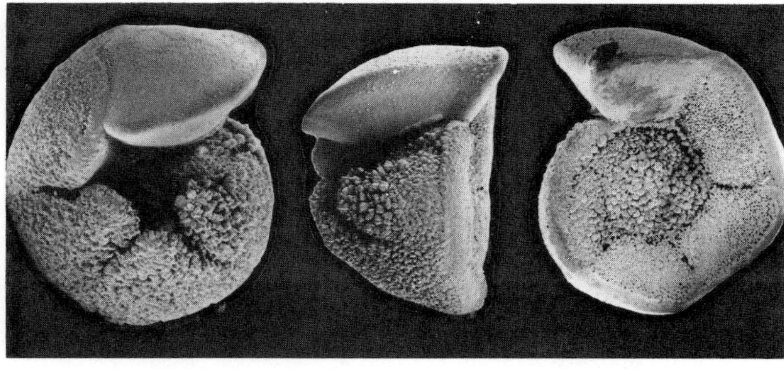

a.

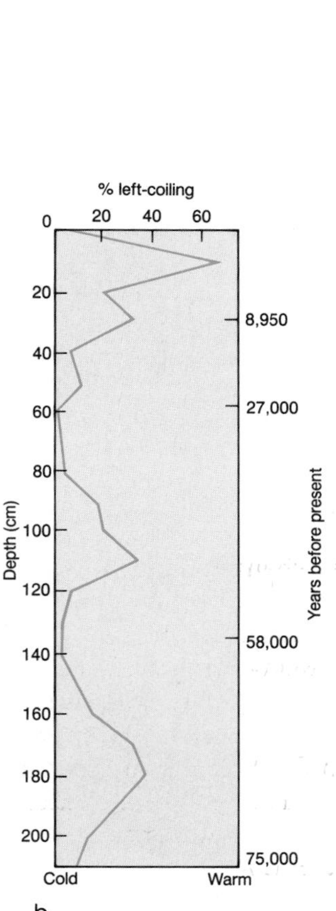


b.

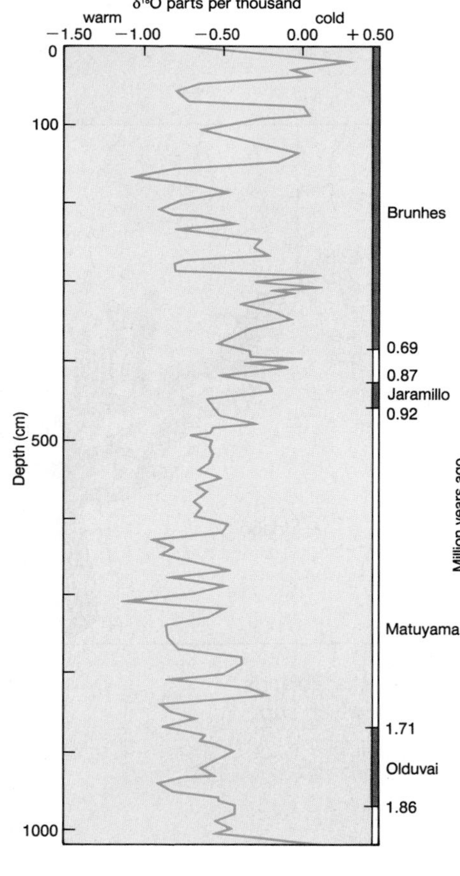


c.

**Figure 17–10** Two methods used to determine changes in ocean suface water temperatures. a. Some planktonic foraminifera species coil in different directions depending on water temperatures. *Globorotalia truncatulinoides* coils to the left in water temperatures below 8 to 10° C. b. Changes in abundance of coiling direction can be used to reconstruct surface water temperatures and hence oceanic climatic conditions. These data, from a core in the Caribbean, show evidence of three intervals of mild climate during the Wisconsin Glaciation. c. Changes in the $0^{18}$ to $0^{16}$ ratio as preserved in planktonic foraminifera shells also reflects surface water temperature changes, and thus climatic changes due to glaciation. (Photo courtesy of Dee Breger, Lamont-Doherty Geological Observatory).

## Perspective 17-1

# The Channeled Scablands of Eastern Washington—A Catastrophic Pleistocene Flood

In 1923 J Harlan Bretz published his "heretical" hypothesis that the channeled scablands of eastern Washington were formed by a single, gigantic flood. Instead of explaining the unusual topography of the channeled scablands as the result of normal stream erosion over a long period of time, Bretz proposed what amounted to a catastrophic hypothesis. As might be expected, his unorthodox explanation was rejected by almost all geologists, who preferred a more traditional interpretation based on normal-stream action. Nonetheless, Bretz persisted and some 40 years later was vindicated.

The term *scabland* is used in the Pacific Northwest to describe areas where the overlying soil covering has been removed or not allowed to form, exposing the underlying rock (Fig. 1). Underlying

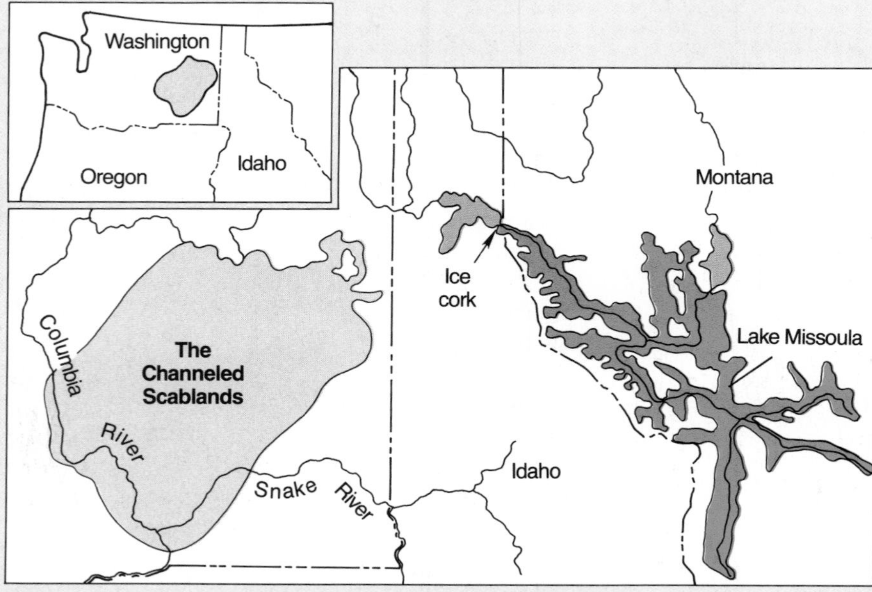


**Figure 1** Outline map showing channeled scablands of eastern Washington and Lake Missoula which supplied the water to form the scablands.

Another method for determining climatic events involves changes in the $O^{18}$ to $O^{16}$ ratio in planktonic foraminifera shells. The abundance of these two oxygen isotopes in the calcareous ($CaCO_3$) shells of foraminifera is related to the amount of dissolved oxygen in seawater when the shell is secreted. The exact ratio of these two isotopes reflects the amount of ocean water stored in glacier ice. When water is evaporated from the ocean and precipitated on land to form glaciers, water containing the light $O^{16}$ isotope is more easily evaporated than water containing the heavier $O^{18}$ isotope. As a result, Pleistocene glaciers contained more of the light isotope, while the oceans became enriched in the heavier isotope. The decreasing percentage of $O^{16}$ and consequent gain of $O^{18}$ in ocean wa-

most of eastern Washington are numerous basalt flows. Most of this lava erupted from fissures during the Miocene Epoch as flow after flow filled in the valleys and then covered most of the landscape. Most of this lava field was then covered by a thick layer (up to 70 m thick in places) of windblown silt or loess during the Pleistocene Epoch. In the region bounded by the Spokane, Snake, and Columbia rivers (Fig. 1), and covering more than 5,100 sq km, are many deep, elongate, subparallel channels that cut through the loess and into the underlying basalt flows. These anastomosing channels are over 70 m deep, have floors covered by gigantic ripple marks up to 10 m high and 70 to 100 m apart (Fig. 2), and have a number of high hills of loess that were not stripped away and are arranged as if they were at one time islands in a huge braided stream. Such are the channeled scablands of eastern Washington.

**Figure 2** Giant ripple marks formed when powerful currents flowed across this area near Crab Creek in eastern Washington. The height of these ripple marks is about 2 meters, and the distance from crest to crest is about 58 meters. (Photo courtesy of Victor R. Baker, University of Arizona).

What Bretz proposed in 1923 was that a single, gigantic flood of glacial meltwater, lasting at most a few days, carved these spectacular channels sometime between 18,000 to 20,000 years ago.

The problem with Bretz's hypothesis was it identified no adequate source for the floodwater. He knew that the glaciers had advanced as far south as Spokane, but he could theorize no mechanism to melt so much water so rapidly. The answer to Bretz's dilemma came from western Montana, where an enormous ice-dammed lake (Lake Missoula) had formed with its spillway leading right into the channeled scablands. Geologists had known about Lake Missoula since the 1880s, but Bretz had not made the connection between the lake as a source of water and the channeled scablands resulting from the draining of the lake.

ters during times of glaciation is recorded in the shells of planktonic foraminifera. Therefore, oxygen isotope fluctuations in planktonic foraminifera accurately reflect surface water temperature changes, and thus climatic changes due to glaciation (Fig. 17–10).

Unfortunately, geologists have not yet been able to correlate these detailed climatic changes to corresponding changes recorded in the terrestrial sedimentary record. The time lag between the onset of cooling and the resulting glacial advance produces discrepancies between the marine and terrestrial records. Thus, it is unlikely that all the minor climatic fluctuations recorded in deep-sea sediments will ever be correlated with the continental stratigraphic units.

## Perspective 17–1 (continued)

Lake Missoula formed when the advancing glacier plugged the Clark Fork Valley at Ice Cork, causing water to fill the valley and create the lake (Fig. 1). At its highest level, the lake covered an area of 7,770 square km and contained an estimated 2,090 cubic kilometers of water (half the volume of present-day Lake Michigan). It was nearly 700 m deep at its deepest point. One can still see the shorelines of Lake Missoula in the mountains around Missoula, Montana.

When the ice dam at Ice Cork finally failed, the water drained to the south and southwest into the scablands at a rate unmatched by any known flood. The maximum rate of flow is estimated to have been 10.8 million cubic meters per second, or about 10 times the combined flow of all the rivers of the world. For comparison, the present-day Columbia River averages 7,220 cubic meters per second, while the Amazon, the world's largest river, flows at 168,000 cubic meters per second. When the flood reached eastern Washington, it carved out huge valleys in the loess, exposing the basalt below, thus creating the scablands. The currents were so turbulent and forceful they eroded the basalt, plucking out pieces up to 10 m across.

While Bretz originally believed that only one massive flood occurred, geologists now know Lake Missoula formed and reformed at least four times and perhaps as many as seven times as the glacier advanced and retreated. The largest lake formed 18,000 to 20,000 years ago, and its draining produced the last great flood.

How long did the flood last, and did early humans witness it? It has been estimated that from the time when the ice dam first broke, letting water rush out onto the scablands, to the time scabland streams returned to normal flow was about one month. No one knows for sure if anyone was around to witness the great flood. The earliest known evidence of humans in the region is from the Marmes Man site in southeastern Washington dated at 10,130 years ago, or several thousand years after the flood.

## THE EFFECTS OF GLACIATION

The direct and indirect effects of glaciation have been many. Movement of glaciers over the Earth's surface has altered the landscape. The formation and melting of glaciers has changed sea level, which in turn, affected the margins of continents. Glaciers have also affected the world's climate, causing cooler and wetter conditions in some areas that today are arid to semiarid. In addition to the usual evidence of glacial activity (Fig. 17–7), one of the largest floods in history was caused by the breaking of a glacier dammed lake resulting in huge ripple marks in eastern Washington (See Perspective 17–1).

### Continental Shelves and Submarine Canyons

During maximum glacial coverage, more than 70 million cu km of snow and ice covered the continents. This not only had a major effect on the glaciated areas, but also on regions far removed from the glaciers. The storage of ocean waters in the glaciers resulted in a 130 m lowering of sea level. The drop in sea level exposed extensive tracts of the present-day continental shelves, which were quickly covered by grasslands and forests.Indeed, a land bridge stretched over the Bering Strait from Siberia to Alaska, allowing the migration of animals between the two continents. Because the shallow North Sea area was above sea level, the British Isles were joined to Europe during these glacial intervals. When temperatures rose and the glaciers retreated, these areas were again flooded, drowning the trees and plants and forcing the animals inland.

Lowering of sea level during times of glacial maxima also affected the base level of most major rivers. When sea level dropped, downcutting accelerated as rivers sought to achieve equilibrium with the new lower sea level. River channels in coastal areas were extended and deepened along the emergent continental shelves. When sea level

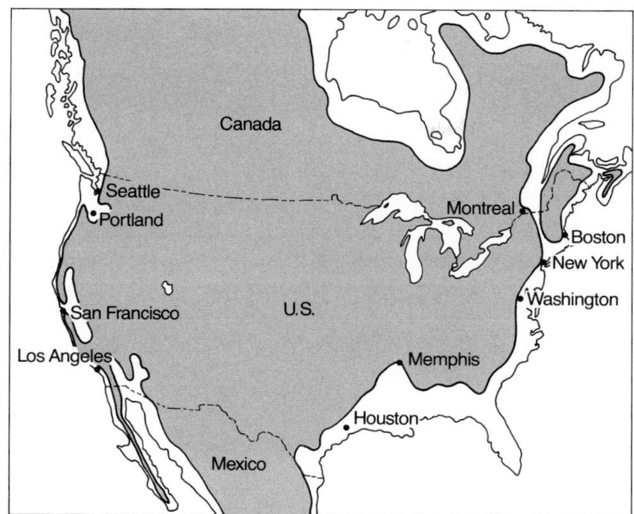


**Figure 17–11** Paleogeography of North America if the present ice caps should melt. Sea level would rise at least 65 m, flooding all the coastal areas including most of the major cities of the world.

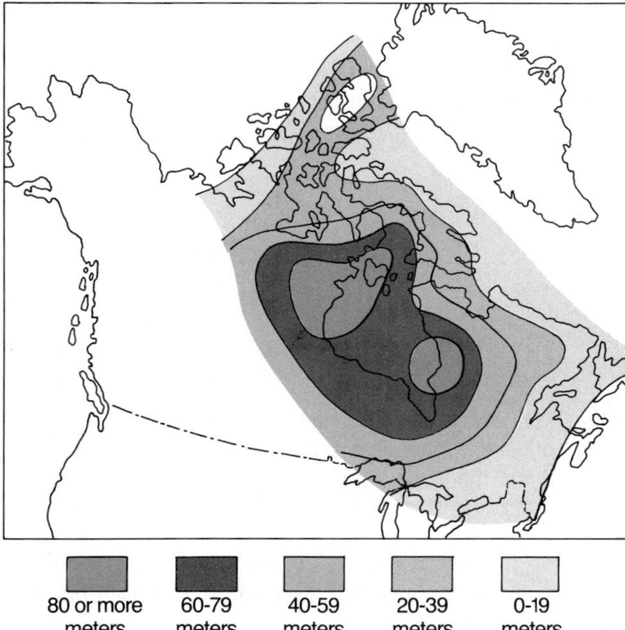


**Figure 17–12** Evidence of isostatic rebound. Areas of North America have risen more than 80 m since the last Pleistocene ice cap melted. The amount of rebound has been determined by measuring elevations of marine sediments 6,000 years old.

rose with the melting of the glaciers, the channels became submarine canyons. In addition, great amounts of sediment were transported to the sea during the Pleistocene Epoch due to accelerated erosion, contributing to the growth of submarine fans along the base of the continental slope.

A tremendous amount of water is still stored in present-day glaciers (25 million cu km). Should these glaciers completely melt, sea level would rise by about 65 m, flooding all of the coastal areas, including most of the major cities of the world such as New York, San Francisco, Los Angeles, London, and Tokyo (Fig. 17–11).

## Isostatic Rebound

Where sufficient sediment or ice accumulates, the Earth's crust gradually sinks under the burden of this weight. Conversely, when the weight is removed by erosion or melting, the crust slowly rises, or rebounds. Such a phenomenon is known as **isostasy** and helps explain some of the features seen in formerly glaciated areas.

When an ice sheet increases its size, the weight of the ice causes the crust to slowly sink. Some places have been depressed as much as 200 to 300 m below their preglacial elevation. As the ice sheet retreats by melting, the downwarped areas of the crust gradually rebound to their former positions. Evidence of isostatic rebound can be found in many parts of the Baltic, the Arctic, and the North American Great Lakes region where former coastal features are now elevated high above sea level and are still rising (Fig. 17–12).

## Pluvial Lakes

The climate of the southwestern United States today is hot and dry, and the area has few major rivers or lakes. However, during the Wisconsin Glacial Stage, there were many large lakes in what are now dry basins. These lakes formed as a result of overall cooler temperatures (especially summer temperatures) that lowered the evaporation rate. At the same time, increased precipitation and runoff helped maintain high water levels. Lakes that formed during those times are called **pluvial lakes** and they correspond to the expansion of glaciers elsewhere (Fig. 17–13). Ancient shoreline features such as wave-cut cliffs, beaches, and deltas found today on the sides of the enclosing slopes are evidence of these former lakes (Fig. 17–13). The largest was Lake Bonneville, which attained a maximum size of 50,000 sq km and a depth of at least 335 m. The Great Salt Lake in Utah is simply the shrunken remnant of this once great lake.

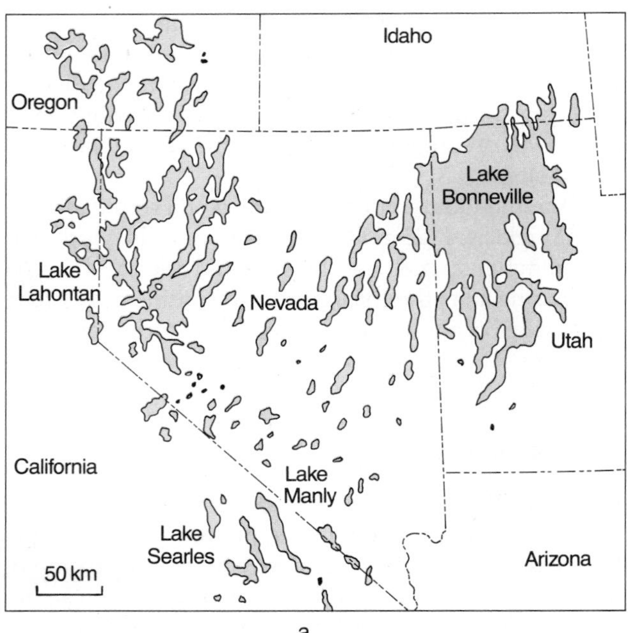


a.

b.

**Figure 17–13** a. Pluvial lakes in the southwestern United States during the last Pleistocene (Wisconsian) Ice Age. Lake Bonneville was one of the largest pluvial lakes, and the Great Salt Lake, Utah is a remnant of it. b. Evidence of the extent of Lake Bonneville can be seen in the ancient shorelines cut into the slope of the Oquirrh Range, Utah.

## History of the Great Lakes

The Great Lakes area of North America provides us with an excellent example of the direct effects of glaciation. Prior to the Pleistocene Epoch, only generally flat lowlands existed in this part of North America. The northern segment of the Missouri River drained northward into Hudson Bay, and the northern part of the Ohio River flowed northeastward into the Gulf of St. Lawrence. As the Laurentide Ice Sheet advanced, those parts of the Missouri and Ohio that once flowed north were turned aside by the ice sheet and forced to flow along the edge of the glacier until they found a southern outlet. The present east-west trend of the northern segment of the Missouri and Ohio Rivers approximates the southern boundary of that ice sheet.

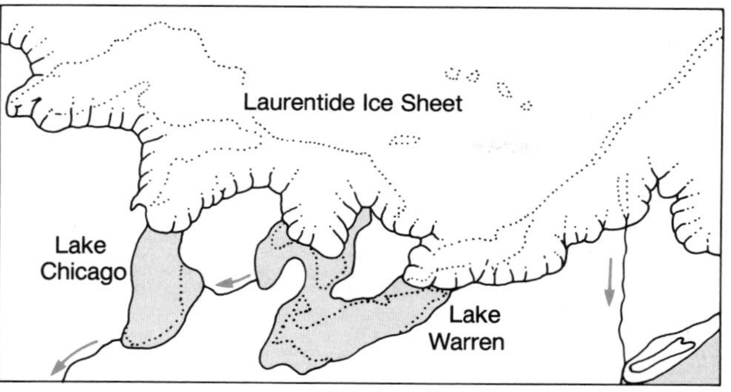


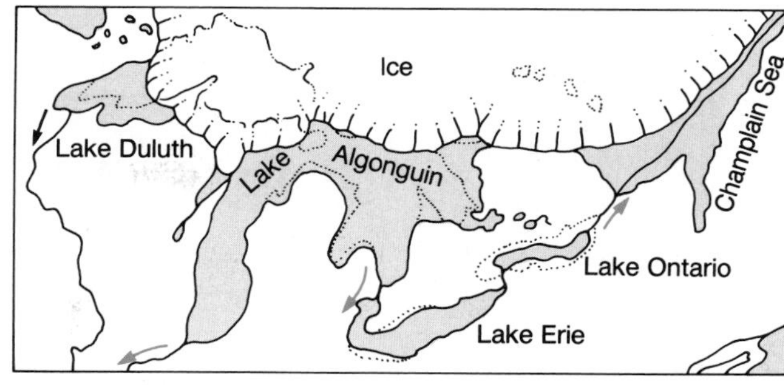


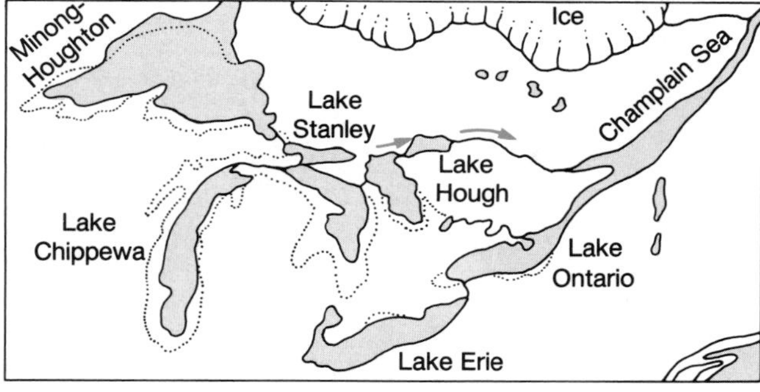


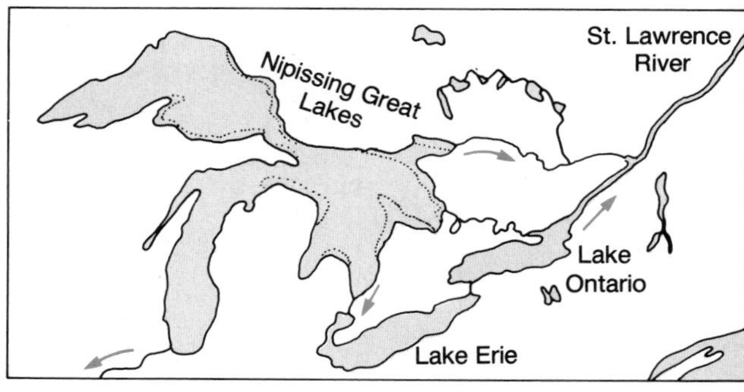


**Figure 17–14** Glacial history of the Great Lakes region from erosion by glaciers 13,000 years ago to subsequent filling by meltwater.

As the glaciers advanced southward, they scoured out the lowlands, and when they retreated, their meltwaters filled the vacated depressions (Fig. 17–14). As the ice melted, isostatic rebound raised the southern parts of the Great Lakes, greatly altering the previous southward-flowing drainage system. Niagara Falls, between Lake Ontario and Lake Erie, came into existence when the retreating glacier uncovered a resistant escarpment of gently dipping Silurian Lockport Dolomite, a remnant of a time when this area was located in warm low latitudes and large coral reefs were growing rather than cold glaciers! Water from the Niagara River flowed over the edge of this escarpment, and Niagara Falls was born.

## CAUSES OF PLEISTOCENE GLACIATION

The evidence for glaciation and the effects of Pleistocene glaciation, still leave the central question of what causes large-scale glaciation and why isn't it more common in the geologic past? The answer to that question is not simple. Because climates change on different time scales, ranging from decades to millions of years, different mechanisms are likely responsible for these changes (Fig. 17–15). These mechanisms interact in complex ways, making it all the more difficult to understand the causes of climatic variability.

Periods of glaciation, some encompassing millions of years, and separated by long intervals of mild climate, can be identified in the geologic record (Fig. 17–15). Climatic changes on this scale are probably the result of slow geographic changes that affected the Earth's crust. These changes resulted from plate tectonic activity that included (1) the movement of continents as they were carried along by plates; (2) the creation of mountain chains at convergent plate margins; (3) the opening and closing of ocean basins and seaways between continents; and (4) the resulting changes in atmospheric and oceanic circulation patterns caused by the changing shapes, positions, and altitudes of the various landmasses through time.

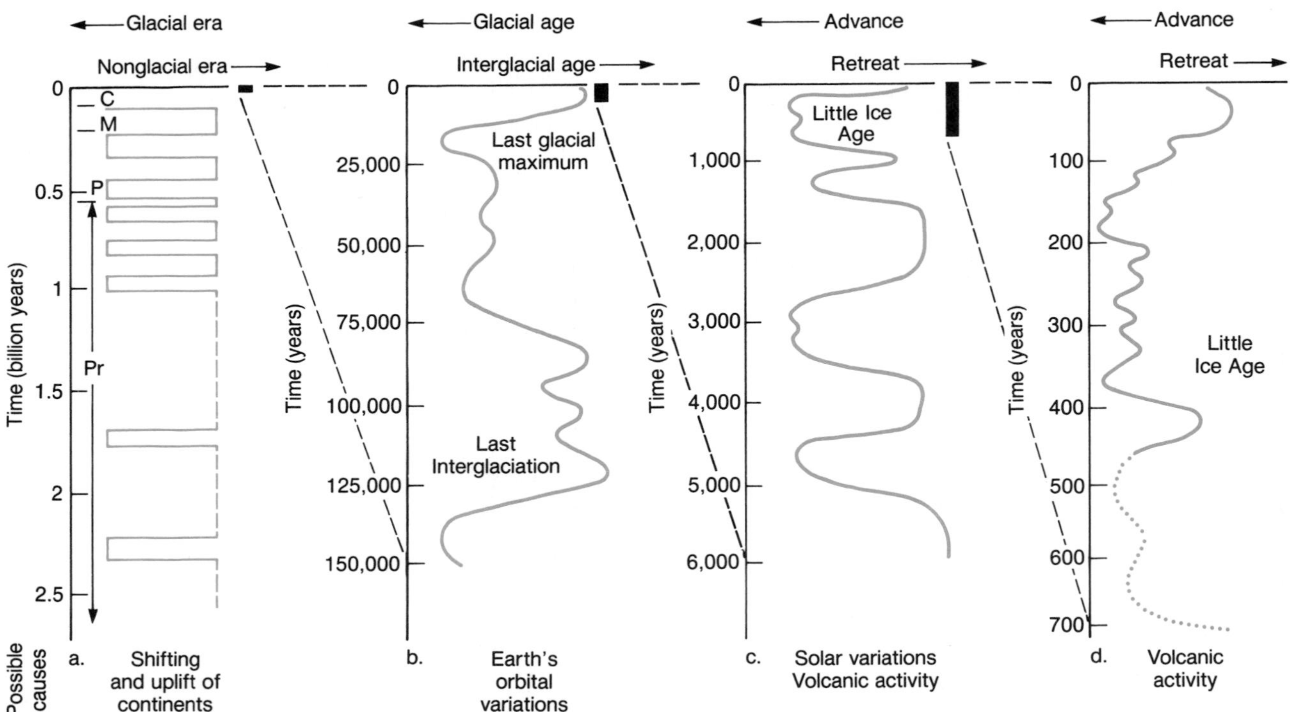


**Figure 17–15** Time scales of climatic variations and possible causes. a. Major glacial and interglacial periods encompassing millions of years and probably resulting from slow geographic changes due to plate movement. Pr = Proterozoic, P = Paleozoic, M = Mesozoic, and C = Cenozoic. b. Glacial and interglacial periods on the order of tens of thousands of years due to variation in the Earth's orbit. c. and d. Short-term climatic events of only a few hundred years duration. They are probably due to variations in the sun's energy output or to volcanic activity.

While the changing geography brought about by plate movement influenced long-term climatic changes, geologists must still explain what controls the glacial-interglacial cycles, the little ice ages, and the even shorter climatic changes we have evidence for in the geologic record. The answer to these questions is of more than academic interest. Understanding the cause of geologically short climate changes could provide the key to long-range weather forecasting and even the potential to affect local weather patterns to increase crop production.

## Ice-Age Periodicity and the Milankovitch Theory

The major glacial periods in Earth history resulting primarily from plate tectonic activity can be viewed as first-order climatic events. The glacial-interglacial cycles are considered second-order events, and superimposed on them are a series of third-order events having durations of only several hundred years, such as the Little Ice Age (Fig. 17–15). As initially discovered from terrestrial glacial deposits and later verified by studies of deep-sea sediments, glacial and interglacial ages have alternated during the past 2 million years. The cause of this cyclicity has long been a problem in developing a comprehensive theory of climatic change.

A particularly interesting hypothesis for these second-order climatic events of glacial-interglacial cyclicity was put forth by the Yugoslavian astronomer Milutin Milankovitch in the 1940s. He proposed that minor irregularities in the Earth's rotation and orbit around the sun are sufficient to alter the amount of solar radiation the Earth receives at any given latitude and hence affect climatic changes (Fig. 17–16). Called the **Milankovitch theory,** it was largely ignored because it was considered untestable because the Ice Age chronology was insufficiently precise to correlate with orbital variations. Recently, variations in oxygen isotope ratios from deep-sea cores have provided the necessary data for finer resolution of Pleistocene climatic events. These data demonstrate a correlation between Ice Age events

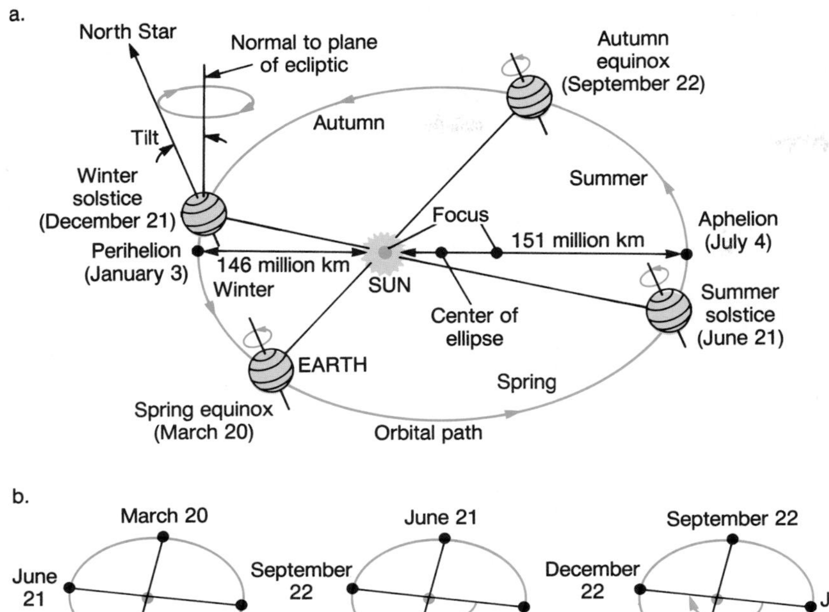


**Figure 17–16** a. Geometry of the sun-Earth system. The Earth's orbit, an ellipse with the sun at one focus, defines the plane of the ecliptic. The Earth moves around its orbit in the direction of the arrows, while spinning about its axis, which is tilted to the plane of the ecliptic at 23.5° and points toward the North Star. b. Precession of the equinoxes causes the position of the equinoxes and solstices to shift slowly around the Earth's elliptical orbit.

and orbital variation and have revived interest in the Milankovitch theory.

Milankovitch attributed the onset of the Ice Age to variations in three parameters of the Earth's orbit (Fig. 17–17). The first parameter— the shape, or eccentricity, of the orbit—is the degree to which the orbit departs from a perfect circle. Calculations reveal a roughly 100,000-year-cycle between times of maximum eccentricity. This 100,000-year-cycle corresponds closely to the 20 warm-cold climatic cycles indicated by paleontologic and isotope data from deep-sea cores. The second parameter, the tilt of the Earth's axis, is the angle between the Earth's axis and a line perpendicular to the plane of the ecliptic (Fig. 17–16). This angle, which is now 23.5° shifts about 1.5° one way or the other during a 41,000-year-cycle. The third parameter is the precession of the equinoxes, which causes the position of the equinoxes and solstices to shift slowly around the Earth's elliptical plane (Fig. 17–16). These positions complete one full cycle about every 23,000 years. Each of these parameters operates on a different time scale, so when they interact, the resulting pattern is quite complicated (Fig. 17–17). The amount of solar heat received at any latitude fluctuates continuously by a small amount due to the interaction of these three parameters. However, the total heat received by the planet remains little changed.

The close correlation between climatic changes and the amount of solar radiation received by the Earth has provided considerable support for the theory that astronomical factors control the timing of glacial-interglacial events.

## Solar Variation, Volcanic Activity, and Third-Order Climatic Events

The third-order climatic events have a time duration of only a few hundred years and are too short to be affected by either continental movement or the three astronomical parameters. For these we must seek other explanations. Numerous hypotheses have been proposed, two of which have received special attention.

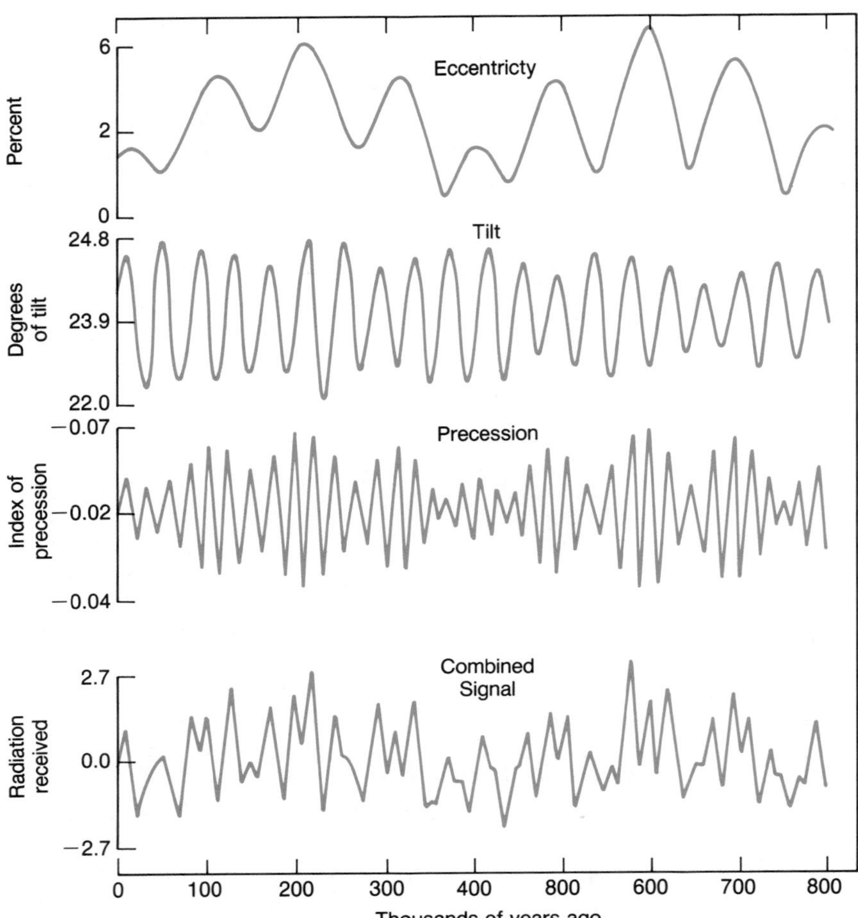


**Figure 17–17** Pattern of orbital eccentricity, axial tilt, and precession during the last 800,000 years. When combined, these factors produce an irregular curve that portrays variations in the amount of radiation reaching the Earth as a function of time.

The first is based on the idea that the sun's energy output has fluctuated through time (Fig. 17–15). This idea is appealing because it could explain climatic variations on many time scales. While measurements of incoming solar radiation have been made for the last 75 years, they show only a slight change during that time. Fluctuations in the production of carbon 14 in the upper atmosphere are known to be related to solar activity. These variations are revealed by isotopic measurements of tree rings and do not show any significant correlation between climatic and solar radiation changes for the past 1,000 years. While the sun may influence climate over a short time frame, a relationship has yet to be clearly established.

The second hypothesis involves volcanic eruptions (Fig. 17–15). Some volcanos erupt tremendous amounts of ash into the atmosphere. The ash reflects incoming solar radiation and thus reduces atmospheric temperatures. It can remain in the atmosphere for several years as it circles the globe, producing spectacular sunsets and resulting in a slight cooling at the Earth's surface. Following the 1883 eruption of Krakatoa in Indonesia, worldwide temperatures dropped about 2° C. The 1815 eruption of Mount Tambora, also in Indonesia, resulted in one of the coldest summers (1816) on record in the Northern Hemisphere. It was known in North America as "eighteen hundred and froze to death." Killer frosts and even snows in June, July, and August virtually wiped out the summer New England crops, leading to higher food prices, lower meat prices, and more farm failures.

While volcanic eruptions definitely affect climates, no correlation between episodes of volcanic activity and periods of glaciation has been demonstrated. This may be due in part to the fact we still know so little about the history of volcanic activity prior to this century. Although geologists can relate, with varying degrees of confidence, different mechanisms of climatic change to the types of glacial events recorded in the rock record, it is the climate of the immediate future we are most concerned about. This uncertainty poses a major challenge to geologists, whose vast array of information about past climates and surface conditions will, it is hoped, provide clues to predicting future climatic changes and their consequences.

## HUMAN EVOLUTION

**Primates** are difficult to characterize as an order since they lack the strong specializations seen in most other mammalian orders. We can point to some trends such as larger brain size, stereoscopic vision, and a grasping hand, all of which are related to an arboreal, or tree-dwelling, existence. However, in a number of other respects, primates remain primitive. For example, primates retain five fingers and toes, while in carnivorous and herbivorous mammals the trend has been toward a reduction of these elements. Furthermore, the teeth of carnivores and herbivores are specialized for their respective diets, yet primates lack such specializations because they are omnivorous, meaning that they include both meat and vegetable matter in their diets.

The primates are divided into two suborders (Fig. 17–18). The **prosimians** are primitive primates that include the tree shrews, lemurs, and tarsiers. The **anthropoids** descended from ancestors of prosimians and include monkeys, apes, humans, and their ancestors.

## THE PROSIMIANS

Extant prosimians retain the primitive characteristics of the primates in that they are arboreal, have five-digits on each hand and foot with either claws or nails, have at least partial stereoscopic vision, and are typically omnivorous. As a group, prosimians were abundant, diversified, and geographically widespread during the Late Paleocene and Eocene (Fig. 17–19). However, as the continents continued drifting northward from the warm tropical climates into the cooler middle latitudes, the prosimian population decreased in both numbers and distribution. Tarsiers and lemurs, abundant in North America during the Eocene Epoch, were virtually gone by the Oligocene Epoch. The widespread Paleocene-Eocene prosimian populations of Europe and Asia began migrating southward during the Oligocene Epoch to the warmer latitudes of Africa, Asia, and the East Indies. Today, prosimians are found only in the tropical regions of Asia, India, Africa, and Madagascar.

## THE ANTHROPOIDS

The anthropoids can be divided into three superfamilies: the New World monkeys, the Old World monkeys, and the hominoids. The New World monkeys, identified by their prehensile tail, flattish faces, and widely separated nostrils, include the howler, spider, and squirrel monkeys (Fig. 17–18). The New World monkeys evolved independently from the Old World monkeys and hominoids, and their earliest fossil record is from the Oligocene Epoch of South America.

The Old World monkeys are characterized by close-set downward-directed nostrils (like those of humans), grasping hands, and a non-prehensile tail. They include the macaque, baboon, and proboscis monkeys (Fig. 17–18). The oldest fossils of this group come from African Oligocene beds. Today Old World monkeys are widely distributed in the tropical regions of Africa and Asia.

The **hominoids** currently consist of three families: the apes, which include the chimpanzees, orangutans, and gorillas; the gibbons and siamangs; and the **hominids,** which are humans and their extinct relatives. The divergence of the hominoids from the Old World monkeys occurred after the Middle Oligocene and before the Early Miocene.

Presently, fossil evidence indicates that the human family (Hominidae) did not evolve from the modern ape family (Pongidae), but rather followed an independent line of evolution. Whether or not they evolved from a single family of primitive apes is still not clear.

Two families of ape-like species that evolved during the Miocene Epoch have both been considered potential ancestors of humans and apes. The first family, the **dryopithecines,** originated in Africa near the beginning of the Miocene Epoch about 20 million years ago and

**Figure 17–18** Primates are divided into two suborders, the prosimians (a) and anthropoids (b through d). The anthropoids are divided into three superfamilies: New World monkeys (b), Old World monkeys (c), and Great apes (d) and humans.

**Figure 17–19** The Eocene lemur *Notharctus.*

spread to Eurasia about 14 million years ago (m. y. a.), approximately 6 million years after Africa collided with Asia. While no complete skull or skeletons of dryopithecines have been found, partial skulls, jaws, and fragmentary limb-bone fossils reveal they had a relatively small brain, ape-like teeth and jaw, and an ape-like face (Fig. 17–20). Their limb structure suggests a four-legged arboreal existence with limited activity on the ground.

The second possible ancestors of modern humans and apes are the **ramapithecines,** which lived from 17 m.y.a. to about 7 m.y.a. Members of the ramapithecines were small creatures, a little over a meter tall, ranging in weight from 20 to 70 kg. Their face was foreshortened and more hominid than ape-appearing. They had rather small canine teeth, and molars with thick enamel and flat chewing surfaces (Fig. 17–21).

Currently, insufficient fossil evidence exists to establish the relationship between the dryopithecines and ramapithecines, or to establish their relationships to modern apes or humans. The fossil evidence clearly indicates that early primate evolution was not a case of orderly change along a single path, but instead a complex of parallel and diverging branches.

**Figure 17–20** Probable life appearance of a dryopithecine. The fossil record of dryopithecines is poor. Based on fragmentary limb-bone fossils, it is suggested they led a four-legged arboreal existence with limited activity on the ground.

## The Australopithecines

The fossil record of the hominids extends back to only about 4 m.y.a., while that of the dryopithecines and

a.

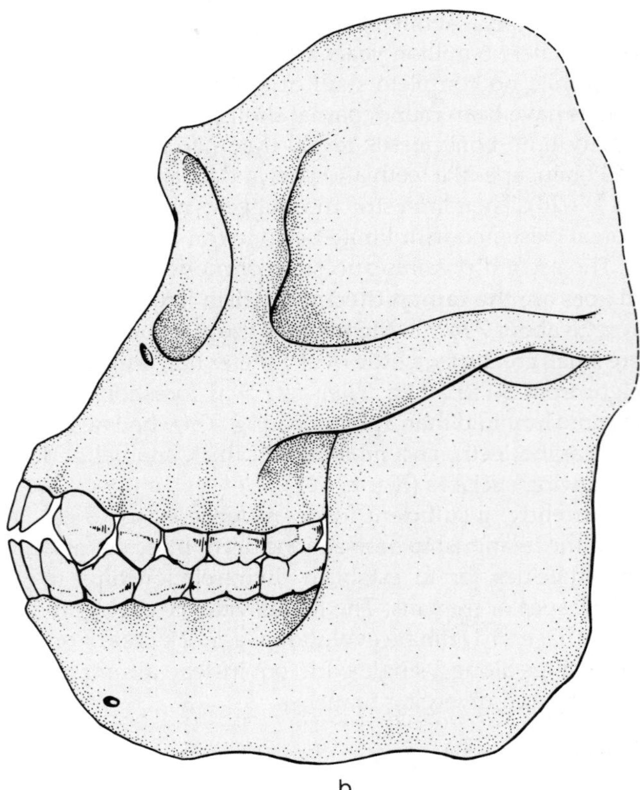

b.

**Figure 17–21** a. Molars and upper jawbone of *Ramapithecus*. b. A reconstruction of the skull of *Ramapithecus*.

ramapithecines, considered by some to be potential ancestors of modern apes and humans, ends about 7 m.y.a. This approximately 3- to 4-million-year interval is believed to be the time when hominids diverged from their hominoid ancestors. Hominids differ from other hominoids in two significant features. The first is an upright posture (hominids are bipeds), with its modification of the pelvis and limbs (Fig. 17–22), and the second is a trend toward a large and internally reorganized brain (Fig. 17-23). Other changes include a reduced face and canine teeth, manual dexterity, and the associated use and construction of sophisticated tools.

The oldest hominids belong to the genus *Australopithecus.* The first specimens of this genus were found in South Africa in the 1920s and 1930s by Raymond Dart and Robert Broom. Since then, other specimens have been found in Tanzania, Kenya, and Ethiopia by a number of different workers, the most famous being Louis and Mary Leakey and their son Richard. To date, four species of *Australopithecus (A. afarensis, A. africanus, A. robustus,* and *A. boisei)* are generally recognized.

Although debate still abounds, especially in light of a recent find of *Australopithecus boisei,* many paleontologists accept the evolutionary scheme shown in Figure 17–24 in which *Australopithecus afarensis* is considered the ancestor to both the later australopithecines and the genus *Homo,* which is the human lineage. We will consider this evolutionary scheme as well as an alternative one following a short discussion of the fossils themselves.

The earliest and most primitive of the **australopithicines** is *Australopithecus afarensis,* which lived from

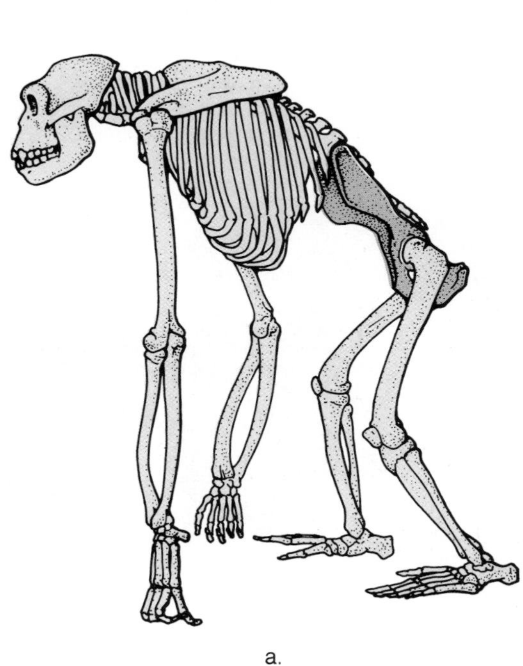

a.

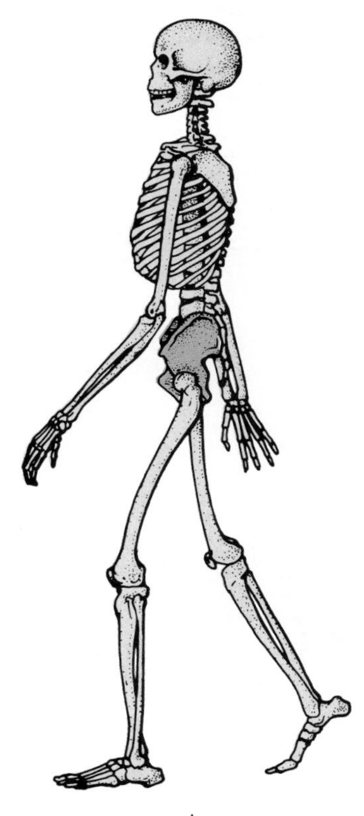

b.

**Figure 17–22** Comparison between quadrupedal and bipedal locomotion in gorillas and humans. a. In gorillas, the ischium bone is long, and the whole pelvis is tilted toward the horizontal. b. In humans the ischium bone is much shorter, and the pelvis is vertical.

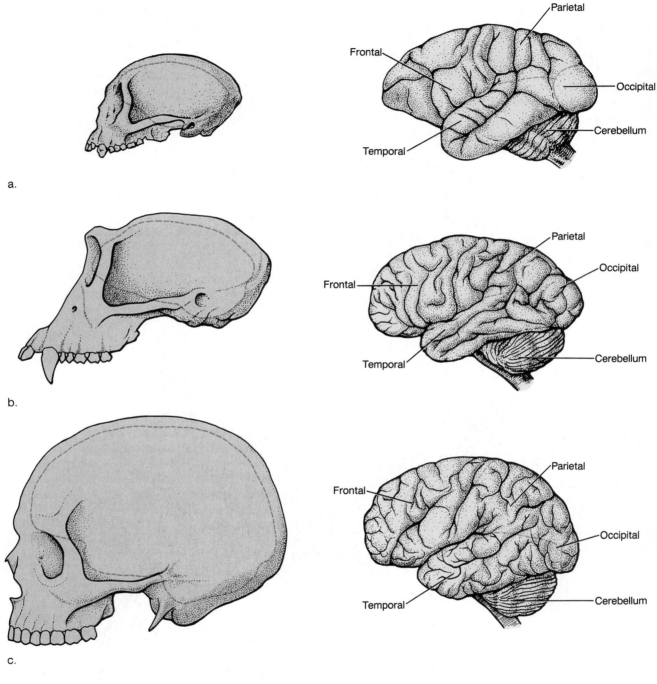

**Figure 17–23** Increase in brain size and organization is apparent in comparing the skulls and brains of a New World monkey (a), a great ape (b), and a modern human (c). All three brains shown at same scale.

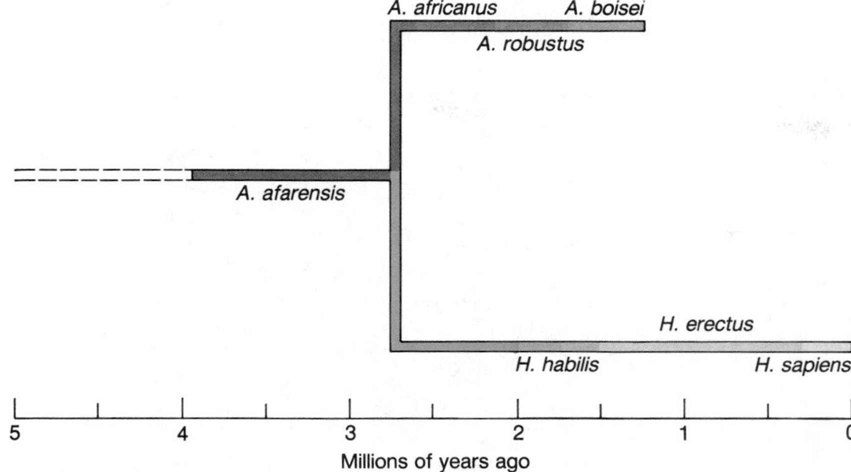


**Figure 17–24** The evolutionary scheme proposed by Don Johanson and Tim White in which humans and the late australopithecines split from a common ancestor less than 3 million years ago.

about 4 to 2.75 m.y.a. (Fig. 17–25). Members of this species were fully bipedal and exhibited great variability in size and weight, particularly between males and females. They ranged from just over 1 m to about 1.7 m tall and weighed between 25 and 60 kg. Their average brain capacity was larger than that of a chimpanzee—380 to 450 cubic centimeters (cc) vs. 300 to 400 cc for a chimpanzee—but much smaller than that of a modern human (1,300 cc average).

The skull of *A. afarensis* retained many ape-like features, including massive brow ridges, a low forehead and a forward-jutting jaw. Its teeth were intermediate between those of apes and humans with relatively smaller incisors than apes, relatively larger canines than humans, and relatively larger molars than apes. Their heavily enameled molars were probably an adaptation to chewing fruits, seeds, and roots (Fig. 17–26).

About 3 million years ago, *A. afarensis* was succeeded by another species, *Australopithecus africanus*, which lived from 3 to about 1.6 m.y.a. The distinction between the two species is relatively minor and is mostly a matter of degree. *A. africanus* was slightly taller on average (1.4 m versus 1.2 m), and had a slightly larger cranial capacity (400 to 600 cc versus 380 to 450 cc). Furthermore, its face was slightly flatter and its incisor teeth were relatively smaller. While no evidence shows that *A. africanus* modified stones into primitive tools, they probably used sticks, leaves, or stones for food gathering or processing much as chimpanzees do today. The use of

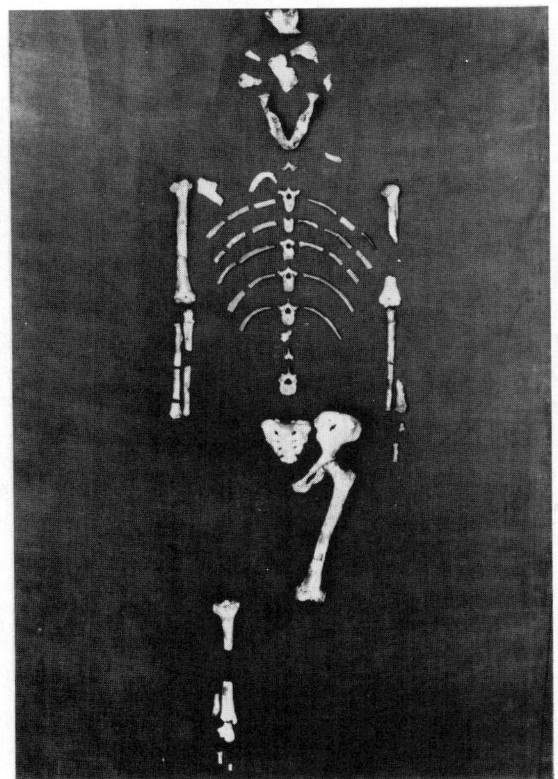

**Figure 17–25** The skeletal remains of "Lucy," a nearly complete specimen of a 3.5-million-year-old *Australopithecus afarensis* female.

**Figure 17–26** Recreation of a Pliocene landscape showing members of *Australopithecus afarensis* gathering and eating various fruits, seeds, and roots.

these types of tools would be impossible to decipher from the fossil record (Fig. 17–27).

Both *A. afarensis* and *A. africanus* differ markedly from the two latter so-called robust species of *Australopithecus*. The first of these robust species, *A. robustus* lived from about 2.3 to 1.3 m.y.a., partially overlapping the geologic and geographic range of *A. africanus*. *A. robustus* was heavier and taller (an average 1.5 m tall) than *A. africanus*, had a larger brain (500 to 600 cc), a flat face, and strong jaws with very broad, flat molars. The crown of its skull was elevated into a bony crest much like that of a gorilla, thus providing an additional area for the attachment of strong jaw muscles (Fig. 17–28). It appears from fossil-teeth evidence that *A. robustus* was primarily a vegetarian.

The fourth species of *Australopithecus*, *A. boisei* is also a robust form found in eastern Africa. It lived from about 2.2 to 1.2 m.y.a. However, a recent specimen found near Lake Turkana, Kenya, is from beds dated at 2.5 m.y.a. *A. boisei* is very similar to *A. robustus* in possessing a ridge on the top of its skull, having a flat face, and large, broad, flat molar teeth. However, in *A. boisei* these features are much exaggerated (Fig. 17–29).

In the evolutionary scheme proposed by Don Johanson and Tim White, *A. afarensis* is the single ancestral species that gave rise to all later hominids. This species split into two lineages about 2.5 m.y.a. (Fig. 17–24). One lineage contains all the australopithecines, arranged in a progressing increase of robusticity through time, while the second lineage leads directly to humans.

The discovery of a 2.5-million-year-old skull of *A. boisei* casts doubt on such a simple evolutionary scheme. This skull displays many primitive features (found in the braincase and jaw joint) in common with *A. afarensis*, yet its derived features, mostly in the face and teeth are shared with later specimens of *A. boisei*, the most specialized of the australopithecines. With these new facts, Richard Leakey and Alan Walker have proposed an alternative evolutionary scheme in which there are two lineages of australopithecines, with *A. africanus* a contemporary of *A. robustus* and *A. boisei*, and not an ancestor (Fig. 17–30). Furthermore, in this scheme the two australopithecine lineages and the human lineage evolve independently from ancestors whose fossils have not yet been discovered.

**Figure 17–27** An *Australopithecus africanus* male digs with a stick, while a female harvests berries.

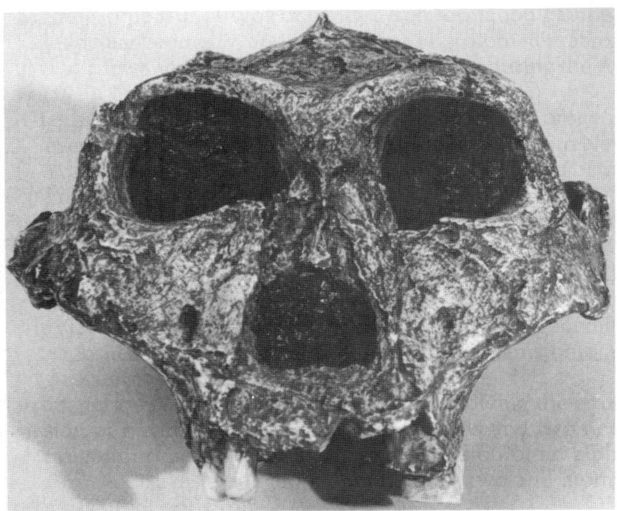

**Figure 17–28** The skull of *Australopithecus robustus*. This species had a massive jaw, powerful chewing muscles, and large, broad, flat chewing teeth apparently used for grinding up coarse plant food.

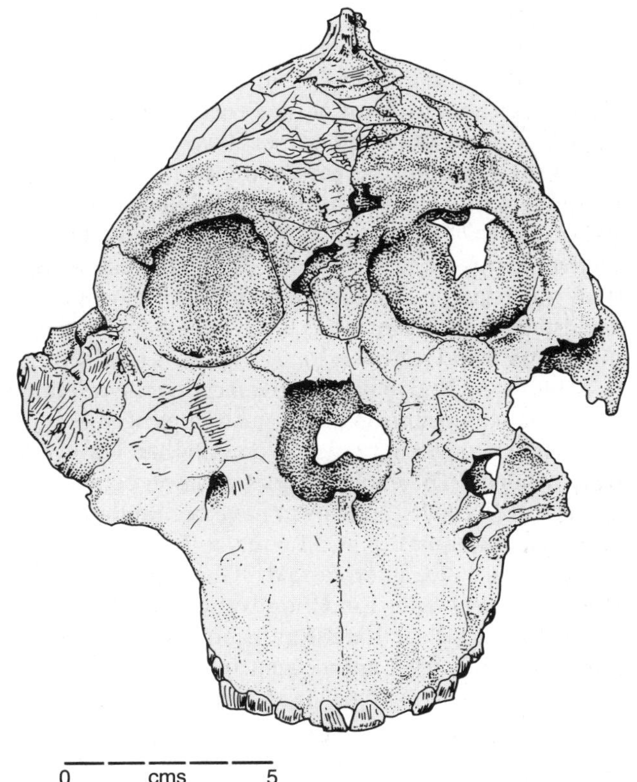


**Figure 17–29** The cranium of *Australopithecus boisei*. This specimen is 1.8 million years old and comes from Olduvai Gorge, Tanzania.

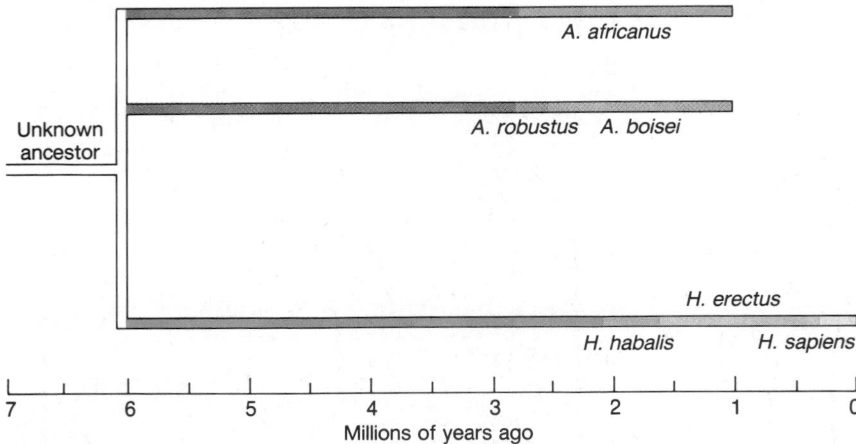


**Figure 17–30** The evolutionary scheme proposed by Richard Leakey and his colleagues has humans and the two main australopithecine groups evolving from creatures that have not yet been discovered.

## The Human Lineage

### Homo habilis

The earliest representative of our own genus *Homo* is a species named *Homo habilis.* Its remains were first found at Olduvai Gorge but are now known from South Africa, Kenya, and Ethiopia. Depending on which evolutionary scheme is accepted (Fig. 17–24 or 17–30), *H. habilis* either evolved from *A. afarensis,* or from an as yet undiscovered early hominid about 2 m.y.a. and survived as a species until about 1.4 m.y.a. A recent discovery of a 1.8-million-year-old skull and partial skeleton of *H. habilis* has cast new light on this earliest member of the human lineage (Fig. 17–31). The head of *H. habilis* is rather like that of *Australopithecus* in that it has a flattish face and relatively large molars. However, its brain size is definitely much larger, averaging about 700 cc. What is most surprising about *Homo habilis* is how ape-like in appearance it was; it had relatively long arms and a small body.

This recent find highlights the fact that the primitive body form characteristic of the australopithecines continued much longer in human evolution than was previously believed. The evolutionary transition between *H. habilis* and *H. erectus* appears to have occurred in a very short time, between 1.8 and 1.6 m.y.a. As Tim White of the University of California said, "Given the primitiveness we see in *H. habilis* and the advanced characteristics in *H. erectus,* it's clear that the transition was much more abrupt than has been appreciated."

### Homo erectus

In contrast to the australopithecines and *H. habilis,* which are unknown outside Africa, *H. erectus* was a widely distributed species and the first to penetrate the temperate climatic zone. Specimens have been found not only in Africa, but in Europe, India, China ("Peking Man") and Indonesia ("Java Man"). *H. erectus* first appeared in Africa 1.6 m.y.a. and by 1 m. y. a. was present in southeastern and eastern Asia, where it survived until about 250,000 years ago.

*H. erectus* differed from modern humans in several ways. The cranial capacity (800 to 1,300 cc), while much larger than that of *H. habilis,* was still less than in *Homo sapiens* (about 1,300 cc on the average). *H. erectus* possessed a long, low skull, sharply angled at the back. Its face was massive, and the forehead receded sharply behind strongly developed brow ridges (Fig. 17–32). The teeth of *H. erectus* were smaller than those of australopithecines but bigger than the teeth of modern humans. A recent discovery in Africa of a 1.6-million-year-old skeleton of a twelve-year-old boy indicates that *H. erectus* rivaled modern humans in body size.

The archeological record indicates that *H. erectus* were tool makers. Their stone culture, known as the *Acheulean,* included the production of handaxes, flakes, and pointed scrapers. Further evidence indicates that *H. erectus* used fire and lived in caves (Fig. 17–33).

### Neanderthals

The evolution of modern humans from *H. erectus* is not completely clear from the fossil record (See Perspective 17–2). Nonetheless the most famous of all fossil humans were the **Neanderthals,** who inhabited Europe and the Near East between about 150,000 and 32,000 years ago. Some paleontologists regard the Neanderthals as a vari-

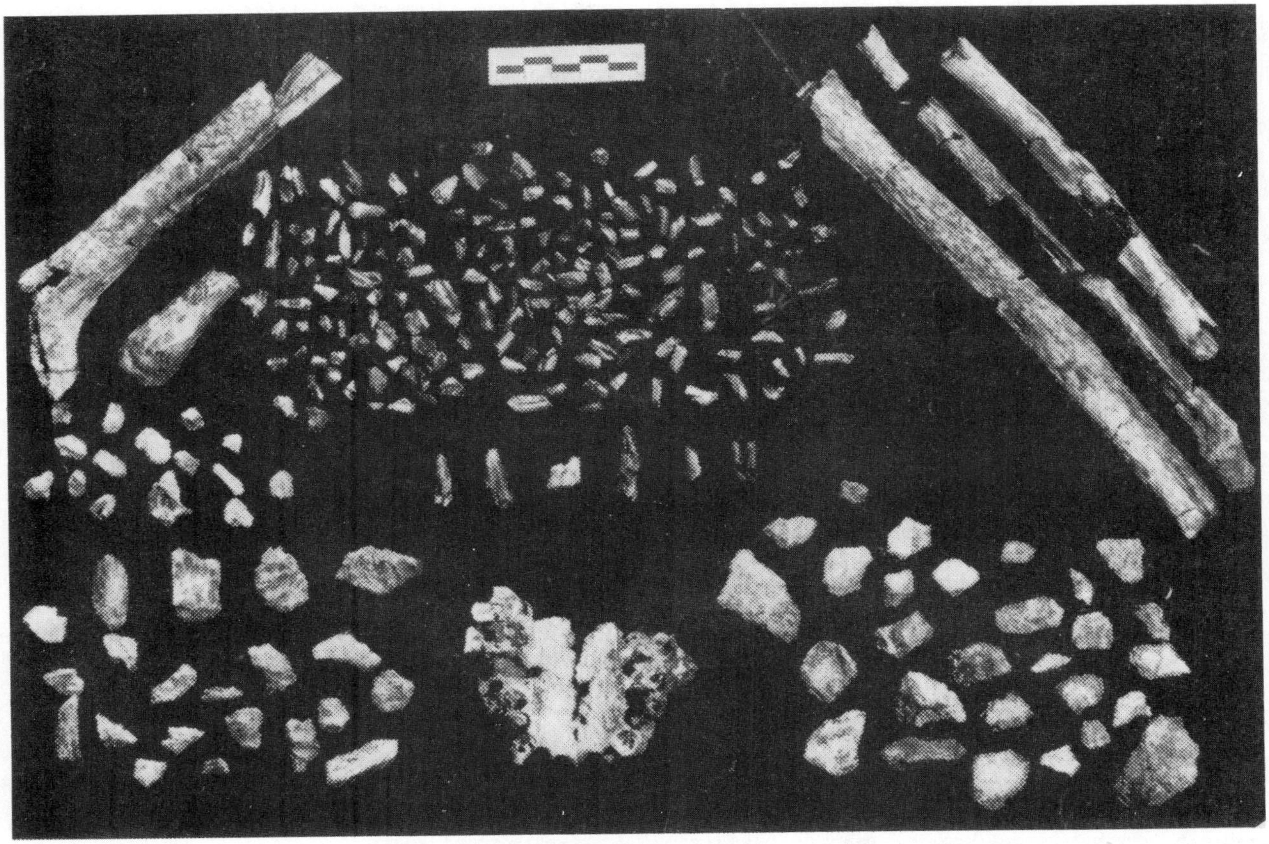

Figure 17–31 Fossil skull and skeleton fragments from a 1.8-million-year-old specimen of *Homo habilis*.

ety or subspecies (*Homo sapiens neanderthalensis*) of our own species, while others regard them as a separate species (*Homo neanderthalensis*). In any case, their name comes from the first specimens found in the Neander Valley near Dusseldorf, Germany in 1856.

The Neanderthal's skeleton differs slightly from our own, with the most notable differences occurring in the features of the skull. The Neanderthal skull was long and low with prominent heavy brow ridges, a projecting mouth, and a weak, receding chin (Fig. 17–34). Its brain was large, actually slightly larger on average than our own, and somewhat differently shaped. The body of the Neanderthals was somewhat more massive and heavily muscled than ours, with rather short lower limbs, much like Eskimos and other cold-adapted people of today.

Of all the early hominids, none have been so maligned and misunderstood as the Neanderthals. The name alone conjures up images of brutishness and low intelligence. This unfortunate image has been fostered from reconstructions based on a nearly complete skeleton found in 1908 in the Carreze region of southwestern

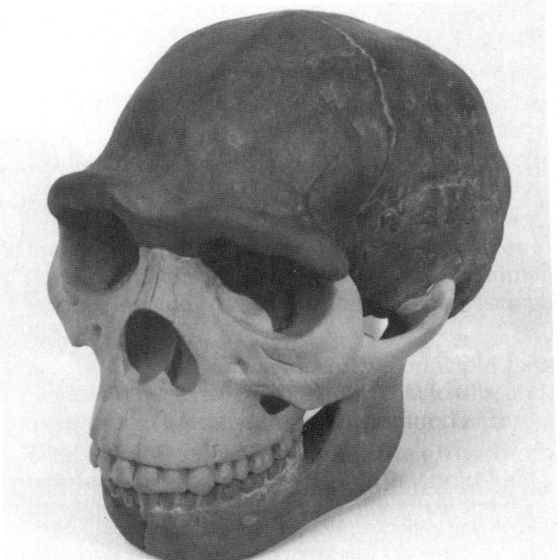

**Figure 17–32** A reconstruction of the skull of *Homo erectus*, a widely distributed species whose remains have been found in Africa, Europe, India, China, and Indonesia.

**Figure 17–33** Reconstruction of a Pleistocene setting in which *Homo erectus* is using fire.

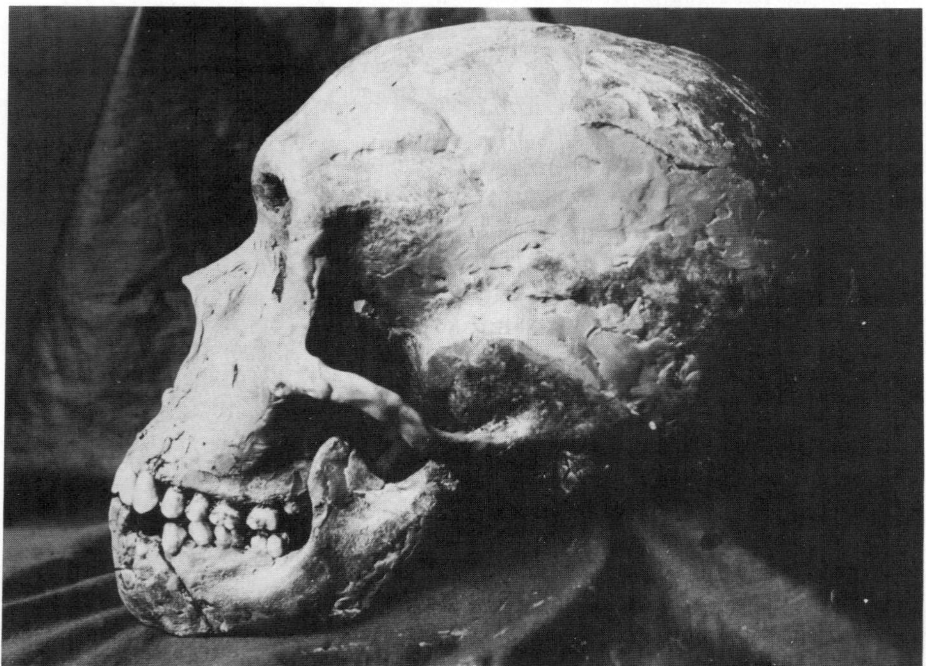

**Figure 17–34** Reconstructed Neanderthal skull. The Neanderthals were characterized by prominent heavy brow ridges and a weak chin. Their brain was also slightly larger on average than that of modern humans.

**Figure 17–35** Members of a Neanderthal living group prepare their kill after a successful hunt. Archeological evidence indicates that Neanderthals used tools skillfully.

France. From this skeleton has come an image of a hulking, dim-witted brute who shuffled along with the bent-knee walk of an ape. At the time of its discovery, however, people failed to realize that this was the skeleton of an individual who was stooped over because of severe hip arthritis, a broken rib, and diseased vertebrae. Also it was missing most of its teeth due to the effects of gum disease.

Based on specimens from more than a hundred sites, we now know that the Neanderthals were not much different from us, only more robust. Europe's Neanderthals were the first humans to move into truly cold climates, enduring miserably long winters and short summers as they pushed north into tundra country (Fig. 17–35).

The remains of Neanderthals are found chiefly in caves and hut-like rock shelters, which also contain a variety of specialized stone tools and weapons. Evidence such as the arthritic individual just mentioned above indicates that Neanderthals commonly took care of their injured and buried their dead, frequently with such grave items as tools, food, and perhaps even flowers.

While the Neanderthals may have been socially and technologically less advanced than modern humans, they were certainly not the dim-witted, brutish boors they are frequently portrayed as.

## Cro-Magnons

Around 35,000 years ago, humans closely resembling modern Europeans moved into the region inhabited by the Neanderthals and completely replaced them. When and how this happened is still unknown. **Cro-Magnon**, the name given to the successors of the Neanderthals in France, lived from about 35,000 to 10,000 years ago, and during this period the development of art and technology far exceeded anything the world had seen before.

Highly skilled nomadic hunters, Cro-Magnons followed the herds in their seasonal migrations. They used a variety of specialized tools in their hunts, including perhaps the bow and arrow (Fig. 17–36). They sought refuge in caves and rock shelters and formed groups of 50 to 75 people. Cro-Magnons were also cave painters. Using paints made from manganese and iron oxides,

**Figure 17–36** Recreation of a Cro-Magnon camp in Europe. Cro-Magnons were highly skilled hunters that formed living groups of 50 to 75 people.

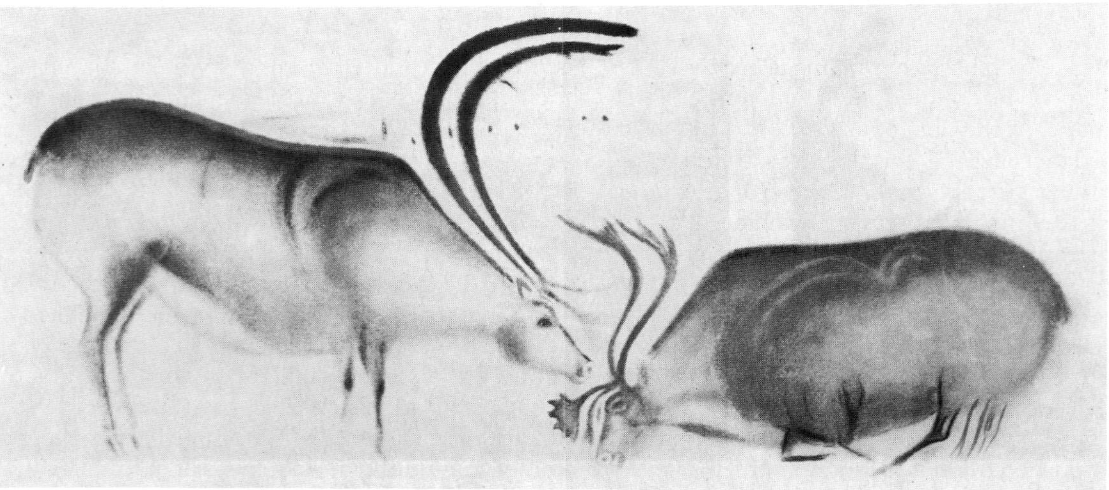

**Figure 17–37** Cro-Magnons were very skilled cave painters. Shown are a male and female reindeer from Fonte de Gaume, France.

Cro-Magnon people painted hundreds of scenes on the ceilings and walls of caves in France and Spain, where many of them are still preserved today (Fig. 17–37). About 10,000 years ago, modern humans succeeded Cro-Magnon, and about 5000 years ago, writing was invented and the era of recorded history began.

## Perspective 17-2

# The Search for the Elusive Eve—A Controversial Theory Concerning the Origin of Humans

A controversial new theory on the origin and evolution of modern humans has sparked a bitter debate between traditional paleoanthropologists who work with fossil bones and a new breed of anthropologists who are trained in molecular biology. The molecular anthropologists claim to have found our common ancestor, a woman they call Eve, who lived in the hot savanna of Africa 200,000 years ago and left as her legacy, resilient genes that are carried by all living humans. What makes this theory so controversial is that it challenges the traditional view based on fossils that the human family tree began around 1.8 million years ago. Furthermore, the molecular anthropologists believe that modern humans did not evolve slowly in different parts of the world, as believed by many traditional anthropologists, but instead evolved from Eve's family in Africa. Then, sometime between 180,000 and 90,000 years ago, a group of her offspring migrated from their homeland and, apparently endowed with some special advantage, replaced every tribe of early humans they encountered until they settled the entire world.

What makes the molecular anthropologists so sure they are right? To answer that question we need to look at an earlier battle between geneticists and paleoanthropologists concerning the time of divergence between the ancestral hominid and the chimpanzee. Prior to 1967, most paleoanthropologists believed the split between the chimpanzees and hominids occurred at least 15 million years ago, based on bones of an ape-like creature who seemed to be ancestral to humans but not apes. Then, in 1967, geneticists began studying the molecular structure of a blood protein believed to change at a slow, steady rate as a species evolved. When comparing this blood protein among baboons, chimpanzees, and humans, the geneticists discovered that there were major differences between the blood protein molecules of chimpanzees and baboons. This was expected, since the two species had been evolving separately for 30 million years. However, the difference between chimpanzees and humans was very small, leading them to conclude the two species separated no more than 8 million years ago, which was at least 6 million years later than the fossil evidence suggested, much to the chagrin of paleontologists. However, as more fossils turned up, paleontologists realized that the 15-million-year-old bones did not belong to a human ancestor and that the split between chimpanzees and humans occurred more recently than they had believed (Fig. 1).

The current, controversial Eve hypothesis arose from an examination of the DNA from the mitochondria of cells. This mitochondrial DNA is useful for tracing family trees because, unlike the DNA from a cell's nucleus, it is inherited only from the mother and so preserves a family record (Fig. 2). Mi-

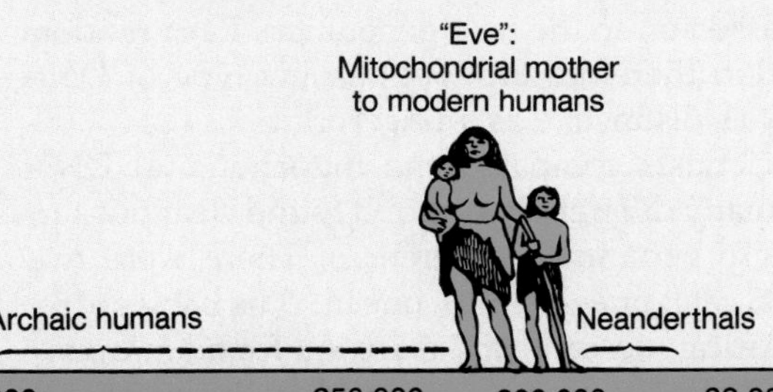


**Figure 1** The evolutionary lineage of humans based on fossil bones and molecular biology.

## Perspective 17–2 (continued)

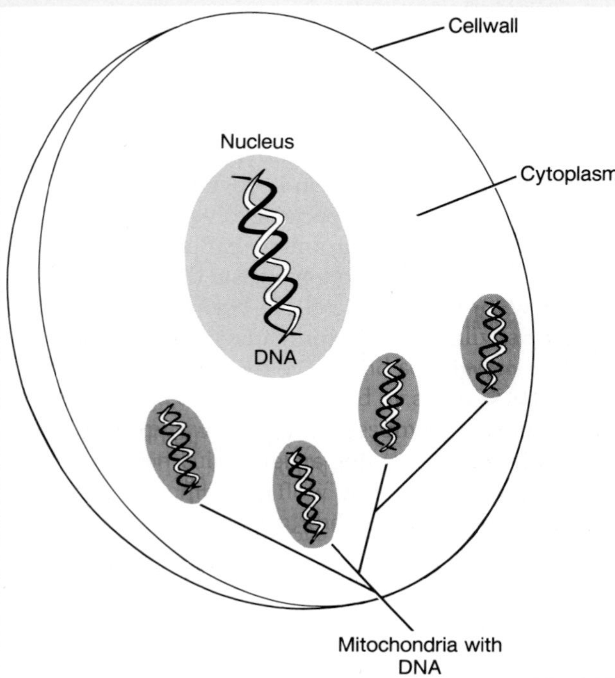


**Figure 2** Source of the genetic code contained in all humans. Nucleic DNA is a mixture of the genes from both parents. Mitochondrial DNA is inherited only from the mother and therefore is not a mixture of parental genes.

tochondrial DNA is altered only by mutations that are passed on to the next generation. Each random mutation therefore produces a new type of DNA that is as distinctive as a fingerprint.

Scientists compared the mitochondrial DNA from many different babies and found clear but surprisingly very small differences. There were two general categories of DNA; one in some babies of recent African descent, and a second found in everyone else and the other Africans. What the geneticists interpret this to mean is that modern humans began in Africa and then at some point a group of Africans split off, forming a second branch of DNA that they carried and spread throughout the rest of the world.

All of the babies' DNA could be traced back to a single woman from whom we are all descended. This in itself is not surprising, considering the statistics of genetic inheritance; and even most traditional paleoanthropologists do not argue the existence of an ancestral Eve. What surprises them is how recently modern humans may have evolved. This age is calculated by counting the mutations that have occurred back to Eve's DNA. By looking at the DNA types that are most different from one another, and assuming mutations occurred at a regular rate, the geneticists have calculated how many steps it would take for Eve's original DNA to mutate into three different types. The molecular calculations indicate Eve lived around 200,000 years ago.

Where does this leave the traditional paleoanthropologists who base their evolutionary schemes on evidence from the fossil record? Most are skeptical of the genetic data. After all, the geneticists' molecular clock may be way off. By changing a few basic assumptions, one might use their calculations to show that Eve evolved many hundreds of thousands of years ago, which would be more in line with the fossil data. Certainly the fossil hunters will continue to keep digging, perhaps eventually vindicating the geneticists, or perhaps making a discovery that will lead to a completely different interpretation of our ancestry. As Alan Mann of the University of Pennsylvania stated:

> "We are dealing with a dramatic jump. Maybe the origin of creatures like us occurred very recently. Certainly the mitochondrial data is a significant advance. But there really isn't any good fossil evidence from that period to back it up. If you look at the fossils, the good evidence on Africa can be placed on the palm of your hand. In this field a person kicks over a stone in Africa, and we have to rewrite the textbooks."

## CHAPTER SUMMARY

1. The Quaternary Period consists of two unequal epochs: the Pleistocene Epoch, from 2 m.y.a. to 10,000 years ago, and the Holocene Epoch, from 10,000 years ago to the present.
2. About 20 warm-cold Pleistocene climatic cycles are recognized from paleontologic and oxygen isotope data derived from deep-sea cores. Terrestrial stratigraphy indicates that at least four major glacial and interglacial episodes occurred in North America, and six or seven such episodes occurred in Europe.
3. During periods of maximum glacial development, huge ice sheets directly altered the landscape as they advanced and retreated. Such ice sheets also caused lower sea level and climatic changes.
4. Climates change on different time scales, so different mechanisms are likely responsible for the changes. First-order periods of glaciation lasting millions of years are probably caused by plate tectonic activity. Second-order events, such as occurred during the Pleistocene Epoch and involving cyclical advances and retreats of glaciers, can be explained by astronomical causes. The third-order, and shortest-duration events may be the result of variation in solar energy received or volcanic eruptions.
5. The primates probably evolved during the Late Cretaceous. Two major features developed that had profound consequences: a grasping hand with opposable thumb and the development of stereoscopic vision.
6. The primates are divided into two suborders, the prosimians and anthropoids. The prosimians include the lemurs, tarsiers, and tree shrews. The anthropoids include monkeys, apes, and hominids (humans, and their ancestors).
7. The australopithecines were a group of fully bipedal hominids that evolved in Africa. Two hypotheses exist concerning their evolutionary place. One holds that the oldest species, *Australopithecus afarensis*, gave rise to the rest of the australopithecines and the human lineage, while the other states the australopithecine and human lineage evolved independently and were derived from a still to be discovered ancestor.
8. The human lineage began about 2 million years ago in Africa with the evolution of *Homo habilis*. By 1.6 million years ago *Homo erectus* evolved and quickly migrated to Europe, Asia, and Indonesia. This species used fire and made tools.
9. *Homo sapiens* evolved around 150,000 years ago (Neanderthals) and from then on, cultural and technological changes occurred with increasing rapidity.

## IMPORTANT TERMS

anthropoids
australopithecines
Cro-Magnon
dryopithecines
hominids
hominoids
isostasy
Milankovitch theory
Neanderthal
pluvial lakes
pollen analysis
primates
prosimians
ramapithecines

## REVIEW QUESTIONS

1. What is the direct evidence for Pleistocene glaciation?
2. What is the evidence that glaciation occurred in areas far removed from the ice sheets?
3. Why do the Pleistocene and Holocene chronologies based on deep-sea cores fail to correlate with those based on terrestrial stratigraphy?
4. What are the three ways foraminifera can be used to determine past climatic conditions?
5. Explain how the Milankovitch theory accounts for glacial-interglacial cyclicity.
6. Why was the development of a grasping hand with opposable thumb and the development of stereoscopic vision so important for the ultimate evolution of humans?
7. Discuss both evolutionary schemes for the human lineage.

## ADDITIONAL READINGS

Covey, C. 1984. The Earth's Orbit and the Ice Ages. *Scientific American,* 250, no. 2: 58–67.

Johanson, D. C., and M. A. Edey. 1981. *Lucy, the Beginnings of Humankind.* New York: Simon and Schuster, Inc.

Leakey, R. E. 1981. *The Making of Mankind.* New York: E. P. Dutton.

Nilsson, T. 1983. *The Pleistocene: Geology and Life in the Quaternary Ice Age.* Holland: D. Reidel Publishing Co.

Pilbeam, D. 1984. The Descent of Hominoids and Hominids. *Scientific American,* 250, no. 3: 84–96.

Shipman, P. 1986. Baffling Limb on the Family Tree. *Discover,* 7, no. 9: 86–93.

Stringer, C. B., and P. Andrews. 1988. Genetic and Fossil Evidence for the Origin of Modern Humans. *Science,* 239: 1263–1268.

Weaver, K. F. 1985. The Search for our Ancestors. *National Geographic,* 168, no. 5: 560–623.

# APPENDIX A
# English–Metric Conversion Chart

| English Unit | Conversion Factor | Metric Unit | Conversion Factor | English Unit |
|---|---|---|---|---|
| *Length* | | | | |
| Inches (in) | 2.54 | Centimeters (cm) | 0.39 | Inches (in) |
| Feet (ft) | 0.305 | Meters (m) | 3.28 | Feet (ft) |
| Miles (mi) | 1.61 | Kilometers (km) | 0.62 | Miles (mi) |
| *Area* | | | | |
| Square inches ($in^2$) | 6.45 | Square centimeters ($cm^2$) | 0.16 | Square inches ($in^2$) |
| Square feet ($ft^2$) | 0.093 | Square meters ($m^2$) | 10.8 | Square feet ($ft^2$) |
| Square miles ($mi^2$) | 2.59 | Square kilometers ($km^2$) | 0.39 | Square miles ($mi^2$) |
| *Volume* | | | | |
| Cubic inches ($in^3$) | 16.4 | Cubic centimeters ($cm^3$) | 0.061 | Cubic inches ($in^3$) |
| Cubic feet ($ft^3$) | 0.028 | Cubic meters ($m^3$) | 35.3 | Cubic feet ($ft^3$) |
| Cubic miles ($mi^3$) | 4.17 | Cubic kilometers ($km^3$) | 0.24 | Cubic miles ($mi^3$) |
| *Weight* | | | | |
| Ounces (oz) | 28.3 | Grams (g) | 0.035 | Ounces (oz) |
| Pounds (lb) | 0.45 | Kilograms (kg) | 2.20 | Pounds (lb) |
| Short tons (st) | 0.91 | Metric tons (t) | 1.10 | Short tons (st) |
| *Temperature* | | | | |
| Degrees Fahrenheit (°F) | −32° x 0.56 | Degrees Celsius (Centigrade) (°C) | x 1.80 + 32° | Degrees Fahrenheit (°F) |

Examples: 10 inches = 25.4 centimeters; 10 centimeters = 3.9 inches
100 square feet = 9.3 square meters; 100 square meters = 1080 square feet
50°F = 10.08°C; 50°C = 122°F

# APPENDIX B
# Classification of Organisms

Any classification is an attempt to make order out of disorder and to group similar items into the same categories. All classifications are schemes that attempt to relate items to each other based on current knowledge and therefore are progress reports on the current state of knowledge for the items classified. Because classifications are to some extent subjective, classification of organisms may vary among different texts.

The classification that follows is based on the five-kingdom system of classification of Margulis and Schwartz.* We have not attempted to include all known life forms, but rather major categories of both living and fossil groups.

**Kingdom MONERA**—Prokaryotes
- *Phylum Anaerobic Photosynthetic Bacteria*—(Archean-Recent)
- *Phylum Cyanobacteria*—Blue-green algae or blue-green bacteria (Archean-Recent)

**Kingdom PROTOCTISTA**—Solitary or colonial unicellular eukaryotes
- *Phylum Acritarcha*—Organic unicellular algae of unknown affinity (Proterozoic-Recent)
- *Phylum Bacillariophyta*—Diatoms (Jurassic-Recent)
- *Phylum Charophyta*—Stoneworts (Silurian-Recent)
- *Phylum Chlorophyta*—Green algae (Proterozoic-Recent)
- *Phylum Chrysophyta*—Golden-brown algae, silicoflagellates and coccolithophorids (Jurassic-Recent)
- *Phylum Euglenophyta*—Euglenids (Cretaceous-Recent)
- *Phylum Myxomycophyta*—Slime molds (Proterozoic-Recent)
- *Phylum Phaeophyta*—Brown algae, multicellular, kelp, seaweed (Proterozoic-Recent)
- *Phylum Protozoa*—Unicellular heterotrophs (Cambrian-Recent)
  - *Class Sarcodina*—Forms with pseudopodia for locomotion (Cambrian-Recent)
    - *Order Foraminifera*—Benthonic and planktonic sarcodinids most commonly with calcareous tests (Cambrian-Recent)
    - *Order Radiolaria*—Planktonic sarcodinids with siliceous tests (Cambrian-Recent)
- *Phylum Pyrrophyta*—Dinoflagellates (Silurian?, Permian-Recent)
- *Phylum Rhodophyta*—Red algae (Proterozoic-Recent)
- *Phylum Xanthophyta*—Yellow-green algae (Miocene-Recent)

**Kingdom FUNGI**
- *Phylum Zygomycota*—Fungi that lack cross walls (Proterozoic-Recent)
- *Phylum Basidiomycota*—Mushrooms (Pennsylvanian-Recent)
- *Phylum Ascomycota*—Yeasts, bread molds, morels (Mississippian-Recent)

**Kingdom PLANTAE**—Photosynthetic eukaryotes
- *Division* Bryophyta*—Liverworts, mosses, hornworts (Devonian-Recent)
- *Division Psilophyta*—Small, primitive vascular plants with no true roots or leaves (Silurian-Recent)
- *Division Lycopodophyta*—Club mosses, simple vascular systems, true roots and small leaves, including scale trees of Paleozoic Era (lycopsids) (Devonian-Recent)
- *Division Sphenophyta*—Horsetails, scouring rushes, and sphenopsids such as the Carboniferous *Calamites* (Devonian-Recent)
- *Division Pteridophyta*—Ferns (Devonian-Recent)
- *Division Pteridospermophyta*—Seed ferns (Devonian-Jurassic)
- *Division Coniferophyta*—Conifers or cone-bearing gymnosperms (Carbonaceous-Recent)
- *Division Cycadophyta*—Cycads (Triassic-Recent)
- *Division Ginkgophyta*—Maidenhair tree (Triassic-Recent)
- *Division Angiospermophyta*—Flowering plants and trees (Cretaceous-Recent)

**Kingdom ANIMALIA**—Nonphotosynthetic multicellular eukaryotes (Proterozoic-Recent)
- *Phylum Porifera*—Sponges (Cambrian-Recent)
- *Phylum Archaeocyatha*—Extinct spongelike organisms (Cambrian)
- *Phylum Coelenterata*—Hydrozoans, jellyfish, sea anemones, corals (Cambrian-Recent)
  - *Class Hydrozoa*—Hydrozoans (Cambrian-Recent)
    - *Order Stromatoporoida*—Extinct group of reef-building organisms (Cambrian-Cretaceous)

*Margulis, L., and K. V. S. Schwartz, 1982. *Five Kingdoms.* New York: W. H. Freeman and Co.

*In botany, division is the equivalent to phylum.

*Class Scyphozoa*—Jellyfish (Proterozoic-Recent)
*Class Anthozoa*—Sea anemones and corals (Cambrian-Recent)
*Order Tabulata*—Exclusively colonial corals with reduced to nonexistent septa (Ordovician-Permian)
*Order Rugosa*—Solitary and colonial corals with fourfold symmetry (Ordovician-Permian)
*Order Scleractinia*—Solitary and colonial corals with sixfold symmetry. Most colonial forms have symbionic dinoflagellates in their tissue. Important reef builders today (Triassic-Recent)

*Phylum Bryozoa*—Exclusively colonial suspension feeding marine animals that are useful for correlation and ecological interpretations (Ordovician-Recent)

*Phylum Brachiopoda*—Marine suspension feeding animals with two unequal sized valves. Each valve is bilaterally symmetrical (Cambrian-Recent)
*Class Inarticulata*—Primitive chitino-phosphatic or calcareous brachiopods that lack a hinging structure. They open and close their valves by means of complex muscles (Cambrian-Recent)
*Class Articulata*—Advanced brachiopods with calcareous valves that are hinged (Cambrian-Recent)

*Phylum Mollusca*—A highly diverse group of invertebrates (Cambrian-Recent)
*Class Monoplacophora*—Segmented, bilaterally symmetrical crawling animals with cap-shaped shells (Cambrian-Recent)
*Class Amphineura*—Chitons. Marine crawling forms, typically with 8 separate calcareous plates (Cambrian-Recent)
*Class Scaphopoda*—Curved, tusk-shaped shells that are open at both ends (Ordovician-Recent)
*Class Gastropoda*—Single shelled, general coiled crawling forms. Found in marine, brackish, and fresh water as well as terrestrial environments (Cambrian-Recent)
*Class Bivalvia*—Mollusks with two valves that are mirror images of each other. Typically known as clams or oysters (Cambrian-Recent)
*Class Cephalopoda*—Highly evolved swimming animals. Includes shelled sutured forms as well as non-shelled types such as octopus and squid (Cambrian-Recent)
*Order Nautiloidea*—Forms in which the chamber partitions are connected to the wall along simple, slightly curved lines (Cambrian-Recent)
*Order Ammonoidea*—Forms in which the chamber partitions are connected to the wall along wavy lines (Devonian-Cretaceous)
*Order Coleoidea*—forms in which the shell is reduced or lacking. Includes octopus, squid, and the extinct belemnoids (Mississippian-Recent)

*Phylum Annelida*—Segmented worms. Responsible for many of the Phanerozoic burrow and trail trace fossils (Proterozoic-Recent)

*Phylum Arthropoda*—The largest invertebrate group comprising about 80% of all known animals. Characterized by a segmented body and jointed appendages (Cambrian-Recent)
*Class Trilobita*—Earliest appearing arthropod class. Trilobites had a head, body, and tail and were bilaterally symmetrical (Cambrian-Permian)
*Class Crustacea*—Diverse class characterized by a fused head and body and an abdomen. Included are barnacles, copepods, crabs, ostracodes, and shrimp (Cambrian-Recent)
*Class Insecta*—Most diverse and common of all living invertebrates, but rare as fossils (Silurian-Recent)
*Class Merostomata*—Characterized by four pairs of appendages and a more flexible exoskeleton than crustaceans. Includes the extinct eurypterids, horseshoe crabs, scorpions, and spiders (Cambrian-Recent)

*Phylum Echinodermata*—Exclusively marine animals with fivefold radial symmetry and a unique water vascular system (Cambrian-Recent)
*Subphylum Crinozoa*—Forms attached by a calcareous jointed stem (Cambrian-Recent)
*Class Crinoidea*—Most important class of Paleozoic echinoderms. Suspension feeding forms that are either free-living or attach to sea floor by a stem (Cambrian-Recent)
*Class Blastoidea*—Small class of Paleozoic suspension feeding sessile forms with short stems (Ordovician-Permian)
*Class Cystoidea*—Globular to pear-shaped suspension feeding benthonic sessile forms with quite short stems (Ordovician-Devonian)
*Subphylum Homalozoa*—A small group with flattened, asymmetrical bodies with no stems. Also called carpoids (Cambrian-Devonian)
*Subphylum Echinozoa*—Globose, predominantly benthonic mobile echinoderms (Ordovician-Recent)
*Class Helioplacophora*—Benthonic, mobile forms, shaped like a top with plates arranged in a helical spiral (Early Cambrian)
*Class Edrioasteroidea*—Benthonic, sessile or mobile, discoidal, globular or cylindrical shaped forms with five straight or curved feeding areas shaped like a starfish (Cambrian-Pennsylvanian)
*Class Holothuroidea*—Sea cucumbers. Sediment feeders having calcareous spicules embedded in a tough skin (Ordovician-Recent)

*Class Echinoidea*—Largest group of echinoderms. Globe or disk shaped with movable spines. Predominantly grazers or sediment feeders. Epifaunal and infaunal (Ordovician-Recent)

*Subphylum Asterozoa*—Stemless, benthonic mobile forms (Ordovician-Recent)

*Class Asteroidea*—Starfish. Arms merge into body (Ordovician-Recent)

*Class Ophiuroidea*—Brittle star. Distinct central body (Ordovician-Recent)

*Phylum Hemichordata*—Characterized by a notochord sometime during the life history. Modern acorn worms and extinct graptolites (Cambrian-Recent)

*Class Graptolithina*—Colonial marine hemichordates having a chitinous exoskeleton. Predominantly planktonic (Cambrian-Mississippian)

*Phylum Chordata*—Animals with notochord, hollow dorsal nerve chord, and gill slits in embryo (Cambrian-Recent)

*Subphylum Urochordata*—Sea squirts, tunicates. Larval forms have notochord in tail region.

*Subphylum Cephalochordata*—Small marine animals with notochords and small fish-like bodies (Cambrian-Recent)

*Subphylum Vertebrata*—Animals with a backbone of vertebrae (Cambrian-Recent)

*Class Agnatha*—Jawless fish. Includes the living lampreys and hagfish as well as the extinct armored ostracoderms (Cambrian-Recent)

*Class Acanthodii*—Primitive jawed fish with numerous spiny fins (Silurian-Permian)

*Class Placodermii*—Primitive armored jawed fish (Silurian-Permian)

*Class Chondrichthyes*—Cartilaginous fish such as sharks and rays (Devonian-Recent)

*Class Osteichthyes*—Bony fishes (Devonian-Recent)

*Subclass Actinopterygii*—Ray-finned fishes (Devonian-Recent)

*Subclass Sarcopterygii*—Lobe-finned, air-breathing fish (Devonian-Recent)

*Order Crossoptergii*—Lobe-finned fish that were ancestral to amphibians (Devonian-Recent)

*Order Dipnoi*—Lungfishes (Devonian-Recent)

*Class Amphibia*—Amphibians. The first terrestrial vertebrates (Devonian-Recent)

*Subclass Labyrinthodontia*—Earliest amphibians. Solid skulls and complex tooth pattern (Devonian-Triassic)

*Subclass Salientia*—Frogs, toads, and their relatives (Triassic-Recent)

*Subclass Condata*—Salamanders and their relatives (Triassic-Recent)

*Class Reptilia*—Reptiles. A large and varied vertebrate group characterized by having scales and laying an amniote egg (Pennsylvanian-Recent)

*Subclass Anapsida*—Reptiles whose skull has a solid roof with no openings (Pennsylvanian-Recent)

*Order Cotylosauria*—Captorhinomorphs or earliest reptiles (Pennsylvanian-Triassic)

*Order Chelonia*—Turtles (Triassic-Recent)

*Subclass Euryapsida*—Reptiles with one opening high on the side of the skull behind the eye. Mostly marine (Permian-Cretaceous)

*Order Protorosauria*—Land living ancestral euryapsids (Permian-Cretaceous)

*Order Placodontia*—Placodonts. Bulky, paddle-limbed marine reptiles with rounded teeth for crushing mollusks (Triassic)

*Order Ichthyosauria*—Ichthyosaurs. Dolphin-shaped swimming reptiles (Triassic-Cretaceous)

*Subclass Diapsida*—Most diverse reptile class, characterized by two openings in the skull behind the eye. Includes lizards, snakes, crocodiles, thecodonts, dinosaurs, and pterosaurs (Permian-Recent)

*Infraclass Lepidosauria*—Primitive diapsids including snakes, lizards, and the large Cretaceous marine reptile group called mosasaurs (Permian-Recent)

*Infraclass Archosauria*—Advanced diapsids (Triassic-Recent)

*Order Thecodontia*—Thecodonts were a diverse group that was ancestral to the crocodilians, pterosaurs, and dinosaurs (Permian-Triassic)

*Order Crocodilia*—Crocodiles, alligators, and gavials (Triassic-Recent)

*Order Pterosauria*—Flying and gliding reptiles called pterosaurs (Triassic-Cretaceous)

*Infraclass Dinosauria*—Dinosaurs (Triassic-Cretaceous)

*Order Saurischia*—Lizard-hipped dinosaurs (Triassic-Cretaceous)

*Suborder Theropoda*—Bipedal carnivores (Triassic-Cretaceous)

*Suborder Sauropoda*—Quadrupedal herbivores, including the largest known land animals (Jurassic-Cretaceous)

*Order Ornithischia*—Bird-hipped dinosaurs (Triassic-Cretaceous)

*Suborder Ornithopoda*—Bipedal herbivores, including the duck-billed dinosaurs (Triassic-Cretaceous)

*Suborder Stegosauria*—Quadrupedal herbivores with bony spikes on their tails and bony plates on their backs (Jurassic-Cretaceous)

*Suborder Ceratopsia*—Quadrupedal herbivores typically with horns or a bony frill over the top of the neck (Cretaceous)

*Suborder Ankylosauria*—Heavily armored quadrupedal herbivores (Cretaceous)

*Subclass Synapsida*—Mammal-like reptiles with one opening low on the side of the skull behind the eye (Pennsylvanian-Triassic)

*Order Pelycosauria*—Early mammal-like reptiles including those forms in which the vertebral spines were extended to support a "sail" (Pennsylvanian-Permian)

*Order Therapsida*—Advanced mammal-like reptiles with legs positioned beneath the body and the lower jaw formed largely of a single bone. Many therapsids may have been endothermic (Permian-Triassic)

*Class Aves*—Birds. Endothermic and feathered (Jurassic-Recent)

*Class Mammalia*—Mammals. Endothermic animals with hair (Triassic-Recent)

*Subclass Prototheria*—Egg-laying mammals (Triassic-Recent)

*Order Docodonta*—Small, primitive mammals (Triassic)

*Order Triconodonta*—Small, primitive mammals with specialized teeth (Triassic-Cretaceous)

*Order Monotremata*—Duck-billed platypus, spiny anteater (Cretaceous-Recent)

*Subclass Allotheria*—Small extinct early mammals with complex teeth for grinding food (Jurassic-Eocene)

*Order Multituberculata*—The first mammalian herbivores and the most diverse of Mesozoic mammals (Jurassic-Eocene)

*Subclass Theria*—Mammals that give birth to live young (Jurassic-Recent)

*Order Symmetrodonta*—Small, primitive Mesozoic therian mammals (Jurassic-Cretaceous)

*Order Pantotheria*—Trituberculates (Jurassic-Cretaceous)

*Order Creodonta*—Extinct ancient carnivores (Cretaceous-Paleocene)

*Order Condylartha*—Extinct ancestral hoofed placentals (ungulates) (Cretaceous-Oligocene)

*Order Marsupialia*—Pouched mammals. Opossum, kangaroo, koala (Cretaceous-Recent)

*Order Insectivora*—Primitive insect-eating mammals. Shrew, mole, hedgehog (Cretaceous-Recent)

*Order Xenungulata*—Large South American mammals that broadly resemble pantodonts and uintatheres (Paleocene)

*Order Taeniodonta*—Includes some of the most highly specialized terrestrial placentals of the Late Paleocene and Early Eocene (Paleocene-Eocene)

*Order Tillodontia*—Large, massive placentals with clawed, five-toed feet (Paleocene-Eocene)

*Order Dinocerata*—Uintatheres. Large herbivores with bony protuberances on the skull and greatly elongated canine teeth (Paleocene-Eocene)

*Order Pantodonta*—North American forms are large, sheep to rhinoceros-sized. Asian forms are as small as a rat (Paleocene-Eocene)

*Order Astropotheria*—Large placental mammals with slender rear legs and stout forelimbs and elongate canine teeth (Paleocene-Miocene)

*Order Natoungulata*—Largest assemblage of South American ungulates with a wide range of body forms (Paleocene-Pleistocene)

*Order Liptoterna*—Extinct South American hoofed-mammals (Paleocene-Pleistocene)

*Order Rodentia*—Rodents. Squirrel, mouse, rat, beaver, porcupine, gopher (Paleocene-Recent)

*Order Lagomorpha*—Hare, rabbit, pika (Paleocene-Recent)

*Order Primates*—Lemur, tarsier, loris, monkey, human (Paleocene-Recent)

*Order Edentata*—Anteater, sloth, armadillo, glyptodont (Paleocene-Recent)

*Order Carnivora*—Modern carnivorous placentals. Dog, cat, bear, skunk, seal, weasel, hyena, raccoon, panda, sea lion walrus (Paleocene-Recent)

*Order Pyrotheria*—Large mammals with long bodies and short columnar limbs (Eocene-Oligocene)

*Order Chiroptera*—Bats (Eocene-Recent)

*Order Dermoptera*—Flying lemur (Eocene-Recent)

*Order Cetacea*—Whale, dolphin, porpoise (Eocene-Recent)

*Order Tubulidentata*—Aardvark (Eocene-Recent)

*Order Perissodactyla*—Odd-toed ungulates (hoofed placentals). Horse, rhinoceros, tapir, titanothere, chalicothere (Eocene-Recent)

*Order Artiodactyla*—Even-toed ungulates. Pig, hippo, camel, deer, elk, bison, cattle, sheep, antelope, entelodont, oredont (Eocene-Recent)

*Order Proboscidea*—Elephant, mammoth, mastadon (Eocene-Recent)

*Order Sirenia*—Sea cow, manatee, dugong (Eocene-Recent)

*Order Embrithopoda*—Known primarily from a single locality in Egypt. Large mammals with two gigantic bony processes arising from the nose area (Oligocene)

*Order Desmostyla*—Amphibious or seal-like in habit. Front and hind limbs well developed, but hands and feet somewhat specialized as paddles (Oligocene-Miocene)

*Order Hyracoidea*—Hyrax (Oligocene-Recent)

*Order Pholidota*—Scaly anteater (Oligocene-Recent)

# Glossary

**Absaroka sequence** A widespread sequence of Pennsylvanian and Permian sedimentary rocks bounded above and below by unconformities; deposited during a transgressive-regressive cycle of the Absaroka Sea.

**absolute dating** The process of assigning an actual age to geologic events. Various radioactive decay dating techniques yield absolute ages. (*See relative dating*)

**Acadian orogeny** A Devonian orogeny in the northern Appalachian mobile belt resulting from a collision of Baltica with Laurentia.

**acanthodian** Any of the fish first having a jaw or jaw-like mechanism; a class of fishes (class Aconthodii) appearing during the Early Silurian and becoming extinct during the Permian.

**acondrite** A type of stony meteorite lacking condrules; composition similar to that of terrestrial basalt.

**acritarch** Organic walled microfossils that probably represent the cysts of planktonic algae; appeared in the fossil record about 1.4 billion years ago and became abundant during the Late Proterozoic through Devonian.

**adaptive radiation** The adaptation of species of related ancestry to various aspects of the environment; the branching out of organisms of related ancestry into new habitats.

**Algoma-type BIF** A type of banded iron formation having a lenticular geometry, and measuring a few kilometers across and a few meters thick; most were deposited during the Archean Eon. (*See Superior-type BIF*)

**Alleghenian orogeny** Pennsylvanian to Permian orogenic event during which the Appalachian mobile belt from New York to Alabama was deformed; occurred in the area of the present-day Appalachian Mountains.

**allele** Alternative form of a gene controlling the same trait.

**allopatric speciation** Model for the origin of a new species from a small population that has become geographically isolated from its parent population.

**alluvial fan** A cone-shaped deposit of alluvium; generally deposited where a stream flows from mountains onto an adjacent lowland.

**alpha decay** A type of radioactive decay involving the emission of a particle consisting of two protons and two neutrons from the nucleus of an atom; emission of an alpha particle decreases the atomic number by two and atomic mass number by four.

**Alpine-Himalayan belt** A major linear belt of deformation extending from the Atlantic Ocean eastward across southern Europe and northern Africa, through the Middle East, and into southeast Asia; one of two major Mesozoic-Cenozoic orogenic belts. (*See circum-Pacific belt*)

**Alpine orogeny** A Late Mesozoic-Cenozoic episode of mountain building affecting southern Europe and northern Africa.

**altered remains** Fossil remains that have been changed from their original composition or structure or both. (*See unaltered remains*)

**anaerobic** A term referring to organisms that are not dependent on oxygen for respiration.

**analogous organs** Body parts, such as wings of insects and birds, that serve the same function, but differ in structure and development. (*See homologous organ*)

**Ancestral Rockies** Late Paleozoic uplift in the southwestern part of the North American craton.

**angiosperm** Vascular plants having flowers and seeds; the flowering plants.

**angular unconformity** An unconformity below which older strata dip at a different angle (usually steeper) than the overlying younger strata. (*See disconformity and nonconformity*)

**anthropoid** Any member of the primate suborder Anthropoidea; includes New World and Old World monkeys, apes, and humans.

**Antler orogeny** A Late Devonian to Mississippian orogeny that affected the Cordilleran mobile belt; deformation extended from Nevada to Alberta, Canada.

**Appalachian mobile belt** A mobile belt located along the eastern margin of the North American craton; extends from Newfoundland to Georgia; probably continuous to the southwest with the Ouachita mobile belt.

**arc orogen** An area of deformation such as an island arc, that results from subduction of an oceanic plate; characterized by deformation and igneous activity.

**Archean Eon** A part of Precambrian time beginning 3.8 billion years ago, corresponding to the age of the oldest known rocks on Earth, and ending 2.5 billion years ago. (*See Proterozoic Eon*)

**argillite** Fine-grained sedimentary rock such as shale that has been slightly altered by metamorphism, but does not have the properties of slate.

**artificial selection** The practice of selective breeding of plants and animals for desirable traits.

**artiodactyl** Any member of the order Artiodactyla, the even-toed hoofed mammals. Living artiodactyls include swine, sheep, goats, camels, deer, bison, and musk oxen.

**aseismic ridge** A submarine ridge with no seismic activity; commonly oriented at a high angle to a spreading center.

**asthenosphere** Part of the upper mantle lying below the lithosphere; behaves plastically and flows.

**aulacogen** A sediment-filled, inactive rift of a triple junction that formed above a rising mantle plume.

**australopithecine** A term referring to several extinct species of the genus *Australopithecus* that existed in South and East Africa during the Pliocene and Pleistocene epochs.

**autotrophic** Describes organisms that synthesize their organic nutrients from inorganic raw materials; photosynthesizing bacteria and plants are autotrophs. (*See heterotrophic*)

**backarc marginal basin** A basin formed on the continent-ward side of a volcanic island arc; thought to form by backarc spreading; the site of a marginal sea, e.g., the Sea of Japan.

**Baltica** One of six major Paleozoic continents; comprised of Russia west of the Ural Mountains, Scandinavia, Poland, and northern Germany.

**banded iron formation (BIF)** Sedimentary rocks consisting of alternating thin layers of silica (chert) and iron minerals (mostly the iron oxides hematite and magnetite). (*See Algoma-type and Superior-type BIF*)

**barrier island** An elongate sand body oriented parallel to a shoreline, but separated from the shoreline by a lagoon.

**Basin and Range province** An area centered on Nevada but extending into adjacent states; characterized by Cenozoic block-faulting.

**bed** A sedimentary layer greater than 1 cm thick. (*See laminae*)

**Bedding plane** The bounding surface that separates one layer of strata from another.

**benthos** Any organism that lives on the bottom of seas or lakes; may live upon the bottom or within bottom sediments.

**beta decay** A type of radioactive decay during which a fast-moving electron is emitted from a neutron and thus is converted to a proton; results in an increase of one atomic number, but no change in atomic mass.

**Big Bang** A model for the evolution of the Universe from a dense, hot state followed by expansion, cooling, and a less dense state.

**biogenic sedimentary structure** Any structure in sedimentary rocks produced by the activities of organisms, e.g., tracks, trails, burrows. (*See trace fossil*)

**biostratigraphic unit** A unit of sedimentary rock defined by its fossil content.

**bioturbation** The process of churning or stirring of sediments by organisms.

**bipedal** Walking on two legs as a means of locomotion.

**body fossil** The actual remains of any prehistoric organism; includes shells, teeth, bones, and, rarely the soft parts of organisms. (*See trace fossil*)

**bony fish** A class of fishes (class Actinopterygii) that evolved during the Devonian; the most common fishes; characterized by an internal skeleton of bone; divided into two subgroups, the ray-finned fishes and lobe-finned fishes.

**Caledonian orogeny** A Silurian-Devonian orogeny that occurred along the northwestern margin of Baltica resulting from the collision of Baltica with Laurentia.

**Canadian shield** The Precambrian shield of North America; exposed mostly in Canada, but outcrops occur in Minnesota, Wisconsin, Michigan, and New York.

**captorhinomorph** The oldest known reptiles; evolved during the Early Pennsylvanian; ancestors of all other reptiles, thus commonly called the stem reptiles.

**carbon 14 dating** An absolute dating method that relies upon determining the ratio of $C_{14}$ to $C_{12}$ in a sample; useful back to about 60,000 years ago; can be applied only to organic substances.

**carbonaceous chondrite** A type of stony meteorite; same as ordinary chondrites except they contain about 5 percent organic compounds including inorganically produced amino acids.

**carnassials** A pair of specialized upper and lower shearing teeth in members of the order Carnivora.

**carnivore-scavenger** Any animal that depends on other animals, living or dead, as a source of nutrients.

**cartilaginous fish** Fishes such as living sharks, rays, and skates, and their extinct relatives that have a skeleton composed of cartilage.

**cast** A replica of an object such as a shell or bone formed when a mold of that object is filled by sediment or mineral matter.

**catastrophism** The concept proposed by Baron Georges Cuvier that explained the physical and biological history of the Earth by a series of sudden, widespread catastrophes.

**Catskill Delta** The Devonian clastic wedge that was deposited adjacent to the highlands that formed during the Acadian orogeny.

**China** One of six major Paleozoic continents; comprised of all of southeast Asia, including China, Indochina, part of Thailand, and the Malay Peninsula.

**chondrite** A stony meteorite that contains chondrules.

**chondrule** A small, round mineral body formed by rapid cooling; found in chondritic meteorites.

**chordate** All members of the phylum Chordata; characterized by a notochord, a dorsal, hollow nerve cord, and gill slits at some time during the animal's life cycle.

**chromosomes** Complex, double-stranded, helical molecules of deoxyribonucleic acid (DNA); specific segments of chromosomes are genes.

**circum-Pacific belt** One of two major Mesozoic-Cenozoic orogenic belts; located around the margins of the Pacific Ocean basin; includes the orogens of South and Central America, the Cordillera of western North America, the Aleutian, Japan, and Philippine arcs. (*See Alpine-Himalayan belt*)

**clastic texture** Sedimentary rocks consisting of the broken particles of pre-existing rocks or organic structures such as shells are said to have a clastic texture.

**clastic wedge** An extensive accumulation of mostly clastic sediments eroded from and deposited adjacent to an uplifted area; clastic wedges are coarse-grained and thick near the uplift and become finer-grained and thinner away from the uplift, e.g., the Queenston Delta.

**collision orogen** An orogen produced as a result of a collision of two continents, e.g., the collision of India with Asia that resulted in the formation of the Himalayas.

**colonial organism** Any organism that lives in close association with other members of the same species, thus forming a colony.

**Colorado-Central Plains province** A part of Laurentia that was accreted to the continent during the Proterozoic Eon.

**concurrent range zone** A type of biozone established by plotting the overlapping ranges of fossils that have different geologic ranges; the first and last occurrences of fossils are used to establish concurrent range zone boundaries.

**condylarth** Members of an archaic order of placental mammals (order Condylartha). Condylarths were ancestors of several other mammalian orders including artiodactyls and perissodactyls.
**conformable** The relationship between beds in a sedimentary sequence containing no discontinuities. (*See unconformity*)
**continental-continental plate boundary** A type of convergent plate boundary along which two continental lithospheric plates collide (e.g., the collision of India with Asia).
**continental drift** The theory that the continents were once joined into a single landmass which broke apart with the various fragments (continents) moving with respect to one another; the theory of continental movement proposed by Alfred Wegener in 1912.
**convergent evolution** The development of similar structures or general appearance in two or more distantly related organisms as a consequence of adapting to a similar life style, e.g., ichthyosaurs and porpoises. (*See parallel evolution*)
**convergent plate boundary** The boundary located between two plates that are moving toward one another; three types of convergent plate boundaries are recognized. (*See continental-continental plate boundary, oceanic-continental plate boundary, and oceanic-oceanic plate boundary*)
**coprolite** Fossilized feces.
**Cordilleran orogeny** A protracted episode of deformation affecting the western margin of North America from Jurassic to Early Cenozoic time; typically divided into three separate phases called the Nevadan, Sevier, and Laramide orogenies.
**Cordilleran mobile belt** A mobile belt in western North America bounded on the west by the Pacific Ocean and on the east by the Great Plains; extends north-south from Alaska into central Mexico. (*See North American Cordillera*)
**core** The interior part of the Earth beginning at a depth of about 2900 km; probably composed mostly of iron and nickel; divided into an outer liquid core, and an inner solid core.
**correlation** Demonstration of the physical continuity of rock units or biostratigraphic units in different areas, or demonstration of time equivalence as in time-stratigraphic correlation.
**craton** A name applied to the relatively stable part of a continent; consists of a Precambrian shield and a platform, a buried extension of a shield; the ancient nucleus of a continent.
**cratonic sequence** A widespread sequence of sedimentary rocks bounded above and below by unconformities; deposited during a transgressive-regressive cycle of an epeiric sea, e.g., the Sauk sequence.
**creodont** Early Tertiary order of archaic carnivorous mammals (order Creodonta).
**Cretaceous Interior Seaway** An interior seaway that existed during the Late Cretaceous; formed as northward-transgressing waters from the Gulf of Mexico joined with southward-transgressing water from the Arctic; effectively divided North America into two large land masses.
**Cro-Magnon** A race of *Homo sapiens* that lived mostly in Europe from 35,000 to 10,000 years ago.
**cross-bedding** Strata containing beds that were deposited at an angle to the surface upon which they were accumulating, as in desert dunes, are said to be cross-bedded.
**cross-cutting relationships** An important principle in determining the relative ages of events; holds that an igneous intrusion or fault must be younger than the rocks that it intrudes or cuts.
**crossopterygian** A specific type of lobe-finned fish; possessed lungs; ancestral to amphibians.
**crust** The outer part of the Earth; the upper part of the lithosphere; separated from the mantle by the Moho; divided into continental and oceanic crust.
**crystalline texture** A texture of rocks consisting of an interlocking mosaic of mineral crystals.
**Curie point** The temperature at which iron-bearing minerals in a cooling magma attain their magnetism.
**cyclothem** A vertical sequence of cyclically repeated sedimentary rocks resulting from alternating periods of marine and nonmarine deposition; commonly contain a coal bed.
**cynodont** A type of therapsid (advanced mammal-like reptile); the probable ancestors of mammals.

**daughter element** An element formed by the radioactive decay of another element, e.g., argon 40 is the daughter element of potassium 40.
**delta** An alluvial deposit at the mouth of a river.
**depositional environment** Any area in which sediment is deposited; a depositional site that differs in physical aspects, chemistry, and biology from adjacent environments.
**desiccation crack** A crack formed in clay-rich sediments in response to drying and shrinkage.
**disconformity** An unconformity above and below which the strata are parallel. (*See angular unconformity and nonconformity*)
**divergent evolution** The diversification of a species into two or more descendant species. (*See adaptive radiation*)
**divergent plate boundary** The boundary located between two plates that are moving apart; new oceanic lithosphere forms at the boundary; characterized by volcanism and seismicity.
**DNA (deoxyribonucleic acid)** The substance of which chromosomes are composed; the genetic material of all organisms except bacteria.
**Doppler effect** The apparent change in frequency of a wave resulting from motion of the source of the wave, the receiver, or both.
**drift** A collective term for all sediment deposited by glacial activity including material deposited directly by glacial ice (till), and material deposited in streams derived from the melting of ice (outwash).
**dryopithecine** Any of the members of a Miocene family of apelike primates; possible ancestors of apes and humans. (*See ramapithicine*)

**ectotherm** An animal having a variably body temperature. Ectotherms acquire heat from their environment. (*See endotherm*)

**Ediacaran faunas** A collective name for all Late Proterozoic faunas containing animal fossils similar to those of the Ediacara fauna of Australia.

**electromagnetic force** A combination of electricity and magnetism into one force; binds atoms into molecules.

**electron capture decay** A type of radioactive decay involving the capture of an electron by a proton and its conversion to a neutron; results in a loss of one atomic number, but no change in atomic mass.

**Ellesmere orogeny** A Mississippian orogeny affecting the northern margin of Laurentia.

**encounter theory** A theory for the evolution of the solar system involving a star passing close to the sun thus pulling away gaseous filaments that later accreted into solid bodies.

**endotherm** An animal whose body temperature is maintained by the animal's metabolic processes. (*See ectotherm*)

**epeiric sea** A broad shallow sea that covers part of a continent; six epeiric seas covered parts of North America during the Phanerozoic Eon, e.g., the Sauk Sea.

**eukaryotic cell** A type of cell with a membrane-bounded nucleus containing chromosomes; also contains such organelles as plastids and mitochondria which are absent in prokaryotic cells. (*See prokaryotic cell*)

**exotic terrane** Microplate that shows little in common in terms of fossils, stratigraphy, structural trends, and paleomagnetic properties with other plates with which it is associated; assumed to have formed elsewhere and because of plate motions was carried to its present position

**Farallon plate** A Late Mesozoic-Cenozoic oceanic plate that was largely subducted beneath North America; remnants of the Farallon plate are the Juan de Fuca and Cocos plates.

**fermentation** An anaerobic metabolic process during which organic compounds are decomposed and energy is made available to cells.

**fission track dating** The process of dating samples by counting the number of small linear tracks (fission tracks) that result from damage to a mineral crystal by rapidly moving alpha particles generated by radioactive decay of uranium.

**fluvial** A term referring to rivers, river action, and the deposits of rivers.

**formation** The basic lithostratigraphic unit; a unit of strata that is mappable and that has distinctive upper and lower boundaries.

**fossil** The remains or traces of prehistoric life preserved in rocks of the Earth's crust. (*See body fossil and trace fossil*)

**Franklin mobile belt** The most northerly mobile belt in North America; extends from northwestern Greenland westward across the Canadian Arctic islands.

**gene** The basic unit of inheritance; a specific segment of a chromosome. (*See allele*)

**gene pool** The sum total of all alleles of all genes available to an interbreeding population.

**geologic record** The record of past events preserved in rocks.

**geologic time scale** A vertical geologic chart arranged such that the designation for the earliest part of geologic time appears at the bottom, and progressively younger designations appear in their proper chronologic sequence.

**geology** The science concerned with the study of the Earth; includes studies of Earth materials (minerals and rocks), surface and internal processes, and Earth history.

***Glossopteris* flora** A Late Paleozoic flora found only on the Southern Hemisphere continents and India, and named after its best known genus, *Glossopteris*.

**Gondwana** One of six major Paleozoic continents; comprised of the present-day continents of South America, Africa, Antarctica, Australia, and India, and parts of other continents such as southern Europe, Arabia, and Florida; began fragmenting during the Triassic Period.

**graded bedding** A type of sedimentary bedding in which an individual bed is characterized by a decrease in grain size from bottom to top.

**grain size** A term relating to the size of the particles making up sediment or sedimentary rock.

**granite-gneiss complex** One of the two main types of rock bodies characteristic of areas underlain by Archean rocks.

**granite-rhyolite provinces** Areas in the present-day north-central and southwestern United States underlain by intrusive and extrusive rocks; formed during the Proterozoic Eon.

**gravity** The attractive force that acts between all objects, e.g., the Earth and the Moon.

**graywacke** A type of sandstone with abundant fine-grained matrix; also commonly contains chemically unstable minerals and rock fragments.

**grazer** Any animal that crops low-growing vegetation, especially grasses.

**greenstone belt** A linear or podlike association of volcanic and sedimentary rocks particularly common in Archean terranes; typically synclinal and consists of lower and middle volcanic units and an upper sedimentary rock unit.

**Grenville province** An area in the eastern United States and Canada that was accreted to Laurentia during the Late Proterozoic.

**guide fossil** Any easily identifiable fossil that has a wide geographic distribution and short geologic range; used to determine the geologic ages of strata and to correlate strata of the same age.

**gymnosperm** The flowerless, seed-bearing land plants.

**half-life** The time required for one-half of the original number of atoms of a radioactive element to decay to a stable daughter product, e.g., the half-life of potassium 40 is 1.3 billion years.

**herbivore** An animal that is dependent on plants as a source of nutrients.

**Hercynian orogeny** Pennsylvanian to Permian orogeny in the Hercynian mobile belt of southern Europe.

**heterotrophic** Describes organisms such as animals that depend on preformed organic molecules from the environment as a source of nutrients. (*See autotrophic*)

**hiatus** The interval of geologic time not represented by strata in a sequence of strata containing an unconformity.

**hominid** Abbreviated form of Hominidae; the family of hominoids to which humans belong. Such bipedal primates as *Australopithecus* and *Homo* are hominids. (*See hominoid*)

**Hominoid** Abbreviated form of Hominoidea, the superfamily of primates to which apes and humans belong. (*See hominid*)

**homogeneous accretion** A model for the differentiation of the Earth into a core, mantle, and crust; holds that the Earth was originally compositionally homogeneous, but heated up, allowing heavier elements to sink to the core.

**homologous organ** Body parts in different organisms that have a similar structure, similar relationships to other organs, and similar development, but do not necessarily serve the same function, e.g., the wing of a bird and the forelimbs of whales and dogs. (*See analogous organs*)

**hot spot** Localized zone of melting below the lithosphere; detected by volcanism at the surface.

**Hudsonian orogeny** An episode of Proterozoic mountain building in Laurentia; occurred between 1.9 and 1.8 billion years ago.

***Hyracotherium*** A small Early Eocene mammal; the oldest genus assigned to the horse family (family Equidae).

**Iapetus Ocean** A Paleozoic ocean basin that separated North American from Europe; the Iapetus Ocean began closing when North America and Europe began moving toward one another, and it was eliminated when the continents collided during the Late Paleozoic.

**ichthyosaur** Any of the porpoise-like, Mesozoic marine reptiles.

**igneous rock** Any rock formed by cooling and crystallization of magma, or formed by the accumulation of volcanic ejecta such as ash.

**inheritance of acquired characteristics** A mechanism proposed by Jean Baptiste de Lamarck to account for evolution stating that characteristics acquired during an organism's lifetime could be passed on to its offspring.

**inhomogeneous accretion** A model for the differentiation of the Earth into a core, mantle, and crust by the sequential condensation of core, mantle, and crust from hot nebular gases.

**insectivore** Any mammal, such as moles, shrews, and hedgehogs, belonging to the placental order Insectivora. Insectivores appeared during the Late Cretaceous and were ancestors of all other placental orders.

**intracontinental rift** A linear zone of deformation within a continent produced by tensional forces; characterized by normal faults and volcanism, e.g., the East African Rift System.

**iron meteorite** A type of meteorite consisting of iron and nickel.

**isostasy** Theoretical concept of the Earth's crust "floating" on a dense underlying layer; areas of less dense crust, such as continental crust, rise topographically above more dense oceanic crust.

**isotope** All atoms of a chemical element have the same number of protons in the nucleus, but may have variable numbers of neutrons; those with different numbers of neutrons are isotopes of that element, e.g., carbon 12 and carbon 14.

**Kaskaskia sequence** A widespread sequence of Devonian and Mississippian sedimentary rocks bounded above and below by unconformities; deposited during a transgressive-regressive cycle of the Kaskaskia Sea.

**Kazakhstan** One of six major Paleozoic continents; a triangular-shaped continent centered on Kazakhstan (part of Asia).

**Kenoran orogeny** The last major event of deformation during the Archean Eon in what is now North America.

**Klamath Arc** A Middle Paleozoic island arc that formed off the western margin of the North American craton; collided with the craton during the Late Devonian and Early Mississippian causing the Antler orogeny.

**labyrinthodont** Any of the amphibians, from the Devonian to the Triassic, characterized by the labyrinthine wrinkling and folding of their teeth.

**laminae** Layers of sedimentary rock less than 1 cm thick. (*See bed*)

**Laramide orogeny** The Late Cretaceous to Early Cenozoic phase of the Cordilleran orogeny; responsible for many of the structural features of the present-day Rocky Mountains. In contrast to the preceding Nevadan and Sevier orogenies, the Laramide orogeny deformed the margin of the craton.

**lateral continuity** A principle developed by Nicolas Steno that holds that sediment layers extend outward in all directions until they terminate.

**Laurasia** A Late Paleozoic, northern hemisphere continent comprised of the present-day continents of North America, Greenland, Europe, and Asia.

**Laurentia** The name given to a Proterozoic continent that was comprised mostly of North America and Greenland, and parts of northwestern Scotland, and perhaps parts of the Baltic shield of Scandinavia.

**lithology** The physical characteristics of rocks.

**lithosphere** The outer, rigid part of the Earth consisting of the upper mantle, oceanic crust, and continental crust; lies above the asthenosphere.

**lithostratigraphic unit** A unit of sedimentary rock, such as a formation, defined by its lithologic characteristics rather than its biologic content or time value.

**living fossil** A living organism that has descended from ancient ancestors with little apparent change.

**Llano province** An area located in the southwestern United States that was accreted to Laurentia during the Late Proterozoic.

**lobe-finned fish** A type of fish in which the fin contains a series of articulating bones, and the fin is attached to the body by a fleshy shaft; one of the two major subgroups of bony fish.

**magnetic anomaly** Any change, such as a change in average strength, of the Earth's magnetic field.

**magnetic reversal** The phenomenon of the complete reversal of the north and south magnetic poles.

**marine regression** The withdrawal of the sea from a continent or coastal area resulting in the emergence of land as sea level falls or the land rises with respect to sea level.

**marine transgression** The invasion of coastal areas or much of a continent by the sea resulting from the rise in sea level or subsidence of the land.

**marsupial** Any of the pouched mammals such as opossums, kangaroos, and wombats. The order Marsupialia appeared during the Late Cretaceous. At present, marsupials are common only in Australia.

**mass extinction** A time during which extinction rates are greatly accelerated thus resulting in a marked decrease in the diversity of organisms, e.g., the terminal Cretaceous extinction event.

**meiosis** A type of cell division during which the number of chromosomes is reduced by one-half. The cell division process that yields sex cells, sperm and eggs in animals, and pollen and ovules in plants (*See mitosis*)

**metamorphic rock** Any rock type altered by high temperature and pressure, and the chemical activities of fluids is said to have been metamorphosed, e.g., slate, gneiss, marble.

**meteorite** A mass of matter of extraterrestrial origin that has fallen to the Earth's surface.

**miacid** Small, short-limbed, Early Tertiary carnivorous mammals; ancestors of all later members of the order Carnivora.

**micrite** Microcrystalline calcium carbonate mud; a limestone composed of micrite.

**microplate** Small lithospheric plate.

**midcontinent rift** A Late Proterozoic intracontinental rift within Laurentia; contains thick accumulations of extrusive igneous rocks and detrital sedimentary rocks.

**Milankovitch theory** A theory that explains cyclic variations in climate as a consequence of irregularities in the Earth's rotation and orbit.

**mineral** A naturally occurring, inorganic, crystalline solid having characteristic physical properties and chemical composition.

**mitosis** A type of cell division during which the two cells resulting from division receive the same number of chromosome as possessed by the parent cell. (*See meiosis*)

**mobile belt** Elongated area of deformation as indicated by folds and faults; generally located adjacent to a craton, e.g., the Appalachian mobile belt.

**modern synthesis** A synthesis of the ideas of geneticists, paleontologists, population biologists and others to yield a neo-Darwinism view of evolution; includes chromosome theory of inheritance, mutation as a source of variation, and gradualism.

**molarization** An evolutionary trend in the hoofed mammals (ungulates) during which the premolars became more molarlike.

**molars** The teeth of mammals used for grinding and chewing.

**mold** A cavity or impression of an organism or part thereof in sediment or sedimentary rock, e.g., a mold of a clam shell.

**monomer** A comparatively simple organic molecule, such as an amino acid, that is capable of linking with other monomers to form polymers. (*See polymer*)

**monotreme** The egg-laying mammals; only two types of monotremes now exist, the platypus and spiny anteater of the Australian region.

**morphology** The form or structure of an organism or any of its parts.

**mosaic evolution** The concept that all features of an organism do not evolve at the same rate; all organisms are mosaics of characteristics some of which are retained from the ancestral condition, and some of which are more recently evolved.

**multicellular organism** Any organism consisting of many cells as opposed to a single cell; possesses cells specialized to perform specific functions such as reproduction and respiration.

**mutation** Any change in the genetic determinants, genes, of organisms; some of the inheritable variation in populations upon which natural selection acts arises from mutations in sex cells.

**natural selection** A mechanism proposed by Charles Darwin and Alfred Russell Wallace to account for evolution; natural selection results in the survival to reproductive age of those organisms best adapted to their environment.

**Neanderthal** A type of human that inhabited Europe and the Near East from 150,000 to 32,000 years ago; considered by some to be a variety or subspecies (*Homo sapiens neanderthalensis*) of *Homo sapiens*, and by some as a separate species (*Homo neanderthalensis*).

**nebular theory** A theory for the evolution of the solar system from a rotating cloud of gas.

**nekton** Actively swimming organisms, e.g., fishes, whales, and squids. (*See plankton*)

**Neptunism** The concept held by Abraham Gottlob Werner and others that all rocks were precipitated from a worldwide ocean.

**Nevadan orogeny** Late Jurassic to Cretaceous phase of the Cordilleran orogeny; most strongly affected the western part of the Cordilleran mobile belt.

**nonconformity** An unconformity in which stratified rocks above the erosion surface overlie igneous or metamorphic rocks. (*See angular unconformity and disconformity*)

**North American Cordillera** A complex mountainous region in western North America extending from Alaska into central Mexico. (*See Cordilleran mobile belt*)

**oceanic-continental plate boundary** a type of convergent plate boundary along which oceanic lithosphere and continental lithosphere collide; characterized by subduction of the oceanic plate beneath the continental plate and by volcanism and seismicity.

**oceanic-oceanic plate boundary** A type of convergent plate boundary along which two oceanic lithospheric plates collide.

**Oolite** A spherical grain of sand size consisting of concentric layers of calcium carbonate; a sedimentary particle common in some limestones.

**ophiolite suite** A sequence of igneous rocks thought to represent a fragment of oceanic lithosphere; composed of peridotite overlain successively by sheeted basalt and pillow basalts.

**ordinary chondrite** The most abundant type of stony meteorite; composed of high-temperature ferromagnesium minerals such as olivine and some pyroxenes.

**organic evolution** *See theory of evolution.*

**organic reef** A wave resistant limestone structure with a structural framework of animal skeletons, e.g., stromatoporoid reef or coral reef.

**original horizontality** A principle developed by Nicolas Steno that holds that sediment layers when deposited were laid down horizontally or very nearly so.

**ornithischian** Any dinosaur belonging to the order Onithischia; characterized by a birdlike pelvis. (*See saurischian*)
**orogen** A linear part of the Earth's crust that was deformed during an orogeny. (*See orogeny*)
**orogeny** The process of forming mountains, especially by folding and thrust faulting; an episode of mountain building, e.g., the Acadian orogeny.
**ostracoderm** The "bony-skinned" fish; first appeared during the Late Cambrian and thus are the oldest known vertebrates; characterized by a lack of jaws and teeth, and presence of bony armor.
**Ouachita mobile belt** A mobile belt located along the southern margin of the North American craton; probably continuous with the Appalachian mobile belt.
**Ouachita orogeny** An orogeny that deformed the Ouachita mobile belt during the Pennsylvanian Period.
**outgassing** The process whereby gases derived from the Earth's interior are released into the atmosphere by volcanic activity.
**outwash** A term for the sediment deposited in streams that issue from glaciers. (*See drift*)

**paleoclimate** A climate that existed during the past.
**paleocurrent** A term referring to an ancient current direction; determined by measuring the orientations of various sedimentary structures such as cross-bedding.
**paleogeography** The study of the Earth's geography throughout geologic time; paleogeography may be determined for the entire Earth, such as the position of continents through time, or for a local area.
**Pangaea** The name proposed by Alfred Wegener for a supercontinent that existed at the end of the Paleozoic Era, and that consisted of all landmasses of the Earth.
**Panthalassa Ocean** The Late Paleozoic worldwide ocean that surrounded the supercontinent Pangaea.
**pantothere** Any member of an order (order Pantotheria) of mammals thought to be ancestors of both marsupial and placental mammals.
**parallel evolution** The development of similar structures in two or more closely related but separate lines of descent as a consequence of similar adaptations. (*See convergent evolution*)
**parent element** An unstable element that by radioactive decay is changed into a stable daughter element (*See daughter element*)
**pelycosaur** Pennsylvanian to Permian "finback reptiles"; possessed some mammalian characteristics.
**period** The fundamental unit in the hierarchy of time units; a part of geologic time during which a particular sequence of rocks designated as a system was deposited.
**perissodactyl** Any member of the order Perissodactyla, the odd-toed hoofed mammals; living perissodactyls include horses, rhinoceroses, and tapirs. (*See artiodactyl*)
**photochemical dissociation** A process whereby water molecules in the upper atmosphere are disrupted by ultraviolet radiation thus yielding oxygen ($O_2$) and hydrogen (H).
**photosynthesis** The metabolic process of synthesizing organic molecules from water and carbon dioxide, using the radiant energy of sunlight captured by chlorophyll-containing cells.
**phyletic gradualism** An evolutionary concept holding that a species evolves gradually and continuously through time to give rise to new species. (*See punctuated equilibrium*)
**placental** Any of the mammals that have a placenta to nourish the embryo; fusion of the amnion of the amniote egg with the walls of the uterus forms the placenta; most mammals, living and fossil, are placentals.
**placoderm** The "plate-skinned" fish; Late Silurian through Permian; characterized by jaws and bony armor especially in the head-shoulder region.
**planetesimal** An asteroid-sized body that along with other planetesimals aggregated to form protoplanets.
**plankton** Animals and plants that float passively, e.g., phytoplankton and zooplankton. (*See nekton*)
**plate tectonics (plate tectonic theory)** The theory that large segments of the outer part of the Earth (lithospheric plates) move relative to one another; lithospheric plates are rigid and move over the asthenosphere which behaves much like a very viscous fluid.
**playa lake** Broad, shallow, temporary lake that forms in an arid region and quickly evaporates to dryness.
**plesiosaurs** A group of Mesozoic, marine reptiles.
**pluvial lake** Any lake that formed beyond the areas directly affected by glaciation during the Pleistocene as a result of increased precipitation and lower evaporation rates, e.g., Lake Bonneville.
**point bar** The sediment body deposited on the inside of a meander loop.
**pollen analysis** Identification and statistical analysis of pollen from sedimentary rocks; such analyses provide information about ancient floras and climates.
**polymer** A comparatively complex organic molecule, such as nucleic acids and proteins, formed by monomers linking together. (*See monomer*)
**Precambrian** A widely used informal term referring to all rocks stratigraphically beneath strata of the Cambrian System, and for all geologic time preceding the Cambrian Period.
**Precambrian shield** A vast area of exposed Precambrian rocks on a continent; areas of relative stability for long periods of time, e.g., the Canadian Shield.
**primary producer** Those organisms in a food chain, such as green plants and bacteria, upon which all other members of the food chain depend directly or indirectly; those organisms not dependent on an external source of nutrients. (*See autotrophic*)
**Primates** The order of mammals that includes prosimians (lamurs and tarsiers), monkeys, apes, and humans; characteristics include large brain, stereoscopic vision, and grasping hand.
**principle of fossil succession** The principle based on the work of William Smith that holds that fossils, and especially assemblages of fossils, succeed one another through time in a regular and determinable order.
**principle of inclusions** A principle that holds that inclusions, or fragments, in a rock unit are older than the rock unit itself, e.g., granite fragments in a sandstone are older than the sandstone rock unit.
**proboscidian** The elephants and their extinct relatives (order Proboscidea); characterized by an elongate or snoutlike feeding organ (a trunk).

**progradation** The seaward (or lakeward) advance of the shoreline as a result of nearshore sedimentation.

**prokaryotic cell** A type of cell having no nucleus, and lacking such organelles as plastids and mitochondria; cells of bacteria and cyanobacteria (blue-green algae) are prokaryotic. (*See eukaryotic cell*)

**prosimian** Any member of the primate suborder Prosimii; includes tree shrews, lemurs, tarsiers, and lorises; commonly called lower primates.

**Proterozoic Eon** That part of Precambrian time beginning 2.5 billion years ago and ending 570 million years ago. (*See Archean Eon*)

**pseudosuchian** Members of a group of thecodont reptiles; probable ancestors of both orders of dinosaurs.

**pterosaur** Any of the Mesozoic flying reptiles.

**punctuated equilibrium** An evolutionary concept that holds that a new species evolves rapidly, perhaps in a few thousands of years, then remains much the same during its several millions of years of existence.

**quadrupedal** Referring to locomotion on all four legs; as opposed to bipedal.

**Quaternary** A term for a geologic period or system comprising all geologic time or rocks from the end of the Tertiary to the present; consists of two epochs or series, the Pleistocene and the Recent (Holocene).

**quartzite-carbonate-shale assemblage** A suite of sedimentary rocks characteristic of passive continental margins, but also known from intracratonic basins and backarc basins.

**Queenston Delta** The clastic wedge resulting from the erosion of the highlands formed during the Taconic orogeny; deposited on the west side of the Taconic highlands.

**radioactive decay** The spontaneous decay of an atom by emission of a particle from its nucleus (alpha and beta decay) or by electron capture, thus changing the atom to an atom of a different element.

**ramapithecine** A family of Miocene ape-like primates; possible ancestors of humans. (*See dryopithecine*)

**range zone** Biostratigraphic unit defined by the occurrence of a single type of organism such as a species or genus.

**red beds** Sedimentary rocks, mostly sandstone and shale, with red coloration due to the presence of ferric oxides.

**reef** *See organic reef.*

**relative dating** The process of determining the age of an event relative to other events; involves placing geologic events in their correct chronologic order, but involves no consideration of when the events occurred in terms of numbers of years ago. (*See absolute dating*)

**ripple mark** Wavelike (undulating) structure produced in granular sediment such as sand; formed by wind, unidirectional water currents, or wave currents.

**rock** A consolidated aggregate of minerals or particles of other rocks; although an exception to this definition, coal, natural glass, and aggregates of shells are also considered rocks.

**rock cycle** A sequence of processes through which Earth materials may pass as they are transformed from one rock type to another.

**ruminant** Any of the cud-chewing placental mammals such as deer, cattle, camels, bison, sheep, and goats.

**sabkha** The term applied to coastal supratidal flats, expecially those in arid regions.

**salt dome** A structure resulting from the upward movement of a mass of salt through overlying layers of sedimentary rocks. Oil and gas fields are commonly associated with salt domes.

**San Andreas transform fault** A major transform fault extending through part of California; connects with spreading centers in the Gulf of Mexico and in the Pacific Ocean off the northwest coast of the United States.

**Sauk sequence** A widespread sequence of sedimentary rocks bounded above and below by unconformities; deposited during a latest Proterozoic to Early Ordovician transgressive-regressive cycle of the Sauk Sea.

**saurischian** Any dinosaur belonging to the order Saurischia; characterized by a lizardlike pelvis. (*See ornithischian*)

**sea-floor spreading** The theory that the sea-floor moves away from spreading centers and is eventually subducted and consumed at convergent plate margins.

**sediment-deposit feeder** Any animal that ingests sediment and extracts the nutrients from it.

**sedimentary facies** Any aspect of a sedimentary rock unit that makes it recognizably different from adjacent sedimentary rocks of the same, or approximately same, age, e.g., a sandstone facies.

**sedimentary rock** Any rock composed of sediment. The sediment may be particles of various sizes such as gravel or sand, the remains of animals or plants as in coal and some limestones, or chemicals in solution that are extracted by organic or inorganic processes.

**seedless vascular plant** A type of land plant with vascular tissues for transport of fluids and nutrients throughout the plant; reproduces by spores rather than seeds, e.g., ferns and horsetail rushes.

**sedimentary structure** Any structure in sedimentary rock such as cross-bedding, desiccation cracks, and animal burrows.

**Sevier orogeny** The Cretaceous phase of the Cordilleran orogeny that affected the continental shelf and slope areas of the Cordilleran mobile belt.

**Siberia** One of six major Paleozoic continents; comprised of Russia east of the Ural Mountains, and Asia north of Kazakhstania and south of Mongolia.

**solar nebular theory** A modification of the nebular theory. (*See nebular theory*)

**Sonoma orogeny** A Permian-Triassic orogeny caused by the collision of an island arc with the southwestern margin of North America.

**sorting** In sediment or sedimentary rock a term referring to the degree to which all particles are of about the same size; e.g., well-sorted, poorly sorted.

**Southwestern province** A part of Laurentia that was accreted to the southern margin of the continent during the Proterozoic Eon.

**species** A population of similar individuals which in nature can reproduce and produce fertile offspring.

**stony meteorite (stone)** Meteorite composed of silicate minerals.

**stony-iron meterorite (stony-iron)** Meteorite composed of iron-nickel and silicate minerals.

**stratigraphy** A branch of geology; concerned with the composition, origin, and areal and age relationships of stratified rocks.

**stromatolite** A structure in sedimentary rocks, especially limestones, produced by entrapment of sediment grains on sticky mats of photosynthesizing bacteria; a biogenic sedimentary structure.

**strong nuclear force** The force that holds protons together within the nucleus of an atom.

**subduction zone** An elongate, narrow zone at a convergent plate boundary where an oceanic plate descends relative to another plate, e.g., the subduction of the Nazca plate beneath the South American plate.

**submarine fan** A cone-shaped sedimentary deposit that accumulates on the continental slope and rise.

**Sundance Sea** A wide seaway that existed in western North America during the Middle Jurassic Period.

**Superior-type BIF** A type of banded-iron formation composed of alternating layers of silica (chert) and iron oxides; most were deposited during the Early Proterozoic Eon. (*See Algoma-type BIF*)

**superposition** A principle developed by Nicolas Steno that holds that younger layers of strata are deposited on top of older layers of strata.

**supracrustal sequence** A sequence of rocks deposited upon a pre-existing basement of crust.

**suspect terrane** A microplate of probable but not demonstrated exotic origin.

**suspension feeder** An animal that consumes microscopic plants, animals, or dissolved nutrients from water.

**system** The fundamental unit in the time-stratigraphic hierarchy of units; the Devonian System refers to rocks deposited during a specific interval of geologic time, the Devonian Period.

**Taconic orogeny** An Ordovician orogeny that resulted in deformation of the Appalachian mobile belt.

**Tejas epeiric sea** A Cenozoic epeiric sea that was largely restricted to the Atlantic and Gulf Coastal plains and parts of coastal California, but did extend into the continental interior in the Mississippi Valley.

**Tertiary** A term for a geologic period or system comprising all geologic time or rocks from the end of the Cretaceous to the beginning of the Quaternary. The Tertiary consists of five epochs or series; Paleocene, Eocene, Oligocene, Miocene, and Pliocene.

**thecodont** Any of the Triassic reptiles assigned to the order Thecodonta; pseudosuchian theocodonts are the probable ancestor of both orders of dinosaurs.

**theory** An explanation for some natural phenomenon that has a large body of supporting evidence; to be considered scientific, a theory must be testable (e.g., plate tectonic theory).

**theory of evolution** The theory that all living things are related, and that they descended with modification from organisms that lived during the past.

**therapsid** Permian to Triassic reptiles that possessed mammalian characteristics and thus are called mammal-like reptiles; one group of therapsids, the cynodonts, gave rise to mammals.

**thermal convection cell** In plate tectonics a type of circulation of material in the asthenosphere during which hot material rises, moves laterally, cools and sinks, and is reheated and reenters the cycle.

**till** All sediment deposited directly by glacial ice.

**tillite** Lithified glacial till.

**time-stratigraphic unit** A unit of strata that was deposited during a specific interval of geologic time, e.g., the Devonian System, a time-stratigraphic unit, was deposited during that part of geological time designated as the Devonian Period.

**time transgressive** Refers to a rock unit that was deposited in an environment that shifted with time, as during a marine transgression. Thus, the age of the rock unit varies over its geographic extent.

**time unit** Any of the units such as eon, era, period, epoch, and age used to refer to specific intervals of geologic time.

**Tippecanoe sequence** A widespread sequence of sedimentry rocks bounded above and below by unconformities; deposited during an Ordovician to Early Devonian transgressive-regressive cycle of the Tippecanoe Sea.

**trace fossil** Any indication of prehistoric organic activity such as tracks, trails, burrows, borings, or nests. (*See body fossil*)

**Transcontinental Arch** An area consisting of several large islands extending from New Mexico to Minnesota that was above sea level during the Cambrian transgression of the Sauk Sea.

**transform-fault plate boundary** Plate boundary along which plates slide past one another, and crust is neither produced nor destroyed; a type of strike-slip fault along which oceanic ridges are offset.

**tree-ring dating** The process of determining the age of a tree or wood in structures by counting the number of annual growth rings.

**triple junction** A point where three plates meet, e.g., the triple junction formed at the intersection of the East African Rift System and the rifts in the Red Sea and Gulf of Aden.

**unaltered remains** Fossil remains that retain their original composition and structure. (*See altered remains*)

**unconformity** An erosion surface that separates younger strata from older rocks. (*See angular unconformity, disconformity, and nonconformity*)

**ungulate** An informal term referring to the hoofed mammals, especially the orders Artiodactyla and Perissodactyla.

**uniformitarianism** The principle that holds that we can interpret past events by understanding present-day processes, and the assumption that natural laws have not changed through time.

**varve** A couplet of sedimentary laminae representing deposition that occurred in one year, e.g., the dark (winter) and light (summer) layers in varved sediments deposited in glacial lakes.

**vertebrate** Any animal having a segmented vertebral column; members of the subphylum Vertebrate; includes fishes, amphibians, reptiles, mammals, and birds.

**vestigial structure** A body part that serves little or no function, e.g., the dewclaws of dogs; a vestige of a structure or organ that was well developed and functional in some ancestor.

**volcanic island arc** A curved chain of volcanic islands parallel to a deep-sea trench where oceanic lithosphere is subducted causing volcanism and the origin of volcanic islands.

**Walther's Law** The observations by Johannes Walther serve as the basis for a principle that holds that facies deposited in adjacent environments will be superposed one upon the other when the environments migrate laterally; in a conformable vertical sequence of facies, the same facies that occur vertically will replace one another laterally.

**weak nuclear force** A force responsible for the breakdown of atomic nuclei, thus producing radioactive decay.

**Zuni sequence** An Early Jurassic to Early Paleocene sequence of sedimentary rocks bounded above and below by unconformities; deposited during a transgressive-regressive cycle of the Zuni Sea.

# Index

printed with the permission of Cambridge University Press; p. 64 Diagram from *Giants of Geology* by Carroll Lane and Mildred Adams. Copyright 1945, 1952 by Carroll Lane Fenton and Mildred Adams Fenton. Reprinted by permission of Double-day, a division of Bantam, Doubleday, Dell Publishing Group, Inc.; p. 64 From L.W. Mintz, *Historical Geology*, 1981, 1977, 1972; Charles E. Merrill Publishing Company; p. 65 From John W. Harbaugh, *Stratigraphy and the Geologic Time Scale*, 2d ed. Copyright © 1974 Wm. C. Brown Publishers, Dubuque, IA. All Rights Reserved. Reprinted with permission; p. 67 Reprinted from Gregory, H.E. 1950. Geology and Geography of the Zion Park Region Utah and Arizona. U.S.G.S. Professional Paper 220, Fig. 24; p. 69 *From History of the Earth*, 2/E by Bernhard Kummel. Copyright © 1961, 1970 by W.H. Freeman and Company. Reprinted with permission; p. 71 Reproduced, with permission, for Rider, M.H. 1986. *The Geological Interpretation of Well Logs.* Blackie and Son. Ltd, Glasgow and London; p. 71 B. Rascoe Jr., 1962, reprinted by permission of American Association of Petroleum Geologists; p. 72 From *Elements of Petroleum Geology* by Richard C. Selley. Copyright © 1985 W.H. Freeman and Company. Reprinted with permission; p. 72 O.R. Berg and R.J. Hubbard, J. Pape and R.G. Wolverton (eds.), 1985; reprinted by permission of American Association of Petroleum Geologists; p. 75 K.J. Hsu, K. Kelts, and J.W. Valentine, 1980, reprinted by permission of American Association of Petroleum Geologists.

**CHAPTER 4** p. 80 Modified from figure on page 36, "When the Mediterranean Dried Up" by K.J. Hsu, *Scientific American*, 227, # 6, 1972; p. 80 W.L. Fisher, L.F. Brown, Jr., A.J. Scott, and J.H. McGowen, 1969, reprinted with permission of Texas Bureau of Economic Geology; p. 81 Reprinted with permission from M.B. Cita, Geodynamics Scientific Report No. 7, 1973, Elsevier Science Publishers; p. 81 From Hsu, Kenneth J. "When the Mediterranean Dried Up." Copyright © 1972 by Scientific American, Inc. All rights reserved; p. 82 R. C. Selley, 1978, reprinted by permission of American Association of Petroleum Geologists; p. 87 From Charles C. Plummer and David McGreary, *Physical Geology*, 4th ed. Copyright © 1988 Wm. C. Brown Publishers, Dubuque, Iowa. All Rights Reserved. Reprinted by permission; p. 87 Photograph by G. Shanmugan. Reprinted with permission from Shanmugam, Moiola, McPherson, and O'Connell, *Sedimentary Geology*, © 1988, Elsevier Science Publishers; p. 88 Photograph by R.G. Walker. Reproduced with the permission of the Geological Association of Canada; p. 89 J.C. Harms, J.B. Southward, D.R. Spearing, R.G. Walker, 1975, reprinted by permission of the Society of Economic Palentologists and Mineralogists; p. 89 J.C. Harms, Southward, D.R. Spearing, and R.G. Walker, 1975, reprinted by permission of the Society of Economic Palentologists and Mineralogists; p. 91 From Pettijohn, et al., Sand and Sandstone, © 1972, Springer-Verlag. Reprinted with permission; p. 91 Hamblin, The Earth's Dynamic System © 1982. Reprinted with permission of Burgess International; p. 92 From Allen, J.R.L., "Review of the Origin and Characteristics of Recent Alluvial Sediments." *Sedimentology* 5, (1965). Reprinted with permission of Blackwell Scientific Publications Ltd.; p. 92 From Allen, J.R.L., "Review of the Origin and Characteristics of Recent Alluvial Sediments." *Sedimentology* 5, (1965). Reprinted with permission of Blackwell Scientific Publications Ltd.; p. 93 R.C. Selley, *Ancient Sedimentary Environment*, 2d ed., 1978, Chapman & Hall Ltd. Reprinted with permission; p. 93 Reproduced with permission of the Geological Association of Canada; p. 93 Reproduced with permission of the Geological Association of Canada; p. 95 Reprinted with permission form R.W. Kelley and W.R. Farrand, "The Glacial Lakes Around Michigan," 1967 Geological Survey Bulletin 4. Geological Survey Division, Michigan Department of Natural Resources; p. 95 Reprinted with permission from R.W. Kelley and W.R. Farrand, "The Glacial Lakes Around Michigan," 1967 Geological Survey Bulletin 4. Geological Survey Division, Michigan Department of Natural Resources; p. 96 Coleman, Deltas: Processes of Deposition and Models for Exploration, © 1976, Continuing Education Publication Company, Inc. Reprinted with permission; p. 96 H.N. Fisk, E. McFarlan, Jr., C.R. Kolb, and L.J. Wilbert Jr., 1954, reprinted with permission of the Society of Economic Paleontologists and Mineralogists; p. 97 C.R. Kolb and J.R. Van Lopik, 1966, reprinted by permission of the Houston Geological Society; p. 98 a. Photo courtesy of NASA; b. Reprinted from Fisher, W.L., Brown, L.F. Jr., Scott, A.J., and McGowen, J.H. 1969. Delta Systems in the Exploration for Oil and Gas—A Research Colloquium; p. 99 P.C. Scruton, 1960, reprinted with permission of American Association of Petroleum Geologists; p. 102 Reproduced from *Evolution of Sedimentary Rocks* by Robert M. Garrels and Fred T. Mackenzie, by permission of W.W. Norton & Company, Inc. Copyright © 1971 by W.W. Norton & Company, Inc.; p. 102 From Galloway and Hobday, *Terrigenous Clastic Depositional Systems* © 1983 Springer-Verlag. Reprinted with permission; p. 103 From Selley, *Introduction to Sedimentology* ©1976 Academic Press. Reprinted with permission; p. 103 From Selley, *Introduction to Sedimentology* © 1976 Academic Press. Reprinted with permission; p. 103 From Davies, T.A. and Gorsline, D.S. "Oceanic Sediments and Sedimentary Processes." In *Chemical Oceanography*, ed. J.P. Riley and R. Chester, 1976. Reprinted with permission of Academic Press; p. 104 From Wilson, *Carbonate Facies in Geologic History*, © 1975, Springer-Verlag. Reprinted with permission; p. 105 Blatt, Middleton, Murray, *Origin of Sedimentary Rocks*, © 1980, 2e, p. 498. Reprinted by permission; p. 107 (both) Found, Fiddler, Jones, 1972, reprinted with permission of American Association of Petroleum Geologists; p. 108 Courtesy of Rocky Mountain Association of Geologists; p. 109 (all) Weimer, 1960, reprinted with permission of American Association of Petroleum Geologists; p. 110 (both) Courtesy of Rock Mountain Association of Geologists.

**CHAPTER 5** p. 117 Neg. #326697, Courtesy Department of Library Services, American Museum of Natural History; p. 122 E. Mayr and C. Vaurie, *Evolution*, v. 3, 1948. Reprinted with permission; p. 123 From W.R. Atchley and D.S. Woodruff, 1981, *Evolution and Speciation.* Reprinted with the permission of Cambridge University Press; p. 124 Steffins, G. Ledyard and Franbcisco J. Ayala, "The Evolution of Darwinism." Copyright © 1985 by Scientific American, Inc. All rights reserved; p. 125 From *Principles of Paleontology*, 2/E by David M. Raup and Steven M. Stanley. Copyright © 1971, 1978 W.H. Freeman and Company. Reprinted with permission; p. 128 From Olson and Robinson, *Concepts of Evolution*, © 1975 Charles E. Merrill Publishers. Reprinted with permission; p. 129 Stanley, *Evolution*, v. 28, 1974. Reprinted with permission; p. 132 Courtesy Department of Library Services, American Museum of Natural History; p. 141 From O.C. March, *American Journal of Science*, 1982. Reprinted by permission of American Journal of Science; p. 141 From O.C. Marsh, *American Journal of Science*, 1982. Reprinted by permission of American Journal of Science; p. 133 d. Seaphot Limited: Planet Earth Pictures; p. 135 Ruse, The *Darwinian Revolution*, 1979, the University of Chicago Press; p. 139 The National Audubon Society Collection/PR; p. 142 Smithsonian Institution, photograph #15771.

**CHAPTER 6** p. 147 Reprinted with permission of Bildarchiv Preussischer Kulturbersitz; p. 148 From L.W. Mintz, *Historical Geology*, 1981, 1977, 1972; after R.J. Foster, *General Geology*, 1983, 1978, 1973, 1969; Charles E. Merrill Publishing Company; p. 149 Modified from A. Hallam, 1972, Continental Drift and the Fossil Record, *Scientific American*, v. 227, no. 5, p. 63; p. 150 Reprinted with permission from Geological Survey of America Bulletin v. 71, A. Cox and R.R. Doell, © 1960, the Geological Survey of America; p. 151 From A. Cox, "Geomagnetic Reversals," *Science*, vol. 163, pp. 237-245. Copyright 1969 by the AAAS; p. 151 Aluminum Company of America; p. 153 From *The Bedrock Geology of the World* by R.L. Larson et al. Copyright © 1984. Reprinted with the permission of W.H. Freeman and Company; p. 154 Woods Hole Oceanographic Institution; p. 160 Modified from F. Press and R. Siever, *Earth*, 4th Edition, p. 513, fig. 20-12; p. 62, 1972; p. 165 M. Kay, Geological Society of America Memoir 48, © 1951, the Geological Society of America; p. 166 Modified from E.H. Colbert, 1973, *Wandering Lands and Animals*, p. 72, fig. 31; p. 167 b. Scientific American, v. 227 #5, p. 62, 1972.

**CHAPTER 7** p. 171 Lick Observatory Photograph; p. 172 *The Restless Universe* © Nigel Henbest and Heather Couper. Published by George Philip & Son Ltd.; p. 174 Reprinted with permission. © West Publishing Co., Snow, *The Dynamic Universe*, 1985; p. 176 Reprinted with permission. © West Publishing Co., Snow, *The Dynamic Universe*, 1985; p. 176 NASA; p. 178 Reprinted with permission. © West Publishing Co., Snow, *The Dynamic Universe*, 1985; p. 179 Reprinted with permission. © West Publishing Co., Snow, *The Dynamic Universe*, 1985; p. 180 D.L. Eicher and A.L. McAlester, *History of the Earth*, © 1980, p. 11. Reprinted by permission of Prentice-Hall, Inc.; p. 182 both NASA; p.183 Science Graphics, Bend, Oregon; p. 184 Reprinted with permission. ©West Publishing Co., Snow, *The Dynamic Universe*, 1985; p. 188 Eicher and McAlester, *History of the Earth*, © 1980, p. 24. Reprinted by permission of Prentice-Hall, Inc., Englewood Cliffs, New Jersey; p. 189 Lick Observatory Photograph.

**CHAPTER 8** p. 194 James, 1972, reprinted by permission of American Association of Petroleum Geologists; p. 194 Harrison and Peterman, 1982, reprinted by permission of American Association of Petroleum Geologists; p. 197 W.R. Dickinson and W.C. Luth, "A Model for Plate Tectonic Evolution of Mantle Layers," *Science*, vol. 174, pp. 400–404. Copyright 1971 by the AAAS; p. 198 From Goodwin, *The Geology of Continental Margins*, © 1974, Springer-Verlag. Reprinted with permission; p. 199 Stockwell, 1965, reprinted by permission of American Association of Petroleum Geologists; p. 201 From A.N. Strahler, *Physical Geography*, 4th ed. John Wiley & Sons, Inc. Copyright © 1975 by Arthur N. Strahler. Used by permission; p. 203 From Condie, *Archean Greenstone Belts*, © 1981 Elsevier Science Publishers. Reprinted with permission; p. 205 GGU (A.P. Nutman); p. 205 Reproduced with permission of the Geological Association of Canada; p. 206 Reprinted with permission from Condie, Plate *Tectonics and Crustal Evolution.* Copyright 1982, Pergamon; p. 206 Reprinted by permission, *American Scientist*, Journal of Sigma Xi, P. Cloud, "Evolution of Ecosystems," v. 62, no. 1, 1974; p. 207 Photo by R.W. Oakangas. From Ojakangas and Matsch, *Minnesota's Geology*, 1982, University of Minnesota Press; p. 208 Photo by P.F. Hoffman, Geological Survey of Canada; p. 208 Reprinted with permission from Erikson, *Sedimentary Geology*, © 1977, Elsevier Science Publishers; p. 210 Reprinted with permission from Condie and Hunter, *Earth and Planetary Science Letters*, © 1976, Elsevier Science Publishers; p. 211 B.F. Windley, *The Evolving Continents*, 2nd Ed., 1984, John Wiley & Sons, Ltd.; p. 211 Reprinted with permission from Condie, *Plate Tectonics and Crystal Evolution*, 2d ed., copyright

1982, Pergamon Books Ltd.; p. 211 From Landford, F.F. and Morin, J.A., American Journal of Science, 1976. Reprinted by permission of American Journal of Science; p. 213 From Ahrens, *Meteorology Today,* ©1988, West Publishing Company. Reprinted with permission; p. 216 Reprinted with permission from *Modern Ideas on Spontaneous Generation,* v. 69, Miller, "The Formation of Organic Compounds on the Primitive Earth," copyright 1957, Annals of the New York Academy of Sciences; p. 219 J.B. Corliss, J.A. Baross, and S.E. Hoffman, *Oceanological Acta,* special issue no. 2 "Geology of the Oceans," supplement to vol. 4, 1981; p. 220 (all) R.L. Anstey and T./L. Chase, *Environments Through Time,* 1974. Reprinted with permission of Burgess International.

**CHAPTER 9** p. 225 Reprinted with permission of Jones & Barlett Publishers, Boston, from *Early Life,* by L. Marqulis, 1982, p. 29; p. 226 P. Cloud, "The Ediecarian Period and System Metazoa Inherit the Earth," *Science,* vol. 217, pp. 783–792. Copyright 1988 by the AAAS; p. 228 R.A. Kerr, "Tectonics Goes Back 2 Billion Years," *Science,* vol. 230, pp. 1364–1367. Copyright 1985 by the AAAS; p. 230 Reprinted with permission from Ray and Wanless, "The Age and Geological History of the Wollaston, Peter Lake, and Rottenstone Domains in Northern Saskatchewan," copyright 1980, the *Canadian Journal of Earth Sciences;* p. 231 P.F. Hoffman, Geological Survey of Canada; p. 232 Reprinted with permission from *Geology,* v. 10, Condie, ©1982, the Geological Society of America; p. 233 Reprinted with permission from *Geology,* v. 14, Bickford, et al., © 1986, the Geological Society of America; p. 233 Courtesy of W.B. Hamilton, U.S.G.S.; p. 234 From Green, *Tectonophysics,* © 1983, Elsevier Science Publishers. Reprinted with permission; p. 235 Reprinted with permission from *Geology,* GSA Memoir 156, Daniels, © 1982, the Geological Society of America; p. 236 From Porter, *Precambrian Plate Tectonics,* © 1981. Reprinted with permission of Elsevier Science Publishers; p. 236 Reprinted with permission from Condie, *Plate Tectonics and Crustal Evolution,* copyright 1982, Pergamon Journals Ltd.; p. 237 (top) Young, G.M., Tectono-Sedimentary History in Geology and Tectonics of the Lake Superior Basin, Wold, R.J. and Heinze, W.J. (eds.), GSA mem. 156, 1982. Reprinted with permission from the Geological Society of America Memoir; p. 239 From Kontinen, *Precambrain Research,* © 1987, Elsevier Science Publishers. Reprinted with permission; p. 240 Reprinted from Harrison, J.E., Griggs, A.B., and Wells, J.D., 1974. Tectonic Features of the Precambrian Belt Basin and Their Influence on Post-Belt Structures, U.S.G.S. Professional Paper 866, Fig. 1; p. 242 From Lindsey. *Palaeogeography, Palaeoclimatology, Palaeoecology,* © 1971. Reprinted with permission of Elsevier Science Publishers; p. 242 From Young and Nesbitt, *Precambrian Research,* © 1985. Reprinted with permission of Elsevier Science Publishers; p. 242 (Figs. 9–20c and 9–20d) Geological Survey of Canada; p. 243 From Frakes, *Climates Throughout Geologic Time,* © 1979. Reprinted with permission of Elsevier Science Publishers; p. 243 From Frakes, *Climates Throughout Geologic Time,* © 1979. Reprinted with permission of Elseiver Science Publishers; p. 245 From Eichler, *Handbook of StrataBound and StrataForm Ore Deposits,* © 1976. Reprinted with permission of Elsevier Science Publishers; p. 245 From A.M. Goodwin, *Rensta Brasileira de Geociencias,* 1982; p. 246 From K.A. Eriksson and R.G. Vos, *Precambrain Research,* © 1979, Elsevier Science Publishers; p. 248 From *Symbiosis in Cell Evolution* by Margulis. Copyright © 1981 W.H. Freeman and Company. Reprinted with permission; p. 250 From "Symbiosis and Evolution" by Margulis. Copyright © 1971 by Scientific American, Inc. All rights reserved.

**CHAPTER 10** p. 257 Photo courtesy of National Park Service, U.S. Department of the Interior; p. 260, 261 Reprinted by permission *American Scientist,* Journal of Sigma Xi, R.K. Bambach, C.R. Scotese, A.M. Ziegler, "Before Pangea: The Geographies of the Paleozoic World", v. 68, no. 1, 1980; p. 262 Reprinted with permission from *Geological Society of America Bulletin,* v. 74, © 1963, the Geological Society of America; p. 263 Modified from Cooper, Miller, and Patterson, 1986; p. 264 Reprinted with permission from *Geological Society of America Bulletin,* v. 83, Stewart, © 1972, the Geological Society of America; p. 268 Photo Courtesy of Wisconsin Dells Visitor & Convention Bureau; p. 269 Photo courtesy of Wisconsin Department of Natural Resources; p. 270 Modified from Cooper, Miller, and Patterson, 1986; p. 272 Reprinted with permission. Dorr and Eschman, *Geology of Michigan,* 1970, University of Michigan Press; p. 273 Mesolella, et al., 1974, reprinted by permission of American Association of Petroleum Geologists; p. 275 Reprinted with permission. Dorr and Eschman, *Geology of Michigan,* 1970, University of Michigan Press; p. 276 Reprinted with permission from *Geological Society of America Bulletin,* v. 81, J.M. Bird and J.F. Dewey, © 1970, the Geological Society of America.

**CHAPTER 11** p. 281 Photo courtesy of Field Museum of Natural History; Chicago. Photos No. 81442-A, 81011; p. 282, 283 Reprinted by permission *American Scientist.* Journal of Sigma Xi, R.K. Bambach, C.R. Scotese, A.M. Ziegler, "Before Pangea: The Geographies of the Paleozoic World", v. 68, no. 1, 1980; p. 285 Modified from D.L. Eicher and A.L. McAlester, 1980, *History of the Earth,* p. 233, fig. 7.31; p. 289 Modified from Cooper, Miller, and Patterson, 1986; p. 290 Modified from Cooper, Miller, and Patterson, 1986; p. 292 Modified from R.H. Dott, Jr. and R.L. Batten, *Evolution of the Earth,* 4th Edition, p. 397, fig. 14-14; p. 296 Modified from Cooper, Miller and Patterson, 1986; p. 297 (top left) From a diorama by T.L. Chase at the Permian Basin Petroleum Museum, Midland, Texas; (bottom) The Permian Basin Petroleum Museum; p. 299 Courtesy of James Gilluly, U.S.G.S. p. 304 Modified from C.K. Seyfert and L.A. Sirkin, 1979, *Earth History and Plate Tectonic,* 2nd Edition, p. 117, fig. 7-9.

**CHAPTER 12** p. 309 Smithsonian Institution, Transparency No. 86-13471A; p. 311 Reprinted by permission, *American Scientist,* journal of Sigma Xi, S. Conway Morris, "The Search for the Precambrian - Cambrian Boundary," v. 75, no. 2, 1987. Steroscan micrographs courtesy of Simon Conway Morris and Stefan Bengtson; p. 312 Modified from S.M. Stanley, American Journal of Science, v. 276, p. 63; p. 313 Geological Survey of Canada. Photo 202109 - S; p. 315 From Tappan, Proceedings of the North American Paleontological Convention, 1970 and 1971. Reprinted with permission of Allen Press Inc.; p. 316 (bottom) Carnegie Museum of Natural History; p. 317 Courtesy of Smithsonian Institution; p. 318 Field Museum of Natural History (Geo80820c), Chicago; p. 321 From *Lethaia,* v. 16, Briggs, Clarkson, and Aldridge, 1983. Reprinted with permission; p. 322 a. Modified from P. Cooper, 1975, Proceeding of Second International Coral Reef Symposium, 1: 365-386; b. Field Museum of Natural History (Geo80821c) Chicago; p. 323 (top) Field Museum of Natural History (Geo80819c), Chicago; (bottom) Courtesy of Smithsonian Institution; p. 325 Modified from S.M. Stanley, 1987. *Extinctions;* p. 327 Neg. #K10257, Courtesy Department of Library Services, American Museum of Natural History; p. 327 Neg. #K10269, Courtesy Department of Library Services, American Museum of Natural History; p. 328 *Leptotriticities tumidus* (Skinner) x10, Neva Limestone, Wolfcampian, Cowley County, Kansas. (Specimen furnished by G.A. Sanderson); p. 328 *Triticities* spp. Amazonias Limesone, Pennsylvanian, Elk County, Kansas. (Specimen furnished by G.A. Sanderson); p. 328 *Robustoschwagerian stanislavi* (Dunbar), Lower Bone Spring Limestone, Leonardian, Sierra Diablo, Culberson County, Texas (speciman furnished by G.A. Sanderson); p. 329 D.M. Raup and J.J. Sepkoski, "Mass Extinctions in the Marine Fossil Record," *Science,* vol. 215, Fig. 2, pp. 1501–1502. Copyright 1982 by the AAAS; p. 329 From *Living Invertebrates,* Pearse/Buchsbaum, by permission; p. 330 From Romer, *The Vertebrate Story,* 1959, University of Chicago Press. Reprinted by permission; p. 331 Reproduction from *Alcheringa,* by permission of Geological Society of Australia; p. 331 Reprinted with permission from *Science News,* the weekly news magazine of Science, copyright 1988 by Science Service, Inc.; p. 331 Photo courtesy of Pierre-Yves Gagnier, Institut de Paleontologie, Paris. Drawing by Henri Lavina; p. 332 John E. Repetski, "A Fish From The Upper Cambrian of North America," *Science,* vol. 200, pp. 529–531. Copyright 1978 by the AAAS; p. 335 McAlester, *The History of Life,* 2nd ed., Prentice-Hall Inc., 1977; p. 337 b. Modified from H. Levin, *The Earth through Time,* 3rd Edition, p. 400, fig. 11–68; p. 337 b. Neg. #321683, Courtesy Department of Library Services, American Museum of Natural History; p. 340 From McAlester, *The History of Life,* 1977, Prentice-Hall Inc. Reprinted by permission; p. 342 C.W. Stearn, R.L. Carroll, T.H. Clark, *Geological Evolution of North America,* 3rd ed., 1979, John Wiley & Sons, Inc.; p. 344 From Robert T. Baker, "Dinosaur Renaissance," Copyright © 1975 by Scientific American, Inc. All rights reserved; p. 345 Reprinted by permission, *American Scientist,* journal of Sigma Xi, L. Graham, "The Origin of the Life Cycle of Land Plants," v. 73, no. 2, 1985; p. 346 Reprinted by permission from *Nature,* vol. 333, p. 770, copyright © 1988, Macmillan Magazines Ltd.; p. 347 Reprinted by permission, *American Scientist,* journal of Sigma Xi, P.G. Gensel and H.N. Andrews, "The Evolution of Early Land Plants," 1987; p. 349 a. Modified from S. Stanley, *Earth and Life through Time,* p. 372; b. Modified from A. L. McAlester, 1977, *The History of Life,* 2nd Edition, p. 101, fig. 5–9.

**CHAPTER 13** p. 357 This item is reproduced by permission of *The Huntington Library,* San Marino, California; p. 359 Dietz, R.S. and J.C. Holden. Journal of Geophysical Research. Reconstruction of Pangaea: Breakup and Dispersion of Continents, Permian to Present. v. 75, no. 26, pp. 4939–4956, 1970. © American Geophysical Union; p. 360 R.H. Dott and R.H. Shaver (eds.), 1974, reprinted with permission of the Society of Economic Paleontologists and Mineralogists; p. 361 (top) Dietz, R.S. and J.C. Holden. Journal of Geophysical Research. Reconstruction of Pangaea: Breakup and Dispersion of Continents, Permian to Present. v. 75, no. 26, pp. 4939–4956, 1970. © American Geophysical Union; Reprinted with permission from *Geology,* v. 3, Burke, © 1975, the Geological Society of America; p. 362 Dietz, R.S. and J.C. Holden. Journal of Geophysical Research. Reconstruction of Pangaea: Breakup and Dispersion of Continents, Permian to Present. v. 75, no. 26, pp. 4939–4956, 1970. © American Geophysical Union; p. 364 Reproduced by permission of the author and the Palaeontological Association; p. 364 Reprinted with permission from Geological Society of America Bulletin v. 88, W.L. Donn and D.M. Shaw, © 1977, the Geological Society of America; p. 366 Modified from Cooper, Miller, and Patterson, 1986; p. 369 Palisades Interstate Park Commission; p. 369 Photo courtesy of the Peabody Museum of Natural History; p. 370 From M.S. Petersen, J.K. Rigby, and L.F. Hintze, *Historical Geology of North America.* Copyright © 1973 Wm. C. Brown Publishers, Dubuque,

IA. All Rights Reserved. Reprinted by permission; p. 371 From Morris S. Petersen, J. Keith Rigby, and Lehi F. Hintze, HISTORICAL GEOLOGY OF NORTH AMERICA, 2d ed. Copyright © 1973, 1980 Wm. C. Brown Publishers, Dubuque, Iowa. All Rights Reserved. Reprinted by permission; p. 372 Bebout and Loucks, 1983, reprinted by permission of American Association of Petroleum Geologists; p. 373, 375 Photo courtesy of National Park Services, U.S. Department of Interior; p. 376 Reprinted with permission from Geological Society of America Bulletin, v. 86, Schweickert and Cowan, © 1975, the Geological Society of America; p. 377 B.M. Page, San Joaquin Geological Society Short Course 1977, with permission of San Joaquin Geological Society; p. 379 (top right) Reprinted from W.R. Dickinson, Cenozoic Plate Tectonic Setting of the Cordilleran Region in the United States, from Cenozoic Paleogeography of the Western United States: SEPM Pacific Section, Pacific Coast Paleogeography Symposium 3, ©1979; p. 381 Photo by John S. Shelton; p. 383 Modified from S.M. Stanley, *Earth and Life through Time*, p. 473, fig. 15–44; p. 384 Carnegie Museum of Natural History; p. 386 Reprinted with permission from *Geological Society of America Bulletin*, v. 79, Armstrong, © 1968, the Geological Society of America; p. 386 Modified from G.O. Williams and C.R. Stelck, 1975, Geol. Soc. Canada Spec. Paper 13; p. 387 Reprinted by permission, *American Scientist*, Zvi Ben-Avraham, "The Movement of Continents," v. 69, no. 3, 1981.

**CHAPTER 14** p. 394 Smithsonian Institution Photo No. 106677; p. 396 Colbert, *Evolution of the Vertebrates*, 3rd ed., John Wiley & Sons, Inc. 1980; p. 396 Reprinted by permission from Botanical Review Vol. 43 (1), p. 2 Front piece, Fig. 57 by L.J. Hickey. Copyright 19, The New York Botanical Garden; p. 397 Colbert, *Evolution of the Vertebrates*, 3rd ed., John Wiley & Sons, Inc., 1980; p. 398 From Banks, *Evolution and Plants of the Past*, © 1970, Wadsworth Publishing Company. Reprinted with permission; p. 404 Copyright, 1980, Smithsonian Institution. Reprinted with the permission of the Smithsonian Institution and John Gurche; p. 408 Reprinted with permission of Hamlyn Publishing Group Ltd.; p. 409 Claude Babin, Elements of Paleontology, 1980, John Wiley & Sons, Ltd.; p. 409 From Wann Langston, Jr., "Pterosaurs," Copyright © 1981 by Scientific American, Inc. All rights reserved; p. 413 Modified from p. 84, Scientific American, v. 255, no. 4, 1986; p. 415 (left) British Museum (Natural History), London; (right) Colbert, *Evolution of the Vertebrates*, 3rd ed., John Wiley & Sons, Inc. 1980; p. 417 (top) Reprinted with permission. Hopson, "The Mammal-Like Reptiles: A Study of Transitional Fossils," *The American Biology Teacher*, National Association of Biology Teachers, Reston, VA, vol. 49, #1, January 1987, p. 18; p. 417 (bottom) Illustration from SYNAPSIDA by John C. McLoughlin. Copyright © 1979 by John C. McLoughlin. All rights reserved. Reprinted by permission of Viking Penguin, Inc.; p. 418 Reprinted with permission. Hopson, "The Mammal-Like Reptiles: A Study of Transitional Fossils." *The American Biology Teacher*, National Association of Biology Teachers, Reston, VA, vol. 49, #1, January 1987, p. 22; p. 419 Reprinted with permission of University of California Press; p. 419 Reprinted with permission of MacMillan Publishing Company from *Ancient Environment and the Interpretation of* by Fichter and Poche. Copyright © 1979 by Lynn S. Fichter and David J. Poche; p. 419 Reprinted with permission. Hopson, "The Mammal-Like Reptiles: A Study of Transitional Fossils." *The American Biology Teacher*, National Association of Biology Teachers, Reston, VA, vol. 49, #1, January 1987, p. 23; p. 420 Reprinted with permission of University of California Press; p. 421 Reproduced from *Early Mammals* (eds,) D.M. and K.A. Kermack, *Zoo J. Linn. Soc.* London SO *Suppl. 1* p. 103-15; by kind permission of the Linnean Society and the authors; p. 421 courtesy of the British Museum (Natural History); p. 422 (bottom) Courtesy of the British Museum (Natural History); p. 423 MONTANARI et, al., 1983 *Geology*, vii, p. 668-631; p. 425 D.M. Raup and J.J. Sepkiski, "Testing for periodicity of Extinction," *Science*, vol. 241, pp. 94-96. Copyright 1988 by the AAAS.

**CHAPTER 15** p. 473 Modified from Figures 1 and 6, Seeland, D. 1985. Oligocene Paleogeography of the Northern Great Plains and Adjacent Mountains. *In* Flores, R.M. and Kaplan, S.S. (eds.). Cenozoic Paleography of the West-Central United States. Rocky Mountain Paleography Symposium 3. p. 431 From Gordon and Jurdy, Cenozoic Global Plate Motions, v. 91, pp. 12,394-5, 1986, copyright by the American Geophysical Union; p. 432 Baker, Mohr, and Williams, Geological Society of America, Special Paper 136, 1972; p. 433 From P. Guennoc and Y. Thisse, Bureau de Recherches Geologigues et Minieres, 1982, reprinted by permission of the author from *Sedimentary Environments and Facies*, 19, Blackwell Scientific Publications Ltd.; p. 433 From Hempton, *Tectonics*, v. 6, p. 688, 1987, copyright by the American Geophysical Union; p. 434 Reproduced by permission of the Geological Society from A.M. Spencer ed., Mesozoic-Cenozoic Orogenic Belts, 1974; p. 435 From Smith and Woodcock, *Alpine-Mediterranean Geodynamics*, Geodynamic Series 7, 1982, copyright by the American Geophysical Union; p. 436 B.E. Hobbs, W.D. Means, P.F. Williams, *An Outline of Structural Geology*, 1976, John Wiley & Sons, Inc.; p. 436 Dewey and Bird, Journal of Geophysical Research, v. 75, p. 2627, 1970, copyright by the American Geophysical Union; p. 437 Reprinted by permission, *American Scientist*, journal of Sigma Xi, P. Molnar, "The Geologic History and Structure of the Himalaya," v. 74, no. 2, 1986; p. 438 Graham, Dickinson, and Ingersoll, *Himalayan-Bengal Model for Flysch Dispersal in the Appalachian-Quachita System*, v. 86, 1975; p. 440 Reprinted with permission from Geological Society of America Bulletin, v. 83, Malfait and Dinkelman, © 1972, the Geological Society of America; Modified from unnumbered figure showing Peruvian Andes on page 77. "The Structure of Mountain Ranges" by P. Molner, *Scientific American*, 255, #1, 1986; p. 443 C.W. Stearn, R.L. Carroll, T.H. Clark, *Geological Evolution of North America*, 3rd Ed., 1979, John Wiley & Sons, Inc.; p. 444 From Dickinson and Snyder, GSA Mem. 151. 1978, the Geological Society of America. Reprinted with permission; p. 445 W.R. Dickinson, 1979, reprinted with permission of the Society of Economic Paleontologists and Mineralogists; p. 446 W.R. Dickinson, 1979, reprinted with permission of the Society of Economic Paleontologists and Mineralogists; p. 447 Philip B. King, *The Evolution of North America*, rev. ed. Copyright ©1959, 1977 by Princeton University Press. Fig. 74 reprinted with permission of Princeton University Press; p. 449 Photo by Mel Bersch used with permission from Marie Huizing, Managing Editor, *Rocks and Minerals*; p. 450 Origin of the Carbonate Sediments in the Wilkins Peak Member of the Lacustrine Green River Formation (Eocene), Wyoming, U.S.A. *In Modern and Ancient Lake Sediments*, Matter, A. and Tucker, M.E. (eds.) Proceedings of a symposium held at the H.C. orsted Institute, University of Copenhagen, 12-13 August, 1977; p. 451 From M. Suter, *Geological Society of America Bulletin*, v. 98, 1987, the Geological Society of America. Reprinted with permission; p. 452 Diagrams by Dr. S.H. Knight, *in* Blackstone (1988); p. 453 R.L. Christiansen and P.W. Lipman, Philosophical Transaction Royal Society of London, v. A271, 1972. Reprinted with permission; p. 454 P.R. Hooper, "The Columbia River Basalts," *Science*, vol. 215, pp. 1563-1568. Copyright 1982 by the AAAS; p. 457 From J.H. Stewart, GSA Mem. 152, 1978, the Geological Society of America. Reprinted with permission; p. 458 From J.H. Stewart, GSA Mem. 152, 1978, the Geological Society of America. Reprinted with permission; p. 460 W.R. Dickinson, 1979, reprinted with permission of the Society of Economic Paleontologists and Mineralogists; p. 461 Courtesy of D. L. Durham, U.S.G.S.; p. 461 International Association of Sedimentologists; p. 462 From *Natural Regions of the United States and Canada* by Charles B. Hunt. Copyright © 1967, 1974 W.H. Freeman and Company. Reprinted with permission; p. 463 R.M. Flores and S.S. Kaplan (eds.), 1985, reprinted with permission of the Society of Economic Paleontologists and Mineralogists; p. 464 By Courtesy of the British Museum (Natural History); p. 465 Lowman, 1949, reprinted by permission of American Association of Petroleum Geologists; p. 466 Fisher and McGowen, 1969, reprinted by permission of American Association of Petroleum Geologists; p. 467 From Hanna, 1959, reprinted by permission of the Oil and Gas Journal; p. 467 John Suppe, *Principles of Structural Geology*, © 1985, figure 7-35 p. 246. Reprinted by permission of Prentice-Hall, Inc., Englewood Cliffs, New Jersey; p. 468 G.M. Griedman and J.E. Sanders, *Principles of Sedimentology*, 1978, John Wiley & Sons, Inc.; p. 469 Schlee and Klitgord, *Journal of Geological Education*, 1986. Reprinted by permission of the National Association of Geology Teachers.

**CHAPTER 16** p. 475 Furnished by B.A. Masters; p. 475 Courtesy of Paul Brenckle; p. 479 Reproduced with permission from Colbert, Edwin H., *An Outline of Vertebrate*; p. 485 From Gingerich 1984, *in* Mammels; Notes for a Short Course, University of Tennessee Studies in Geology 8; p. 486 Reprinted from the July-August Issue of Geotimes by permission from the American Geological Institute; p. 486 Figure 48 from *The Vertebrate Body*, Third Edition, by Alfred S. Romer, copyright © 1962 by Saunders College Publishing, a division of Holt, Rinehart and Winston, Inc., reprinted by permission of the publisher; p. 487 From Evolution of the Vertebrates by E.H. Colbert. Illustrations drawn and copyrighted by Lois Darling. p. 488 Figure 224 from The Vertebrate Body, Third Edition, by Alfred S. Romer, copyright © 1962 by Saunders College Publishing, a division of Holt, Rinehart, and Winston, Inc., reprinted by permission of the publisher; p. 488 The Carnegie Museum of Natural History; p. 489 K.D. Rose, "Skeleton of Diacodexis, Oldest Known Artiodactyl," *Science*, vol. 216, pp. 621-623, Fig. 1. Copyright 1982 by the AAAS; p. 489 E.H. Colbert, *Evolution of the Vertebrates*, 3rd Ed., 1980, John Wiley & Sons, Ltd. Illustrations drawn and copyrighted by Lois Darling; p. 491 Courtesy of the British Museum (Natural History); p. 493 © 1962 by Saunders College publishing, a division of Holt, Rinehart and Winston, Inc., reprinted by permission of the publisher; p. 493 Courtesy of the British Museum (Natural History); p. 494 Colbert, *Evolution of the Vertebrates* 3rd ed., John Wiley & Sons, Inc., 1980; p. 495 B.J. MacFadden, Paleobiology, v. 11, no. 3, 1985, the Paleontological Society. Reprinted with permission (right) Reprinted from J.S. Monroe, Basic Created Kinds and the Fossil Record of Perissodactyls, *Creation/Evolution* Issue XVI, v. 5, no. 2, 1985; p. 497 Courtesy of the British Museum (Natural History); p. 499 By courtesy of the British Museum (Natural History); p. 500 Reprinted by permission, *American Scientist*, journal of Sigma Xi, J.A. Wolfe, "A Paleobotanical Interpretation of Tertiary Climates in the Northern Hemisphere," v. 66, no. 6, 1978; p. 501 S.D. Webb, 1984, *in Mammals*, University of Tennessee Studies in Geology 8; p. 502 From Fritz, *Geology*, v. 8, 1980, the Geological Society of America. Reprinted with permission; p. 502 (both) From Yuretich, *Geology*, v. 12, 1984, the Geological Society of America. Reprinted with permission; Reprinted from R.F. Yuretich. "Yellowstone Fossil Forests: New Evidence for Burial in Place" Geology v. 12, 1984; p. 503 Photo courtesy of Lyn Topinka, U.S.G.S. David A. Johnston Cascades Volcano Observatory, Vancouver, Wash.; p. 504 Reproduced with permission from Tarling, D.H., *Continental Drift & Biological Evolution* 1980. Carolina Biology Reader Series. Copyright Carolina Biological Supply Company, Burlington, North Carolina.

**CHAPTER 17** p. 511 Reprinted with permission of Cambridge University Press; p. 511 Photo by John S. Shelton; p. 513 B.J. Skinner and S.C. Porter, *Physical Geology*, 1987, John Wiley & Sons, Inc.; p. 514a. Modified from B.J. Skinner and S.C. Porter, 1987, *Physical Geology*, p. 564, fig. 20.12a; p. 515 B.J. Skinner and S.C. Porter, *Physical Geology*, 1987. John Wiley & Sons, Inc.; p. 518 Adapted from H.B. Willman and J.C. Frye, 1970, "Pleistocene Stratigraphy of Illinois," *Illinois Geological Survey Bulletin 94* in *North America and the Great Ice Age*, C.L. Matsch, 1976, McGraw-Hill Book Co.; p. 519 a. Reprinted from Geological Society of America Memoir 145, van Donk, 1976, the Geological Society of America; b. Reprinted from G. Wollin, D.B. Ericson, and M. Ewing, The Late Cenozoic Glacial Ages - K.K. Turekian (ed.), 1971, Yale University Press; p. 523 simplied and reprinted with permission from Andrews, 1969, Canadian Geological Survey. p. 524 H.E. Wright, Jr., David G. Frey, eds., *Quaternary of the United States*. Copyright © 1965 by Princeton University Press. Fig. 1, p. 266. Reprinted with permission of Princeton University Press; p. 524b Courtesy of U.S. Geological Survey; p. 525 Reprinted with permission from Prest, 1970, Canadian Geological Survey; p. 526 B.J. Skinner and S.C. Porter, *Physical Geology*, 1987, John Wiley & Sons, Inc.; p. 527 B.J. Skinner and S.C. Porter, *Physical Geology*, 1987, John Wiley & Sons, Inc.; p. 528 B.J. Skinner and S.C. Porter, *Physical Geology*, 1987, John Wiley & Sons, Inc.; p. 532 Photograph #334A from the Peabody Museum of Natural History, Yale University New Haven, Conn. 06511; Modified from unnumbered figure (b) on page 29. "Ramapithecus" by E.L. Simons, *Scientific American*, 236, #5, 1977; p. 533 Modified from unnumbered figure on page 63. "The Antiquity of Human Walking" by J. Napier, *Scientific American*, 216, #4, 1967; p. 534 Modified from unnumbered figure on pages 108 and 109. "The Casts of Fossil Hominid Brains" by R.L. Holloway, *Scientific American*, 231, #1, 1974; p. 535 The Cleveland Museum of Natural History; p. 537 From Clark Spencer Larson and Robert M. Matter, *Human Origins: The Fossil Record*. Copyright © 1985 by Waveland Press, Inc., Prospect Heights, Il; p. 537 (Fig. 17–28) Property of the Wenner-Gren Foundation for Anthropological Research, Inc. New York, New York; p. 539 R. Lewin, "The Earliest 'Humans' Were More Like Apes," *Science*, vol. 236, p. 1062. Copyright 1987 by the AAAS; p. 539 Carolina Biological Supply Company replica; p. 540 Neg. no. 125964. Courtesy Department of Library Services American Museum of Natural History; p. 542 Neg. no. 15509. Courtesy Department of Library Service American Museum of Natural History.